Mkomazi: the Ecolog
Biodiversity and Con
of a Tanzanian Savanna

Edited by
Malcolm Coe, Nicholas McWilliam,
Graham Stone & Michael Packer

Editorial Assistant Jo Lyas

Published by the Royal Geographical Society
(with The Institute of British Geographers), 1999

Mkomazi Ecological Research Programme Collaboration

The Mkomazi Ecological Research Programme was organised by the Royal Geographical Society (with The Institute of British Geographers), the Government of Tanzania's Department of Wildlife, and the University of Oxford.

The UK Darwin Initiative funded the invertebrate biodiversity assessment and the habitat mapping projects of the Programme. Friends of Conservation was the Programme's launch sponsor, providing funds for the renovation of the Ibaya Field Centre in the Mkomazi Game Reserve. The British Council provided funding and administrative support for the Programme in Tanzania.

The Programme benefited from the support and collaboration of six major Benefactors. The main sponsor of this book is Friends of Conservation and further funds were provided by the George Adamson Wildlife Preservation Trust.

SUGGESTED CITATION Coe, M.J., McWilliam, N.C., Stone, G.N. & Packer, M.J. (eds.) (1999) *Mkomazi: the Ecology, Biodiversity and Conservation of a Tanzanian Savanna*. Royal Geographical Society (with The Institute of British Geographers), London.

ISBN 0-907649-75-0

EDITORIAL ASSISTANT Jo Lyas
DESIGN & PRODUCTION Nicholas McWilliam
PRINTER Page Bros. (Norwich) Ltd.

FRONT COVER picture by Jonathan Kingdon: "This painting maps the Mbono Valley, Mkomazi, stage for a daily drama of life and death in which zebras and lions are major players". See *Mkomazi Mind Memory Maps*, page 587.

BACK COVER Top left: Ndea peak and Kenya beyond, seen from Vitewini ridge (CC). Top right: *Eulophia speciosa* (Orchidaceae), an African ground orchid which appears on seasonally inundated grassland as it dries out (MC). Bottom left: Measuring ant activity on an *Acacia* bush (NMcW). Bottom right: well-fed lion and the remains of a zebra, with Ibaya Camp beyond—a similar story is told on the front cover painting (NMcW).

COLOUR PLATE PHOTOGRAPHS Chris Caldicott 40; Susan Canney 1, 2, 7, 14, 31; Malcolm Coe 8–13, 15, 23, 25–30, 32, 34–37, 39; Tom Craig 38; Angus Jackson 18; George McGavin 19, 21; Nick McWilliam 6, 17, 41, 42; Tony Russell-Smith 20; Graham Stone 3, 16, 22, 24; Nigel Winser 33.

The opinions expressed in this volume are those of the authors and do not necessarily reflect those of the Royal Geographical Society (with The Institute of British Geographers) or the Department of Wildlife, Tanzania.

Royal Geographical Society (with The Institute of British Geographers)
1 Kensington Gore, London SW7 2AR, UK
Tel. 020 7591 3000, fax 020 7591 3031
E-mail info@rgs.org
Website www.rgs.org

Contents

continued

Preface

Malcolm Coe & Bakari N.N. Mbano
Directors, Mkomazi Ecological Research Programme

When wildlife managers are faced with administrative and environmental problems, it is easy to forget that there is nothing new about most of these difficulties: they have been experienced almost continuously over the last 60 or more years by those that have been allocated the task of conserving our animal and plant resources. Without the foresight of the Government of the United Republic of Tanzania, much of the current land that is protected from the ecologically damaging effects of human husbandry practices would long since have disappeared. Sayers (1930) pointed out that in the late 1920s the "...provision in Kenya law prohibiting the killing of game was not embodied in Tanganyika legislation". In fact at that time, although there were 15 quite small areas that were designated as Game Reserves, Sayers outlines massive areas around these Reserves which are described as the "best general shooting areas"[1].

Tanzania covers an area of 930,700 km^2 and exhibits a range of physiographic and biotic diversity hardly matched by any other single country in Africa. These unique environments include the staggering species-rich coral reefs of the Indian Ocean fringe, the borders of the Congo Basin rain forests in the west, the montane forests and frozen alpine wastes of the high mountains, the grasslands with their vast herds of mammalian herbivores of the Rift Valley and its savanna borders, alkaline lakes pink with flamingos, active volcanoes, and the extensive southern Miombo woodlands. This impressive physical range and the vast age-span of the landscape are to a large degree responsible for the country's wealth of biological diversity, many species of which are endemic, or unique, to Tanzania, especially on the archipelago-like Mountains of the Eastern Arc, the more recent volcanic masses and the coastal forests. Indeed, Polhill (1968) estimated that the flora of

[1] The area surrounding the Selous Reserve measured 530 km on a SW–NE axis with an average width of 155 km, encompassing in all an area almost 258,200 km^2 or five times the area of the Selous Game Reserve today.

Tanzania probably comprises 10,000 plant species, of which up to 11% are endemic.

An important contributory factor to the service of conservation in Tanzania is the allocation of about 28% of its surface to protected area status, which is made up of National Parks (12), Game Reserves (31), Conservation Area (1), Forest Reserves (5), Biosphere Reserves (2) and the internationally designated UNESCO World Heritage Sites (4) (IUCN 1987, Wildlife Division, pers. comm.). Areas gazetted primarily as wildlife protected areas comprise four conservation categories, each of which has a different level of legal status, administration and protection: National Parks such as Serengeti represent areas where all consumptive exploitation (e.g. hunting, grazing, tree felling) is prohibited; Game Reserves such as Mkomazi and Selous, where access and activities are by special permission only; Game Controlled Areas where restrictions apply specifically to hunting animals in the area; and the Ngorongoro Conservation Area which is managed for both wildlife conservation and resident Maasai herders (Homewood & Rodgers 1991).

The national census of 1921 enumerated a human population of 4,107,000, a figure less than that estimated from the German census of 1913 which totalled 4,145,000 (Sayers 1930). Since this time, however, human numbers have risen consistently, to 7.5 million in 1948, 8.7 million in 1957, 10.3 million in 1964, 12.2 million in 1967, 17.5 million in 1978, 19.7 million in 1983 and 23.2 million in 1988 (Blacker 1969, Morgan 1973, Lyogello 1991). The figure for 1988 represents a 210% increase over the population of 1948, indicating the expanding pressures that have been exerted on the natural resources of Tanzania over a period of 40 years and especially on rangelands, which Peberdy (1969) estimated occupy two thirds of the country's surface.

The resource demands of rising populations have resulted in the increased migration of pastoralists into Mkomazi Game Reserve, which in consequence of its higher level of protection has previously suffered far less from exploitative environmental damage than its surroundings. This intensification of conflict over biodiversity conservation and resource utilisation is reflected by a high national priority being attributed to the development of management plans that balance the protection of biodiversity with use of the resource base, wherever practicable, for local development.

In February 1996, the Tanzanian Parliament ratified the Convention on Biological Diversity, committing the country to the conservation of biodiversity, the sustainable utilisation of biological resources and the equitable sharing of the benefits of this utilisation. Thus Tanzania aims to conserve biodiversity in all its forms through ecosystems, communities, species and their genetic variation, in order that environments may be used in a sustainable manner for the benefit of both the indigenous flora and fauna, and the human population. Conservation has taken a long time to shake itself free of the purely preservationist dogma to this new one, in which the needs of the human exploiters are considered alongside those of the natural world. In coming to terms with this modified conservation philosophy, we must all recognise that ultimately, without control of rising human numbers and changes in patterns of consumption, accommodation between the needs of mankind and the wild becomes increasingly difficult.

The Wildlife Sector, recognising the need to re-formulate policy, published *The Wildlife Policy of Tanzania* in March 1998, which takes account of indigenous population needs in determining how natural resources are to be managed (Ministry of Natural Resources and Tourism 1998). Game Reserves, the direct management responsibility of the Wildlife Division of the Ministry of Natural Resources and Tourism, provide the only opportunity for directly attending the needs of both conservation and development.

The costs of managing the natural resources of protected areas are becoming increasingly high, especially where the areas do not lie on the mainstream tourist routes that are geared to the most spectacular wildlife. Yet specialist tourism, which is capable of generating income from less invasive pursuits, is expanding steadily. The *Integrated Tourism Master Plan* (Ministry of Natural Resources & Tourism 1996) recommends a Tourist Circuit Development which would encompass a route from the coast at Tanga through the biologically rich Usambara Mountains and the Mkomazi Game Reserve. Such a development would not only benefit the Mkomazi Game Reserve but provide essential financial benefit for local communities around the reserve's periphery.

Co-operative programmes like the Mkomazi Ecological Research Programme (MERP) help the development of an infrastructure that can contribute significantly to the conservation of Tanzania's unique biodiversity. To these ends, MERP has provided technical staff with training in insect and plant taxonomy, the curation of taxonomic collections and skills in information management. It has also provided new storage facilities for the insects collected by MERP as a contribution to the establishment of the National Insect Collection, has added 1,100 savanna plant species to the National Herbarium's collection and has established research facilities within the reserve that would allow continued research and monitoring activities.

At present, although Mkomazi is almost certainly of great biological significance, representing species and ecosystems not commonly found elsewhere in Tanzania, national economic benefits such as tourism are currently limited. Internationally Mkomazi has unwittingly become a *cause célèbre* for both those championing the rights of local people to resources and those seeking to preserve and restore ecosystems. These issues require us to ask ourselves how we define and value biodiversity, wilderness, conservation and development. The process of addressing these questions needs to be well informed and it is in this context that the research in Mkomazi, presented in this book, was conducted. Whether the expense and intellectual effort of this research will lead to the conservation of these unique resources will require a high level of good will on all sides.

References

Blacker, J.G.C. (1969) Demography. In: Morgan, W.T.W. (ed.) *East Africa: its peoples and resources*. Oxford University Press, Nairobi and London. pp. 41-58.

Homewood, K.M. & Rodgers, W.A. (1991) *Maasailand Ecology: Pastoralist Development and Wildlife Conservation in Ngorongoro, Tanzania*. Cambridge University Press, Cambridge.

IUCN (1987) *IUCN Directory of Afrotropical Protected Areas*. IUCN, Gland Switzerland, and Cambridge, UK.

Lyogello, L.N. (1991) *A Guide to Tanzania National Parks*. Tourist Publishing Consortium. Dar es Salaam, Tanzania.

Morgan, W.T.W. (ed.) (1973) *East Africa*. Longman, London.

Peberdy, J.R. (1969) Rangeland. In: Morgan, W.T.W. (ed.) *East Africa: its peoples and resources*. Oxford University Press, Nairobi and London. pp. 153-176.

Polhill, R.M. (1968) Tanzania. In: Hedberg, I. & Hedberg, O. (eds.) *Conservation of vegetation in Africa south of the Sahara*. Acta Phytogeographica Suecica 54. pp. 166-178.

Sayers, G.F. (ed.) (1930) *The Handbook of Tanganyika*. Macmillan & Co., London.

Ministry of Natural Resources & Tourism (1996) *Integrated Tourism Master Plan. Final Report: Executive Summary*. Report for Ministry of Natural Resources & Tourism, United Republic of Tanzania, by CHL Consulting Group, Dublin, Ireland.

Ministry of Natural Resources and Tourism (1998) *The Wildlife Policy of Tanzania*. United Republic of Tanzania, Dar es Salaam.

CHAPTER 1

Introduction

Malcolm Coe

Biologically, the Mkomazi Game Reserve[1] is a component of two species-rich and characteristic biogeographic zones. First, Mkomazi represents the southern-most extent of the Sahel in east Africa. Second, the ancient mountains along the southern borders of the reserve are home to a highly diverse and regionally-specific fauna and flora.

The Sahel or Sahelian region of Africa is a belt of semi-arid savanna which borders the southern Sahara, extending from Mauritania eastwards in a broad arc across the northern borders of Ghana, Nigeria, Cameroon and the Central African Republic (Allaby 1977), until it sweeps through southern Sudan and Ethiopia to Somalia and the Indian Ocean. The *Acacia-Commiphora* savanna of eastern Kenya and north-eastern Tanzania represents the southern most extension of the Sahel, which is now under increasing pressure from rapidly expanding human populations and their domestic stock.

White (1979) identified the eastern extension of the Sahel as an important centre for regionally-specific plants (and associated animals). Named the Somalia-Masai Regional Centre of Endemism (RCE), the area covers 1.87 million km^2 and includes south-western Sudan, southern and eastern Ethiopia, Somalia, north-eastern Uganda, most of Kenya and the north-eastern arid regions of Tanzania. 50% of the region's plant species (around 2,500 species) are found nowhere else (i.e. are endemic). The coastal areas from southern Somalia to South Africa represent Regional Transition Zones, within which less than 50% of the species are endemic and many species extend their distribution into adjacent regions. Within the Somalia-Masai RCE, isolated areas of equatorial highlands (including Mount Kenya and Kilimanjaro) support highly characteristic floras, including many endemic species. The distribution of many of these highland plants, like the mountain enclaves themselves, are discontinuous.

The separate identity of the Somalia-Masai RCE persists at taxonomic levels above the species, with up to 50 endemic genera, pointing to a relatively long

[1] The names 'Mkomazi Game Reserve', 'MGR' and 'Mkomazi' are used in this book to refer to the area formally designated the 'Mkomazi/Umba Game Reserves'. The village of Mkomazi lies about 40 km south of the reserve.

history of local radiations and isolation from other regions. The endemic genera do however show strong affinities with the flora of the island of Socotra and south-western Arabia, indicating the transient land connections that must have been present across the southern end of the Red Sea during the major glaciations. This north-eastern corner of the Horn of Africa represents an evolutionary 'hot spot' within which a large number of arid-adapted genera have radiated dramatically. Most notable among these is the genus *Acacia* which includes 38 species in Somalia, 44 species in Kenya and 51 in Tanzania; these three countries individually contain 11, 2 and nine endemic species (Coe & Skinner 1993). The tree genus *Commiphora* is similarly centred in the Somalia-Masai RCE, with 66 species recorded for the area covered by the *Flora of Tropical East Africa* (i.e. excluding Somalia), 51 species in Kenya and 35 in Tanzania (Gillett 1991). The *Acacia-Commiphora* savannas of eastern Kenya extend over the border into north-east Tanzania and the region of Mkomazi, between Kilimanjaro and the coastal forests. Many animal and plant species exhibit more or less continuous distributions between here and at least southern Somalia. Many consider this to make the area one of great biogeographical and biodiversity significance.

In addition to affinities with the Sahel biota, the flora and fauna of the Somalia-Masai RCE show some similarities to those found in areas of southern and south-western Africa. The explanation of this similarity lies in corridors of semi-arid habitats which are thought to have linked the two regions during past glacial and inter-glacial episodes (Chapin 1923, Coe & Skinner 1993, Hamilton 1982, Verdcourt 1969, Winterbottom 1967).

The Mkomazi Game Reserve is unique among reserves and national parks in east Africa: it not only marks the southern limit of the Sahel biota but has striking affinities with adjacent montane regions. Mkomazi is surrounded on its western and southern boundaries by the North and South Pare Mountains and in the south-east by the Usambara Mountains, which rise respectively to 2,111 m (Kindoroko), 2,463 m (Shangema) and 2,304 m (Chambolo) above sea level. Although these peaks are low when compared with the adjacent peaks of Kilimanjaro (5,896 m) and Mount Meru (4,567 m), their proximity to the reserve's southern boundary has a profound influence on the local climate, leading to increased precipitation in some areas and 'rain shadows' in others.

The Pare and Usambara Mountains are formed of extremely ancient Precambrian rocks (dated at between 2,000 and 600 million years old), but were probably only thrust and folded into their present ranges about 25 million years ago. Since then they have been weathered and eroded into their present forms (Rodgers & Homewood 1982). Because of this great age, these mountains, despite their comparatively low elevation, have had much longer to recruit and evolve a locally characteristic assemblage of plants and animals than the far better known but much younger volcanic peaks of equatorial Africa. The Usambara Mountains probably have more endemic animal and plant species than any other comparable mountain range in Africa, as a result of (i) their great age, (ii) their relative isolation and (iii) occasional arrivals of colonising plants and animals from central Africa, west Africa, eastern Uganda and central Kenya along the 'stepping stones' provided by the other isolated peaks of the Eastern Arc Mountains (Kingdon 1981 & 1990).

The Mkomazi Game Reserve

History of protected status

The Mkomazi Game Reserve was established in 1951 in accordance with the Fauna Conservation Ordinance of that year (IUCN 1987). The area gazetted was about 2,850 km^2 (Anstey 1958, Mduma 1986) and was intended to replace the old Ruvu Game Reserve on the northern side of the South Pare Mountains—much of which had been degazetted as a hunting area. The boundaries of the reserve were retracted twice, in 1955 and 1966. In 1973 (just before the enactment of the Wildlife Conservation Act of 1974) the administration of the reserve was split into the Mkomazi Game Reserve in the west and the Umba Game Reserve in the east. The two reserves were administered respectively by the Kilimanjaro and Tanga regional administrations as part of the Government's policy of decentralisation. Both areas were later brought back under the administrative responsibility of the Department of Wildlife. The basic features of the reserve were described by Anstey (1958). Harris conducted the earliest significant ecological field work in the reserve, between late 1964 and mid-1967 (Harris 1970 & 1972, Harris & Fowler 1975). At this time he recorded the area of the MGR as 3,276 km^2. This is almost the same as the area calculated from current maps, although some parts of the boundary are uncertain. The reserve lies along the Kenya border, with a maximum length of 130 km, a maximum width of 41 km, and co-ordinates extending 3°47'–4°33' south and 37°45'–39°32' east (see Map, inside back cover).

Physiographic outline

The basal terrain of the MGR slopes from 760 m in the north-west to 230 m on the Umba steppe in the south-east. Within the reserve, outlying hills of the Pare and Usambara ranges form outcrops rising above the general surface of the MGR. The highest points in the reserve are Maji Kununua (1,620 m) and Kinondo Hill (1,595 m). Many other isolated hills are scattered over the plains, with decreasing elevations from west to east. In some areas of the reserve, such outcrops have been completely eroded away to leave large oval or round areas of red soil. Smith (1894) provides the first written description of the large natural water-filled potholes of Ngurunga[1] (which he called Gurungani) in the Ngurunga valley. This site was

[1] "A large rock" (Krapf 1882).

dammed by Anstey in the early 1950s, and remains an important source of water in the dry season for both wild animals and the domestic stock of local herders.

The extensive central plain slopes slightly in a south-easterly direction, dissected by fairly regular alluvial valleys. In the central and western areas of the reserve in particular, water flow from hill slopes has led to the formation of a series of large basins containing accumulations of eroded alluvial material. These 'black-cotton soils' or *mbugas* are rich in clay, and as a consequence often remain waterlogged throughout the rainy season. These soils also often develop deep gullies or *korongos* which carry storm waters during the rains. In the east, water from the Kisima and Tussa hills drains either southwards towards the Kisiwani River or eastwards towards the Umba River (Tanganyika Ministry of Lands, Forests and Wildlife, Survey Division 1963), the only permanent flowing water within the MGR. To the north-west, the outlying hills of the North Pare range largely drain north-westwards to Lake Jipe or into the extensive basin of the Mbono and Mzukune Valleys. Harris (1972) points out that in many of the valley bottoms, erosion processes have frequently exposed ancient gneisses, schists and crystalline limestones. Such an area may be observed in an extensively eroded valley north-east of the Mandi Hills and Ngurunga. The geochemistry of the MGR soils is described by Abrahams & Bowell (Chapter 3).

The only major sign of faulting in the MGR is in the Pangaro Valley in the north-west (Harris 1972). This elevated valley is bordered by Kinondo Hill and the steep escarpment of Maji Kununua, and its floor lies up to 100 m above that of the Dindira and Mbono valleys below. Earth tremors are still felt in the region, suggesting that movement is still taking place.

Rainfall and climate

Generally, Mkomazi falls within a region with two periods of rainfall each year—the 'short rains' in October to December and the 'long rains' in March to May. The first thunderstorms usually occur in mid-October, followed by sporadic (and sometimes very heavy) showers through November and December. These are followed by further low intensity rains in February and March with the ground remaining damp for a period of 6–8 weeks.

As in most semi-arid environments, the timing and extent of rainfall varies enormously between years; during the present studies conditions in the autumn 'short rains' have varied from drought between November and January to intense rainfall early in the New Year. In addition to seasonal patterns, rainfall generally increases with altitude, with maxima in the Pare and Usambara Mountains. Towns at the foot of the northern slopes of the South Pare Mountains (Same, Mnazi and Gonja) have annual rainfall averages of around 570 mm, 770 mm and 890 mm respectively. Elevated areas of the Usambara Mountains have higher averages of around 1,260 mm at Lushoto (1,371 m) and 1,910 mm at Amani (911 m). Spatial and temporal patterns of rainfall in Mkomazi are described in more detail by McWilliam & Packer (Chapter 2).

Mean annual temperature is largely related to altitude and was recorded by Harris (1972) in the west of the MGR as 23.1°C, with mean minima of between 9.4 and 17.5°C and mean maxima of between 29.0 and 37.8°C.

Vegetation types

Mkomazi contains a wide variety of vegetation types, which are structured largely by variations in altitude and rainfall (Harris 1972). Most of the hills are found in the western part of the reserve, generating a concentration of habitat diversity in the region. This, in turn, is associated with a concentration of animal and plant species. The principal vegetation types in the reserve are introduced briefly below, and discussed in more detail by Coe *et al.*, Chapter 7.

1. *Bushland*: woody vegetation dominated by *Commiphora* and *Acacia*, with other less dominant shrubs and trees, and grassy spaces.
2. *Bushed and wooded grassland*: this vegetation type is found on more freely drained soils on mountain and hill slopes. The trees are usually from 10–12 m in height but are widely spaced, usually not exceeding 20% canopy cover. The trees in this community are dominated by *Acacia, Platycelyphium, Boscia* and *Melia.* The well-drained but generously watered ground supports tall grasses such as *Themeda triandra*, *Heteropogon contortus*, *Digitaria* spp., and *Bothriochloa radicans*.
3. *Grassland*: this vegetation type occurs in valley bottoms where it receives ample (seasonal) rainfall and may be seasonally waterlogged. The dominant grass is *Pennisetum mezianum* but may also include other species.
4. *Upland dry forest*: virtually all upland areas >1,000 m are (or were) covered with a closed canopy forest, whose height varies from 15–20 m. The common development of regular cloud cover on these hilltops and uplands is responsible for the development of profuse epiphyte growth on the forest trees. The commonest trees in the upper canopy include *Calodendrum* and *Albizia*, with a species-rich shrubby understorey. Open glades are colonised by a characteristic set of grass species.
5. *Riparian forest*: river habitats, groundwater forest and the rather narrower margins of seasonal watercourses support good stands of tall trees, with a relatively rich diversity of shrubs in the understorey and at the forest margins.
6. *Rock outcrops and boulder slopes*: these habitats support highly characteristic drought adapted succulent and xerophytic plants, particularly apparent on Kisima hill. Rocky cracks are often occupied by stunted acacias and other shrubs (e.g. *Boswellia neglecta)* or perennial woody herbs.
7. *Permanent and seasonal river margins*: these habitats are often associated with highly saline or alkaline soils, and have their own characteristic floras.

Harris (1972) and Harris & Fowler (1975) concluded that the bushland habitats covered up to 70% of the surface of the MGR, although this percentage has probably been reduced to below 50% in recent years. Human activities have almost certainly played a part in the changing distribution of vegetation types within the reserve, particularly through changes in the distribution and abundance of large herbivores and the incidence of fire. Both have significant effects on vegetation structure. In general, fires started either deliberately or accidentally by people have become more abundant in recent times and large wild herbivores have become less abundant. As a result, the distribution of the vegetation types described above is almost certainly in a period of rapid change.

Abundance of large herbivores in Mkomazi

Since Harris's early work on the vegetation of the MGR in the 1960s, there have been major changes in the large herbivores whose activities have a major impact on vegetation structure. There has been a drastic reduction in the number of elephants, the extinction of the black rhinoceros and a huge increase in the numbers of domestic animals entering the reserve. A large part of the MGR's northern boundary is shared with the southern boundary of Kenya's vast 21,000 km^2 Tsavo National Park. The two protected areas comprise an indivisible Tsavo ecosystem, estimated to cover about 40,000 km^2 (Cobb 1976).

In the recent past, this ecosystem supported a huge elephant population; as recently as the late 1960s the elephant population of the Tsavo National Park and its environs was estimated at 35,000 (Laws 1969). By 1978, the population was reported to have fallen to about 20,000 through the effects of drought induced mortality and poaching. Subsequent poaching had a devastating effect, and by the late 1980s the population had crashed to between 5,000 and 6,000 (Olindo *et al.* 1988, Ottochilo 1986). In less than 20 years the elephant population fell by over 80%. An aerial count carried out by the Kenya Wildlife Service in 1991 estimated the total number of elephants in a 38,000 km^2 region including the Tsavo National Park, the Galana Ranch, the Taita Hills and the MGR to be 6,763 (Kenya Wildlife Service 1991). Although far smaller than earlier populations, this actually represents an increase of 8.6% over estimates from the previous count in June 1989. The total elephant population in Mkomazi at this time was estimated to be 131. The count was carried out in December, at the end of the 'short rains', representing the season when elephants are expected to use the MGR as a wet season dispersal area (see Eltringham *et al.* on large mammals, Chapter 31).

Frequency of fires in Mkomazi

It is thought that there has been a recent increase in the number of fires started by people either in the reserve or spreading into the reserve. This has had the effect of suppressing the growth of woody vegetation and maintaining much of the west part of the reserve as open grassland. There is very tentative evidence that stands of bushland may have been opened up and fragmented (Cox 1994). As a result, over much of the reserve the vegetation has become fragmented into a fire-induced mosaic, while many of the hill slopes in the western and biologically-richest segment of the reserve are now almost completely devoid of woody vegetation. Damage to the hilltop forests is a cause of particular concern in the future management of Mkomazi if it can be shown (as it is in the following chapters) that these hills represent islands of diversity for species of plants and animals absent from other habitats in the reserve. Ferguson (Chapter 9) considers the causes of fire in Mkomazi and investigates the effects in detail.

Conservation and development

The potential for conflicts of interest between people and wildlife has long been recognised: Sayers, the General Editor of *The Handbook of Tanganyika*, wrote in 1930 that "There are two extremist schools of thought in regard to the preservation of African fauna, the first which favours the enactment of stringent laws for the protection of game and their rigorous observance, even at the expense of native interests and economic development, and the second which holds that as the progress of native and non-native agriculture and game preservation cannot go hand in hand, the game is inevitably doomed to extinction in the course of the next few generations, so that protective steps may as well be abandoned. To neither of these policies does the Government of Tanganyika subscribe and the problem before it is to reconcile the reactionary views of the one with the pessimism of the other."

Ecologists working in Africa have long recognised that if conservation areas are to survive the impacts of rising human numbers, it is essential for the managing authorities to recognise that income derived from such areas must be shared with the local community. Semi-arid savannas such as Mkomazi have suffered in the past from a great shortage of infrastructural finance. This has left Mkomazi in a backward state of development (Mduma 1986) and the income generated by the reserve fails to pay even the cost of its administration let alone produce a surplus which could be shared with the communities surrounding the reserve.

An understanding of the needs of these surrounding communities is required in assessing how biodiversity conservation in Mkomazi can be achieved at the same time as enabling local development. Research to gain such an understanding within the cultivator and pastoralist communities is reported by Brockington & Homewood (Chapter 33) and Kiwasila & Homewood (Chapter 34).

The Mkomazi Ecological Research Programme

Most of the studies presented in this book arose from the Mkomazi Ecological

Research Programme (MERP). In 1988, the Tanzanian Government's Director of Wildlife invited the Royal Geographical Society in London to consider undertaking an ecological inventory study of the Mkomazi Game Reserve, with a view to providing the information that the Department of Wildlife could use to plan for the area's future management and utilisation. This invitation was accepted, leading to the establishment of the Mkomazi Ecological Research Programme with the overall aim of providing high quality scientific data and research, and training opportunities for overseas and local scientists.

The scientific aim of the programme has been to describe the floral and faunal diversity of Mkomazi's habitats and to work towards an understanding of observed patterns of distribution, abundance and species diversity. Additionally the programme has attempted to generate baseline data against which any effects of human-induced change inside and around the reserve might be studied. The background and activities of the programme are elaborated in the MERP section later in this book.

The studies undertaken during the active field phase of the MERP between 1993 and 1997 have demonstrated just how much still awaits discovery and understanding in the African savanna. Whether in the long run it will be thought of as worthwhile will depend entirely on the manner in which the information contained in this book is used to sustainably protect and utilise these superb and shrinking wilderness areas for the benefit of all Tanzanians, both now and in the future.

References

Allaby, M. (1977) *A Dictionary of the Environment*. Macmillan Press, London.

Anstey, D. (1958) Mkomazi Game Reserve. *Tanganyika Notes and Records* 50: 68-70.

Chapin, J.P. (1923) Ecological aspects of bird distribution in tropical Africa. *American Naturalist* 57: 106-125.

Cobb, S.M. (1976) The distribution and abundance of the large mammal herbivore community of Tsavo (East) National Park, Kenya. D.Phil. Thesis. University of Oxford.

Coe, M.J. & Skinner, J.D. (1993) Connections, disjunctions and endemism in the Eastern and Southern African mammal faunas. *Transactions of the Royal Society of South Africa* 48: 233-255.

Cox, J. (1994) *Mkomazi Ecological Research Programme: Pilot Study (August-September 1993): Remote Sensing Habitat Survey*. Report to Royal Geographical Society, London.

Gillett, J.B. (1991) *Flora of Tropical East Africa: Burseraceae*. Balkema, Rotterdam.

Hamilton, A.C. (1982) *Environmental History of East Africa: a study of the Quaternary.* Academic Press, London.

Harris, L.D. (1970) *Some structural and functional attributes of a semi-arid East African ecosystem.* PhD Thesis, Michigan State University.

Harris, L.D. (1972) An ecological description of a semi-arid East African ecosystem. *Colorado State University Range Science Series* 11: 1-80.

Harris, L.D. & Fowler, N.K. (1975) Ecosystem analysis and simulation of the Mkomazi Reserve, Tanzania. *East African Wildlife Journal* 13: 325-346.

IUCN (1987) *IUCN Directory of Afrotropical Protected Areas.* IUCN, Gland, Switzerland and Cambridge, UK.

Kenya Wildlife Service (1991) *Tsavo Elephant Count.* Kenya Wildlife Service.

Kingdon, J. (1981) Where have all the colonists come from? A zoogeographical examination of some mammalian isolates in eastern Africa. *African Journal of Ecology* 19: 115-124.

Kingdon, J. (1990) *Island Africa: The evolution of Africa's rare animals and plants.* Collins, London.

Krapf, L. (1882) *Suahili-English Dictionary.* Trübner and Co., London.

Laws, R.M. (1969) The Tsavo Research Project. *Journal of Reproduction and Fertility*, Supplement 6: 495-531.

Mduma, S.R. (1986) Mkomazi Game Reserve: Design and reconstruction measures for its survival. *Miombo* 1: 17-19.

Olindo, P., Douglas-Hamilton, I. & Hamilton, P. (1988) *The Tsavo Elephant Count 1988.* Report to Kenya Wildlife Service.

Ottochilo, W.K. (1986) Population estimates and distribution patterns of the elephant in the Tsavo Ecosystem, Kenya, in 1980. *African Journal of Ecology* 24: 53-57.

Rodgers, W.A. & Homewood, K.M. (1982) Species richness and endemism in the Usambara Mountain forests, Tanzania. *Biological Journal of the Linnean Society* 18: 197-242.

Sayers, G.F. (ed.) (1930) *The Handbook of Tanganyika.* Macmillan & Co., London.

Smith, C.S. (1894) The Anglo-German boundary in East Equatorial Africa: Proceedings of the British Commission, 1892. *Geographical Journal* 4: 424-437.

Tanganyika Ministry of Lands, Forests and Wildlife, Survey Division (1963) Geological Survey maps. Dodoma, Tanganyika.

Verdcourt, B. (1969) The arid-corridor between the north-east and south-west areas of Africa. *Palaeoecology of Africa* 7: 151-181.

White, F. (1979) The Guineo-Congolian region and its relationship with other phytochoria. *Bulletin Jardin Botanique Nationale Belgique* 49: 11-55.

Winterbottom, J.M. (1967) Climatological implications of avifaunal assemblages between south-western Africa and Somaliland. *Palaeoecology of Africa* 2: 77-79.

SECTION 1

Ecology and biodiversity

This section, by far the largest in the volume, presents the results of a great variety of field research activities that aimed to inventory the biodiversity of the Mkomazi Game Reserve. Inventorying is not merely the drawing up of species lists but involves surveying, sorting, cataloguing, quantifying and mapping of a variety of components of ecosystems as well as the integration of resulting information for analysis of ecological patterns and processes. Inventorying, therefore, provides the science for understanding biodiversity.

A wide range of sampling methods have been used to study different groups of animals and plants, from vegetation sweep netting and direct observation to insecticide fogging and satellite remote sensing, and wherever possible the advantages and limitations of each method are described by those using them. An appreciation of what can and cannot reliably be concluded from field data is essential to their meaningful application in management decisions.

Correct identification of sampled 'material' is critical to sound interpretation of studies. For larger animals and plants, species can be identified by reference to existing identification or taxonomic keys. For many of the smaller invertebrates, however, identification to the species level was impossible, and many of the organisms collected belong to species yet to receive formal scientific names. The common practice for such groups is to identify each individual sampled as a member of an identifiably distinct group, termed a 'morphospecies'. In this way, notional species, that is those without formal scientific names, still contribute to diversity assessments (usually expressed as an index) and assessments of abundance of individuals 'species' found in particular habitats. Identifications and naming of the thousands of invertebrates sampled in Mkomazi is still in progress, and the collections made during this work will remain a valuable resource for savanna ecologists for many years to come.

The following 31 chapters describe in detail the biophysical characteristics of the Mkomazi locality, and the diversity and ecology of particular groups of organisms. A common theme is to consider variations in patterns of species richness and abundance across different habitat types, and at a variety of spatial scales, as well as how they may be explained. Such an understanding is clearly important for determining conservation priorities and management actions.

CHAPTER 2

Climate of Mkomazi: variability and importance

Nicholas C. McWilliam & Michael J. Packer

Ecology and climate in Mkomazi

Rainfall in semi-arid areas such as Mkomazi is a critical factor driving ecological processes and determining the distribution of vegetation and animals. Plant productivity and mammalian herbivore biomass, for instance, have been closely correlated with mean annual rainfall in semi-arid areas (Coe *et al.* 1976). Plant species richness in semi-arid habitats was found to be related to climatic heterogeneity (Cowling *et al.* 1994). Elsewhere in this book, rainfall is considered in relation to plant phenology (Chapter 8), hangingfly distribution (Chapter 15), beetle richness (Chapter 14) and large mammal distribution (Chapter 4).

Temperature is also a significant factor in ecological patterns. For instance, tsetse fly distribution in Africa is markedly affected by temperature (Rogers *et al.* 1996). In a biological context, temperature may operate directly by limiting plant and animal metabolic processes, although conditions in Mkomazi are not considered critical in this respect. Using temperatures from Same as representative of western Mkomazi, Harris (1970) reported a mean annual temperature for 1965–66 of 23.1°C, with a mean annual minimum of 17.5°C and maximum of 29.0°C. The absolute minimum recorded at Ibaya was 9.4°C, although lower, possibly more critical, temperatures may occur at the higher points (up to 1,600 m). The indirect effects of temperature are far more significant. Harris estimated the potential evapotranspiration in western Mkomazi to be 1,450 mm/year, increasing in the lower, warmer parts of the reserve. This is well in excess of Mkomazi's annual rainfall, so has a dramatic effect on the water available to plants and animals.

In this context, rainfall is critical to water availability and ecology in Mkomazi. The rainfall regime is characterised by complex spatial and temporal patterns: of Pratt & Gwynne's six ecoclimatic zones in east Africa, Mkomazi lies in or close to four zones (Pratt & Gwynne 1977). This chapter investigates the seasonality, variability and distribution of rainfall in Mkomazi, and compares it with that of Tsavo National Park in Kenya and east Africa as a whole.

Rainfall data for Mkomazi

Published rainfall maps of east Africa and Tanzania (e.g. Department of Lands & Surveys 1956, Nicholson *et al.* 1988) cover Mkomazi but their isohyets are interpolated from a very sparse raingauge network and do not have sufficient accuracy or resolution to show local variations. Our study of rainfall patterns in Mkomazi, although constrained by a lack of long-term and spatially representative raingauge data, has been made possible by several reliable datasets from within or near Mkomazi, and these are outlined now.

We have used monthly raingauge records from Same (since 1935) and Voi, Kenya (since 1904) (Figure 2.1). Further records for 1941–86 came from raingauges at Gonja and Mnazi, although they are very sporadic. Within Mkomazi, Larry Harris established eight raingauges during 1965 and operated them until June 1967 (Harris 1970). The Mkomazi Ecological Research Programme (MERP) re-established raingauges at four of these sites (Ibaya, Kisima, Maore and Kamakota) plus new ones at eight other sites, operating them between January 1995 and February 1997 (Figure 2.1). The 12 MERP gauges were locally made using a simple funnel design and were set into the ground for protection, with the rim extending c. 10 cm above ground level. A thin oil layer minimised evaporation between readings. Whenever possible they were read monthly, although removal, damage and impassable roads have left gaps in the record. To study Mkomazi's spatial rainfall variations in more detail, we have also used satellite data.

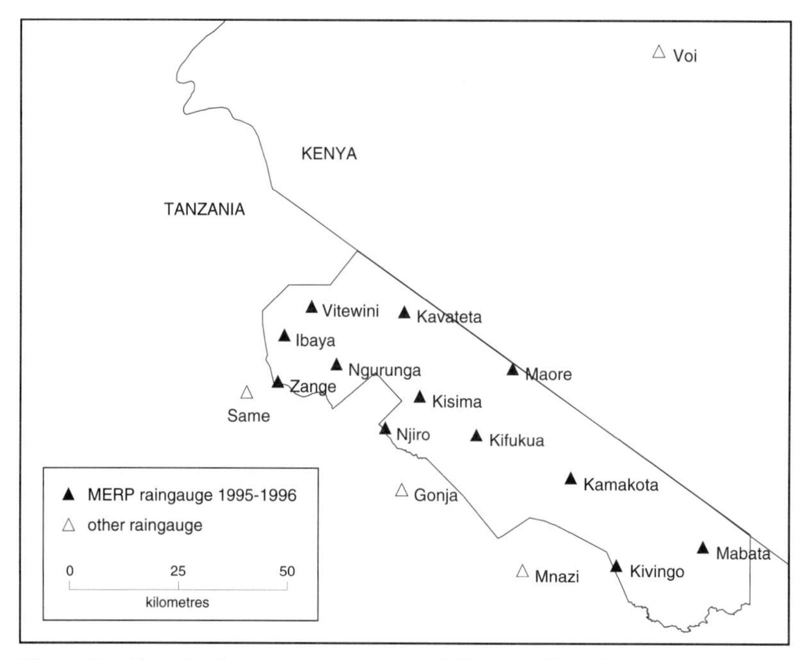

Figure 2.1 Map of raingauges in and around Mkomazi Game Reserve.

Using Same and Voi rainfall data for Mkomazi

Rainfall in east Africa varies greatly from place to place and from year to year; however, inter-annual rainfall variations are remarkably similar across the region. This is mainly because the early rains—the source of most inter-annual rainfall variability—are driven by large-scale processes which affect the region as a whole (Washington 1998). This pattern potentially lets us use long-term rainfall records from outside Mkomazi as indicators of fluctuations within the reserve itself.

Figure 2.2 shows that there is a statistically-significant correlation between the seasonal rainfall of Same and Voi, for both early and late rains. Figure 2.3 uses the few records available from inside the reserve to indicate that the pattern extends into Mkomazi. Given this pattern of spatial coherence, we can use the records from Same and Voi as good indicators of rainfall variations within Mkomazi. It is worth noting that we found no evidence for long-term upward or downward trends in the amount of rain falling in either Same or Voi.

Figures 2.2. and 2.3, and other analyses in this chapter, use October to December to represent the early rains and March to May to represent the late rains. This gives a useful seasonal measure, while providing consistency with other published studies and helping to diminish the impact of extreme rainfall months.

Temporal patterns and variability of rainfall

Seasonality

The dominant characteristic of rainfall in east Africa is its marked seasonality. This is a result of the north–south movement of the inter-tropical convergence

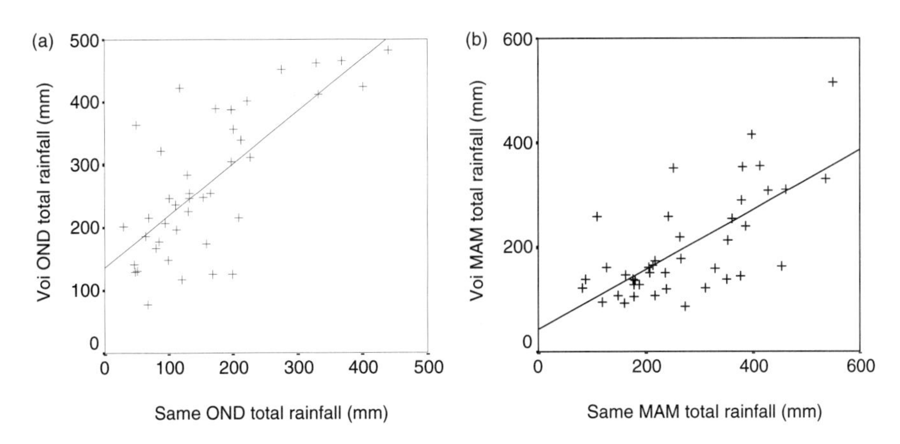

Figure 2.2 The relationships between seasonal rainfall in Same and Voi, 1950–90. Both are statistically significant. (a) Early rains. OND = October, November, December. Pearson's correlation co-efficient $r = 0.732$, $p<0.001$. (b) Late rains. MAM = March, April, May. Pearson's $r = 0.692$, $p<0.001$.

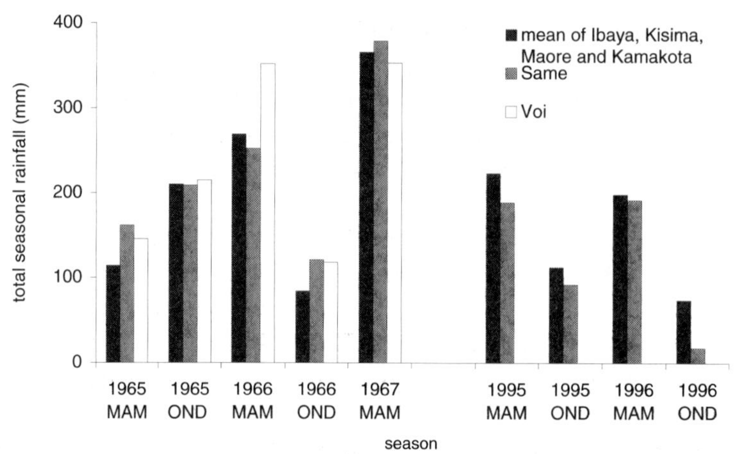

Figure 2.3 Comparison between seasonal rainfall inside Mkomazi and at Same and Voi, for 1965 to 1967 (Harris 1970) and 1995–96 (MERP). MAM = March, April, May (late rains); OND = October, November, December (early rains). Data from Voi for 1995–96 are not yet available.

zone (ITCZ) and its associated rains. In July, the ITCZ is at its northernmost extent and east Africa is generally very dry. As it moves south, it brings the early rains to Mkomazi in October and November, before reaching its southernmost extent and moving back north bringing the late rains to Mkomazi in March to May. The resulting bimodal pattern of rainfall in Mkomazi is very apparent in Figure 2.4. The effects on vegetation can be seen in Plates 1 and 2.

In east Africa generally, the majority of annual rain falls during the late rains. This is clearly true of Same, with 27% falling in October to December and 48% in March to May. Our very limited raingauge data from inside Mkomazi (representing at most three years of readings) suggest a more complicated local pattern, even over distances of tens of kilometres. Raingauges lying closer to the Pare and Usambara mountains appear to have had greater early rains (Zange, Njiro, Gonja and Mnazi). This is supported by longer records from Gonja (45% early, 32% late) and Mnazi (43%, 38%). By contrast, greater late rains were recorded further away from the mountains at Ibaya, Kisima, Maore and Kivingo, in keeping with prevailing east African conditions. Tyrell & Coe (1974) also suggest a reversal of the east African norm around Tsavo, although with rather different zonation.

Variability

The preceding discussion has looked at mean rainfall conditions. Ecologically, it is also useful to look at the variability of seasonal rainfall. In the late rains of 1995 and the early rains of 1996, for example, rainfall at most raingauges in Mkomazi and at Same was almost certainly well below average. Many studies presented in this book were undertaken during this period, and the low rainfall is likely to have affected vegetation phenology as well as the occurrence of insects, birds and large mammals.

Variability can be measured by the co-efficient of variation, defined as the standard deviation divided by the mean, and expressed here as a percentage. A high co-efficient of variation thus indicates a greater variability.

Generally in east Africa, the amount of rain falling in the early rains is far more variable than in the late rains (co-efficient of variability for October to December = 74%, for March to May = 35%, Washington 1988). Same accords with this pattern (61% and 45%). Again, however, Tyrrell & Coe (1974) found contrary local patterns in much of the Tsavo region. Their evidence suggests that a zone which includes Mkomazi has more variable late rains. Results from around Mkomazi broadly match this: Voi (42%, 50%), Gonja (46%, 53%) and Mnazi (40%, 48%) are all slightly more variable in March to May. Unfortunately our data from within Mkomazi cover too short a period to give meaningful averages. As in the preceding section, there may well be a greater deviation from the east African pattern towards the Pare and Usambara mountains.

Causes of variability

The causes of year-to-year rainfall variations in east Africa are increasingly being studied in relation to large-scale atmospheric and oceanic processes. The effect of these large-scale processes is seen in the spatially coherent pattern of rainfall variation in east Africa, which, as noted earlier, is evident around Mkomazi.

In particular, the El Niño Southern Oscillation (ENSO) has been associated with periodic climate fluctuations worldwide and with above-average rainfall in east Africa (for example, see Ogallo 1989 and Washington 1998).

We have therefore examined Same's rainfall in relation to an indicator of ENSO activity, the Southern Oscillation Index (SOI). The SOI compares the sea-level

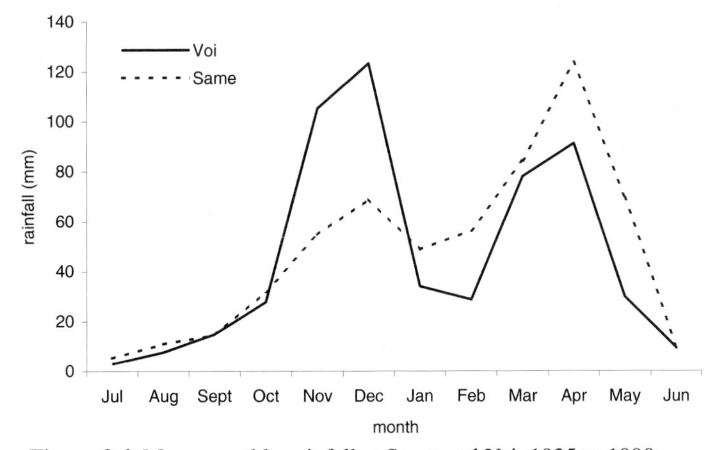

Figure 2.4 Mean monthly rainfall at Same and Voi, 1935 to 1990.

atmospheric pressure in the western Pacific (Darwin, Australia) to that of the mid-Pacific (Tahiti). Negative values tend to indicate ENSO activity. Monthly SOI data were obtained free from http://nic.fb4.noaa.gov/data/cddb/.

We found no statistically-significant correlation between SOI and rainfall during the late rains at Same and Voi. As noted earlier, the variability in the late rains is generally less than that in the early rains, and is likely to be related to local climatic influences. In contrast, SOI is highly correlated with the early rains at Same (Figure 2.5); the same is also true of Voi. Thus it seems likely that much of the inter-annual variation in the early rains around Mkomazi is caused by very large-scale phenomena related to ENSO, although the mechanisms involved in this connection are beyond the scope of this chapter.

Spatial patterns of rainfall in Mkomazi

We have seen the marked spatial unevenness of rainfall in Mkomazi. In trying to map this, straightforward interpolation between raingauges is potentially the easiest method. However, there were not enough raingauges to represent the many local, yet still significant, rainfall variations in the reserve. The altitude-rainfall relationship described for east Africa by Trapnell & Griffiths (1960) could also be used at a broad scale to model rainfall patterns in Mkomazi. In this case, the relationship between rainfall at the 12 raingauges and their altitude was found to be not statistically significant—probably due to local rainfall variations not directly determined by altitude.

A less direct measure of rainfall comes from satellite imagery. Data from the NOAA-AVHRR is used to generate the normalised difference vegetation index (NDVI, see Table 4.1 of Chapter 4 for details), which varies with plant photosynthetic activity (Tucker 1979) and has been closely linked to rainfall patterns in semi-arid African environments (e.g. Davenport & Nicholson 1993). We there-

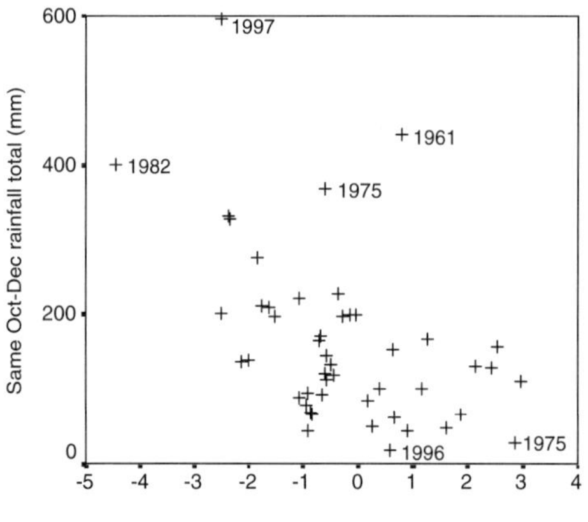

Figure 2.5 Total rainfall for October to December (early rains) at Same, plotted against the Southern Oscillation Index anomaly (an indicator of El Niño activity), for 1951 to 1998. The relationship is statistically significant (Pearson's $r = -0.491$, $n = 48$, $p<0.001$). Exceptionally wet and dry year years are labelled.

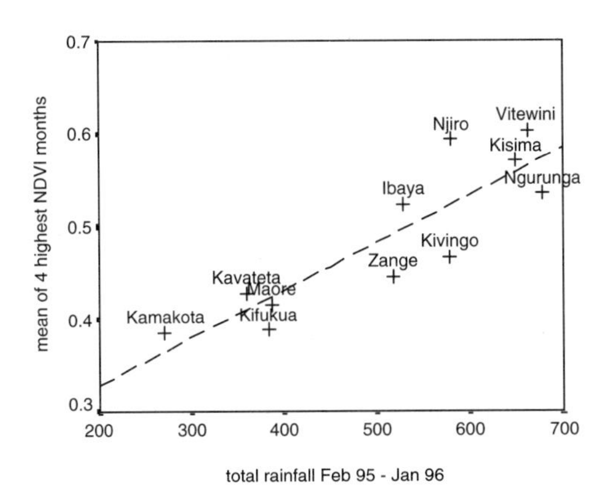

Figure 2.6 Total rainfall plotted against the average of the four highest monthly NDVI values at 11 raingauges in Mkomazi, for the period Februray 1995 to January 1996. The relationship is statistically significant (Pearson's correlation co-efficient $r = 0.878$, $p<0.001$). There were insufficient readings from Mabata to be used.

fore obtained monthly NDVI data for a 12-month period, February 1995 to January 1996, fortunately coinciding with raingauge records from within the reserve. Although this is a relatively short period, the dataset has the key advantages of providing coverage of the whole reserve, and at frequent intervals. Its resolution of 1 km is appropriate for showing local variations. We thus aimed to establish a relationship between NDVI and rainfall measured at the raingauges in Mkomazi to provide the basis for a map of estimated rainfall across the entire reserve.

For each 1 km grid cell which contained a raingauge, we extracted the 12 monthly NDVI values. From these we calculated the mean NDVI of the highest two, three, four and so on months, up to the mean of all 12 months. Each of these mean values was regressed on the 12-month rainfall total for 11 of the raingauges. The best relationship with total rainfall was found to be with the mean of the four highest monthly NDVI values (see Figure 2.6). The regression equation linking NDVI to rainfall was used to estimate rainfall for all 1 km grid cells in Mkomazi.

The resulting map (Figure 2.7) is presented as an approximate indication of rainfall in Mkomazi during 1995. As expected, the main determinant of high estimated rainfall appears to be topography. This is seen partly in the overall increase between the south-east and the north-west of around 300 to 600 mm, but is more marked in relation to the hilly areas of Kinondo, Maji Kununua, Gulela and Kisima-Tussa, and even the smaller hills of Mzara and Maore. In these areas, the estimated rainfall was 600 to 775 mm, although special caution is needed with these figures since higher altitudes were represented by only one raingauge, Vitewini. Between the Tussa hills and the Umba River, rainfall increased slightly towards the edge of the reserve (over 400 mm), probably due to the proximity of the Usambara Mountains. Relatively high raingauge readings from Gonja and Mnazi, between the reserve and the mountains, support this trend. There is some indication of an increase in the extreme east of the reserve, possibly related to increasing rainfall towards the coast, 60 km away (Department of Lands & Surveys 1956, Nicholson *et al.* 1988).

The driest parts appear to be in east-central Mkomazi, with estimates as low as 250 mm. Again, such areas were not represented by raingauges, so the figures are

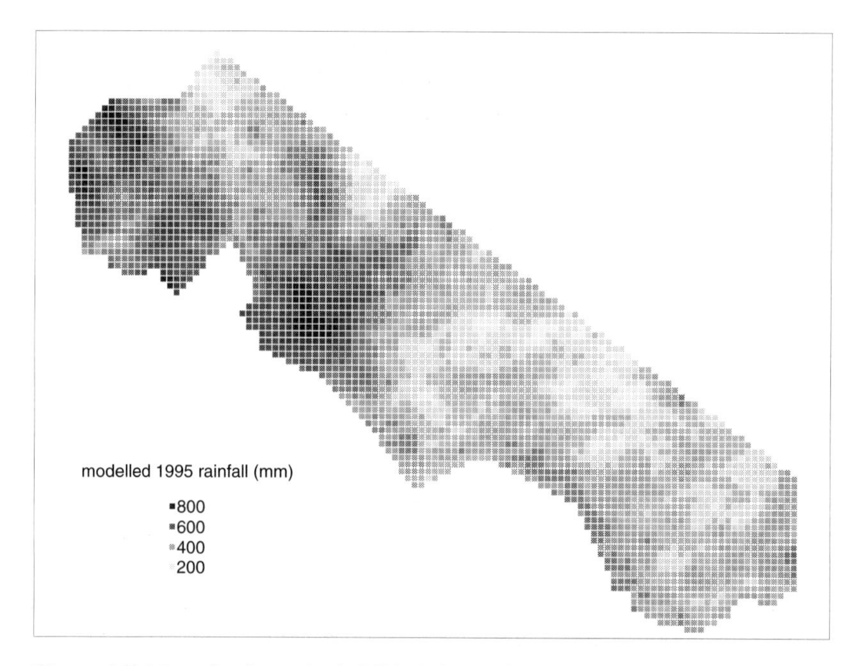

Figure 2.7 Map of estimated rainfall in Mkomazi, February 1995 to January 1996, modelled according to the relationship between rainfall at 11 raingauges and the mean of the four highest monthly NDVI values at each 1 km grid cell during this period.

very tentative. The unexpectedly arid area north-east of Ndea is discussed below. The areas around Ibaya camp, where much of MERP's research took place, have estimates of around 400 to 600 mm. The patterns and figures described here tally well with those of Harris (1970: 15–16).

The relationship we have found between rainfall and the four highest NDVI months accords with other vegetation studies. Chapter 8 discusses how most plant growth in Mkomazi takes place in response to high-intensity rainfall, which occurs for around four months of the year (see Figure 2.4). Likewise, Deshmukh (1984) found a widespread positive relationship between rainfall and peak biomass in east Africa, a finding he supports with Harris's data from Mkomazi. By contrast, rain which falls early in the wet seasons, and any in the drier months, is usually insufficient to reach plant roots.

The limitations of this map must be recognised. We have used a function of NDVI as a surrogate measure for rainfall, and while NDVI in Mkomazi is influenced mainly by rainfall, it is significantly affected by other factors. These include soil fertility, soil moisture availability, fire and herbivory. For example, the map shows unexpectedly low values of estimated rainfall in the extreme north of the reserve, between Ndea and the Kenyan border. This area is known to have experienced high levels of cattle grazing during 1995, resulting in reduced vegetation cover and, almost certainly, in reduced NDVI values.

The map should not be used to infer typical rainfall conditions, as its source data are for 1995 only. Also, the raingauge network did not fully sample the reserve's range of rainfall, so the particularly low and high rainfall estimates remain untested. Ideally, we would use data from a longer period and from more raingauges.

Finally, the regression methods used could be improved, as we are effectively attempting to predict an independent variable (rainfall) from a dependent variable (NDVI), rather than, as is usual, the other way round (Sokal & Rohlf 1981).

Conclusion

Potential evapotranspiration is greater than rainfall in Mkomazi, making water availability a key factor in the distributions and dynamics of the reserve's plants and animals—the ecological significance is high-lighted by other chapters in this book. As temperature is relatively uniform in Mkomazi, our main concern is to understand spatial and temporal patterns of rainfall. Seasonality is the dominant rainfall characteristic, although this chapter has shown how the amount of seasonal rain varies greatly from year to year and from place to place. Data are sparse and we cannot draw a detailed picture of Mkomazi's rainfall, but some underlying patterns can be suggested. Proximity to the Pare and Usambara mountains appears to locally alter and even reverse some seasonal characteristics seen across east Africa as a whole, while smaller spatial variations appear dominated by altitude and local topography. There is evidence that much of the year-to-year rainfall variability is related to the El Niño Southern Oscillation. The resulting extreme wet and dry years undoubtedly shape important aspects of habitat structure and dynamics. Further research on the periodicity of ENSO events and long-term trends in their magnitude provides the prospect of investigating the nature of rainfall as a long-term driving factor in east African savanna dynamics.

Acknowledgements

We are most grateful to Tim Morgan who against the odds developed and operated the raingauge network in Mkomazi for over two years; to the staff of Same Meteorological Station who have provided their latest rainfall records; to Richard Washington at the School of Geography, University of Oxford, for providing invaluable guidance for this chapter; and to the Climate Research Unit at the University of East Anglia and Dan Brockington for providing rainfall data.

References

Coe, M.J., Cumming, D.H. & Phillipson, J. (1976) Biomass and production of large African herbivores in relation to rainfall and primary productivity. *Oecologia* 22: 341-354.

Cowling, R.M., Esler, K.J., Midgley, G.F. & Honig, M.A. (1994) Plant functional diversity, species diversity and climate in arid and semi-arid southern Africa. *Journal of Arid Environments* 27: 141-158.

Davenport, M.L. & Nicholson, S.E. (1993) On the relation between rainfall and the normalized difference vegetation index for diverse vegetation types in East Africa. *International Journal of Remote Sensing* 14: 2369-2389.

Department of Lands & Surveys (1956) *Atlas of Tanganyika, East Africa.* 3rd edition. Department of Lands and Surveys, Dar es Salaam.

Deshmukh, I.K. (1984) A common relationship between precipitation and grassland peak biomass for East and southern Africa. *African Journal of Ecology* 22: 181-186.

Harris, L.D. (1970) *Some structural and functional attributes of a semi-arid East African ecosystem.* Ph.D. thesis, Michigan State University.

Nicholson S.E., Kim J. & Hoopingarner J. (1988) *Atlas of African rainfall and its interannual variability*. Department of Meteorology, Florida State University, Tallahassee, Florida, USA.

Nicholson, S.E. (undated) *A Review of Climate Dynamics and Climate Variability in Eastern Africa.* Unpublished manuscript, Department of Meteorology, Florida State University.

Ogallo, L.J. (1989) Relationships between seasonal rainfall in East Africa and the Southern Oscillation. *International Journal of Climatology* 8: 31-43.

Pratt, D.J. & Gywnne, M.D. (1977) *Rangeland Management and Ecology in East Africa.* Hodder & Staughton, London.

Rogers, D.J., Hay, S.I. & Packer, M.J. (1996) Predicting the distribution of tsetse flies in West Africa using temporal Fourier processed meteorological satellite data. *Annals of Tropical Medicine and Parasitology* 90: 225-241.

Sokal, R.R. & Rohlf, F.J. (1981) *Biometry.* Second edition.W.H. Freeman, New York.

Trapnell, C.G. & Griffiths, J.F. (1960) The rainfall-altitude relation and its ecological significance in Kenya. *East African Agricultural Journal.* 25: 207-213.

Tucker, C.J. (1979) Red and photographic infrared linear combinations for monitoring vegetation. *Remote Sensing of Environment* 8: 127-150.

Tyrrell, J. G. & Coe, M. J. (1974) The rainfall regime of Tsavo National Park, Kenya and its potential phenological significance. *Journal of Biogeography* 1: 187-192.

Washington, R. (1998) *Interannual and interdecadal variability of African rainfall.* Ph.D. thesis, School of Geography, University of Oxford.

CHAPTER 3

Soil geochemical mapping of Mkomazi

Peter W. Abrahams & Robert J. Bowell

As the discipline is viewed today, environmental geochemistry and health is concerned with the applications of geochemistry, and in particular geochemical mapping, to plant, animal and human health (Thornton 1993). For example, the sampling and analysis of stream sediments and soils from the south-west of England led to the production of geochemical maps which delineated extensive areas of copper and arsenic enrichment attributable to mineralisation and pollution from mining activity (Abrahams & Thornton 1987). These maps identified areas for further study, where the implications of the soil pollution for the health of cattle could be evaluated (Abrahams & Thornton 1993).

Although early research highlighted links between the geochemistry of soils and the nutritional status of plants and animals (e.g. Patterson 1938, Ferguson *et al.* 1943), the foundations of geochemistry and health were really established in the 1960s (Webb 1964). With the evolution of the discipline, increasing attention has focused on the links between geochemistry and health in the developing countries (Appleton *et al.* 1996), and some recent studies have included a consideration of the influence of geochemistry on wildlife nutrition. For example, working in Mole National Park (Ghana), Bowell & Ansah (1993) suggested a potential problem of cobalt deficiency which may affect wildlife. Similar research by Maskall & Thornton (1991) in Lake Nakuru National Park (Kenya) demonstrated the low blood copper status of impala. In this latter study, a reconnaissance geochemical survey identified the low soil copper concentrations of the Rift Valley floor which are likely to account, at least in part, for the depressed blood copper levels.

With the increasing importance of wildlife conservation and associated nature-oriented tourism in developing countries, there will be a continued need for managing protected areas which will include gaining further understanding of the mineral nutrient requirements of wildlife and identifying nutritional imbalances. In this respect, geochemical mapping will relatively rapidly and cheaply assess the nutrient status of wildlife areas and it is likely that such surveys will become more common in the future.

Our work in the Mkomazi Game Reserve comprised such a rapid survey and here, in order to illustrate the nature of the work, we report some results obtained from the analysis of soil samples collected throughout the reserve. The total concentrations of the major elements calcium (Ca), potassium (K), magnesium (Mg), sodium (Na) and phosphorus (P) are described, and their spatial distributions are mapped. In addition, the soil pH and organic matter content of the soil samples are reported.

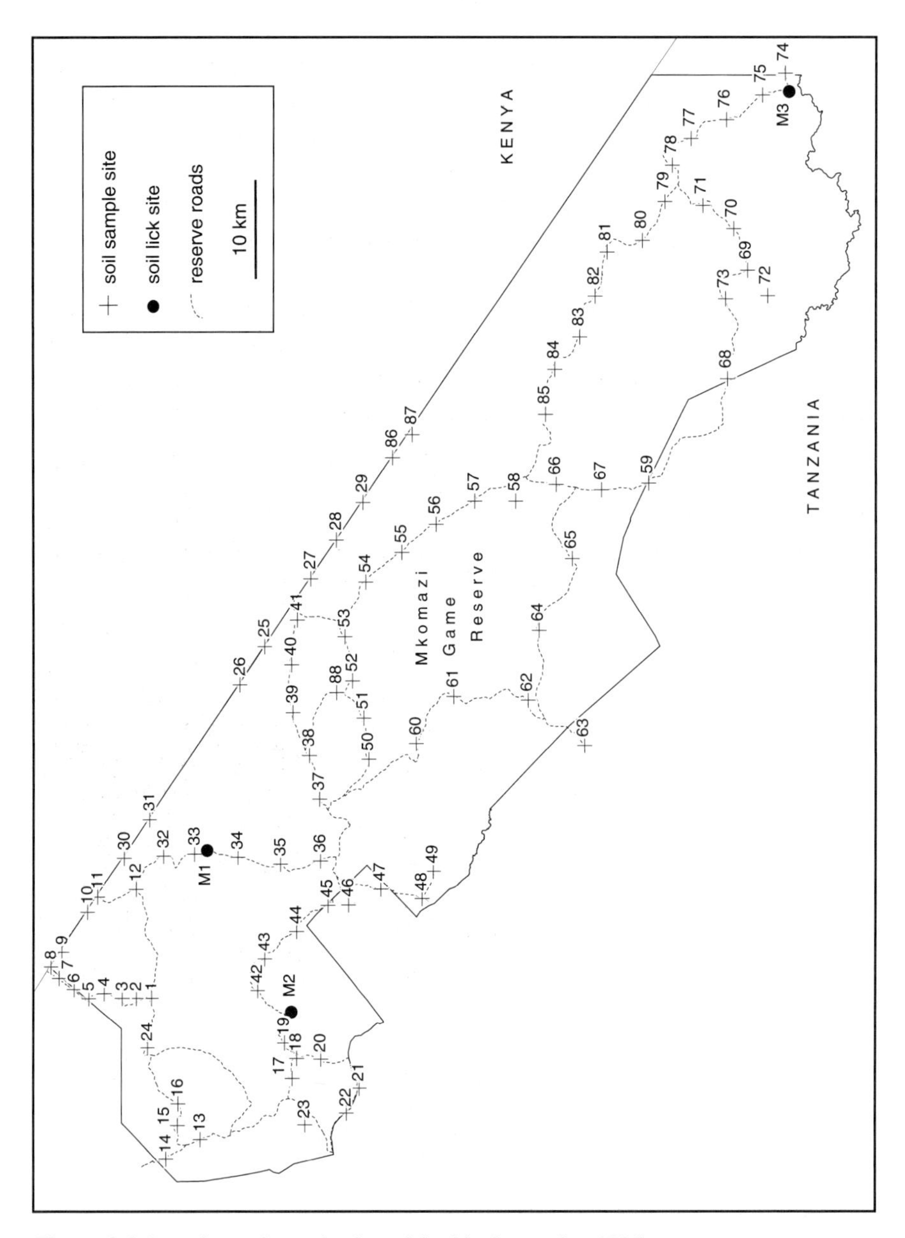

Figure 3.1 Locations of sample sites visited in September 1995.

Methodology

Field sampling

One of us (RJB) undertook an early survey of the reserve in September 1992 and April 1993, collecting water, vegetation, sediment, soil and rock samples. The results from much of this work, including details of the rocks and soils found within the reserve, will be reported elsewhere. As part of this survey, 80 surface soils (0–30 cm) were collected throughout the reserve. At each site, a composite of nine sub-samples taken from a 3 x 3 m grid was collected using a 5 cm Teflon-coated soil auger. Because of the lack of a detailed map of the reserve at the time of sampling, the location of these sites is not precisely known.

In September 1995, PWA undertook a rapid soil sampling programme. Using the reserve road network, 88 locations throughout the reserve were visited for sampling. A representative topsoil sample (0–15 cm) was collected from the surrounding savanna scrub at each location. This was achieved by establishing a 6 x 6 m plot at every location, and collecting and bulking the topsoil from the corners, mid-points and centre of each plot. Samples were made using a 5 cm diameter Edelman auger. During the survey, the 88 samples were made systematically at 2 or 5 km intervals along the reserve roads. In addition, sampling was undertaken whenever it was considered appropriate (Figure 3.1). The use of a global positioning system allowed grid co-ordinates to be established for every sample location.

In addition to the collection of the 88 topsoils in 1995, soil licks (otherwise known as mineral, natural or salt licks; Kreulen 1985) were sampled using a trowel. Perhaps surprisingly, bearing in mind the c. 3,250 km^2 area of the reserve, only three soil licks were known to the reserve authorities, all of which were sampled (Figure 3.1). Obvious use of these soils by game animals such as elephant was evident by tusk marks and 'mined' excavations (Figure 3.2).

Figure 3.2 *A bank of a seasonal river channel. Daniel Mafunde is standing in an excavation made by wildlife for geophagical (i.e. soil eating) purposes.*

Laboratory analysis

Soil samples were stored in Kraft paper bags in the field and on arrival at the UK laboratory were air-dried, disaggregated and passed through a 2 mm nylon mesh sieve. With the < 2 mm fine earth fraction, a 1:2.5 soil/water and soil/0.01M $CaCl_2$ suspension was used for the determination of pH by electrode and meter (Avery & Bascombe 1974).

An estimate of soil organic matter content was made following the ignition (375°C for 16 hours) of < 2 mm oven-dried (105°C for 16 hours) soil (Ball 1964).

Some selected soils were assessed for calcium carbonate content by observing the reaction of soil with a few drops of 10% hydrochloric acid (Hodgson 1976). The salinity of these and other soils was rapidly appraised by measuring the electrical conductance of soil solutions extracted from saturated soil pastes (Rowell 1994).

For elemental analysis, a portion of the < 2 mm soil was crushed to a fine powder in a ball mill. Following a nitric-perchloric acid digestion (Thompson & Wood 1982), soil concentrations of Ca and Mg were determined by atomic absorption spectrometry, K and Na by flame emission spectrometry, and P by spectrophotometric measurement of the vanadomolybdophosphoric yellow colour (Jackson 1962). Concentrations were recorded in units of mg/kg (i.e. parts per million). In addition to the samples collected by PWA, the surface soil samples collected by RJB were also subjected to these elemental analyses.

Because of the number of samples analysed, and laboratory limitations, each sample for elemental analysis was determined in one of six separate batches. Analytical quality control procedures were followed in order to appreciate the standard of these laboratory results and to ensure consistency in measurement between the batches of samples (Thompson 1983). All samples were randomised and anonymous to the analyst. The analysis of 'blanks' indicated negligible contamination problems.

Repeat analysis of a sample selected at random allowed for the determination of within-batch and overall variation. The coefficient of variation was the index used to assess these measurements of precision (Table 3.1). The inclusion of an appropriate (i.e. laterite) soil reference material into each batch of samples al-

Table 3.1 Average (median) within-batch variation and overall variation assessed by repeat analysis of a randomly chosen soil sample[1].

	element				
	Ca	K	Mg	Na	P
within batch variation[2]	11.2	6.4	5.6	10.2	9.0
overall variation[3]	11.8	23.1	14.1	15.3	57.8

[1] variation is assessed by computation of the coefficient of variation.
[2] the values presented are medians determined from the six batches of samples analysed.
[3] obtained from the repeat analysis of the random sample which was included in all six batches of samples.

lowed for the assessment of accuracy (Table 3.2). Reference to this table indicates that, excepting P, the nitric-perchloric acid digestion used for the determination of the elements underestimates to a varying degree the values recorded. The variable recovery of elements ranges from 41.9% (Na) to 79.3% (K). For P, the concentrations recorded from the soil reference material are approximately twice those of the certified value. The inclusion of another soil reference material (National Council for Certified Reference Materials (China), GSS-1: podzol) yielded a more satisfactory recovery of some 120% (J. Parsons, pers. comm.). There is some uncertainty, therefore, regarding the accuracy of analysis for P.

Results and discussion

Table 3.3 provides a summary of the data. Because there is a difference in the sampling procedures employed by PWA and RJB for the collection of the surface soils, the data are separated accordingly. Table 3.3 also provides an overall summary of the data by considering all of the samples.

A feature of the data is the positively skewed distribution of the measured variables. This is a common characteristic of data generated by geochemists. Excepting soil pH, the mean values of variables are appreciably higher than their respective median values. A relative measure of skewness can be made by calculation of Pearson's coefficient of skewness (*Sk*) for each variable. *Sk* will yield a value of zero if perfect symmetry is present. As long as *Sk* does not exceed ± 3, the skewness may be considered moderate (Croxton *et al.* 1967, cited in Earickson & Harlin 1994). In this study, the highest value of *Sk* equals 1.44 (Table 3.3). Whilst this index shows that none of the variables are strongly skewed, clearly the median is the most appropriate measure of central tendency to use in an appreciation of the data. Similarly, the inter-quartile range is an appropriate measure of dispersion, although Table 3.3 also presents the maximum and minimum values obtained for each variable.

Table 3.2 Certified values for the soil reference material and the mean (and standard deviation) concentrations determined from this sample following a nitric-perchloric acid digestion.

	element				
	Ca	K	Mg	Na	P
certified value	1144	1660	1568	549	1150
mean	569	1317	837	230	2491
standard deviation	143.9	105.1	101.6	22.8	496
% recovery	49.7	79.3	53.4	41.9	216.6

Concentrations are mg/kg. Soil reference material used: National Council for Certified Reference Materials (China), GSS-7: laterite.

Table 3.3 Descriptive statistics summarising the variables considered in this study. Major element concentrations are in mg/kg; na = not analysed.

		pH (water)	pH ($CaCl_2$)	% organic matter	Ca	K	Mg	Na	P
PWA's	*n*	88	88	88	88	88	88	88	86
samples	mean	6.5	5.8	2.4	5,522	3,487	3,980	442	1,335
	median	6.5	5.9	2.0	2,150	3,200	2,400	182	1,135
	interquartile range	6.0–7.1	5.1–6.4	1.7–2.5	730–5,475	2,440–4,335	1,281–4,350	140–326	800–1,625
	range	3.6–8.5	3.5–7.9	1.2–8.7	190–73,600	200–9,600	240–58,000	85–5,900	240–4,600
	Sk	0.10	-0.21	0.85	0.91	0.50	0.71	0.89	0.72
RJB's	*n*	na	na	na	80	80	80	80	78
samples	mean	–	–	–	9,032	3,009	12,768	885	2,237
	median	–	–	–	4,250	2,450	3,200	325	1,300
	interquartile range	–	–	–	1,950–10,250	1,115–4,780	1,962–6,400	220–691	737–3,663
	range	–	–	–	354–98,000	194–7,000	280–400,000	90–9,000	100–8,000
	Sk	–	–	–	1.05	0.85	0.54	0.95	1.44
combined	*n*	88	88	88	168	168	168	168	164
samples	mean	6.5	5.8	2.4	7,193	3,259	8,165	653	1,764
	median	6.5	5.9	2.0	2,880	3,080	2,950	250	1,200
	interquartile range	6.0–7.1	5.1–6.4	1.7–2.5	1,115–6,525	2,000–4,485	1,563–5,000	160–449	753–2,230
	range	3.6–8.5	3.5–7.9	1.2–8.7	190–98,000	194–9,600	240–400,000	85–9,000	100–8,000
	Sk	0.10	-0.21	0.85	1.03	0.29	0.42	0.84	1.10

Soil pH

Soil pH was measured only on PWA's freshly collected, air-dried samples. In ecological studies, pH measurements are generally carried out on slurries of the soil sample in distilled water (Allen *et al.* 1974). A measured pH value depends to some extent, however, upon the soluble salt content of a soil. In many soils, a high soluble salt content will be reflected by a lower apparent pH reading when measured in water. Consequently, differences in pH values between samples can be due to variations of soluble salts in soil-water mixtures, the so-called salt effect (for further discussion, see Cresser *et al.* 1993). To overcome any salt effect variations, it is common practice when measuring soil pH to add a dilute calcium chloride solution (i.e. a salt solution), rather than distilled water, to the soil. Typically, in many soils, pH measurements made in such a salt solution are some 0.5 pH units lower than those measurements made in a water suspension (Rowell 1994).

Bearing in mind the semi-arid nature of the Mkomazi environment, where rainfall may be insufficient to remove soluble salts, it may be important in the pH analysis of soils to take account of a salt effect. In this study, therefore, the soil pH was measured in both distilled water and in a 0.01 M $CaCl_2$ suspension, which standardises pH measurement by providing a constant salt concentration. This was done in spite of the fact that measurement of the electrical conductance (EC) of selected Mkomazi soil solutions extracted from soil pastes indicated that only one sample (from site number 69) can actually be described as saline (with an EC > 4 mS/cm).

Reference to Table 3.3 indicates that the average soil in Mkomazi is slightly acidic (e.g. mean pH in water suspension = 6.5 units; in $CaCl_2$ suspension = 5.8 units). There is, however, considerable variability in the pH of soils within the reserve, with soils ranging from what can be classified as strongly acidic to alkaline in reaction (Figure 3.3). Table 3.4 shows that slightly acid soils are the most abundant, and that overall more than 50% of the soils can be described as acidic. The remaining soils are classed as either neutral or alkaline. For a soil to have a

Table 3.4 Interpretation of soil pH analysis, and frequency of pH measurements made in distilled water and 0.01M $CaCl_2$ suspensions

soil pH measurement			
in water		in $CaCl_2$	
interpretation	*n*	interpretation	*n*
strongly acid (<4.5)	2	strongly acid (<4.0)	2
moderately acid (4.5–5.5)	11	moderately acid (4.0–5.0)	18
slightly acid (5.6–6.5)	32	slightly acid (5.1–6.0)	34
neutral (6.6–7.5)	30	neutral (6.1–7.0)	24
alkaline (>7.5)	13	alkaline (>7.0)	10

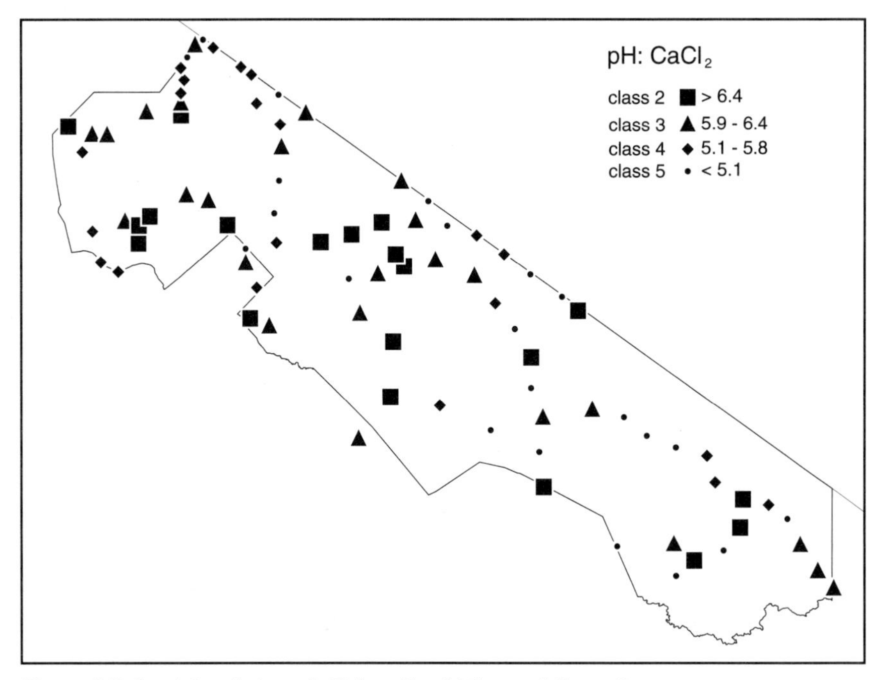

Figure 3.3 Spatial variation of pH in soils of Mkomazi Game Reserve.

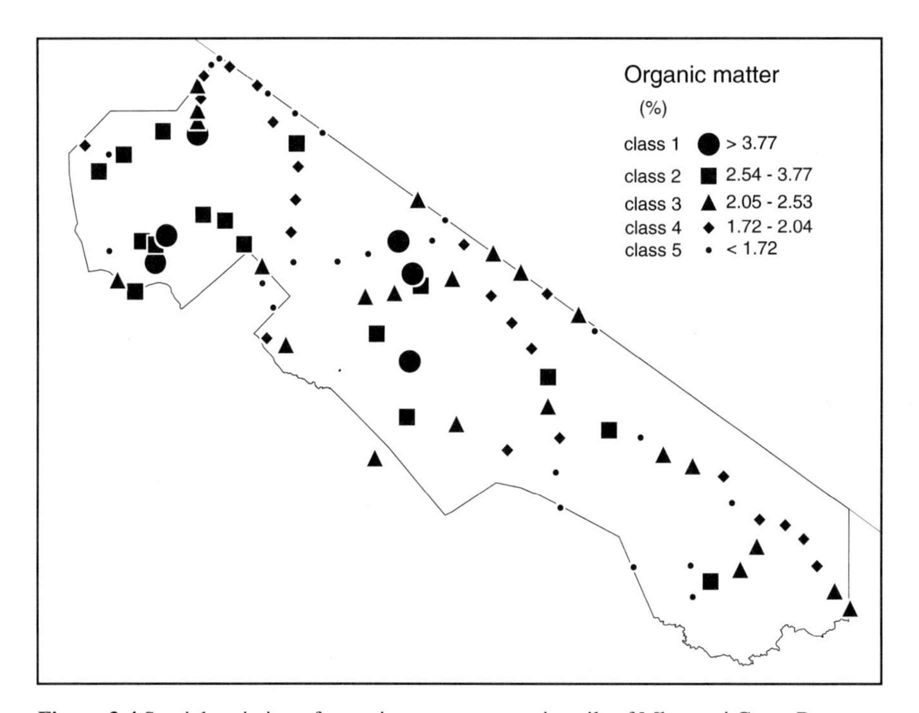

Figure 3.4 Spatial variation of organic matter content in soils of Mkomazi Game Reserve.

pH above 7, it must either be calcareous (containing calcite, $CaCO_3$), dolomitic (containing dolomite, $CaCO_3.MgCO_3$) or sodic (containing Na_2CO_3) (Rowell 1994). According to Brady (1990), the pH of sodic soils is always above 8.5, which is greater than any pH measurement determined in this study. An estimate of calcium carbonate content can be determined by observing the reaction of soil with dilute HCl (note: this method is less reliable for dolomite). All the alkaline soils found in this study reacted with HCl, and they can be classified as ranging from very slightly calcareous to very calcareous soils. Clearly, this information indicates the importance of carbonate minerals in maintaining the alkaline reaction of these soils.

Soil organic matter

As with the pH measurements, the soil organic matter content was only determined on those samples collected by PWA. The loss-on-ignition procedure used to assess the organic matter content of the soil samples yields an approximate measure of this variable. With a median content of 2% organic matter (and an inter-quartile range of 1.7%–2.5%; Table 3.3), the majority of the soil samples have a very low amount of this important soil constituent. These amounts are, however, typical of soils in a savanna environment. The spatial variation in organic matter content throughout the reserve (Figure 3.4) indicates that there are particular areas of higher content. The amount and the distribution of organic matter in savanna soil is related both to the total quantity of organic residues available for decomposition and the nature and source of those residues. Ahn (1993) argues that relative to tropical forest areas, in savanna regions the amount of plant material that is available to form humus is much less because:

- the above-ground plant parts are grazed by animals;
- dead plant material in the dry season is subjected to fire and is therefore converted to ash rather than being available for humus formation; and
- the total weight of savanna vegetation is usually only a fraction of the weight of forest vegetation.

Ahn (1993) also observes the importance of roots and their contribution to humus in savanna soils. This typically results in a more or less even spread of organic matter down through the upper 100 cm of the soil, in contrast to any topsoil concentration as found in forest soils where the main source of organic matter is the surface litter layer.

Major element concentrations

In an initial appreciation of the major element concentrations determined from the soil samples, it is apparent from Table 3.3 that there is a difference in the concentrations recorded from PWA's and RJB's samples. Taking Ca as an example, a median concentration of 2,150 mg/kg is evident from PWA's samples, which contrasts with the median of 4,250 mg/kg determined from the samples collected by RJB. To investigate this further, the major element concentrations from the two sets of data were subjected to the non-parametric Mann-Whitney *U* test. The null hypothesis, that there is no significant difference between the median major element concentrations determined from PWA's and RJB's samples, is rejected at a 95% confidence level for the elements Ca, Mg and Na (concentrations are significantly different) but it is accepted for K and P (concentrations do not differ significantly). Based on these results, it can be concluded that PWA's and RJB's samples come from different sample populations of soil. It should be borne in mind that the two sets of samples were collected using slightly different protocols (e.g. sample depth). Also, in an attempt to survey the reserve as fully as possible, PWA's sampling tended to be biased towards areas not covered in detail by RJB. Such factors may contribute to the significant differences detected.

Ideally, in an appreciation of the geochemical data, a comparison with previously published studies undertaken in a similar environment to Mkomazi is desirable. There are in reality many problems in making such comparisons. Reasons for this are highlighted by Plant *et al.* (1996). First, modern geochemical data are rarely available for such environments, or may be inappropriate for comparative environmental purposes having been collected principally for mineral exploration surveys. Second, those *ad hoc* studies undertaken have used a range of methodologies, which vary according to short-term goals and the practices established in the country or organisation doing the research. Third, previous studies have often been limited in the number of samples collected and the range of elements determined with much emphasis being centred on the so-called potentially harmful elements such as arsenic, cadmium, lead and mercury.

With these difficulties in mind, the data generated in this study are considered without reference to other data sets derived from work undertaken in a savanna environment. Reference is made in the following paragraphs, however, to the data generated for the compilation of *The Soil Geochemical Atlas of England and Wales* (McGrath & Loveland 1992). This publication is based on the analysis of a considerable number of soils (5,692 in total) collected from 0–15 cm depth. A large number of major and trace elements were determined, and the analytical and statistical methods used are similar to those employed in the Mkomazi research.

A useful technique for summarising geochemical data is the production of box-and-whisker plots (also known as boxplots; McGrath & Loveland 1992). This procedure divides the distribution of each element into quartiles, firstly by finding the median, and then doing the same for each of the remaining halves. These upper and lower points or 'hinges', respectively known as the upper quartile and lower quartile, define the central box shown in Figure 3.5. 'Whiskers' are drawn from the ends of the box, each extending 1.5 times the so-called interquartile range, towards the maximum and minimum values. Any values outside the whisk-

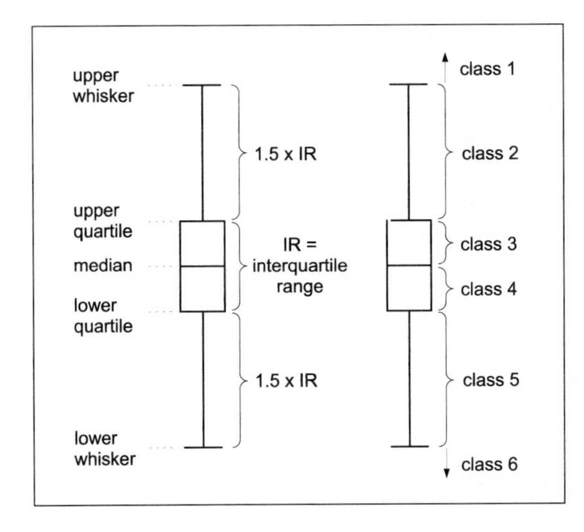

Figure 3.5 Boxplot analysis used for (left) summarising the results for concentrations of elements in soils, and (right) to set classes for use in producing maps of the data.

ers are defined as outliers. This procedure allows the concentrations of each element to be divided into six classes (Figure 3.5). The technique was applied to the major elements discussed in this paper (and also the soil pH and organic matter data). Since PWA's samples are the only ones with a precise known geographical location, the boxplot procedure was applied only to the data derived from these samples. Accurate maps showing the spatial distribution of the elements, soil pH and organic matter content following boxplot transformation could then be produced (Figures 3.3 to 3.4 and 3.6 to 3.10).

For the five major elements considered in this appraisal, no anomalous low values were recorded (i.e. no concentration was placed into Class 6 by the boxplot procedure). In comparison, some element concentrations were assigned to the Class 1 category, and in this respect they may be regarded as anomalous and high relative to the remaining sample concentrations. For example, for Na (Fig 3.6), some 13 soil samples have concentrations that fall within the Class 1 category. Most of these sites are located several kilometres east of Zange Gate (Figure 3.6). For the remaining elements, three K, eight Ca, six Mg and five P concentrations were placed into the Class 1 category (Figures 3.7 to 3.10).

It is of interest to note that when the boxplot classifications are looked at overall, the limits that divide the concentrations of elements into their respective classes are broadly similar to those derived from the comprehensive soil geochemical survey of England and Wales undertaken by McGrath & Loveland (1992). Such a finding is perhaps surprising bearing in mind the contrasting environments (and soils) of Mkomazi and the UK, but it seems reasonable to conclude that these major element concentrations recorded within Mkomazi from PWA's samples can be regarded as typical for soils. The exception to this finding, however, is P, where the classification limits are higher for the Mkomazi data. One factor which may be contributing to this observation is that the quality control procedures used in the

Table 3.5 Spearman rank correlation coefficients determined for the paired variables investigated in this study.

	Na	K	Ca	Mg	P	pH ($CaCl_2$)	% OM
Na	–						
K	0.093	–					
Ca	0.777†	0.214*	–				
Mg	0.709†	0.242*	0.879†	–			
P	0.221*	0.141	0.233*	0.195	–		
pH ($CaCl_2$)	0.701†	0.357†	0.915†	0.830†	0.164	–	
% OM	0.348†	0.089	0.460†	0.383†	0.204	0.421†	–

* significant at 95% confidence level

† significant at 99% confidence level; OM = organic matter

analysis indicate that the P concentrations are being possibly overestimated (see *Methodology*).

To assess the similarity of the geochemical maps for the five major elements, Spearman's rank correlation coefficients were determined. This statistic provides a measure of the strength of the relationship between paired rankings of the chemical variables (Earickson & Harlin 1994). The correlation matrix compiled following the computation of the coefficients is presented in Table 3.5. This matrix indicates three groups of elements. Firstly, for the paired elements Na/Ca, Na/Mg and Ca/Mg, moderately strong correlation coefficients were determined. These are significant at the 99% confidence level. Secondly, for the paired elements Na/P, K/Ca, K/Mg and Ca/P, significant though weak correlation coefficients were determined. These are significant at the 95% confidence level. Thirdly, for the paired elements Na/K, K/P and Mg/P, no significant correlation was evident. The implications of this statistical treatment are that the geochemical maps for the elements Ca, Mg and Na should look broadly similar. This is indeed the case with, for example, many of the highest concentrations of these three elements (i.e. Class 1 and Class 2 concentrations) being found several kilometres east of Zange Gate. In comparison, a cluster of Class 4 and Class 5 concentrations of these three elements are found in the north-west corner of Mkomazi, and many such concentrations dominate the centre and eastern half of the reserve.

The estimates of soil pH and organic matter content determined here were also included in the correlation analysis (Table 3.5). A particularly strong correlation ($r_s = 0.915$) is evident between Ca and the pH of soils. The importance of calcium carbonate in maintaining alkaline soil conditions in parts of the game reserve has already been indicated. Moderately strong correlation coefficients also exist, however, between the variables Mg/pH and Na/pH, whilst a moderately weak (though significant at the 99% confidence level) correlation exists between K/pH. In exchangeable form, the ions Ca^{2+}, Mg^{2+}, K^+ and Na^+ (collectively referred to as base cations) do have a positive association with pH. The evidence of the correlation analysis suggests that Ca is the dominant cation in this positive association within the soils of the reserve. As with any correlation analysis, care is needed in interpreting correlation coefficients. The existence of a high correlation between two variables does not necessarily mean that one is causing the other, and sometimes correlations occur because the variables are linked through a third variable. For example, although Na correlates moderately strongly with soil pH, this may be because both variables are associated with Ca. Because of the significant positive correlations observed especially between the variables Na/pH, Ca/pH and Mg/pH, it is not surprising that the maps displaying the spatial variations of these soil chemical constituents are similar (Figures 3.3, 3.6, 3.8 and 3.9).

The map showing the spatial variations of soil pH in the reserve (Figure 3.3) indicates that no anomalous pH values were recorded in this study, since no pH values were assigned to Class 1 or Class 6 categories. In contrast, the map displaying the spatial variations of soil organic matter within the reserve (Figure 3.4) indicates that six soils have organic matter contents within the Class 1 category. These values are only anomalous in the context of the reserve, since the highest organic matter content (8.7%, sample 39) can hardly be considered unusual if considered from a global perspective. Reference to the correlation matrix (Table 3.5) indicates a number of significant (though moderately weak) correlation coefficients linking the organic matter content of soil to the total concentrations of various elements and the soil pH.

Soil licks

The soil licks have a distinctive and variable composition relative to the 168 soils collected for the reconnaissance geochemical survey (Table 3.6). Lick soil M2 was the most alkaline soil found within the reserve, and had the highest Na and second-highest Ca (but lowest K and P) concentrations (the soil can be described as highly calcareous and moderately saline). Only one sample collected for the

Table 3.6 Soil pH, organic matter content and major element concentrations determined from the three soil licks. Major element concentrations are in mg/kg. OM=organic matter.

site	pH (water)	pH ($CaCl_2$)	% OM	Ca	K	Mg	Na	P
M1	4.3	3.8	1.6	280	2000	940	120	1250
M2	11.0	10.7	1.0	91000	160	8000	12600	<100
M3	8.2	8.0	1.1	7450	3300	5400	5040	1600

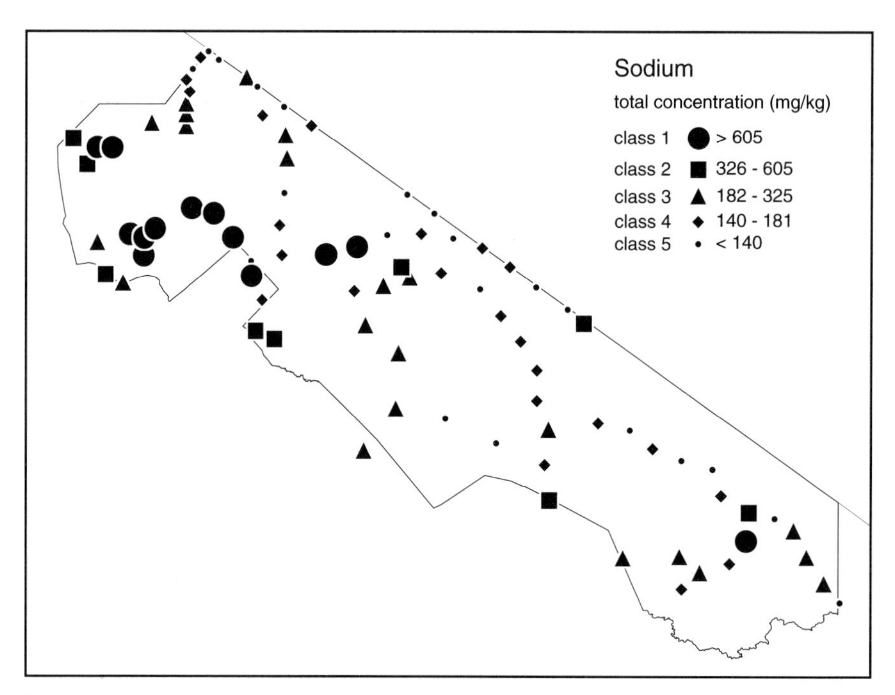

Figure 3.6 Spatial variation of sodium (Na) in soils of Mkomazi Game Reserve.

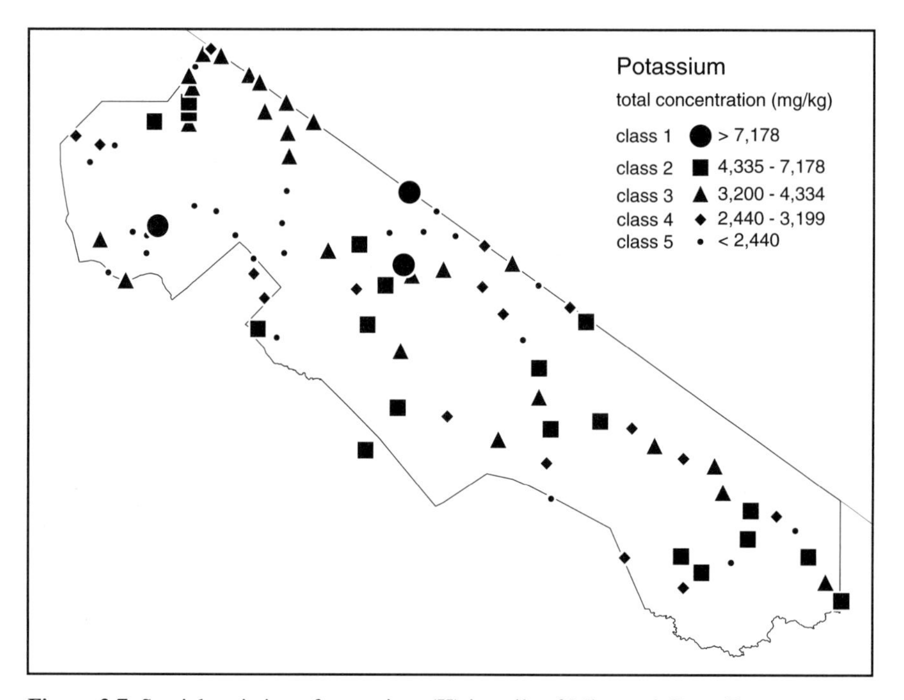

Figure 3.7 Spatial variation of potassium (K) in soils of Mkomazi Game Reserve.

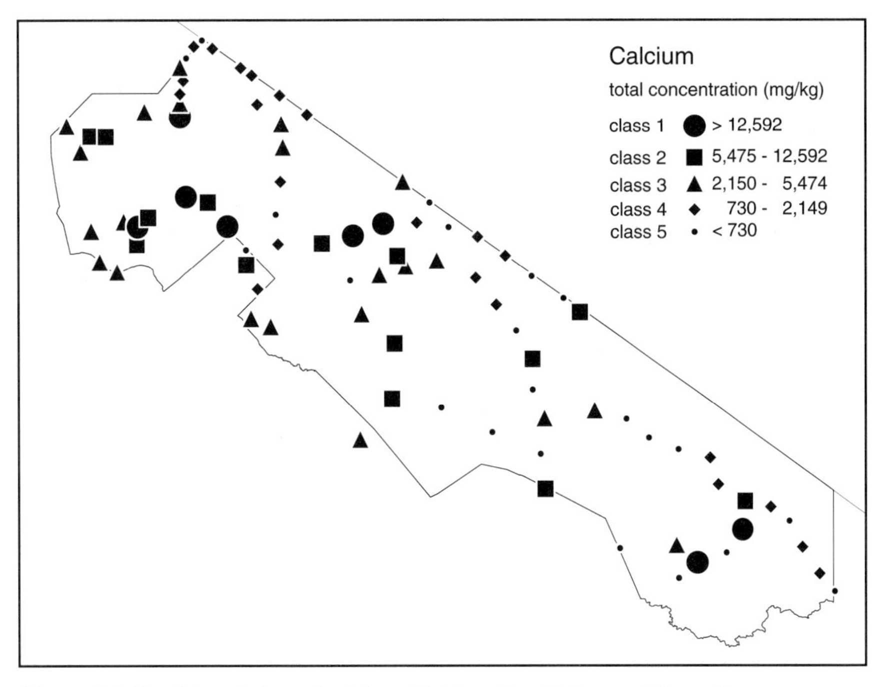

Figure 3.8 Spatial variation of calcium (Ca) in soils of Mkomazi Game Reserve.

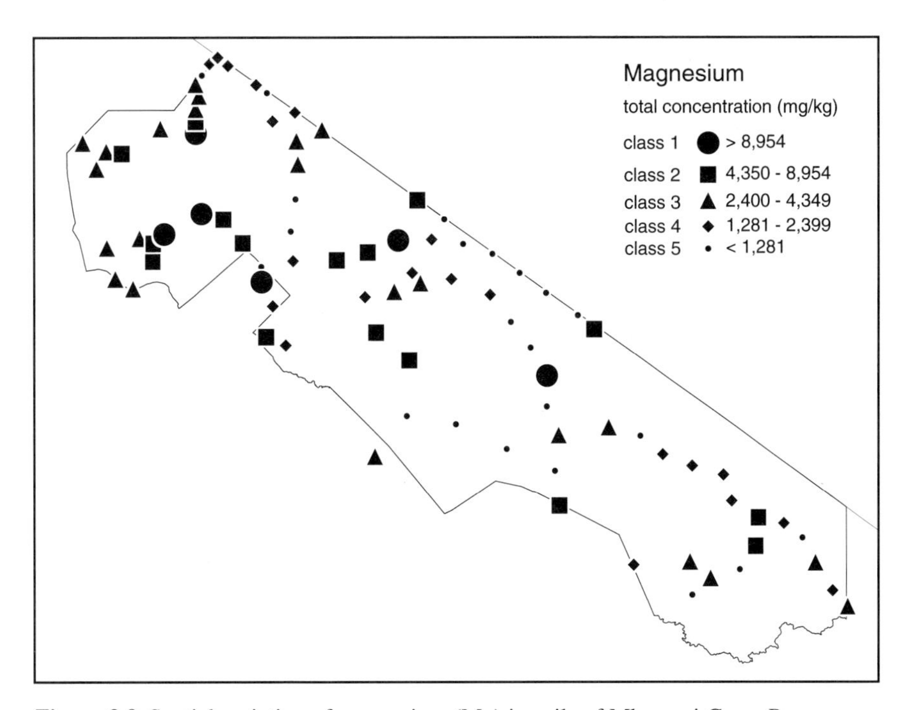

Figure 3.9 Spatial variation of magnesium (Mg) in soils of Mkomazi Game Reserve.

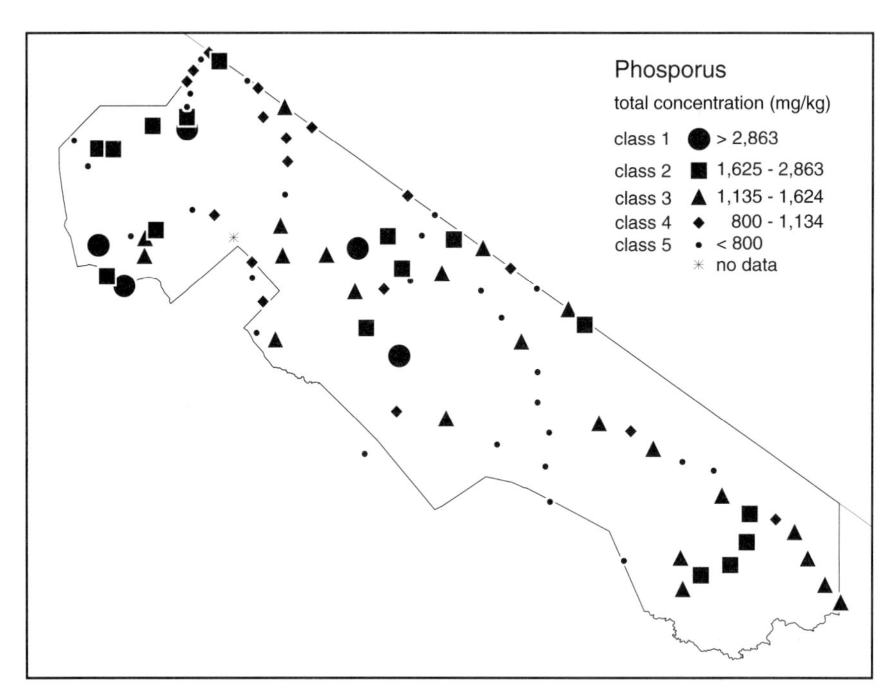

Figure 3.10 Spatial variation of phosphorus (P) in soils of Mkomazi Game Reserve.

geochemical survey was more acidic than M1, and the latter has comparatively low concentrations of the bases Ca, K, Mg and Na. M3 can be described as slightly calcareous, and was one of the most alkaline soils found in Mkomazi. Only two reconnaissance samples contained more Na than M3.

It is not immediately obvious why the lick soils are so distinctive in their composition, although the very low amounts of organic matter associated with all three samples can be explained by the fact that the animals were exploiting, to a large extent, subsoil material (whilst the soils collected for the geochemical survey were topsoils collected from 0–15 cm depth.). The reason why animals are deliberately exploiting these lick soils may be because of their distinct chemical composition.

The classical view is that the function of soil ingestion is to provide essential nutrients to the animal. In particular, Na is the element associated with this argument. However, while M2 and M3 are enriched in this element relative to the majority of soils found in the reserve, the low Na concentration associated with M1 indicates that this element cannot uniquely explain soil ingestion. A comprehensive review of the literature by Kreulen (1985) concludes that lick soils can vary greatly in their composition, can be ingested deliberately by animals for a multiplicity of purposes and can supply animals with different substances (e.g. essential nutrients, clay minerals) to satisfy these purposes. The reasons why animals within Mkomazi are exploiting these soils remains unclear but further research is currently being undertaken on this important aspect of animal behaviour.

Conclusions

The topsoils of Mkomazi vary considerably in terms of their reaction (from strongly acid to alkaline) but typically contain low amounts of organic matter. A comparison of the total major nutrient concentrations of soils from Mkomazi and England/Wales indicates that the soils are essentially similar despite the contrasting environments. In this respect the nutrient concentrations of Mkomazi soils may be regarded as typical. Statistically significant associations are evident between the determined variables, which can be recognised visually by comparison of the geochemical maps.

The purpose of this paper is primarily to describe the nature of our work undertaken at Mkomazi. The soil samples have also been analysed for a number of potentially important micro-nutrients and trace elements. For virtually all elements, both total and bioavailable (extractable) concentrations have been determined. When details of this complete data set are published, some 27 maps will be available for use by research workers interested in the mineral nutrient status of soils in Mkomazi. These data will provide important baseline environmental information in a subject area which has received little attention to date. Such information may highlight potential problems of nutritional imbalance in the plants and animals of the reserve. For example, preliminary consideration of the data for Cu indicates that many soils within Mkomazi may contain insufficient amounts of this element to meet the dietary requirements of some animals. Geochemical maps may also be of use in explaining any observed patterns of biodiversity within Mkomazi.

References

Abrahams, P.W. & Thornton, I. (1987) Distribution and extent of land contaminated by arsenic and associated metals in mining regions of southwest England. *Transactions of the Institution of Mining and Metallurgy (Section B: Applied Earth Science)* 96: B1-B8.

Abrahams, P.W. & Thornton, I. (1993) The contamination of agricultural land in the metalliferous province of southwest England: implications to livestock. *Agriculture, Ecosystems and Environment* 48: 125-137.

Ahn, P.M. (1993) *Tropical Soils and Fertiliser Use.* Longman, Essex.

Allen, S.E., Grimshaw, H.M., Parkinson, J.A. & Quarmby, C. (1974) *Chemical Analysis of Ecological Materials.* Blackwell, Oxford.

Appleton, J.D., Fuge, R. & McCall, G.J.H. (eds.) (1996) *Environmental Geochemistry and Health.* The Geological Society, London.

Avery, B.W. & Bascomb, C.L. (1974) *Soil Survey Laboratory Methods.* Bartholomew Press, Dorking.

Ball, D.F. (1964) Loss-on-ignition as an estimate of organic matter and organic carbon in non-calcareous soils. *Journal of Soil Science* 15: 84-92.

Bowell, R.J. & Ansah, R.K. (1993) Trace element budget in an African savannah ecosystem. *Biogeochemistry* 20: 103-126.
Brady, N.C. (1990) *The Nature and Properties of Soils.* Macmillan, New York.
Cresser, M., Killham, K. & Edwards, A. (1993) *Soil Chemistry and its Applications.* Cambridge University Press, Cambridge.
Earickson, R.J. & Harlin, J.M. (1994) *Geographic Measurement and Quantitative Analysis.* Macmillan, New York.
Ferguson, W.S., Lewis, A.H. & Watson, S.J. (1943) The teart pastures of Somerset. 1. The cause of teartness. *Journal of Agricultural Science, Cambridge* 33: 44-51.
Hodgson, J.M. (ed.) (1976) *Soil Survey Field Handbook.* Bartholomew Press, Dorking.
Jackson, M.L. (1962) *Soil Chemical Methods.* Constable, London.
Kreulen, D.A. (1985) Lick use by large herbivores: a review of benefits and banes of soil consumption. *Mammal Review* 15: 107-123.
Maskall, J.E. & Thornton, I. (1991) Trace element geochemistry of soils and plants in Kenyan conservation areas and implications for wildlife nutrition. *Environmental Geochemistry and Health* 13: 93-107.
McGrath, S.P. & Loveland, P.J. (1992) *The Soil Geochemical Atlas of England and Wales.* Blackie, London.
Patterson, J.B.E. (1938) Some observations on a disease of sheep on Dartmoor. *Empire Journal of Experimental Agriculture* 6: 262-267.
Plant, J.A., Baldock, J.W. & Smith, B. (1996) The role of geochemistry in environmental and epidemiological studies in developing countries: a review. In: Appleton, J.D., Fuge, R. & McCall, G.J.H. (eds.) *Environmental Geochemistry and Health.* The Geological Society, London. pp. 7-22.
Rowell, D.L. (1994) *Soil Science: Methods and Applications.* Longman, Essex.
Thompson, M. & Wood, S.J. (1982) Atomic absorption methods in applied geochemistry. In: J.E. Cantle (ed.) *Atomic Absorption Spectrometry.* pp. 261-284.
Thompson, M. (1983) Control procedures in geochemical analysis. In: Howarth, R.J. (ed.) *Statistics and Data Analysis in Geochemical Prospecting.* pp. 39-58.
Thornton, I. (1993) Environmental geochemistry and health in the 1990s: a global perspective. *Applied Geochemistry*, Supplementary Issue No. 2: 203-210.
Webb, J.S. (1964) Geochemistry and life. *New Scientist* 23: 504-507.

CHAPTER 4

Ecological mapping of a semi-arid savanna

Michael J. Packer, Susan M. Canney, Nicholas C. McWilliam & Raphael Abdallah

Ecological mapping is largely concerned with describing plant and animal distributions. It is also concerned with understanding the biological, environmental and anthropogenic driving factors and processes that shape observed distribution patterns. In this chapter we place our ecological mapping work in a biodiversity conservation context, outline our methodological approach to mapping the biodiversity of Mkomazi and discuss examples of on-going research to map vegetation cover and large mammal distributions in Mkomazi.

Ecological mapping in context

Mapping biodiversity is an increasingly important part of the inventory process, defined by Stork & Samways (1995) as "the surveying, sorting, cataloguing, quantifying and mapping of entities such as genes, individuals, populations, species, habitats, biotopes, ecosystems and landscapes or their components, and the synthesis of the resulting information for the analysis of processes". Inventorying provides basic data needed for understanding the distribution and dynamics of biodiversity. Thorough inventories are used to provide baseline information for studying change, which subsequent monitoring (repeated and more limited inventories) seeks to assess. Inventorying and monitoring are therefore essential in efforts to conserve and sustainably use biodiversity.

There are many policy frameworks that offer guidance for research into biodiversity that is relevant to its management. The international *Convention on Biological Diversity* (see Johnson 1993) and the national *Wildlife Policy of Tanzania* (MNRT 1998) are especially relevant here.

Convention on Biological Diversity

The *Convention on Biological Diversity* is, despite its shortcomings, considered to be the most comprehensive and useable framework available for tackling issues of conservation and development. Tanzania became a Party to the Convention in

February 1996, thereby making a long term commitment to its objectives, which are defined in Article 1 as "...the conservation of biological diversity, the sustainable use of its components, and the fair and equitable sharing of the benefits arising from...utilisation". The Convention sets out some 20 core Articles for achieving its objectives. Of particular relevance to this chapter is Article 7, which commits Parties to the *identification and monitoring* of key components of biodiversity. Activities under Article 7 are relevant to other key Articles: developing strategies (Article 6); *in situ* conservation (Article 8); *ex situ* conservation (Article 9); sustainable use of biodiversity (Article 10); environmental assessment (Article 14); and, access to genetic resources (Article 15) (Johnson 1993). The Convention is considered further in Chapter 35.

Wildlife Policy of Tanzania

The Wildlife Policy of Tanzania, published in March 1998, sets out an agenda for conserving biodiversity in ways which attend also to national and local development needs. According to the Policy, achieving its goals requires the development of the protected area (PA) network and the involvement of all interested groups (notably local communities and the private sector) in management planning and action. Developing the PA network will involve the establishment of a new category of PA called Wildlife Management Areas: within these, conservation activities will be managed entirely by and for the benefit of local people through community-based conservation programmes. It might also entail an increase in the number of PAs or a change of management practice in particular PAs.

Achievement of the Policy's goals and international co-operation objectives will involve implementing 16 strategies. These range from "Protecting biological diversity", through "Integrating wildlife conservation and rural development" to "Wildlife conservation education awareness". The Policy recognises that wildlife management planning and action must be based on sound scientific knowledge. This need is captured in its "Wildlife research and monitoring" strategy (MNRT 1998: 3.3.13) which promotes both basic socio-economic and ecological research, and research focused on providing answers to management questions.

Large area ecological mapping

Ecological mapping uses ecological, environmental and anthropological data. The reliability of the derived maps depends on the quality and suitability of these data for a particular purpose. More commonly than might be expected, data used in such mapping exercises are unreliable for a variety of reasons, and can result in suboptimal management decision-making (Conroy & Noon 1996).

Mapping at the scale of a typical protected area in Tanzania presents both logistical and scientific challenges. Constraints of time, expertise, finance and scientific techniques limit the spatial and temporal extent of such studies, as well as the nature of activities that can be undertaken. Recent advances in geographical information system (GIS) and satellite remote sensing technologies provide, however, great potential for developing suitable methodologies. The major constraint has been the lack of reliable field data (known as 'reference' or 'ground-truth' data) with which to test and refine these capabilities (Davis *et al.* 1991).

Using spatial data

A GIS is a database system in which data are spatially referenced to particular geographical co-ordinate systems. GISs are computer-based tools for the storage,

Box 1—Essentials of geographical information systems

Collecting and using information about the environment requires an appropriate representation (model) of the real world. The nature of the model determines the types of data that can be used and the spatial analyses that can be performed.

Spatial features on Earth can be viewed as *discrete entities* (having specific properties including a location) or *continuous fields of variation* (having no clearly located boundaries). Such features may be characterised in a GIS by either a series of points, lines or polygons, which are termed *vector* representations; or as regular grids containing values of continuous spatial variables, which are termed *raster* data.

Spatial data can exist in *analogue* or *digital* formats (paper maps, tables, aerial photographs, digital vector data, satellite images) and are likely to be of variable accuracy. Data gathered for a study are increasingly created in a digital format. Secondary data (those available prior to the start of a study) are commonly in analogue format, typically as maps (for instance of vegetation, animal species distribution, elevation, rivers). Analogue data need to be digitised for integration into a GIS. Increasingly, secondary data are available in digital format.

GIS data needs to be registered to a common geodetic co-ordinate system (known as *georeferencing*). Georeferencing may involve the use of ground control points in conjunction with accurate maps and global positioning systems (GPSs).

The completed database can be used simply for management, retrieval and display of data or in more complex ways—for example for deriving new data (such as interpolating point data to produce elevation models or climate surfaces) and for spatial modelling. Various capabilities of GIS are particularly useful in analysis: geometric conversion; proximity analysis; overlay analysis; vector to raster conversion; network analysis; classification.

The results can be presented as maps. The model used at the outset to represent reality influences the nature of data gathered and types of analysis performed, and is influenced by user discipline and culture. This bias in the description or explanation of spatial phenomena or processes ultimately influences the resulting graphical depiction of a new aspect of reality, a consideration which must be borne in mind when interpreting results.

integration, manipulation, analysis and display of both spatial and associated non-spatial data sets. An outline of the essentials of GIS is given in Box 1, while Burrough & McDonnell (1998) provide a detailed treatment of the subject.

The rapid, user-driven evolution of GISs has resulted in powerful tools for assisting efforts to understand spatial problems particularly those relating to managing natural resources, such as in the identification of areas for conservation (e.g. see Miller 1994, Freitag *et al.* 1998). There are still, however, many analytical challenges to be addressed in the further development and application of GIS technology. For example, there is a need to define methods for the combination of data of different resolutions so that, for instance, fine spatial/coarse temporal satellite imagery can be combined with that of coarse spatial but fine temporal resolution. There is also a need to be able to combine data from very different sources such as the natural and social sciences (Burrough & McDonnell 1998), as well as to develop intrinsic analytical and modelling capabilities.

Table 4.1 Temporal, spatial and spectral characteristics of imagery from operational satellite-borne sensors. *Temporal resolution* refers to the frequency with which a particular site is imaged. *Spatial resolution* indicates the smallest area for which data can be gathered (with the width of the image swath in parentheses). *Spectral resolution* is given as the number of wavebands (sensor channels) sensed and the extent of the EMS sensed (wavelength range in μm, where 0.3 is at the visible and 14 the thermal infrared part of the spectrum).

	Resolution characteristics		
Satellites & sensors	Temporal	Spatial	Spectral
National Oceanic and Atmospheric Administration (NOAA)			
Advanced Very High Resolution Radiometer (AVHRR)	12 h	1,000 m (3,600 km)	Ch 1–5 (0.58–11.50)
Landsat series [a]			
Multi-Spectral Scanner (MSS)	18/16 d	79/82 m [b] (185 km)	Ch 4–7 (0.50–1.10)
Thematic Mapper (TM)	16 d	30 m (185 km)	Ch 1–5 & 7 (0.45–2.35)
		120 m	Ch 6 (10.40–12.50)
Meteosat series			
High Resolution Radiometer (HRR)	0.5 h	2.5 km ([c])	Ch 1 (0.40–1.10)
		5 km	Ch 2 & 3 (5.70–12.50)
Earth Observing System (EOS) [d]			
MODerate-resolution Imaging Spectroradiometer (MODIS)	1-2 d	250 m (2,330 km)	Ch 1 & 2 (0.620–0.876)
		500 m	Ch 3–7 (0.459–2.115)
		1,000 m	Ch 8–36 (0.405–14.385)

[a] MSS - on Landsat 1 to 5; TM - on Landsat 4 and 5.

[b] Spatial resolution is 79 m for Landsats 1–3 and 82 m for Landsats 4–5 with temporal resolutions shown.

[c] This satellite is geostationary, viewing Africa and Europe, so the swath width is not applicable.

[d] TERRA (EOS-AM1), the first EOS satellite, with the MODIS is due for launch in August 1999.

Box 2—Principles of satellite remote sensing

Remote sensing concerns the acquisition of data about distant objects. As we search for a familiar face among a nearby crowd, our eyes are sensing light of different wavelengths and intensities that has been reflected from all the objects in our field of view. Similarly, satellite-borne multispectral scanners measure radiation in various regions of the electromagnetic spectrum (EMS) which emanates from surface features on Earth. Because different objects reflect and emit radiation differentially throughout the EMS, we can characterise them with a distinctive (although not always unique) 'spectral response pattern' or 'spectral fingerprint'.

'Passive' remote sensing is based on the detection of sunlight (typically from the visible to thermal infra-red part of the EMS, 0.4–14 μm) reflected from Earth. This radiation can be recorded onboard a satellite, either using film sensitive to a particular wavelength or using filters and arrays of electronic sensors that respond to particular values that tell us something about conditions at specific times and locations on the surface of Earth. This process takes into account the relative movement of the satellites and Earth; optical and geometric effects; incoming solar radiation; atmospheric scattering and distortion; and, finally, properties of the surface of Earth itself.

Satellite data may be statistically clustered on the basis of the spectral similarity of pixels across an area of interest. Classes determined by such an 'unsupervised classification' often relate to meaningful features on the ground (although they may not). The resulting classified image can be used to guide fieldwork. Finally, when reliable, spatial information about the distribution of natural resources is available for part of an area, it may be used to inform or 'train' the classification process in order to produce a 'supervised classification' of the whole area for which satellite imagery is available; in effect it can be used to predict the distribution of the resource (vegetation cover, for example).

Satellite remote sensing

Satellite remote sensing generates data about features on the surface of Earth. The usefulness of these data depends on the properties of the sensor. Table 4.1 details key characteristics of the imagery available from selected satellite-borne sensors. An outline of the principles of satellite remote sensing, including how satellite data are generated and used, is given in Box 2. Lillesand & Kiefer (1994) provide a detailed treatment of the subject.

The potential environmental information content of satellite imagery is vast, and the major challenge is to make ecologically meaningful interpretations that are relevant to mapping, monitoring and managing biodiversity. Only recently, however, has research begun to elaborate methodologies for using such data in reliable ways (see Ehrlich *et al.* 1994, Rogers *et al.* 1996, 1997, Lambin 1997).

Satellite-derived data

Various data 'products' can be derived from satellite imagery. For the purposes of ecological mapping, there are two broad categories: spectral vegetation indices

(SVIs) and meteorological surrogates. SVIs are proving particularly useful in ecological studies of land cover, vegetation phenology and classification, and are becoming central to many monitoring methodologies such as detecting drought, monitoring climate impacts on vegetation, and estimating crop yields (see Justice *et al.* 1986). Meteorological surrogates are proving very useful in studies of animal species distributions (see Hay *et al.* 1996 and Rogers *et al.* 1996).

Satellite imagery used in our studies was primarily provided by the NOAA-AVHRR. The AVHRR was designed for meteorological purposes and therefore has a relatively coarse spatial resolution of 1 km but high temporal resolution which provides daily (visible) and twice-daily (thermal infra red) coverage of Earth's surface (see Table 4.1 for details). The availability of daily imagery allows the 'contaminating' influence of atmospheric moisture (clouds in the extreme) to be removed through a process of image compositing. This uses images taken over relatively short periods of time and selects the most cloud-free image to represent the given time period. Compositing is usually carried out over a ten-day period resulting in the production of three 'cloud-free' images per month (Holben 1986, Eidenshink & Faundeen 1994). Further technical features of AVHRR data are described by Kidwell (1995), while Hay *et al.* (1996) review their characteristics and their application in studies of biological distributions.

AVHRR data are available free of charge from the USGS website (http://edcwww.cr.usgs.gov/landdaac/1KM/comp10d.html) for the periods September 1992–August 1993 and February 1995–January 1996 (Eidenshink & Faundeen 1994). Data for the Mkomazi region were downloaded, rescaled into geophysically meaningful values, and further composited to provide monthly images—an appropriate temporal resolution for our study purposes.

These data were then processed into SVIs and meteorological surrogates. SVIs exploit the fact that healthy vegetation has a low reflectance in the visible red (detected by AVHRR Channel 1) because photosynthetic pigments in plant tissues absorb such light; and a high reflectance in the near-infrared (detected by Channel 2) because the structure of mesophyll tissue reflects radiation of these wavelengths (Tucker & Sellers 1986). The indices therefore measure rates of photosynthesis at the pixel scale, and when integrated over time are related to green biomass. The normalised difference vegetation index (NDVI) is the most commonly used SVI and is defined for the AVHRR as:

$$\text{NDVI} = \frac{(\text{Channel 2} - \text{Channel 1})}{(\text{Channel 2} + \text{Channel 1})}$$

The resulting ratio has values between -1 and +1. Negative values indicate water, while values of 0.0–0.2 correspond to bare ground and values of 0.2–0.7 indicate the presence of actively photosynthesising vegetation (Tucker 1979).

Other SVIs derived from Channels 1 and 2 data (the soil-adjusted vegetation index (SAVI) and the global environment monitoring index (GEMI)) were calculated from NOAA-AVHRR using similarly straight-forward formulae (see Hay *et al.* 1996). Several surrogates for 'standard' meteorological variables were calculated for inclusion in our predictive modelling. These were Price's thermal temperature index, a surrogate for surface temperature, and vapour pressure deficit, a surrogate for humidity (although this varies with the time of day).

The environmental variables derived from AVHRR data were all georeferenced to the same co-ordinate system as other environmental and ecological data in the Mkomazi GIS (see below). At this stage each environmental variable comprised two sets of 12 consecutive months of data. The sets of data were time-series analysed, using temporal Fourier processing (see Rogers *et al.* 1996), to derive a few uncorrelated and ecologically meaningful variables which could be used in our descriptive and predictive analyses.

Field survey

The constraints mentioned earlier mean that, in general, our knowledge of ecological pattern and process is limited to relatively small areas and short time periods, and the data used in mapping exercises tend to be 'patchy'. Many attempts to deal with such data 'patchiness' in conservation planning have resorted to the use of biodiversity surrogates. This approach, however, has significant short-comings (van Jaarsveld *et al.* 1998), which highlight the importance of field research.

Basic ecological data are gathered through various types of field survey. These typically involve one or more of the following approaches: systematic and direct recording and measurement of features; sampling of representative habitats; opportunistic recording of entities as part of other research activities; and, observations made by amateurs naturalists or local individuals. The choice of sampling design has important implications for how the data may be subsequently used and is often constrained by logistical considerations such as time available for gathering field data (see Conroy & Noon 1996, Freitag & van Jaarsveld 1998).

Rapid biodiversity assessments are unlikely to yield adequate amounts of appropriate kinds of data for reliable large-scale mapping. It should also be apparent, however, that the effort needed to compile spatially extensive data (such as Tanzania-wide bird distributions) is a major constraint to timely progress with ecological mapping, and severely limits possibilities for monitoring change. In planning fieldwork, a balance needs to be struck between survey effort (sampling intensity) and geographical extent, so that taxonomic information is maximised at known spatial and temporal scales. A thorough treatment of methodological approaches to field survey and issues of designing survey protocols is provided by Southwood (1978).

A variety of sampling approaches was used in Mkomazi, depending on the taxonomic group of interest (see relevant Chapters in this volume, and below). The inventorying of taxa which are patchily distributed and which require labour intensive sampling, such as reptiles, amphibians and arthropods, was organised to ensure that all habitat types were included, with little attempt to be either spatially or temporally systematic. Inventorying of more spatially extensive taxa which require less labour intensive sampling, such as vegetation and large mammals, included systematic, representative and opportunistic survey methods.

Ecological mapping of Mkomazi

Our approach to the ecological mapping of Mkomazi has been to use standard multi-variate statistical techniques to define relationships between environmental variables derived primarily from satellite imagery, and ecological phenomena such as species, communities, habitats, described by field survey. These relationships have then been used to spatially predict patterns across the whole reserve, thus assuming a causal link between the 'measured' environmental variables and the presence or absence of surveyed phenomena. In other words, we develop an understanding of why species and habitats are where we have recorded them and use this understanding to predict where else they may occur using environmental 'predictor' variables.

Data, analysis and mapping

The first step in our mapping activities was to create a base GIS of Mkomazi. The database consisted of information digitised from 1:50,000 topographic sheets, including protected area and administrative boundaries, communications (roads, railways), rivers, human settlements, agriculture/forestry, and elevation contours. Subsequent steps involved the compilation of various 'themes' of environmental and ecological data from a wide variety of sources. As with many parts of Africa, few environmental data are available and particularly not at resolutions appropriate for the relatively small area and heterogeneous environment of Mkomazi.

The reliability of data sources is sometimes relatively straightforward to verify, as in the case of geo-referenced point biological data in original publications or in museum collections of plant and animal species (e.g. Inamdar 1996). The quality of data from thematic and topographic paper and digital maps may require more effort to verify as the source data are usually unavailable to the map reader.

A particular constraint to ecological mapping in Mkomazi and similar areas is the availability of environmental data of a high enough spatial resolution to 'capture' the relatively high habitat heterogeneity across its relatively small area. Although the spatial resolution—about 1 km^2—of the satellite imagery we used is relatively coarse, it was considered adequate for our attempts to map extensive patterns of vegetation and large mammal distribution. Field surveys were planned to provide reference data at appropriate spatial resolutions to explore the utility of satellite-derived variables for mapping distribution patterns (see below). Our use of hand-held global positioning systems in the field was critical to ensure that reference data were spatially registered to the satellite-derived data.

Once the data set is compiled, it can be analysed using multi-variate statistical techniques. In effect, these form a model with two main components. First is the distribution of the species (or other ecological phenomenon) which we are trying to explain or predict. In terms of a model, this constitutes the dependent variable. In the case where ground observations have been made in a sample area as the basis for making predictions in other areas, these are regarded as 'training data'. Second are environmental variables, particularly those obtained from satellites, which are expected to be associated with the dependent variable; these constitute the independent or predictor variables. The aim of the statistical techniques is to describe the association between the dependent and independent variables. This association can then be interpreted in terms of ecological relationships and can further be used to predict the dependent variable, resulting in distribution maps. Multiple regression techniques tend to be used when the relationship among variables is to be examined, while discriminant analysis (and logistic regression) is used for developing predictive equations. The applicability of the various analytical methods for particular purposes is discussed in Morrison *et al.* (1992).

In the following sections we detail our preliminary work in mapping the vegetation and the large mammal species of Mkomazi.

Mapping vegetation cover of Mkomazi

As the primary producer of energy in an ecosystem, vegetation composition, distribution and dynamics underlie the ecosystem patterns we observe. A reliable description of the distribution of vegetation types throughout Mkomazi is therefore an important first step in understanding the ecology of this semi-arid savanna.

Vegetation diversity is commonly described in terms of extent, pattern, composition and structure, and these facets of vegetation can be assessed using satellite remotely-sensed environmental data. Different vegetation types manifest particular spectral reflectance and emittance properties, which result in distinctive spectral response patterns. The present study was designed to exploit this behaviour. Its aim is to use vegetation data collected at geo-referenced locations on the ground in conjunction with satellite-derived data to produce a map showing the contemporary distribution and extent of different vegetation types.

The marked seasonality of savannas is well known. Another characteristic is the relative dominance of woody and herbaceous species. Variation in the proportion of these types depends on the effects of disturbance (fire, herbivory) superimposed on the effects of climate and soil (Huntley & Walker 1982). Changes in the disturbance regime and variations in climate are thus reflected in a change in the ratio of woody to herbaceous vegetation. As a result, the available descriptions of the vegetation in east Africa are based mostly on the structure of the vegetation, sometimes qualified by the dominant genera or species (Pratt *et al.* 1966, Pratt & Gwynne 1977). As well as having ecological meaning, these are criteria that can be easily recognised in the field and are thus important for natural resource managers.

Data, analyses and results

This study used 1 km AVHRR imagery from September 1992 to August 1993, which had been processed to provide NDVI data. As described earlier, NDVI is directly related to the photosynthetic activity of plants and has been found to be a sensitive indicator of the presence and condition of green vegetation. The maximum NDVI value in each of the 12 months (the value least affected by atmospheric moisture) was selected and these were combined to give a 12 band image of Mkomazi in which each band was associated with a monthly NDVI value.

This image was subjected to an unsupervised classification (using ERDAS Imagine) in which a clustering algorithm was used to assign each 1 km pixel to one of 22 classes on the basis of a user-prescribed statistical rule ('unsupervised' means that the classification procedure is largely computer-automated). Thus each class is composed of pixels with similar spectral characteristics in terms of both absolute NDVI values and its seasonal variation throughout the year.

The result is shown in Plate 4. The colours follow the colours of the rainbow with the red class representing the highest values of NDVI, through orange, yellow, green, blue, to violet which represents the lowest values. In this classification, the boundaries between classes generally follow gradients of rainfall and altitude, which is itself closely related to rainfall and temperature. As vegetation activity in a semi-arid savanna is largely determined by moisture availability, these variables are important determinants of photosynthetic activity in this region.

Thus the areas of highest NDVI are located on the mountain tops, with the two large patches to the south of the reserve representing the South Pare and the Usambara Mountains. This is where significant patches of evergreen hill-summit forest are likely to be found. Areas represented by the indigo-violet classes represent the dry plains environment covered with sparse grassland or sparse dry bushland. The coastal influence on climate can be detected in the lower right hand corner where the orange and yellow classes signify a denser type of wooded bushland and woodland composed of larger trees and shrubs.

One part of the reserve, however shows a greater NDVI than would be expected according to rainfall and altitude influences. This is the Kifukua 'seasonal swamp' which shows up as a dark green area in the southern central region of the reserve. Although this area receives relatively little rain and is relatively low-lying, the high NDVI is probably the result of the rain water that drains into the Kifukua 'bowl' from hills to the west in the centre of the reserve (see the map of Mkomazi inside the back cover) and provides moisture for vegetation activity. In a notably dry year for which we have subsequently obtained NDVI data (1995–96), very little water would have been falling on the central hills, and the Kifukua area had a very low NDVI.

The availability of monthly values of NDVI means that seasonal variation in green plant activity of the vegetation at any site (pixel) also exerts a strong influence in the classification process. This variation is an important feature of vegetation in a highly seasonal environment such as the savanna and can be used to distinguish vegetation types. Examples of the seasonal variation in the NDVI of Mkomazi vegetation types are given in Figure 4.1 together with their corresponding colour category. Figure 4.1a shows the high NDVI values and relatively low seasonality associated with dry upland evergreen forest. Figure 4.1b represents a more deciduous dense wooded bushland, which attains a slightly lower maximum NDVI but shows much greater seasonal variation.

Categories dominated by grass such as the seasonally-inundated grassland (Figure 4.1c) and the bushed grassland (Figure 4.1d) show two NDVI peaks. This is because grasses are shallow rooted and die back soon after the rain ceases, whereas vegetative activity of the deeper-rooted woody vegetation continues for longer

Figure 4.1 Graphs of NDVI through the year September 1992–August 1993 for different vegetation types (with their Plate 4 classification category in brackets). The vertical axis is an arbitrary scale of NDVI values.

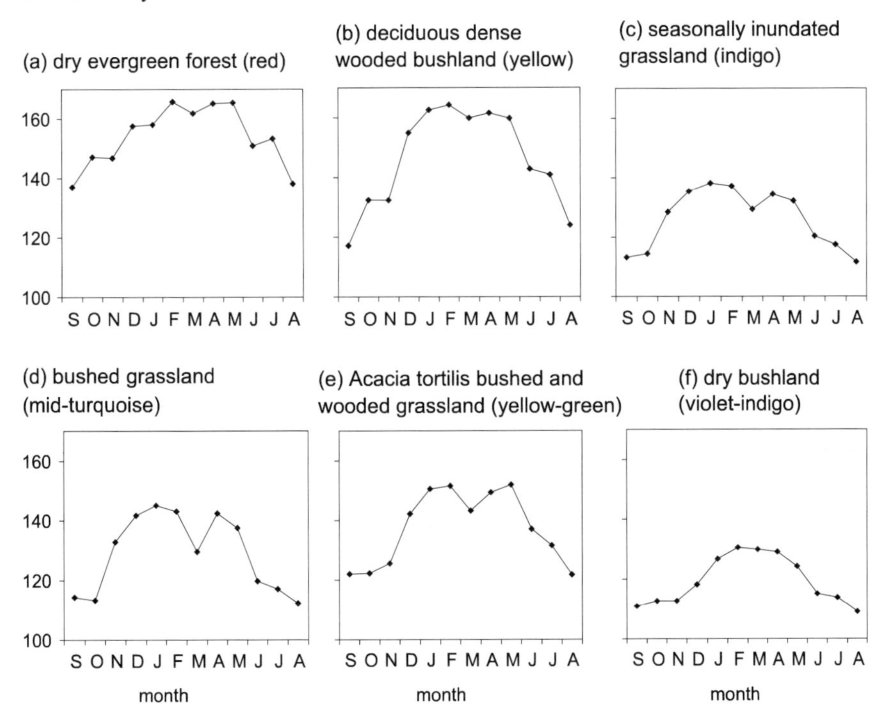

into the dry seasons. Thus vegetation types dominated by grasses will show a double peak representing the die-back between rainy seasons. Similarly, the double peak in NDVI of the *Acacia tortilis* wooded and bushed grassland (Figure 4.1e) indicates the influence of the grass, although it is less pronounced because of the influence of the large trees with high woody biomass associated with this vegetation type. By contrast the violet category in Plate 4 represents areas of dry bushland with little grass cover and low values of NDVI (Figure 4.1f).

Discussion

The advantage of an unsupervised classification, as used here, is that minimal input is required from the user. The classification algorithm, however, only identifies clusters of pixels with similar *spectral* characteristics and these do not necessarily correspond to *ecologically* meaningful features on the ground. The more difficult and critical task is to interpret these classes.

As indicated above, we can interpret these classes qualitatively within the reserve (outside the reserve there is the added component of human derived land covers such as clearing forest for agricultural plots). A quantitative interpretation of vegetation composition would, however, require ground data collected at a similar scale to that of a pixel, i.e. 1 km^2. Data at a smaller resolution would not necessarily be representative of the pixel within which it was located, and so aerial photographs or ground-truthed finer-resolution satellite imagery aggregated to c. 1 km^2 would therefore be required (see below).

An improved classification procedure would use supervised techniques which incorporate such reference data into the analysis. In this case, the analyst selects pixels that represent vegetation classes that have been previously identified and characterised through 'ground-truth' vegetation sampling. The training pixels are then used to define a 'statistical signature' for each vegetation type in the image, which may include spatial data derived from satellite imagery or other sources. The signature set is then used to classify each pixel in the rest of the image, using a particular statistical decision rule such as maximum likelihood classifier, into the vegetation class to which it is statistically most similar.

Further development of this work

This work is very much ongoing and has two components:

- Producing a vegetation map using 'supervised' procedures.
- Identifying vegetation change from the early 1970s to the present.

Satellite-based mapping of different vegetation types requires that information about species composition and vegetation structure can be sensed at a resolution which is similar to that of the ground data. Landsat-TM senses radiation in wavelength bands that are particularly finely tuned for the discrimination of vegetation types, according to species differences (Reys-Benayas & Pope 1995). Radar imagery, such as JERS-Synthetic Aperture Radar (SAR), is unaffected by cloud cover and is particularly sensitive to vegetation structure (Waring *et al.* 1993, Nezry *et al.* 1993). This high resolution imagery (30 m and 22 m respectively) will be combined with ground-collected vegetation data to produce a vegetation map of Mkomazi.

Vegetation surveys were conducted in Mkomazi throughout 1997 and are almost complete. Sites for vegetation sampling were spread as widely as possible throughout the vegetation types, and located by GPS. For the woody vegetation, data were collected on species, canopy cover and height class, while for the herbaceous vegetation species and percentage cover were recorded (as well as percentage cover of bare ground). Analysis of these data enabled the identification of vegetation types described in terms of structure and species. Species associations were determined using two-way indicator species cluster analysis (TWINSPAN; Hill 1979), while vegetation structure was described using the Kenya Soil Survey classification (Weg & Mbuvi 1975), based on the percentage cover by trees and shrubs.

A range of environmental variables such as rainfall, soil type, slope, aspect and altitude were also determined for each sample site to allow the analysis of how these might influence the distribution of vegetation types or individual plant species, and to assist in the interpretation of change.

As the ground data were collected at a similar scale to the resolution of Landsat-TM and JERS-SAR imagery they can be used directly to inform a supervised classification of the satellite imagery. This, in conjunction with techniques such as linear mixture modelling, will be used to produce a vegetation map, whose accuracy will subsequently be tested by visiting sites located at random and comparing the observed vegetation type with that predicted by the map. This map can then be used to provide a quantitative analysis of the AVHRR imagery.

By using a variety of satellite data (Landsat-TM, -MSS and NOAA-AVHRR) in combination with aerial photography and ground-gathered vegetation and environmental data from various periods, it is hoped that an analysis of vegetation/habitat change over the period from early 1970s to mid-1990s will be possible. Imagery from different dates will be compared to identify where vegetation has changed over time. Further analyses will attempt to identify the nature of the changes, and to correlate them with historical data on rainfall and disturbances such as fire, herbivore abundance and human activity.

It is hoped that the results will improve our understanding of how such factors influence the distribution of vegetation and provide insights to processes shaping savanna ecosystems. These studies will also give an indication of the type of data that needs to be collected on a regular basis for the future monitoring of the reserve as well as of a framework within which additional studies can contribute to our understanding of ecological pattern and process.

Mapping large mammal distributions in Mkomazi

In this section, we describe how geo-referenced observations of large mammals are being used together with environmental data to model large mammal distribution in Mkomazi, resulting in maps of predicted distributions. Large mammals have an important ecological role in semi-arid ecosystems, particularly in terms of vegetation, and they relate closely to management issues such as tourism, fire, hunting and water provision. More generally, maps of predicted distributions are important for understanding biodiversity patterns and dynamics in areas like Mkomazi where extensive distributional data are scarce.

The large extent of Mkomazi (and even larger extents of many other Tanzanian protected areas) means that fieldwork can only 'sample' a relatively small proportion of the whole area. Ideally such a sample is systematic and representative of all environmental conditions, typically using aerial survey techniques. Cost and logistics, however, make aerial survey inappropriate for most east African protected areas, particularly where regular monitoring is desirable. If sampling is not representative and systematic, then a simple map of observations is likely to say more about where observations were made than about the actual distribution of a species. This is particularly true in Mkomazi where habitats, and therefore mammal distributions, are far from uniform. We have therefore aimed to develop methods to map distributions in a way which is appropriate for Mkomazi in terms of needs, techniques and costs.

Modelling approach

The approach is based on the association between sightings of a species and environmental factors which might determine its distribution: in effect, defining its habitat or environmental niche.

Modelling this habitat association requires a 'classifier' or rule, which uses the environmental variables as predictors in order to 'decide' whether or not a species is likely to be present in each part of the study area. We chose logistic regression as a classifier. It uses a set of predictor variables to describe the variance in a dependent variable which has two states—in this case the presence (1) or absence (0) of a species. Logistic regression has the advantages of allowing the predictor variables to have non-normal distributions and to contain categorical data (such as vegetation class) as well as numerical data (such as slope angle). It results in a probability, ranging from 0 to 1, of a species being present in each part of the study area. The probability value can itself be mapped; or more usefully, a suitable threshold can be applied, above which the probability value indicates a presence, and below which it indicates an absence.

This approach depends on there being an ecological relationship between a species' distribution and its habitat. In Mkomazi, two major factors likely to determine large mammal distributions are each species' level of dependence on drinking water, and its feeding habit (e.g. grazer, browser or mixed) (van Wijngaarden 1985). Both of these characteristics can be related to environmental conditions which vary over space according to the availability of drinking water and the availability of vegetation suitable for grazers or browsers. For example, Coe (Chapter 7), observes that in Mkomazi "the greater aridity in the east is responsible for a low standing crop of grasses, most of which are relatively unpalatable, making the area largely unsuitable for wild and domestic grazers but ideal for browsers".

This observation also hints at one of the potential problems in multivariate modelling. While a grazer's presence or absence might be most immediately *determined* by the standing crop of grasses, it is also likely to be *correlated* with rainfall. However, rainfall also helps to determine the grass crop. This correlation among predictor variables complicates the interpretation of the models' results.

Large mammal data

The 'training data' for our model consist of 'presence' and 'absence' observations. The presences are points where large mammal species were observed in Mkomazi during 1994–97. The majority of observations were made from vehicles using existing tracks. Essential data recorded were: species, date and location (using a GPS), and additional data were: time, number of individuals, distance from vehicle (estimated by eye), and compass bearing to the group. The bearings and distances were later combined with the GPS co-ordinates to calculate the coordinate for the location of each group. The large mammals mapping project considered records of all mammal species from dikdik size (c. 60 cm head and body height) and above, and which were observed at least 20 times during the study period. These species and their details are given in Chapter 31.

The observations were made during three main types of survey, each representing different sampling effort. Full explanations and maps are also given in Chapter 31, but summary details are provided here and in Table 4.2.

- A series of *periodic surveys* from January to November 1996 on four routes in north-west Mkomazi aimed to represent temporal changes in distribution across the northern third of the reserve.
- Systematic, *dry season surveys* (four, each lasting 3–4 days in July to September 1996) were conducted along roads throughout the reserve, providing a spatially extensive although seasonally limited representation of Mkomazi.
- *Opportunistic records*, from *ad hoc* sightings made during the course of other research in Mkomazi, were made between 1994 and 1997 and were in no way controlled in terms of temporal or spatial coverage.

Sampling distributions using road surveys involves numerous biases (Wilson 1996). The methods used here present two particular biases. First, the 'envelope' of environmental conditions to be represented was not well known at the outset.

Instead, the survey routes were chosen according to field observations of habitat heterogeneity. Second, the survey routes were essentially confined to existing road tracks and to more accessible parts, both limiting the sample area and possibly introducing complications due to the influence of roads. As mentioned earlier, however, the modelling methods developed here expressly aim to use data collected with such sub-optimal methods. A particular problem does remain: we cannot validly use logistic regression to make predictions in areas whose environmental conditions are not represented in the training data. Further consideration of sampling biases in this study is given by McWilliam (in prep.).

All the mammal observations were collated in a spreadsheet. The co-ordinates were converted to the same sinusoidal projection as the environmental variables (based on the AVHRR satellite data). Simple maps of observations could then be readily produced using MapInfo software (such as those in Chapter 31). To prepare the training data for the logistic regression, they were consolidated on a pixel-by-pixel basis, rather than on a sighting-by-sighting basis. Thus for each 1 km grid square in Mkomazi, the number of sightings were grouped according to species, survey method and season. An example is given in Table 4.3.

In the logistic regression model, the 'absences' are important because they define those environmental conditions which are *not* suitable for a species. In general, the survey methods used here were not intensive enough indicate with certainty that the *absence of an observation* of a species represented an *actual absence* of that species in a given area. We therefore had to infer absences, using rule-of-thumb criteria. These are detailed in McWilliam (in prep.), but in general they estimate whether sufficient survey effort was expended in a given area for a species to have been seen were it actually present there. The estimates aimed to be conservative: a few inferred absences, attributed with a high degree of certainty, are far better as training data than a large number of less-certain absences.

Table 4.2 Summary of field surveys used to collect large mammal data in Mkomazi. The right-hand column refers to the number of sightings of a group or an individual, rather than the numbers of individual animals.

characteristics	location	length (km)	survey timing	sightings made
spatially intensive, temporally extensive sample	Kavateta Ngurunga Vitewini Zange	27.8 11.2 37.6 15.0	twice each month, Jan.–Nov. 1996	875
spatially extensive, temporally intensive sample	long route	393	4 times during July–Sept. 1996	669
opportunistic sightings	all routes	–	*ad hoc*, 1994–97	2,125

Table 4.3 Sample of observations classified by species, grid cell, survey method and season. This indicates, for example, that of all the opportunistic records, two dikdik observations were made in cell (4194000, -441000) during wet season months.

		dikdik									
		periodic survey			opportunistic			dry season survey			dikdik
grid X	grid Y	wet	dry	total	wet	dry	total	wet	dry	total	total
4194000	-442000	1	1	2							2
	-441000				2		2				2
	-440000										
4195000	-447000				2		2				2
	-446000										
...	...	...	...	...	...	...	...	...	...	...	...
total		56	78	134	248	35	283	0	318	318	735

Environmental variables

Environmental variables are factors likely to determine species' distributions. To be useful as predictors in our model, they need to cover the entire study area at spatial and temporal resolutions appropriate to the distribution patterns we are trying to describe. An important source of predictor variables is therefore the AVHRR satellite imagery.

The derivation of the AVHRR data for this study was described earlier, in the section on *Satellite-derived data*. Of the large number of potential AVHRR variables, those describing NDVI were selected. NDVI is widely used and is readily obtainable. Also, tests for co-linearity between the various AVHRR products showed that most were significantly correlated with NDVI: including them all would therefore not add significantly to the model.

Several NDVI products were used for the model: the 1995 mean; the November–June wet season mean; the July–October dry season mean; and seven variables which describe seasonal patterning, derived from Fourier processing of the NDVI time series. Our model also used four environmental predictor variables derived from field data and map data held in the Mkomazi GIS.

- *Elevation* is likely to be correlated with environmental gradients which we could not readily measure, particularly rainfall. Elevation data, at 1 km resolution, were interpolated using Erdas Imagine from 100 m contours digitised from 1:50,000 maps.
- *Slope* may directly influence some species (such as klipspringer, *Oreotragus oreotragus*, which favours rocky hillslopes) but is more likely to influence distribution indirectly, through vegetation types. The slope angle for each 1 km pixel was derived from the elevation data, again using Erdas Imagine.

- *Distance to open water:* as discussed, this is a key determinant of the distribution of large mammal species which depend on regular access to drinking water. Two variables, expressed as the distance from a given pixel to the closest water source, were produced. One measured the distance to wet season water sources, and one the distance to dry season water sources, which are far fewer. The locations of water sources in Mkomazi came from GPS survey and the course of the Umba river was digitised from 1:50,000 maps.
- *Vegetation* was available digitally from a recent 1:250,000 vegetation map of Tanzania (Surveys and Mapping Division 1996). The maps were derived from Landsat Thematic Mapper images (1994–96), classified on the basis of ground survey (Hunting Technical Services, undated). Despite classification errors due to the limited ground survey in Mkomazi, the data remain valuable as a predictor variable in being able to differentiate between vegetation types.

All of the environmental predictor variables were prepared as raster images and registered to the same sinusoidal projection as the AVHRR data. They were combined with the training data, matched according to pixel, and imported into the statistical programme SPSS 9.0 for Windows for analysis.

Analysis

The modelling procedure was developed using zebra distribution, which is presented here to illustrate the analytical process. Zebra were chosen because they have a relatively large set of training data and, being water-dependent, their distribution is expected to change seasonally.

Our aim was to develop a logistic regression model which combines accurate prediction (particularly of observed presence), robustness (which is desirable so the model can be applied to different species, time periods and locations) and simplicity (using the fewest number of input variables commensurate with the model's predictive power).

In summary, the procedure involved: formulating and comparing different distribution models; choosing the best one; generating presence probability values for each cell in Mkomazi; identifying a presence/absence threshold for probability values; and finally mapping the results. These steps are now described in more detail.

Twelve different logistic regression models were performed. There were six wet season and six dry season models. For each season, we used different combinations of predictor variables and different definitions of zebra absence. In each case, the regression generated a value between 0 and 1 for each 1 x 1 km cell in Mkomazi, indicating the probability of zebra being present.

Two techniques were used to assess and compare the different models (both described by Fielding & Bell 1997). The first, a 'confusion matrix', measures the performance of a model according to its ability to predict its own training data. It cross-tabulates the observed and predicted number of presences and absences. Figure 4.3 in the *Results* section is an example.

Although easy to construct and interpret, confusion matrices are specific to whichever threshold was used in deciding whether a given probability value indicates a presence or an absence. They do not provide a good overall assessment of a model's performance. A technique just starting to be used in ecology provides a better assessment: the 'receiver operating characteristics' (ROC) plot.

In an ROC plot, the vertical axis effectively measures the ability of a model to predict presences (the true positive fraction), while the horizontal axis measures the inverse of the model's ability to predict absences (the false positive fraction). These measures are plotted for all possible threshold values, ranging from 0 to 1, resulting in a curve. The area under this curve (AUC) then provides a useful overall measure of a model's performance. The AUC is used here to compare models.

Figure 4.2 shows a sample plot. The diagonal line results from a hypothetical prediction based entirely on chance. It has an AUC of 0.5. By contrast, the upper curve comes from data in this study and shows a much-better-than-chance result, with an AUC of 0.852. The maximum possible AUC is 1.0—this is the ideal.

For the zebra distribution model, we calculated the AUC for each of the six wet season and six dry season logistic regressions. This allowed us to compare models and select the best one for each season. The results of these two models were then mapped by exporting the predicted probability values for all grid cells from SPSS to the MapInfo GIS.

The final step was to identify suitable thresholds for assigning presence/absence from the probability values. ROC plots were used again. A threshold of 0.5 is often arbitrarily chosen: predicted probabilities of over 0.5 are considered presences, while below 0.5 are absences. However, as the sample ROC plot shows, different thresholds (i.e. different points along the curve) can result in greatly different assignations of presence/absence. To find a suitable threshold value, we chose a point on the ROC curve with as high as possible a true positive fraction while corresponding to a relatively low false positive fraction. The procedure is detailed in McWilliam (in prep.) and is explained further in Fielding & Bell (1997).

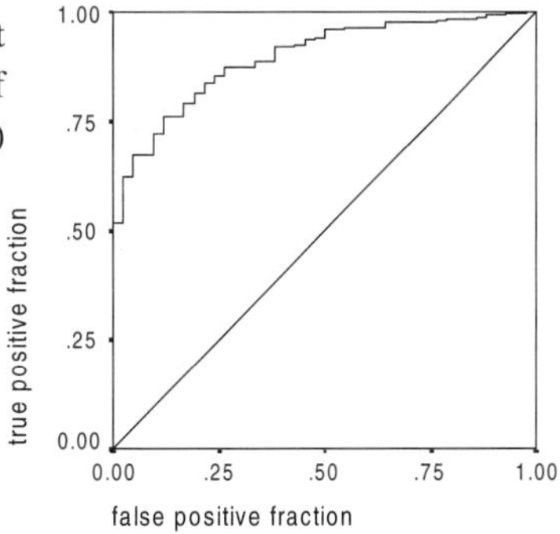

Figure 4.2 Sample ROC plot. Although not apparent here because of the stepped nature of the upper curve, all curves pass through (0,0) and (1,1).

Results of zebra distribution model

Among the six wet season models of zebra distribution, the AUC values varied from 0.581 to 0.892 . The dry season values varied from 0.816 to 0.908. Details of the two models with the highest AUC values are shown in Table 4.4. In both cases, the high AUC values suggest well above-chance predictions. The high AUC values for all of the dry season models are probably a result of the strong influence of the distance to water at a time when water sources are scarce.

ROC plots of the two 'best' models were then used to choose suitable probability thresholds, as described earlier. For the wet season model, the threshold was 0.6, i.e. if the predicted probability for a given cell in Mkomazi was above 0.6, then zebra was predicted to be present in that cell. For the dry season model, the threshold was 0.1. In the dry season case, 0.1 did not in fact provide the best *overall* prediction of both presence and absence—but it did provide the best classification for observed presence, which is considered more important here.

Confusion matrices were then calculated according to the chosen threshold values (Figure 4.3). These give a more readily-interpreted indication of the predictive performance of the models.

Finally, Plates 5(a) and 5(b) show the resulting predicted distribution maps. As the threshold for the wet season model is 0.6, the red and grey areas are classified as zebra presence. In the dry season model the threshold is 0.1, so only the bright red areas are zebra presence.

Table 4.4 Characteristics of the wet and dry season zebra distribution models selected from the original 12 models on account of their ROC plots having the highest area-under-curve (AUC) values.

	season	
	wet	dry
model reference number	3	15
presence data used	all surveys	all surveys
absence data used	1996 NW Mkomazi survey routes	1996 extensive survey routes
number of presence observations	94	50
number of inferred absences	41	345
variables used in logistic regression (the NDVI variables are different products of the temporal Fourier analysis)	vegetation class NDVI 1992 mean NDVI_92amp1 NDVI_92phase1 NDVI_92amp2 NDVI_92amp3	distance to water (dry season) vegetation class NDVI 1992 mean NDVI_92amp1 NDVI_92amp2 NDVI_92amp3
area under ROC curve (AUC)	0.892	0.908

Figure 4.3 Confusion matrices for predicted zebra distribution models: wet season (left) and dry season (right). '0' = absences; '1' = presences. The thresholds used to classify a presence are 0.6 and 0.1 respectively. On the right are the percentages of successful predictions.

		predicted 0	predicted 1	
observed	0	28	13	68.3%
	1	9	85	90.4%
			overall	83.7%

		predicted 0	predicted 1	
observed	0	288	57	83.5%
	1	11	39	78.0%
			overall	82.8%

Discussion

One particular aspect of the results is explored here: the predicted dry season 'presences' in central Mkomazi shown in Plate 5(b). These presences demonstrate the sensitivity of the model to its predictor variables; they show some important limitations to the model; and they give directions for further development.

The training data used by the model to make these predictions comprised zebra sightings made in the vicinity of Maore waterhole (eight sightings) and Kavateta waterhole (two sightings) between July and mid-August—early in the period defined as the dry season (July–October). It is known that water persisted at both Maore and Kavateta into July and this is likely to explain the presence of zebra in these locations when water was not generally available in Mkomazi.

However, the only water sources included in the dry season model as predictor variables are at Dindira and the Umba River, at opposite ends of the reserve. The model assumed over-simplistically that Maore and Kavateta were dry during the dry season. Thus the successful prediction of zebra presences in central Mkomazi was due to the inclusion of vegetation class and NDVI as predictor variables, rather than the more likely reason of remaining water at Maore and Kavateta.

We tested this with a model which included only two predictor variables: distance to water and slope, but neither vegetation nor NDVI. This model was used first with the full training data set, then with the 'anomalous' presence cells in central Mkomazi temporarily removed from the training data. The effect of removal was to considerably improve the prediction from AUC = 0.816 to AUC = 0.912.

We then added vegetation class and NDVI as predictors in the revised model, resulting in a small further improvement to AUC = 0.938 (Plate 5(c)). This improvement might be expected, as the model could then account for habitat variations within a radius of the water sources.

Thus the differences between Plates 5(b) and 5(c) are due simply to the removal of the seven central Mkomazi presences from the training data. Large areas of presence, predicted on account of NDVI and vegetation in Plate 5(b), disappear from Plate 5(c). The distribution becomes far more as expected for the dry season without water at Maore and Kavateta.

Our original model successfully described known distribution patterns. It subsequently appeared, however, that its combination of training data and predictor variables was not as ecologically accurate nor as parsimonious as possible. One practical consequence of this is that the model is liable to make false predictions in parts of the reserve where no training data exist. More generally, this example shows the importance of interpreting and improving the model using existing ecological knowledge of the reserve. In this case, the model could be improved by better temporal partitioning, specially for water-dependent species such as zebra. Thus, for example, both the training data and the distance to water predictor data could be divided into months rather than simply wet and dry seasons. This in turn would require better monitoring of water availability at each water point.

Further work

This study has highlighted directions for future research. Priorities include:

- Improved temporal partitioning and sensitivity to changing water availability, particularly for water-dependent species in dry season months.
- Comparison of models using different definitions of absence.
- Examination of mis-classified or unexpectedly classified pixels to help refine models.
- Combining the training data for species likely to have similar habitat requirements and producing predictions for 'functional types' rather than species (e.g. lesser kudu and gerenuk, as water-independent browsers).
- Examination of limitations in the predictive capacity of the technique, particularly in relation to:
  - the thematic, spatial and temporal resolution of the predictor variables
  - whether the predictor variables used here actually represent important habitat variations
  - the size of the training data sets
  - behaviour or ecological characteristics of the species.
- Description of patterns of predicted species richness by overlaying predicted distributions for all species.

Conclusion

Semi-arid savannas are highly dynamic ecosystems. Marked temporal and spatial heterogeneity in the many factors and processes shaping such savannas make mapping their biodiversity a particular challenge. The need for such information, which must be both reliable and robust, is critical to attempts to conserve and sustainably utilise biological resources.

Vegetation dynamics are the most apparent manifestation of savanna heterogeneity. Having set ourselves the objective of understanding these dynamics at a variety of temporal and spatial scales in Mkomazi, our preliminary attempts at mapping vegetation have pointed to the utility of satellite imagery in extrapolating 'patchy' ground data collected at point locations over the whole area of Mkomazi. Much further work is needed to gather comprehensive ground data sets and to process imagery that will provide multi-resolution environmental data. The prospects are good for reliable mapping of vegetation patterns, and for detecting change in these patterns. A likely major challenge will be in attributing these patterns and changes in them to particular factors and processes, be they ecological, environmental or anthropogenic.

We have also used ecological mapping to study the distribution of large mammals in Mkomazi. The results, although for just one species so far, are promising: predicted distributions have a good match with actual observations and are statistically strong. What do we learn from this?

First, we expect the methods to be applicable to other large mammal species in Mkomazi. The results from this exercise should provide seasonal maps that can help the management and monitoring of one of the reserve's key components. Second, we tentatively show that useful distribution models can be built with relatively sparse and crude data, even in an area like Mkomazi whose habitat varies over small distances. We hope that this might contribute to developing appropriate methodologies for studying mammal distributions in other parts of east Africa.

The predictive approaches to mapping biodiversity presented in this chapter are based on statistical models, and the use of satellite remotely-sensed data within a geographical information system has been critically important. Further improvement in such predictive approaches to mapping biodiversity is largely dependent on the gathering of spatially and temporally more extensive and higher quality reference and other relevant data. In the meantime, results of the kinds of modelling described here, whether used for mapping or monitoring, must be viewed with some caution.

Acknowledgements

We acknowledge financial support from: the UK Darwin Initiative for the Survival of Species (MJP); the Natural Environment Research Council for a CASE studentship (SMC) in co-operation with the RGS-IBG; GlaxoWellcome support (SMC); the British Airways Assisting Conservation programme (MJP); and, the Friends' Provident Stewardship Joanna Lumley Research Fellowship at Green College, Oxford (MJP). Many thanks to the following: Azmilen bin Ramlee (RGS-IBG) for assistance building the GIS base-map; the field survey 'crew' (Omari Mohammed, Firimini Saidi and Yona Zawadi) for field assistance to SMC and MJP and for making the camps work despite the best efforts of the 1997–98 ENSO event; Mr Paul Marenga, Project Manager of Mkomazi Game Reserve, for providing ranger assistance during the vegetation surveys; Ken Ferguson for helping to initiate and design the systematic large mammal surveys; the rangers and all the members of MERP, particularly the 1996 Oxford University expedition team, who contributed to the large mammal surveys; and the British Council in Dar es Salaam for excellent logistical support outside of Mkomazi. NMcW is grateful to Dr Malcolm Coe for all his help with the large mammal mapping work and Dr Robert Whittaker, Oxford University School of Geography, for co-supervising the project with MJP. Finally, many thanks to Andy Anderson for allowing us to use the illustration detail of giraffe in the Dindira basin.

References

Burrough, P.A. & McDonnell, R.A. (1998) *Principles of geographical information systems*. Oxford University Press.

Conroy, M.J. & Noon, B.R. (1996) Mapping species richness for conservation of biological diversity: conceptual and methodological issues. *Ecological Applications* 6: 763-773.

Davis, F.W., Quattrochi, D.A., Ridd, M.K., Lam, N.S.N., Walsh, S.J., Michaelsen, J.C., Franklin, J., Stow, D.A., Johannsen, C.J. & Johnston, C.A. (1991) Environmental analysis using integrated GIS and remotely sensed data: some research needs and priorities. *Photogrammetric Engineering and Remote Sensing* 57: 689-697.

Ehrlich, D., Estes, J.E. & Singh, A. (1994) Applications of NOAA-AVHRR 1 km data for environmental monitoring. *International Journal of Remote Sensing* 15: 145-161.

Eidenshink, J.C. & Faundeen, J.L. (1994) The 1km AVHRR global land data set—1st stages in implementation. *International Journal of Remote Sensing* 15: 3443-3462.

Fielding, A.H. & Bell, J.F. (1997) A review of methods for the assessment of prediction errors in conservation presence/absence models. *Environmental Conservation* 24: 38-49.

Freitag, S., Nicholls, A.O. & van Jaarsveld, A.S. (1998) Dealing with established reserve networks and incomplete distribution data sets in conservation planning. *South African Journal of Science* 94: 79-86.

Freitag, S. & van Jaarsveld, A.S. (1998) Sensitivity of selection procedures for priority conservation areas to survey extent, survey intensity and taxonomic knowledge. *Proceedings of the Royal Society of London Series B—Biological Sciences* 265: 1475-1482.

Hay, S.I., Tucker, C.J., Rogers, D.J. & Packer, M.J. (1996) Remotely sensed surrogates of meteorological data for the study of the distribution and abundance of arthropod vectors of disease. *Annals of Tropical Medicine and Parasitology* 90: 1-19.

Hill, M.O. (1979) *TWINSPAN – a FORTRAN program for arranging multivariate data in an ordered two way table by classification of the individuals and the attributes.* Cornell University, Department of Ecology and Systematics, Ithaca, New York.

Holben, B.N. (1986) Characteristics of maximum-value composite images from temporal AVHRR data. *International Journal of Remote Sensing* 7: 1417-1434.

Huntley, B.J. & Walker, B.H. (1982) *Ecology of tropical savannas.* Springer-Verlag.

Hunting Technical Services (undated) *Forest resources mapping project: national reconnaissance level land use and natural resources mapping project.* Hunting Technical Services Ltd., Hemel Hempstead, UK. (www.htsconsult.co.uk/profile/tanzania/mainpage.htm).

Inamdar, A. (1996) *The ecological consequences of elephant depletion.* PhD thesis, University of Cambridge.

Johnson, S.P. (ed.) (1993) *The Earth Summit: the United Nations Conference on Environment and Development (UNCED).* Graham & Trotman/Martinus Nijhoff, London.

Justice, C.O., Holben, B.N. & Gwynne, M.D. (1986) Monitoring East African vegetation using AVHRR data. *International Journal of Remote Sensing* 7: 1453-1474.

Kidwell, J.B. (1995) *NOAA Polar Orbiter Data Users Guide.* National Oceanic and Atmospheric Administration, Washington DC.

Lambin, D.F. (1997) Land cover changes in sub-Saharan Africa (1982-1991): application of a change index based on remotely sensed surface temperature and vegetation indices at a continental scale. *Remote Sensing of Environment* 61: 181-200.

Lillesand, T.M. & Kiefer, R.W. (1994) *Remote Sensing and Image Interpretation.* John Wiley & Sons, Inc.

McWilliam, N.C. (in prep.) *Modelling the distribution of large mammals in Mkomazi Game Reserve, Tanzania.* MSc Thesis, University of Oxford.

Miller, R.I. (ed.) (1994) *Mapping the diversity of nature.* Chapman & Hall, London.

MNRT (1998) *The Wildlife Policy of Tanzania.* Ministry of Natural Resources and Tourism, Dar es Salaam, Tanzania.

Morrison, M.L., Marcot, B.G. & Mannan, R.W. (1992) *Wildlife-habitat relationships: concepts and applications.* University of Wisconsin Press.

Nezry, E., Mougin, E., Lopes, A., Gastelluetchegorry, J. P. & Laumonier, Y. (1993) Tropical vegetation mapping with combined visible and SAR spaceborne data. *International Journal of Remote Sensing* 14: 2165-2184.

Pratt, D.J. & Gwynne, M.D. (1977). *Rangeland management and ecology in East Africa.* Hodder and Stoughton, London.

Pratt, D.J., Greenaway, P.J. & Gwynne, M.D. (1966). A classification of east African rangeland with an appendix on terminology. *Journal of Applied Ecology* 3: 369-382.

Reys-Benayas, J.M. & Pope, K.O. (1995) Landscape ecology and diversity patterns in the seasonal tropics from Landsat TM imagery. *Ecological Applications* 5: 386-394.

Rogers, D.J., Hay, S.I. & Packer, M.J. (1996) Predicting the distribution of tsetse flies in West Africa using temporal Fourier processed meteorological satellite data. *Annals of Tropical Medicine and Parasitology* 90: 225-241.

Rogers, D.J., Hay, S.I., Packer, M.J. & Wint, G.R.W. (1997) Mapping land-cover over large areas using multispectral data derived from the NOAA-AVHRR: a case study of Nigeria. *International Journal of Remote Sensing* 18: 3297-3303.

Stork, N.E. & Samways, M.J. (1995) Inventorying and monitoring. In: Heywood, V.H. & Watson, R.T. (eds.) *Global biodiversity assessment*. UNEP, Cambridge University Press. pp. 453-543.

Southwood, T.R.E.S. (1978) *Ecological Methods* (2nd ed.). Chapman and Hall, London. 524 pp.

Surveys and Mapping Division (1996) 1:250,000 map series of Tanzania: Land Cover and Land Use. Sheets SA-37-14 (Voi) [northern Mkomazi] and SB-37-2 (Lushoto) [central and southern Mkomazi]. Prepared by Hunting Technical Services for the National Reconnaissance Level Land Use and Natural Resources Mapping Project, Ministry of Natural Resources and Tourism, Tanzania.

Tucker, C.J. (1979) Red and photographic infrared linear combinations for monitoring vegetation. *Remote Sensing of Environment* 8: 127-150.

Tucker, C.J. & Sellers, P.J. (1986) Satellite remote sensing of primary production. *International Journal of Remote Sensing* 7: 1395-1416.

van Jaarsveld, A.S., Freitag, S., Chown, S.L., Muller, C., Koch, S., Hull, H., Bellamy, C., Kruger, M., Endrody Younga, S., Mansell, M.W. & Scholtz, C.H. (1998) Biodiversity assessment and conservation strategies. *Science* 279: 2106-2108.

Van Wijngaarden, W. (1985) *Elephants—trees—grass—grazers: relationships between climate, soil,* vegetation *and large herbivores in a semi-arid ecosystem (Tsavo, Kenya)*. ITC Publication number 4.

Waring, R. H., Way, J., Hunt, E. R. Jnr., Morrissey, L., Ranson, K. J., Weishampel, J. F. & Franklin, E. (1995) Imaging radar for ecosystem surveys. *Bioscience* 45: 715-723.

Weg, R. F. & Mbuvi, J.F. (1975) *Soil of the Kindaruma Area*. Reconnaissance Soil Survey Report No. R2, Government Printer, Nairobi.

Wilson, D.E. (ed.) (1996) *Measuring and Monitoring Biological Diversity: Standard Methods for Mammals*. Biological diversity handbook series. Washington & London, Smithsonian Institution Press.

CHAPTER 5

The flora of Mkomazi and its regional context

Malcolm Coe, Kaj Vollesen, Raphael Abdallah & Emmanuel I. Mboya

Introduction

In spite of great international interest in biodiversity and its conservation we still have very few detailed floral checklists for much of eastern Africa. Those that do exist have been made by biologists who have spent varying periods in a particular locality. Sadly, the collecting effort varies considerably between them, making it extremely difficult to compare the flora of one geographical area with another. Where regular sampling efforts have been carried out, the accumulation of species new to the area with time in relation to the cumulative species total makes it fairly simple to estimate a realistic species total, although such records of a collecting sequence are difficult to obtain. In spite of the differences in sampling effort, the present study combined with those in other savanna areas in eastern Africa make it possible for us to look for distributional patterns that might tell us about the location of arid glacially induced refugia and other potential centres of speciation.

The Mkomazi Game Reserve lies at the southern end of the sahelian *Acacia-Commiphora* savanna, which is an important vegetation type within White's (1979) Somalia-Masai regional centre of endemism (RCE). The local MGR flora is however enriched by the area's high physiographic diversity, its proximity to the isolated coastal vegetation of the Zanzibar-Imhambane regional mosaic in the east, the elevated Usambara and Pare Mountains around the MGR's southern and western boundary, and the Taita Hills just over the border in Kenya. The rangeland surrounding the MGR (except for the small quadrant of the Tsavo West National Park in Kenya in the north-west) is becoming degraded at a frightening rate as human populations and their domestic stock continue to rise. The first casualty from such grazing, browsing and fire pressures is the steady loss of woody plant species, many of which have a very limited geographical range. More particularly, these species comprise important and often as yet unstudied items in the daily lives of the local people (see Chapter 34, Kiwasila & Homewood) and of significance perhaps to the whole world.

The Mkomazi Game Reserve flora

Although it is difficult to estimate the potential plant species total for a remote area like the MGR, Coe (1995) used the records of Harris (1972) and Greenway (1969) to examine the percent species overlap between the two areas and concluded that the potential species total was at least 1,026 species (Table 5.1).

Table 5.1 Predicted plant species numbers in the MGR and its environs. Sources: Greenway (1969) (Tsavo East National Park) and Harris (1972) (MGR).

group	total potential MGR species	Harris species not found in Greenway	Harris species found in Greenway	total species recorded by Harris
Ferns	11	0	0	0
Dicotyledons	773	107	58	165
Monocotyledons	242	49	46	95
total	1,026	156	104	260

We were able to call on a small number of records using plant specimens formerly lodged in the National Herbariums of Tanzania (Arusha) and Kenya (Nairobi); the 260 species recorded by Harris (1972) in the MGR; and Greenway's (1969) checklist for the Tsavo East National Park, whose Tsavo West salient is adjacent to the MGR. These data were incorporated into our tentative flora of probable and possible plant species (Coe & Windsor 1993), in advance of the first major period of field work in the MGR in 1994. At the time of going to press, the flora of the MGR totals 1,307 taxa, a figure made up of 1,148 taxa whose identity has been confirmed and a further 159 taxa which await confirmation of their species or are as yet undescribed. Bearing in mind that it was not possible to collect during the torrential rains of late 1997 and early 1998, it does not seem unrealistic that the eventual species total will exceed 1,500, or 15% of the 10,000 species estimated for the whole of Tanzania[1] by Polhill (1968).

The MGR flora is represented by 120 families/subfamilies (including 21 monocotyledons and 97 dicotyledons). 993 species are distributed between the 25 families which contain 15 or more species, or 76% of the total observed taxa (Table 5.2). It is worth noting here that 37 of the families (21%) are represented by only a single species.

The 21 largest genera in the MGR flora (Table 5.3) bear a striking numerical resemblance to those recorded in the Kora National Reserve, Kenya (KNR)[2] between 1982 and 1985 (Kabuye, Mungai & Mutangah 1986). These resemblances

[1] The MGR represents 0.43% of the total land area of Tanzania.

[2] The KNR covers an area of 1,700km^2 or 46% of the MGR.

Table 5.2 Plant families containing 15 or more species in the MGR.

family	species	family	species
Poaceae	130	Convolvulaceae	29
Fabaceae-Papilionoideae	99	Capparaceae	28
Euphorbiaceae	79	Cucurbitaceae	25
Acanthaceae	76	Amaranthaceae	22
Asteraceae	67	Boraginaceae	22
Rubiaceae	54	Verbenaceae	20
Lamiaceae	46	Commelinaceae	20
Fabaceae-Mimosoideae	37	Sterculiaceae	19
Cyperaceae	40	Combretaceae	18
Liliaceae	37	Tiliaceae	18
Malvaceae	37	Fabaceae-Caesalpinoideae	16
Asclepiadaceae	36	Ferns	19

do not have any great ecological or phytogeographical significance, for they are mainly very large genera in African and cosmopolitan floras at large; although in the case of *Maerua* and *Grewia* the species recorded respectively represent 20% and 8% of their total. The estimated species total for each genus (after Mabberley 1990) is included.

It is interesting to consider the MGR plant species in a local geographical rather than a Pan-African context, for there are striking patterns of changing species composition with distance. Kabuye *et al.* (1986) compared the flora of the Kora National Reserve (KNR) with that of the Meru National Park (MNP) (Ament & Gillett 1975), both in eastern Kenya. The arid south-eastern corner of the MNP is

Table 5.3 MGR plant genera containing 10 or more species. The first column gives the number of species recorded in the MGR with the projected total (i.e. including unnamed and provisionally named species) in brackets. The second column gives the approximate world total, after Mabberley (1990).

genus	MGR species	world total	genus	MGR species	world total
Euphorbia	24 (29)	1,600	Grewia	11 (12)	150
Acacia	22 (23)	1,200	Combretum	11 (12)	250
Justicia	17 (19)	420	Solanum	10 (11)	1,400
Ipomoea	15 (18)	500	Cassia	10 (11)	535
Crotalaria	13 (18)	600	Commiphora	9 (11)	190
Indigofera	16 (17)	700	Tephrosia	9 (10)	400
Hibiscus	15 (16)	200	Chlorophytum	9 (10)	300
Cyperus	13 (16)	600	Mariscus (= Cyperus)	9 (10)	
Plectranthus	8 (15)	300			
Barleria	12 (13)	250	Maerua	8 (10)	50
Eragrostis	12 (13)	350	Sporobolus	8 (10)	160

separated from the north-western corner of the KNR by the width of the Tana River, while the conservation areas of the Rahole and Bisanadi National Reserves abut the northern bank of the river. Kabuye *et al.* (1986) pointed out that Ament & Gillett (1975) recorded 536 species in the MNP while they had listed 720 species in the KNP. Of the MNP total only 312 species (58%) were shared with the KNR, while 224 species (42%) found in the MNP were not observed in Kora. Some caution is necessary when discussing such distributional data, but for the more conspicuous elements of at least the woody flora such presence or absence data may be considered to be of some biogeographic significance. In this case are some striking faunal examples of an almost inexplicable but genuine barrier posed by the Tana River, which in its middle reaches is seldom more than 50–100 m wide, yet isolates the Grevy zebra (*Equus grevyi*) and the reticulate giraffe (*Giraffa camelopardalis reticulata*) (Kingdon 1979) and sub-species of the flat-headed bat (*Platymops setiger macmillani* and *P. s. setiger*) to the north and south of the river respectively (Coe 1986).

We are fortunate in also being able to draw on botanical checklists for Tsavo East National Park (TNP), probably also applicable to most of lowland Tsavo West, adjacent to the MGR (Greenway 1969); the MGR (this volume); and the low-lying but diverse 50,000 km^2 Selous Game Reserve (SGR) (Vollesen 1980) in southern Tanzania. The approximate distance between central MNP and the SGR is 1,000 km, although the largest proportion of the distance separating these five conservation areas is that between the MGR and SGR (540 km). The MNP and KNR are adjacent; the KNR and TNP are separated by 210 km of largely lowland savanna; while Tsavo West National Park and the MGR are adjacent. In spite of the problems posed by comparing collecting programmes of different intensity and duration, it is instructive to compare their floral composition.

Since the SGR essentially falls within the more typically central african 'Miombo' woodland, we will initially examine the floral composition of the four northern conservation areas (Table 5.4). It is clear that numerically the number of species recorded in individual families is not closely correlated with the area being sampled. In many families the numbers recorded in the MGR, TNP and KNR are very similar and the number collected in the MNP are consistently lower. Bearing in mind that the MNP slopes from dense arid *Acacia-Commiphora* bush in the east (c. 370 m) to well-watered *Combretum* wooded grassland on the west (c. 800 m) at the foot of the Nyambeni Mountains (2,513 m) (Ament & Gillett 1975), it seems likely that this 1,813 km^2 National Park is almost certainly much richer than these records suggest. Interestingly some families (Rubiaceae, Malvaceae, Fabaceae-Mimosoideae, Cyperaceae, Tiliaceae, Combretaceae and Fabaceae-Caesalpinoidea) exhibit similar species numbers in the TNP, KNR and MNP, suggesting that it is perhaps not only a question of collecting intensity.

At the species level we can observe how the compositions of the genera *Acacia*, *Commiphora* and *Grewia* change with distance. Between the MNP in the north and the SGR in the south, we have data on the distributions of 37 taxa (36 species) (Table 5.5), about 27% of the African *Acacia* species. The species recorded within the MGR represent 61% of the total, while the other four conservation areas exhibit very similar levels, 13–17 taxa or 36–47%. Ross's (1979) study of the African *Acacia* species reveals that the Somalia-Masai regional centre of endemism represents the African centre of the species radiation and/or the location of a glacial refuge. If, however, we examine *Acacia* occurence in the five conser-

Table 5.4 Comparison of the MGR plant families represented by 15 or more species with those of other adjacent semi-arid conservation areas. Study areas: MGR–Mkomazi Game Reserve; TNP–Tsavo National Park, Kenya (Greenway 1969); KNR–Kora National Reserve, Kenya (Kabuye *et al.* 1986); MNP–Meru National Park, Kenya (Ament & Gillett 1975).

	location and area (km^2)			
family	MGR 3,250	TNP 20,821	KNR 1,700	MNP 1,813
Poaceae	130	106	71	72
Fabaceae-Papilionoideae	99	61	49	45
Acanthaceae	76	62	46	25
Euphorbiaceae	79	58	39	30
Asteraceae	67	45	36	13
Rubiaceae	54	21	15	16
Lamiaceae	46	36	16	13
Cyperaceae	40	26	28	21
Malavaceae	37	22	19	16
Fabaceae-Mimosoideae	37	20	17	23
Liliaceae	37	24	20	14
Asclepiadaceae	36	25	13	8
Convolulaceae	29	30	22	17
Capparaceae	28	24	29	12
Cucurbitaceae	25	28	19	5
Amaranthaceae	22	18	19	7
Boraginaceae	22	18	19	9
Verbenaceae	20	14	13	9
Commelinaceae	20	15	9	9
Sterculiaceae	19	14	13	7
Tiliaceae	18	13	18	10
Combretaceae	18	12	13	12
Fabaceae-Caesalpinoidea	16	15	14	13
Ferns	19	13	18	10
total species (*all* families)	1,307	955	734	566
total families	120	88	77	75

Table 5.5 Distribution of *Acacia* species (Fabaceae–Mimosaceae) in five east African conservation areas. Taxonomy after Beentje (1994), Coe (1991) and Ross (1979).

Acacia species	MNP*	KNR	TNP	MGR	SGR	total
A. adenocalyx					+	2
A. (faidherbia) albida	+?			+	+?	1(3)
A. ancistroclada				+		1
A. ataxacantha	+	+				2
A. brevispica	+		+	+		3
A. bussei	+	+	+	+		4
A. clavigera			+			1
A. dolichocephala				+		1
A. drepanolobium	+			+		2
A. elatior elatior	+	+	+			3
A. etbaica				+		1
A. gerrardii				+	+	2
A. goetzii					+	1
A. hamulosa		+				1
A. hockii	+			+	+	3
A. horrida	+	+	+			3
A. kirkii		+				1
A. (macrothyrsa) amythethophylla					+	1
A. mellifera	+	+	+	+		4
A. nigrescens					+	1
A. nilotica ssp. *kraussiana*	+		+	+	+	4
A. paolii		+				1
A. pentagona				+		1
A. polyacantha			+	+	+	3
A. reficiens	+	+	+	+		4
A. robusta var. *clavigera*					+	1
A. robusta var. *usambarensis*	+		+	+	+	4
A. rovumae				+	+	2
A. senegal	+	+	+	+	+	5
A. seyal var. *fistula*	+			+		2
A. sieberana					+	1
A. stuhlmannii	+	+	+	+		4
A. taylorii					+	1
A. thomasii			+	+		2
A. tortilis	+	+	+	+		4
A. xanthophloea	+	+	+?			2(3)
A. zanzibarica	+		+	+	+	4
total	17	13	16	22	15	

* MNP–Meru National Park; KNR–Kora National Reserve; TNP–Tsavo National Park; MGR–Mkomazi Game Reserve; SGR–Selous Game Reserve.

vation areas, 15 species (41.7%) are recorded in only one of the areas; eight, five and eight species in two, three and four areas respectively (13.9–22.2%); while only a single species (*A. senegal*) is found in all five conservation areas. Geographically widespread species such as *A. nilotica and A. tortilis* should be present in all five areas or at least in the vicinity. The presence of 15 species whose distribution is truly restricted is also of interest, although this is not always true, for the presence of *A. nigrescens* in the Selous Game Reserve indicates a strong southern/central African influence in southern Tanzania. *A. adenocalyx* is limited to southern Tanzania and central Mozambique and *A. taylorii* is restricted to southern Tanzania. *A. hamulosa*, *A. horrida* and *A. paolii* are all representatives of the strictly Somalian flora and extend their range into southern Kenya.

The genus *Commiphora* was recently reviewed for the *Flora of Tropical East Africa* (FTEA) (Gillett 1991), which enables us to look at its distribution across our five conservation areas with a reasonable level of taxonomic reliability. About 190 species are known to be centred in the Somalian-Arabian area, although the species extend to western India and South America. Up to 65 species (about 35% of the total) occur within the FTEA area (Kenya, Tanzania and Uganda). Table 5.6 lists the 28 species found between the MNP and the SGR and demonstrates that these species are largely arid-adapted forms that are concentrated in the east of these countries, spreading southwards from the Somalian region. Gillett (1991) underlines this pattern when he points out that the species (taxa) recorded in the FTEA area were 6(7) in Uganda, 51(64) in Kenya and 35(48) in Tanzania. Indeed by the time we reach southern Africa, only 33 species are recorded in the whole region (Palgrave 1981) and of those only five also occur in the FTEA area.

Of the *Commiphora* species recorded in the five conservation areas, 17 (63%) are found in only one area, five (18.5%) are observed in two, one in three, three in four, and only a single species (the widespread *C. africana*) is recorded in all five. Although a small numbers of species are widespread, the majority exhibit relatively restricted distributions (Gillett 1991). Like so many African tree genera, it is surprising that they are so poorly collected over much of their range, despite the potential importance of the oleo-gum resins that they contain (Waterman 1986).

The third genus included in this examination of species distribution patterns is *Grewia,* whose species are widespread in savanna Africa and of considerable ecological significance as sources of pollen and nectar for arthropods and for their nutrient-rich seeds which are significant in the diet of many mammals and birds and in many cases are an important source of famine food for indigenous human populations (Beentje 1994). 22 species[1] (plus two potential species) are recorded between the five conservation areas (Table 5.7), out of a total of 150 species in the Old World tropics. There are significantly more species in the MGR and SGR than in those to the north. As in the cases of *Acacia* and *Commiphora* above, only

[1] The family Tiliaceae has not yet been published in the FTEA making the nomenclature provisional only.

Table 5.6 Distribution of *Commiphora* species (Burseraceae) in five east African conservation areas. Taxonomy after Gillett (1991).

Commiphora species	MNP*	KNR	TNP	MGR	SGR	total
C. africana	+	+	+	+	+	5
C. baluensis				+		1
C. campestris	+	+	+	+		4
C. confusa		+		+		2
C. edulis	+	+	+	+		4
C. eminii				+		1
C. engleri	+					1
C. erosa		+				1
C. fulvotomentosa					+	1
C. habessinica				+		1
C. holtziana	+	+	+	+		4
C. incisa		+				1
C. kua	+					1
C. longipedicillata		+				1
C. madagascariensis	+		+		+	3
C. merkeri			+	+		2
C. mildbraedii		+	+			2
C. mollis	+					1
C. mombassensis					+	1
C. pedunculata					+	1
C. pteleifolia					+	1
C. rostrata	+	+				2
C. sambarensis	+					1
C. schimperi	???		+	+	???	2
C. serrata					+	1
C. unilobata		+				1
C. zanzibarica					+	1
total	10	11	8	10	8	

* MNP – Meru National Park; KNR – Kora National Reserve; TNP – Tsavo National Park; MGR – Mkomazi Game Reserve; SGR – Selous Game Reserve.

one species (*G. bicolor*) has been recorded in all five conservation areas, while a further five species (*G. fallax, G. forbesii, G. tembensis, G. tenax* and *G. villosa*) are found in four out of five. Seven of these species are also recorded by Palgrave (1981) in southern Africa although only three (*G. bicolor, G. tenax* and *G. villosa*) are found from the MNP to the south. Of special interest is that four of these species (*G. lepidopetala, G. micrantha, G. microcarpa* and *G. sulcata*) are limited to the SGR, another example of the botanical affinity of this area with the southern African flora. 13 of the *Grewia* species are found in only one of the five conservation areas and as above, four of these are southern African forms.

Discussion

In our attempts to understand the importance of species richness and abundance in tropical environments, it is increasingly apparent that the type of field work that has been carried out in the MGR, largely by groups of unpaid volunteers, is vital if

Table 5.7 Distribution of *Grewia* species (Tiliaceae) in five east African conservation areas. Taxonomy after Beentje (1994) and Vollesen (1980).

Grewia species	MNP*	KNR	TNP	MGR	SGR	total
G. bicolor	+	+	+	+	+	5
G. calymmatosepala					+	1
G. conocarpa					+	1
G. densa		+				1
G. fallax		+	+	+	+	4
G. forbesii		+	+	+	+	4
G. goetzeana				+		1
G. herbacea					+	1
G. holstii					+	1
G. lepidopetala					+	1
G. lilacina	+	+	+			3
G. micrantha					+	1
G. microcarpa					+	1
G. plagiophylla		+				1
G. stulmannii					+	1
G. sulcata				+	+	2
G. tembensis (includes var. *nematopus & kakothamnos*)	+	+	+	+		4
G. tenax	+	+	+	+		4
G. tephrodermis				+		1
G. tristis			+	+		2
G. truncata				+		1
G. villosa	+	+	+	+		4
total	5	9	8	11	12	

* MNP – Meru National Park; KNR – Kora National Reserve; TNP – Tsavo National Park; MGR – Mkomazi Game Reserve; SGR – Selous Game Reserve.

we are to understand what we have on the ground and what we likely to lose if land degradation continues at its present rate. It is not enough for those studying biodiversity to sit in the world's herbaria and produce lists of what has been collected and recorded to date: in many cases large areas remain entirely unstudied, with as many as 15 or 20% of the species being either little known or even entirely new and un-named.

The brief consideration of the distribution of just three genera above has demonstrated that a small number of species are widespread, while the majority have much more limited distributions. Of special interest though are the plants which show disjunct patterns between the area under consideration and the southern African region—exemplified by the euphorbiaceous tree species *Heywoodia lucens*, which is isolated in the Igire forest of the MGR and coastal regions of Natal and Mozambique with virtually no records in between.

It is hoped that the Division of Wildlife, in conjunction with the National Herbarium at TPRI, will ensure that futher extensive collections are made over the whole MGR, especially during the wet seasons, in order that we may have a complete understanding of plant biodiversity on this very rich environment.

Acknowledgements

This checklist represents a programme of intensive collecting which was carried out by the authors and many other members of the MERP between 1994 and 1998. In addition, periods spent by RA, EM and KV in the National Herbaria in Arusha (TPRI) and Nairobi were generously facilitated by Dr William Mziray in Tanzania and Geoffrey Mungai and Christine Kabuye in Kenya. We must also record our grateful thanks to Professor Sir Ghillean Prance, Professor Grenville Lucas and Dr Roger Polhill, all of the Royal Botanic Gardens, Kew for their help and support. Above all, though, we must all thank the Department of Wildlife for allowing us to contribute to our understanding of the very high plant species richness exhibited by the Mkomazi Game Reserve.

References

Ament, J.G. & Gillett, J.B. (1975) The vascular plants of Meru National Park, Kenya. Part 2. Checklist of the vascular plants recorded. *Journal East African Natural History Society and National Museum* 154: 11-34.

Beentje, H. (1994) *Kenya Trees, Shrubs and Lianes*. National Museums of Kenya, Nairobi.

Coe, M. (1986) The ecology of rock outcrops in the Kora National Reserve, Kenya. In: Coe, M. & Collins, N.M. (eds.) *Kora: an Ecological Inventory Study of the Kora National Reserve, Kenya*. Royal Geographical Society, London. pp. 159-171.

Coe, M. (1995) Botanical studies in the Mkomazi Game Reserve. In: Coe, M. and Stone, G.N. (eds.) *Mkomazi Ecological Research Programme: Progress Report, July 1995*. Royal Geographical Society, London. pp. 7-12.

Coe, M. & Windsor, N. (eds.) (1993) *Introduction, Flora & Fauna Species Lists and Bibliography for Mradi wa Utafiti, Mkomazi (Mkomazi Ecological Research Programme) 1993-1996*. Department of Wildlife, Tanzania; Royal Geographical Society, London; University of Oxford.

Greenway, P.J. (1969) A checklist of plants recorded in Tsavo National Park, East. *Journal East African Natural History Society and National Museum* 27(3): 162-209.

Gillett, J.B. (1991) Burseraceae. In Polhill, R.M. (ed.) *Flora of Tropical East Africa*. A.A. Balkema, Rotterdam.

Harris, L.D. (1972) *An ecological description of a semi-arid East African ecosystem*. Science Series No. 11, Colorado University Range Science Department.

Kabuye, C.H.S., Mungai, G.M. & Mutangah, J.G. (1986) Flora of Kora National Reserve. In: Coe, M. & Collins, N.M. (eds.) *Kora: an Ecological Inventory Study of the Kora National Reserve, Kenya*. Royal Geographical Society, London. pp. 57-104.

Kingdon, J. (1979) *East African Mammals (An Atlas of Evolution in Africa)*. Vol. IIIB. Academic Press, London.

Mabberley, D.J. (1990) *The Plant-Book: A Portable Dictionary of Higher Plants*. Cambridge University Press, Cambridge.

Palgrave, K.C. (1981) *Trees of Southern Africa*. C. Struik, Cape Town.

Polhill, R.M. (1968) Tanzania. In: Hedberg, I. & Hedberg, O. (eds.) *Conservation of Vegetation in Africa South of the Sahara*. Acta Phytogeographica Suecica 54. pp. 166-178.

Ross, J.H. (1979) *A Conspectus of the African Acacia Species*. Memoirs of the Botanical Survey of South Africa 44.

Vollesen, K. (1980) *Annotated check-list of the vascular plants of the Selous Game Reserve, Tanzania*. Opera Botanica 59.

Waterman, P.G. (1986) Resins and other exudates from the flora of the Kora National Reserve. In: Coe, M. & Collins, N.M. (eds.) *Kora: an Ecological Inventory Study of the Kora National Reserve, Kenya*. Royal Geographical Society, London.

White, F. (1979) The Guineo-Congolian region and its relationship with other phytochoria. *Bulletin Jardin Botanique Nationale Belgique*. 49(1&2), 11-55.

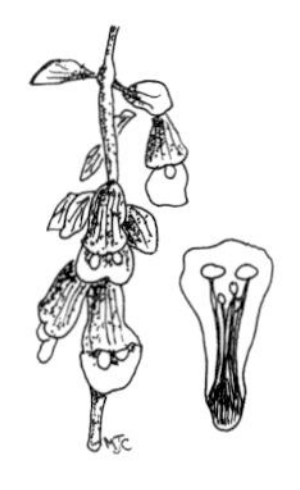

CHAPTER 6

Checklist: Vascular plants and pteridophytes of Mkomazi

Kaj Vollesen, Raphael Abdallah, Malcolm Coe & Emmanuel Mboya

Species marked '*' are as yet unnamed or only provisionally named. These are *not* included in the species total given after each family name. Abbreviations:
sp. nov. — *species nova* (new species)
sp. nov. aff. — *species nova affinis* (new species with a strong affinity to)
SR — sight record
s.n. — *sine numero* (without number)

GYMNOSPERMS

Cycadaceae (1)
Encephalartos kisambo Faden & Beentje. Rocky hilltops: rare (only on Kisima Hill). Abdallah 3000; Abdallah & Coe s.n.

ANGIOSPERMS, DICOTYLEDONS

Acanthaceae (70)
Asystasia charmian S. Moore. Seasonally inundated grassland, Acacia and Acacia-Commiphora bushland; locally common. Abdallah 2846, 3100; Abdallah & Mboya 3995; Abdallah & Vollesen 95/49; Vollesen 96/46.
Asystasia gangetica (L.) T. Anderson. Montane forest, riverine forest, Acacia-Commiphora bushland; occasional. Abdallah & Mboya 3926, 3991, 3997.
Asystasia mysorensis (Roth) T. Anderson. Seasonally inundated grassland, Acacia-Commiphora bushland (disturbed places); occasional. Abdallah 2880, 3063, 3152.
Asystasia sp. nov., not matched at Kew. Seasonally inundated grassland, rocky hilltops; rare. Abdallah & Vollesen 95/66; Vollesen 96/39.
Barleria acanthoides Vahl. Acacia-Commiphora bushland; occasional. Abdallah 2613; Bally 259.
Barleria argentea Balf. f. Acacia-Commiphora bushland; common. Abdallah 3141, 3260, 3309; Abdallah & Mboya 4032; Abdallah & Vollesen 95/19.
Barleria eranthemoides C.B. Clarke. Acacia-Commiphora bushland; common. Abdallah 3006, 3312; Abdallah & Mboya 3453, 3619; Abdallah & Vollesen 95/56; Cox & Abdallah 2304, Greenway 4579.
Barleria holstii Lindau. Montane forest, riverine forest; rare. Abdallah & Mboya 3459; Abdallah & Vollesen 95/110.
Barleria marginata Oliv. Seasonally inundated grassland, Acacia bushland; occasional. Abdallah 3242; Abdallah & Vollesen 95/17.
Barleria mucronata Lindau. Acacia-Combretum bushland; rare. Abdallah, Mboya & Vollesen 96/62; Abdallah & Vollesen 95/71.
Barleria ramulosa C.B. Clarke. Acacia-Commiphora bushland; occasional. Abdallah 3221, 3352; Abdallah & Vollesen 95/47; Faulkner 4428; L.D. Harris 116.
Barleria submollis Lindau. Montane forest, rocky hilltops; occasional. Abdallah & Mboya 3400; Abdallah & Vollesen 95/29, 95/68.
Barleria taitensis S. Moore. Acacia-Commiphora bushland (eastern); locally common. Abdallah 3207; Abdallah & Mboya 4178; Abdallah, Mboya & Vollesen 96/171.
Barleria ventricosa Nees. Montane forest; rare. Abdallah, Mboya & Vollesen 96/74.

Barleria sp. nov. aff. *B. acanthoides* Vahl (= Luke et al TPR700, etc.). Acacia-Commiphora bushland (eastern); rare. Abdallah, Mboya & Vollesen 96/172.

Barleria sp. nov. aff. *B. spinisepala* E.A. Bruce (= Vollesen MRC4642). Seasonally inundated grassland, Acacia bushland; rare. Abdallah & Mboya 3703; Abdallah, Mboya & Vollesen 96/114.

Blepharis edulis (Forssk.) Pers. Acacia-Commiphora bushland; locally common. Abdallah 3209.

Blepharis hildebrandtii Lindau subsp. *hildebrandtii*. Acacia-Commiphora bushland (eastern); locally common. Abdallah & Mboya 4050, 4051; Semsei 2122.

Blepharis integrifolia (L. f.) Schinz var. *integrifolia*. Seasonally inundated grassland, Acacia-Combretum bushland; occasional. Abdallah & Mboya 3949; Abdallah & Vollesen 95/136; L.D. Harris 65

Blepharis maderaspatensis (L.) Roth. Montane forest, riverine forest, Acacia-Commiphora bushland; occasional. Abdallah 3354; Abdallah & Mboya 4009, 4059; Abdallah & Vollesen 95/58.

Brillantaisia pubescens Oliv. var. *pubescens*. Riverine forest; rare. Abdallah & Vollesen 95/176.

Crabbea velutina S. Moore. Acacia-Combretum bushland, Acacia-Commiphora bushland; common. Abdallah 2719, 2869; Abdallah & Mboya 3853, 3879.

Crossandra friesiorum Mildbr. Montane forest; locally common. Abdallah & Mboya 3834; Abdallah, Mboya & Vollesen 96/81.

Crossandra mucronata Lindau. Riverine forest, rocky hilltops; occasional. Abdallah 2614; Abdallah & Mboya 3696, 3915; Abdallah & Vollesen 95/177; L.D. Harris 57A, 183.

Crossandra stenostachya Lindau. Seasonally inundated grassland; locally common. Abdallah 3287; Abdallah & Vollesen 95/131; Greenway 6496.

Crossandra subacaulis C.B. Clarke. Acacia-Combretum bushland, seasonally-inundated grassland; common. Abdallah 2615; Abdallah & Mboya 3653; Vollesen 96/37.

Dicliptera umbellata (Vahl) Juss. Montane forest; rare. Abdallah & Vollesen 96/259.

Duosperma kilimandscharicum (Lindau) Dayton. Acacia-Commiphora bushland (eastern), rocky outcrops; common. Abdallah & Mboya 3871; Abdallah, Mboya & Vollesen 96/162.

Duosperma sp. nov. aff. *D. eremophilum* (Milne-Redh.) Brummitt (= Mwasumbi 10841). Seasonally inundated grassland; locally common. Abdallah, Mboya & Vollesen 96/150; Abdallah & Vollesen 95/133.

Dyschoriste hildebrandtii (S. Moore) S. Moore. Acacia-Commiphora bushland, Acacia-Combretum bushland; common. Abdallah 2612, 3012, 3320, 3439; Abdallah & Mboya 3869, 4057; Abdallah & Vollesen 95/4, 95/163; Cox & Abdallah 2301.

Dyschoriste radicans Nees. Dry montane forest, Acacia-Combretum bushland; rare. Vollesen 96/19.

Dyschoriste sp. nov. (= L.D. Harris 129). Dry montane forest, Acacia-Combretum bushland, rocky hilltops; locally common (western part only). Abdallah & Vollesen 95/32, 95/54; Greenway 6497A; L.D. Harris 129; Vollesen 96/21.

Ecbolium subcordatum C.B. Clarke var. *glabratum* Vollesen. Acacia-Commiphora bushland; rare. Abdallah 3104; Greenway 3959, Semsei 2126.

Hygrophila auriculata (Schumach.) Heine. Seasonally inundated grassland; common. Greenway 3940; Semsei 2337.

Hypoestes aristata (Vahl) Roem. & Schult. Montane forest, rocky hilltops; occasional. Abdallah, Mboya & Vollesen 96/135.

Hypoestes forskaolii (Vahl) R. Br. Acacia-Combretum bushland, Acacia-Commiphora bushland; rare. Abdallah & Mboya 3944.

Justicia betonica L. Acacia-Combretum bushland; rare. Vollesen SR.

Justicia calyculata Defl. Acacia-Commiphora bushland; rare. Abdallah 3090; Abdallah & Vollesen 95/118.

Justicia capensis Thunb. Montane forest; rare. Abdallah & Mboya 3839; Abdallah, Mboya & Vollesen 96/87.

Justicia coerulea Forssk. Acacia-Commiphora bushland, seasonally-inundated grassland; locally common. Abdallah 2626, 2762, 2839; Abdallah & Mboya 3580; B.J. Harris 792; L.D. Harris 1; Semsei 2119.

Justicia cordata (Nees) T. Anderson. Acacia bushland, Acacia-Commiphora bushland; locally common. Abdallah & Mboya 3943, 4078; Abdallah & Vollesen 95/18.

Justicia diclipteroides Lindau. Acacia-Combretum bushland, Acacia-Commiphora bushland, rocky hilltops, montane forest; common. Abdallah 2654, 3315; Abdallah, Mboya & Vollesen 96/78; Abdallah & Vollesen 95/57, 95/84; Greenway 2128.

Justicia engleriana Lindau. Montane forest, riverine forest; locally common. Abdallah & Mboya 4239; Abdallah & Vollesen 95/104.

Justicia flava (Vahl) Vahl. All types of bushland; common. Abdallah 2617, 2747, 3371; Cox & Abdallah 2324; L.D. Harris 1A.

Justicia heterocarpa T. Anderson. Acacia-Commiphora bushland, rocky hilltops; occasional. Abdallah 3329, 3379; Abdallah & Mboya 4026.
Justicia matammensis (Schweinf.) Oliv. Acacia-Combretum bushland, Acacia-Commiphora bushland; occasional. Abdallah 2625, 3231; Abdallah & Vollesen 95/173.
Justicia nyassana Lindau. Montane forest, riverine forest; locally common. Abdallah 2755, 2765, 2766; Abdallah & Mboya 3702, 3874; Abdallah & Vollesen 95/78.
Justicia odora (Forssk.) Vahl. Acacia-Commiphora bushland; occasional. Abdallah 3311.
Justicia ornatopila Ensermu. Seasonally inundated grassland; locally common. Abdallah 2845; Abdallah & Vollesen 95/167.
Justicia pseudorungia Lindau. Montane forest; rare. Abdallah, Mboya & Vollesen 96/99.
Justicia scandens Vahl. Riverine forest, riverbeds; locally common. Abdallah 3316; Abdallah & Vollesen 95/148.
Justicia stachytarphetoides C.B. Clarke. Riverine forest (Umba River); rare. Abdallah, Mboya & Vollesen 96/195.
Justicia striata (Klotzsch) Bullock. Montane forest, rocky hilltops; locally common. Abdallah 3384; Abdallah & Mboya 3842; Abdallah & Vollesen 95/67.
Lepidagathis scabra (Lindau) C.B. Clarke. Acacia-Commiphora bushland, Acacia-Combretum bushland; occasional. Abdallah 3359; Abdallah & Vollesen 95/22, 95/72.
Lepidagathis scariosa Nees. Acacia-Commiphora bushland; locally common. Abdallah 3189, 3255, 3310; Greenway 2038, 4576.
Megalochlamys revoluta (Lindau) Vollesen subsp. *revoluta*. Acacia bushland; locally common. Abdallah 3319; Abdallah & Vollesen 95/15; Greenway 4566, Semsei 2123.
Metarungia pubinervia (T. Anderson) C. Baden. Montane forest; locally common. Abdallah 3520; Abdallah, Mboya & Vollesen 96/94; Vollesen 96/14.
Monechma debile (Forssk.) Nees. All types of bushland, rocky hilltops; common. Abdallah & Vollesen 95/63; B.J. Harris 791.
Neuracanthus tephrophyllus Bidgood & Brummitt subsp. *conifer* Bidgood & Brummitt. Acacia bushland; locally common. Abdallah & Mboya 4070; Abdallah & Vollesen 95/16; Greenway 6497B; Mbano CAWM 5790; Vollesen 96/49.
Peristrophe paniculata (Forssk.) Brummitt. Acacia-Combretum bushland; occasional. Vollesen SR.
Phaulopsis imbricata (Forssk.) Sweet subsp. *imbricata*. Montane forest; locally common. Abdallah & Vollesen 95/94; Vollesen 96/9.
Pseuderanthemum hildebrandtii (Lindau) C.B. Clarke. Montane forest, riverine forest; locally common. Abdallah 3223; Abdallah & Mboya 3966; L.D. Harris 32.
Rhinacanthus gracilis Klotzsch. Riverine forest, rocky riverbeds; rare. Abdallah & Vollesen 95/45.
Ruellia bignoniiflora S. Moore. Acacia-Commiphora bushland, rocky hilltops; occasional. Abdallah 2652, 3442; Abdallah & Mboya 3832; Abdallah & Vollesen 95/158.
Ruellia patula Jacq. All types of bushland; common. Vollesen SR.
Ruellia prostrata Poir. Montane forest, riverine forest; occasional. Vollesen SR.
Thunbergia alata Sims. Montane forest, Acacia-Commiphora bushland; occasional. Abdallah & Mboya 3564, 3649.
Thunbergia holstii Lindau. Rocky hilltops and ravines, Acacia-Commiphora bushland; occasional. Abdallah 2803; Abdallah & Mboya 3783; Abdallah & Vollesen 95/14; Greenway 6486.
Thunbergia guerkeana Lindau. Acacia-Commiphora bushland (eastern); rare. Abdallah 2750; Abdallah & Mboya 4106; Greenway 2144.
Thunbergia sp. nov. D of Kew. Acacia-Commiphora bushland; locally common. Abdallah & Mboya 3446, 4109; Abdallah, Mboya & Vollesen 96/153, 96/212; Abdallah & Vollesen 95/21.
**Barleria* sp. L.D. Harris 66.
**Duospermum crenatum*. Abdallah & Mboya 4052.
**Dyschoriste clinopodioides*. Abdallah 3390; Abdallah & Mboya 3988.
**Justicia anselliana*. Abdallah & Mboya 3723.
**Justicia uncinulata*. Abdallah 3027.
**Thunbergia* sp. Abdallah 2616.

Aizoaceae (1)
Zaleya pentandra (L.) C. Jeffrey. Acacia-Combretum bushland, Acacia-Commiphora bushland; occasional. Vollesen SR.

Amaranthaceae (20)
Achyranthes aspera L. var. *pubescens* (Moq.) C.C. Townsend. All types of bushland, montane forest; common. Abdallah 3170, 3377; Abdallah & Mboya 3517.
Aerva lanata (L.) Schult. All types of bushland; occasional. Abdallah 3036; Abdallah & Mboya 3704.
Aerva persica (Burm. f.) Merr. All types of bushland; occasional. Greenway 1989, 3994; L.D. Harris 179; Semsei 2373.

Alternanthera pungens Kunth. Weed; occasional. Vollesen SR.

Alternanthera sessilis (L.) DC. Seasonally inundated grassland, waterholes; occasional. Abdallah & Mboya 3503.

Amaranthus graecizans L. Seasonally inundated grassland, waterholes, weed; occasional. Abdallah 3058.

Celosia schweinfurthiana Schinz. Rocky hilltops, riverine forest; occasional. Vollesen SR.

Celosia trigyna L. Acacia-Combretum bushland, rocky hilltops; occasional. Abdallah 3262.

Centemopsis kirkii (Hook. f.) Schinz. Acacia-Commiphora bushland; occasional. Abdallah 2599, 3175; L.D. Harris 47; Mwamba 47.

Centrostachys aquatica (R. Br.) Moq. Temporary waterholes, dams; locally common. Abdallah & Mboya 3421; Vollesen 96/23.

Cyathula cylindrica Moq. Rocky hilltops; rare. Abdallah 3374.

Cyathula orthacantha (Aschers.) Schinz. Seasonally inundated grassland, Acacia-Commiphora bushland; common. Vollesen SR.

Dasysphaera tometosa Lopr. All types of bushland; common. Abdallah 3324; Ibrahim 635; Vollesen 96/33.

Digera muricata (L.) Mart. subsp. *trinervis* C.C. Townsend. All types of bushland; common. Abdallah 3014; Greenway 2134; L.D. Harris 7.

Gomphrena celosioides Mart. Weed; occasional. Vollesen SR.

Psilotrichum elliotii Bak. Acacia-Combretum bushland, rocky hilltops; common. Abdallah 2784, 3382; Abdallah & Mboya 3530.

Pupalia lappacea (L.) A. Juss. var. *velutina* (Moq.) Hook. f. All types of bushland, rocky outcrops; common. Abdallah 3259; L.D. Harris 25; Semsei 2133.

Sericocomopsis hildebrandtii Schinz. All types of bushland; common. L.D. Harris 32A, 59; Vollesen 96/34.

Sericocomopsis pallida (S. Moore) Schinz. All types of bushland; common. Abdallah 3065; Greenway 1991, 3960; Gillman 745; L.D. Harris 75; Semsei 2136.

Volkensinia prostrata (Gilg) Schinz. All types of bushland; rare. Semsei 2156.

**Cyathula prostrata*. Abdallah 2621.

**Cyathula* sp. Abdallah & Mboya 3454.

Anacardiaceae (9)

Lannea alata (Engl.) Engl. Acacia-Commiphora bushland; locally common. Abdallah 2758, 3084; Abdallah & Mboya 3759, 3803, 3896; Abdallah & Vollesen 95/24; Bally 145; Greenway 9672.

Lannea rivae (Chiov.) Sacleux. Acacia-Combretum bushland; occasional. Abdallah 2637.

Lannea schweinfurthii (Engl.) Engl. var. *stuhlmannii* (Engl.) Kokwaro. Acacia-Combretum bushland, rocky hilltops; common. Abdallah 2537, 2689, 2735; Abdallah & Mboya 3637, 3669, 3670, 3689, 3691, 3908; L.D. Harris 128.

Lannea triphylla (A. Rich.) Engl. All types of bushland, rocky hilltops; common. Abdallah 2636, 2871, 3222; Abdallah & Mboya 3693, 3774, 3812, 3909.

Ozoroa insignis Del. subsp. *reticulata* (Bak. f.) Gillett. Acacia-Combretum bushland, rocky hilltops; occasional. Abdallah 2825.

Rhus natalensis Krauss. Acacia-Combretum bushland, rocky hilltops and ravines, riverine forest; occasional. Abdallah & Mboya 3494, 3569; Abdallah, Mboya & Vollesen 96/55.

Rhus tenuinervis Engl. Rocky hilltops, montane forest; rare. Abdallah, Mboya & Vollesen 96/138.

Sclerocarya birrea (A. Rich.) Hochst. subsp. *caffra* (Sond.) Kokwaro. Acacia-Combretum bushland; occasional. Greenway SR; Vollesen SR.

Sorindeia madagascariensis DC. Riverine forest; locally common. Vollesen SR.

Annonaceae (5)

Annona senegalensis Pers. Acacia-Combretum bushland; occasional. Vollesen SR; Greenway SR.

Uvaria acuminata Oliv. Montane forest, rocky hilltop thickets; occasional. Abdallah 3123, 3376.

Uvaria leptocladon Oliv. Dry montane forest; locally common. Vollesen SR.

Uvaria scheffleri Diels. Montane forest; occasional. Vollesen SR.

Uvariodendron anisatum Verdc. Montane forest; occasional. Abdallah, Mboya & Vollesen 96/86; Vollesen 96/15.

Apiaceae (3)

Heteromorpha trifoliata (Wendl.) Eckl. & Zeyh. Acacia-Combretum bushland, rocky hilltops; occasional. Abdallah 3389.

Steganotaenia araliacea Hochst. Rocky hilltops and ravines; rare. Abdallah & Mboya 4063; Hughes 242.

Trachyspermum aethusifolium Chiov. var. *aethusifolium*. Seasonally inundated grassland, Acacia bushland; locally common. Abdallah 3280; Abdallah, Mboya & Vollesen 96/154.

Apocynaceae (4)

Adenium obesum (Forssk.) Roem. & Schult. Acacia-Combretum bushland, rocky outcrops; occasional. Vollesen SR.

Carissa edulis Vahl. Dry montane forest, ravines; occasional. Abdallah & Mboya 3477, 3585.

Landolphia buchananii (Hall. f.) Stapf. Montane forest; occasional. Abdallah & Mboya 3848.

Landolphia kirkii Dyer. Acacia-Combretum bushland, rocky hilltops; occasional. Vollesen SR.

**Apocynaceae/Asclepiadaceae*. Abdallah & Mboya 4000, 4049, 4135.

Araliaceae (1)

Cussonia holstii Engl. var. *holstii*. Montane forest, rocky hilltops; rare. Vollesen SR.

Aristolochiaceae (1)

Aristolochia bracteolata Lam. Seasonally inundated grassland; occasional. Abdallah & Mboya 4037, 4136; Abdallah, Mboya & Vollesen 96/147; Greenway 1973, 3951.

Asclephiadaceae (32)

Baseonema gregorii Schltr. & Rendle. Acacia-Commiphora bushland; rare. Cox & Abdallah 2028.

Brachystelma sp., material insufficient. Acacia-Commiphora bushland; rare. Abdallah & Mboya 3948.

Caralluma gracilipes K. Schum. Acacia-Commiphora bushland; rare. Greenway 6492.

Caralluma priogonium K. Schum. Acacia-Commiphora bushland; rare. Greenway 3939, 4066.

Caralluma speciosa (N. E. Br.) N. E. Br. Acacia-Combretum bushland, rocky hilltops; occasional. Abdallah & Mboya 3473; Greenway 3938, 4067, Richards 21927.

Ceropegia meyeri-johannis Engl. Montane forest; rare. Abdallah & Mboya 4176.

Ceropegia stenantha K. Schum. Montane forest; rare. Abdallah & Mboya 4062.

Ceropegia sp. ? nov. aff. *C. brevirostris* Bally & Field, not matched at Kew. Montane forest; rare. Abdallah, Mboya & Vollesen 96/67.

Cryptolepis sp. ? nov. aff. *C. obtusa* N. E. Br., not matched at Kew. Riverine forest; rare. Abdallah & Mboya 3979.

Curroria volubilis (Schltr.) Bullock. Acacia-Commiphora bushland; rare. Burtt 5325.

Cynanchum altiscandens K. Schum. Montane forest, ravines; rare. Abdallah & Mboya 4084.

Diplostigma canescens K. Schum. Acacia-Commiphora bushland, rocky outcrops; occasional. Abdallah 3139; Abdallah & Mboya 3960; Abdallah, Mboya & Vollesen 96/203.

Edithcolea grandis N. E. Br. Acacia-Commiphora bushland; rare. Greenway 2160.

Fockea sp. nov. Acacia-Commiphora bushland, Acacia-Combretum bushland; occasional. Abdallah & Mboya 3683, 3884, 4031; Abdallah & Cox 1948; Abdallah & Vollesen 95/23; Vollesen 96/22.

Gomphocarpus kaessneri N. E. Br. Acacia-Commiphora bushland; rare. Greenway 3931.

Gomphocarpus physocarpus E. Mey. Seasonally inundated grassland; rare. Abdallah & Mboya 3518, 4038.

Kanahia laniflora (Forssk.) R. Br. Riverbanks, rocky riverbeds; occasional. Abdallah & Mboya 3623.

Orbea semota (N. E. Br.) Leach. Rocky outcrops; rare. Greenway 4592.

Pachycymbium dummeri (N. E. Br.) M.G. Gilbert. Acacia-Commiphora bushland, rocky hilltops; rare. Abdallah & Mboya 3876.

Pentarrhinum insipidum E. Mey. Rocky hilltops, montane forest; occasional. Abdallah 2830, 3362.

Pentatropis nivalis (J.F. Gmel.) D. Field & J.R.I. Wood. Seasonally inundated grassland, Acacia bushland; rare. Greenway 4554, Semsei 2154.

Pergularia daemia (Forssk.) Chiov. Acacia-Combretum bushland, Acacia-Commiphora bushland; occasional. Abdallah & Mboya 3573; Greenway 2042; Semsei 2140.

Periploca Iineariifolia Dillon & A. Rich. Rocky hilltops, montane forest; occasional. Vollesen SR.

Raphionacme jurensis N. E. Br. Acacia-Commiphora bushland; rare. Abdallah 2733; Faulkner 4558.

Raphionacme sp. nov. aff. *R. splendens* Schltr., not matched at Kew. Acacia-Commiphora bushland; rare. Abdallah & Mboya 3623.

Sacleuxia newii (Benth.) Bullock. Rocky hilltops; rare. Abdallah & Vollesen 95/64.

Sarcostemma viminale (L.) R. Br. Rocky hilltops and outcrops, ravines; occasional. Semsei 2157.

Secamone attenuifolia Goyder. Acacia-Combretum bushland; rare. Abdallah & Vollesen 96/234; Greenway 4569.

Secamone parvifolia (Oliv.) Bullock. Acacia-Commiphora bushland; rare. Abdallah & Mboya 4086; Abdallah & Vollesen 95/160

Secamone punctulata Decne. Rocky hilltops; rare. Abdallah 2754.

Stathmostelma pedunculatum (Decne.) K. Schum. Seasonally inundated grassland; rare. Abdallah 2729, 2849.

Tenaris rostrata N. E. Br. Seasonally inundated grassland; rare. Coe SR.

**Dregea schimperi*. Abdallah & Mboya 4135.

**Dregea* sp. Abdallah 2704.

**Oxystelma bornuense*. Abdallah 2560.

**Raphionacme* sp. Abdallah 2692.

Asteraceae (58)

Achyrothalamus marginatus O. Hoffm. Montane forest; occasional. Abdallah 3112; Abdallah & Mboya 3617; Abdallah & Vollesen 95/81.

Anisopappus holstii (O. Hoffm.) Wild. Acacia-Combretum bushland, rocky hilltops; rare. Abdallah & Mboya 3603; Abdallah & Vollesen 96/237.

Aspilia kotschyi (Hochst.) Oliv. Seasonally inundated grassland, Acacia-Commiphora bushland; occasional. Vollesen SR.

Aspilia monocephala Baker. Seasonally inundated grassland, Acacia-Combretum bushland; rare. Abdallah & Vollesen 95/146.

Aspilia mossambicensis (Oliv.) Wild. All types of bushland; common. Abdallah 2806, 3134, 3148, 3240; Abdallah & Mboya 3612, 3647.

Athroisma gracile (Oliv.) Mattf. subsp. *psyllioides* (Oliv.) T. Eriksson. Acacia-Commiphora bushland, seasonally-inundated grassland; occasional. Abdallah 3181; Abdallah & Vollesen 95/157, 95/165.

Bidens pilosa L. Montane forest, weed; occasional. Vollesen SR.

Blainvillea gayana Cass. Acacia-Combretum bushland; rare. Abdallah & Vollesen 95/12.

Blepharispermum zanguebaricum Oliv. & Hiern. Acacia-Combretum bushland, dry montane forest, ravines; occasional. Abdallah & Mboya 3613; Abdallah & Vollesen 95/10.

Bothriocline argentea (O. Hoffm.) Wild & Pope var. *argentea*. Montane forest; rare. Semsei 2110.

Bothriocline longipes (Oliv. & Hiern) N. E. Br. Montane forest; rare. Abdallah, Mboya & Vollesen 96/85.

Brachylaena huillensis O. Hoffm. Dry montane forest; locally common. L.D. Harris 39.

Conyza aegyptiaca (L.) Ait. Seasonally inundated grassland, Acacia-Combretum bushland; rare. Abdallah & Mboya 3470; Abdallah, Mboya & Vollesen 96/145.

Conyza pyrrhopappa A. Rich. subsp. *oblongifolia* (O. Hoffm.) H. Wild. Acacia-Combretum bushland, Acacia-Commiphora bushland, rocky hilltops; common. Abdallah 2558, 2586, 3325; Abdallah & Mboya 3788; Abdallah & Vollesen 95/2; L.D. Harris 18.

Conyza stricta Willd. Acacia-Combretum bushland, rocky hilltops; rare. Abdallah & Vollesen 96/240.

Dicoma tomentosa Cass. Acacia-Commiphora bushland; rare. Abdallah & Vollesen 95/120.

Eclipta prostrata (L.) L. Seasonally inundated grassland, temporary waterholes; occasional. Vollesen SR.

Emilia coccinea (Sims) G. Don. Acacia-Commiphora bushland; rare. Abdallah 3239.

Erlangea alternifolia (O. Hoffm.) S. Moore. Acacia and Acacia-Commiphora bushland, seasonally-inundated grassland; locally common. Abdallah & Vollesen 95/25; Vollesen 96/47.

Ethulia angustifolia DC. Seasonally inundated grassland, rare. Peter 54015.

Gutenbergia rueppellii Sch. Bip. var. *rueppellii*. Seasonally inundated grassland, Acacia-Combretum bushland; common. Abdallah 3236; Abdallah, Mboya & Vollesen 96/58; Abdallah & Vollesen 95/142.

Helichrysum globosum (Sch.Bip.) A. Rich. Montane forest; rare. Abdallah 3125.

Helichrysum glumaceum DC. Acacia-Commiphora bushland, Acacia-Combretum bushland; occasional. Abdallah 3238; Abdallah & Mboya 3433; L.D. Harris 16, 76; Semsei 2137.

Helichrysum traversii Chiov. Dry montane forest; rare. Abdallah, Mboya & Vollesen 96/71.

Hirpicium diffusum (O. Hoffm.) Roess. Acacia-Combretum bushland, Acacia-Commiphora bushland, weed; common. Vollesen SR.

Kleinia abyssinica (A. Rich.) A. Berger var. *abyssinica*. Rocky hilltops and outcrops; occasional. Abdallah & Mboya 3486; Semsei 2127.

Kleinia grantii (Oliv. & Hiern) Hook. f. Acacia-Commiphora bushland; occasional. Abdallah 3269; Abdallah & Mboya 3872.

Kleinia gregorii (S. Moore) C. Jeffrey. Acacia-Commiphora bushland; rare. Abdallah 3135.

Kleinia odora (Forssk.) DC. Acacia-Commiphora bushland; rare. Semsei 2148.

Kleinia squarrosa Cuf. Rocky hilltops and outcrops; rare. Abdallah & Mboya 3405; Greenway 4574.

Lactuca inermis Forssk. Acacia-Commiphora bushland; rare. L.D. Harris 20.

Launaea cornuta (Oliv. & Hiern) C. Jeffrey. Acacia-Commiphora bushland; occasional. Abdallah 3085, 3200.

Osteospermum vaillantii (Decne.) Norl. Acacia-Commiphora bushland; occasional. Abdallah 2740; Abdallah & Mboya 4129.

Pluchea bequaertii Robyns. Seasonally inundated grassland, riverbeds; rare. Abdallah 2811; Abdallah, Mboya & Vollesen 96/100.

Pluchea dioscoridis DC. Riverbanks, riverbeds; locally common. Vollesen SR.

Psiadia punctulata (DC.) Vatke. Montane forest; occasional. Abdallah & Mboya 4171; Abdallah, Mboya & Vollesen 96/75.

Senecio hadiensis Forssk. Montane forest, rocky hilltops and outcrops, ravines; locally common. Abdallah, Mboya & Vollesen 96/137; Doughty 13, Semsei 2093.

Solanecio angulatus (Hook. f.) C. Jeffrey. Montane forest, rocky hilltops and ravines; rare. Abdallah, Mboya & Vollesen 96/56.

Solanecio cydoniifolius (O. Hoffm.) C. Jeffrey. Montane forest, rocky hilltops and ravines; rare. Abdallah & Mboya 3483, 3852; Abdallah, Mboya & Vollesen 96/53.

Solanecio mannii (Hook. f.) C. Jeffrey. Montane forest, rocky hilltops; occasional. Abdallah & Mboya 3462.

Sphaeranthus greenwayi Ross-Craig. Seasonally inundated grassland, temporary waterholes; rare. Abdallah & Vollesen 95/53.

Sphaeranthus kirkii Oliv. & Hiern var. *cyathuloides* (O. Hoffm.) Beentje. Seasonally inundated grassland, temporary waterholes; common. Abdallah & Mboya 3411, 3468, 4098; Faulkner 4429; Vollesen 96/25.

Tridax procumbens L. Weed; occasional. Vollesen SR.

Triplocephalum holstii O. Hoffm. Seasonally inundated grassland, Acacia bushland; locally common. Abdallah & Mboya 4205; Abdallah & Vollesen 95/51; Semsei 3975.

Vernonia aemulans Vatke. Acacia-Commiphora bushland (eastern); rare. Abdallah, Mboya & Vollesen 96/170.

Vernonia albocinerascens C. Jeffrey. Acacia-Combretum bushland; occasional. Abdallah, Mboya & Vollesen 96/52; L.D. Harris 178, 246.

Vernonia anthelmintica (L.) Willd. Dry montane forest; occasional. Abdallah & Mboya 4154; Abdallah & Vollesen 95/187.

Vernonia brachycalyx O. Hoffm. Montane forest, rocky hilltops and ravines; locally common. Abdallah 3121; Abdallah & Mboya 3604, 3838; Abdallah, Mboya & Vollesen 96/68; Abdallah & Vollesen 95/30, 96/233.

Vernonia cinerascens Sch. Bip. Acacia-Commiphora bushland, Acacia-Combretum bushland; occasional. Abdallah 3232, 3273; Abdallah & Mboya 3830, 3873, 3971; Greenway 1986, 3989.

Vernonia galamensis (Cass.) Less. Acacia-Commiphora bushland; occasional. Abdallah 3294.

Vernonia hildebrandtii Vatke. Montane forest; rare. Abdallah & Mboya 3450.

Vernonia hochstetteri Sch. Bip. Acacia-Combretum bushland, montane forest; rare. Abdallah & Vollesen 96/245.

Vernonia holstii O. Hoffm. Montane forest, rocky hilltops; occasional. Abdallah & Vollesen 95/70; Vollesen 96/16.

Vernonia karaguensis Oliv. & Hiern. Acacia-Combretum bushland; rare. Abdallah & Mboya 3408; Peter 54019.

Vernonia lasiopus O. Hoffm. subsp. *iodocalyx* (O. Hoffm.) C. Jeffrey. Montane forest edges, Acacia-Combretum bushland; locally common. Abdallah & Mboya 3413; Abdallah & Vollesen 96/242.

Vernonia usambarensis O. Hoffm. Montane forest, rocky hilltops and ravines; occasional. Abdallah, Mboya & Vollesen 96/63; Abdallah & Vollesen 96/241; Vollesen 96/17.

Vernonia wakefieldii Oliv. Dry montane forest, rocky hilltops and ravines; rare. Abdallah & Vollesen 95/186; Cox & Abdallah 2313.

Vernonia wollastonii Sch. Bip. Seasonally inundated grassland; rare. Vollesen 96/45.

**Aspilia pluriseta.* Abdallah & Mboya 3891.

**Aspilia* sp. Abdallah 3291.

**Conyza* sp. Abdallah 4091.

**Erythrocephalum zambezianum.* Abdallah 3177.

**Erythrocephalum* sp. Abdallah & Mboya 3568.

**Pluchea* sp. Abdallah & Mboya 3469.

**Pseudoconyza viscosa (Blumea aurita).* Abdallah & Mboya 3407.

**Sonchus oleraceus.* Abdallah & Mboya 3464.

**Vernonia* sp. Abdallah 2827, 3321; Abdallah & Mboya 4168.

Balanitaceae (2)

Balanites aegyptiaca (L.) Del. var. *aegyptiaca.* Acacia-Combretum bushland; occasional. Abdallah & Mboya 3680.

Balanites pedicellaris Mildbr. & Schltr. subsp. *pedicellaris.* Acacia-Commiphora bushland; occasional. Greenway 4058; Mgaza 54.

Balsaminaceae (1)

Impatiens walleriana Hook. f. Montane forest; rare. Abdallah & Mboya 3708; Abdallah & Vollesen 95/85.

Basellaceae (1)

Basella paniculata Volkens. Riverine forest; rare. Abdallah & Mboya 3984; Milne-Redhead & Taylor 7244

Bignoniaceae (3)

Kigelia africana (Lam.) Benth. Acacia-Combretum bushland, Acacia-Commiphora bushland, riverine forest, seasonally-inundated grassland; occasional. Vollesen SR.

Markhamia obtusifolia (Bak.) Sprague. Acacia-Combretum bushland; occasional. Greenway SR; Vollesen SR.

Stereospermum kunthianum Cham. Acacia-Combretum bushland; rare. Vollesen SR.

Bombacaceae (1)

Adansonia digitata L. Acacia-Combretum bushland, Acacia-Commiphora bushland; occasional. Vollesen SR.

Boraginaceae (17)

Bourreria teitensis (Gürke) Thulin. Acacia-Commiphora bushland; locally common. Abdallah 2593; Abdallah & Mboya 4033; Abdallah & Vollesen 95/161.

Coldenia procumbens L. Temporary waterholes; rare. Abdallah & Mboya 4228.

Cordia goetzei Gürke. Riverine forest, montane forest (?); rare. Abdallah & Mboya 3977, 4152; Greenway 218.

Cordia monoica Roxb. Acacia-Combretum bushland, Acacia-Commiphora bushland; common. Abdallah 2685, 3099; Abdallah & Mboya 3789, 3808, 4182; L.D. Harris 60.

Cordia quercifolia Klotzsch. Seasonally inundated grassland, Acacia-Commiphora bushland; locally common. Abdallah 2594; Abdallah & Mboya 4230; Greenway 2124, 3963.

Cordia sinensis Lam. Riverine forest; rare. Semsei 2108.

Ehretia amoena Klotzsch. Acacia-Commiphora bushland; occasional. Abdallah 2746; Abdallah & Mboya 3701; L.D. Harris 56.

Ehretia bakeri Britten. Acacia-Commiphora bushland, Acacia-Combretum bushland; occasional. Abdallah 3002; Abdallah & Mboya 3697, 3951.

Heliotropium indicum L. Temporary waterholes; common. Abdallah 2797, 2859, 2874, 3062.

Heliotropium longiflorum (A. DC.) Jaub. & Spach subsp. *undulatifolium* (Turrill) Verdc. Acacia-Commiphora bushland; rare. Abdallah 3274; Greenway 3978; Semsei 2142.

Heliotropium ovalifolium Forssk. Riverbeds, temporary waterholes; rare. Abdallah & Mboya 3513; Abdallah, Mboya & Vollesen 96/103.

Heliotropium rariflorum Stocks subsp. *hereroense* (Schinz) Verdc. Acacia-Commiphora bushland; rare. Abdallah 2748.

Heliotropium steudneri Vatke subsp. *steudneri* var. *steudneri*. All types of bushland; common. Abdallah 2764, 2774, 2889, 3018, 3023, 3034, 3157, 3162, 3347; L.D. Harris 40.

Heliotropium strigosum Willd. All types of bushland; common. Abdallah 2714.

Heliotropium supinum L. Riverbeds, temporary waterholes; rare. Abdallah & Mboya 3511; Cox & Abdallah 2347.

Heliotropium zeylanicum (Burm. f.) Lam. All types of bushland; occasional. Abdallah 2796, 3035, 3214, 3350; Abdallah & Mboya 3648, 3686; B.J. Harris 789.

Trichodesma zeylanicum (Burm. f.) R. Br. Acacia-Combretum bushland, weed; occasional. Vollesen SR.

**Cordia ovalis*. Abdallah & Mboya 3681, 3726.

**Cordia* sp. Abdallah & Mboya 3773.

**Ehretia caerulea*. Abdallah 2715; Abdallah & Mboya 3764.

**Ehretia petiolaris*. Abdallah 2534.

**Heliotropium "faitimatum"*. Abdallah 2751.

Brassicaceae (2)

Farsetia stenoptera Hochst. subsp. *stenoptera*. All types of bushland; occasional. Abdallah 2627, 2861; Abdallah & Mboya 3608, 3674; Abdallah & Vollesen 95/155.

Rorippa micrantha (Roth) Jonsell. Riverbeds, temporary waterholes; rare. Abdallah, Mboya & Vollesen 96/106; Semsei 2336.

**Erucastrum arabicum*. Abdallah & Mboya 3420.

Burseraceae (10)

Boswellia neglecta S. Moore. Acacia-Commiphora bushland (eastern); locally common. Abdallah 3194; Greenway 1970.

Commiphora africana (A. Rich.) Engl. var. *africana*. Acacia-Commiphora bushland, Acacia-Combretum bushland; common. Abdallah 2770, 2820, 3333, 3342; Abdallah & Mboya 3445, 3811, 3900; Greenway 4724.

Commiphora baluensis Engl. Dry montane forest, rocky hilltops and ravines; occasional. Greenway 4586.

Commiphora campestris Engl. subsp. *campestris* var. *heterophylla* (Engl.) Gillett. Acacia-Commiphora bushland; common. Abdallah 3307, 3331; Abdallah & Mboya 3588, 3870; Abdallah & Vollesen 95/27; Procter 3276.

Commiphora confusa Vollesen. Acacia-Commiphora bushland; rare. Greenway 1968.

Commiphora edulis (Klotzsch) Engl. subsp. *boiviniana* (Engl.) Gillett. Acacia-Commiphora bushland, rocky riverbanks; rare. Abdallah & Mboya 3823.

Commiphora eminii Engl. subsp. *zimmermannii* (Engl.) Gillett. Dry montane forest, rocky hilltops and ravines; occasional. Abdallah & Mboya 3768, 3888; Abdallah & Vollesen 95/190.

Commiphora habessinica (O. Berg) Engl. subsp. *habessinica*. Acacia-Commiphora bushland, Acacia-Combretum bushland; common. Abdallah & Mboya 3624, 3883, 3912.

Commiphora holtziana Engl. subsp. *holtziana*. Acacia-Commiphora bushland; common. Abdallah & Mboya 3828; Leippert 6204.

Commiphora schimperi (O. Berg) Engl. Acacia-Commiphora bushland, Acacia-Combretum bushland; common. Abdallah 3172, 3335; Abdallah & Mboya 3793, 3865.

Commiphora merkeri. Abdallah & Mboya 4067.

**Commiphora* sp. L.D. Harris 34A.

Campanulaceae (1)

Cyphia glandulifera A. Rich. Acacia-Commiphora bushland, rocky hilltops; rare. Abdallah 2708; Abdallah & Mboya 4165.

Capparaceae (24)

Boscia angustifolia A. Rich. var. *angustifolia.* Acacia-Commiphora bushland; occasional. Abdallah, Mboya & Vollesen 96/169; L.D. Harris 108.

Boscia coriacea Pax. Acacia-Commiphora bushland; occasional. Greenway 4061; Kisena & Msanga 343; Semsei 2134, 2328.

Boscia mossambicensis Klotzsch. Acacia-Commiphora bushland; occasional. Abdallah 3317; Abdallah & Mboya 3914.

Boscia salicifolia Oliv. Acacia-Combretum bushland; occasional. Abdallah & Mboya 4212, 4218; L.D. Harris 100.

Cadaba farinosa Forssk. subsp. *adenotricha* (Gilg & Bened.) R.A. Graham. All types of bushland, rocky outcrops; common. Abdallah 2546, 2604, 2606; Abdallah & Mboya 3438, 3650, 3656, 3786, 4219; Abdallah & Vollesen 95/41; L.D. Harris 111.

Cadaba glandulosa Forssk. All types of bushland; common. Abdallah 2605; Abdallah & Mboya 4210, 4224; Cox & Abdallah 2327; Greenway 4065; Semsei 2153.

Cadaba ruspolii Gilg. Acacia-Commiphora bushland; occasional. Abdallah & Mboya 3930, 4213; Burtt 5346; L.D. Harris 122.

Cadaba stenopoda Gilg & Bened. Acacia-Commiphora bushland; rare. Greenway 4059.

Capparis sepiaria L. var. *fischeri* (Pax) DeWolf. Acacia-Commiphora bushland; rare. Abdallah & Mboya 3945.

Capparis tomentosa Lam. Riverine forest, dry forest edges; occasional. Ruffo 2468.

Capparis viminea Oliv. var. *viminea.* Montane forest; rare. Abdallah & Mboya 4236; Milne-Redhead & Taylor 7250.

Cleome hirta (Klotzsch) Oliv. Acacia-Commiphora bushland; occasional. Abdallah 3055, 3077; Abdallah, Mboya & Vollesen 96/218.

Cleome monophylla L. Acacia-Combretum bushland, Acacia-Commiphora bushland, seasonally-inundated grassland; occasional. Abdallah 2522, 2850, 3024.

Cleome stenopetala Gilg & Bened. Acacia-Commiphora bushland; rare. Abdallah & Vollesen 95/119.

Cleome usambarica Pax. Acacia-Commiphora bushland, rocky outcrops; occasional. Abdallah 2607, 2608; Abdallah, Mboya & Vollesen 96/206.

Maerua angolensis DC. Acacia-Combretum bushland; occasional. Abdallah & Mboya 4008.

Maerua crassifolia Forssk. Acacia-Commiphora bushland, rocky hilltops, occasional. Abdallah & Mboya 3791, 4061.

Maerua edulis (Gilg & Bened.) De Wolf. Acacia-Commiphora bushland; rare. Abdallah & Mboya 3684.

Maerua endlichii Gilg & Bened. Acacia-Commiphora bushland; rare. Abdallah & Mboya 4221.

Maerua grantii Oliv. Acacia-Combretum bushland, Acacia-Commiphora bushland, seasonally-inundated grassland; common. Abdallah 2545, 2603, 3357; Abdallah & Mboya 3611, 3660, 3685; Abdallah & Vollesen 95/156; Bally 16389; Drummond & Hemsley 1330; Greenway 2187; L.D. Harris 15, 64.

Maerua holstii Pax. Riverine forest (Umba River); rare. Abdallah, Mboya & Vollesen 96/167.

Maerua subcordata (Gilg) De Wolf. Acacia-Commiphora bushland; occasional. Greenway 4062; Richards 21915; Semsei 2145.

Maerua triphylla A. Rich. Acacia-Combretum bushland, rocky hilltops and rocky outcrops, montane forest; common. Abdallah 2602; Abdallah & Mboya 3432, 3463, 3529, 3666, 3717, 3855, 3861, 3958; Abdallah, Mboya & Vollesen 96/136; Bally 16385; Ruffo 2475; Semsei 2161.

Thilachium africanum Lour. Acacia-Combretum bushland; common. Abdallah & Mboya 3436, 3561, 3659, 3678, 3679, 4199; L.D. Harris 92, 126A; Richards 21909.

**Cadaba* sp. Abdallah & Mboya 3622; L.D. Harris 121.

**Capparis* sp. Abdallah & Mboya 4028.

Maerua eminii. Abdallah 2672.

Maerua kirkii. Abdallah & Mboya 3419, 4237.

Caryophyllaceae (2)

Drymaria cordata (L.) Roem. & Schult. Montane forest; rare. Abdallah & Vollesen 96/258.

Polycarpaea eriantha A. Rich. All types of bushland; occasional. Vollesen SR.

Celastraceae (7)

Hippocratea (Loeseneriella) africana (Willd.) Loes. Riverine forest; rare. Abdallah & Mboya 3690; Abdallah, Mboya & Vollesen 96/196.

Maytenus drummondii Robson & Sebsebe. Montane forest; locally common. Abdallah, Mboya & Vollesen 96/93.

Maytenus heterophylla (Eckl. & Zeyh.) Robson. Acacia-Combretum bushland, rocky hilltops; occasional. Abdallah, Mboya & Vollesen 96/142.

Maytenus senegalensis (Lam.) Exell. Montane forest, rocky hilltops; common. Abdallah & Mboya 3664, 4169.

Maytenus undata (Thunb.) Blakelock. Montane forest, rocky hilltops; locally common. Abdallah & Mboya 3606; Abdallah & Vollesen 95/153.

Mystroxylon aethiopicum (Thunb.) Loes. Montane forest, rocky hilltops; occasional. Abdallah & Vollesen 95/152.

Salacia stuhlmanniana Loes. Montane forest, riverine forest; occasional. Abdallah & Mboya 3968, 4235.

Maytenus sp. Abdallah 2664.

Ceratophyllaceae (1)

Ceratophyllum submersum L. Temporary waterholes, dams; occasional. Abdallah & Mboya 3424, 4197.

Chenopodiaceae (2)

Chenopodium opulifolium Koch & Ziz. Acacia-Commiphora bushland, weed; rare. Semsei 2342.

Suaeda monoica J.F. Gmel. Acacia bushland, seasonally-inundated grassland; rare. Greenway 1981, 3955.

Clusiaceae (1)

Garcinia volkensii Engl. Montane forest; locally common. Abdallah, Mboya & Vollesen 96/98; Abdallah & Vollesen 96/246.

Combretaceae (16)

Combretum aculeatum Vent. Acacia-Combretum bushland; common. Abdallah 2638; Abdallah & Mboya 3731; Greenway 1975B.

Combretum adenogonium A. Rich. Acacia-Combretum bushland; occasional. Vollesen SR.

Combretum apiculatum Sond. Acacia-Combretum bushland; rare. Abdallah & Mboya 4183.

Combretum contractum Engl. & Diels. Combretum-Terminalia bushland (extreme east); locally common. Abdallah, Mboya & Vollesen 96/189; Greenway 1975A.

Combretum exalatum Engl. Dry montane forest, Acacia-Combretum bushland, Acacia-Commiphora bushland, rocky hilltops and ravines; common. Abdallah & Mboya 3761, 3787; Abdallah & Vollesen 95/33; Greenway 1977, 4571.

Combretum hereroense Schinz subsp. *grotei* (Exell) Wickens. Acacia-Commiphora bushland; rare. Jussif bin Mohamed 4.

Combretum hereroense Schinz subsp. *volkensii* (Engl.) Wickens. Acacia-Commiphora bushland, rocky hilltops; locally common. Abdallah 3353; Abdallah & Mboya 3760, 3816, 4083, 4162; Gillman 778; Greenway 3987.

Combretum molle G. Don. Acacia-Combretum bushland, rocky hilltops, rocky outcrops; common. Abdallah 2591, 2676, 3349; Abdallah & Mboya 3628, 3663; Greenway 4580; L.D. Harris 52.

Combretum padoides Engl. & Diels. Dry montane forest, rocky hilltops and ravines; occasional. Abdallah 3133, 3391; Abdallah & Vollesen 95/8; Semsei 2113.

Combretum xanthothyrsum Engl. & Diels. Riverine forest; rare. Abdallah & Vollesen 95/178.

Combretum zeyheri Sond. Acacia-Combretum bushland; common. Abdallah & Mboya 4007.

Terminalia brownii Fresen. Rocky hilltops and ravines; occasional. Abdallah 2892.

Terminalia kilimandscharica Engl. Acacia-Combretum bushland, Acacia-Commiphora bushland, rocky riverbeds; common. Abdallah 2674; Abdallah & Vollesen 95/26, 95/44.

Terminalia prunioides Laws. Acacia-Commiphora bushland; common. Abdallah & Mboya 4147; Abdallah & Vollesen 95/43, 95/162; Greenway 3936; Semsei 2078, 2081, 2088, 2109, 2112.

Terminalia sambesiaca Engl. & Diels. Riverine forest; locally common. L.D. Harris 95.

Terminalia spinosa Engl. Combretum-Terminalia bushland (extreme east); locally common. L.D. Harris 125.

**Combretum* sp. Abdallah & Mboya 3640, 4003, 4005.

**Terminalia* sp. Abdallah 2587.

Convolvulaceae (25)

Astripomoea hyoscyamoides (Vatke) Verdc. Acacia-Combretum bushland, Acacia-Commiphora bushland, weed; common. Abdallah & Mboya 3409; Greenway 2184; Ibrahim 655; Semsei 2371.

Evolvulus alsinoides (L.) L. Acacia-Combretum bushland, Acacia-Commiphora bushland; common. Abdallah 2657, 2785, 3143.

Ipomoea aquatica Forssk. Temporary waterholes, dams; locally common. Greenway 3983.

Ipomoea blepharophylla Hall. f. Seasonally inundated grassland, temporary waterholes;

common. Abdallah & Mboya 4054.

Ipomoea bullata Oliv. Acacia-Commiphora bushland; occasional. Semsei 2125.

Ipomoea ficifolia Lindl. Acacia-Combretum bushland; occasional. Abdallah 3500; Doughty 14.

Ipomoea hildebrandtii Vatke subsp. *hildebrandtii*. Acacia-Combretum bushland; common. Vollesen SR.

Ipomoea involucrata P. Beauv. Acacia-Combretum bushland, Acacia-Commiphora bushland, seasonally-inundated grassland; occasional. Abdallah 3030, 3250; Abdallah & Mboya 4095.

Ipomoea kituiensis Vatke var. *kituiensis*. Acacia-Combretum bushland, rocky hilltops; occasional. Abdallah & Mboya 4190.

Ipomoea lapathifolia Hall. f. var. *lapathifolia*. Acacia-Commiphora bushland, seasonally-inundated grassland; rare. Abdallah 3304; Abdallah & Mboya 4079.

Ipomoea mombassana Vatke. Seasonally inundated grassland; rare. Abdallah 2847.

Ipomoea obscura (L.) Ker-Gawl. All types of bushland; common. Abdallah 2717, 3187, 3254; Mgaza 48.

Ipomoea ochracea (Lindl.) G. Don var. *ochracea*. Acacia-Commiphora bushland (eastern part); rare. Abdallah, Mboya & Vollesen 96/182.

Ipomoea pes-tigridis L. var. *longibracteata* Vatke. Acacia-Combretum bushland, Acacia-Commiphora bushland, seasonally-inundated grassland; common. Abdallah 2789, 3042, 3195; Abdallah & Mboya 3615, 4096; Greenway 2135; L.D. Harris 3; Ibrahim 659.

Ipomoea sinensis (Desr.) Choisy subsp. *blepharosepala* (A. Rich.) Meeuse. Acacia-Combretum bushland, Acacia-Commiphora bushland, seasonally-inundated grassland; common. Abdallah 3159; Greenway 3985; Peter 51538.

Ipomoea tenuipes Verdc. Rocky riverbanks; rare. Abdallah & Mboya 3589; Abdallah, Mboya & Vollesen 96/105.

Ipomoea transvaalensis Meeuse subsp. *orientalis* Verdc. Acacia-Commiphora bushland (eastern part); rare. Abdallah, Mboya & Vollesen 96/186.

Merremia ampelophylla Hall. f. Acacia-Commiphora bushland; occasional. Abdallah 2883, 3227; Greenway 2012; Vollesen 96/31.

Merremia palmata Hall. f. Rocky hilltops; rare. Abdallah 2813.

Seddera hirsuta Hall. f. var. *gracilis* (Chiov.) Verdc. Acacia-Commiphora bushland; common. Abdallah & Mboya 3662, 3775, 4112; Abdallah, Mboya & Vollesen 96/184; Cox & Abdallah 1958; Greenway 3991.

Seddera humilis Hall. f. var. *humilis*. Acacia-Commiphora bushland; occasional. Abdallah & Vollesen 95/116; Faulkner 4551.

Seddera suffruticosa (Schinz) Hall. f. var. *suffruticosa*. Acacia-Commiphora bushland (eastern part); rare. Abdallah, Mboya & Vollesen 96/185.

Stictocardia incompta (Hall. f.) Hall. f. Acacia-Commiphora bushland; rare. Semsei 2117.

Turbina stenosiphon (Hall. f.) Meeuse var. *stenosiphon*. Dry montane forest, rocky hilltops and ravines; occasional. Abdallah & Vollesen 96/232.

Xenostegia tridentata (L.) Austin & Staples. All types of bushland; common. Abdallah & Mboya 3928, 3985.

**Ipomoea coscinosperma*. Abdallah 3253.

**Ipomoea wightii*. L.D. Harris 71.

**Ipomoea* sp. Abdallah 3228; Abdallah & Mboya 3435.

**Merremia* sp. Abdallah 3031.

Crassulaceae (8)

Crassula expansa Dryand. subsp. *fragilis* (Bak.) Tölken. Rocky hilltops and ravines; occasional. Abdallah & Vollesen 95/9.

Crassula schimperi Fisch. & Mey. subsp. *phyturus* (Mildbr.) R. Fernandes. Rocky hilltops and ravines; rare. Abdallah, Mboya & Vollesen 96/141.

Crassula volkensii Engl. subsp. *volkensii*. Rocky hilltops and ravines; occasional. Abdallah & Mboya 3455, 3769; Abdallah, Mboya & Vollesen 96/124.

Kalanchoe glaucescens Britten. Rocky hilltops and ravines; rare. Abdallah & Mboya 3476.

Kalanchoe lanceolata (Forssk.) Pers. Rocky hilltops and ravines; occasional. Abdallah & Mboya 3474.

Kalanchoe lateritia Engl. var. *lateritia*. All types of bushland; common. Vollesen SR.

*Kalanchoe nyikae*Engl. subsp. *nyikae*. Rocky hilltops and ravines; rare. Abdallah & Mboya 3482; Greenway 6695.

Kalanchoe rotundifolia (Haw.) Haw. Rocky hilltops and ravines; rare. Abdallah, Mboya & Vollesen 96/119.

**Kalanchoe* sp. Abdallah & Mboya 3466.

Cucurbitaceae (22)

Coccinia grandis (L.) Voigt. Riverine forest; rare. Abdallah & Mboya 3798.

Coccinia microphylla Gilg. Acacia-Combretum bushland, Acacia-Commiphora bushland; common. Abdallah & Mboya 3634; Abdallah & Vollesen 95/198; Archbold 1339; Faulkner 4561.

Coccinia trilobata (Cogn.) C. Jeffrey. Acacia-Combretum bushland; occasional. Abdallah 2736, 2782; Abdallah & Mboya 3785.
Corallocarpus epigaeus (Rottl.) C.B. Clarke. Acacia-Combretum bushland; occasional. Abdallah 2781.
Cucumis anguaria L. Acacia-Commiphora bushland; rare. Abdallah & Mboya 3829.
Cucumis dipsaceus Spach. All types of bushland; common. Milne-Redhead & Taylor 7252.
Cucumis hirsutus Sond. Rocky hilltops; rare. Abdallah 2823.
Cucumis prophetarum L. supsp. *dissectus* (Naud.) C. Jeffrey. Acacia-Combretum bushland; rare. Abdallah 3119.
Cyclantheropsis parviflora (Cogn.) Harms. Dry montane forest, rocky hilltops and ravines; occasional. Abdallah & Mboya 4153; Abdallah & Vollesen 95/189; Semsei 2107.
Diplocyclos palmatus (L.) C. Jeffrey. Dry montane forest, rocky hilltops; occasional. Abdallah 2668; Abdallah & Vollesen 95/98.
Gerrardanthus lobatus (Cogn.) C. Jeffrey. Dry montane forest, riverine forest, rocky hilltops and ravines; occasional. Abdallah & Mboya 3491, 4159; Abdallah & Vollesen 95/184; Milne-Redhead & Taylor 7243.
Kedrostis foetidissima (Jacq.) Cogn. Acacia-Combretum bushland, Acacia-Commiphora bushland, rocky hilltops; occasional. Abdallah 3032; Abdallah & Mboya 3597.
Lagenaria siceraria (Molina) Standley. Acacia-Commiphora bushland; occasional. Abdallah & Mboya 3507, 4121.
Momordica anigosantha Hook. f. Montane forest; rare. Abdallah & Mboya 3492.
Momordica boivinii Baill. Acacia-Commiphora bushland, Acacia-Combretum bushland; common. Abdallah 3174; Abdallah & Mboya 3654, 3780, 3809; Abdallah & Vollesen 95/195; Greenway 4555; Vollesen 96/32.
Momordica rostrata A. Zimm. Acacia-Commiphora bushland; occasional. Abdallah & Mboya 3646, 3866; Milne-Redhead & Taylor 7247.
Momordica spinosa (Gilg) Chiov. Acacia-Commiphora bushland; rare. Abdallah & Mboya 3437, 4074.
Momordica trifoliolata Hook. f. Riverine forest, Acacia-Commiphora bushland, seasonally-inundated grassland; occasional. Abdallah 3064; Abdallah & Mboya 3895.
Peponium vogelii (Hook. f.) Engl. Acacia-Commiphora bushland; rare. Abdallah 2882.
Zehneria oligosperma C. Jeffrey. Riverine forest; rare. Abdallah & Vollesen 95/149.
Zehneria pallidinervia (Harms) C. Jeffrey. Rocky hilltops and ravines; rare. Abdallah & Mboya 4157.
Zehneria scabra (L. f.) Sond. Montane forest; occasional. Abdallah & Vollesen 95/99.
**Coccinia* sp. Cox & Abdallah 1950.
**Momordica cissoides*. Abdallah 2876.
**Momordica* sp. Abdallah & Mboya 3645.

Cyclocheilaceae (1)
Asepalum eriantheumum (Vatke) Marais. Acacia-Commiphora bushland; locally common. Abdallah 3093; Abdallah & Mboya 3939, 4132.

Ebenaceae (8)
Diospyros abyssinica (Hiern) F. White. Montane forest; locally common. Vollesen SR.
Diospyros consolatae Chiov. Acacia-Commiphora bushland; rare. Abdallah & Mboya 4097.
Diospyros mespiliformis A. DC. Riverine forest; rare. Mgaza 50.
Diospyros natalensis (Harv.) Brenan. Dry montane forest, rocky hilltops; occasional. Abdallah, Mboya & Vollesen 96/125.
Diospyros squarrosa Klotzsch. Dry montane forest; rare. Abdallah & Mboya 3932.
Diospyros zombensis (B.L. Burtt) F. White. Riverine forest; occasional. Abdallah & Vollesen 95/106; Milne-Redhead & Taylor 7242, 7251.
Euclea divinorum Hiern. Acacia-Commiphora bushland; rare. Abdallah & Mboya 4196.
Euclea racemosa Murray subsp. *schimperi* (A. DC.) F. White. Dry montane forest, ravines; rare. Vollesen SR.
**Diospyros* sp. Abdallah & Mboya 3586, 3980, 4225.

Erythroxylaceae (1)
Erythroxylum emarginatum Thonn. Montane forest; occasional. Vollesen SR.

Euphorbiaceae (69)
Acalypha ciliata Forssk. Acacia-Combretum bushland, Acacia-Commiphora bushland; rare. Abdallah 3039.
Acalypha fruticosa Forssk. var. *fruticosa*. All types of bushland, rocky hilltops and ravines; common. Abdallah 2530; Abdallah & Mboya 3818, 4200; Abdallah, Mboya & Vollesen 96/101; Greenway 2129.
Acalypha indica L. Acacia-Commiphora bushland; rare. Abdallah & Mboya 4119.
Acalypha neptunica Muell. Arg. var. *neptunica*. Riverine forest; rare. Abdallah & Vollesen 95/108.
Acalypha ornata A. Rich. Montane forest, riverine forest; occasional. Vollesen SR.

Acalypha racemosa Baill. Montane forest; rare. Abdallah & Mboya 3457.

Acalypha psilostachya Hochst. var. *psilostachya*. Montane forest, Acacia-Combretum bushland; rare. Vollesen 96/8.

Acalypha volkensii Pax. Rocky hilltops, rocky outcrops; occasional. Abdallah & Mboya 4187; Abdallah, Mboya & Vollesen 96/127.

Antidesma venosum Tul. Rocky outcrops; rare. Abdallah 3215.

Bridelia cathartica Bertol. f. Riverine forest; rare. Greenway 4582.

Cephalocroton cordofanus Hochst. Seasonally inundated grassland; locally common. Abdallah 2759, 3051; Abdallah & Mboya 3661, 3996; Abdallah & Vollesen 95/168.

Cleistanthus schlechteri (Pax) Hutch. var. *pubescens* (Hutch.) J. Léon. Riverine forest (Umba River); rare. Abdallah, Mboya & Vollesen 96/193.

Croton dichogamus Pax. Acacia-Combretum bushland; rare. Hedberg TMP294.

Croton megalocarpus Hutch. Montane forest; locally common. Vollsen SR.

Croton polytrichus Pax. Montane forest; occasional. Abdallah 3114; Abdallah & Mboya 3714; Abdallah & Vollesen 95/101.

Croton pseudopulchellus Pax. Montane forest, rocky hilltops and ravines; common. Abdallah 2589, 3381; Abdallah & Mboya 3578, 3885; Abdallah & Vollesen 95/92.

Dalechampia scandens L. var. *cordofana* (Webb) Muell. Arg. Seasonally inundated grassland; locally common. Abdallah 2588, 3293; Abdallah & Mboya 3735; Abdallah & Vollesen 95/130.

Drypetes parvifolia (Muell. Arg.) Pax & K. Hoffm. Montane forest; occasional. Abdallah & Vollesen 95/74.

Erythrococca bongensis Pax. Riverine forest; rare. Abdallah 2705.

Erythrococca fischeri Pax. Montane forest; locally common. Abdallah & Vollesen 96/249.

Erythrococca kirkii (Muell. Arg.) Prain. Montane forest; rare. Greenway 5878.

Euphorbia acalyphoides Boiss. subsp. *acalyphoides*. Seasonally inundated grassland; occasional. Abdallah 3281; Abdallah & Vollesen 95/170; Greenway 6493.

Euphorbia breviarticulata Pax var. *beviarticulata*. Rocky hilltops, rocky outcrops; common. Vollesen SR.

Euphorbia buruana Pax. Rocky hilltops; occasional. Abdallah & Mboya 3875, 3952; Balslev 237.

Euphorbia bussei Pax var. *bussei*. Rocky hilltops, rocky outcrops; common. Abdallah, Mboya & Vollesen 96/120; Greenway 6488.

Euphorbia candelabrum Kotschy var. *bilocularis* (N. E. Br.) S. Carter. All types of bushland; occasional. Greenway 6481.

Euphorbia crotonoides Boiss. subsp. *crotonoides*. Acacia-Commiphora bushland, rocky hilltops; occasional. Abdallah 3278, 3338; Abdallah & Mboya 3722.

Euphorbia cuneata Vahl subsp. *spinescens* (Pax) S. Carter var. *pumilans* S. Carter. Acacia-Combretum bushland, Acacia-Commiphora bushland; common. Abdallah & Vollesen 95/1.

Euphorbia cuneata Vahl subsp. *spinescens* (Pax) S. Carter var. *spinescens*. Rocky hilltops; occasional. Abdallah 2673; Greenway 4060.

Euphorbia espinosa Pax. Acacia-Combretum bushland, Acacia-Commiphora bushland; occasional. Abdallah 3068; Abdallah & Mboya 3402; Abdallah, Mboya & Vollesen 96/107.

Euphorbia furcata N. E. Br. Acacia-Commiphora bushland; rare. Abdallah & Mboya 3931; Cox & Abdallah s.n.

Euphorbia goetzei Pax. Rocky hilltops; rare. Greenway 6491.

Euphorbia gossypina Pax var. *gossypina*. Acacia-Commiphora bushland; rare. Greenway SR.

Euphorbia heterochroma Pax subsp. *heterochroma*. Acacia-Commiphora bushland; rare. Balslev 231, 241; Greenway 2216; Richards 21926.

Euphorbia hirta L. Acacia-Combretum bushland, weed; occasional. Vollesen SR.

Euphorbia kilwana N. E. Br. Seasonally inundated grassland; rare. Abdallah & Vollesen 95/139.

Euphorbia lissosperma S. Carter. Seasonally inundated grassland; locally common. Abdallah, Mboya & Vollesen 96/146.

Euphorbia platycephala Pax. Seasonally inundated grassland; rare. Abdallah 2843; Abdallah & Mboya 3993.

Euphorbia polyantha Pax. Acacia-Commiphora bushland; rare. Abdallah 3360; Abdallah & Mboya 3825, 3913.

Euphorbia prostrata Ait. Weed; rare. Abdallah & Mboya 4126.

Euphorbia quinquecostata Volk. Rocky hilltops, rocky outcrops; occasional. Balslev 230, 238; Greenway 4593, 6489.

Euphorbia scheffleri Pax. Rocky hilltops and ravines; locally common. Abdallah & Vollesen 95/36; Greenway 2127, 6487.

Euphorbia systyloides Pax. Acacia-Combretum bushland, Acacia-Commiphora bushland, rocky hilltops; common. Abdallah 2711, 3266; Abdallah & Mboya 3484.

Euphorbia tenuispinosa Gilli. Acacia-Commiphora bushland, rocky hilltops; rare. Abdallah & Mboya 3942; Abdallah, Mboya & Vollesen 96/116.
Euphorbia tirucalli L. Acacia-Combretum bushland; rare. Greenway 2151.
Excoecaria madagascariensis (Baill.) Muell. Arg. Montane forest; occasional. Abdallah 3111; Abdallah, Mboya & Vollesen 96/77.
Flueggea virosa (Willd.) Voigt. Acacia-Combretum bushland; occasional. Abdallah 2523.
Heywoodia lucens Sim. Montane forest; locally common. Abdallah & Mboya 3620, 3715; Abdallah, Mboya & Vollesen 96/97; Abdallah & Vollesen 95/80.
Jatropha spicata Pax. All types of bushland; widespread but scattered. Abdallah 2888; Abdallah, Mboya & Vollesen 96/180; Greenway 4567.
Manihot glaziovii Muell. Arg. Acacia-Combretum bushland; escaped from cultivation. Vollesen SR.
Meineckia phyllanthoides Baill. subsp. *somalensis* (Pax) Webster. Seasonally inundated grassland; rare. Abdallah, Mboya & Vollesen 96/148.
Mildbraedia carpinifolia (Pax) Hutch. var. *carpinifolia.* Riverine forest; rare. Abdallah & Vollesen 95/179.
Monadenium crispum N. E. Br. Rocky hilltops; rare. Faulkner 4557.
Monadenium heteropodum (Pax) N. E. Br. var. *heteropodum.* Acacia-Commiphora bushland; rare. Vollesen SR.
Phyllanthus leucanthus Pax. Rocky hilltops; rare. Abdallah, Mboya & Vollesen 96/122.
Phyllanthus maderaspatensis L. All types of bushland; common. Abdallah 2590, 2786, 2826B, 2872, 2895, 3098, 3201, 3249, 3355; Abdallah & Mboya 3562; L.D. Harris 30.
Phyllanthus reticulatus Poir. Dry montane forest, Acacia-Combretum bushland; occasional. Vollesen SR.
Phyllanthus rotundifolius Willd. Rocky hilltops; rare. Abdallah & Mboya 3667.
Phyllanthus sepialis Muell. Arg. All types of bushland, montane forest; occasional. Abdallah 2684, 3203; Abdallah & Mboya 3843.
Phyllanthus welwitschianus Muell. Arg. var. *beillei* (Hutch.) A. Radcl.-Sm. Montane forest; occasional. Abdallah & Vollesen 95/95.
Ricinus communis L. Ravines, weed; rare. Abdallah & Mboya 3497; L.D. Harris 106.
Spirostachys africana Sond. Dry montane forest, Acacia-Combretum bushland; common. Abdallah 2680; Shabani 202.
Spirostachys venenifera (Pax) Pax. Riverine forest (Umba River); locally common. Abdallah, Mboya & Vollesen 96/191; Drummond & Hemsley 2345.
Suregada zanzibariensis Baill. Riverine forest (Umba River); locally common. Vollesen SR.
Synadenium glaucescens Pax. Rocky hilltops and ravines, montane forest; occasional. Greenway 6494.
Tragia hildebrandtii Muell. Arg. Seasonally inundated grassland; occasional. Abdallah & Mboya 4144; Abdallah, Mboya & Vollesen 96/157.
Tragia subsessilis Pax. Seasonally inundated grassland; common. Abdallah & Vollesen 95/144; Vollesen 96/44.
Tragia ukambensis Pax var. *ukambensis.* Rocky outcrops; rare. Abdallah, Mboya & Vollesen 96/205.
Tragiella natalensis (Sond.) Pax & K. Hoffm. Montane forest; rare. Abdallah & Mboya 3835.
**Euphorbia agowensis.* Abdallah 3196; Abdallah & Mboya 3590, 3905.
**Euphorbia cuneata.* Abdallah & Mboya 3672.
**Euphorbia hyssopifolia.* Abdallah & Mboya 3594.
**Euphorbia mossambicensis.* Abdallah 2833; Abdallah & Mboya 3894, 3990.
**Euphorbia* sp. Abdallah 3102; Abdallah & Mboya 4025.
**Mallotus* sp. Abdallah 2601.
**Phyllanthus leucocalyx.* Abdallah 2822, 2873, 3075.
**Phyllanthus paxii.* Abdallah 3076.
**Phyllanthus* sp. Abdallah 2655.
**Tragia subsessilis.* Abdallah 2791, 3244; Abdallah & Mboya 3986.

Fabaceae – Caesalpinioideae (15)
Afzelia quanzensis Welw. Acacia-Combretum bushland, riverine forest; occasional. L.D. Harris 105; Milne-Redhead & Taylor 7061.
Cassia abbreviata Oliv. subsp. *beareana* (Holmes) Brenan. Acacia-Combretum bushland; rare. L.D. Harris 89.
Cassia abbreviata Oliv. subsp. *kassneri* (Bak. f.) Brenan. Acacia-Commiphora bushland; occasional. Abdallah 2568; Abdallah & Mboya 4020; Greenway 1971.
Cassia absus L. All types of bushland; occasional. Abdallah 3145, 3272; Abdallah & Mboya 4113.
Cassia fallacina Chiov. Seasonally inundated grassland; rare. Abdallah & Vollesen 95/147.
Cassia hildebrandtii Vatke. Rocky hilltops; rare. Abdallah & Vollesen 95/65.
Cassia kirkii Oliv. var. *kirkii.* Acacia-Commiphora bushland; rare. Peter 49186.
Cassia longiracemosa Vatke. Acacia-Comm-

iphora bushland; occasional. Abdallah & Mboya 3965, 4076; Abdallah & Vollesen 95/135.

Cassia mimosoides L. All types of bushland; common. Abdallah 3144, 3346; Abdallah & Mboya 3964.

Cassia siamea Lam. Acacia-Commiphora bushland; originally introduced, rare. Abdallah & Mboya 3496.

Cassia singueana Del. Acacia-Combretum bushland; common. Abdallah 2569; Abdallah & Mboya 3498; L.D. Harris 82.

Delonix elata (L.) Gamble. Acacia-Commiphora bushland; occasional. Abdallah 3256; Abdallah & Mboya 3827; Greenway 2005; L.D. Harris 81; Richards 21914; Semsei 2092.

Piliostigma thonningii (Schumach.) Milne-Redh. Riverine forest, seasonally-inundated grassland; rare. Vollesen SR.

Tamarindus indica L. Acacia-Combretum bushland, riverine forest; common. L.D. Harris 84.

Tylosema fassoglensis (Schweinf.) Torre & Hillc. Acacia-Combretum bushland; common. Abdallah 2535, 2622; Greenway 2146; L.D. Harris 70.

**Cassia* sp. Abdallah 3216.

Fabaceae – Mimosoideae (36)

Acacia albida Del. Acacia bushland, seasonally-inundated grassland; rare. Abdallah & Mboya 4234; Greenway 2164; Semsei 2080; Wilson 25.

Acacia ancistroclada Brenan. Acacia-Combretum bushland; rare. Abdallah & Mboya 3471; L.D. Harris 118.

Acacia brevispica Harms. All types of bushland; common. L.D. Harris 109.

Acacia bussei Sjöstedt. Acacia and Acacia-Commiphora bushland; locally common. Abdallah 3193; Abdallah & Mboya 3929, 4238; Greenway 2048, 2157, 4558; L.D. Harris 113A; Hedberg TMP289; Richards 21931.

Acacia dolichocephala Harms. Acacia-Combretum bushland; rare. Abdallah & Vollesen 95/150.

Acacia drepanolobium Sjöstedt. Seasonally inundated grassland; locally common. Abdallah & Mboya 4029, 4194.

Acacia etbaica Schweinf. subsp. *australis* Brenan. Acacia-Combretum bushland, Acacia-Commiphora bushland; common. Abdallah & Mboya 3526, 3732, 3831, 4231; Greenway 6498; L.D. Harris 113.

Acacia etbaica Schweinf. subsp. *platycarpa* Brenan. Acacia- Commiphora bushland on sand; occasional. Abdallah 2538; L.D. Harris 97, 107.

Acacia gerrardii Benth. var. *gerrardii*. Acacia-Commiphora bushland; occasional. Abdallah & Mboya 4215; Abdallah & Vollesen 95/183.

Acacia hockii De Wild. Acacia-Combretum bushland, rocky hilltops; common. Abdallah 2681, 3398; Abdallah & Vollesen 95/87.

Acacia mellifera (Vahl) Benth. subsp. *mellifera*. Acacia-Commiphora bushland; common. Abdallah & Mboya 4002, 4060; Greenway 6495; L.D. Harris 15A, 80; Semsei 2080.

Acacia nilotica (L.) Del. subsp. *subalata* (Vatke) Brenan. Acacia-Commiphora bushland; common. Abdallah 3302; Abdallah & Mboya 4184 ; Gillman 791, 892; Greenway 2136.

Acacia pentagona (Schumach. & Thonn.) Hook. f. Montane forest; locally common. Abdallah & Mboya 3890.

Acacia polyacantha Willd. subsp. *campylacantha* (A. Rich.) Brenan. Riverine forest, seasonally-inundated grassland; occasional. Gillman 788.

Acacia reficiens Wawra subsp. *misera* (Vatke) Brenan. Acacia-Commiphora bushland; occasional. Abdallah & Mboya 3817, 3892.

Acacia robusta Burch. subsp. *usambarensis* (Taub.) Brenan. Riverine forest (Umba River); locally common. Abdallah, Mboya & Vollesen 96/173.

Acacia rovumae Oliv. Riverine forest (Umba River); locally common. Abdallah, Mboya & Vollesen 96/194; Gillman 889.

Acacia senegal (L.) Willd. var. *senegal*. All types of bushland; very common. Abdallah 2743; Abdallah & Mboya 3629, 3736; L.D. Harris 86, 88; Leippert 6112; Richards 21910.

Acacia seyal Del. var. *fistula* (Schweinf.) Oliv. Acacia-Combretum bushland, Acacia-Commiphora bushland; occasional. Abdallah & Mboya 4030.

Acacia stuhlmannii Taub. Acacia and Acacia-Commiphora bushland; locally common. Abdallah & Mboya 3582, 3936, 4034; L.D. Harris 115; Semsei 2089.

Acacia thomasii Harms. Acacia-Combretum bushland, Acacia-Commiphora bushland; locally common. Abdallah & Mboya 4150, 4201.

Acacia tortilis (Forssk.) Hayne subsp. *spirocarpa* (A. Rich.) Brenan. Acacia-Commiphora bushland on sand. Greenway 2125; L.D. Harris 98.

Acacia zanzibarica (S. Moore) Taub. var. *zanzibarica*. Acacia bushland, seasonally-inundated grassland; locally common. Abdallah 2573, 3074; Abdallah & Mboya 3625, 3643, 3651, 3934; Greenway 2049; L.D. Harris 114A.

Adenopodia rotundifolia (Harms) Brenan. Riverine forest; rare. Greenway 4559.
Albizia amara (Roxb.) Boiv. subsp. *sericocephala* (Benth.) Brenan. Acacia-Combretum bushland; Acacia-Commiphora bushland; occasional. Abdallah 2679; Abdallah & Mboya 3598, 3797; Abdallah, Mboya & Vollesen 96/59.
Albizia anthelmintica Brongn. Acacia-Combretum bushland, Acacia-Commiphora bushland; occasional. Abdallah & Mboya 3584, 4217; L.D. Harris 87; Hedberg TMP293; Leippert 6111.
Albizia glaberrima (Schumach. & Thonn.) Benth. var. *glabrescens* (Oliv.) Brenan. Riverine forest; locally common. Hedberg TMP288; Milne-Redhead & Taylor 7248, 7248A; Semsei 2079.
Albizia harveyi Fourn. Acacia-Combretum bushland; rare. Gillman 783.
Albizia petersiana (Bolle) Oliv. Dry montane forest, rocky hilltops and ravines; locally common. Vollesen SR.
Albizia schimperiana Oliv. var. *schimperiana*. Montane forest; common. Vollesen SR.
Albizia zimmermannii Harms. Acacia-Commiphora bushland; rare. Abdallah, Mboya & Vollesen 96/102; Semsei 2085.
Dichrostachys cinerea (L.) Wight & Arn. All types of bushland; very common. Abdallah 2753, 3343; Abdallah & Mboya 3733, 4048.
Entada leptostachya Harms. Acacia-Commiphora bushland; rare. Abdallah & Mboya 4010; Milne-Redhead & Taylor 7254.
Mimosa pigra L. Riverbanks and riverbeds; occasional. Vollesen SR.
Newtonia hildebrandtii (Vatke) Torre var. *hildebrandtii*. Acacia-Combretum bushland, Acacia-Commiphora bushland, in areas with high ground water level; locally common. Abdallah & Mboya 4016; Abdallah, Mboya & Vollesen 96/144; Greenway 2120, 4564; L.D. Harris 90, 96.
Parkia filicoidea Oliv. Riverine forest; rare. Peter 49063.
**Acacia* sp. L.D. Harris 112.

Fabaceae – Papilionoideae (87)
Abrus schimperi Bak. subsp. *africanus* (Vatke) Verdc. Acacia-Combretum bushland, rocky hilltops; occasional. Abdallah 3364; Greenway 2172.
Aeschynomene indica L. Seasonally inundated grassland, temporary waterholes; common. Abdallah 3106, 3186; Abdallah & Mboya 3504, 3999, 4080, 4160.
Aeschynomene schimperi A. Rich. Seasonally inundated grassland; rare. Abdallah 3137.
Alysicarpus glumaceus (Vahl) DC. subsp. *glumaceus* var. *glumaceus*. All types of bushland, temporary waterholes; common. Abdallah 3211, 3245, 3271, 3330, 3340; Abdallah & Mboya 3443, 3514, 3998; Abdallah & Vollesen 95/164, 96/223.
Alysicarpus glumaceus (Vahl) DC. subsp. *hispidicarpus* (Fiori) J. Léon. var. *hispidicarpus*. Seasonally inundated grassland; common. Abdallah & Vollesen 95/141.
Clitoria ternatea L. All types of bushland; occasional. Abdallah 2531, 3147.
Craibia brevicaudata (Vatke) Dunn. Montane forest, rocky hilltops; common. Abdallah 3127, 3386; Abdallah & Mboya 3720, 3860, 4203; L.D. Harris 85; Vollesen 96/12, 96/13.
Crotalaria barkae Schweinf. subsp. *cordisepala* Polhill. Acacia-Commiphora bushland; rare. Abdallah, Mboya & Vollesen 96/152; Abdallah & Vollesen 95/175.
Crotalaria bernieri Baill. Seasonally inundated grassland; rare. Abdallah & Vollesen 95/143.
Crotalaria bogdaniana Polhill. Seasonally inundated grassland; rare. Vollesen 96/43.
Crotalaria burttii Bak. f. Seasonally inundated grassland; rare. Abdallah & Vollesen 95/132.
Crotalaria deserticola Bak. f. subsp. *deserticola* var. *deserticola*. Acacia-Combretum bushland; rare. Abdallah & Mboya 3572; Abdallah & Vollesen 96/221.
Crotalaria goodiiformis Vatke. Dry montane forest, Acacia-Combretum bushland, rocky hilltops; occasional. Abdallah 3365; Abdallah & Vollesen 95/73.
Crotalaria greenwayi Bak. f. Acacia-Combretum bushland; occasional. Abdallah & Vollesen 95/196.
Crotalaria kirkii Bak. Acacia-Commiphora bushland; rare. Abdallah 3356.
Crotalaria laburnifolia L. All types of bushland on sand; occasional. Abdallah 2635.
Crotalaria polysperma Kotschy. Acacia-Commiphora bushland; rare. Abdallah & Mboya 3854.
Crotalaria tsavoana Polhill. Acacia-Commiphora bushland; rare. Abdallah, Mboya & Vollesen 96/217.
Crotalaria uguenensis Taub. Rocky hilltops and ravines; rare. Abdallah & Mboya 3441.
Crotalaria ukambensis Vatke. Acacia-Commiphora bushland, rocky outcrops; occasional. Abdallah, Mboya & Vollesen 96/208; Abdallah & Vollesen 95/185.
Dalbergia melanoxylon Guill. & Perr. Acacia-Combretum bushland; common. L.D. Harris 104.
Dalbergia microphylla Chiov. Acacia-Combretum bushland, montane forest; rare. Abdall-

ah 2571; Abdallah & Mboya 3677.

Dolichos oliveri Schweinf. Acacia-Commiphora bushland; rare. Abdallah & Mboya 3444.

Dolichos sericeus E. Mey. subsp. *glabrescens* Verdc. Montane forest; locally common. Abdallah & Mboya 3490, 3994; Abdallah, Mboya & Vollesen 96/70.

Dolichos trilobus L. var. *trilobus.* Seasonally inundated grassland; rare. Abdallah & Mboya 3631, 3987.

Erythrina abyssinica DC. Acacia-Combretum bushland; occasional. Vollesen SR.

Erythrina melanacantha Harms. Acacia-Commiphora bushland; occasional. Vollesen SR.

Indigastrum parviflorum (Wight & Arn.) Schirire. Seasonally inundated grassland; rare. Vollesen 96/38.

Indigofera arrecta A. Rich. Acacia-Commiphora bushland; occasional. Abdallah 3092; Abdallah & Mboya 3521; Abdallah & Vollesen 95/38.

Indigofera atriceps Hook. f. subsp. *kaessneri* (Bak. f.) Gillett. Acacia-Combretum bushland; rare. Vollesen 96/20.

Indigofera brachynema Gillett. Acacia-Combretum bushland; rare. Abdallah, Mboya & Vollesen 96/60.

Indigofera colutea (Burm. f.) Merr. All types of bushland, temporary waterholes; occasional. Abdallah & Mboya 3423; Abdallah & Vollesen 96/222; Mwamba 35.

Indigofera garckeana Vatke. Acacia-Combretum bushland; common. Abdallah 2677, 2728, 2732, 3124, 3387; Abdallah & Mboya 3712, 3851, 3923; Abdallah & Vollesen 95/5; Cox & Abdallah 2315.

Indigofera lupatana Bak. f. Acacia-Combretum bushland; common. Abdallah 3129; Abdallah & Vollesen 95/7.

Indigofera schimperi Jaub. & Spach var. *baukeana* (Vatke) Gillett. Acacia-Commiphora bushland; occasional. Abdallah 3314; Abdallah & Mboya 3523; Abdallah & Vollesen 95/124; L.D. Harris 2.

Indigofera sisalis Gillett. Acacia-Commiphora bushland; occasional. Abdallah 3206; Abdallah & Vollesen 95/125.

Indigofera spicata Forssk. Acacia-Combretum bushland; occasional. Vollesen SR.

Indigofera spinosa Forssk., forma vel sp. aff. Acacia-Commiphora bushland (eastern part); locally common. Abdallah 3217; Abdallah, Mboya & Vollesen 96/163; Faulkner 4334.

Indigofera swaziensis Bolus var. *perplexa* (N. E. Br.) Gillett. Acacia-Combretum bushland; rare. Abdallah & Vollesen 95/88.

Indigofera tinctoria L. var. *tinctoria.* Acacia bushland on alluvial riverbanks (Umba River); locally common. Abdallah, Mboya & Vollesen 96/174.

Indigofera trita L. f. Acacia-Commiphora bushland, seasonally-inundated grassland; occasional. Abdallah 3289; Abdallah & Mboya 3877.

Indigofera vohemarensis Baill. All types of bushland; common. Abdallah 3019; Abdallah, Mboya & Vollesen 96/204; Abdallah & Vollesen 95/6; Vollesen 96/28.

Indigofera volkensii Taub. Acacia-Commiphora bushland, rocky hilltops; occasional. Abdallah 3348; Abdallah & Mboya 3440, 3947; Abdallah & Vollesen 95/115.

Indigofera zenkeri Bak. f. Acacia-Combretum bushland; rare. L.D. Harris 35.

Lablab purpureus (L.) Sweet subsp. *uncinatus* Verdc. Acacia-Combretum bushland, Acacia-Commiphora bushland, seasonally-inundated grassland; occasional. Abdallah 3073, 3251; Abdallah & Mboya 3567.

Lonchocarpus eriocalyx Harms. Acacia-Combretum bushland; common. Abdallah 2693; Abdallah & Mboya 3630; L.D. Harris 38, 79.

Macrotyloma axillare (E. Mey.) Verdc. Seasonally inundated grassland; rare. Abdallah 2837.

Macrotyloma maranguense (Taub.) Verdc. Acacia-Combretum bushland; occasional. Abdallah 3327; Abdallah & Mboya 3610.

Macrotyloma uniflorum (Lam.) Verdc. var. *stenocarpum* (Brenan) Verdc. Acacia-Combretum bushland; common. Abdallah & Vollesen 95/3.

Millettia usaramensis Taub. subsp. *usaramensis* var. *usaramensis.* Acacia-Commiphora bushland (eastern part); occasional. Abdallah 2572; Abdallah, Mboya & Vollesen 96/177; Mgaza 43.

Mundulea sericea (Willd.) A. Chev. Acacia-Commiphora bushland (eastern part); occasional. Abdallah 3180; Abdallah & Mboya 3954.

Neonotonia wightii (Wight & Arn.) Lackey subsp. *wightii* var. *longicauda* (Schweinf.) Lackey. Montane forest; common. Abdallah, Mboya & Vollesen 96/76.

Neorautanenia mitis (A. Rich.) Verdc. All types of bushland; common. Greenway 2011; L.D. Harris 54.

Ormocarpum kirkii S. Moore. Acacia-Combretum bushland, Acacia-Commiphora bushland; common. Abdallah 2691.

Ormocarpum trachycarpum (Taub.) Harms. Acacia-Combretum bushland; occasional. Abdallah, Mboya & Vollesen 96/57.

Platycelyphium voënse (Engl.) Wild. Acacia-Combretum bushland, Acacia-Commiphora bushland (western part); locally common.

Abdallah & Mboya 3587; Abdallah & Vollesen 95/50.

Pseudarthria hookeri Wight & Arn. Acacia-Combretum bushland; occasional. Abdallah 3363; Abdallah & Vollesen 95/76.

Rhynchosia densiflora (Roth) DC. subsp. *chrysadenia* (Taub.) Verdc. Acacia-Combretum bushland; rare. Abdallah & Vollesen 95/151.

Rhynchosia malacophylla (Spreng.) Boj. Seasonally inundated grassland; rare. Abdallah 3323.

Rhynchosia minima (L.) DC. var. *nuda* (DC.) O. Ktze. Acacia-Combretum bushland; rare. Faulkner 4333.

Rhynchosia minima (L.) DC. var. *prostrata* (Harv.) Meikle. Acacia-Combretum bushland, Acacia-Commiphora bushland, seasonally-inundated grassland; common. Abdallah 3295, 3322; Abdallah & Mboya 3429; Abdallah, Mboya & Vollesen 96/65.

Rhynchosia pulchra (Vatke) Harms. Acacia-Combretum bushland, Acacia-Commiphora bushland; occasional. Abdallah 2670, 3922, 4045; Abdallah, Mboya & Vollesen 96/118.

Rhynchosia usambarensis Taub. subsp. *usambarensis* var. *usambarensis*. Acacia-Combretum bushland, montane forest; local. Abdallah & Vollesen 96/244; Vollesen 96/18.

Rhynchosia verdcourtii Thulin (R. sp. C of FTEA). Acacia-Combretum bushland; rare. Abdallah & Vollesen 96/220.

Sesbania quadrata Gillett. Acacia-Commiphora bushland; locally common. Abdallah 3069; Bally 16388; Greenway 2190.

Sesbania sesban (L.) Merr. var. *nubica* Chiov. Temporary waterholes, riverbanks; occasional. Abdallah 2570; Abdallah & Mboya 4088, 4120.

Sesbania subalata Gillett. Seasonally inundated grassland; locally common. Abdallah & Vollesen 95/197.

Spathionema kilimandscharicum Taub. Acacia-Commiphora bushland; rare. Abdallah 3230, 3332; Abdallah & Mboya 4233.

Stylosanthes fruticosa (Retz.) Alston. All types of bushland; occasional. Abdallah 2550; Abdallah & Mboya 3906.

Tephrosia elata Defl. var. *tomentella* Brummitt. Acacia-Combretum bush; rare. Abdallah & Mboya 3566.

Tephrosia hildebrandtii Vatke. Acacia-Combretum bushland, Acacia-Commiphora bushland, montane forest, rocky hilltops; occasional. Abdallah 2816, 3028; Abdallah & Mboya 3596; Abdallah & Vollesen 95/82.

Tephrosia noctiflora Bak. All types of bushland, temporary waterholes; common. Abdallah 3049, 3108, 3306; Abdallah & Mboya 3427.

Tephrosia pentaphylla (Roxb.) G. Don. Seasonally inundated grassland; rare. Abdallah & Vollesen 95/138.

Tephrosia pumila (Lam.) Pers. var. *pumila*. All types of bushland; common. Abdallah 2540, 3142; L.D. Harris 53.

Tephrosia purpurea (L.) Pers. subsp. *leptostachya* (DC.) Brummitt var. *leptostachya*. Acacia-Combretum bushland, rocky hilltops, seasonally-inundated grassland; common. Abdallah 2817, 3277; Abdallah & Mboya 3426; Abdallah & Vollesen 95/140.

Tephrosia subtriflora Baker. Acacia-Commiphora bushland; common. Abdallah 2792, 3345; Abdallah & Vollesen 95/126, 95/174.

Tephrosia uniflora Pers. All types of bushland; common. Abdallah 2650, 2885, 3050, 3079, 3110, 3163; Abdallah & Mboya 4177.

Tephrosia villosa (L.) Pers. subsp. *ehrenbergiana* (Schweinf.) Brummitt. All types of bushland; common. Abdallah 2539, 3038, 3229; Abdallah & Mboya 3430; L.D. Harris 162.

Teramnus labialis (L. f.) Spreng. subsp. *arabicus* Verdc. Montane forest and riverine forest; occasional. Abdallah & Mboya 3506.

Vatovea pseudolablab (Harms) Gillett. Seasonally inundated grassland; occasional. Abdallah & Mboya 3757, 3973; Abdallah & Vollesen 95/52; Vollesen 96/42.

Vigna frutescens A. Rich. subsp. *frutescens* var. *frutescens*. Acacia-Commiphora bushland, Acacia-Combretum bushland; occasional. Abdallah 2658; Abdallah & Mboya 4056; L.D. Harris 53A.

Vigna membranacea A. Rich. subsp. *caesia* (Chiov.) Verdc. Acacia-Commiphora bushland; rare. Abdallah 3237; Abdallah & Mboya 3946; Abdallah & Vollesen 95/127.

Vigna pubescens Wilczek. Acacia-Commiphora bushland; rare. Abdallah 3305.

Vigna vexillata (L.) A. Rich. Acacia-Commiphora bushland; occasional. Abdallah 3080, 3270.

Xeroderris stuhlmannii (Taub.) Mendonça & Sousa. Acacia-Commiphora bushland; locally common. Abdallah 3149, 3527; Abdallah & Mboya 3802, 3897; L.D. Harris 120.

Zornia capensis Pers. subsp. *tropica* Milne-Redh. Acacia-Commiphora bushland (eastern part); rare. Abdallah 3213.

Zornia glochidiata DC. Acacia-Commiphora bushland (eastern part); locally common. Abdallah, Mboya & Vollesen 96/160.

**Crotalaria comanestiana*. Abdallah 2768.

**Crotalaria distantiflora*. Abdallah & Mboya 4166.
**Crotalaria grata*. Abdallah 3212.
**Crotalaria* sp. *aff. massaiensis*. Abdallah 3184.
**Crotalaria* sp. Abdallah & Mboya 4027.
**Dolichos kilimandscharicus*. Abdallah & Mboya 3633.
**Dolichos* sp. Abdallah & Mboya 3652.
**Indigofera* sp. Abdallah 2557.
**Phaseolus* sp. Abdallah & Mboya 4089.
**Tephrosia* sp. Abdallah & Mboya 3524.
**Teramnus uncinatus* subsp. *ringoeti*. Abdallah & Mboya 3605, 3920.
**Vigna* sp. Abdallah & Mboya 4094.

Flacourtiaceae (2)
Dovyalis macrocalyx (Oliv.) Warb. Montane forest; locally common. Abdallah & Mboya 3709.
Rawsonia lucida Harv. & Sond. Montane forest; occasional. Vollesen 96/6.
**Rawsonia* sp. Abdallah & Mboya 4011.
**Flacourtiaceae*. Abdallah & Mboya 4021.
**Flacourtiaceae/Celastraceae*. Abdallah & Mboya 4023.

Gentianaceae (2)
Enicostema axillare (Lam.) A. Raynal subsp. *axillare*. Acacia-Commiphora bushland; occasional. Abdallah 3043; Drummond & Hemsley 1335; Haarer 1570.
Sebaea microphylla (Edgew.) Knobl. Montane forest; rare. Abdallah & Vollesen 96/267.

Geraniaceae (4)
Monsonia angustifolia A. Rich. Acacia-Commiphora bushland; occasional. Abdallah 3025, 3033; Abdallah & Mboya 3902.
Monsonia ovata Cav. subsp. *glauca* (Knuth) Bowden & T. Müller. Acacia-Commiphora bushland; rare. Abdallah & Mboya 3963.
Pelargonium alchemilloides (L.) Ait. subsp. *multibracteatum* (A. Rich.) Kokwaro. Rocky hilltops; occasional. Abdallah 2829, 3279.
Pelargonium quinquelobatum A. Rich. Acacia-Commiphora bushland; occasional. Vollesen SR.

Gisekiaceae (1)
Gisekia pharnaceoides L. var. *pharnaceoides*. Acacia-Commiphora bushland; occasional. Abdallah 3011.

Hamamelidaceae (1)
Trichocladus ellipticus Eckl. & Zeyh. subsp. *malosanus* (Bak.) Verdc. Montane forest; rare. Abdallah & Vollesen 96/264.

Hernandiaceae (1)
Gyrocarpus americanus Jacq. subsp. *americanus*. Rocky hilltops and ravines; rare. Greenway 4714.

Hydnoraceae (1)
Hydnora johannis Beccari. Acacia-Combretum bushland; rare. Abdallah 2722.

Icacinaceae (2)
Apodytes dimidiata Arn. var. *acutifolia* (A. Rich.) Boutique. Montane forest, rocky hilltops; common. Abdallah & Mboya 3595; Vollesen 96/5.
Pyrenacantha malvifolia Engl. Acacia-Commiphora bushland (eastern part); rare. Abdallah 2671.

Lamiaceae (33)
Basilicum polystachion (L.) Moench. Temporary waterholes, seasonally-inundated grassland; occasional. Abdallah & Mboya 3425; L.D. Harris 6.
Becium filamentosum (Forssk.) Chiov. Acacia-Commiphora bushland; rare. Abdallah & Vollesen 95/59.
Becium obovatum (Benth.) N.E. Br. Acacia-Commiphora bushland; rare. Abdallah 2610.
Capitanya otostegioides Gürke. Acacia-Commiphora bushland; rare. Abdallah & Mboya 3911; Greenway 4562.
Endostemon camporum (Gürke) Ashby. Acacia-Combretum bushand; occasional. Abdallah & Vollesen 95/55.
Endostemon tenuiflorus (Benth.) Ashby. Acacia-Commiphora bushland; occsional. Abdallah 2609; Abdallah & Vollesen 95/117.
Endostemon tereticaulis (Poir.) Ashby. Acacia-Commiphora bushland, Acacia-Combretum bushland; occasional. Abdallah 2665; Abdallah & Vollesen 95/171; Faulkner 4550.
Fuerstia africana Th. C. E. Fries. Montane forest, Acacia-Combretum bushland; rare. Abdallah & Vollesen 96/261.
Geniosporum hildebrandtii (Vatke) Ashby. Acacia-Combretum bushland; rare. Abdallah & Vollesen 96/236.
Hoslundia opposita Vahl. All types of bushland, montane forest, rocky hilltops; common. Abdallah 3156; Abdallah & Mboya 3641, 3729, 3969; L.D. Harris 57.
Hyptis pectinata Poit. Temporary waterholes, seasonally-inundated grassland; rare. Abdallah & Mboya 3487; Vollesen 96/26.
Leucas densiflora Vatke. Montane forest, Acacia-Combretum bushland; occasional. Abdallah 3308; Abdallah & Mboya 3844; Abdallah & Vollesen 96/251.

Leucas glabrata (Vahl) R. Br. All types of bushland, seasonally-inundated grassland, rocky hilltops; common. Abdallah 2835, 3282; Abdallah & Mboya 3642, 4040; L.D. Harris 69.
Leucas neuflizeana Courbon. All types of bushland; common. Abdallah 3067, 3301; Abdallah & Mboya 3410; Abdallah, Mboya & Vollesen 96/104.
Leucas tsavoensis Sebald. Acacia-Commiphora bushland; occasional. Abdallah & Mboya 3910, 4044; Abdallah & Vollesen 95/128.
Ocimum americanum L. var. *americanum*. All types of bushland, seasonally-inundated grassland; common. Abdallah 3161; Abdallah & Mboya 3528; Greenway 2143.
Ocimum fischeri Gürke. Acacia-Commiphora bushland (eastern part); rare. Abdallah, Mboya & Vollesen 96/183; Faulkner 4332.
Ocimum gratissimum L. subsp. *gratissimum* var. *gratissimum*. Acacia-Combretum bushland; occasional. Abdallah 2738; Haarer 1479.
Ocimum kilimandscharicum Gürke. Montane forest; rare. Abdallah & Mboya 3609.
Orthosiphon pallidus Benth. Acacia-Combretum bushland, seasonally-inundated grassland; common. Abdallah 2767; Abdallah & Vollesen 95/169.
Orthosiphon parvifolius Vatke. All types of bushland, seasonally-inundated grassland; common. Abdallah 2651; Abdallah & Mboya 3698, 3781, 3805, 3959; Abdallah & Vollesen 95/40; Drummond & Hemsley 1329; Faulkner 4555; Greenway 4560; L.D. Harris 44.
Orthosiphon suffrutescens (Thonn.) J.K. Morton. Acacia-Combretum bushland, rocky hilltops; occasional. Abdallah 2725, 2739.
Plectranthus caninus (Roth) Vatke. All types of bushland; common. Abdallah 2773; Abdallah & Mboya 3499, 3599.
Plectranthus kamerunensis Gürke. Montane forest; rare. Abdallah, Mboya & Vollesen 96/73.
Plectranthus longipes Bak. Rocky hilltops; occasional. Abdallah 2819; Abdallah & Mboya 3621; Abdallah & Vollesen 95/31; Greenway 2145, 4583.
Plectranthus tetensis (Baker) Agnew. Acacia-Commiphora bushland; rare. Abdallah, Mboya & Vollesen 96/109.
Plectranthus tetragonus Gürke. Riverine forest; rare. Abdallah & Vollesen 95/182.
Plectranthus sp. nov. aff. *P. cyaneus* Gürke (= Burtt 5176, etc). Rocky hilltops; occasional. Abdallah, Mboya & Vollesen 96/129.
Plectranthus sp. nov. aff. *P. laxiflorus* Benth., not matched at Kew. Riverine forest; rare. Abdallah & Vollesen 95/181.
Plectranthus sp. nov. aff. *P. tenuiflorus* (Vatke) Agnew (= Drummond & Hemsley 3686, etc.). Rocky hilltop and ravines; rare. Abdallah & Vollesen 96/227.
Solenostemon silvaticus (Gürke) Agnew. Montane forest; occasional. Abdallah & Vollesen 96/263.
Tinnea aethiopica Hook. f. subsp. *aethiopica*. Acacia-Commiphora bushland; occasional. Abdallah 2544, 3009; Abdallah & Mboya 4013; Grenway 2180.
Tinnea aethiopica Hook. f. subsp. *litoralis* Vollesen. Acacia-Commiphora bushland (eastern part); occasional. Abdallah, Mboya & Vollesen 96/168.
**Becium capitatum*. Abdallah & Mboya 3655.
**Becium grandiflorum*. Abdallah 3297.
**Englerastrum scandens*. Abdallah & Mboya 3458.
**Ocium angustifolium*. Abdallah 2611.
**Ocimum hadiense*. Abdallah 3263.
**Orthosiphon suffrutescens*. Abdallah & Mboya 3699.
**Plectranthus assurgens*. Abdallah & Mboya 3602.
**Plectranthus cylindraceus*. Abdallah & Mboya 3481.
**Plectranthus decumbens*. Abdallah & Mboya 3456.
**Plectranthus kapatensis*. Abdallah & Mboya 3475.
**Plectranthus lanuginosus*. Abdallah 3300.
**Plectranthus pauciflorus*. Abdallah & Mboya 4191.
**Plectranthus* sp. Abdallah 2532; Abdallah & Mboya 3493, 4172.

Lentibulariaceae (1)
Utricularia inflexa Forssk. Temporary waterholes, dams; locally common. Vollesen 96/29.

Loganiaceae (5)
Nuxia floribunda Benth. Montane forest; locally common. Abdallah, Mboya & Vollesen 96/83.
Strychnos henningsii Gilg. Rocky hilltops and ravines, rocky outcrops; common. Abdallah 2703; Abdallah & Mboya 3563; Abdallah, Mboya & Vollesen 96/140; Cox & Abdallah 2318.
Strychnos madagascariensis Poir. Riverine forest; rare. Vollesen SR.
Strychnos panganensis Gilg. Montane forest; occasional. Faulkner 4335.
Strychnos usambarensis Gilg. Montane forest; occasional. Abdallah & Mboya 4042; Abdallah & Vollesen 95/103; L.D. Harris 37
**Strychnos* sp. L.D. Harris 101, 103.

Loranthaceae (8)
Erianthemum dregei (Eckl. & Zeyh.) Tieghem. Riverine forest; rare. Wiens 6523.
Helixanthera kirkii (Oliv.) Danser. Acacia-Commiphora bushland; occasional. Abdallah & Mboya 3976.
Oliverella hildebrandtii (Engl.) Tieghem. Acacia-Commiphora bushland; occasional. Abdallah & Mboya 4103.
Oncocalyx fischeri (Engl.) M.G. Gilbert. Acacia-Commiphora bushland; occasional. Wiens 6522, 6524.
Oncocalyx kelleri (Engl.) M.G. Gilbert. Acacia-Commiphora bushland; rare. Wiens 6521.
Plicosepalus curviflorus (Oliv.) Tieghem. Acacia-Commiphora bushland; occasional. Cox & Abdallah 1987; Gill 23.
Plicosepalus sagittifolius (Engl.) Danser. Acacia-Commiphora bushland; rare. Abdallah & Mboya 4195; Tweedie 1680.
Spragueanella rhamnifolia (Engl.) Balle. Montane forest, rocky hilltops; common. Abdallah 3368; Abdallah, Mboya & Vollesen 96/95; Abdallah & Vollesen 95/100.
**Englerina holstii*. Abdallah & Mboya 3974.
**Phragmanthera* sp. Abdallah & Mboya 4220.

Lythraceae (3)
Ammannia prieuriana Guill. & Perr. Waterholes, seasonally-inundated grassland; occasional. Vollesen 96/27.
Lawsonia inermis L. Acacia-Commiphora bushland (eastern part), riverbanks (Umba River); occasional. Abdallah 2576; Abdallah & Mboya 4017.
Nesaea volkensii Koehne. Riverbeds; rare. Abdallah & Mboya 4130.

Malpighiaceae (2)
Caucanthus auriculatus (Radlk.) Niedenzu. Acacia-Combretum bushland, Acacia-Commiphora bushland, dry montane forest, rocky hilltops and ravines; occasional. Abdallah 2886, 3226.
Triaspis mozambica A. Juss. Dry montane forest, rocky hilltops and ravines; occasional. Abdallah & Mboya 3956, 3967; Abdallah & Vollesen 95/11; Ibrahim 658.

Malvaceae (36)
Abutilon figarianum Webb. Acacia bushland, seasonally-inundated grassland; locally common. Vollesen 96/48.
Abutilon fruticosum Guill. & Perr. Acacia-Commiphora bushland; common. Abdallah 3257; Abdallah & Mboya 3862, 3940.
Abutilon guineense (Schumach.) Bak. f. & Exell. Seasonally inundated grassland; locally common. Abdallah, Mboya & Vollesen 96/117; L.D. Harris 61A.
Abutilon hirtum (Lam.) Sweet. Seasonally inundated grassland; rare. Greenway 3950.
Abutilon mauritianum (Jacq.) Medic. Acacia-Commiphora bushland; rare. Abdallah & Mboya 3525.
Abutilon pannosum (G. Forst.) Webb. Seasonally inundated grassland; locally common. Abdallah 2854; Abdallah & Mboya 3417, 4141.
Abutilon rehmannii Bak. f. Acacia-Commiphora bushland (eastern part); rare. Abdallah 2598.
Azanza garckeana (F. Hoffm.) Exell & Hillc. Acacia-Combretum bushland; rare. Vollesen SR.
Hibiscus calyphyllus Cav. Riverine forest; rare. Abdallah & Mboya 3451, 3626; Greenway 2167.
Hibiscus cannabinus L. Seasonally inundated grassland; common. Vollesen SR.
Hibiscus dongolensis Del. Seasonally inundated grassland; locally common. Abdallah 3292; Abdallah & Mboya 3962; Abdallah, Mboya & Vollesen 96/151.
Hibiscus flavifolius Ulbr. Seasonally inundated grassland; common. Vollesen SR.
Hibiscus fuscus Garcke. Montane forest, Acacia-Combretum bushland; occasional. Vollesen SR.
Hibiscus greenwayi Bak. f. Rocky hilltops; occasional. Abdallah & Mboya 3434; Abdallah & Vollesen 95/69.
Hibiscus lobatus (Murr.) O. Ktze. Riverine forest, montane forest; occasional. Abdallah & Mboya 4155; Abdallah & Vollesen 95/96.
Hibiscus mastersianus Hiern. Acacia-Combretum bushland; rare. Abdallah & Vollesen 95/199.
Hibiscus micranthus L. f. All types of bushland; common. Abdallah 2559, 2600A, 2690; Abdallah & Mboya 3755, 3933.
Hibiscus palmatus Forssk. All types of bushland; occasional. Abdallah 2667, 3045, 3096, 3105; Abdallah & Mboya 3614; Vollesen 96/50.
Hibiscus panduriformis Burm. f. Seasonally inundated grassland, temporary waterholes; rare. Abdallah & Mboya 4229.
Hibiscus sidiformis Baill. Acacia-Combretum bushland; rare. Abdallah 3299.
Hibiscus vitifolius L. Seasonally inundated grassland, temporary waterholes; occasional. Abdallah & Mboya 4039; Greenway 2041.
Hibiscus sp. nov. aff. *H. flavifolius* Ulbr. (= Leippert 6337, etc). Acacia-Commiphora bushland, seasonally-inundated grassland; occasional. Abdall & Mboya 3810; Abdall-

ah, Mboya & Vollesen 96/111; L.D. Harris 207.

Hibiscus sp. nov. aff. *H. vitifolius* L. (sp. D of UKWF). Acacia-Commiphora bushland; occasional. Abdallah & Vollesen 95/61.

Pavonia arabica Boiss. Acacia-Commiphora bushland; occasional. Abdallah 3190; Abdallah & Mboya 3819, 3935; Abdallah & Vollesen 95/20; Cox & Abdallah 1965.

Pavonia burchellii (DC.) Dyer. All types of bushland; common. Abdallah 2524, 2547, 2737, 2760, 3091, 3146; Abdallah & Mboya 3592, 3740, 4082.

Pavonia elegans Garcke. Acacia-Combretum bushland, rocky hilltops; common. Abdallah 3087; Abdallah & Mboya 3616, 3953, 4072; Abdallah & Vollesen 95/13.

Pavonia ellenbeckii Gürke. Seasonally inundated grassland; occasional. Abdallah 2818; Abdallah, Mboya & Vollesen 96/110.

Pavonia glechomifolia (A. Rich.) Garcke. Seasonally inundated grassland; occasional. Abdallah 3303; Abdallah, Mboya & Vollesen 96/158; Semsei 2118.

Pavonia procumbens (Wight & Arn.) Walp. Acacia-Commiphora bushland, seasonally-inundated grassland;; occasional. Abdallah 2744; Abdallah & Mboya 3807; Abdallah & Vollesen 95/172.

Pavonia propinqua Garcke. Acacia-Commiphora bushland, seasonally-inundated grassland; common. Abdallah 3173, 3285; Abdallah & Mboya 3903, 4071; Abdallah, Mboya & Vollesen 96/149.

Pavonia zeylanica Cav. Acacia-Commiphora bushland; common (more so in the east). Abdallah 2600, 3103, 3191; Abdallah & Mboya 3522, 4140; Abdallah, Mboya & Vollesen 96/161; Greenway 1997.

Sida alba L. Acacia-Combretum bushland, seasonally-inundated grassland; common. Abdallah 3169; Abdallah & Mboya 3415, 3918, 3970.

Sida chrysantha Ulbr. Acacia-Commiphora bushland (eastern part); locally common. Abdallah, Mboya & Vollesen 96/210.

Sida cordifolia L. All types of bushland; common. L.D. Harris 33; Kisena 551.

Sida javensis Cav. Riverine forest; rare. Mgaza 53.

Sida ovata Forssk. All types of bushland; common. Abdallah 2810, 3171; Abdallah & Mboya 3416.

**Hibiscus* sp. Abdallah & Mboya 4081.

Melastomataceae (1)

Memecylon, sterile. Montane forest; rare. Vollesen SR.

Meliaceae (5)

Melia volkensii Gürke. Acacia-Combretum bushland, Acacia-Commiphora bushland; common (western part). Abdallah & Mboya 3581; Greenway 8584; L.D. Harris 77; Semsei 2077.

Trichilia emetica Vahl. Riverine forest; occasional. Vollesen SR.

Turraea floribunda Hochst. Riverine forest; locally common. Abdallah & Vollesen 95/109.

Turraea mombassana C. DC. subsp. *cuneata* (Gürke) Styles & F. White. Rocky hilltops; occasional. Abdallah 2828; Abdallah & Mboya 4185; Abdallah & Vollesen 95/62.

Turraea robusta Gürke. Montane forest; rare. Abdallah & Mboya 3570.

**Trichilia* sp. L.D. Harris 83.

**Turraea* sp. Abdallah 3392.

Menispermaceae (1)

Cissampelos pareira L. var. *orbiculata* (DC.) Miq. All types of bushland; common. Abdallah 3130; Abdallah & Mboya 3401, 3576, 4108, 4134.

Molluginaceae (5)

Corbichonia decumbens (Forssk.) Exell. Acacia-Commiphora bushland, seasonally-inundated grassland; common. Abdallah 2731, 3298; Abdallah & Mboya 4077.

Glinus lotoides L. Temporary waterholes; occasional. Abdallah & Mboya 3502.

Glinus oppositifolius (L.) A. DC. Temporary waterholes; occasional. Vollesen SR.

Limeum viscosum (J. Gay) Fenzl. Acacia-Commiphora bushland (eastern part); occasional. Abdallah 3219.

Mollugo nudicaulis Lam. Acacia-Commiphora bushland, seasonally-inundated grassland; common. Abdallah 2840, 3089.

Moraceae (12)

Dorstenia hildebrandtii Engl. var. *hildebrandtii.* Rocky riverbeds and riverbanks; rare. Drummond & Hemsley 3726.

Dorstenia zanzibarica Oliv. Montane forest; rare. Abdallah, Mboya & Vollesen 96/80.

Ficus bubu Warb. Dry montane forest, ravines, rocky riverbeds; single trees, widespread. Abdallah & Mboya 3737; van Noort s.n.

Ficus bussei Mildbr. & Burret. Acacia-Commiphora bushland, rocky riverbeds; single trees, widespread. van Noort s.n.

Ficus glumosa Del. Acacia-Commiphora bushland, dry montane forest, rocky riverbeds, rocky hilltops and ravines, rocky outcrops; common. Abdallah & Mboya 3488, 4222; Greenway 2159; van Noort s.n.

Ficus ingens (Miq.) Miq. Rocky hilltops, rocky outcrops, rocky riverbeds; common. Abdallah 2618; Abdallah & Mboya 3767; Abdallah, Mboya & Vollesen 96/211; van Noort s.n.
Ficus lutea Vahl. Dry montane forest, ravines, Acacia-Commiphora bushland; single trees, rare. Abdallah & Mboya 3814; van Noort s.n.
Ficus natalensis Hochst. Montane forest; rare. van Noort s.n.
Ficus sansibarica Warb. subsp. *sansibarica*. Riverine forest, rocky riverbeds; single trees, widespread. van Noort s.n.
Ficus stuhlmannii Warb. Rocky riverbeds, ravines, Acacia-Combretum bushland; single trees, widespread. Abdallah 2749; Abdallah & Mboya 3887, 4148; van Noort s.n.
Ficus sycomorus L. Riverbeds, riverine forest, ravines; common. van Noort s.n.
Ficus thonningii Blume. Rocky outcrops; rare. Abdallah 2619.

Myrtaceae (1)
Syzygium guineense (Willd.) DC. Acacia-Combretum bushland; rare. Vollesen SR.

Nyctaginaceae (6)
Boerhavia coccinea Mill. Acacia-Commiphora bushland; occasional. Abdallah & Mboya 3975.
Boerhavia diffusa L. Acacia-Commiphora bushland, weed; common. Abdallah & Mboya 4093.
Boerhavia erecta L. All types of bushland, weed; common. Vollesen SR.
Commicarpus grandiflorus (A. Rich.) Standl. Acacia-Combretum bushland, Acacia-Commiphora bushland; occasional. Abdallah 2533; Abdallah & Mboya 3428, 3515; Abdallah, Mboya & Vollesen 96/54.
Commicarpus helenae (J.A. Schultes) Meikle. Acacia-Commiphora bushland; occasional. Abdallah & Vollesen 95/123.
Commicarpus plumbagineus (Cav.) Standl. All types of bushland; common. Abdallah 3029; Abdallah & Mboya 3668, 3796; Milne-Redhead & Taylor 7063.
**Commicarpus pedunculosus*. Abdallah 2620.

Ochnaceae (2)
Ochna holstii Engl. Montane forest; occasional. Abdallah & Mboya 3840, 3841.
Ochna ovata F. Hoffm. Acacia-Commiphora bushland; occasional. Abdallah 2541, 2565; Abdallah & Mboya 3644, 3856.

Olacaceae (1)
Ximenia americana L. Acacia-Combretum bushland; rare. Vollesen SR.

Oleaceae (5)
Chionanthus battiscombei (Hiern) Stearn. Montane forest; locally common. Vollesen 96/4.
Jasminum fluminense Vell. Acacia-Commiphora bushland, rocky hilltops and ravines; common. Abdallah 2566, 2834; Abdallah & Mboya 3479, 3867; Abdallah, Mboya & Vollesen 96/123; Kisena 408.
Jasminum schimperi Vatke. Rocky hilltops; rare. Abdallah & Mboya 3782; Abdallah, Mboya & Vollesen 96/143.
Jasminum stenolobum Rolfe. Montane forest; rare. Kisena 407.
Jasminum streptopus E. Mey. Acacia-Commiphora bushland; rare. Abdallah 2875, 2887.

Onagraceae (3)
Ludwigia jussiaeoides Desr. Seasonally inundated grassland, temporary waterholes, riverbanks; occasional. Greenway 3948.
Ludwigia octovalvis (Jacq.) Raven subsp. *brevisepala* (Brenan) Raven. Seasonally inundated grassland, temporary waterholes, riverbanks; occasional. Peter K662.
Ludwigia stenorrhaphe (Brenan) Hara. Seasonally inundated grassland, temporary waterholes, riverbanks; occasional. Drummond & Hemsley 3728; Semsei 2159.

Opiliaceae (1)
Opilia celtidifolia (Guill. & Perr.) Walp. Riverine forest; rare. Abdallah & Mboya 3730; Abdallah, Mboya & Vollesen 96/197.

Papaveraceae (1)
Argemone mexicana L. Temporary waterholes, weed; rare. Vollesen SR.

Passifloraceae (5)
Adenia globosa Engl. subsp. *globosa*. Rocky hilltops and ravines; occasional single plants. Abdallah & Mboya 4202; L.D. Harris 117; Semsei 2087.
Adenia keremanthus Harms. Acacia-Commihpora bushland; rare. Volllesen SR.
Adenia wightiana (Wight & Arn.) Engl. subsp. *africana* de Wilde. Rocky hilltops and ravines; rare. Abdallah & Vollesen 96/230.
Basananthe hanningtoniana (Mast.) de Wilde. Acacia-Combretum bushland, Acacia-Commiphora bushland, rocky hilltops; occasional. Abdallah 2775, 2896, 3154; Abdallah & Vollesen 95/134, 96/224; Faulkner 4564.
Basananthe lanceolata (Engl.) de Wilde. Rocky outcrops; rare. Abdallah, Mboya & Vollesen 96/201.
**Adenia* sp. Abdallah & Mboya 3447, 3571.

Pedaliaceae (2)

Pedaliodiscus macrocarpus Ihlenfeldt. Acacia-Commihpora bushland, seasonally-inundated grassland; rare. Abdallah 2730; Abdallah, Mboya & Vollesen 96/219.

Sesamothamnus rivae Engl. Acacia-Commiphora bushland; occasional. Abdallah & Mboya 3820, 4035; Abdallah & Vollesen 95/46; Greenway 4565.

Piperaceae (2)

Peperomia blanda (Jacq.) Kunth. Dry montane forest, rocky hilltops and ravines; rare. Abdallah & Mboya 3461, 3495; Abdallah & Vollesen 95/188; Greenway 2176, 6485.

Peperomia rotundifolia (L.) Kunth. Montane forest; rare. Abdallah & Vollesen 96/250.

Plumbaginaceae (2)

Plumbago dawei Rolfe. Riverine forest; rare. Greenway 2182; Milne-Redhead & Taylor 7249; Semsei 2120.

Plumbago zeylanica L. Riverine forest; rare. L.D. Harris 93.

Polygalaceae (5)

Polygala erioptera DC. Acacia-Commiphora bushland; occasional. Abdallah & Vollesen 95/122.

Polygala kilimandjarica Chod. Montane forest, Acacia-Combretum bushland, rocky hilltops; occasional. Abdallah & Mboya 3725; Vollesen 96/1.

Polygala sphenoptera Fresen. All types of bushland; common. Abdallah 2567, 2898, 3118, 3284; Abdallah & Mboya 3658, 3919, 3989.

Polygala sadebeckiana Gürke. Seasonally inundated grassland; rare. Abdallah 2836; Drummond & Hemsley 1327.

Polygala sp. nov., not matched at Kew. Combretum-Terminalia bushland (eastern); rare. Abdallah, Mboya & Vollesen 96/190.

Polygonaceae (5)

Oxygonum sinuatum (Meisn.) Dammer. All types of bushland; common. Abdallah 2771, Greenway 2149; L.D. Harris 21.

Oxygonum sp. nov. aff. *O. sinuatum* (Meisn.) Dammer, not matched at Kew. Acacia-Combretum bushland; locally common. Abdallah & Vollesen 96/238.

Polygonum salicifolium Willd. Temporary waterholes, dams; occasional. Abdallah & Mboya 3418.

Polygonum senegalense Meisn. Temporary waterholes, dams; occasional. Vollesen 96/24.

Rumex abyssinicus Jacq. Montane forest; rare. Abdallah, Mboya & Vollesen 96/72.

Portulacaceae (7)

Calyptrotheca taitensis (Pax & Vatke) Brenan. Acacia-Commiphora bushland, rocky hilltops and ravines; occasional. Abdallah & Mboya 3868, 3950; Greenway 2039; L.D. Harris 50.

Portulaca kermesina N. E. Br. Acacia-Commiphora bushland; occasional. Vollesen SR.

Portulaca oleracea L. All types of bushland, seasonally-inundated grassland; common. Abdallah 3048; Greenway 2154, 4563.

Portulaca peteri von Poelln. Acacia-Commiphora bushland; rare. Abdallah, Mboya & Vollesen 96/115.

Portulaca quadrifaria L. Acacia-Commiphora bushland; occasional. Vollesen SR.

Talinum caffrum (Thunb.) Eckl. & Zeyh. Acacia-Combretum bushland; occasional. Abdallah 2799.

Talinum portulacifolium (Forssk.) Schweinf. Acacia-Combretum bushland, rocky hilltops and ravines; common. Abdallah 2649, 2694; Bally 16387.

**Portulaca foliosa.* Abdallah & Mboya 4139.

Ptaeroxylaceae (1)

Ptaeroxylon obliquum (Thunb.) Radlk. Montane forest; rare. Abdallah, Mboya & Vollesen 96/96.

Ranunculaceae (1)

Clematis simensis Fresen. Montane forest; locally common. Abdallah & Vollesen 96/243.

Rhamnaceae (7)

Berchemia discolor (Klotzsch) Hemsl. Riverine forest, rocky riverbeds; occasional. Abdallah & Mboya 3815, 3941; Greenway 2165; Semsei 2084.

Helinus integrifolius (Lam.) O. Ktze. Dry montane forest, Acacia-Commiphora bushland, rocky hilltops and ravines; common. Abdallah & Mboya 3864.

Helinus mystacinus (Ait.) Steud. Montane forest; rare. Abdallah & Mboya 3607.

Lasiodiscus mildbraedii Engl. Montane forest; rare. Abdallah & Vollesen 96/248.

Scutia myrtina (Burm. f.) Kurz. Montane forest; occasional. Abdallah & Vollesen 95/91.

Ziziphus mucronata Willd. subsp. *mucronata.* Acacia-Combretum bushland, Acacia-Commiphora bushland; occasional. Abdallah 3235; Abdallah & Mboya 3657, 4133; L.D. Harris 127; Semsei 2090.

Ziziphus pubescens Oliv. Riverine forest; rare. Greenway 2174.

Rhizophoraceae (1)

Cassipourea malosana (Baker) Alston. Dry

montane forest, Acacia-Combretum bushland; locally common. Abdallah & Vollesen 95/28; L.D. Harris 31, 36A.

Rubiaceae (43)

Breonadia salicina (Vahl) Hepper & Wood. Riverine forest (Umba River); rare. Abdallah & Mboya 4015.

Canthium keniense Bullock. Montane forest; locally common. Abdallah & Vollesen 96/255.

Canthium mombazense Baill. Montane forest, rocky hilltops; common. Abdallah & Mboya 3601, 4156; Abdallah, Mboya & Vollesen 96/126; Abdallah & Vollesen 95/75, 95/97; Greenway 2179 (TYPE of C. greenwayi), 4587; Mgaza 45.

Carphalea glaucescens (Hiern) Verdc. subsp. *glaucescens*. Acacia-Commiphora bushland; rare. Abdallah 3821.

Catunaregam nilotica (Stapf) Tirvengadum. Acacia-Commiphora bushland, rocky riverbanks; rare. Abdallah & Mboya 4087.

Gardenia volkensii K. Schum. subsp. *volkensii*. Acacia-Commiphora bushland; occasional. Vollesen SR.

Gardenia ternifolia Schumach. & Thonn. subsp. *jovis-tonantis* (Welw.) Verdc. Acacia-Combretum bushland; occasional. Abdallah & Mboya 3978.

Hymenodictyon parvifolium Oliv. subsp. *parvifolium*. Rocky hilltops, rocky outcrops; common. Abdallah 2624; Greenway 4577.

Kohautia aspera (Roth) Bremek. Seasonally inundated grassland, Acacia-Combretum bushland; rare. Abdallah 3264; Vollesen 96/40.

Oldenlandia corymbosa L. All types of bushland; common. Abdallah 3210; Abdallah & Mboya 4161.

Oldenlandia fastigiata Bremek. var. *fastigiata*. All types of bushland, temporary waterholes; common. Abdallah 2574, 2815, 3138; Abdallah & Mboya 3414.

Oldenlandia wiedemannii K. Schum. var. *wiedemannii*. Acacia-Commiphora bushland; locally common. Abdallah & Mboya 4075, 4111; Abdallah & Vollesen 95/114.

Pachystigma loranthifolium (K. Schum.) Verdc. subsp. *salaense* Verdc. Acacia-Commiphora bushland; rare. Abdallah & Mboya 3824, 3938.

Pavetta crebrifolia Hiern var. *crebrifolia*. Montane forest; occasional. Abdallah & Mboya 3849; Abdallah & Vollesen 95/77.

Pavetta dolichantha Bremek. Rocky hilltops and ravines; rare. Abdallah 2891.

Pavetta gardeniifolia A. Rich. var. *gardeniifolia*. Montane forest, riverine forest, rocky hilltops; common. Abdallah 2756, 3397; Abdallah & Mboya 4014; Abdallah, Mboya & Vollesen 96/69; Abdallah & Vollesen 95/180.

Pavetta gardeniifolia A. Rich. var. *subtomentosa* K. Schum. Rocky hilltops; rare. Ibrahim 637.

Pavetta holstii K. Schum. Montane forest; rare. Abdallah, Mboya & Vollesen 96/82.

Pavetta sepium K. Schum. var. *merkeri* (K. Krause) Bridson. Acacia-Commiphora bushland; rare. Greenway 4568.

Pentanisia ouranogyne S. Moore. Acacia-Combretum bushland, Acacia-Commiphora bushland; common. Abdallah 2721, 3066; Faulkner 4554; Greenway 2009; L.D. Harris 4.

Pentas parvifolia Hiern. Montane forest, riverine forest, rocky hilltops; occasional. Abdallah 3204; Abdallah & Mboya 3460, 3763; Greenway 2178, 4585; L.D. Harris 33A.

Pentas zanzibarica (Klotzsch) Vatke var. *zanzibarica*. Acacia-Combretum bushland; rare. Abdallah & Vollesen 95/89.

Pentodon pentandrus (Schumach. & Thonn.) Vatke var. *pentandrus*. Temporary waterholes; occasional. Vollesen SR.

Polysphaeria braunii K. Krause. Riverine forest; occasional. Abdallah & Vollesen 95/111.

Polysphaeria lanceolata Hiern subsp. *lanceolata* var. *lanceolata*. Riverine forest; rare. Greenway 2168.

Polysphaeria macrantha Brenan. Montane forest; rare. Abdallah & Vollesen 96/260.

Psychotria alsophila K. Schum. Montane forest; occasional. Abdallah, Mboya & Vollesen 96/84; Abdallah & Vollesen 96/256.

Psychotria kirkii Hiern var. *nairobiensis* (Bremek.) Verdc. Montane forest, Acacia-Commiphora bushland; occasional. Abdallah & Mboya 2805, 3008, 3784, 3937; Abdallah, Mboya & Vollesen 96/66.

Psychotria kirkii Hiern var. *swynnertonii* (Bremek.) Verdc. Rocky hilltops; rare. Abdallah 2821.

Psychotria sp. nov. aff. *P. alsophila* K. Schum., not matched at Kew. Montane forest; rare. Abdallah, Mboya & Vollesen 96/88.

Psydrax schimperiana (A. Rich.) Bridson. Montane forest, rocky hilltops; occasional. Abdallah, Mboya & Vollesen 96/139; Abdallah & Vollesen 96/252.

Rothmannia fischeri (K. Schum.) Bullock subsp. *verdcourtii* Bridson. Montane forest; occasional. Vollesen SR.

Rytigynia celastroides (Baill.) Verdc. var. *celastroides*. Montane forest, Acacia-Combretum bushland; occasional. Abdallah 3126; Abdallah & Mboya 3859; Greenway 4589.

Rytigynia decussata (K. Schum.) Robyns. Rocky hilltops; rare. Abdallah 3388.

Rytigynia eickii (K. Schum. & K. Krause) Bullock. Montane forest, rocky hilltops; rare. Abdallah 3369.

Rytigynia uhligii (K. Schum. & K. Krause) Verdc. Montane forest, riverine forest; occasional or locally common. Abdallah 3004, 3115, 3202; Abdallah & Mboya 3847; Abdallah & Vollesen 95/86, 96/254.

Spermacoce sp. A of FTEA. Acacia-Commiphora bushland (eastern part); rare. Abdallah, Mboya & Vollesen 96/159.

Tarenna graveolens (S. Moore) Bremek. subsp. *graveolens* var. *graveolens*. Rocky hilltops and ravines; occasional. Greenway 4591.

Tennantia sennii (Chiov.) Verdc. & Bridson. Acacia-Commiphora bushland (eastern part); locally common. Abdallah, Mboya & Vollesen 96/166.

Tricalysia ovalifolia Hiern var. *glabrata* (Oliv.) Brenan. Montane forest, rocky hilltops, Acacia-Commiphora bushland; occasional. Abdallah 2634, 3344; Abdallah & Mboya 3480; Mgaza 51.

Tricalysia ovalifolia Hiern var. *taylori* (S. Moore) Brenan. Montane forest, Acacia-Combretum bushland; rare. Vollesen 96/2.

Vangueria madagascariensis J.F. Gmel. Acacia-Commiphora bushland; rare. Abdallah & Mboya 4024.

Vangueria randii S. Moore subsp. *acuminata* Verdc., vel aff. Montane forest; occasional. Abdallah & Mboya 3718, 3836, 4173.

**Canthium phyllanthoideum.* Abdallah 3399.

**Gardenia* sp. L.D. Harris 119C.

**Lamprothamnus zanguebaricus.* Abdallah & Mboya 3983.

**Pavetta abyssinica.* Abdallah & Mboya 3880.

**Pavetta oliveriana.* Abdallah & Mboya 4004.

**Psychotria amboniana.* Abdallah & Mboya 4085.

**Rothmannia urcelliformis.* Abdallah & Mboya 3719.

**Rytigynia* sp. Abdallah & Mboya 4043.

**Tarenna* sp. Abdallah & Mboya 3449.

**Vangueria tomentosa.* Abdallah & Mboya 3713.

**Rubiaceae indet.* Abdallah & Mboya 4117.

Rutaceae (6)

Clausena anisata (Willd.) Benth. Montane forest; occasional. Vollesen SR.

Vepris glomerata (F. Hoffm.) Engl. Rocky hilltops, Acacia-Combretum bushland; occasional. Abdallah & Mboya 3893.

Vepris nobilis (Del.) Mziray. Montane forest; rare. Vollesen SR.

Vepris uguenensis Engl. Rocky hilltops; occasional. Abdallah & Vollesen 95/34; L.D. Harris 94.

Zanthoxylum chalybeum Engl. Acacia-Commiphora bushland; occasional. Abdallah 3367; Abdallah & Mboya 4092.

Zanthoxylum deremense (Engl.) Kokwaro. Montane forest; occasional. Abdallah 3128.

**Vepris* sp. Abdallah & Mboya 4047.

Salvadoraceae (3)

Azima tetracantha Lam. Acacia bushland on clay riverbanks (Umba River); rare. Abdallah & Mboya 3889.

Dobera loranthifolia (Warb.) Harms. Acacia bushland; occasional. Greenway 1972, 4553; L.D. Harris 124A; Leippert 6110.

Salvadora persica L. var. *persica*. Acacia bushland, riverine forest; occasional. L.D. Harris 91, 123A.

Santalaceae (1)

Osyris lanceolata Hochst. & Steud. Acacia-Commiphora bushland; rare. Abdallah & Mboya 4226.

Sapindaceae (6)

Allophylus rubifolius (A. Rich.) Engl. var. *dasystachys* (Gilg) Verdc. Montane forest, Acacia-Commiphora bushland; occasional. Abdallah 3339; Abdallah & Mboya 3846, 3925, 3927, 3981, 4124; Abdallah & Vollesen 95/102.

Cardiospermum halicacabum L. var. *microcarpum* (Kunth) Blume. Seasonally inundated grassland; occasional. Abdallah 3296; Abdallah, Mboya & Vollesen 96/156; Abdallah & Vollesen 95/129.

Haplocoelum foliolosum (Hiern) Bullock. Dry montane forest, rocky hilltops and ravines; common. Abdallah & Mboya 4022; L.D. Harris 102.

Haplocoelum inopleum Radlk. Riverine forest (Umba River); rare. Vollesen SR.

Lecaniodiscus fraxinifolius Bak. subsp. *vaughanii* (Dunkley) Friis. Riverine forest; common. Vollesen SR.

Pappea capensis Eckl. & Zeyh. Dry montane forest, rocky hilltops and ravines; common. Vollesen SR.

**Allophylus griseotomentosus.* Abdallah & Mboya 3826.

Sapotaceae (3)

Manilkara discolor (Sond.) Hemsl. Montane forest, rocky hilltops; occasional. Abdallah & Mboya 4227; Abdallah, Mboya & Vollesen 96/134.

Manilkara mochisia (Baker) Dubard. Acacia-Combretum bushland, Acacia-Commiphora bushland; rare. Abdallah & Mboya 3771.

Mimusops fruticosa A. DC. Riverine forest; rare. Drummond & Hemsley 3712.

Scrophulariaceae (9)

Buchnera hispida D. Don. Seasonally inundated grassland; rare. Abdallah & Mboya 3583.

Craterostigma pumilum Hochst. Acacia-Combretum bushland; rare. Abdallah 2706.

Cycnium herzfeldianum (Vatke) Engl. Acacia-Combretum bushland; rare. Abdallah & Vollesen 96/247.

Cycnium tubulosum (L. f.) Engl. subsp. *montanum* (N. E. Br.) O.J. Hansen. Acacia-Commiphora bushland, rocky hilltops; rare. Abdallah & Mboya 3766, 4180.

Cycnium veronicifolium (Vatke) Engl. subsp. *veronicifolium*. Rocky hilltops; rare. Abdallah, Mboya & Vollesen 96/133.

Harveya obtusifolia (Benth.) Vatke. Rocky hilltops and ravines; rare. Abdallah & Vollesen 96/228.

Stemodiopsis buchananii Skan. Rocky outcrops; rare. Cox & Abdallah 2353.

Striga asiatica (L. f.) O. Ktze. All types of bushland; occasional. Vollesen SR.

Striga latericea Vatke. All types of bushland; occasional. Abdallah 2683, 3176, 3341; Abdallah & Mboya 3724; Cox & Abdallah 2331; L.D. Harris 9; Vollesen 96/41.

**Veronica crassifolia*. Abdallah & Mboya 3792.

Simaroubaceae (1)

Harrisonia abyssinica Oliv. All types of bushland; occasional. Abdallah & Mboya 3957.

Solanaceae (13)

Lycium shawii Roem. & Schult. Acacia-Commiphora bushland; rare. Vollesen SR.

Physalis angulata L. Weed in seasonally-inundated grassland; rare. Abdallah & Mboya 4041.

Solanum campylacanthum A. Rich. All types of bushland, weed; occasional. Vollesen SR.

Solanum coagulans Forssk. Acacia-Commiphora bushland, seasonally-inundated grassland; rare. Abdallah 3153; Greenway 2010.

Solanum hastifolium Dunal. All types of bushland; common. Abdallah 2575, 2666, 3001, 3047, 3094; Abdallah & Vollesen 95/37.

Solanum incanum L. All types of bushland, weed; common. Abdallah 2659, 2794, 3117, 3122, 3158, 3160; Abdallah & Mboya 3734, 3845; L.D. Harris 12A.

Solanum lanzae Lebrun & Stork. Acacia-Commiphora bushland; rare. Abdallah, Mboya & Vollesen 96/216.

Solanum renschii Vatke. All types of bushland; common. Abdallah 2542, 3086, 3313; Abdallah & Mboya 3961, 4179; Abdallah & Vollesen 96/235.

Solanum terminale Forssk. Montane forest; occasional. Abdallah & Vollesen 96/257.

Solanum zanzibarense Vatke var. *vagans* (Wright) Bitter. Montane forest; rare. Abdallah & Vollesen 95/79.

Solanum zanzibarense Vatke var. *zanzibarense*. Acacia-Commiphora bushland; rare. Abdallah & Vollesen 95/159.

Solanum sp. aff. *S. renschii* Vatke, not matched at Kew. Acacia-Commiphora bushland; rare. Abdallah, Mboya & Vollesen 96/108.

Withania somnifera (L.) Dunal. Acacia-Commiphora bushland; rare. Abdallah & Mboya 3516, 3618.

**Solanum setaceum*. Abdallah 3164; Abdallah & Mboya 3776, 3804.

Sterculiaceae (15)

Byttneria fruticosa K. Schum. Riverine forest (Umba River); rare. Abdallah, Mboya & Vollesen 96/192.

Dombeya kirkii Mast. Dry montane forest, rocky hilltops and ravines; occasional. Abdallah & Mboya 3403; Abdallah & Vollesen 95/192.

Dombeya rotundifolia (Hochst.) Planch. Acacia-Combretum bushland; rare. Abdallah 2536.

Dombeya shupangae K. Schum. subsp. *shupangae*. Riverine forest; rare. Hedberg TMP287.

Hermannia fischeri K. Schum. Acacia-Commiphora bushland (eastern part); common. Abdallah, Mboya & Vollesen 96/164.

Hermannia oliveri K. Schum. Acacia and Acacia-Commiphora bushland; common. Abdallah 3150, 3192; Abdallah, Mboya & Vollesen 96/175, 96/213; Greenway 2035; Vollesen 96/51.

Hermannia uhligii Engl. Acacia-Commiphora bushland, seasonally-inundated grassland; occasional. Abdallah 2548, 2597, 2742; Abdallah, Mboya & Vollesen 96/113, 96/188.

Melhania ferruginea Forssk. All types of bushland; occasional. Abdallah 2592, 3007, 3178, 3318; L.D. Harris 61.

Melhania ovata (Cav.) Spreng. Acacia-Combretum bushland, Acacia-Commiphora bushland; occasional. Abdallah 2595, 2596, 3132, 3140, 3167; Abdallah & Mboya 4138; Abdallah & Vollesen 95/121.

Melhania rotundata Mast. Acacia-Commiphora bushland (eastern part); rare. Abdallah, Mboya & Vollesen 96/165.

Pterygota sp. nov. Riverine forest (Umba River); rare. Abdallah & Mboya 4018; Semsei 2130.
Sterculia quinqueloba (Garcke) K. Schum. Acacia-Commiphora bushland; occasional. Coe SR.
Sterculia rhynchocarpa K. Schum. All types of bushland; common. Abdallah & Mboya 4232; Greenway 2006, 2155, 4064; Semsei 2083.
Sterculia stenocarpa H. Winkl. Acacia-Commiphora bushland; occasional. Cox & Abdallah 2350.
Waltheria indica L. All types of bushland; occasional. Abdallah 3188; Abdallah & Mboya 4046.
Hermannia boranensis. Abdallah 2741.
Hermannia exappendiculata. Abdallah 2881; L.D. Harris 110.
Melhania angustifolia. Abdallah & Mboya 3727.
Melhania parviflora. Abdallah 3183.

Thymelaeaceae (2)
Gnidia latifolia (Oliv.) Gilg. Dry montane forest, Acacia-Combretum bushland, ravines; occasional. Abdallah 2682, 2802; Abdallah & Vollesen 95/194.
Synaptolepis alternifolia Oliv. Montane forest; occasional. Abdallah & Mboya 3452; Abdallah & Vollesen 95/93.

Tiliaceae (15)
Corchorus fascicularis Lam. Seasonally inundated grassland, temporary waterholes; rare. Abdallah 3283.
Corchorus olitorius L. Seasonally inundated grassland, temporary waterholes; rare. Abdallah & Mboya 4053.
Corchorus trilocularis L. Acacia-Commiphora bushland; occasional. Abdallah 3199; L.D. Harris 34.
Grewia bicolor Juss. All types of bushland; common. Abdallah 2529, 2584, 3197; Abdallah & Mboya 3822; L.D. Harris 17.
Grewia fallax K. Schum. All types of bushland; common. Abdallah 2585; Abdallah & Mboya 3758, 3806, 3972, 4012; Abdallah & Vollesen 95/39; Greenway 2156.
Grewia forbesii Mast. All types of bushland, rocky hilltops and outcrops; common. Abdallah 2528, 2656, 3218, 3395; Abdallah & Mboya 3431, 3921, 4102; Abdallah, Mboya & Vollesen 96/178, 96/215; Greenway 2177; Ibrahim 661.
Grewia goetzeana K. Schum. Montane forest; rare. Abdallah & Mboya 3916.
Grewia kakothamnos K. Schum. All types of bushland; common. Abdallah & Mboya 3694, 3857, 3863, 3882; Abdallah & Vollesen 95/60; Cox & Abdallah 2358.
Grewia tembensis Fresen. All types of bushland; common. Abdallah 2525, 2526, 2527, 2577, 2582, 2675, 3337; Abdallah & Mboya 3665, 4090; Abdallah & Vollesen 95/35, 95/42; L.D. Harris 10.
Grewia tenax (Forssk.) Fiori. Acacia-Commiphora bushland, rocky hilltops; common. Abdallah 2581, 3179; Abdallah & Mboya 3639, 3721, 3799; Abdallah & Vollesen 95/48; Greenway 2139, 4556; L.D. Harris 182, 208.
Grewia tephrodermis K. Schum. Acacia-Commiphora bushland; rare. Greenway 1976, 4575.
Grewia tristis K. Schum. Acacia-Commiphora bushland (eastern part); locally common. Abdallah 2578, 2580, 2583, 3168; Abdallah, Mboya & Vollesen 96/176, 96/187.
Grewia truncata Mast. Montane forest; rare. Abdallah & Mboya 3917; Greenway 2173.
Grewia villosa Willd. All types of bushland; common. Abdallah 2543; Greenway 2040, 2126; L.D. Harris 240; Ruffo 2561.
Triumfetta rhomboidea Jacq. Acacia-Commiphora bushland; rare. Abdallah 3358.
Grewia sulcata. Abdallah 2579.
Triumfetta cordifolia. Abdallah 3373.
Triumfetta flavescens. Abdallah & Mboya 3728.

Ulmaceae (1)
Celtis africana Burm. f. Riverine forest; rare. Abdallah & Mboya 3695.

Urticaceae (7)
Didymodoxa caffra (Thunb.) Friis & Wilmot-Dear. Montane forest; rare. Abdallah, Mboya & Vollesen 96/92.
Girardinia diversifolia (Link) Friis. Montane forest; occasional. Abdallah, Mboya & Vollesen 96/89.
Laportea aestuans (L.) Chew. Rocky hilltops and ravines; rare. Abdallah & Vollesen 96/231.
Laportea alatipes Hook. f. Riverine forest; rare. Abdallah & Vollesen 95/112.
Laportea lanceolata (Engl.) Chew. Riverine forest; rare. Abdallah & Vollesen 95/113.
Obetia radula (Bak.) B.D. Jackson. Rocky outcrops and ravines; rare. Vollesen SR.
Pilea johnstonii Oliv. subsp. *johnstonii*. Montane forest; rare. Abdallah & Vollesen 96/266.

Vahliaceae (1)
Vahlia digyna (Retz.) O. Ktze. Temporary waterholes; rare. Vollesen 96/35.

Verbenaceae (17)

Chascanum hildebrandtii (Vatke) Gillett. All types of bushland, rocky hilltops, seasonally-inundated grassland; common. Abdallah 2752, 2814; Abdallah & Vollesen 95/137.

Chascanum laetum Walp. Acacia bushland; rare. Abdallah, Mboya & Vollesen 96/214.

Clerodendrum eriophyllum Gürke. Acacia-Commiphora bushland; rare. Abdallah 3234; Abdallah & Mboya 3593.

Clerodendrum hildebrandtii Vatke var. *hildebrandtii.* Seasonally inundated grassland, Acacia-Commiphora bushland; occasional. Abdallah 2633, 2800; Ibrahim 708.

Clerodendrum makanjanum H. Winkler. Acacia-Commiphora bushland; rare. Abdallah & Mboya 4058.

Clerodendrum myricoides (Hochst.) Vatke subsp. *myricoides* var. *discolor* (Klotzsch) Bak. Rocky hilltops; rare. Abdallah 3385.

Lantana trifolia L. Acacia-Combretum bushland; occasional. Abdallah 2769, 2777.

Lantana ukambensis (Vatke) Verdc. Acacia-Combretum bushland; occasional. Abdallah 3131; L.D. Harris 24A.

Lantana viburnoides (Forssk.) Vahl. Acacia-Combretum bushland, Acacia-Commiphora bushland; occasional. Abdallah 3268; Abdallah & Mboya 3579, 4142.

Lippia javanica (Burm. f.) Spreng. Seasonally inundated grassland, riverbanks; occasional. Abdallah 3328.

Lippia kituiensis Vatke. Montane forest, Acacia-Combretum bushland; common. Abdallah & Vollesen 95/90.

Phyla nodiflora (L.) Greene. Riverbanks (Umba River); rare. Vollesen SR.

Premna hildebrandtii Gürke. Riverine forest; occasional. Abdallah 3265; Abdallah & Mboya 3982; Abdallah & Vollesen 95/107.

Premna oligotricha Baker. Dry montane forest, all types of bushland; occasional. Abdallah 2745, 2893; Abdallah, Mboya & Vollesen 96/179; Abdallah & Vollesen 95/193; Mbano CAWM5750.

Premna resinosa (Hochst.) Schauer subsp. *resinosa.* Acacia-Commiphora bushland (eastern part); rare. Abdallah 3208; Abdallah & Mboya 3955.

Stachytarpheta urticifolia Sims. Riverbeds, temporary waterholes; rare. Abdallah 3101.

Vitex strickeri Vatke & Hildebrandt. Dry montane forest, Acacia-Combretum bushland; occasional. Abdallah & Mboya 4127, 4175; Cox & Abdallah 1986.

**Lantana viburnoides.* Abdallah 3005; Abdallah & Mboya 4100, 4101.

**Premna senensis.* Abdallah & Mboya 3682.

**Premna* sp. Abdallah & Mboya 4006; L.D. Harris 103A.

Violaceae (3)

Hybanthus enneaspermus (L.) F. Muell. var. *enneaspermus.* Seasonally inundated grassland, all types of bushland; common. Abdallah 3182; Abdallah & Mboya 3472, 3901, 4104; Abdallah & Vollesen 95/166.

Rinorea angustifolia (Thouars) Baill. subsp. *ardisiiflora* (Oliv.) Grey-Wilson. Montane forest; rare. Abdallah & Mboya 3711; Abdallah & Vollesen 95/83.

Rinorea elliptica (Oliv.) O. Ktze. Riverine forest (Umba River); rare. Abdallah & Mboya 4019.

Viscaceae (2)

Viscum chyuluense Polhill & Wiens. Rocky hilltops; rare. Abdallah & Vollesen 96/239.

Viscum hildebrandtii Engl. Rocky hilltops; rare. Abdallah & Mboya 3770.

Vitaceae (10)

Ampelocissus africana (Lour.) Merr. var. *africana.* Rocky hilltops; rare. Abdallah & Mboya 4068.

Cayratia gracilis (Guill. & Perr.) Suesseng. Montane forest, Acacia-Combretum bushland; occasional. Abdallah 2686; Abdallah & Mboya 3406.

Cissus aphyllantha Gilg & Brandt. Rocky outcrops; rare. Abdallah, Mboya & Vollesen 96/207.

Cissus cactiformis Gilg. Acacia-Commiphora bushland; rare. Abdallah & Mboya 3501.

Cissus quadrangularis L. Rocky hilltops and outcrops, riverine forest; occasional. Vollesen SR.

Cissus rotundifolia (Forssk.) Vahl. Rocky hilltops and outcrops, riverine forest; occasional. Abdallah 2801; Abdallah & Mboya 3632.

Cyphostemma adenocaule (A. Rich.) Wild & Drummond subsp. *adenocaule.* All types of bushland; occasional. Abdallah 2688, 2702.

Cyphostemma buchananii (Planch.) Wild & Drummond. Montane forest, rocky hilltops and ravines; occasional. Abdallah 2890, 3267; Abdallah & Mboya 4158.

Cyphostemma engleri (Gilg) Descoings. Acacia-Commiphora bushland; rare. Greenway 4063.

Rhoicissus revoilii Planch. Rocky hilltops and ravines, rocky outcrops, riverine forest; occasional. Abdallah 2899, 3366; Abdallah & Mboya 3577, 3673, 4055.

**Cyphostemma* sp. Abdallah 2865; Abdallah & Mboya 4001.

Zygophyllaceae (1)
Tribulus terrestris L. All types of bushland; common. Abdallah 3155; Abdallah & Mboya 4073; Greenway 2142.

Dicot. Indet. (or numbers not used)
Abdallah 2653, 2734, 2831, 2842, 2860, 2866, 2877, 2879, 3275, 3276, 3334, 3351, 3361, 3375, 3378; Abdallah & Mboya 4216, 4223, 4240.

ANGIOSPERMS, MONOCOTYLEDONS

Aloaceae (4)
Aloe ballyi Reynolds. Rocky hilltops and outcrops, rockfaces; rare. Greenway 4573.
Aloe desertii Berger. Rocky hilltops and outcrops, rockfaces; common. Carter 2238; Greenway 6385, 6490; Reynolds 8772; Volkens 2378 (TYPE).
Aloe flexilifolia Christian. Rocky hilltops and outcrops, rockfaces; common. Abdallah, Mboya & Vollesen 96/121; Abdallah & Vollesen 96/225.
Aloe rabaiensis Rendle. Rocky hilltops and outcrops, rockfaces; rare. Greenway 6427.
**Aloe* sp. Abdallah & Mboya 3478, 3638.

Amaryllidaceae (3)
Crinum kirkiii Baker. Seasonally inundated grassland; occasional. Abdallah 2561.
Cyrtanthus sanguineus (Lindl.) Walp. subsp. *wakefieldii* (Sealy) Nordal. Acacia-Commiphora bushland; rare. Abdallah & Mboya 3508.
Scadoxus multiflorus (Martyn) Raf. Acacia-Combretum bushland; occasional. Vollesen SR.
**Amaryllidaceae.* Abdallah & Mboya 4128.

Anthericaceae (9)
Chlorophytum angustissimum (Poelln.) Nordal. Rocky outcrops; rare. Abdallah, Mboya & Vollesen 96/202.
Chlorophytum cameronii (Bak.) Kativu var. *pterocaulon* (Bak.) Kativu. All types of bushland; common. Abdallah 2695, 2699; Abdallah & Mboya 3801.
Chlorophytum comosum (Thunb.) Jacq. Montane forest; occasional. Abdallah 2798; Abdallah & Mboya 3687, 4143; Abdallah & Vollesen 96/262.
Chlorophytum macrophyllum (A. Rich.) Aschers. Rocky hilltops, Acacia-Commiphora bushland; common. Abdallah 2562; Abdallah & Mboya 3688, 3795, 3924.
Chlorophytum silvaticum Dammer. Acacia-Combretum bushland, Acacia-Commiphora bushland; occasional. Abdallah 2707, 3013.
Chlorophytum somaliense Bak. Rocky hilltops; rare. Abdallah & Mboya 3790.
Chlorophytum suffruticosum Baker. Rocky hilltops; rare. Burtt 6421.
Chlorophytum tuberosum Baker. Acacia-Combretum bushland, rocky hilltops; rare. Abdallah 2563; Abdallah & Mboya 3800; L.D. Harris 160.
Chlorophytum viridescens Engl. Seasonally inundated grassland. Abdallah 2838; Abdallah & Mboya 4122.
**Chlorophytum affine.* Abdallah & Mboya 3600.
**Chlorophytum micranthum.* Abdallah & Mboya 4069.
**Chlorophytum* sp. Abdallah 2648, 2650A, 2700.
**Ornithogalum* sp. L.D. Harris 14, 61A.

Aponogetonaceae (1)
Aponogeton abyssinicus A. Rich. var. *abyssinicus.* Seasonal waterholes; rare. Greenway 1978.

Araceae (3)
Gonatopus boivinii (Decne.) Engl. Acacia-Commiphora bushland; rare. Abdallah & Mboya 3636.
Stylochaeton bogneri Mayo. Acacia-Commiphora bushland (eastern); rare. Drummond & Hemsley 7245.
Stylochaeton borumensis N. E. Br. Acacia-Commiphora bushland (eastern); rare. Abdallah, Mboya & Vollesen 96/181.

Arecaceae (1)
Hyphaene compressa H. Wendl. Acacia-Commiphora bushland, seasonally-inundated grassland; occasional. Greenway 2122.

Asparagaceae (6)
Asparagus africanus Lam. Rocky hilltops; occasional. Abdallah 3372; L.D. Harris 130.
Asparagus aspergillus Jessop. All types of bushland; common. Vollesen 96/30.
Asparagus falcatus L. Rocky hilltops and outcrops, Acacia-Commiphora bushland; occasional. Abdallah & Mboya 4198.
Asparagus flagellaris (Kunth) Baker. Riverine forest; rare. Milne-Redhead & Taylor 7062.
Asparagus racemosus Willd. Acacia-Combretum bushland; rare. L.D. Harris 12.
Asparagus setaceus (Kunth) Jessop. Montane forest; occasional. Abdallah 3116.

Asphodelaceae (1)
Trachyandra saltii (Bak.) Oberm. var. *secunda* (Krause & Dinter) Oberm. Acacia-Combretum bushland, Acacia-Commiphora bush-

land, rocky hilltops; occasional. Abdallah & Mboya 3813, 3850, 3881, 3886.

Commelinaceae (18)

Aneilema aequinoctiale (P. Beauv.) Kunth. Montane forest; occasional. Abdallah & Mboya 3858; Vollesen 96/3.

Aneilema hockii De Wild. Acacia-Combretum bushland, seasonally-inundated grassland; common. Greenway 2185.

Aneilema leiocaule K. Schum. Montane forest; occasional. Abdallah 2808; Abdallah, Mboya & Vollesen 96/90.

Aneilema rendlei C.B. Clarke. Dry montane forest, rocky hilltops and ravines; occasional. Abdallah 2646; Abdallah & Vollesen 95/191, 96/229.

Aneilema taylori C.B. Clarke. Riverine forest; rare. Abdallah & Vollesen 95/105.

Anthericopsos sepalosa (C.B. Clarke) Engl. Acacia-Combretum bushland; rare. Abdallah 2663.

Commelina africana L. All types of bushland, rocky outcrops; common. Abdallah 2644, 2645, 2696, 2697, 2804; Abdallah & Mboya 3898.

Commelina albescens Hassk. Seasonally inundated grassland; occasional. Abdallah 2701.

Commelina benghalensis L. Montane forest; occasional. Abdallah 2687.

Commelina bracteosa Hassk. Seasonally inundated grassland; occasional. Vollesen SR.

Commelina diffusa Burm. f. All types of bushland, seasonally-inundated grassland; common. Abdallah 3020, 3290.

Commelina erecta L. subsp. *livingstonei* (C.B. Clarke) J.K. Morton. Acacia-Commiphora bushland, seasonally-inundated grassland; common. Abdallah 2727, 2855, 3078.

Commelina forskalaei Vahl. Acacia-Commiphora bushland; common. Abdallah 3205.

Commelina imberbis Hassk. Seasonally inundated grassland; occasional. Vollesen SR.

Commelina lugardii Bullock. Seasonally inundated grassland; occasional. Abdallah & Vollesen 95/145.

Cyanotis foecunda Hassk. Rocky outcrops; common. Abdallah 3383.

Cyanotis longifolia Benth. Rocky outcrops; common. Abdallah 2661.

Murdannia simplex (Vahl) Brenan. All types of bushland, rocky hilltops; common. Abdallah 2824; Abdallah & Mboya 4189.

**Aneilema johnstonii*. Abdallah 2857.

**Aneilema sericeum*. Abdallah & Mboya 4065.

Cyperaceae (33)

Cyperus alopecuroides Rottb. Seasonally inundated grassland, temporary waterholes; occasional. L.D. Harris 304.

Cyperus alternifolius L. subsp. *flabelliformis* (Rottb.) Kuek. Riverbeds, temporary waterholes, seasonally-inundated grassland; occasional. Abdallah & Mboya 3575.

Cyperus articulatus L. Seasonally inundated grassland, temporary waterholes; common. Abdallah & Mboya 3412.

Cyperus bulbosus Vahl. Acacia-Combretum bushland, Acacia-Commiphora bushland; occasional. Abdallah 3015; Mwamba 10.

Cyperus difformis L. Seasonally inundated grassland, temporary waterholes; occasional. Peter 12134.

Cyperus distans L. f. Acacia-Commiphora bushland; common. Abdallah 2894, 3185; L.D. Harris 29.

Cyperus exaltatus Retz. Seasonally inundated grassland, temporary waterholes; occasional. Abdallah & Mboya 3512; Greenway 3962.

Cyperus immensus C.B. Clarke. Seasonally inundated grassland, temporary waterholes; occasional. Greenway 2162.

Cyperus obtusiflorus Vahl. Acacia-Combretum bushland, common. Abdallah 2555; Greenway 2147; L.D. Harris 67.

Cyperus pseudoleptocladus Kük., forma vel sp. aff. Montane forest; occasional. Vollesen 96/7.

Cyperus rotundus L. Acacia- Commiphora bushland, seasonally-inundated grassland; common. Abdallah 3166, 3243; L.D. Harris 22.

Cyperus rubicundus Vahl. Acacia-Commiphora bushland; occasional. Abdallah & Mboya 4118.

Cyperus undulatus Kuek. Seasonally inundated grassland; occasional. Greenway 2161.

Fimbristylis hispidula (Vahl) Kunth. All types of bushland; common. Abdallah 3095.

Kyllinga alata Nees. Acacia-Combretum bushland; occasional. Abdallah & Mboya 3467.

Kyllinga crassipes Boeck. Acacia bushland, Acacia-Combretum bushland, common. Abdallah & Mboya 3706, 3779.

Kyllinga polyphylla Kunth. Seasonally inundated grassland, temporary waterholes; common. Abdallah 2723.

Kyllinga pumila Michx. Rocky hilltops and outcrops; rare. Abdallah 3396.

Kyllinga squamulata Vahl. Montane forest; rare. Abdallah 3070.

Kyllinga triceps Rottb. All types of bushland; common. Abdallah 2632, 2718, 2878.

Kyllingiella microcephala (Steud.) R. Haines & K. Lye. Rocky hilltops and outcrops; occasional. Abdallah 2757.

Mariscus amauropus (Steud.) Cuf. Acacia-Combretum bushland, rocky outcrops; common. Abdallah 2678, 2716, 2793; Abdallah & Mboya 3778; Mwamba 40, 46.

Mariscus circumclusus C.B. Clarke. Acacia-Combretum bushland, common. L.D. Harris 25A.

Mariscus cyperoides (L.) O. Ktze. Acacia-Commiphora bushland; common. Abdallah 2709, 2710.

Mariscus dubius (Rottb.) Hutch. var. *dubius*. Rocky outcrops; common. Abdallah 2712.

Mariscus hemisphaericus (Boeck.) C.B. Clarke. All types of bushland; common. Vollesen SR.

Mariscus macrocarpus Kunth. Acacia-Combretum bushland, Acacia-Commiphora bushland; common. Abdallah & Mboya 4107.

Mariscus mollipes C.B. Clarke. All types of bushland; common. Abdallah & Mboya 3565; L.D. Harris 114; Mwamba 6.

Mariscus obsoletenervosus (Peter & Kuek.) Greenway. Rocky outcrops; rare. Abdallah 2647.

Mariscus pseudovestitus C.B. Clarke. Rocky outcrops; rare. Abdallah 2662; Abdallah & Mboya 3756; L.D. Harris 20A.

Pycreus macrostachyos (Lam.) J. Raynal. Seasonally inundated grassland, waterholes; occasional. Abdallah 3151; Peter 12143.

Pycreus polystachyos (Rottb.) P. Beauv. Acacia-Combretum wooded grassland, riverbanks; occasional. Abdallah 2631, 3120.

Schoenoplectus articulatus (L.) Palla. Seasonally inundated grassland, waterholes; occasional. Greenway 3992.

**Cyperus dilatatus*. Abdallah & Mboya 4170.

**Cyperus fulgens*. Abdallah 2623.

**Cyperus* sp. Abdallah 2669, 2778, 3022; Abdallah & Mboya 3485, 3671.

**Kyllinga alba*. Abdallah 2726.

**Kyllinga richardii*. Abdallah & Mboya 3700.

**Kyllinga* sp. Abdallah & Mboya 3675.

**Mariscus* sp. Abdallah 2832; Abdallah & Mboya 3635, 3676.

Dioscoreaceae (1)

Dioscorea quartiniana A. Rich. var. *quartiniana*. Rocky hilltops; rare. Abdallah & Vollesen 96/226.

Dracaenaceae (5)

Dracaena laxissima Engl. Montane forest; occasional. Vollesen SR.

Sansevieria conspicua N. E. Br. Rocky hilltops and outcrops; occasional. Abdallah & Mboya 3448, 4192.

Sansevieria ehrenbergii Schweinf. Rocky hilltops and outcrops; occasional. Greenway 2148.

Sansevieria gracilis N. E. Br. Rocky hilltops and outcrops; occasional. Abdallah & Mboya 3627.

Sansevieria volkensii Gürke. Rocky hilltops and outcrops; occasional. Vollesen SR.

**Sansevieria* sp. Abdallah 3220; Abdallah & Mboya 3489.

Eriospermaceae (1)

Eriospermum triphyllum Baker. Seasonally inundated grassland; rare. Abdallah & Mboya 4214.

Flagellariaceae (1)

Flagellaria guineensis Schumach. Riverine forest (Umba River); rare. Abdallah, Mboya & Vollesen 96/198.

Hyacinthaceae (3)

Albuca abyssinica Murr. Acacia-Commiphora bushland; rare. Abdallah 2564, 2841.

Dipcadi viride (L.) Moench. Acacia-Combretum bushland, Acacia-Commiphora bushland; rare. Abdallah & Mboya 3465, 3794; Greenway 4572.

Ledebouria revoluta (L. f.) Jessop. Acacia-Combretum bushland; rare. Abdallah 2698.

**Albuca wakefieldii*. Abdallah & Mboya 3878.

Hypoxidaceae (1)

Hypoxis angustifolia Lam. Acacia-Combretum bushland, rocky hilltops; rare. Abdallah & Mboya 3765.

**Hypoxis urceolata*. Abdallah & Mboya 3710.

Iridaceae (3)

Dietes iridioides (L.) Klatt. Montane forest; rare. Abdallah & Vollesen 96/253.

Gladiolus candidus (Rendle) Goldblatt. Acacia-Combretum bushland; rare. Abdallah 2776.

Gladiolus dalenii van Geel. Rocky hilltops, Acacia-Combretum bushland; rare. Abdallah & Mboya 4186.

Orchidaceae (10)

Aerangis kirkii (Reichb. f.) Schltr. Rocky hilltops; epiphytic on Euphorbia; locally abundant. Abdallah & Mboya 4193..

Angraecopsis breviloba Summerh. Rocky hilltops; epiphytic on Euphorbia; locally abundant. Abdallah, Mboya & Vollesen 96/130.

Angraecum cultriforme Summerh. Rocky hilltops, epiphytic; occasional. Moreau 724.

Cyrtorchis praetermissa Summerh. Rocky hilltops; epiphytic; occasional. Vollesen SR.

Eulophia cucullata (Sw.) Steud. Acacia-Combretum bushland; rare. Abdallah 2660.

Eulophia petersii Reichb. f. Acacia-Commiph-

ora bushland; rare. Greenway 2158.
Microcoelia exilis Lindl. Montane forest, rocky hilltops, epiphytic; locally abundant. Cox & Abdallah 2321; Drummond & Hemsley 3727.
Oberonia disticha (Lam.) Schltr. Rocky hilltops; epiphytic on Euphorbia; locally abundant. Abdallah, Mboya & Vollesen 96/131.
Polystachya dendrobiiflora Reichb. f. Rocky hilltops and outcrops, epiphytic on Xerophyta; locally common. Abdallah & Mboya 3404; Abdallah & Vollesen 95/154; Cox & Abdallah 2303.
Polystachya isochiloides Summerh. Rocky hilltops; epiphytic on Euphorbia; locally abundant. Abdallah, Mboya & Vollesen 96/132.
**Eulophia wakefieldii*. Abdallah 2812; Abdallah & Mboya 3707.
**Eulophia* sp. Abdallah 2858.

Poaceae (116)
Acrachne racemosa (Roem. & Schult.) Ohwi. Acacia-Commiphora bushland; rare. Greenway 1987.
Alloteropsis cimicina (L.) Stapf. Seasonally inundated grassland; occasional. Abdallah 3252.
Andropogon chinensis (Nees) Merr. Acacia-Commiphora bushland, rocky outcrops; occasional. Abdallah 2553; Mwamba 41.
Andropogon schirensis A. Rich. Acacia-Combretum bushland; common. Vollesen SR.
Aristida adscensionis L. All types of bushland, rocky outcrops; common. Greenway 2208; L.D. Harris 13, 78.
Aristida barbicollis Trin. & Rupr. Acacia-Commiphora bushland; occasional. L.D. Harris 283.
Aristida mutabilis Trin. & Rupr. Acacia-Commiphora bushland; occasional. Greenway 2001.
Bothriochloa bladhii (Retz.) S.T. Blake. Seasonally inundated grassland; common. Mwamba 3.
Bothriochloa radicans (Lehm.) A. Camus. All types of bushland, weed, seasonally-inundated grassland; common. Abdallah 3017; Abdallah & Mboya 3509; L.D. Harris 8, 72.
Brachiaria deflexa (Schumach.) Robyns. Acacia-Commiphora bushland, riverine forest; occasional. Greenway 1992, 2131; Mwamba 23.
Brachiaria dictyoneura (Fig. & De Not.) Stapf. Acacia-Commiphora bushland, rocky hilltops; occasional. Abdallah 2897.
Brachiaria eruciformis (J.E. Smith) Griseb. Seasonally inundated grassland; common. Abdallah 3241; Abdallah & Mboya 3992; Mwamba 1.
Brachiaria leersioides (Hochst.) Stapf. Acacia-Commiphora bushland, disturbed places; occasional. Abdallah 3097; Leippert 6207.
Brachiaria leucacrantha (K. Schum.) Stapf. Acacia-Commiphora bushland; occasional. Mwamba 50.
Brachiaria rugulosa Stapf. Acacia-Combretum bushland; rare. Abdallah 2724.
Brachiaria serrifolia (Hochst.) Stapf. Acacia-Commiphora bushland; occasional. Greenway 2141; Mwamba 22, 28.
Cenchrus ciliaris L. All types of bushland, weed; common. Abdallah 3057; Greenway 1998; L.D. Harris 27, 60A.
Chloris mossambicensis K. Schum. Acacia-Combretum bushland; common. L.D. Harris 5, 231; Mwamba 2.
Chloris pycnothrix Trin. All types of bushland, weed; occasional. Vollesen SR.
Chloris roxburghiana Schult. All types of bushland, weed; common. Abdallah 2556, 2628, 2863; Greenway 1996, 2196; L.D. Harris 28, 124; Ruffo 2487, 2488.
Cleistachne sorghoides Benth. Seasonally inundated grassland; common. Greenway 3258.
Cymbopogon caesius (Hook. & Arn.) Stapf. Acacia-Combretum bushland; occasional. Abdallah & Mboya 3591.
Cymbopogon commutatus (Steud.) Stapf. Acacia-Commiphora bushland; rare. L.D. Harris 123.
Cynodon dactylon (L.) Pers. All types of bushland, weed; common. L.D. Harris 198.
Cynodon nlemfuensis Vanderyst var. *robustus* W.D. Clayton & Harlan. Seasonally inundated grassland; common. Greenway 1999.
Dactyloctenium aegyptium (L.) Willd. All types of bushland, weed; common. Abdallah 3041; L.D. Harris 41.
Dactyloctenium giganteum Fisher & Schweick. All types of bushland, weed; occasional. Abdallah 2870; Greenway 2138.
Dichanthium annulatum (Forssk.) Stapf var. *annulatum*. All types of bushland, seasonally-inundated grassland; common. Vollesen 96/36.
Digitaria macroblephara (Hack.) Stapf. Seasonally inundated grassland; occasional. L.D. Harris 46, 203.
Digitaria milanjiana (Rendle) Stapf. All types of bushland; common. Abdallah 2780, 3044, 3056, 3081, 3224; Abdallah & Mboya 4110; Greenway 2003, 2130, 2189; L.D. Harris 26, 56A, 294; Mwamba 5.
Digitaria pennata (Hochst.) T. Cooke. Acacia-Commiphora bushland; occasional. Vollesen SR.
Digitaria rivae (Chiov.) Stapf. Acacia-Commiphora bushland; rare. L.D. Harris 35A.
Digitaria velutina (Forssk.) P. Beauv. All types

of bushland, often disturbed, seasonally-inundated grassland, weed; common. Abdallah 2848, 3037, 3248; Abdallah & Mboya 4123, 4137; Greenway 1993; Mwamba 27.

Dinebra polycarpha S.M. Phillips. Seasonally inundated grassland; occasional. Abdallah & Mboya 4145; Abdallah, Mboya & Vollesen 96/155; Mwamba 14.

Dinebra retroflexa (Vahl) Panzer var. *condensata* S.M. Phillips. Acacia bushland, Acacia-Commiphora bushland; rare. Abdallah 3198; Greenway 2116.

Diplachne fusca (L.) Stapf. Seasonally inundated grassland; rare. Semsei 3964.

Drake-Brockmania haareri (Stapf & Hubbard) S.M. Phillips. Seasonally inundated grassland; common. Abdallah, Mboya & Vollesen 96/112; Greenway 3974, 3982; L.D. Harris 233.

Echinochloa colona (L.) Link. Seasonally inundated grassland, swamps; common. Abdallah 3136.

Echinochloa haploclada (Stapf) Stapf. Seasonally inundated grassland; common. Abdallah 2852; Greenway 2210; L.D. Harris 11.

Echinochloa pyramidalis (Lam.) Hitch. & Chase. Seasonally inundated grassland; rare. Greenway 2054.

Eleusine indica (L.) Gaertn. Acacia-Commiphora bushland, disturbed places; occasional. Abdallah 3165, 3336.

Enneapogon cenchroides (Roem. & Schult.) Hubbard. Acacia-Commiphora bushland, rocky outcrops; occasional. Abdallah & Mboya 3899; Abdallah, Mboya & Vollesen 96/200; L.D. Harris 24.

Enneapogon persicus Boiss. Acacia-Combretum bushland; occasional. L.D. Harris 302; Mwamba 43.

Enteropogon macrostachus (A. Rich.) Benth. Acacia-Combretum bushland; common. Abdallah 2809; Greenway 1995, 2152; L.D. Harris 36, 199.

Enteropogon rupestris (J.A. Schmidt) A. Chev. Acacia-Commiphora bushland; rare. Mbano CAWM5748.

Eragrostiella bifaria (Vahl) Bor. Acacia-Commiphora bushland, rocky outcrops; common. Abdallah & Mboya 4211.

Eragrostis aethiopica Chiov. Seasonally inundated grassland, disturbed; occasional. Abdallah & Mboya 4167; Greenway 2007, 2218; Mwamba 58.

Eragrostis aspera (Jacq.) Nees. All types of bushland, often disturbed, weed; occasional. L.D. Harris 272.

Eragrostis caespitosa Chiov. Acacia-Commiphora bushland (eastern); common. Abdallah 2643; Abdallah, Mboya & Vollesen 96/209; Mwamba 54.

Eragrostis cilianensis (All.) Lut. All types of bushland, often disturbed, weed; common. Abdallah 2884; Greenway 1988, 2209; L.D. Harris 63.

Eragrostis cylindriflora Hochst. Acacia-Commiphora bushland, often disturbed; common. Abdallah 3107.

Eragrostis exasperata Peter. Acacia-Commiphora bushland, rocky outcrops; occasional. Abdallah 3052.

Eragrostis heteromera Stapf. Acacia-Commiphora bushland; occasional. Abdallah 3054, 3082.

Eragrostis inamoena K. Schum. Acacia-Commiphora bushland; rare. Abdallah 2554.

Eragrostis racemosa (Thunb.) Steud. All types of bushland; common. Abdallah 3072; Abdallah & Mboya 4105.

Eragrostis rigidior Pilg. Acacia-Combretum bushland; rare. Mwamba 19.

Eragrostis superba Peyr. All types of bushland; common. Greenway 2195, 4588; L.D. Harris 26A, 45.

Eragrostis tenella (L.) Roem. & Schult. All types of bushland, weed; common. Abdallah & Mboya 3692.

Eriochloa fatmensis (Hochst. & Steud.) W.D. Clayton. Seasonally inundated grassland; occasional. Mwamba 13

Eriochloa meyeriana (Nees) Pilg. Seasonally inundated grassland; occasional. Mwamba 4.

Eriochloa parvispiculata Hubbard. Acacia-Commiphora bushland; rare. Abdallah 2641.

Eustachys paspaloides (Vahl) Lanza & Mattei. Acacia-Combretum bushland, Acacia-Commiphora bushland; occasional. Abdallah 3021; L.D. Harris 55.

Harpachne schimperi A. Rich. Acacia-Commiphora bushland, temporary waterholes; common. Abdallah & Mboya 3422.

Heteropogon contortus (L.) Roem. & Schult. All types of bushland; common. Abdallah 2788; Abdallah & Mboya 3510; L.D. Harris 23.

Hyparrhenia filipendula (Hochst.) Stapf. All types of bushland; common. Vollesen SR.

Hyparrhenia rufa (Nees) Stapf. All types of bushland; common. Vollesen SR.

Ischaemum afrum (J.F. Gmel.) Dandy. Seasonally inundated grassland; occasional. Abdallah 2844, 2851; Greenway 2191; L.D. Harris 229.

Leptocarydion vulpiastrum (De Not.) Stapf. Dry montane forest, Acacia-Commiphora bushland; occasional. Abdallah & Mboya 4114, 4181; Mwamba 51.

Leptochloa obtusiflora Hochst. Acacia-Combretum bushland, Acacia-Commiphora bushland; common. Abdallah 2783, 2864, 3225; L.D. Harris 193; Mwamba 26.

Leptochloa uniflora A. Rich. Seasonally inundated grassland; rare. Greenway 2170.

Leptothrium senegalense (Kunth) W.D. Clayton. Acacia-Combretum bushland; common. Abdallah 2867; Greenway 1990.

Lintonia nutans Stapf. Acacia-Combretum bushland, Acacia-Commiphora bushland; common. Abdallah & Mboya 3904; L.D. Harris 206.

Microchloa kunthii Desv. Acacia-Combretum bushland; common. Mwamba 11.

Neyraudia arundinacea (L.) Henr. Seasonally inundated grassland; occasional. Semsei 2322.

Oplismenus hirtellus (L.) P. Beauv. Montane forest, rocky hilltops and ravines; common. Abdallah 3394.

Panicum atrosanguineum A. Rich. Acacia-Combretum bushland, Acacia-Commiphora bushland; common. Abdallah 3016, 3060, 3061; Greenway 2197.

Panicum coloratum L. var. *minus* Chiov. Acacia-Combretum bushland, Acacia-Commiphora bushland; common. Abdallah 2642; Mwamba 7, 55.

Panicum deustum Thunb. Acacia-Combretum bushland; common. Abdallah 2790; L.D. Harris 197; Mwamba 29.

Panicum infestum Peters. Acacia-Combretum bushland, Acacia-Commiphora bushland; common. Mwamba 16, 45.

Panicum issongense Pilg. Rocky hilltops and ravines; rare. Abdallah 3380.

Panicum maximum Jacq. Acacia-Combretum bushland, seasonally-inundated grassland; common. Greenway 2123, 2194; L.D. Harris 30A.

Panicum trichocladum K. Schum. Montane forest, Acacia-Commiphora bushland; rare. Abdallah 2868.

Pennisetum mezianum Leeke. Seasonally inundated grassland; common. Abdallah 2630, 2772, 2853; Abdallah & Mboya 3519, 3705; Drummond & Hemsley 1325; Greenway 2188; L.D. Harris 39A, 68.

Pennisetum polystachion (L.) Schult. subsp. *polystachion.* Acacia-Combretum bushland, seasonally-inundated grassland; common. Abdallah 2549, 3059.

Phragmites mauritianus Kunth. Seasonally inundated grassland, swamps; occasional. Semsei 2139.

Rhynchelytrum nerviglume (Franch.) Chiov. Acacia-Combretum bushland; rare. L.D. Harris 38A.

Rhynchelytrum repens (Willd.) C.E. Hubbard. Acacia-Combretum bushland, Acacia-Commiphora bushland, weed; common. Abdallah 2795, 3046; L.D. Harris 37A.

Rottboellia cochinchinensis (Lour.) W.D. Clayton. Seasonally inundated grassland; occasional. Abdallah 3247; Mwamba 17.

Schmidtia pappophoroides J.A. Schmidt. Acacia-Commiphora bushland; occasional. Abdallah 3026, 3233; Abdallah & Mboya 3762; Mwamba 42.

Schoenefeldia transiens (Pilg.) Chiov. Seasonally inundated grassland; common. Abdallah 2629; Abdallah & Mboya 4131; L.D. Harris 228.

Setaria incrassata (Hochst.) Hack. Seasonally inundated grassland; common. Abdallah 2856; Greenway 2211; L.D. Harris 19.

Setaria megaphylla (Steud.) Th. Dur. & Schinz. Montane forest; common. L.D. Harris 202.

Setaria pumila (Poir.) Roem. & Schult. Acacia-Commiphora bushland; rare. Abdallah 3083.

Setaria sagittifolia (A. Rich.) Walp. Riverine forest, Acacia-Commiphora bushland; common. Abdallah 3258; Greenway 2119; Leippert 6205; Semsei 2160.

Setaria sphacelata (Schumach.) Moss var. *sphacelata.* Seasonally inundated grassland; common. Vollesen SR.

Setaria verticillata (L.) P. Beauv. Acacia-Combretum bushland; common. Vollesen SR.

Sorghum arundinaceum (Desv.) Stapf. Seasonally inund. grassland; occasional. Vollesen SR.

Sorghum versicolor Anderss. Seasonally inundated grassland; occasional. Abdallah 3286, 3326.

Sporobolus africanus (Poir.) Robyns & Tournay. Acacia-Combretum bushland; occasional. Abdallah 2862.

Sporobolus consimilis Fresen. Seasonally inundated grassland; occasional. L.D. Harris 119.

Sporobolus cordofanus (Steud.) Coss. Acacia-Combretum bushland; occasional. Greenway 2137.

Sporobolus festivus A. Rich. Acacia-Combretum bushland, Acacia-Commiphora bushland; common. Abdallah 2713; Greenway 2004; L.D. Harris 18A.

Sporobolus fimbriatus (Trin.) Nees. Acacia-Combretum bushland, Acacia-Commiphora bushland; common. Abdallah 2779; Greenway 1994; L.D. Harris 59A, 191.

Sporobolus helvolus (Trin.) Th. Dur. & Schinz. Acacia-Commiphora bushland, seasonally-inundated grassland; occasional. Abdallah 3288; Abdallah & Mboya 3907, 4036; Greenway 4557.

Sporobolus ioclados (Trin.) Nees. Acacia-Combretum bushland, Acacia-Commiphora bushland; common. Vollesen SR.

Sporobolus pyramidalis P. Beauv. Seasonally inundated grassland; common. Abdallah 2551, 2761; L.D. Harris 126.
Tetrachaete elionuroides Chiov. Acacia-Commiphora bushland (eastern), rocky outcrops; rare. Abdallah, Mboya & Vollesen 96/199.
Tetrapogon bidentatus Pilg. Acacia-Combretum bushland, Acacia-Commiphora bushland; common. Abdallah 2639, 3040; L.D. Harris 172; Mwamba 44.
Tetrapogon tenellus (Roxb.) Chiov. Acacia-Combretum bushland, Acacia-Commiphora bushland; common. Ibrahim 657; Mwamba 31, 52.
Themeda triandra Forssk. Acacia-Combretum bushland, seasonally-inundated grassland; common. Abdallah 2720; Greenway 2212; L.D. Harris 43, 51.
Tragus berteronianus Schult. Acacia-Combretum bushland, Acacia-Commiphora bushland, weed; common. Abdallah 2640, 3010, 3088; Greenway 2133; L.D. Harris 42, 119A.
Tragus heptaneuron W.D. Clayton. Acacia-Combretum bushland; rare. L.D. Harris 119B.
Tripogon minimus (A. Rich.) Steud. Acacia-Commiphora bushland, rocky outcrops; occasional. Mwamba 8.
Urochloa mossambicensis (Hack.) Dandy. Acacia-Combretum bushland, seasonally-inundated grassland; common. Greenway 2153; Mwamba 21, 48.
Urochloa panicoides P. Beauv. Acacia-Combretum bushland, Acacia-Commiphora bushland; common. Mwamba 49.
Urochloa trichopus (Hochst.) Stapf. Acacia-Combretum bushland, Acacia-Commiphora bushland; common. Greenway 2132.
**Andropogon amethystinus*. Abdallah 2807.
**Cymbopogon pospischilii*. Abdallah 3053, 3071.
**Dichanthium* sp. Abdallah 3246.
**Digitaria* sp. Abdallah 3261.
**Dignathia hirtella*. Abdallah & Mboya 4151.
**Dinebra* sp. Abdallah & Mboya 4115.
**Eragrostis heteromera*. Abdallah 3109.
**Eriochloa* sp. Abdallah & Mboya 4099.
**Hyparrhenia* sp. Abdallah & Mboya 4146.
**Panicum poaeoides*. Abdallah & Mboya 3505.
**Pennisetum thunbergii*. Abdallah & Mboya 3777.
**Sporobolus confinis*. Abdallah 2763.
**Sporobolus myrianthus*. Abdallah 2787.
**Poaceae indet*. Abdallah & Mboya 4116, 4125.

Typhaceae (1)
Typha domingensis Pers. Swamps, seasonally-inundated grassland; rare. Greenway 2163.

Velloziaceae (1)
Xerophyta spekei Bak. Rocky outcrops and hilltops; common. Abdallah 2552; Abdallah, Mboya & Vollesen 96/128; L.D. Harris 239.

PTERIDOPHYTES

Ferns (15)
Actiniopteris radiata (Sw.) Link. Rocky hilltops and outcrops; common. Abdallah & Mboya 3772; Cox & Abdallah 2300.
Actiniopteris semiflabellata Pichi Serm. Rocky outcrops; rare. Greenway 4590.
Adianthum incisum Forssk. Montane forest; common. Vollesen SR.
Asplenium blastophorum Hieron. Montane forest; rare. Vollesen 96/11.
Asplenium mannii Hook. f. Montane forest, rocky hilltops; rare. Abdallah 2826.
Asplenium rutifolium (Berg.) Kunze. Montane forest, rocky hilltops and ravines; common. Abdallah 3393; Cox & Abdallah 2316; Vollesen 96/10.
Doryopteris concolor (Langsd. & Fisch.) Kuhn. Montane forest, riverine forest; occasional. Abdallah 3113.
Drynaria volkensii Hieron. Montane forest, rocky hilltops; occasional. Abdallah & Mboya 4174, 4188.
Ophioglossum petiolatum Hook. Acacia-Combretum bushland; rare. Abdallah, Mboya & Vollesen 96/64.
Pellaea doniana Hook. Rocky hilltops and outcrops; common. Abdallah 3370.
Pellaea viridis (Forssk.) Prantl. Rocky hilltops and outcrops; common. Abdallah & Mboya 4064; Abdallah, Mboya & Vollesen 96/61.
Pleopeltis macrocarpa (Willd.) Kaulf. Montane forest; rare. Abdallah & Mboya 3833.
Pteris catoptera Kunze. Montane forest; rare. Abdallah & Vollesen 96/265.
Pteris sp. ? nov. aff. *P. atrovirens* Willd., not matched at Kew. Montane forest; common. Abdallah & Mboya 3837; Abdallah, Mboya & Vollesen 96/91.
Selaginella abyssinica Spring. Montane forest; locally common. Abdallah, Mboya & Vollesen 96/79.
**Asplenium buettneri*. Abdallah & Mboya 3716.
**Blechnum* sp. Abdallah & Mboya 3574.
**Pellaea involuta*. Abdallah 3003.
**Pellaea longipilosa*. Abdallah & Mboya 4066.

Total (4/4/99): 1,147 taxa.

CHAPTER 7

Vegetation and habitats of Mkomazi

Malcolm Coe, Kaj Volleson, Raphael Abdallah & Emmanuel I. Mboya

Introduction

A habitat may be broadly defined as "…a place where an animal or plant normally lives, often characterised by a dominant plant form or physical characteristic (e.g. the stream habitat, the forest habitat)" (Ricklefs 1973). Basically the climate, substrate and local physiographic features are the primary determinants of plant species distribution, abundance and association; while the animals are almost entirely constrained in similar terms, by the distribution and abundance of such plant associations. Thus in savanna environments like Mkomazi the mammalian fauna of higher rainfall and elevated areas dominated by woody vegetation are characterised by primates, bushbabies, the rare eastern tree hyrax (*Dendrohyrax validus*), bush pig (*Pomatochoerus larvatus*), cephalophine antelopes (*Sylvicapra grimmia* and *Cephalophus* spp.) and bushbuck (*Tragelaphus scriptus*); hill slopes by Bohor reedbuck (*Redunca redunca*) and klipsringer (*Oreotragus oreotragus*); the lower lying seasonally-inundated or fire-derived grasslands and their fringes are occupied by buffalo (*Syncerus cafer*), Burchell's zebra (*Equus burchelli*), eland (*Taurotragus oryx*), Grant's gazelle (*Gazella granti*); kongoni (*Alcelaphus buselaphus*), and steinbuck (*Raphicerus campestris*); while the surviving fire-free *Acacia-Commiphora* bushlands still contain the diminutive Kirk's dikdik (*Madoqua kirkii*) and the increasingly rare lesser kudu (*Tragelaphus imberbis*) and gerenuk (*Litocranius walleri*), which are entirely restricted to these woodlands and their fringes. Passing from the west to the east along a gradient of increasing aridity, tree stature and canopy cover decreases steadily and we begin to encounter the shy Beisa oryx (*Oryx beisa*) (see Eltringham *et al.* on large mammal distribution, Chapter 31, and Lack on bird species distribution, Chapter 26).

The altitudinal range of the MGR (230–1,600 m) imposes a powerful climatic gradient from south-north and west-east, which is to a predominant degree the primary determinant of the species composition and physiognomy of the vegetational associations. Greenway (1969) published the first comprehensive flora of adjacent areas when he gathered together both his own botanical field records and those of other botanists that have collected in and on the fringes of the Tsavo

(East) National Park, a list that in all comprised 955 species in an area of up to 13,000 km^2, although relatively little is known of the far northern areas of Tsavo East, beyond the Tiva River. Tsavo West, which lies adjacent to the Mkomazi Game Reserve not only covers an area of 8,000 km^2 (approximately twice that of the MGR) but includes the peaks of Ngulia (1,824 m) in the east and the Chyulu Hills (2,170 m) in the north-west, a very recent (around 22,000 years before present) volcanic range whose slopes are clothed by montane forest, and its peaks by afro-montane moorland. This high level of physiographic diversity and the consequent fragmentation of its habitats makes it clear that the flora of the whole Tsavo National Park (21,000 km^2) could well exceed 1,500 and perhaps even approach 2,000 species.

Tsavo and Mkomazi both lie within the Eco-climatic Zone V of Pratt, Greenway and Gwynne (1966), which is characterised by a moisture index of -42 to -51[1]. Rainfall in this zone seldom exceeds evaporation and although it may rise to 700 mm, it is more frequently at or below 600 mm. Precipitation is mostly bimodal, with maximum rainfall in April and November though in central Tanzania this tends to contract into a single season between December and April. The soils are usually sandy, often derived from basement rocks or are colluvial in origin (Pratt and Gwynne 1977).

The vegetation of this zone is characterised by *Acacia-Commiphora* complexes, although *Acacia* tends to predominate in the west and *Commiphora* in the east, a feature that adequately describes the Mkomazi bush, over much of which *Commiphora* consistently outnumbers *Acacia*; while in the Kora National Reserve, Kenya to the north and west *Acacia* is a very important generic component of the woody vegetation (Agnew, Payne & Waterman 1986). Pratt and Gwynne (1977) suggest that the dominance of *Acacia* in the west is closely related to overgrazing by domestic stock, while the frequency of *Commiphora* in the eastern and more arid environments is more closely related to edaphic factors. It is clear that the greater degree of aridity in the east is responsible for a low standing crop of grasses, most of which are relatively unpalatable, making the area largely unsuitable for wild and domestic grazers but ideal for browsers. These authors suggest that woodlands dominated by *Commiphora* may be considered as a cycle of quite distinct vegetation types, in which senescent trees are replaced by grassland until the region once more enters a new woodland phase. In recent years a great deal has been written about this phenomenon, in which woodland clearance has been largely attributed to the high elephant densities (Laws 1970, Wijngaarden 1985) although observation in many of these environments (MJC) indicate that long-term rainfall cycles seem to be involved in some aspects of woodland regeneration and cyclicity; while anthropogenically-generated fires must also be playing a significant role in such a succession-like cycle.

[1] An expression derived from monthly rainfall and evaporation. An index of -60 is the maximum possible and equates with no rainfall (Pratt & Gwynne 1977).

The major vegetation associations identified by Greenway (1943) and Pratt, Greenway and Gwynne (1966) within this eco-climatic zone are:

WOODLAND with *Commiphora* and areas of tall *Acacia tortilis*, *A. etbaica* and *Balanites aegyptiaca*. Within the Commiphora woodlands we also commonly observe *Boswellia hildebrandtii*, *Delonix elata*, *Melia volkensii*, *Lannea* spp., and *Sterculia africana* (=*S. rhynchocarpa*), while the Baobab (*Adensonia digitata*) is locally common. The grasslands are rich in grass (and sedge) species but are often dominated by *Panicum* spp., and/or *Chloris roxburghiana*.

BUSHLAND AND SHRUBLAND comprise important components of this arid zone, with distinct vegetation types associated with *Commiphora*- and *Acacia*-dominated types, respectively. Overgrazing frequently appears to be responsible for the inclusion of *Lannea alata*, *Terminalia*, *Combretum*, *Cordia* and *Grewia* in these associations. Shrub-*Commiphora* may be associated with *Terminalia orbicularis*, while in central Tanzania an *Commiphora-Cordyla* bushland is well established. Shrub thickets of *Acacia mellifera*, *A. nubica* or *A. reficiens* are quite common. Grasses are dominated by annual species although the palatable perennials *Cenchrus ciliaris*, *Chloris roxburghiana* and *Chrysopogon plumulosus* are frequent.

GRASSLANDS, most commonly associated with drainage impeded soils or flood plains, though on dry soils they are almost certainly part of a natural grassland phase of periodic oscillations in *Commiphora* woodland. Fire-derived grasslands may intervene in any of these associations. The most productive species are *Cynodon* sp. and *Sporobolus helvolus* on seasonally-flooded areas. Also locally common are *Cenchrus* sp., *Chrysopogon* sp., *Pennisetum mezianum* and *Themeda triandra*.

Greenway's (1969) study of the Tsavo National Park flora classifies the plant associations within the Park, in accordance with his *Provisional Classification of East African Vegetation Types* written for the Vegetation Committee of the Pasture Research Conference (Greenway 1943) and later incorporated into Pratt, Greenway & Gwynne (1966). These include:

GROUND-WATER FOREST[1]. The woody flora of this forest commonly comprises *Acacia* spp., *Dobera glabra*, *Albizia* spp., *Newtonia hildebrandtii*, *Kigelia africana*, *Ficus* spp., *Terminalia kilimandscharica* and *Tamarindus indica*. Like Tsavo East this association occurs in the MGR at a small number of restricted localities (e.g. the Umba River, Tossa, Kisima Hills, Kinondo and Maji Kununua).

WOODLAND. Described as comprising open cover trees whose crowns do not form an interlaced canopy, and which are leafless for several (up to six) months of the

[1] His description of Tsavo East Swamp Vegetation is omitted since, except for a small area on the Umba River and around a number of small springs, it is absent in the MGR.

year. Comprising quite dense stands of *Commiphora–Lannea–Boswellia* and three species of *Sterculia* (*africana*, *rhynchocarpa* (=*africana*) and *stenocarpa*). Other constituent species include *Acacia reficiens*, *A. thomasii*, *A. tortilis*, *Adansonia digitata* (as an emergent), *Cassia abbreviata*, *Delonix elata, Melia volkensii, Platycelyphium voense*. When intact, the open canopies of these woodlands allow the development of a rich ground flora of shrubs, creepers, herbaceous perrenials and grasses. At the time of Greenway's study (1969) much of the woody cover of this vegetation type and its regeneration had been almost completely suppressed by tree damage from high elephant densities (and fire) in Tsavo.

WOODED GRASSLAND comprises a variety of vegetation types which are characterised by low stature and canopy cover of the woody species. They are often of compact habit (even clump-forming), although many of the species are shared with the true woodlands. These are classified by Greenway (1969) as Grouped-trees grassland, Scattered-trees grassland and Shrub or Dwarf-trees grassland. The last two named associations grade into one another and the more open phases often develop into open grassland under elephant or fire disturbance regimes. The first named may have good stands of *Acacia* species, *Commiphora* species, *Delonix elata, Dobera glabra, Melia volkensii, Platycelyphium voense* and a rich grass cover. The other two associations are similar but the Scattered-trees grassland is often characterised by *Melia volkensii, Platycelyphium voense* and *Commiphora* species. The larger woody plants comprise *Boscia coriacea, Dobera glabra, Cadaba heterotricha, Terminalia parvula, T. spinosa, Platycelyphium voense* and *Commiphora* species. The low stature and density of these woody species favours a dense shrub layer of *Sericocomopsis hildebrandtii, Premna resinosa, Thylachium thomasii, Calyptrotheca somalensis, C. taitensis, Cordia ovalis, C. gharaf, Ehretia teitensis, Premna hildebrandtii, P. holstii* and *Combretum aculeatum*. These associations have much in common with similar stands in Mkomazi, although in a number of cases species common in Tsavo are quite absent or at least very rare in Mkomazi.

GRASSLANDS. Trees or shrubs in these associations seldom covering more than 10% of the ground, though there are virtually no true grasslands, devoid of small trees or shrubs in Tsavo. Greenway (1969) mentions over 20 grass species being common on these scattered (often anthropogenically-derived or edaphic) grasslands. The genera include *Aristida, Brachiaria, Cenchrus, Chloris, Chrysopogon, Cymbopogon, Digitaria, Enneapogon, Ischaemum, Latipes, Leptochloa, Panicum, Schoenfeldia* and *Tetrapogon*.

BUSHLAND. Land with 50% small tree or shrub cover. Many of the trees evergreen or deciduous, armed or unarmed. Greenway (1969) recognises that this was a very rich and mixed mosaic of plant associations with groups *of Acacia bussei, A.*

mellifera, A. nilotica, A. reficiens, Adansonia digitata, Delonix elata, Euphorbia robecchii, Lonchocarpus eriocalys, Melia volkensii, Terminalia species and many other species of shrubs and low trees on different soil types. These include *Acacia bussei, A. mellifera, A. nilotica, A. reficiens, Adenia globosa, Bauhinia taitensis, Boswellia hildebrandtii, Bridelia taitensis, Cadaba heterotricha, Caesalpinia trothae, Calytrotheca somalensis, Combretum aculeatum, Commiphora riparia* (=*C. mildbraedii*), *Dobera glabra, Ehretia taitensis, Erythrochlamys spectabilis, Euphorbia* spp., *Grewia fallax, Grewia. tembensis, G. villosa, Hymenodictyon parvifolium, Lannea alata, Premna resinosa, Sesamothamnus rivae, Strychnos decussata, Terminalia orbicularis, T. spinosa, T. parvula* and *Thunbergia guerkeana*.

ROCKY HILLS AND PAVEMENTS. These environments are common in both Tsavo and the MGR where basement inselbergs rise clear of the general surface, these rocky hills often bear eroded platforms of flat rocky pavement where as a consequence of the shallow soils they bear a characteristic flora. Greenway (1969) lists *Acacia tortilis, Adensonia digitata, Boscia coriacea, Boswellia neglect, Bridelia taitensis, Combretum aculeatum, C. exalatum, Cordia ovalis, Delonix elata, Carphalea (Dirichletia) glaucescens, Euphorbia heterochroma, E. quinquecostata, E. scheffleri, Ficus populifolius, F. sonderi, Grewia villosa, Haplocoelium foliosum, Melia volkensii, Premna resinosa, Sarcostemma viminale, Strychnos decussata, Tarenna graveoloens, Thylachium thomasii* and *Vepris eugenifolia* amongst the common or occasional trees and shrubs; while the liane and herb flora include *Cissus quadrangularis, C. rotundifolius, Cardiospermum halicacabum, Ipomoea bullata, Sansevieria ehrenbergiana, S. singularis, Ruellia patula, Barleria prionitis, Amorphophallus galaensis, Stylochiton angustifolius, Crossandra mucronata* and a relatively rich bulbous flora.

The primary aims of the vegetation programme were to prepare a reasonably complete flora of the area as a subsequent aid to management and plant biodiversity conservation; to describe the main habitat types in relation to their biotic and physical correlates; and to demarcate a number of re-locatable habitat plots that can be visited in the future by members of the Department of Wildlife in the execution of the routine monitoring of vegetation change and research.

Methods

Plant collections were made by the authors in all parts of the reserve between 1993 and 1997, although for logistic reasons the central and western end localities were visited more frequently, particularly since the greatest physiographic and geomorphological diversity was located in this region. In addition to our own collections and those of Harris (1972), these were supplemented by collections in the earlier phases of the study by Julie Cox (Cox 1994) and after the conclusion of the main studies in 1997 by Susan Canney. For further details of the flora see Volleson *et al.* (Annotated Checklist: Plants) and for the GIS recording of botanical information for the vegetation map see Packer *et al.* (Chapter 4). Transects were first installed in 1993 and the one hectare plots in 1994. Measurements of canopy cover were carried out on road transects during the course of an *Acacia* distribution study with Yvetes Kalema.

Transects

Initial surveys were carried out by driving through all the major physiognomic vegetation types and siting transects beyond the influence of an ecotone, its origin being identified using a random stopping point determined by the vehicle odometer and its position with a GPS. Each transect was 100 m long and observation points were located every 10 m. Woody plant density was estimated using the point centred quarter (PCQ) method of Curtis (1959). The origins of the transects were marked and each 10 m observation point identified. At each of these points the terrain is divided into four quadrants, the transect being taken as the divisor for two 180° sections. The four plants to be measured at each observation point on the transect were grouped into those greater than and those less than 1.5 m. The distance to nearest plant in each quadrant is measured, the plant identified, its diameter at breast height (dbh) measured with a tape and its canopy height (and maximum and minimum diameter) measured with a graduated pole. Only the species and its height were recorded for plants <1.5 m. When each plant was measured, its four nearest neighbours were also recorded. After gathering data for the transect the observers walked along the transect and recorded other woody plant species that were not recorded in the point centred quarter observations. In spite of the justifiable criticisms of this method (Agnew *et al.* 1986) it does have the advantage of allowing rapid assessment in difficult bush conditions. The locations of these transects are indicated in Figure 7.1.

Plots

Based upon our accumulated knowledge of physiognomic vegetation types within the MGR and the need for the scientists to be able to use re-locatable plots for collection and observation, we selected ten sites on which we demarcated 1 hectare plots. We selected favourable locality and then sited the plot using the vehicle odometer to determine our stopping point. The origin and its altitude, at the SE corner was located by means of the Garmin GPS installed in the vehicle and this position entered into the 'proximity alarm' for purposes of subsequent relocation. The sides of the plot were oriented north–south and east–west, and the corners marked with termite-resistant wooden posts which were then painted white. At the time of handing over the FOC Ibaya Conservation Centre to the Department of Wildlife in 1997 the posts had virtually all survived intact in spite of their visibility. Plant species found on the plots were recorded at the time of initial siting and during subsequent visits by members of the botanical team. The locations of these plots are also indicated on Figure 7.1.

Canopy cover

During the course of a study of *Acacia* species distribution in the MGR (Kalema 1996) we took the opportunity to measure canopy cover along all major roads and tracks in central and western MGR. During these drives the vehicle stopped every kilometre and the canopy cover was estimated by the Bitterlich Stick method (Cooper 1963, Agnew 1968, Agnew *et al.* 1986). This technique assumes that at their widest point the trees' canopies are approximately circular and would appear as a series of circles on the ground. From a fixed observation point, the diameter of a tree will get progressively smaller the further it is away from the observer (i.e. for a fixed diameter the smaller the angle it subtends with the eye, the further it is away). Thus the apparent diameter of any one canopy (or bole) observed has a

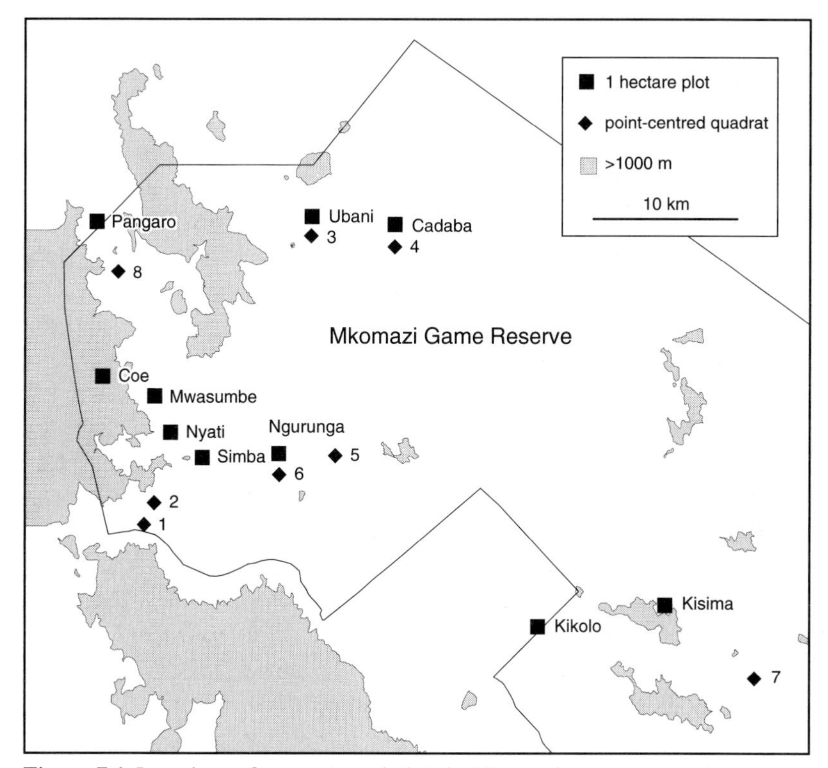

Figure 7.1 Locations of transects and plots in Mkomazi.

direct relationship with the proportion of the ground plan area that it occupies. The simple equipment used in this technique consists of a simple sighting stick on which is constructed a cross-bar at its end such that when the sighting stick is held to the cheek and the observer turns through 360° they can identify and count the number of trees whose canopy diameters exceed that of the cross-bar. Thus we are measuring trees that are occluded by the cross-bar and hence the sampling points using this method are frequently referred to as occlusion quadrats (Agnew 1968). In order to cope with the low stature and density of trees over most of the MGR areas a Bitterlich Stick 76.2 cm long with a cross-bar of 10.75 cm was used in this study. The relationship of area of the circle turned by the sighting stick and that of the cross-bar allows us to calculate the correction factor which must be applied to the sum of tree canopies exceeding the cross-bar diameter at each sampling point. In this case the number of individuals counted through 360° were halved to obtain percentage cover.

Results

Transects

The transects sited at the central and western end of the MGR represent all of the major vegetation types found in the valley bottoms and the lower slopes. Table 7.1 lists their co-ordinates, altitude, physical location, vegetation type, soil colour/texture and the commoner woody and herbaceous perrenial plant species (>1.5 m and <1.5 m) observed during siting of the transects. All these locations are in areas with woody vegetation cover in the lower valley sides and bottoms, and occur within about 150 m of one another. Most of the transects have a number of species in common, although their species richness varies considerably. 36 woody species >1.5 m in height were encountered, of which *Commiphora* represented 133 (41.6%) out of the 320 specimens measured. The number of species encountered on the PCQ Transects 1–8 varied from 3 to 16 (Table 7.2). The richest site with 16 species is that of the Ngurunga Valley (Transect 5) with poor, shallow quartz-gravel soils but a certain amount of retained ground water encourages a rich woody and herbaceous flora, while Transect 7 with 11 species lies on the slope down to Kisima from the Njiro (east of the Kisiwani) which exhibits a dense woodland of large trees and a frequently dense under-story. The lowest number of species (3) was recorded on Transect 8 where frequent anthropogenically induced fires have led to a dense, almost monospecific stand of *Combretum aculeatum* developing on the lower valley wall, at the entrance to the climatically ameliorated Pangaro Valley. Transect 3 is also species-poor (4) and lies in a more arid location at the foot of Ndea and is dominated by open *Commiphora* bush.

Table 7.1 Transect sites 1–8: locations, outline descriptions and species greater than and less than 1.5 m high. Co-ordinates are given in latitude/longitude and, for use with the 1:50,000 maps of Mkomazi, as UTM co-ordinates.

Transect 1

LOCATION	Crossing Same–Zange–Kisiwani road 1a. North of road, recently burnt. Low tree cover 1b. South side of road. Pare Forest Reserve, unburnt 1. Combined data from 1a & 1b
CO-ORDS.	4° 3.59' S, 37° 48.34' E; E367334, N9551467 ALTITUDE 840 m
HABITAT	Open *Acacia etbaica* woodland
SOILS	Dark brown with small gravel particles
>1.5 m	*Acacia etbaica, A. mellifera, Adensonia digitata, Albizia anthelmintica, Astripomoea hyoscyamoides, Balanites aegyptiaca, Capparis sepiaria, Combretum molle, Commiphora campestris, Grewia bicolor, Grewia villosa, Maerua endlichii, Salvadora persica, Sterculia rhynchocarpa*
<1.5 m	*Barleria ramulosa, Maytenus mossambicensis, Asparagus racemosus, Sida cordifolia, Tephrosia* sp.

Transect 2

LOCATION	Ibaya road, 3 km from Zange Gate. 50 m east & west of road
CO-ORDS.	4° 2.88' S, 37° 48.68' E; E367959, N9552771 ALTITUDE 820 m
HABITAT	Fire derived grassland with scattered trees and low capparaceous shrubs
SOILS	Brown, fine grained
>1.5 m	*Acacia etbaica, A. senegal, A. tortilis, Adensonia digitata, Albizia anthelmintica, Commiphora campestris, Balanites aegyptiaca, Dombeya (umbraculifera) kirkii, Grewia bicolor, Salvadora persica*
<1.5 m	*Adenium obesum, Maytenus mossambicensis, Tephrosia* sp., *Justicia flava, Asparagus racemosa, Solanum* sp., *Maerua* sp., *Grewia bicolor, Sida cordifolia*

Transect 3

LOCATION	20 km along Ibaya–Ndea road via Viteo Valley
CO-ORDS.	3° 53.45' S, 37° 53.93' E; E377639, N9570169 ALTITUDE 840 m
HABITAT	Plateau below Ndea, predominantly occupied by fairly dense *Commiphora* bush
SOILS	Dark red, iron stained; fine grained sandy
>1.5 m	*Acacia tortilis, Commiphora campestris, Combretum aculeatum, Xanthoxylum (Fagara) chalybea, Sterculia rhynchocarpa, Lonchocarpus eriocalyx*
<1.5 m	*Indigofera* sp., sapling *Commiphora campestris*, sapling *Combretum aculeatum*

Transect 4

LOCATION	25 km along the Ibaya–Ndea road, on the northern edge of the plateau *Commiphora campestris*
CO-ORDS.	3° 53.65' S, 37° 56.64' E; E382653, N9569793 ALTITUDE 780 m
HABITAT	Dense *Commiphora* scrub with a rich understory. Little sign of recent fire damage, but even-aged stands of young trees suggest that the last severe fire occurred about 10–15 years ago
SOILS	Dark red, iron stained coarse quartz gravel

continued

>1.5 m *Acacia senegal, Capparis tomentosa, Commiphora campestris, Grewia bicolor, Grewia fallax, Boscia salicifolia*

<1.5 m *Cadaba* sp., *Cyathula cylindrica, Euphorbia heterochroma, Justicia flava, Sida cordifolia, Tephrosia* sp.

Transect 5

Location Ngurunga valley, above the Ngurunga falls and pot holes. Flanked by linear rock outcrops. Rock-fall with large specimens of *Ficus sycomorus, Gardenia volkensii, Tamarindus indica* and *Terminalia prunioides*

Co-ords. 4° 1.36' S, 37° 54.68' E; E379046, N9555587 Altitude 760 m

Habitat Open mixed woodland with some tall trees and rich understory

Soils Red, iron stained with coarse quartz gravel and mica

>1.5 m *Acacia ancistroclada, A. etbaica, A. senegal, Astripomoea hyoscyamoides, Capparis* sp., *Combretum hereroense, Commiphora campestris, C. africana, Cordia monoica, Diospyros* sp., *Grewia bicolor, Grewia villosa, Salvadora persica, Terminalia prunioides*

<1.5 m *Acacia etbaica, Abutilon* (*guineense*?), *Cordia monoica, Xanthoxylum* (*Fagara*) *chalybea, Sida cordifolia, Grewia bicolor, Grewia villosa, Cyathula cylindrica, Indigofera* sp.

Transect 6

Location 17.5 km from Ibaya at foot of Mandi Hill on track to Ngurunga

Co-ords. 3° 53.45' S, 37° 53.93'E; E377639, N9570169 Altitude 780 m

Habitat Open grassland with scattered trees of *Acacia tortilis* and *Adensonia digitata* along the slightly elevated seep-line. Recently burnt; little evidence of re-sprouting on the larger trees. *Maerua* sp. regenerating vigorously

Soils Pale fawn, very fine dust-like texture

>1.5 m 45% of all tree sampled were dead to the base with no sign of sap or adventitious growth. *Acacia mellifera, A. zanzibarica, Asparagus* sp., *Capparis* sp., *Commiphora campestris, C. schimperi, Cordia* (*monoica*?), *Indigofera* sp., *Salvadora persica*

<1.5 m All ground vegetation burnt. Remains not identifiable

Transect 7

Location Rich mixed bush on steep slope 3 km down road from Njiro Gate to Kisima

Co-ords. 4° 9.07' S, 38° 0.70'E; E390200, N9541400 Altitude 640 m

Habitat Tree cover includes *Acacia tortilis, A. etbaica, Adensonia digitata* and *Delonix elata*. Ground and shrub cover of dense *Panicum maximum* grass and robust clumps of *Sansevieria ehrenbergii* and *Sansevieria conspicua*. Dense clumps of low shrubs, including *Grewia bicolor, G. tenax, G. villosa, Lannaea triphylla* and *Premna oligotricha* often hide the massive rock-like tubers *of Pyrenacantha malvifolia*

Soils Dark red, iron stained, with fine grained quartz gravel

>1.5 m Trees dominated by *Acacia ancistroclada, A. etbaica, A. senegal, Adensonia digitata, Balanites aegyptiaca*, frequent *Cordia monoica, Combretum aculeatum, C. hereroensis, Croton dichogamus, Grewia bicolor, G. fallax, G. lilacina, Jatropha spicata, Sterculia rhynchocarpa* and *Terminalia prunioides*

<1.5 m Good grass cover (see above). Profuse growths of *Cissus quadrangularis* covering vegetation and trees. A robust woody perennial *Plectranthus* sp. very common in the shrub layer

---

Transect 8

Location	Western route to Pangaro from Ibaya. Ground slopes gently and then steeply from track to the Igire and Kinondo ridges
Co-ords.	3° 55.38' S, 37° 47.47'E; E365689, N9566589 Altitude 900 m
Habitat	Regularly burnt fire-derived grassland, with some dense thickets of *Dichrostachys cinerea*, open patches of *Maerua-Capparis* shrublands regenerating after burning and dense stands of 2–3 m high fire-coppiced *Combretum aculeatum*
Soils	Fawn-reddish, fine grained
>1.5 m	*Acacia nilotica*, *Boscia salicifolia*, *Commiphora schimperi*, *Combretum aculeatum*, *Grewia* (*tembensis*?)
<1.5 m	Although the *Combretum* is quite tall (c. 3 m) its narrow growth-form allows a rich grass understory to develop. These are mainly represented by *Heteropogon contortus*, *Hyparrhenia rufa*, *Panicum maximum* and some *Themeda triandra*. Shrubs quite common including *Astripomoaea hyoscyamoides*, *Indigofera* sp., *Sesbania sesban* and *Solanum* sp. Herbs include *Barleria* sp, *Justicia flava*, *J.* sp. and *Tephrosia* sp.

Table 7.2 shows the results of the PCQ measurements on the above transects. There is a massive difference between densities on these transects, measured by the number of trees over 1.5 m tall per hectare. Those of Transects 1(1a) and 3 are very low and represent the effects of recent fires in the former and the open *Commiphora* bush of the latter which has also been subject to regular burning by wild fires, resulting in the progressive thinning of the trees. Intermediate densities (215–334) are represented by the fire-derived 'parkland' of Transects 6 (Mandi), 5 (Ngurunga Valley) and the unburnt *Acacia tortilis* woodland adjacent to 1b (Kisiwani Road). The highest densities were recorded in the fire-derived *Combretum aculeatum* woodland on Transect 8 (Pangaro road, density 816), the open fire-thinned *Commiphora* woodland on Transect 4 (Ibaya–Ndea road, density 995) and the species-rich woodlands on the slope down to Kisima from Njiro Gate.

Canopy cover measurements similarly reflect the recent history of disturbance of the site: the lowest canopy covers are recorded on 1a, 2, 3, 6 and 8, which have all been subject to fire in the recent past and in most cases repeatedly. Intermediate canopy cover measurements are found on Transects 1b and 5, both of which have been relatively little influenced by recent fire being in the case of the former on the edge of the South Pare Forest Reserve and in the latter, in a fire-protected valley. The high canopy covers recorded on the Ibaya–Ndea road (45.4%) and the Njiro–Kisima slope (93.1%) have both been protected from fire by the lack of water (and herders) in the case of the former and the proximity of the administration in the latter.

The evident consequence of recent fire regimes on both tree density and canopy cover indicate how important it is for the Project Manager and his staff to monitor the ecological effects of pre and post fire ecology, particularly in relation to the resultant reduction in plant diversity.

Table 7.2 Point-centred quarter data for trees >1.5 m on 100 m transects (for co-ordinates and descriptions see Table 7.1)

transect	tree density (per hectare)	% canopy cover	% bole cover x 10^4	mean height (m)	no. tree species in transect
1a	31.9	0.8	0.05	2.8	8
1b	333.7	26.2	0.40	3.4	9
1	70.3	3.6	0.90	3.1	9
2	151.0	7.2	0.23	3.0	8
3	70.6	6.1	1.15	3.5	4
4	995.0	73.2	3.68	3.4	8
5	296.0	45.4	2.06	4.0	16
6	214.7	9.9	0.30	2.6	9
7	1,298.7	93.1	8.45	3.5	11
8	816.3	9.5	0.70	2.7	3

Plots

Outline descriptions of the physical and biological attributes of the study plots 1–10 are provided in Table 7.3. It will be noted that their altitudes are clumped between 720 and 840 m, the only elevated sites being that on the lower slope of Kisima (plot 8, 920 m) and the much more elevated site on the Ibaya–Igire ridge (plot 10, 1,360 m). Of particular interest is the plant species associations recorded for these sites, for although there are strikingly unique characteristics on plot 10 and even plot 8, the other plots bear a large number of species in common on the grassland (either primary or secondary) and the present or former woodland/bushland locations. Thus, while plots 4, 6, 7, and 9 all have a number of *Commiphora* and *Grewia* species in common, the other components of their plant associations are close to being unique.

This observation raises one of the major problems in attempting to designate habitats in these species-rich African savanna environments. Over quite wide altitudinal ranges the 'natural' plant associations are a continuum, within which we may discern species mosaics which may lead to a false impression of habitat fragmentation, such as those observed by Cox (1994) in her earlier studies in the MGR. To a degree such observations of mosaics are derived from modern analytical techniques such as correspondence analysis which can detect minor differences in composition on a very local scale, which may well to say more about the numerical technique than it does about the biological processes.

Local variations in vegetational associations are well illustrated by the *Acacia* species recorded within the MGR. We have observed 22 species within the reserve although it is possible that two or three more may exist. In her study of the

Table 7.3 Plot sites 1 – 10: co-ordinates, location, local physical characteristics and commoner plant species. Plots are listed in the order that they were initially demarcated.

1 Simba

Location	Edge of a rocky outcrop on the Ibaya-Zange road crossing ecotone between surrounding seasonally-inundated grassland and stony substrate. Juvenile lion skull at origin.
Co-ords.	4° 1.26'S, 37° 50.22'E; E370805, N9555767. Altitude 840 m
Habitat	*Acacia thomasii* scattered-tree shrubland.
Soil	Pale brown with quartz gravel and larger boulders
Species	Top capped with large boulders. *A. nilotica, A. senegal, A. thomasii, A. tortilis, Adansonia digitata, Commiphora africana, Cyphostemma adenocaule, Grewia bicolor, G. villosa, Sida chrysantha, Stylochiton bogneri, Xerophyta spekei.* Common grasses include *Cenchrus ciliaris, Pennisetum mezianum* and *Themeda triandra.*

2 Nyati

Location	On border of seasonally-inundated grassland on Ibaya-Zange road. Close to the ecotone between the seasonally wet area and the lower hill slope. Fresh buffalo kill by lions in centre of plot.
Co-ords.	4° 0.44'S, 37° 49.17'E; E368854 N9557270 Altitude 840 m
Habitat	Seasonally inundated grass cover, with good grass cover and a rich herbaceous flora.
Soil	Dark brown-black, fine grained and dust-like with virtually no stony material.
Species	Dense grass cover dense, up to 1.5 m consisting mainly of *Cenchrus ciliaris, Heteropogon contortus, Panicum maximum, Pennisetum mezianum* and *Themeda triandra.* Seasonal herbaceous flora rich, with *Crinum kirkii, Dolichos oliveri, Indigofera* spp., *Orthosiphon parvifolius, Rhynchosia densiflora, Vatovea pseudolablab* and *Vigna membranacea.*

3 Mwasumbi

Location	Lower hill slope on margin of seasonally-inundated grassland from Ibaya.
Co-ords.	3° 59.30'S, 37° 48.65'E; E367888, N9559461 Altitude 840 m
Habitat	Open grassland with scattered bushes regenerating after frequent fire. Rich herbaceous flora during the rains.
Soil	Light brown, fine grained but with some small quartz gravel.
Species	Grassland dominated by *Pennisetum mezianum.* Low, regenerating capparaceous bushes (*Cadaba farinosa* and *Maerua* spp.) and *Dicrostachys cinerea.* Patches of *Hibiscus cannabinus* and *Vigna memranacea.*

4 Ngurunga

Location	A sheltered valley to the east of the Mandi hills fed by a seasonal river which subsequently retains water in pot-holes.
Co-ords.	4° 1.13'S, 37° 52.77'E; E375517, N9556008 Altitude 780 m
Habitat	Dense vegetation, from rocky ridge down to the seasonal river course in valley bottom. *Acacia ancistroclada* (almost entirely restricted to this area), and *Commiphora africana.*
Soil	Pale whiteish sandy soils, with very coarse quartz gravel and abundant mica, weathering from ridges. Surface and herbaceous vegetation greatly disturbed by

Maasai and Pare domestic stock.

SPECIES *Acacia ancistroclada, A. mellifera, Balanites aegyptiaca, Bourreria teitensis, Cadaba* sp., *Capparis tomentosa, Combretum hereroensis, Commiphora africana, C. schimperi, Cordia quercifolia, Grewia mollis, G. fallax, G. tenax, Hymenodictyon, Maerua edulis, Manilkara mochisa, Platycelyphium voense* and *Salvadora persica*

---

5 Pangaro

LOCATION Pangaro, a climatically ameliorated valley west of Ibaya between the fault scarp of Maji ya Kununua to the north and Kinondo to the south.

CO-ORDS. 3° 53.6'S, 37° 46.71'E; E364282, N9569887 ALTITUDE 880 m

HABITAT Badly affected by fire, emanating from agriculuralists above Kinondo and pastoralists intruding from the west. Most of the large *Melia volkensii* are now burnt stumps. Much of the vegetation represents outliers of the hill forest.

SOIL Red sandy soil so familiar over sites of weathered inselbergs.

SPECIES At the eastern end the bush is thick and largely evergreen, comprising *Hoslundia opposita, Rawsonia lucida*, and *Strychnos* sp. while on the more open ground the vegetation rapidly grades into a quite rich open *Acacia* bush containing *Acacia brevispica, A. nilotica, A. senegal, A. stuhlmannii, A. tortilis, Albizia anthelmintica, Balanites aegyptiaca, Boscia salicifolia, B.* sp., *Cadaba farinosa, Capparis tomentosa, Combretum aculeatum, C. hereroense, Commiphora schimperi, G. bicolor, G. fallax, G. mollis, G. villosa, Hymenodictyon parvifolium, Indigofera arrecta, Lannea schweinfurthii, Maerua edulis, M. triphylla, Neorautanenia mitis, Salvadora persica, Solanum hastifolium, Vernonia wakefieldii, Zanthoxylum chalybeum* and *Ziziphus mucronata.*

---

6 Ubani

LOCATION Ibaya-Ndea-Kavateta road open Commiphora bush on SE slope of Ndea.

CO-ORDS. 3° 53.4'S, 37° 53.86'E; E377515, N9570199 ALTITUDE 840 m

HABITAT Open mixed *Acacia-Commiphora* woodland which lies at the northerly edge of the Mbono valley grass fires. Trees mixed with a quite rich woody flora.

SOIL Red sandy soil with abundant quartz gravel.

SPECIES *Acacia nilotica, A. seyal, Commiphora africana, C. campestris, C. schimperi, Lannea triphylla, Lonchocarpus eriocalyx, Terminalia kilimanscharica, Xerroderris stuhlmannii* and *Zanthoxylum chalybeum*

---

7 Cadaba

LOCATION Fairly flat ground on the lower slopes of Ndea, above the seasonally-inundated grasslands of the Mbono Valley bottom.

CO-ORDS. 3° 53.70'S, 37° 56.6'E; E382587, N9569726 ALTITUDE 780 m

HABITAT Dense *Commiphora* bush which has been free of fire for some time but the dbh of the *Commiphora africana* indicates that they all germinated at the same time after a devastating fire (c. 15–20 years before present).

SOIL Red-brown loamy soil, with coarse quartz gravel. Small wallows develop during the rains.

SPECIES Dense *Commiphora-Lannea* bushland. Woody flora includes *A. senegal, A. seyal, Boscia (angustifolia?), Commiphora africana, C. campestris, C. schimperi, Cordia quercifolia, Fockea* sp.*nov., Grewia fallax, G. mollis, G. villosa, Lannea alata, L. triphylla* and *Vernonia* sp.

8 Kisima

LOCATION Dense bush on the stony lower slopes of Kisima Hill. Rich in plant species but on the lower slopes vegetation still very weedy from recent grazing by pastoralists.

CO-ORDS. 4° 6.06'S, 38° 5.58'E; E3992230, N9546954 ALTITUDE 920 m

HABITAT Reasonably undisturbed Acacia-Commiphora bushland

SOIL Red-brown loam with fine gravel.

SPECIES Rich bushland containing *Acacia brevispica, A. senegal, A. seyal, Acalypha fruticosa, Afzelia cuanzensis, Albizia anthelminthica, Blepharispermum zanguebaricum, Bourreria teitensis, Combretum aculeatum, C. exalatum, Commiphora africana, C. campestris, C. habessinica, C. schimperi, Diospyros* sp., *Gnidia latifolia, Grewia kakothamnos, G. mollis, G. villosa, Hoslundia opposita, Hymenodictyon parvifolium, Indigofera lupatana, Lannea schweinfurthii, Lonchocarpus eriocalyx, Newtonia hildebrandtii, Ochnma inermis, Pavetta* sp., *Premna hildebrandtii, P. resinosa, Rhoicissus revoilii, Steganotaenia araliacea, Sterculia rhynchocarpa, Terminalia kilimandscharica, Thylachium africanum, Thunbergia holstii* and *Zanthyoxylum chalybeum.*

9 Kikolo

LOCATION 3 km down the slope from the Njiro Gate close to a small rock outcrop.

CO-ORDS. 4° 6.75'S, 38° 1.35'E; E391406, N9545673 ALTITUDE 720 m

HABITAT *Commiphora campestris* bushland with iopden undergrowth. Frequent bare and weedy patches are almost certainly related to overgrazing.

SOIL Red lateritic soil with iron oxide nodules and some coarse quartz gravel.

SPECIES *Acacia mellifera, A. nilotica, Boscia* sp., *Commiphora africana, C. campestris, C. schimperi, Cordia quercifolia, Grewia mollis, G. villosa, G. sp.* (sterile*), Lannea schweinfurthii, L. triphylla* and *Sterculia rhynchocarpa.*

10 Coe

LOCATION Situated on the Ibaya-Igire ridge above Ibaya camp on the junction of Hill-top mist forest and fire derived ridge forest.

CO-ORDS. 3° 58.6'S, 37° 46.9'E; E364650, N9560625 ALTITUDE 1,360 m

HABITAT Evergreen hill-top forest. Under severe pressure from illegal pole cutters and charcoal burners. At the present rate of damage it will be gone in ten years.

SOIL Dark brown loam with shallow litter. Fine grained, with little gravel.

SPECIES Forest dominated by *Apodytes dimidiata* and *Heywoodia lucens*. Observed in the forest:- *Acalypha fruticosa, Albizia schimperiana, Apodytes dimidiata, Canthium mombazense, Craibia ?brevicaudata, Croton sylvaticus, Dovyalis* sp., *Drypetes parvifolia, Heywoodia lucens, Hoslundia opposita, Myrica salicifolia, Ochna holstii, Pavetta crebrifolia, Pseuderanthemum hildebrandtii, Strychnos usambarensis, Turraea robusta, Uvaria acuminata, U. angolensis, Vangueria randii* subsp. *acuminata, Vepris nobilis.*

factors influencing *Acacia* distribution in the reserve, Kalema (1996) attempted to correlate physical factors with their distribution and association. If we look therefore at the overall patterns of distribution in Mkomazi and much of the rest of savanna Africa, we observe that although species such as *A.* (*Faidherbia*) *albida* (seasonal or permanent water courses), *A. bussei* (the margins of seasonally-inundated grasslands), *A. drepanolobium* (seasonally-inundated grasslands) and *A. thomasii* (rocky hills and slopes) are very specific in their local requirements, many of the other species (*A. brevispica, A. etbaica, A. mellifera, A. nilotica, A. reficiens, A. senegal* and *A. tortilis)* are quite widespread, while the rare *A. dolichocephala* and A. *gerarrdii* are isolated on elevated hillslopes. Species such as A. *hamulosa* and *A. paollii,* absent from the MGR and Tsavo[1], are common in the Kora National Reserve in Kenya (Kabuye *et al.* 1986). In the more arid areas of relatively undisturbed woodland, the *Commiphora* species (*C. africana, C. campestris* and *C. schimperi*) are widespread, varying in local abundance even though the associated species frequently vary greatly from one location to another.

Table 7.4 Summary of canopy cover measurements in central & western MGR

% canopy cover groups	number of observations	% contribution
00.0-09.9	95	24.8
10.0-19.9	76	31.9
20.0-29.9	51	21.4
30.0-39.9	35	14.7
40.0-49.9	11	4.6
50.0-59.9	4	1.7
60.0-69.9	2	0.8
total	238	99.9

Canopy cover

Measurements of canopy cover taken during the study of *Acacia* distribution (Kalema 1996) are presented in Table 7.4 and Figure 7.2. Harris (1972) considered that *Acacia-Commiphora* bushland covered up to 70% of the MGR during his studies in the mid-1960s, but since that time much of that habitat type has been either thinned or suppressed by the effects of fire. The data in this table represent measurements in the central and western segments of the MGR, but bearing in mind that these segments also represent the reserve's more mesic sector, the canopy cover figures are of great interest. The 238 estimates taken 1 km apart indicate that 24.8% of the 'road transects' are under 10%, 31.9% from 10–20% and 21.4%

[1] Only 13 *Acacia* species are listed by Greenway (1969), which suggests that these trees were under-collected in the Tsavo National Park. It should be noted that *A. hamulosa* and *A. paollii* are both found in Somalia as well as in northern and eastern Kenya.

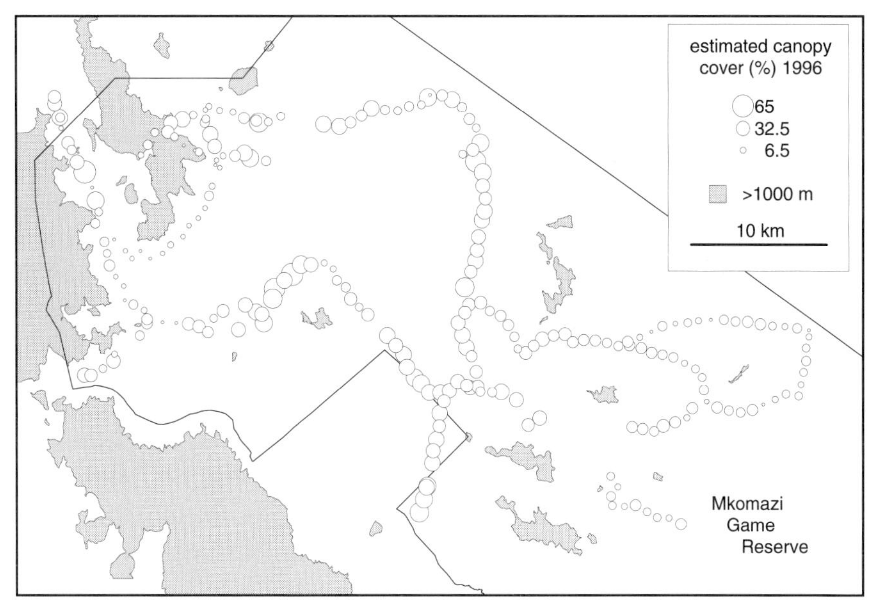

Figure 7.2 Canopy cover measurements made near roads in central and western MGR, 1996. See text for measurement methods.

from 20–30%; or a total of 78.1% of the vegetation has a canopy cover under 30%. By contrast only 6.3% of the road transects exhibited canopy covers of over 50% and none at all over 80%. In the *Commiphora* woodland of the Kora National Reserve in Kenya, nine out of 12 transects exhibited canopy covers in excess of 40% and five over 50%; the highest being 61%, 65% and 75% (Agnew *et al.* 1986). The only high canopy covers in Mkomazi are likely to be in *Albizia* gulley forest at springs and in intact hill-top forest where figures in excess of 80% (even 90%) are to be expected.

Habitats of the MGR

The designation of semi-arid vegetational associations on the east African savannas allows us to compare individual areas in terms of the dominant species present. However, it is soon apparent, as we have observed above, that regional differences are considerable, even when comparing the *Acacia-Commiphora* savanna of the Kora National Reserve in Kenya with that of the MGR in Tanzania, separated by a little over 400 km. These differences are partly related to local differences in physiography, but it seems equally likely that an overiding factor has been speciation events driven by periods of geographic isolation during intervals of glacial advance and retreat in the recent and distant past. The descriptions of plant associations on the transects and plots described above provide us with a valuable picture of species distribution but the overlap in species composition on the lower altitude sites makes it clear that it is unrealistic to attempt to designate a habitat classification in the same way.

In his initial attempt to classify the vegetation types of the MGR, Harris (1972) recognised a preliminary habitat division based on rainfall less than and greater than 500 mm per annum and a vertical pattern based upon the catenary effect. The valuable concept of a catena has its origins in Tanzania, where Milne (1936 & 1947) noted that in the rolling topography of the eroded African surface there was a striking similarity between different soil sequences. He called this succession of soil types along a topographic gradient a 'catena' which has been subsequently widely used in mapping and plant/wildlife ecology (Burtt 1942, Gwynne & Bell 1968, Lind & Morrison 1974, Morison, Hoyle & Hope-Simpson 1948 and Scott 1962). Basically what this concept recognised was that under similar conditions of climate and parent material the topographic sequences are also similar (Harris 1972).

In Mkomazi, conditions are remarkably similar to those described by Milne (1947) at Unyanyambe, Tabora where the rocky surface of eroded granite hills gives way to progressively deeper red earths on the upper slopes, grading to progressively sandier substrates until at the valley bottom we encounter an *mbuga*, an area of often dark seasonally-waterlogged clay soil. These physical processes are often described as the mass wasting creep of deposits (or colluvium) down a slope, terminating in the deposit of fine-grained material (or elluvium) at the bottom. This 'catenary succession' observed down the valley sides is characterised by a soil sequence that results from the sorting processes that take place down the slope, resulting in striking differences not only in soil depth and structure but also in the vegetational associations that they support.

Following Harris (1972) we propose that we should establish a classification of habitat types on the catena principle but widening these definitions to include altitude, aspect, local variations in substrate and their recent history of disturbance. Additionally we propose to institute a series of abbreviations which will allow our colleagues, both now and in the future, to indicate habitat types in relation to the study of animal and plant biodiversity and distribution. These habitats may be described as follows (with the suggested abbreviations in parentheses):

Hill summit forest or 'mist forest' (Mf)

These evergreen forests are now largely located above 1,200 m with canopy covers in excess of 75% and abundant lianes and are characterised by dense early-morning mist over most of the year which produces an additional bonus of 'occult' precipitation that not only favours tree growth but also supports a rich epiphytic flora These forests only still survive on the summits of Ibaya–Igire, Kinondo and Maji ya Kununua. This forest vegetational association is adequately described on the Coe Plot (10) above the FOC Ibaya Conservation Centre.

The Vitewini ridge (c. 1,000 m) has been denuded of much of its tree cover in recent years by fires started in the Viteo and Mbono Valleys but its remaining flora is still quite rich and comprises open *Combretum* bushland with occasional groups of old *Albizia petersiana, Brachylaena huillensis, Cassipourea celastroides* with no undergrowth, suggesting that quite recently there has been dense stands of dry forest. On rocky hilltops and steeper slopes semi-evergreen thickets that have survived fire contain a rich woody flora including *Abrus schimperi, Acalypha fruticosa, Allophylus rubifolius, Blepharospermun zanguebaricum, Combretum exalatum, Commiphora habessinica, Croton dochogamus, C. sylvaticus, Dombeya kirkii, Dyschoriste* sp. *nov., Erythrina burtii, Euphorbia scheffleri, Grewia mollis, G. tembensis, G.* sp. (2–3 sterile species), *Haplocoelium foliosum, Lannea schweinfurthii, Maerua triphylla, Mystroxylon aethiopicum, Pentanisia ouranogyne, Pentas parvifolia, Strychnos henningsii, Tarenna graveolens, Thunbergia holstii, Vepris uguenensis* and *Vitex strickeri.*

Kisima Hill (1,356 m) is an eroded rocky peak which although it has been affected by fire in the recent past, only bears shallow soil on its summit which supports an open shrubby vegetation of *Aloe* spp., *Apodytes dimidiata, Combretum molle, Encephalartos kisambo, Euphorbia* spp., and *Xerophyta* sp. Herbaceous flora includes *Hibiscus greenwayi, Sacleuxia newii, Turraea mombassana* subsp. *cuneata* and *Vernonia holstii.* The summit trees bear a rich flora of lichens and orchids (including *Oberonia disticha*). The vegetation of the slopes are described for the Kisima Plot (8).

Upper hill-slope forest/woodland (Us)

(The hill-slope = footslope of Milne 1947). The vegetation associated with this very distinctive fire-resistant habitat is maintained in its present state by fire, although it appears quite distinct from the 'mist forest' of the summit (c. 1,100–1,300 m). This habitat is typified by that observed just below the ridge above Ibaya, in the Mbono Valley and in the Igire Forest which lies to the south of the ridge and is badly damaged by pole cutters and charcoal burners. It is also found on the upper hill-slopes of Mzara and in small fire-protected pockets and shallow gullies on Vitewini. The species-rich Igire forest will have disappeared in ten years if strict measures are not instituted to control fire and other disturbance. The dominant trees in this forest are the large fire resistant *Brachylaena huillensis* and *Spirostachys africana.* Other species recorded in the Igire forest are *Acacia brevispica, Albizia schimperiana, Allophylus rubifolius, Apodytes dimidiata, Blepharispermum zanguebaricum, Canthium mombazense, Capparis tomentosa, Caucanthus auriculatus, Combretum exalatum, C. padoides, Commiphora eminii* subsp. *zimmermannii, Crotalaria goodiiformis, Croton megalocarpus, C. polytrichus, C. pseudopulchellus, Cussonia* sp., *Dombeya kirkii, Dovyalis* sp., *Euclea schimperi, Euphorbia* sp., *Helinus integrifolius, Landolphia kirkii,*

Mystroxylon aethiopicum, Ochna holstii, Phyllanthus reticulatus, P. welwitschianus var. *beilei, Pseuderanthemum hildebrandtii, Rhoicissis revoilii, Rothmannia* sp., Sapotaceae indet., *Scutia myrtina, Spragueanella rhamnifolia, Steganotaenia araliacea, Strychnos usambarensis, Synaptolepis alternifolia, Tinnea aethiopica, Turraea rubusta, Vitex strickeri* and *Zanthoxylum usambarense.* This habitat is of considerable importance to local human communities, especially in terms of the collection of natural products (see Kiwasila & Homewood, Chapter 34).

Fire-derived ridge woodland (Fd)

This habitat lies on rocky ridges, between surviving patches of Summit and Upper hill-slope forest at about 1,200 m. It is maintained in its present state entirely by fire, and after burning is colonised by dense *Dombeya rotundfolia* whose canopy cover can reach 50%. Between the trees there is a dense sward of tussock grasses.

Middle hill-slope woodland/scrub (Ms)

Probably only occurs as a zone of interdigitation or merging between the Upper and Lower hill-slope woodland ecotones. It does however contain fairly distinctive (but overlapping species), which include *Combretum exalatum, C. fragrans, C. molle, Gnidia latifolia* and *Harrisonia abyssinica.*

Lower hill-slope protected woodland (Lpw)

We have included this habitat category which lies at the eastern end of the Pangaro Valley and represents a quite distinctive lowland vegetation type that bears striking affinities with vegetation found at higher levels on Kinondo. It represents a gentle talus slope whose surrounding hills ameliorate the otherwise dry climate of the region. Its vegetation is dominated by *Acacia tortilis, A. nilotica, A. senegal, Boscia salicifolia, Hoslundia opposita, Melia volkensii* and abundant shrubby evergreens such as *Strychnos usambarensis* and *Rawsonia lucida*. The vegetation is described for the Pangaro Plot (5).

Lower hill-slope woodland (Lsw)

This common habitat is often greatly modified by fire at the bottom of the hill slopes on deep grey-brown soils. It can be observed throughout most of the hilly sections of the central-western sections of the MGR. In a relatively undisturbed state it is a habitat of high plant diversity. The Mwasumbi Plot (3) lies at the foot of this zone as does Transect 2. The FOC Ibaya Conservation Centre lies in this zone where the association may be found between 850 and 1,100 m. Fire has removed many of the larger trees leaving an open severely degraded *Combretum* bushland.

There are patches with tall *Commiphora holtziana*[1] with no undergrowth and scattered specimens of *C. campestris* which suggests that it was originally natural *Commiphora* woodland or wooded grassland.

The woody species of the Combretum woodland are *Acacia hockii, A. nilotica, A. senegal, Afzelia quanzensis, Azanza garckeana, Boscia salicifoliolia, Clerodendrum hildebrandtii, Combretum aculeatum, C. fragrans, C. hereroensis, C. molle, C. zeyheri, Commiphora africana, C. campestris, C. holtziana, C. schimperi, Conyza pyrrhopappa* subsp. *oblongifolia, Dalbergia melanoxylon, Dichrostachys cinerea, Euphorbia cuneata* subsp. *spinescens, Flueggea virosa, Grewia fallax, G. mollis, G. tembensis, G. villosa, Hermannia exappendiculata, Hoslundia opposita, Indigofera garckeana, I. lupatana, Lannea rivae, L. schweinfurthii, Lonchocarpus eriocalyx, Melia volkensii, Ormocarpum trachycarpum, O.* sp., *Platycelyphium voense, Terminalia brownii, T. prunioides, Thunbergia holstii, Zanthoxylum chalybeum* and *Ziziphus mucronata*.

Semi evergreen thicket develops in the gullies of this region, where it is largely protected from the effects of fire and may contain *Blepharispermum zanguebaricum, Cissus quadrangularis, Combretum padoides, C. baluensis, C. eminii* subsp. *zimmermannii, Croton dichogamus, Ficus* sp., *Helinus integrifolius, Hymenodictyon parvifolium, Newtonia hildebrandtii, Sterculia rhynchocarpa* and *Triaspis mozambica*.

Stony north-facing hill-slopes (Ns)

The steep slopes such as those on the north side of Vitawini and the rocky, shallow soiled terrain below the Njiro Gate (and other isolated hills) are, in the absence of regular fire clothed with magnificent stands of the blue-green stemmed *Commiphora holtziana*. This rich habitat is described for Transect 7 at an altitude of 720 m with *Adensonia digitata, Delonix elata* and close to Njiro the superb 20–30 m high green-stemmed *Sterculia appendiculata*. The shallow soil on these slopes and on other outcrops further east are very similar to that of Ngurunga (Plot 4 and Transect 5). Variation between these sites is almost certainly mainly due to differences in exposure, slope and soil depth.

Dichrostachys cinerea thickets (Dt)

In and on the edges of seasonally-inundated grasslands in the Mbono and Viteo Valleys and in the vicinity of Mzara and Hafino, large *Dichrostachys* thickets comprise dense areas of tangled vegetation that may be as much as 100 m x 50 m

---

[1] Note below that *Commiphora holtziana* is particularly common on north-facing hill slopes that have been protected fron fire. These are located below the Njiro Gate (Kikolo Plot (9) and Transect 7) and on the northen side of Vitewini.

(c. 5,000 m^2). Initially we thought that the origins of these thickets lay in old *manyatta* sites that had been evacuated by herders but a detailed examination of those close to Ibaya told a different story. These thickets spread over the plains and even below roads that are being regularly used by traffic. On cutting a path through the centre of such a thicket we observed that the effect of the development of a *Dichrostachys* thicket was to protect many of the enclosed woody plants from fire until they were tall enough to resist the rapid passage of a grass fire. Included in and on the edges of the thickets were healthy (often large) specimens of *Acacia brevispica, Afzelia quanzensis, Cadaba farinacea, Commiphora africana, Combretum aculeatum, C. exalatum, C. hereroensis, Cyphostemma* sp., *Dyschoriste hildebrandtii, Grewia bicolor, Hymenodictyon parvifolium, Maerua edulis, Ormocarpum kirkii,* and *Terminalia prunioides.* The size of the thicket and the absence of most of these woody species in the immediate surroundings leads us to assume that they have grown from saplings since the thicket was established.

Many sub-species and varieties of *Dichrostachys* have been described, whose growth forms vary from 12 m trees to spindly shrubs and thickets. Beentje (1994) refers to *D. cinerea* subsp. *africana* var. *africana* as being in *Combretum* grassland in thickets. Elsewhere in Africa and India this species and other members of the genus (nine species) occur as shrubs or small trees. Examining *Dichrostachys* in the Mbono and Viteo Valleys it was apparent that the spreading, thicket-forming habit may be a result of fire rather than it being a specific thicket-forming variety. When fire sweeps past a thicket it burns all the leaves and the upper stems around the edge of the thicket but the centre (once it is large enough) is virtually unaffected. Subsequently we not only observed adventitious growth originating at the stem bases but around the periphery of the thicket it was clear that subterranean runners have been stimulated by the fire to grow outwards in a very short time. The study of the behaviour of these plants could be conducted in the presence and absence of fire both in nature and artificially (see Ferguson on the effects of fire on vegetation, Chapter 9).

Lowland *Acacia-Commiphora* woodland (Lc)

Most of the lowland MGR (<850 m) was covered with this vegetation type before wild fires drastically reduced its frequency. Harris (1972) considered that it covered 70% of the MGR during the 1960s. The rich flora of these locally variable woodlands are described for Transects 3 and 4 and Ubani (6) and Cadaba (7) Plots on the Ibaya–Ndea–Kavateta road. The rather open nature of this habitat and its canopy cover favours the growth of a very rich bulbous and herbaceous flora during the rains. Some of the commoner components of these areas are *Acacia etbaica, A. mellifera, A. senegal, A. seyal* (on drier ground replaced by *A. reficiens*), *Boscia mossambicensis, Commiphora africana, C. campestris, C. edulis, C. schimperi, Lonchocarpus eriocalyx* and *Zanthoxylum* (*Fagara*) *chalybeum*. The red soils of these habitats usually contain abundant quartz gravel, but out on the plains we observe many large circular or oval areas of bright red soil which are fine grained and during the rains rapidly become waterlogged. They are occupied sparsely by the same vegetation as their surroundings but in some cases contain local endemics (in the Kora National Reserve, Kenya these soils are colonised by *Acacia paollii*). These small sites are important for the formation of small waterholes and wallows during the wet season by elephant, buffalo and formerly by black rhino.

Gully woodland (Gw)

Occurs at the foot of permanent/seasonal watercourses flowing from elevate ground. Though uncommon their presence as a frequent source of seepage during the dry season makes them an important source of water for buffalo, elephant and a variety of other large mammals. Well developed below Ibaya, Kisima, Kinondo, Maji Kununua and Tussa. The forest is composed of a mixed woody flora which includes *Tamarindus indica, Newtonia hildebrandtii, Albizia harveyi* and *A. petersi,* some of which reach a considerable size.

Rock outcrops (Ro)

These are very variable habitats, for their flora is entirely dependent on their location. Within the MGR such outcrops are distributed from the summit of Vitewini and hill-slopes from 1,200 down to 700 m. The outcrop and rock-fall flora of elevated hills are described under Hill summit forest or 'mist forest' above while those of the lower altitudes are included under the Simba Plot (1). Rocky areas with little soil are colonised by a rich variety of succulent, semi-succulent species. These include up to 11 of the arborescent or shrubby *Euphorbia* species; the creeping vitaceous *Cissus* and *Cyphostemma* species; and though not especially common in the MGR, succulent *Caralluma speciosa*, *Adenium obesum*, and four species of *Aloe*, while most of the five species of *Sansevieria* are also commonly found in rocky localities. At higher elevations the most common trees are *Erythrina burtii* while on the lower hill-slopes *Adensonia digitata* and *Acacia thomasii* are regular and conspicuous components of the vegetation. *Adensonia* is remarkably widely distributed but the upper reaches of hill-slopes, where ground water is available for long periods is their commonest locality, whether this be on slopes with deep soil or rocky hillsides.

Kamakota Hill is a good example of a rocky environment in the east of the MGR, and its flora reflects this more arid location. The following were recorded at this location (June 1996): *Duosperma kilimandscharicum, Eragrostis caespitosa, Ficus ingens, Hermannia fischeri, Indigofera spinosa, Melhania rotundata, Pavonia zeylanica, Sida chrysantha, Spermacoce* sp. A of FTEA and *Zornia glochidiata*.

Rock outcrops and rocky pavements clearly grade into each other but there are a number of distinctive plants that seem to be regularly associated with these habitats and their fringes. The most conspicuous are large stands (often up to a metre tall) of *Xerophyta spekei*, while the soil-filled cracks bear a rich herbaceous flora including *Asparagus* sp., *Actinopteris radiata, A. semiflabellata, Brachystelma* sp., *Caralluma gracilipes, C. priogonum, C.* sp., *Craterostigna pumilum, Fimbristylis hispidula* and *Sansevieria* sp.

Seasonally inundated grasslands (Sg)

This habitat is a striking feature of the central to western segments of the MGR and their origins lie entirely with the geomorphology of this area. The large rocky inselbergs both above and below ground have been responsible for the formation of a series of large basins into which eroded minerals and organic matter are washed from the surrounding hills and where subsequent physical and chemical changes form extensive *mbugas* (Figure 7.3). The degree of inundation varies greatly from the centre to the outer edges of these basins. Comparatively few trees are capable of withstanding such a degree of inundation, although *Acacia drepanolobium* commonly forms 1–2 m thickets on the more elevated areas of these basins. This species can withstand long periods of inundation, although there are many regions where

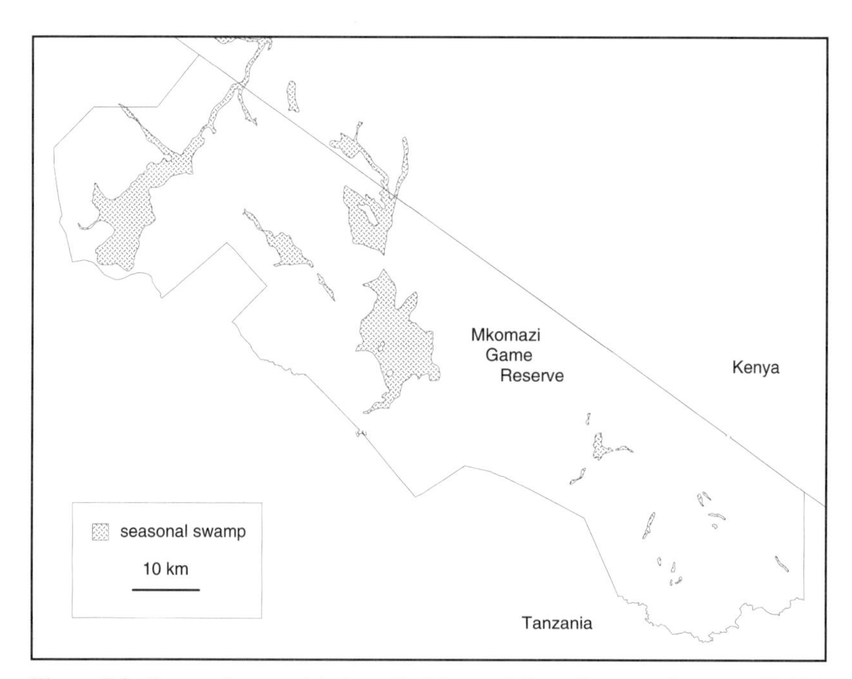

Figure 7.3 'Seasonal swamp' (*mbuga)* in Mkomazi Game Reserve. Source: 1:50,000 maps, 1988.

they occur on much drier ground and the trees reach 3–5 m. Along the grassland ecotone *A. bussei* forms quite dense thickets where it may be mixed with *Grewia bicolor* and *G. fallax*. Frequently these boundaries tend to be alkaline and support thickets of *Salvadora persica* and *Dobera loranthifolia.*

This habitat is dominated by grasses, herbaceous perennials and seasonal herbs, which are outlined under the description of Nyati Plot. The centre of these basins support a number of interesting species, including the rare *Aristolochia brachteata.*

During periods of inundation the swampy surface is suddenly full of the dark red freshwater crabs of the genus *Deckenia*. Their presence here is of great biogeographic interest for they are members of an ancient family (Deckeniidae) of five congeneric species of which three are found in the Seychelles and two in eastern Africa, suggesting that they have persisted since their ancestors were isolated when Gondwanaland broke up 75 million years ago (Haig 1984).

Discussion

We have already suggested that the exceptionally rich flora of the MGR can be largely explained by the high level of physiographic diversity encountered in such a comparatively small area, but it is also important to remember that the Umba River in the east represents the boundary between the Somalia-Masai regional centre of endemism and the coastal Zanzibar-Inhambane regional mosaic (White 1983). In terms of species richness, the 1,148 species currently identified from the MGR may be placed in 120 families, compared with 955, 734 and 566 species respectively in the Tsavo National Park (Greenway 1969), Kora National Reserve and the Meru National Park, all in Kenya. These species numbers are allocated to 88, 77 and 75 families respectively. However, our collections, currently in the National Herbarium in Arusha, include a large number of species (taxa) that have yet to be identified and in some cases described as new species. The number of species yet to be determined is 159, making it quite reasonable to suggest that if the Department of Wildlife continue to monitor the flora, the final list will almost certainly reach c. 1,500 species (see Coe *et al.*, Chapter 5 and Volleson *et al.*, Chapter 6). Coe (1995), in a preliminary calculation based on the differences and similarities observed in the collections of Greenway (1969) and Harris (1972), estimated that the potential flora of the MGR was probably about 1,026 species.

The work currently being undertaken by Mike Packer and Susan Canney will no doubt refine our classification of these habitats during the course of their mapping programme, although we would expect the basic divisions to be quite similar. In the very interesting study of Cox (1994), it may well be that the apparent fragmentation of the Mkomazi habitats, while strongly influenced by anthropogenic fire regimes, has also been influenced by ordination techniques, where the designation of habitat types is often as much a function of the method of analysis as it is of the environment in which they occur.

It is clear though that fire has removed virtually all of the lower hill-slope woodland in the central and western areas of the MGR. The drastic (even complete) removal of the woody plant cover is largely due to fires generated by herders entering the reserve from the north via the Pangaro Valley and along the Mandi Hills, while other fires are undoubtedly due to the activities of poachers. An additional and perhaps more important source of fire-induced habitat change is the conflagrations generated in the vicinity of the Igire, Ibaya and Kinondo hills by agriculturists and charcoal burners, from where fires move down into the Mbono Valley. The joint effect of these fires is to remove the woody cover of the lower, middle and upper hill-slopes. Initially the vegetation of the ridges is denuded but with repeated fires the vegetation of the gullies is penetrated until even these forest/woodland patches are reduced to a much less diverse flora and fauna. The control of fire should be given a high priority in the Department of Wildlife's forthcoming Management Plan.

Acknowledgements

We are all grateful to Dr W.F. Mosha the Director of the Tropical Pesticides Research Institute and Dr W.R. Mziray, the Director of the National Herbarium, also at the TPRI, for all the support they have given us in making RA and EM available to work with us in the field since 1993. Our collections, which are all housed at the National Herbarium in Arusha, will remain as a reminder of the interest of the TPRI and its staff in the need to preserve the rich biodiversity of the Mkomazi savanna. The National Herbarium in Nairobi kindly hosted visits by RA, EM and KV during the course of identifying our collections. The assistance of their staff and especially that of Geoffrey Maina is gratefully acknowledged. We must also thank Dr Roger Polhill of the Royal Botanic Gardens, Kew who arranged for KV to visit Mkomazi during the course of their other studies in Tanzania. Last, but by no means least are our Tanzanian colleagues and friends at Ibaya who shared our enthusiasm, watered our garden and on more than one occasion prevented us from being eaten by the feline guardians of this magical place.

References

Agnew, A.D.Q. (1968) Observations on the changing vegetation of Tsavo National Park (East). *East African Wildlife Journal* 6: 75-80.

Agnew, A.D.Q., Payne, R. & Waterman, P.G. (1986) The structure of Acacia-Commiphora woodland in the Kora National Reserve, Kenya. In Coe, M.J. and Collins, N.M.: *Kora: An ecological inventory study of the Kora National Reserve, Kenya*. Royal Geographical Society, London.

Beentje, H. (1994) *Kenya Trees, Shrubs and Lianas*. National Museums of Kenya, Nairobi.

Burtt, B.D. (1942) Some East African vegetation communities. *Journal of Ecology* 30: 65-146.

Cooper, C.F. (1963) An evaluation of variable plot sampling in shrub and herbaceous vegetation. *Ecology* 44: 565-8.

Coe, M. (1995) Botanical studies in the Mkomazi Game Reserve. In M. Coe & G. Stone (eds.) *Mkomazi Ecological Research Programme: Progress Report July 1995*. Royal Geographical Society, London.

Cox, J. (1994) *Mkomazi Ecological Research Programme Pilot Study: Remote Sensing/Vegetation Survey/Mapping*. Report for the Royal Geographical Society, London.

Curtis, J.T. (1959) *The vegetation of Wisconsin*. Madison.

Faden, R.B. & Beentje, H.J. (1989) *Encephalartos kisambo*, a new cycad from Kenya, with a note on *E. tegulaneus*. *Utafiti* 2: 7-10.

Greenway, P.J. (1943) *Second draft report on vegetation classification for the approval of the Vegetation Committee, Pasture Research Conference*. East African Herbarium, Nairobi, Kenya.

Greenway, P.J. (1969) A check-list of plants recorded in Tsavo National Park, East. *Journal of the East African Natural History Society and National Museum* 27: 161-209.

Gwynne, M.D. & Bell, R.H.V. (1968) Selection of vegetation components by grazing ungulates in the Serengeti National Park. *Nature* 220: 390-393.

Haig, J. (1984) Land and freshwater crabs of the Seychelle and neighbouring islands. In: Stoddart, D.R. (ed.) *Biogeography and Ecology of the Seychelles Islands*. Junk, Hague. pp. 123-139.

Harris, L.D. (1972) An ecological description of a semi-arid East African ecosystem. *Science Series* 11: 1-81. Colorodao State University, Range Science Department.

Kabuye, C.H.S., Mungai, G.M. & Mutangah, J.G. (1986) Flora of Kora National Reserve. In: Coe, M. & Collins, N.M. (eds.) *An ecological inventory of the Kora National Reserve, Kenya*. pp. 57-104.

Kalema, Y. (1996) *Distribution of Acacia species in Mkomazi Game Reserve*. Honours Project Report: University of Dar es Salaam.

Laws, R.M. (1970) The Tsavo Research Project. *Oryx* 10: 355-361.

Lind, E.M. & Morrison, M.E.S. (1974) *East African Vegetation*. Longman, London.

Milne, G. (1936) *A provisional soil map of East Africa. Kenya, Tanzania, Tanganyika and Zanzibar with explanatory memoir*. Amani Memoirs. London: Crown Agents.

Milne, G. (1947) A soil reconnaissance journey through parts of Tanganyika Territory, December 1935 to February 1936. *Journal of Ecology* 35: 191-265.

Morison, C.G.T., Hoyle, A.C. & Hope-Simpson, J.F. (1948) Tropical soil-vegetation catenas and mosaics. A study in the south-western part of the Anglo-Egyptian Sudan. *Journal of Ecology* 36: 1-84.

Pratt, D.J., Greenway, P.J. & Gwynne, M.D. (1966) A classification of East African Rangeland, with an appendix on terminology. *Journal of Applied Ecology* 3: 369-382.

Pratt, D.J. & Gwynne, M.D. (1977) *Rangeland Management and Ecology in East Africa.* Hodder and Stoughton, London.

Ricklefs, R.E. (1973) *Ecology*. Nelson.

Scott, R.M. (1962) The soils of East Africa. In: Russell, E.W. (ed.) *The natural resources of East Africa.* D.A. Hawkins & East African Literature Bureau, Nairobi.

White, F. (1983) *The vegetation of Africa: A descriptive memoir.* Natural Resources Research XX. UNESCO, Paris.

Wijngaarden, W. van (1985) *Elephants–trees – grass – grazers: relationships between climate, soil, vegetation and large herbivores in a semi-arid ecosystem (Tsavo, Kenya).* ITC Publication Number 4.

CHAPTER 8

Phenological observations on the vegetation of Mkomazi

Malcolm Coe, Raphael Abdallah & Emmanuel I. Mboya

Introduction

The Mkomazi Game Reserve lies at the southern extremity of the arid Sahel zone which in its eastern extension is largely typified by *Acacia-Commiphora* savanna. Its structural components have strong affinities with similar vegetation to the north, whose floral composition may be described in terms of White's (1983) Somalia-Masai regional centre of endemism. The proximity of the Usambara and South and North Pare Mountains has a strong influence on the climate on this boundary. Smaller mountains rise from the plains within the MGR and are frequently capped by cloud in the evening and early morning, which must be responsible for the elevation of the recorded annual rainfall by supplementation from dew. The factors expected to influence the growth and phenology of plant species are a combination of local geomorphology/physiography, rainfall, seasonality and the morphological characteristics of individual species.

The primary factor influencing metabolic activity of plants in the MGR is rainfall, although individual species respond in different ways, whether to the actual advent of precipitation or even its anticipation. In savanna environments we expect annual rainfall to be below 300 mm in semi-arid regions and 500–1,000 mm in semi-humid savanna (Harris 1980). Rainfall in the MGR may be expected between early November and mid January (short rains) and March to mid May (long rains) although in recent years the seasonality seems to have been truncated into a late short rains running into an early long rains. Studies in the Tsavo East National Park in Kenya (Tyrrell & Coe 1974) demonstrated that the short rains were more predictable (less likely to fail) than the long rains, although the amount of precipitation falling in each of these two seasons did not differ significantly.

A characteristic of rainfall in most of the more arid African savannas is the tendency for the rain to fall in short sharp showers. This severely limits its effectiveness for plant growth, since the precipitation is likely to exceed the infiltration capacity of the soil—particularly where the soil possesses a high clay fraction, a condition that is quite common on basement-derived soils (Harris 1980). Year-to-year variation of rainfall is also very large, exemplified by the rainfall regime at Voi in Kenya (75 km north of the MGR boundary) where the 64 year mean up to 1974 was 548 mm but the range over the same period was from 184 to 1,201 mm (Anderson and Coe 1974). These patterns of mean variation are repeated locally when we observe that over the whole Tsavo National Park (21,000 km^2) annual rainfall lay between 209 and 1,073 mm in 1968, and between 168 and 694 mm in 1973 (Coe 1990, Glover 1970 & 1973).

The phenological characteristics of the vegetation in tropical arid and semi-arid environments possess many features in common with those of Mediterranean regions, where winter rainfall quite rapidly gives way to very hot summer conditions, placing strong selection pressures on the flora in relation to their reproductive activity. These pressures are observed particularly in respect of the plants' ability to complete their growth and reproduction between a cold, wet winter period and a hot dry summer. The local climate appears to impose severe temporal restrictions on the activity of the plants of eastern African savannas. Tyrrell & Coe (1974) showed that the reproductive rhythms of plants in Tsavo National Park in Kenya appeared to exhibit much greater levels of activity in relation to the more predictable short rains, while they only exhibited sparse or no activity at all during the course of the long rains. These authors predicted that this period of activity coinciding with the short rains may be related to the greater build-up of dead organic matter in the long hot dry season (May–September), making higher levels of decomposer-mediated nutrients available during a period of greater climatic stability. Thus the timing of flowering could be related to the length of the dry season rather than the timing or the amount of rain falling in a particular wet season.

The study area

The climatic characteristics of the MGR, described in Chapter 2, impose severe local restrictions on plant metabolic activity. Apart from sudden brief plant growth surges observed in grasses and herbaceous plants, the major determinant of the onset of phenological activity for most plants is rainfall and the concomitant mobilisation of nutrients from moisture-dependent decomposer activity.

The commencement of the rains in Mkomazi is erratic and during the course of the present studies (1993–97) has varied between below average and complete failure until the torrential downpours and serious flooding of late 1997 and early 1998. The general pattern of precipitation and hence plant phenological activity reveals the highest rainfall on elevated ground in the west and south-west, falling steadily as we progress northwards and eastwards.

As in most semi-arid regions, the location of individual showers is unpredictable. In western Mkomazi, however, there is a clear pattern of directionality in rain showers: little rain appears to originate in the rain-shadow of the Pare Mountains to the south and west, but appears to arise largely over the Kenya border around the Taita Hills (2,210 m). Rain clouds arising on or east of Taita pass southwards, leading to rain over Mzara, the Mandi Hills and finally the Zange area. Rain clouds arising in central Taita pass southwards over Vitewini Ridge and Magunda and deposit rain in the Mbono Valley, Ibaya and on the Igire/Ibaya forest; while clouds from western Taita and Ngulia (1,824 m) deposit rain over Maji Kununua and Pangaro to the west of Ibaya. These shower patterns tend to coalesce during the course of good rains, but during more intermittent rainfall, plant phenological activity is often very erratic and its distribution a mosaic.

Methods

During the planning period of MERP (1991–93) brief visits were made to the MGR when casual observations on leafing, flowering and fruit formation and dispersal were carried out (MJC) as part of the plant diversity species inventory. From July 1993 to January 1997 we visited the MGR and collected flowering/fruiting material which was deposited in the National Herbarium in Arusha and reference material taken to London and Nairobi for identification and taxonomic study.

During the course of making these collections for the floral checklist, we recorded site localities and details of leaf cover, flowering, fruit development and dispersal both on specific plant collecting safaris or during the course of routine travel within the reserve. Many of these observations were made on our study plots, but because of the erratic nature of both rain-dependent and out-of-season flowering activity we used routine visits to various regions of the MGR to make phenological observations.

Although we are interested in the activity of all plants, we have paid particular attention to woody plant species. Their foliage represents significant food items for a wide variety of folivorous animals, and their fruits and seeds are both consumed and dispersed by the avian and mammalian fauna. Indeed, there are indications that the rich vertebrate fauna of these African savannas plays a paramount role in the dispersal of a significant proportion of woody plant propagules and are probably key elements in the maintenance of their rich plant and habitat diversity.

Results

Activity patterns of the savanna flora

Many species exhibit anticipatory flowering before the rains begin, as distinct from those that respond to the first showers or ‘grass rains’ with rapid bursts of activity. These species are spread across a wide spectrum of taxonomic groups represented by both monocotyledons and dicotyledons.

Of the environmental problems faced by plants in these dry habitats, temperature is virtually never a limiting factor on growth, while rainfall and nutrient availability are. There is a strong selection pressure on plants to take advantage of the very narrow 'windows of opportunity' provided by the rains, by growing fast to avoid competition with other species, often larger and more robust. The bulbous and tuberous monocotyledenous species of the savanna have evolved the ability to store nutrients below ground, so they may survive a long dry season and be ready to commence their growth and reproduction while avoiding competition from more vigorous species. *Ammocharis tinneana*, *Crinum kirkii*, *Cyrtanthus sanguineus* subsp. *wakefieldii* (or *ballyi*) and *Scadoxus multiflorus* all flower well before the first showers, although *Cyrtantus* precedes all other by flowering in the middle of bare, sun-baked patches of black cotton soil in August—preceding the first rain by at least two months. In this case we suggest that the advantage of flowering in the middle of the dry season is that small numbers of potential pollinators are active at this time. It would be interesting to find whether pollinators are species-specific to *Cyrtanthus* at this time of year—as we suspect they are for the remarkable *Crinum piliferum:* its single 8 cm long corollas are thrust above the ground in vast mbers over no more than three or four days within the arrival of the first rain showers in the Kora National Reserve in Kenya.

The most striking of all these 'anticipatory' species is undoubtedly *Crinum kirkii* which flowers shortly before the rains commence and during the course of the first few showers. Its flowers are the usual white and pink 'pyjama blooms' but these are followed in a few days by large groups of spherical (rarely elongated or fusiform), inflated pods which may vary in colour from deep crimson to cerise, though at a few sites in the Mbono Valley we have found albino forms. These pods may reach 5 cm in diameter, have thick fleshy alveolated walls which are quite sweet to taste, and number up to over 20 in a single head. Within the pods the seeds swell rapidly after fertilisation into a large hemispherical structures up to 2.5 cm in length, covered with a fibrous net-like epidermis. These seeds already possess a well developed radicle, are pale green in colour and presumably capable of photosynthesising. When ripe these pods are attacked by vertebrates (possibly they are eaten on by rodents and birds) and the heavy seeds are either carried some distance from the parent plant or may simply be dropped an inch or two away. As soon as the rains start they have already pushed their radicle into the soil and shortly afterwards produce their first leaves as they commence a period of rapid growth. We have observed very few really small plants suggesting that during their first season they attain a considerable size, although they may take a lot longer as their outer contractile roots progressively pull them deeper into the soil to depths of up to 0.8 m. First flowering may be delayed for a number of years.

With the arrival of the first rains the soil surface is usually wetted down to a depth of no more than 10 cm, which stimulates activity in the bulbous or tuberous monocotyledons such as *Albuca abyssinica*, *Aneilema johnstonii*, *Anthericopsis sepulosa*, *Anthericum cooperi*, *Chlorophytum bakeri* and shallow-rooted dicotyledons such as annual/perennial grasses and forbs. Succulent species, including *Adenium obesum*, *Aloe ruspoliana*, *Aneilema johnstonii*, *Caralluma speciosa, Talinum caffrum* and *T. portulacifolium,* also respond rapidly to rainfall. Many tuberous species such as *Vovotea pseudolablab* and *Dolichos oliveri* are also able to respond promptly by quickly producing flowers at the expense of leaves, due partly at least to their stored nutrients but also because of the selective advantage of the added visibility of the flowers before they are obscured by foliage. Some trees such as the baobab (*Adensonia digitata*) also exhibit a pulse of early flowering coinciding with the first rains, which may be related to their occurrence on hillsides where ground water is commonly available. However a more likely explanation is that they are mainly pollinated by bushbabies (Coe & Isaac 1965) and fruit bats (Faegri & van der Pijl 1936, Harris & Baker 1960) so that the large white pendulous flowers, dispersed over a massive, almost bare canopy, are more likely to be located and pollinated by these mammals. These flowers are also visited by insects during the day but the availability of their nectar and pollen is probably mainly, if not entirely, a nocturnal phenomenon.

Shallow gravel soils, rocky pavements and outcrops provide the advantages of reduced competition from more robust species but the severe disadvantage of intensely arid conditions during the dry season. Species may respond to these conditions by being an annual which can complete its life history in a very short time, a succulent (*Brachystelma* sp., *Caralluma gracilipes, C. priogonum*, *Edithcolea grandis* etc.), or as one of the most remarkable of all in these arid zones—the resurrection plants. In Mkomazi these are represented by *Xerophyta spekei* (Velloziaceae) which is mainly found on rocky pavements and *Craterostigma pumilum* (Scrophulariaceae) on shallow gravel soils in the Viteo Valley and a number of the larger rock outcrops. The former grows as groups of long fibrous stems which dry out completely in the dry season but after a single shower will absorb water and either produce new or resurrect old leaves and flowers in three or four days. The latter occurs as a procumbent rosette which all but disappears in the dry season when it dries up and shrivels to a few small brown leaf remnants, but within a few hours of rain falling the leaves have unrolled, rehydrated and turned green and will have commenced flowering in the same short interval as *Xerophyta* (see Coe *et al.* on habitats, Chapter 7).

Large tubers are a common feature of savanna plants whether they are small often subterranean structures or massive enlarged structures on the surface of the soil which resemble boulders and whose white surfaces probably impart a high albedo, enabling them to maintain a reduced internal temperature during the dry season. The commonest of these massive species are *Adenia globosa* (up to 2.5 m in diameter) and *Pyrenacantha malvifolia* (to 1.5 m in diameter). However a wide variety of other shrubs and climbers, particularly members of the Cucurbitaceae, also bear either above or below-ground woody tubers, particularly of the genera

Coccinia, *Cucumis*, *Gerrardanthus* and *Momordica*. This effective adaptation for the storage of water and nutrients during the dry season is almost certainly also a valuable pre-adaptation for resisting fire damage. The evolution of massive surface tubers must in part at least be due to the restrictions placed on the size that is achievable below ground in hard quartz gravel soils, so that above a critical size there is no alternative but to develop as storage organs on the surface, 'disguised' as rocks, with a highly reflective surface.

With the arrival of the 'grass rains' many trees respond by rapidly producing flowers and foliage. Some tree species such as *Acacia etbaica, A. nilotica, A. tortilis, A. zanzibarica* and *Lannea schweinfurthii* produce a relatively small number of flowers, some of which may be sterile (Ross 1979), and have no alternative but to abort them if the rain does not continue after four or five days. In the case of *Lannea schweinfurthii,* a single tree at Ibaya produced and aborted its flowers four times in little more than four weeks in November–December 1996 when the start of the rains were delayed. This strategy may be energetically costly but it is a common habit amongst many tree and shrub species in the African savanna, the costs being outweighed by the advantage of being able to utilise the earliest insects emerging as potential pollinators. Gordon-Gray (1975) suggested that the production of groups of sterile flowers and groups of involucellate flowers on the pedicels of *Acacia* species with globose inflorescences has evolved as a device to attract insects in large numbers before the more energetically expensive fertile flowers are produced.

In spite of the variable strategies described above, the commonest is still that of waiting until the rains are fully underway and the soil-water column complete before the plants flower, whether *en masse*, continuously over a period, or in pulses. In some cases flowering is delayed for some time and the plant will be in full, mature leaf before flowering takes place. *Flueggia* (*Securinega*) *virosa* is a common euphorbiaceous shrub in the lower hill-slope habitat and bears small white berries, which are much sought after by fruit-eating birds. It comes into bud quite early in the rains, but flowering is delayed until the mid rains when the flowers open *en masse* in the evening and are moribund by the next afternoon. The flowers are pollinated by a noctuid moth (apparently only one species) which arrives in vast numbers and remains actively feeding on the nectar-laden blooms until the afternoon of the next day. The 20 or more species of *Grewia* in the MGR exhibit a variety of flowering patterns from almost continuous flowering (*G. bicolor* and *G. villosa*) to a single delayed burst in *G. tenax* (like *Flueggia* above).

The 23 species of *Acacia* currently recorded in the MGR have received special attention from both our botanists and zoologists on account of their widespread distribution, the importance of both their foliage and fruits to the mammalian herbivore fauna, providing fruits and nest sites for the avifauna, and the emerging interest of the very diverse insect fauna that is casually or even specifically associated with these tree species.

Phenology of *Acacia* species

The genus *Acacia* is one of most important components of the east African savanna and in Mkomazi comprises 23 species, representing 17% of the African species and 43% of the east African species. We are familiar with 25 *Acacia* taxa in the MGR, although it is possible that two or even three more species (including *A. horrida* and *A. xanthophloea*) may be present. The morphological and habitat characteristics of these species (Table 8.1) show that they are armed with either stipular spines (13 species) or hooks (10 species) as a means of slowing down the rate of feeding by large mammalian herbivores on their foliage (Pellew 1983, Coe & Coe 1987) and—in the case of those with scattered hooks—as scrambling devices. The acacias are distributed widely in all habitats except the hill summit forests, although many species (*A. dolichocephala, gerrardii, laeta, pentagona, polyacantha, rovumae* and *stuhlmannii*) are restricted in their distribution within the MGR. Even for the commoner species, comparatively few of them co-occur in close proximity in the same habitat.

Three of the species currently recorded in the MGR bear 'pseudo-galls', which are normally occupied by crematogastine ants. These galls are referred to as 'pseudo-galls' since they are produced by the plant itself and are not induced as a result of attack by an arthropod. Only those species which possess stipular spines produce galls, which are formed by the swelling of the spine base to form a globular or fusiform structure. Should the gall be swollen to form a large structure, the galls may coalesce to form a single large gall, with the paired stipular spines emerging from the wall *(A. drepanolobium)*. Since only three (19%) out of the 16 MGR species (Table 8.1) with stipular spines bear pseudo-galls, it is tempting to suggest that this may reflect the low density of browsing mammalian herbivores—the number of taxa with pseudo-galls in the whole African *Acacia* flora is 17 (31%) out of about 155 species. These structures provide protection for the ants, while the ants deter excessive feeding by browsers (see Stapley, Chapter 23) which is particularly important for the flowers, as many of these trees are maintained at a low, compact stature by browsing. The protection afforded by the armature and/or galls allows many of these species to produce their flowers and leaves much earlier than would otherwise be possible, particularly since the majority of inflorescences and leaves are borne as bunches in the leaf axils.

The African *Acacia* species may be grouped into those whose flowers are arranged in either globose or spicate inflorescences, of which the former are usually armed with stipular spines while the latter are mostly armed with hooks. This division is fairly distinct but in a limited number of species (e.g. *A. thomasii*), the short, apparently spicate inflorescence resembles a slightly elongated globose form. Stone *et al.* (1996) have recently described their work in the MGR, showing that a number of species which flower at the same time are separated in respect to the time that their flowers are open for pollen and nectar feeding insects. This obser-

Table 8.1 Comparative features of the Mkomazi *Acacia* flora. Infloresence types: S–Spicate, G–Globose. Pod types: ID–Indehiscent, D–Dehiscent.

species	armature	pseudo galls	infl. type	pod type	habitat
A. albida	paired spines	-	S	ID	riverine
A. ancistroclada	spines & hooks	-	G	D	arid bushland
A. brevispica	scattered hooks	-	G	D	grassland fringes widespread
A. bussei	paired spines	+	S	D	flooded grassland ecotone
A. dolichocephala	spines	-	G	D	upper hillslopes
A. drepanolobium	spines	+	G	D	seasonally inundated grassland
A. etbaica subsp. *australis*	spines	-	G	D	lower hillslopes
A. etbaica subsp. *platycarpa*	spines	-	G	D	lower hillslopes
A. gerrardii var. *gerrardii*	spines	-	G	D	upper hillslopes
A. hockii	spines	-	G	D	arid bushland-uncommon
A. laeta?	paired hooks	-	S	D	bushland-rare
A. mellifera subsp. *mellifera*	paired hooks	-	S	D	open bushland
A. nilotica subsp. *subalata*	spines	-	G	ID	lower hillslopes
A. pentagona	scattered hooks	-	G	D	damp gully woodland
A. polyacantha subsp. *campylacantha*	paired hooks	-	S	D	reverie fringe
A. reficiens	paired hooks	-	G	D	arid open bushland-grassland fringe
A. robusta subsp. *usambartensis*	spines	-	G	D	lower hillslopes-uncommon
A. rovumae	paired hooks	-	S	ID	riverine/spring fringe
A. senegal var. *senegal*	triple hooks	-	S	D	arid bushland and grassland fringe
A. seyal var. *fistula*	spines	-	G	D	lower hillslope-arid bushland fringe
A. seyal var. *seyal*	spines	-	G	D	lower hillslope-arid bushland fringe
A. stuhlmannii	spines	-	G	ID	arid bushland on red soils uncommon
A. thomasii	triple hooks	-	S	D	rocky hillslopes-infrequent
A. tortilis subsp. *spirocarpa*	spines & hooks	-	G	ID	lower hillslopes-slightly more mesic
A. zanzibarica var. *zanzibarica*	spines	+	G	D	upper inundated grassland ecotone

vation is of great interest, for the species to which they refer *(A. drepanolobium, A. nilotica, A. senegal, A. tortilis* and *A. zanzibarica*) do not normally occur in the same habitat/sub-habitat (see Table 8.1) and suggests that the insect pollinator guild must be fairly mobile in these savanna environments (see Stone *et al.* in Chapter 20 and Willmer *et al.* in Chapter 21). Doran *et al.* (1983) point out that even among the fertile flowers of African *Acacia* species we can expect no more than two or three pods per 1,000 flowers. Although this is undoubtedly true of most species, those that appear to be pioneers in early successional environments such as *A. karroo* and *A. caffra* in southern Africa very commonly produce much larger numbers. Coe (1989) reported that *A. tortilis heteracantha* only developed an average of 1.53 pods per inflorescence, while even lower numbers were observed *in A. erioloba.* We might also mention the invasive, thicket-forming *Dichrostachys cinerea*, which also produces large bunches of pods from its terminal hermaphrodite inflorescences.

Table 8.2 shows the months of flowering by *Acacia* species in the MGR, together with information from Beentje (1994) on flowering/fruiting in Kenya. Many species have been observed flowering/fruiting in most months of the year, although where there are quite large gaps in reproductive records (*A. ancistroclada, A. bussei, A. etbaica* and *A. reficiens*), this may be as much due to their restricted distribution in arid bushlands as it is to an actual lack of flowering/fruiting. All the other records are solely for flowering activity in the MGR. These indicate that *A. brevispica, A. drepanolobium* and *A. zanzibarica* have been observed flowering in ten months of the year, a factor probably related to their occurrence on or close to the fringes of seasonally-inundated habitats where a high water table persists for some time into the dry season. *A. nilotica* and *A. senegal* appear to flower in up to nine months of the year, the former being largely found in more mesic lower hill-slope habitats, while the latter is widespread in grassland and on seasonally-inundated grassland ecotones. The flowering of the other species (in five to eight months) appears to be largely related to the advent of the long and short rains. *A. ancistroclada, A. bussei, A. stuhlmannii* and *A. thomasii* are all species of the more arid areas of the MGR where water is only available for short periods of growth, and reproductive activity is therefore severely restricted. Combining the MGR and Kenya records, we see that nine out of 13 species have been recorded flowering/fruiting in 10 or more months of the year.

The most complete study of the phenology of African *Acacia* species is that of Milton (1987) who studied seven species (*A. burkei, A. caffra, A. karroo, A. luederitzii, *A. mellifera, *A. nilotica* and **A. tortilis*)[1] on a savanna site at Nylsvley in the Transvaal, South Africa. This site receives mean annual rainfall of 622 mm and she showed that the flowering rhythms of four of these species were spread

---

[1] Species prefixed by * are found in the MGR but even these are considered to be sub-specifically distinct.

Table 8.2 Flowering and fruiting records of MGR *Acacia* species. Data after records by Beentje 1994 (●), Coe 1992–97 (*), Raphael Abdullah 1993–97 (†), Graham Stone 1995–97 (‡) and Linsey Stapley 1996–97 (§). Shading indicates months in which rain can be expected in most years.

	month											
species	1	2	3	4	5	6	7	8	9	10	11	12
A. ancistroclada	●		●*				*	*		●*	●*	●*
A. brevispica	●*	●*	●*	●*	●	●	●*	●*	●*	●*	●*	‡
A. bussei	*		●*	*			*	●*	●	●	●*	●*
A. drepanolobium	●*	●*	●*	●*	§	§	*	●*	●	●	●*	●*
A. etbaica	●*	●				●	*	●*	●	●	●*	●*
A. mellifera	*	●*	●*	●*	●		●	●*	●	●*	●*	*
A. nilotica	●*	●*	●*	●*	●	●	●*	●*	●	●*	●*	●*
A. senegal	●*	●*	●*	●*	●	●	●*	●*	●	●*	●*	●*
A. stuhlmannii	●*	●	●*		●	●	●*	●*	●		*	*
A. reficiens						●	●*	●*	●	†	●*	●*
A. thomasii	*	*	●	●	●	●	●*	●*	●	*	●*	●*
A. tortilis	●	●*	●*	●*	●	●	●*	●*	●	●	●	‡
A. zanzibarica	*	*	*	*	‡	*	*	*			*	*
monthly species total	12	10	11	9	9	10	13	13	11	11	13	13

throughout the rainy season (October–April), while three of them flowered briefly in the spring (August–September) before the summer rainfall in non-Cape areas of southern Africa begins. The success of pod production is probably related to the amount and timing of rainfall. Rainfall also has a profound influence on the availability of pollinators: while the plant cannot start to produce pods without pollination taking place, the development of the pods must be limited by the availability of water and the production of active photosynthate to enable seed formation and maturation to happen. Preece (1971) showed that in Australia production of pods by *A. aneura* was closely related to the amount and the timing of rainfall and was only successful in 10–15% of seasons. The pods of the African *Acacia* species can be divided into those with dehiscent or indehiscent pods, an important distinction: the pods of the former are thin-walled and their seeds are dispersed short distances by wind, while the latter retain their seeds inside the pods and can only be dispersed when the pods are eaten by a herbivore—an effective dispersal method since these pods are a significant food source for large mammalian herbivores in the middle of the dry season when most of them are ripe. There are subtle differences in the flowering and pod ripening in the *Acacia* species, which vary from the slow ripening of the pods spread over an extended period (*A. nilotica*) to the massed synchronised flowering and pod production of species like *A. tortilis*.

If we estimate the flower-pod intervals (the average time in weeks from mid-flowering to mid-pod-ripening), we might expect to find a relationship which is based on pod size: a large indehiscent pod like *A. erioloba* in southern Africa takes longer to develop than a thin-walled indehiscent pod. However, using data for 21 species (15 dehiscent and six indehiscent) from Natal (Ross 1979) and for the seven species listed above for Nylsvlei (Milton 1987), Coe (1989) showed that the flower-pod intervals for dehiscent and indehiscent groups did not differ significantly. The flower-pod intervals were longer in the Transvaal than in Natal for the same species, indicating that the pod development interval is more closely related to climatic conditions and their time of flowering than it is to pod size.

Annual production in all plants and animals is influenced by the energy demands of growth, maintenance and reproduction. Thus the amount of energy (and nutrients) that an *Acacia* can invest in pod growth is finite and is determined by its age and prevailing local conditions. At any one time, the conditions experienced by all the trees in one locality should be much the same, indicating that the plant only has a finite amount of energy and nutrients which it can allocate to pod development. Since we have been unable to show that a relationship exists between the flower-pod interval and pod size, it seems likely that individual species are constrained by the number of pods and their size in relation to the resources invested in the production and maturation of the whole crop. Little information is available on total pod production in relation to both pod and tree size, although observations suggest that far fewer pods are produced by trees with large pods (*A. (Faidherbia) albida, A. erioloba* and *A. sieberana*), compared to those with small pods, irrespective of whether they are dehiscent (*A. caffra, A. karroo* and *A. luederitzii*) or indehiscent (*A. tortilis*) (Coe & Coe 1987). Indeed it may be that *A. tortilis* has opted for a 'dehiscent-like' strategy by producing abundant, relatively small pods and seeds which ripen en masse, thus enhancing their dispersal and reducing seed predation by insects, small mammals and birds. In the Kora National Reserve in Kenya, *A. tortilis* was observed to produce a large crop of pods in August, following the long rains in March–May, which had fallen and been consumed by a variety of wild and domestic herbivores within about 14 days.

The design of the indehiscent pods of some African *Acacia* species in general (21 species[1]) and those of the MGR in particular (5 species) represents one of the best examples of co-evolution between large herbivorous mammals. Not only are the pods eaten by the mammals but the design of their seeds enables them to resist the shearing forces imposed by the mammals' cheek teeth, thus enabling them to be dispersed, often at considerable distance from the parent tree (Coe & Coe 1987). Although passage through the herbivore's gut is often quoted as facilitating germination (Howe & Smallwood 1982, Fenner 1985), a seed being stimulated to germinate in the savanna during the dry season, when most of these pods ripen,

[1] Pods which are both fully indehiscent and tardily dehiscent are included in this category.

would die as soon as its roots extended beyond the influence of the dung fluid at the soil surface. This suggests that dispersal alone is sufficiently important to account for the evolution of these strategies. Chapman *et al.* (1991) however have recently shown that up to 50.9% of *Balanites wilsoniana* seeds recovered from elephant dung in Ghana germinated, compared with only 0.7% of ripe seeds collected directly from the trees.

Discussion

In spite of many detailed studies of large mammal behaviour and ecology that have been conducted in Africa over the last 50 years, comparatively little attention has been paid to the role played by large herbivorous mammals in the establishment and maintenance of plant communities. Laws (1970) and Owen-Smith (1988) have both dealt at length with the importance of the megaherbivores of Africa as natural agents of vegetation change. Sinclair & Fryxell (1985) have drawn attention to the parallel between the behaviour of large migratory herbivores and pastoralists. Both require large areas of land over which they must range, and when their current land is exhausted they must move on if they are not to suffer the devastating effects of habitat degradation imposed by increased numbers, which induces a steady reduction in local productivity. Today we observe the consequences of the success of both wild and domestic large mammalian herbivores as exploiters of the environment, for they now have nowhere to move to unless they can or are allowed to displace adjacent communities, which are likely also to be suffering similar problems of overexploitation.

Howe & Smallwood (1982) in a review of the ecology of seed dispersal have shown that woody plant communities in both temperate and tropical environments are nearly all characterised by a high percentage of vertebrate-dispersed species. These levels of vertebrate-dispersed propagules generally lie above 65%, although there is greater variation when the plant communities are divided into the larger woody species, shrubs and climbers. None the less these data, together with further information for African savanna (Coe & Collins 1986, Coe & Coe 1987) and forest (Gautier-Hion *et al.* 1985) habitats, indicate that a high percentage of woody plant fruits (seeds) are dispersed by birds or mammals, irrespective of the climatic zone in which they are located: examining their dispersal mechanisms in relation to precipitation shows that the percentage of vertebrate-dispersed plants is significantly correlated with mean annual rainfall (Coe 1989).

The animal-dispersed element of the southern African woody flora contributes 59.4% of the total, while a further 25.3% are largely mechanically-dispersed and most of the remaining 15.3% are wind-dispersed. It is probable that many of the dry mechanically-dispersed fruits may also be subsequently dispersed by vertebrates such as seed-hoarding rodents, or eaten whole in the case of indehiscent pods/fruits. If we therefore include these elements in the vertebrate-dispersed category it is likely that the total may approach 80% (Coe & Coe 1987). The same authors compared the southern African woody flora with that of the Kruger National Park and pointed out that the fruits could be divided into several distinct dispersal categories. Of the total vertebrate-dispersed fruits, 37.6% could be classified as succulent (which possess a juicy sugar-rich mesocarp), while 49.4% could be described as dry fruits, whose outer coat is dry and generally less than 2 mm in thickness. The remaining species bore arillate fruits or seeds (13.1%).

The percentage of vertebrate-dispersed fruits that are succulent is low in arid and semi-arid environments compared with their counterparts in more humid ecosystems. The data presented by Howe & Smallwood (1982), supplemented by the information for African habitats detailed above, shows the strong positive relationship between the percentage of the animal-dispersed fruits that are succulent and precipitation. Below 3,000 mm of rain there is a steady decline in the percentage of animal-dispersed fruits that are succulent. In arid environments, plants would find it difficult to retain water, which is well illustrated in Mediterranean habitats where only 5% of the fruits are animal-dispersed (Howe & Smallwood 1982), while in Israel this figure falls to about 3% (Ellner & Shmida 1981).

In Mkomazi a large element of the very high level of floral diversity is almost certainly maintained by the mammalian and avian fauna, and to a lesser extent by the insects. Although the primary determinant of habitat diversity is altitudinal and climatic, the structure and composition of the plant communities is almost certainly much more dependent on the dispersal of plant propagules by vertebrates. The preservation of these natural community attributes depends on our ability to control the effects of anthropogenically induced disturbance, for unless we show the will to do so we will lose a jewel in the crown of Tanzanian savanna.

References

Beentje, H. J. (1994) *Kenya Trees, Shrubs and Lianas*. National Museums of Kenya, Nairobi, Kenya.

Chapman, C.A., Chapman, L.J. & Wrangham, R.W. (1991) Uganda Elephants help maintain a useful tree. *Agroforestry Today* 3(3): 15.

Coe, M. (1989) *Large herbivorous mammals, vegetation and seed dispersal* (Abstract). Zoological Society of Southern Africa Symposium on African Savannas. Etosha National Park, Namibia.

Coe, M. (1990) The Conservation and Management of semi-arid rangelands and their animal resources. In: Goudie, A.S. (ed.) *Techniques for Desert Reclamation*. Wiley, Chichester.

Coe, M. & Coe, C. (1987) Large herbivores, *Acacia* trees and bruchid beetles. *South African Journal of Science* 83: 17-18.

Coe, M. & Collins, N.M. (eds.) (1985) *Kora: an Ecological Inventory Study of the Kora National Reserve, Kenya*. Royal Geographical Society, London.

Coe, M.J. & Isaac, F.M. (1965) Pollination of the baobab (*Adansonia digitata* L.) by the lesser bush baby (*Galago crassicaudatus* E. Geoffroy). *East African Wildlife Journal* 3: 123-124.

Doran, J.C., Turnbull, D.J., Boland, D.J. & Gunn, B.V. (1983) *Handbook of seeds of dry-zone acacias*. FAO, Rome.

Ellner, S. & Shmida, A. (1981) Why are adaptations for long distance dispersal rare in desert plants? *Oecologia* 51: 133-144.

Faegri, K. & van der Pijl, L. (1979) *The principles of pollination ecology.* 3rd edition. Pergamon Press, Oxford.

Fenner, M. (1985) *Seed Ecology*. Chapman and Hall, London.

Gautier-Hion, A., Duplantier, J.-M., Quiris, R., Feer, F., Sourd, C., Decoux, J.-P., Dubost, G., Emmons, L., Erard, C., Heckestweiler, P., Mougazi, A., Roussilhon, C. & Thiollay, J.-M. (1985) Fruit characteristics as a basis for fruit choice and seed dispersal in a tropical forest vertebrate community. *Oecologia* 65: 323-337.

Glover, P.E. (1970) *Tsavo Research Project Report: June 1968–June 1970.* Kenya National Parks, Nairobi, Kenya.

Glover, P.E. (1973) *Tsavo Research Project: Progress Report for the Year 1973.* Kenya National Parks, Nairobi, Kenya.

Gordon-Gray, K.D. & Ward, C.J. (1975) A contribution to the knowledge of floral variation in *Acacia karroo* in eastern South Africa. *Boissiera* 24: 279-284.

Harris, D.R. (1980) Tropical savanna environments: definition, distribution, diversity and development. In: Harris, D.R. (ed.) *Human Ecology in Savanna Environments*). Academic Press, London. pp. 3-27.

Howe, H.F. & Smallwood, J. (1982) Ecology of seed dispersal. *Annual Review of Ecology and Systematics* 13: 201-228.

Laws, R.M. (1970) The Tsavo Research Project. *Oryx* 10: 355-361.

Milton, S.J. (1987) Phenology of seven *Acacia* species in South Africa. *South African Journal of Wildlife Research* 17: 1-6.

Owen-Smith, R.N. (1988) *Megaherbivores: the influence of very large body size on ecology*. Cambridge University Press.

Pellew, R.A. (1983) The impact of elephants, giraffe and fire upon *Acacia tortilis* woodlands of the Serengeti. *African Journal of Ecology* 21: 41-74.

Preece, P.B. (1971) Contributions to the biology of mulga. I. Flowering. *Australian Journal of Botany* 19: 21-38.

Ross, J.H. (1979) A conspectus of African *Acacia* species. *Memoirs of the Botanical Survey of South Africa* 44: 1-155.

Sinclair, A.R.E. & Fryxell, J.M. (1985) *Canadian Journal of Zoology* 63: 987-994.

Stone, G.N., Willmer, P. & Nee, S. (1996) Daily partitioning of pollinators in an African Acacia community. *Proceedings of the Royal Society of London B* 263: 1389-1393.

Tyrrell, J.G. & Coe, M.J. (1974) The rainfall regime of the Tsavo National Park, Kenya and its potential phenological significance. *Journal of Biogeography* 1: 187-192.

White, F. (1983) *The Vegetation of Africa. A descriptive memoir to accompany the UNESCO/AETFAT/UNSO vegetation map of Africa.* UNESCO, Paris.

CHAPTER 9

The effects of fire on shrub survival in Mkomazi

Ken Ferguson

The aetiology of a shrub-grassland fire

Individual fires may rage along the valleys and hill slopes of Mkomazi Game Reserve for a matter of several days during the height of the dry season. Often re-ignitions occur in a localised setting. The main fire front can be several kilometres in breadth, exhibiting variable flame height and intensity. The charred aftermath observed in the valleys can leave no doubt as to the efficiency of burning in these flammable, shrub-grassland communities.

The ignition sources of fire in Mkomazi are two-fold. Firstly, natural lightning can strike at the end of the dry season, when the standing crop of moribund grass is at its greatest. Under these conditions trees may act as lightning rods with the resulting fire igniting the surrounding ground vegetation. Several large trees in the reserve bear obvious heat conductance scars. The second potential source of ignition is anthropogenic. It is highly likely that over a period of many decades, if not longer, most fires in this Reserve have been the result of deliberate ignition. The primary reason for this human manipulation of the productive savanna valleys is to encourage the growth of palatable fire-adapted grasses, in time for the grazing of livestock during the rainy seasons (November – January and March – May). Another commonly cited cause of local ignition is that illegal hunters find it easier to stalk prey and set snares if the accumulated dry season vegetation cover is removed. Arson may also be a cause of ignition.

However, ignitions are only successful if certain climatic and vegetation conditions are favourable (Bond & van Wilgen 1996, Whelan 1995). High humidity and weak winds combined with a high air or fuel moisture content will mitigate the chance of grass fuel igniting. Conversely, dry headwinds will force the flames on to greater heights and speed, allowing fuel consumption to be more complete. The rate of fire spread determines the 'residence time' of lethal fire for plants at any given point. The development of the initial ignition flame into a viable fire depends on the presence of adequate dry fuel which can combust and pre-heat adjacent vegetation. The 'architectural' structure of the vegetation is an important requisite of pre-heating. Vertical grass tussocks and multi-stemmed shrubs offer suitable fire aerated spaces. Dense coarse vegetation, with a closed horizontally arranged canopy, may act as a fire retardant. Plants with the capacity to produce oils or resins may increase the heat yield of the combustion reaction (Bond & van Wilgen 1996, Whelan 1995). Topographic features may assist fire speed especially a fire burning up a slope or into a gully.

Study area and methods

The Mbono Valley floor consists of a gently undulating catenary sequence situated immediately adjacent to the Ibaya camp site. Soils on the valley floor are predominantly seasonally-inundated 'black cotton' lying in the troughs of the catena (*mbugas*) with more sandy loams present on the crests. Drainage lines intersect the catena and originate from the valley walls.

Vegetation composition is a predominantly grass/shrub mixture. Shrubs are arbitrarily defined to be woody plants below 1.5 m in height. Trees up to 6 m in height are sparsely arranged across the catena except along the watercourse where taller tree stands are located.

Four study plots, each of 0.5 hectare, were delineated in November 1995 in the central area of the valley floor, prior to burning by an unplanned fire. Three of the four plots burnt and a fourth remained unburnt (Table 9.1). Additionally ten belt transects (2 x 30 m) were sited, post-burn (after an unplanned October 1995 fire) around the Dindira waterhole. All individual shrubs were allocated coded metal tags and given spatial coordinates with the aid of 50 m measuring tapes. Each shrub was measured for four growth parameters: height in millimetres of the tallest stem; the number of live stems (approximated when this was greater than 20); the diameter of the tallest stem, measured in millimetres with callipers at or as close to the base as possible; and an index of green leaf area, taken by placing a meter-squared quadrat over the shrub crown (an approximate measure of green leaf productivity).

In order to assess the interrelationship between giraffe browsing and fire, wire exclosures were erected around randomly selected individuals of each dominant species. Individuals were then subjected to varying regimes of leaf stripping and clipping, representing the two feeding modes of giraffe (pers. obs.). Results will be presented in a subsequent paper.

Initial measurements were taken in October-November 1995. A random sample of individuals from each plot or transect were measured in the same fashion at approximately two-month intervals. Finally, a full inventory and re-measurement of all tagged plants was undertaken at the end of the study in June 1996. In total the fire survival response of 82 species totalling 1,600 tagged individuals was tracked over a nine-month period.

Results

0.5 hectare plots

82 species of woody plants and forbs were recorded in the study plots and belt transects. Dominant grass species in the general study area were identified. The four 0.5 ha plots were dominated by three shrub species: *Cadaba farinosa, Cadaba ruspolli* (Capparidaceae) and *Acacia drepanolobium* (Mimosaceae). The last named species was not present in the unburnt control plot.

Shrub diversity in the 0.5 ha plots and belt transects was measured using the Simpson Index, a relative index that increases in magnitude with increasing diversity (UNEP 1995). The index varied from 0.0 in a single belt transect consisting of a pure stand of *Dichrostachys cinerea* thicket, to 2.5 in the unburnt control plot (Punda milia plot, *Cadaba* spp. dominant). The three remaining plots measured 2.0 (Kinyonga plot, *Cadaba ruspolli* dominant); 1.3 (Kobe plot, *Cadaba farinosa* dominant) and 1.9 (Twiga plot, *Acacia drepanolobium* dominant). Woody shrub diversity in the central catena is thus low in comparison with other vegetation types (Coe 1995). Individual species were classified into two formal groups according to their fire survival mode: sprouters and non-sprouters (Bond & van Wilgen 1996, Gill 1981). Sprouters can be subjected to near 100% leaf scorch but exhibit resprouting vigour from surviving buds in the post-burn environment. Non-sprouters are killed by fire but are replaced by a new generation of seedlings which often allows them to dominate fire-prone communities (Bond & van Wilgen 1996).

Within-plot comparisons of survival and re-growth of dominant shrubs

The November 1995 fire burnt three of the four 0.5 ha plots. The intensity of the fire was not measured but it is known that the previous fire affecting this study area occurred in October 1993. The study plots were reported to have experienced burning again late in the 1996 season (L. Stapley, pers. comm.). All shrubs below 1 m in height were found to have nearly 100% leaf defoliation caused by the scorching effects of the fire. Fire intensity was probably higher in the troughs of the valley floor catena where taller and more dense grass stands occur. Fire is not a constant phenomenon and the patchiness of a burn will affect the individual persistence of shrubs.

Results of the comparison of survival and resprouting vigour for each dominant species are given in Table 9.1. Comparisons are made for individual species within a plot in relation to pre-burn growth measurements and the final post-burn re-measurement. If a species had not attained the pre-burn average growth at the end of the nine month period, it is described as 'less' in Table 9.1. Similarly, if the average growth was exceeded it is described as 'greater'. An average growth that matched equally between the two time periods was termed 'equal'.

Table 9.1 Intra-plot comparison of growth parameters (see text for explanation).

plot	species	green leaf	height	stem diameter	stem number
Kinyonga	*C. farinosa*	less	less	less	equal
Kinyonga	*C. ruspolli*	less	less	less	equal
Twiga	*C. farinosa*	less	less	less	less
Twiga	*A. drepanolobium*	less	less	less	greater
Kobe	*C. farinosa*	greater	greater	less	equal
Punda milea unburned	*C. farinosa*	equal	greater	equal	less
Punda milea unburned	*C. ruspolli*	equal	greater	greater	less

Mortality was defined as either the presence of a stem 'skeleton' or if an individual tagged plant could not be relocated.

Cadaba farinosa

Cadaba farinosa is a multi-stemmed shrub with an average height of under a metre. It has relatively small leaves born on upright stems. *C. farinosa* was found on slopes, outwith the plots where fire had not penetrated for some time, to be over 2 m in height (pers. obs.). *Cadaba farinosa* was the most numerous tagged plant and was present in all four plots. It was generally found to be at its highest density in the troughs and slopes of the catena valley floor, suggesting a degree of tolerance to both seasonal flooding and intense grass fires.

Density in the burnt plots varied from 0.05–0.09 plants/m^2. Occupancy in the unburnt plot was at an intermediate density. As might be expected, no mortality was observed in the unburnt plot. Mortality was highest in Twiga plot (11.0%) where this species is sub-dominant to *Acacia drepanolobium* and where the lowest density for this species was recorded in any of the plots. Conversely, the highest density was attained in the plot with the lowest recorded mortality (1.3%, Kobe plot). Mortality in the third burnt plot was 8% (Kinyonga plot).

A possible explanation for the high mortality shown may be found in the large disparity in the mean number of pre-burn stems and green leaf area, between Twiga and Kobe plots. In Kobe plot the mean pre-burn stem number was five and the mean green leaf area was 1.3 m^2. In the high-mortality Twiga plot the mean pre-burn stem number was 50 and the mean green leaf area 8.5 m^2. The disparity may represent an older age stand of this species present in Twiga plot or one in which the fire return interval was longer (i.e. last fire was in 1993). Increased stem numbers and a correspondingly larger leaf area may have an important influence on flammability. More densely stemmed individuals, which exhibit greater crown areas, may be more prone to being killed by fire.

Comparison of the four growth parameters within a plot showed that in each plot this species was able to resprout vigourously from root bud sources. However, despite this it was recorded that in two of the three burnt plots there was less stem repsrouting, smaller stem diameters, less height and less green leaf crown area in the post-burnt population in comparison with the pre-burn population. The plot with the highest density (Kobe plot) deviated from the trend by increasing the mean height and leaf area in relation to the pre-burnt population. From Table 9.1 it is noted that post-burn stem numbers did not increase in Kobe plot and there must be some doubt as to how green leaf area was recorded to be greater while stem numbers were less than the pre-burn population. These two parameters should be positively correlated (Whelan 1995). This may be the result of a sampling error or alternatively edaphic/drainage factors in Kobe may induce rapid re-growth in a post-burn environment.

Clearly, a figure of 8-11% mortality per fire, in the case of two of the burnt plots, is unsustainable. If local population extinction is to be avoided for this species then seedling recruitment must at least match mortality. *Cadaba farinosa* was seen to flower in January/February in the burnt plots. Individual persistence attained by vigorous repsrouting can lead to flowering and seeding within a few months of the initial disturbance. But flowering was later than in the unburnt plot. Population stands appear to be mature and even aged within a plot but generations most likely overlap between plots. Juvenile plants tend to be absent. This is a defining factor of the fire life histories of sprouter species under a frequent fire regime (Bond & van Wilgen 1996).

Cadaba ruspolli

Cadaba ruspolli is a multi-stemmed lignotuberous shrub of approximately a metre in height. The lignotuber, a woody knot structure termed a burl, is present semi-submerged in the soil. This structure contains a bud bank from which resprouting stems emerge after a fire (James 1984). *C. ruspolli* also has fewer, larger and more succulent leaves and its woody stems are more robust than its congener. This species was generally situated on the crests of the catena where the soil is more sandy and permeable. The average height was under a metre.

This species was present in one burnt plot (Kinyonga plot) and the unburnt plot (Punda milia plot). The densities in the two plots for this species was similar (0.1 plants/m^2). Post-burn mortality was 7.6% in the burnt plot. No mortality was recorded in the unburnt plot. Comparisons between each of the four regrowth parameters (between the first and last measurements) showed that regrowth vigour had not matched pre-burn levels (Table 9.1).

However, as a rule of thumb, post-burn growth levels were 20–25% of the original measurements. Substantial regrowth had occurred during the fire interval period. An increase in the growth parameters was noted in the unburnt control plot with the exception of stem numbers, which declined over the study period. A possible cause may be the browsing action of giraffes, which were observed to favour this species during, in particular, the mid-dry season (pers. obs.).

This species was not observed to flower in the burnt or unburnt plot during the study period. The explanation for this absence may be that the population is a young cohort. Sprouters may take several years to reach maturity, and in the short term invest in longevity. An alternative explanation is that individuals of this species are not adapted to short fire return intervals and require longer periods between fires for successful reproduction and seedling recruitment. On balance it seems unlikely that this is a young population: the woody lignotubers are large and well established and have most likely acted as bud reserves over a number of generations.

It is suggested tentatively that the survival strategy of this plant, in the presence of annual fires, is to allocate resources to survival and not reproduction, while a frequent fire regime is prevalent, a feature noted in other lignotuberous shrubs (Bond & van Wilgen 1996, James 1984).

Acacia drepanolobium

The third shrub could be more appropriately classified as a tree (Coe & Beentje 1991). *Acacia drepanolobium* reaches a two and a half metre maximum in the plot. The mature plant is usually single stemmed with a brush like canopy situated on the top third of the plant. Small leaflets and dense interconnecting branchlets are a diagnostic feature of several *Acacia* species. Densities were highest in *mbugas*. This species was present only in Twiga plot. The majority of individuals (measured pre-burn) were over 1.3 m in height and post-burn mortality was 7.3%. In one case mortality was not directly fire related but caused by a buffalo. The mean pre-burn height of those that died was 1 m, which is suggestive that a fire escapement height threshold exists for this species.

This species can reach 6 m in height in other locations in east Africa and thus the low stature of individuals measured in this study area must be due to either fire or edaphic factors. *Acacia drepanolobium* has thick corky bark surrounding its long usually single stem. Bark thickness and survival of the stem have been shown to be negatively correlated with the degree of the heat pulse required to kill cambial tissue (Hare 1965). Stem diameter is an approximate index of bark thickness. The plants that died had on average a smaller stem and therefore bark thickness.

Of those that survived, the green leaf area index reveals the extent of the species' reaction to fire. 41% of those re-measured post-burn were technically still considered alive, but had virtually no green leaf production. Re-measurement took place near the end of the growing season. Partial compensation for this loss of photosynthetic area is shown by the observation that 50% of those plants that were recorded to be denuded of leaves had seedling growth recorded at the base of the plant, which presumably arose from a stored seedbank or alternatively was the product of a neighbouring plant. *Acacia drepanaolobium* individuals had a variable rate of flowering and producing seed pods post-fire. The pre and post burn comparison of the growth parameters showed a marked decrease on green leaf area, stem diameter and height in the growth period after the perturbation (Table 9.1). The increase in stem numbers is due to the increase in seedling density. Locally isolated stands of this species were noted elsewhere in the valley, but it is not a predominant shrub on the valley floor (pers. obs.). This low density may be related to the small stature of individuals keeping them below the height of escaping the lethal or growth retarding effects of fire.

The effects of fire on savanna shrubs

Individual and population persistence

Fire directly affects the growth, survival and reproduction of plants. Fire tends to 'prey' selectively on younger/smaller, thinner-barked individuals of a species in a community. The killing of mature plants requires that the live crown scorch is extensive and that protected buds located on or below the plant are destroyed. Those species with high stem growth rates are likely to be vulnerable for shorter periods of time in a frequent fire regime (Bond & van Wilgen 1996).

Savanna trees like mature *Acacia drepanolobium* rely on thick corky bark, long bare trunks and high-held canopies with few large branches near the ground to reduce vulnerability to fire. Savanna trees often lose the ability to resprout when mature but can form multi-stemmed shrubs when young (Coe & Beentje 1991). The low canopy stature and high number of stems of indivdual *Acacia drepanolobium* within the study plots suggests that the local population reaction to fire, in this instance, seems to be to invest in the number of new stems rather than existing stem height as a means of increasing inter-fire productivity. This appears to be a reaction to the frequent fire regime experienced in the study area.

The two *Cadaba* species congeners are identifiable as sprouters. The *Cadaba* species tolerate 100% defoliation and sprout new shoots from subterranean buds. Soil acts as an insulator for buds from critical lethal temperatures during the passage of fire (Gill 1981). Root crown sprouts are the source of the next generation of buds for *Cadaba farinosa* (pers. obs.). *Cadaba ruspolli* buds proliferate from the lignotuber, which is originally formed by a swelling in the axils of cotyledonary leaves (James 1984). The anatomy of the lignotuber is poorly understood but they are known to act as a carbohydrate store (James 1984). Survival of *Cadaba* species is for long periods (perhaps decades) based on the successful persistence of resprouting stems (termed ramets). If a correlation were to be found between burl weight or diameter and plant age then this could act as an indicator of the strength and frequency of past fires.

Sprouter species are less susceptible to fire mortality when juvenile, if sufficient time has passed without fire to allow young individuals to gain the ability to sprout (Bond & van Wilgen 1996). The relatively even age structure (although overlapping generations may have been present between plots) of these shrubs suggests that fire incidence has remained frequent for some time.

In the case of the sprouter *Cadaba ruspolli* flowering was not recorded, in either the burnt or unburnt plots. It is likely that seedling recruitment will only occur in older unburnt vegetation and will not occur under fire regimes with short return intervals (Keeley 1986).

Cadaba farinosa, although a conger of *C. ruspolli,* seems to adopt a different reproductive strategy. Limited and variable post-burn flowering was noted suggesting that this species still has a limited propensity to reproduce under a frequent fire regime. Unfortunately no information is available on seed or seedling survival in this species. However, if the seed or seedlings of *C. farinosa* can remain viable through the dry season and after frequent burns, then in the long-term this species may be set, presuming that edaphic and drainage factors are the same, to outcompete its congener.

Evidence that tends to suggest this hypothesis is that *C. ruspolli* is found almost exclusively on catenary crests where fire intensity is at its lowest. Only a longitudinal study based on the existing tagged plants and seedling density within the plots could prove or disprove this hypothesis. Alternatively, selected plots could be subdivided and protected from fire by the use of fire-breaks.

Why are the dominant shrub species in the study plots sprouters? Bond & van Wilgen (1996) consider resprouting from buds, situated below ground to be the primitive ancestral form of vegetative survival in woody angiosperms. Non-sprouters should have the advantage of flowering and reproducing quickly after a fire when space is freed by the burning of established plants. The chance of seedling predation may be lessened in the post-burn environment (Bond & van Wilgen 1996). Non-sprouters require the vacating of space in order to facilitate population persistence. Non-sprouters tend to be affected more by fire intensity and plant size at the time of the fire than fire frequency. Sprouters survive by initiating repeated regeneration episodes although full reproductive capacity may be temporarily lost. Regenerating after frequent fires ensures that sprouters are closely packed and do not vacate space to competitors, such as non-sprouter seedlings. The cost of sprouting is that resources must be allocated to protective structural tissues such as bark or subterranean carbohydrate stores represented by the lignotuber (Bond & van Wilgen 1996). A recent study conducted on shrubs in Australian heathlands showed that sprouting carries this considerable allocation cost resulting in reduced seedling growth, smaller seed crops and a deferral of reproduction (Pate *et al.* 1990). Deferring the onset of maturity, or some form of reproductive senescence, may be an attempt to maximise the number of reproductive episodes over a longer time scale and under a more favourable fire regime.

Non-sprouters attempt to maximise the probability of being reproductively mature by the time of the next fire disturbance.

Over evolutionary time the frequency of fire in Mkomazi must have been low in comparison with the relatively new phenomenon of anthropogenically induced fire. The fire frequency hypothesis advocated by Bond & van Wilgen (1996) assumes that population dynamics and patterns of survival, growth and reproduction are reset after each fire. However, each fire is a unique event. The effects of an individual fire and the sequence of post-burn events (e.g. resource competition for space) are less predictable than the effects of the fire return interval (Bond & van Wilgen 1996).

With longer fire intervals sprouters undergo thinning which results in increasing the distance between individuals (Keeley 1986). It is at this stage that non-sprouter seedlings could potentially colonise. Sprouters can persist through frequent fires and not vacate space to competitors but at a cost of reduced reproduction (Bond & van Wilgen 1996). Frequent fires, which still have enough seasonally accumulated fuel to be classified as intense, could explain the high plant density of sprouting shrubs recorded in the study plots. Non-sprouters have in some cases a shorter life span than sprouters (Pate *et al.* 1990). During short fire return periods, non-sprouters are out-competed by sprouters and in the course of long fire intervals they can be outlived by sprouters. The optimal fire life history of a non-sprouter appears to be to grow and reproduce at intermediate fire frequencies. Short fire return times may eventually lead to greater variability in individual persistence among sprouters; this can be seen by the relatively high mortalities recorded in some stands but not in others.

Community organisation and fire

Theories of savanna dynamics have in the past largely ignored the role of fire, in favour of resource competition (Bond & van Wilgen 1996, Tilman 1982). The reason why fire has never been given a pre-eminent role may be because fire does not necessarily generate instability in an ecosystem and may only lead to transient effects on species composition. But fire dependent stochasticity can lead to local shifts in composition especially in arid savannas, where the rate of fuel accumulation can be highly variable and vegetation can burn patchily.

Bond & van Wilgen (1996) believe that "fire-induced succession" is a real phenomenon. Community composition in fire-prone systems is altered primarily by changing fire intervals and the effect that this has on the timing of key life history events rather than by elimination of superior competitors by resource competition.

Integration of fire into a general scheme of savanna dynamics has been suggested by Norton Griffiths (1979) and expanded by Dublin (1995). Their theory predicts that competition, fire and herbivory act synergistically. One possible outcome of this is exemplified by the 'Gulliver effect' (Bond & van Wilgen 1996,

Pellew 1983, Dublin 1995). 'Gullivers' are stunted multi-stemmed woody shrubs that dominate communities as adults but struggle as juveniles to emerge from the herb layer. The rapid growth rates of grasses suppress shrub seedlings. Combined with this factor is a slowing of the growth rate of adult shrubs due to competition for resources. The grasses effectively stunt shrub survivors, which are less likely to attain a critical height in order to escape the lethal fire threshold. In addition to this, the rapid fuel accumulation of grass material will ensure the complete burning of these dwarf shrubs. This effect will lead to delays in maturation and reproduction, effectively locking the shrubs into a continuous regeneration cycle and lacking the ability to escape fire and to mature. Large browsing mammals may 'assist' in the stunting process by feeding on stems and keeping the adult shrubs below the fire escapement threshold (Pellew 1983). Dominant shrubs in Mkomazi are seasonally browsed by giraffes (pers. obs.). Release from this stunted form may require either a decrease in the density of browsers or a reduction in fire frequency. An increase in the density of large mammal grazers may also decrease shrub competition with grasses. In this case the composition of woody plants and grasses can remain remarkably stable until a perturbation (e.g. a decrease in giraffe numbers) pushes this state into another stable state (e.g. recruitment of larger trees). Other factors will affect the balance of woody plants to grasses, for example elephant/tree dynamics are particularly well documented (Dublin 1995). Drainage and edaphic factors affect composition more subtly e.g. the development of *Dichrostachys* thickets (Tilman 1982).

Management implications

Fire was once considered a solely destructive force in the early days of African wildlife conservation (Graham 1973). In more recent times, frequent early burns have been advocated as a management tool to encourage large grazer biomass by promoting open savanna and thus facilitating game viewing by tourists (Graham 1973). Botanists may however term the composition and diversity of the vegetation in the central valley as "fire-degraded savanna" (K. Vollensen, pers. comm.).

Patchy prescribed burns may maximise plant biodiversity by using varied fire regimes over time and space to decrease the chance of several species dominating the landscape (UNEP 1995). The extant plant community of the Mbono valley has been subject to a number of perturbations. Currently the plant composition of the savanna would appear to be stable.

The Reserve was gazetted in 1951. It is currently controversial as to whether pastoralist cattle density was higher prior to this date. If it was, then fire almost certainly accompanied grazing and the resultant savanna may have a longer history of frequent burning than is known at present. However, it cannot be disputed that annual burns or a burn every two years are representative of a frequent fire regime.

Palatable grasses such as *Themeda* and *Heteropogon,* the dominant genera at fire locations in Mkomazi (pers. obs.), are known to be intolerant to self-shading, whereby the accumulation of dead litter inhibits growth and reduces the abundance of these species in the herb layer (Everson *et al.* 1988). Frequent fires reduce the quantity of dead organic matter and encourage the dominance of these species. Moderate grazing can have the same effect with the timing and intensity of grazing pressure becoming a critical factor in stimulating grass growth. In these instances "both fire and grazing tend to reduce the chaotic dynamics in native grasslands" (Bond & van Wilgen 1996). Heavy grazing by cattle may lead to a reduction in fire intensity, a consequent reduction in palatable grass species, and perhaps eventually to reversion towards a more closed canopy. A decrease in wild large mammal grazer numbers would then follow. Alternatively exclusive browsing by goats would aid the 'Gulliver effect', acting in concert with fire to keep the area's aspect more open. If in the future a system of pastoralist livestock licence is considered for local people, then a carefully designed management plan which rotates grazing in relation to some knowledge of both the fire regime and the herb layer would be required.

Currently there are no financial resources available to control or manipulate the fire regime and insufficient knowledge of fire behaviour in Mkomazi to warrant any interference with the current fire regime in the valley grasslands.

A more important issue pertains to the conservation of hilltop dry forests, which represent less fire-tolerant communities. Fire can be seen to be creating invasive grassland patches, which enter further in to the remnant forest patches with each subsequent event. A reduction in logging at the edges of remnant patches may prevent further encroachment. Future research into this issue is vital to preserve the integrity of these 'island' forests.

Acknowledgments

This research was funded by the GreenCard Trust and the Royal Geographical Society (with IBG). A debt of gratitude is extended to all Ibaya camp staff who assisted with data collection. Special thanks are extended to TPRI botanical technicians Mr Raphael Abdallah, Mr Emmanuel Mboya and to Dr Keith Eltringham, University of Cambridge, who supervised this research.

References

Bond, W.J. & van Wilgen, B.W. (1996) *Fire and Plants.* Chapman & Hall, London.

Coe M.J. (ed.) (1995) *Mkomazi Research Project Progress Report July 1995.* Royal Geographical Society, London.

Coe, M.J. & Beentje, H. (1991) *A Field Guide to the Acacias of Kenya.* Oxford University Press.

Dublin, H.T. (1995) Vegetation dynamics in the Serengeti-Mara ecosystem: the role of elephants, fire and other factors. In: Sinclair, A.R.E. & Arcese, P. (eds.) *Serengeti II : Dynamics, Management and Conservation of an Ecosystem.* University of Chicago Press. pp. 71-90.

Everson, C.S., Everson, T.M. & Tainton, N.M. (1988) Effects of intensity and height of shading on the tiller initiation of six grass species from the highland sourveld of Natal. *South African Journal of Botany* 54: 315-318.

Gill, A.M. (1981) Adaptive responses of Australian vascular plant species to fires. In: Gill, A.M., Groves R.H. & Noble, I.R. (eds.) *Fire and the Australian Biota.* Australian Academy of Science, Canberra. pp. 243-272.

Graham, A.D. (1973) *Gardeners of Eden.* Allen and Unwin.

Hare, R.C. (1965) Contribution of bark to fire resistance in southern trees. *Journal of Forestry* 63: 248-51.

James, S. (1984) Lignotubers and burls – their structure, function and ecological significance in Mediterranean ecosystems. *Botanical Review* 50: 225-266.

Keeley, J.E. (1986) Resilience of Mediterranean shrub communities to fires. In: Dell, B., Hopkins, A.J.M. & Lamont, B.B. (eds.) *Resilience in Mediterranean-type Ecosystems.* Junk, Dordrecht. pp. 95-112.

Norton Griffiths, M. (1979) The influence of grazing, browsing and fire on the vegetation dynamics of the Serengeti. In: Sinclair, A.R.E. & Norton Griffiths, M. (eds.) *Serengeti: Dynamics of an Ecosystem.* University of Chicago Press. pp. 310-352.

Pate, J.S., Froend, R.H., Bowen, B.J. Hansen, A. & Kuo, J. (1990) Seedling growth and storage characteristics of seeder and resprouter species of Mediterranean-type ecosystems of S.W. Australia. *Annals of Botany* 65: 585-601.

Pellew, R.A. (1983) The impact of elephant, giraffe and fire upon the *Acacia tortilis* woodlands of the Serengeti. *African Journal of Ecology* 21: 41-74.

Tilman, D. (1982) *Resource Competition and Community Structure.* Princeton University Press. Princeton, New Jersey.

Whelan, R.J. (1995) *The Ecology of Fire.* Cambridge University Press, Cambridge.

UNEP (1995) *Global Biodiversity Assessment.* Cambridge University Press.

CHAPTER 10

Invertebrate biodiversity of Mkomazi

Tony Russell-Smith, Graham N. Stone & Simon van Noort

This chapter provides a brief overview of the invertebrate biodiversity research conducted during this programme in Mkomazi. Our purpose is five-fold. First, we introduce the main methods used here to sample invertebrate biodiversity, and outline the ways in which diversity itself is quantified. Second, we ask to what extent the sampling carried out to date provides a full picture of the biodiversity of Mkomazi. Third, we outline the richness of Mkomazi relative to other sites in Africa for those groups in which such a comparison is possible. So few other detailed studies have been made that these comparisons are necessarily limited, and to a large extent the data presented must act as a baseline for comparison with future studies. Fourth, we outline the extent and nature of differences between habitat types in Mkomazi as revealed by studies of particular invertebrate groups. Such differences may be important in assessing conservation priorities in Mkomazi. Lastly, we ask whether any general conclusions can be made about the impact of burning on invertebrate abundance and species richness.

Introduction

Savanna habitats occupy some 40% of the land surface in the tropics (Solbrig *et al.* 1996). Despite this high percentage, very little is known about their invertebrate faunas: mammals, birds and vegetation, rather than invertebrates, have been the usual subjects of studies on tropical savannas (see, for example, Sinclair 1983, Menaut & Cesar 1982, Fry 1983). Much of our knowledge of these small, numerous animals comes from the intensive studies carried out by Lamotte and his colleagues in a humid savanna at Lamto, Côte d'Ivoire (Lamotte 1982, Lamotte & Bourlière 1983), which concentrated on patterns of energy and nutrient flow. Other studies include work in a dry savanna at Nylsvlei, South Africa and, for termites, work in Tsavo East National Park, Kenya (Buxton 1981a & b). It is known that savannas often support characteristic taxa absent from other vegetation types (such as forests) in the same region, and a mosaic of savanna and other habitats may have a higher richness of species than either alone. Many invertebrate groups play a significant role in the functioning of savannas (Lamotte & Bourlière 1983). Among the most abundant of these are the termites and the dung beetles which are fundamental to savanna ecology, as consumers and agents of nutrient transfer. In addition, termites play an important role in nutrient cycling and soil water dynamics (Wood 1976, Elkins, *et al.* 1986), occupying a niche similar to earthworms in more humid savannas. Among the secondary consumers the dominant ecological role is probably played by ants and spiders, although other groups may be important, such as the scorpions and solifuges (sun-spiders) in dry savannas and, as illustrated in this study, the ground beetles and tiger beetles in more moist savannas. The dominant invertebrate herbivores of the grass layer of tropical savannas are grasshoppers (Gillon 1983), while in trees various groups of beetles (grazers) and bugs (sap suckers) are important.

The studies of invertebrates detailed in the following chapters have the following general aims:

- To characterise the species richness of invertebrates present in Mkomazi through collection of baseline data. This is essential if the biodiversity in Mkomazi is to be assessed relative to other regions or protected areas.
- To sample discrete habitat types within the reserve, classified in terms of vegetation and soil type. This is essential in order to establish which habitats are particularly species rich, and also to demonstrate the extent to which particular groups of invertebrates are habitat specific.

  These first two aims have been addressed through detailed surveying of a number of groups in a range of habitats and with an array of sampling methods (described in detail in each chapter). The groups sampled specifically include arachnids (spiders, solpugids, scorpions, pseudoscorpions and harvestmen), beetles, ants and figwasps. Opportunistic collections were also made of cicadas (homopteran bugs), lacewings (Neuroptera) and hangingflies (Mecoptera) and butterflies (Lepidoptera). In particular, we have attempted to assess to what extent the forest remnants on the hills in Mkomazi act as refuges for a characteristic fauna absent from other habitats in the reserve.
- To determine the effects of burning on species richness and diversity in different habitat types. Does burning increase or decrease species richness? Burning (both natural and human-initiated) may have complex (and often counter-intuitive) effects on species richness in a habitat. In Mkomazi, effects of burning were examined by intensive monthly sampling of ground-active beetles and arachnids in burnt and unburnt *Acacia-Commiphora* bushland and foot-slope grassland.
- To identify the ecological links between invertebrates and other organisms (plants and animals) within the reserve. Detailed ecological work on individual species has concentrated on the insects associated with two major tree genera in Mkomazi—*Acacia* and *Commiphora*. How do the diversity and species-richness of the insects in savanna trees compare to data for rainforests and other tropical habitats? To what extent do different *Acacia* and *Commiphora* species support distinctive insect faunas? These issues are addressed by McGavin (Chap-

ter 22) using rapidly acting (and very short-lasting) knock-down pesticide sprays to collect a high percentage of the insects on entire trees. Three papers examine the roles played by particular insect groups in the biology of acacias. Willmer *et al.* (Chapter 21) and Stapley (Chapter 23) examine the links between two highly specialised ant-acacias and the ants which they house in modified spines. Stone *et al.* take a broader view of the pollination biology of Mkomazi acacias (Chapter 20). This overview draws mainly on larger scale comparisons between habitats, and of these detailed ecological studies we will refer only to the *Acacia* and *Commiphora* tree spraying.

Sampling methods and the quantification of biodiversity

To attempt to comprehensively assess the species richness of any group in a given habitat it is necessary to carry out as many different sampling methods as possible. Because of differing behaviour, many species are only susceptible to being collected by one or other method. It is also necessary to implement as intense a sampling programme as possible. The longer the sampling period and the higher the sampling frequency and number of traps the more species will be collected, because more rare species will be represented in the samples (New 1996). However, a compromise must be reached between sampling intensity and the available resources for sorting and identifying the material. For any survey, most of the effort and time goes into the latter—sampling is the easy part.

Interpreting the results obtained in invertebrate surveys requires an understanding of the drawbacks associated with particular sampling techniques and the limitations of given sample sizes. Here we introduce briefly the main sampling methods used in the survey work, with comments on their advantages and disadvantages.

Pitfall trapping

Pitfall trapping is a common sampling technique in which animals fall into small open containers (usually plastic coffee cups) sunk into the soil and partially filled with liquid. The deficiencies of pitfall traps for use in population studies have been widely discussed (Southwood 1973). The numbers of invertebrates trapped in this way are influenced by trap design, climate, and vegetation cover (Curtis 1980, Honek 1988). There is also good evidence that invertebrates differ in their behaviour in relation to traps, resulting in differential 'trapability' (Luff 1975, Topping 1993). The profile of animals caught by pitfalls may thus represent a biased sample of what is actually there. Much discussion has been devoted to what exactly pitfall traps measure, and it is widely accepted that this represents some product of both the population density and the activity levels ('active density') of a given species. Very much less attention has been given to the utility of pitfall traps in invertebrate diversity assessment. It is recognised that some species have physical or behavioural characteristics that render them unlikely to be trapped (e.g. very large species or species that are mostly immobile) and for these, pitfall traps are clearly inappropriate. Set against this, pitfall traps operate continuously, day and night, for as long as the researcher deems appropriate. Most other currently available sampling methods for surface-active invertebrates provide only an instantaneous 'snapshot' of a population at a given moment in time. The other advantage of pitfall traps, particularly in remote areas of the tropics, is their ease of operation and low cost. Despite this, it is recognised that there is an urgent need to calibrate pitfall traps as a method for diversity assessment, preferably in areas where the diversity of the target taxa has been established by some independent method.

Canopy fogging

Work on the invertebrate communities of tree canopies poses some difficulties. You can go up into the canopy using a variety of techniques ranging from climbing and the use of ladders and cranes to net rafts placed on the canopy by means of an airship or balloon. The advantage of being in the canopy is that you can study the fauna and what they do close-up. However these approaches yield results slowly, partly because population densities of many species can be low, especially in rain forest canopies. Direct access also has a relatively high cost and can be dangerous. More simply, fast-acting insecticidal mists, fogs or smokes directed from the ground or generated within the canopy can be used to knock-down material into collecting trays. A history of studies of arthropod diversity in tropical forest canopies is given by Erwin (1995). Stork & Hammond (1997) and Basset *et al.* (1997) discuss fogging and review other canopy collecting techniques respectively. Many factors affect the catch ranging from the height and architectural complexity of the tree being sampled to the type of insecticide being used and application method. Knockdown techniques are quick, cheap and can produce large amounts of material in a short time but it is only possible to infer what the animals collected might be doing using alternative sources of ecological knowledge, based on accumulated ecological knowledge of the taxa to which they belong. Although there has been little uniformity in approach it is generally agreed that samples taken early in the day are more representative of the canopy fauna as a whole. Drop time is also an important factor and most mist blowing or fogging studies have shown that 60–90 minutes is adequate.

No single technique will collect all taxa with equal efficiency. Hand searching in foliage tends to miss small things whereas insecticidal fogging or mist-blowing tends to under-record large things that might be able to fly or leap some distance before the insecticide kills them. Insecticidal knockdown will also fail to sample internal feeders such as leaf miners, stem borers and the occupants of galls and leaf rolls. Sap feeding bugs such as scale insects may die in situ, attached by their mouthparts and it is difficult to estimate how many animals may not drop into the collecting trays by being caught up in bark crevices or flowers. As long as you are aware of these drawbacks, insecticidal knockdown techniques are still the most cost-effective and simple way of sampling tree canopies. Mist blowing rather than fogging was selected for the present study as the majority of the trees are not very high and a blown mist is more easily directed and is not so prone to drift.

Winkler bag leaf litter sampling

A wide range of terrestrial invertebrates are found in leaf litter ranging from the fungus eaters and scavengers through to predators. Most of these species are very small and are not easily collected through conventional collecting methods such as hand collecting and pitfall trapping. Berlese funnels were one of the original methods used for extracting the fauna from leaf litter but they are cumbersome and not easily taken into the field. Winkler bags perform a similar function to Berlese funnels (Besuchet *et al.* 1987, Fisher 1998), but are more appropriate for expedition use in that they are made of material and are light, collapsible and do not need electric lighting to operate. A Winkler bag consists of a cylinder of cloth that tapers at the bottom where a bottle containing alcohol is attached. It is held in a rectangular shape by two horizontal metal rectangular frames and at the top there is loose cloth that can be tied together after the bag has been suspended from a beam and the leaf litter inserted. The leaf litter is first sifted in the field through a coarse (c. 9 mm) sieve and transported back to the base in a bag. The sifted litter is placed in netting bags which are suspended from the upper metal frame within the Winkler bag (3–4 netting bags per Winkler). The disturbance of the litter causes the animals to wander about and fall out of the netting bags on to the interior of the Winkler bag. They then wander about until they fall into the bottle of alcohol at the base. In this study only 1–5 samples were taken from each locality but each sample was large consisting of leaf litter from a number of sites at the locality.

Sweep netting

Sweep netting is a collecting method commonly employed to sample parasitic wasps (Prinsloo 1980, Noyes 1982) and is effective for most groups of Hymenoptera (Noyes 1989). This method also works well for a number of other arthropod groups, such as spiders, grasshoppers, bugs, beetles and flies. The method entails sweeping a net with a robust frame back and forth through vegetation. The sampled insects are then removed from the net, either by hand for larger specimens, or sucked up using an aspirator ('pooter'). The success of sweeping depends on the habitat and type of vegetation that is being sampled. Whether results are representative of the insect fauna associated with a particular vegetation type depends to a large extent on the physical nature of the vegetation. Grassland is most conducive to thorough sampling, whereas thorny bushes are extremely difficult to sample.

Furthermore, the efficiency of any given sweep stroke depends to a large extent on the person carrying out the sampling. Sweep length, speed at which the net is moved and the height that the net is held above the ground are all factors that affect the number of species collected. These factors will vary from person to person, making any direct comparisons between sweep sampling conducted by different people unreliable. Time of day and weather will also affect sweep results. Nevertheless, in a study of five methods of sampling Hymenoptera in tropical rainforest, sweeping was found to be the most efficient single method (Noyes 1989) despite earlier criticism that sweeping underestimates the abundance of small parasitic wasps (Hespenheide 1978).

Malaise trapping

Malaise trapping involves the passive collecting of mostly flying insects using a tent-like structure made of netting, with openings on two or sometimes all four sides. The design is usually based on that of Townes (1972) and consists of a central baffle that guides insects up into a sloped roof-like covering, with a collecting head at the high end that is usually filled with ethanol. The functioning of the trap relies on the principle that most flying insects (except beetles) on hitting an obstruction attempt to fly up towards the light to get around it. The trap is set up so that the collecting head is closest to the sun. Efficiency of Malaise traps as a sampling method can be highly variable, with the siting of the trap, the trap design and mesh size all playing a role in determining the success of the trap (Darling & Packer 1988). A trap moved only a couple of metres can dramatically increase or decrease the number and species of insects collected. As a result efficient sampling of a locality requires that a number of replicate traps are run simultaneously in each habitat type. The drawback of this sampling protocol is that a Malaise trap produces huge quantities of insect material, the sorting of which is very time consuming and labour intensive.

Light trapping

Light trapping is a further method that can be employed effectively to enable comparisons of quantitative species richness data. Light traps usually comprise an ultra-violet light source, such as a blacklight fluorescent tube or mercury vapour bulb, either simply hung in front of a white sheet or mounted in a more complex trap. A trap usually consists of baffles around a centrally mounted light bulb which guide insects down into a funnel, which in turn opens into a collecting container. Light traps are very effective for collecting moths, lacewings and nocturnally flying beetles, bugs, flies and other insects. Open traps need to be manned and specimens of the target insect group hand selected. Efficiency is dependent on weather conditions, with an overcast, moonless and windless night the most productive.

Hand collecting

Hand collecting is an important method for a number of insect groups such as butterflies and cicadas and for any arthropod group supplements quantitative trapping methods. Butterfly and cicada faunas are very poorly sampled in traps or using other quantitative methods such as sweeping. One exception is the butterfly genus *Charaxes* which readily comes to banana baited traps and this method could be used as a quantitative tool to assess species richness of this genus in different habitats. For the rest of the butterflies and the cicadas, hand collecting is the only productive method, and because of their visual or auditory conspicuousness, together with their high levels of diurnal activity, it is easier to assess how much of the fauna has been collected.

An intense collecting programme covering all seasons will produce an inventory for a habitat or locality that can be considered to be representative of the sampled area. This can then be used in assessments of comparative species richness. However, the lack of quantification means that an area must be exhaustively collected before any reliable comparisons may be made with other areas. The variation in collecting intensity and effort between different surveys is a potential shortcoming of this approach.

Quantification

There are many ways of quantifying the biodiversity present in a given location. One way is simply to count the numbers of species present in an area—the 'species richness'. Although useful as a first measure (and sometimes the only measure possible), species richness tells us nothing about the relative abundance of the species sampled. If we were to have a species richness of two, for example, this could represent either two equally common species, or one common one and one very rare one. It is important in describing the biodiversity of a location to bring knowledge of relative abundance into the measure of diversity used, and this is achieved using one of a range of 'diversity indices'. The term 'diversity' in biology has a precise meaning, which differs from species richness. Diversity is a mathematical function of the number of species present *and* of how common each of the species are. In the two-species example given above, diversity is higher when two species are equally abundant than when one is far more common than the other. Intuitively this is because if one species is very rare, then the community is really closer to having only one species than two. There are a number of different measures of diversity, termed indices, which are commonly used, of which Simpson's index and the Shannon-Weaver index are the most common. Further details on the range of diversity indices available and their calculation can be found in ecology texts such as Krebs (1994).

To what extent has the Mkomazi invertebrate fauna been sampled?

It is clear that the estimates of richness for all groups sampled in Mkomazi represent underestimates. One way of assessing the completeness of sampling is to plot numbers of species obtained as a function of sampling effort (number of samples sorted, number of trap nights). Such a species accumulation curve (also called a species sampling curve) typically rises steeply with sampling effort initially, and then levels off at some level of searching effort once the true total is approached. For all the groups sampled, species accumulation curves have yet to level off, showing that many species remain to be discovered (see examples of accumulation curves in Van Noort & Compton, Chapter 18). Furthermore, most of the sampling has been concentrated in the western end of the reserve, and sampling in the east would certainly increase the numbers recorded in all groups.

Species richness of the Mkomazi invertebrate fauna compared to that of other savanna sites

A major difficulty of comparing the data on invertebrate species richness with that from similar sites in Africa is the lack of comparability in sampling methods and sampling effort. Only in the case of ground active spiders does there appear to be an approximately comparable study, that of Griffin from Etosha Pan National Park in Namibia. Here it is clear that Mkomazi is genuinely richer in families, genera and species of ground-active spiders than Etosha. The reasons for this are not clear but may result in part from the much lower rainfall in Etosha National Park. However, the possibility that historical geographical factors also play a role cannot be excluded. Although no other intensive surveys of arachnids have been undertaken in Tanzania, early work in the region of Mt Kilimanjaro indicates that this too has an extremely rich spider fauna and thus that northern Tanzania may prove to be a diversity hotspot for this group.

Although cicadas (Cicadoidea, Hemiptera), lacewings (Neuroptera), hangingflies (Mecoptera) and butterflies (Hesperioidea & Papilionoidea, Lepidoptera) were only collected on an opportunistic basis, comparisons of the resultant species inventories with other east and southern African countries indicate that a rich fauna is protected within Mkomazi Game Reserve. The reserve protects at least 11 of the 32 cicada species known from Tanzania, with a local species richness that is similar to other surveyed African savanna habitats. Mkomazi harbours several endemic cicada species and protects a healthy population of a species last reported in 1914. Several undescribed cicadas were collected in montane forest and hillside bushland. The recorded lacewings are represented by 37 species in 26 genera belonging to seven families, representing a third of the total neuropteran species currently known from Tanzania. Several undescribed species and a number of east African endemic lacewings were recorded from Mkomazi. Five species of hangingfly have been recorded for Mkomazi Game Reserve including four of the nine species previously known to occur in Tanzania. *Bittacus discors* Navás was recorded for the first time from Tanzania. Thus at least 50% of Tanzania's mecopteran fauna is present in Mkomazi and with further collecting it is very likely that the remaining species will also be found to be protected within the reserve. 153 butterfly species of a predicted potential total of 418 species have been recorded for Mkomazi. The predicted total is almost a third of the total number of species recorded for Tanzania and is a reasonably high diversity for a savanna area of this size, a figure that can be attributed to the large geographical and associated habitat and floral diversity within the reserve.

A typical savanna fig tree and fig wasp species richness is protected within Mkomazi, although *Ficus bubu*, a widespread, but rare or overlooked species, is locally abundant in the reserve and has a high associated fig wasp species richness. 88 species of fig wasp have been recorded from the reserve of which around three-quarters are undescribed. This is about half of the fig wasp species expected to be reared from the nine recorded fig tree species in Mkomazi, which represent 23% of Tanzania's fig tree diversity. It is probable that further fig tree species will be recorded from Mkomazi, and because of high host-specificity this will elevate fig wasp species richness beyond comparable sites in southern Africa.

The survey of invertebrates from tree canopies (McGavin, Chapter 22) is the largest so far carried out from a savanna site in Africa and possibly worldwide. While only a fraction of the samples collected have been sorted and identified to morpho-species as yet, it is already clear that the abundance of invertebrates is much higher than those recorded in previous studies. Average abundance of invertebrates from 31 individual *Acacia* trees, representing six species, is more than 10-fold greater than that recorded by West (1986) from 36 trees (half belonging to the genus *Acacia* and half to *Commiphora*) in Kora National Reserve, Kenya. The markedly lower numbers of insects and the somewhat lower species richness in the Kora sample may be partly attributable to the fact that sampling was undertaken during a long period of drought. The cumulative number of insect species recorded in the two surveys is also similar, with 492 from Mkomazi and 496 from Kora. These similarities are perhaps to be expected, given the relative proximity of the two sites and the fact that they form part of the same major vegetation zone (Somalia-Masai *Acacia-Commiphora* deciduous bushland and thicket, *sensu* White 1983) and share a high proportion of tree species.

Factors influencing diversity of invertebrates across habitat types

Different Mkomazi habitats support characteristic invertebrate assemblages. Differences between habitats vary with the group of invertebrates considered, and are more pronounced for some than for others. On a relatively fine scale, species richness of all spiders inhabiting the grass layer was significantly correlated with the proportion of tree and bush cover in each habitat, suggesting that shade is important for these species. This may have resulted from higher availability of prey insects in shaded habitats or, alternatively, could have been associated with differences in the structural characteristics of grasslands in the different habitats. Patterns shown vary between groups, probably as a result of underlying differences in their biology. For example two families of spiders (Gnaphosidae and Salticidae) were more species rich in shaded habitats (woodland and bush) than in open habitats, while crab spiders (Thomisidae) showed the opposite trend, with greatest species richness in open, unshaded grasslands (Russell-Smith, Chapter 11). These differences may explain the observation that a mosaic of different habitat types is often more rich in species than a region of a single habitat type. Superimposed on habitat specific faunas are general patterns in richness. For example, species richness of beetle faunas appeared to respond primarily to moisture availability, being higher in the wetter habitats in the west of the reserve and lower in drier habitats (Davies, Chapter 14).

The presence of montane forest appears to play a critical contributory role to the high insect species richness conserved within Mkomazi. Although these areas of montane forest and associated higher altitude grassland (*Setaria* sp.) comprise under 1% of the reserve, proportionate to their area they contribute the most to spider (Russell-Smith, Chapter 11), cicada (Villet & van Noort, Chapter 12), hangingfly (Londt & van Noort, Chapter 15), lacewing (Mansell & van Noort, Chapter 13), beetle (Davies, Chapter 14), butterfly (van Noort & Stone, Chapter 17), and ant (Robertson, Chapter 19) species richness in Mkomazi. These regions of higher altitude probably also act as a seasonal refuge for many insects, such as species of cicada and hangingflies, during drier periods (Villet & van Noort, Chapter 12, Londt & van Noort, Chapter 15). As such, the geographically restricted montane habitats that are exceptionally prone to degradation require careful management and are a conservation priority.

The effects of burning on invertebrate species diversity

Does burning reduce numbers of species and/or numbers of individuals? The answers to these questions are complex, and depend (a) on the group being studied, and (b) the season in which comparisons between burnt and unburnt habitats are made.

Beetles

For beetles, the answer to the first question appears to be no—there were no statistically significant differences in species richness of beetles between burnt and unburnt habitats. However, there were significant differences in the numbers of individuals caught on burnt and unburnt ground. The relative numbers of individuals caught in burnt and unburnt areas varied with the season. In the hillside *Acacia-Commiphora* bushland more beetles were caught in the unburnt area during the autumn rains in November 1994, while in April 1995 and January 1996 more were captured in the burnt area. This suggests that although burning initially reduces beetle activity, in the longer term it results in significantly higher activity. This could be because beetles are more easy to catch on open burnt ground (and are not necessarily more abundant), or because release of recycled nutrients to the soil through burning actually results in an increase in beetle populations. Seasonal variation in the effects of fire was not consistent in the sampled habitats. In grassland sampling sites at the base of hill slopes, the only major difference between burnt and unburnt habitats was obtained in April 1995. In contrast to the pattern described above for *Acacia-Commiphora* bush at the same time of year, here beetle abundance/activity was significantly higher in the unburnt area. The disparity between the two habitats perhaps relates to the recent history of burning events in the two sites. While there was no recent record of fire in the unburnt hillside, the 'unburnt' grassland had experienced burning 18 months previously. If burning does lead to increases in beetle populations through release of nutrients, then perhaps the 'unburnt' grassland was in fact benefiting from the relatively recent effects of fire. Any impact of fire will depend on frequency and intensity, and it is clear that very frequent fires, with removal of habitat diversity, will reduce beetle numbers and diversity through loss of foodplants and perhaps also increased vulnerability to predation. An optimum environment is likely to exist some months after a burn, particularly if the subsequent wet season brings about a recovery in the vegetation.

Spiders

For spiders, differences in species richness between unburnt and burnt areas of hillside bush varied from month to month. In most months of the year spider species richness was higher in burnt than in unburnt areas—particularly so during the dry season: in October 1995, at the end of the dry season, a maximum difference of 42% more species was obtained in the burnt site. In the wet season months of April, May and December, however, richness was slightly higher in unburnt areas. Species diversity (rather than species richness) was clearly reduced in the burnt habitat in most months and over the whole 12 month period was 26% lower in burnt than unburnt bushland. In the grassland differences between burnt and unburnt areas were far less significant than they were in the bushland sample sites. Unburnt grassland had more spider species than burnt grassland in six out of nine months but the differences in numbers of species between burnt and unburnt were rarely great. Overall, therefore, the effects of burning on invertebrate diversity appeared to be relatively slight, even though the relative abundance of many species varied considerably between burnt and unburnt habitats. A synoptic study such as the present one has obvious limitations in studying the longer term effects of burning on invertebrate diversity. Without information on the frequency and intensity of previous fires in an area, it is impossible to be sure how frequently unburnt 'controls' have themselves been subject to previous burning. Only carefully controlled burning experiments, in which the previous fire history of both burnt and unburnt areas is well documented, could determine long-term changes in diversity resulting from fire. Unfortunately, this was well beyond the resources of the present project.

The implications for conservation management of the present study are limited by the nature of the data and an absence of records of earlier burning events. However, this study does suggest that greatest invertebrate diversity would be maintained by a mosaic of burnt and unburnt patches. What we were unable to determine was the optimal frequency and intensity of burning or the optimal size of burnt areas in different habitats. There is some indication that frequent burning may have a greater impact on biodiversity in bushland habitats, particularly those on steep slopes. However, experience from other African and Australian semi-arid ecosystems suggests that an annual fire is unlikely to have significant long-term effects on either density or diversity of surface-active spiders, which would appear in general to be well adapted to fire as a recurrent natural event in savanna habitats.

Acknowledgements

GNS would like to thank the Apgar Fund of Magdalen College and the Varley-Gradwell Trust of Oxford University Department of Zoology for their funding.

References

Basset, Y., Springate, N.D., Aberlenc, H.P. & Delvare, G. (1997) A review of methods for sampling arthropods in tree canopies. In: Stork, N.E., Adis, J. & Didham, R.K. (eds.) *Canopy Arthropods*. Chapman and Hall. pp. 27-52.

Besuchet, C., Burckhardt, D.H. & Lobl, I. (1987) The 'Winkler/Moczarski' eclector for fungus and litter Coleoptera. *The Coleopterists Bulletin* 41: 393-394.

Buxton, R.D. (1981a) Changes in the composition and activities of termite communities in relation to changing rainfall. *Oecologia* 51: 371-378.

Buxton, R.D. (1981b) Termites and the turnover of dead wood in an arid tropical environment. *Oecologia* 51: 379-384.

Curtis, D.J. (1980) Pitfalls in spider community studies (Arachnida, Araneae). *Journal of Arachnology* 8: 271-280.

Darling, D.C. & Packer, L. (1988) Effectiveness of Malaise traps in collecting Hymenoptera: the influence of trap design, mesh size, and location. *Canadian Entomologist* 120: 787-796.

Elkins, N.Z., Sabol, G.V., Ward, T.J. & Whitford, W.G. (1986) The influence of subterranean termites on the hydrological characteristics of a Chihuahuan desert ecosystem. *Oecologia* 68: 521-528.

Erwin, T.L. (1995) Measuring arthropod biodiversity in the tropical forest canopy. In: Lowman, M.D. & Nadkarni, N.M. (eds.) *Forest Canopies*. Academic Press, London. pp. 109-127.

Fisher, B.L. (1998) Ant diversity patterns along an elevational gradient in the Réserve Spéciale d'Anjanaharibe-Sud and on the Western Masoala Peninsula, Madagascar. *Fieldiana: Zoology* 88, in press.

Fry, C.H. (1983) Birds in savanna ecosystems. In: Bourlière, F. (ed.) *Tropical Savannas*. Ecosystems of the World, Vol. 13. Elsevier, Amsterdam. pp. 337-358.

Gillon, Y. (1983) The invertebrates of the grass layer. In: Bourlière, F. (ed.) *Tropical Savannas*. Ecosystems of the World, Vol. 13. Elsevier, Amsterdam. pp. 289-311.

Hespenheide, H.A. (1978) Are there fewer parasitoids in the tropics? *The American Naturalist* 113: 766-769.

Honek, A. (1988) The effect of crop density and microclimate on pitfall trap catches of Carabidae, Staphylinidae (Coleoptera), and Lycosidae (Araneae) in cereal fields. *Pedobiologia* 32: 233-242.

Krebs, C.J. (1994) *Ecology: the Experimental Analysis of Distribution and Abundance*. 4th edition. Harper Collins.

Lamotte, M. (1982) Consumption and decomposition in tropical grassland ecosystems. In: Huntley, B.J. & Walker, B.H. (eds.) *Ecology of Tropical Savannas* (Ecological Studies, 42). Springer-Verlag, Berlin. pp. 415-430.

Lamotte, M. & Bourlière, F. (1983) Energy flow and nutrient cycling in tropical savannas. In: Bourlière, F. (ed.) *Tropical Savannas*. Ecosystems of the World, Vol. 13. Elsevier, Amsterdam. pp. 583-604.

Luff, M.L. (1975) Some features influencing the efficiency of pitfall traps. *Oecologia* 19: 345-357.

Menaut, J.C. & Cesar, J. (1982) The structure and dynamics of a West African savanna. In: Huntley, B.J. & Walker, B.H. (eds.) *Ecology of Tropical Savannas* (Ecological Studies 42). Springer-Verlag, Berlin. pp. 80-100.

New, T.R. (1996) Taxonomic focus and quality control in insect surveys for biodiversity conservation. *Australian Journal of Entomology* 35: 97-106.

Noyes, J.S. (1982) Collecting and preserving chalcid wasps (Hymenoptera: Chalcidoidea). *Journal of Natural History* 16: 315-334.

Noyes, J.S. (1989) A study of five methods of sampling Hymenoptera (Insecta) in a tropical rain forest, with special reference to the Parasitica. *Journal of Natural History* 23: 285-298.

Prinsloo, G.L. (1980) An illustrated guide to the families of African Chalcidoidea (Insecta, Hymenoptera). *Department of Agriculture and Fisheries Science Bulletin, Republic of South Africa* 395: 1-66.

Sinclair, A.R.E. (1983) The adaptations of African ungulates and their effects on community function. In: Bourlière, F. (ed.) *Tropical Savannas*. Ecosystems of the World, Vol. 13. Elsevier, Amsterdam. pp. 401-426.

Solbrig, O., Medina, E. & Silva, J. (eds.) (1996) *Biodiversity and savanna ecosystem processes: a global perspective*. Ecological Studies 121. Springer, Berlin.

Stork, N.E. & Hammond, P.M. (1997) Sampling arthropods from tree-crowns by fogging with knockdown insecticides: lessons from studies of oak beetle assemblages in Richmond Park (UK). In: Stork, N.E., Adis, J. & Didham, R.K. (eds.) *Canopy Arthropods*. Chapman and Hall. pp. 3-26.

Topping, C.J. (1993) Behavioural responses of linyphiid spiders (Araneae, Linyphiidae) towards pitfall traps. *Entomologia Experimentalis et Applicata* 68: 287-293.

Townes, H. (1972) A light-weight Malaise trap. *Entomological News* 83: 239-247.

White, F. (1983) *The vegetation of Africa.* UNESCO, Paris. pp. 356. 4 maps.

Wood, T.G. (1976) The role of termites (Isoptera) in decomposition processes. In: Anderson, J.M. & Macfadyen, A. (eds.) *The Role of Terrestrial and Aquatic Organisms in Decomposition Processes*. Blackwell, Oxford. pp. 145-168.

1 & 2. Wet and dry season views from Ibaya, 1997. Maji Kununua (1,620 m) lies on the centre horizon. A similar view is published in Harris (1971)—see Chapter 1, this volume.

3. A heavy shower waters the Mbono valley. Hafino Hill lies in the middle distance.

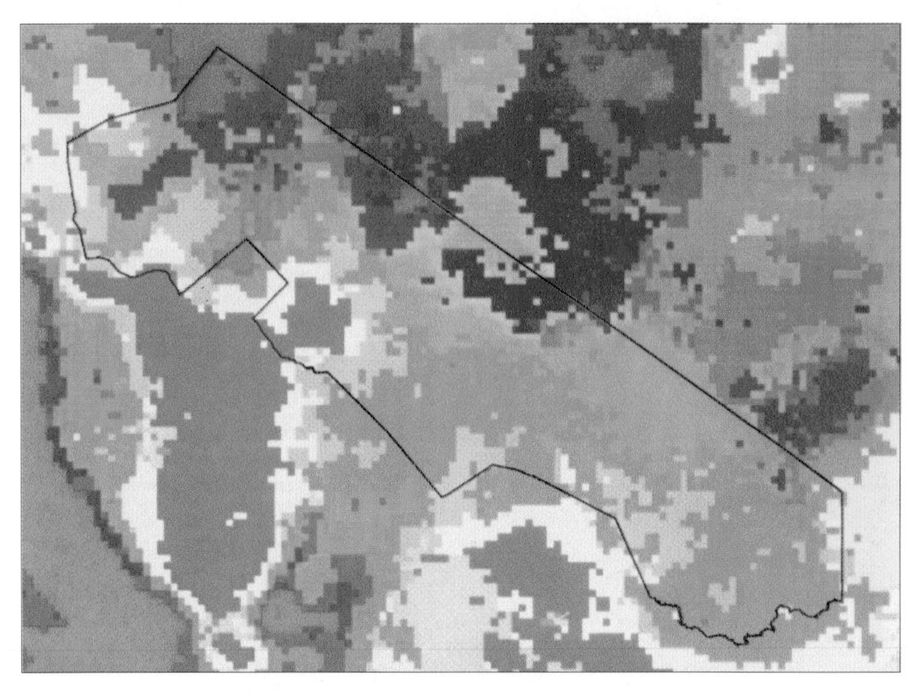

ABOVE **4.** Satellite image of Mkomazi, using a 22-class unsupervised classification of monthly AVHRR data for September 1992 to August 1993, with 1 km resolution (Chapter 4). **5.** Large mammal distribution modelling in Mkomazi: predicted probabilities of zebra presence (Chapter 4). BELOW **5(a).** Wet season model, based on environmental factors (vegetation classes and Fourier-processed NDVI).

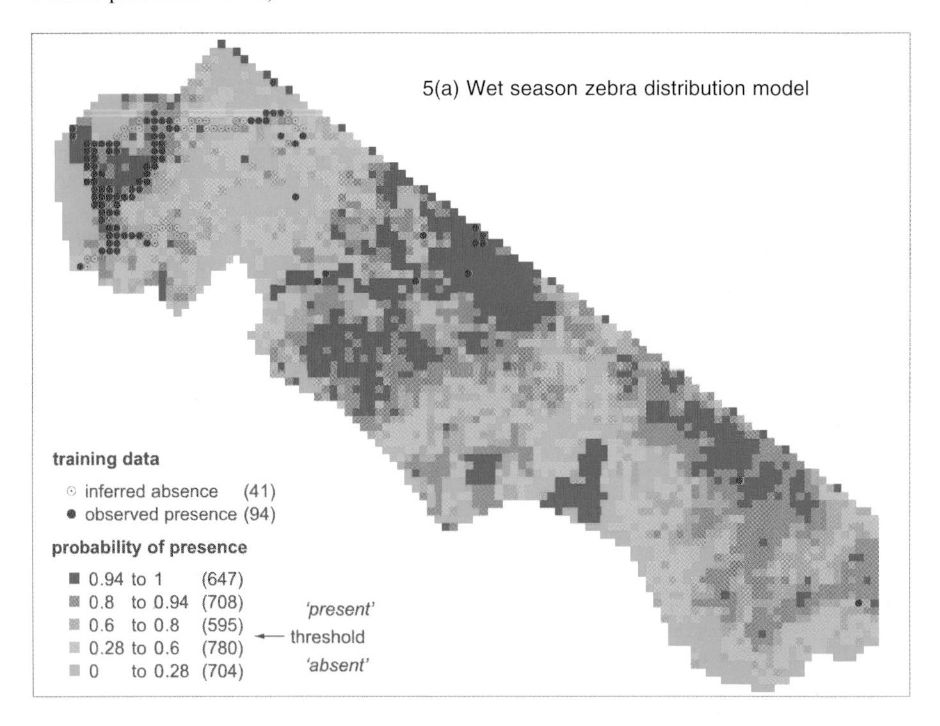

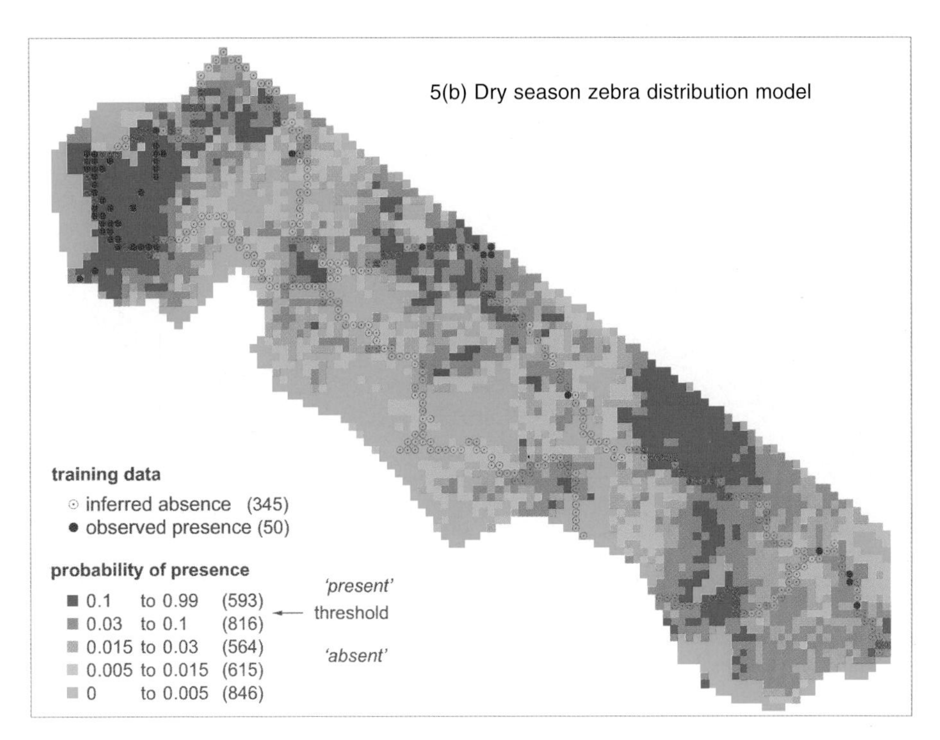

ABOVE **5(b).** Dry season model, which also includes distance to water as a factor. BELOW **5(c).** Modified dry season zebra distribution model. Observations made around ephemeral water holes in central Mkomazi early in the dry season have been removed from the training data, to give a more typical dry season distribution pattern.

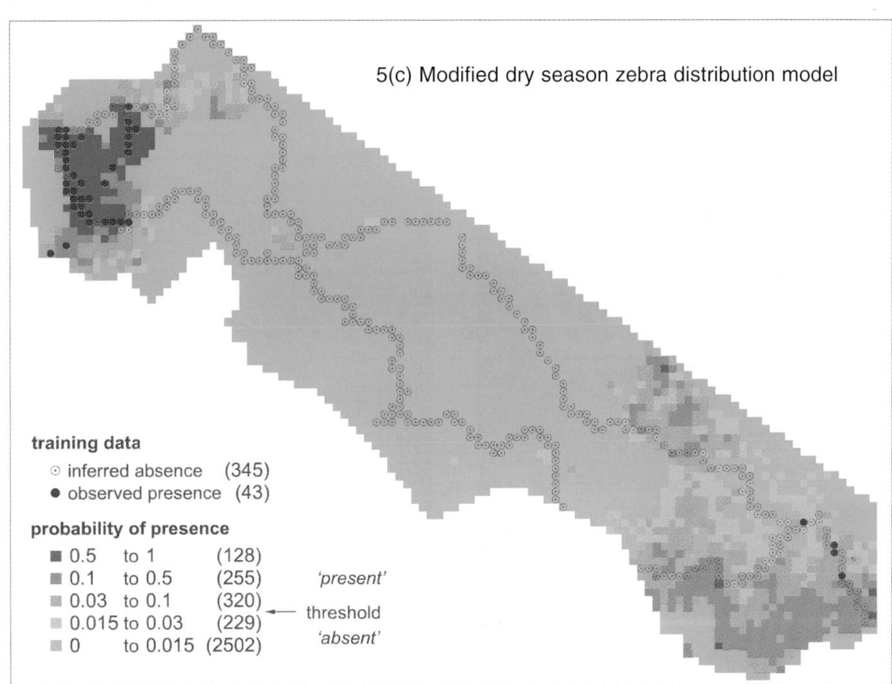

6. Looking south from Ndea Hill with rock outcrop in the foreground, seasonally indundated grassland (*mbuga*) in the valley bottom and the South Pare mountains beyond.

7. Dense *Acacia-Commiphora* woodland (*Lc*—see Chapter 7) in full leaf at the start of the rains. Note the red iron-stained soil.

8. Surveying Nyati Plot in seasonally inundated grassland (*Sg*), with the mist-shrouded Maji Kununua on the horizon.

9. Rock outcrop with scattered boulders (*Ro*) on Vitewini ridge, crowned by an isolated *Erythrina abyssinica.*

10. A fig tree climbs on a cliff face above Ngurunga dam. Even before the dam, this was one of the few permanent water sources in Mkomazi.

11. Stony hillslope woodland (*Ns*) below Mandi Hill with baobab 'parkland'.

12. *Commiphora campestris* trees in lowland *Acacia-Commiphora* woodland (*Lc*) close to Ndea. The intermediate vegetation has been fairly extensively fire-cleared.

13. Mature *Acacia tortilis* in the lower hillslope protected woodland (*Lpw*) at Pangaro.

14. Open *Acacia-Commiphora* woodland (*Lc*) below Kamakota, with rocky pavement (*Rp*) at the base of the footslope.

15. Ibaya-Igire Hill summit forest (*Mf*) above Ibaya Camp. Note the almost closed canopy evergreen forest on the summit and the fire-cleared ridge vegetation (*Fd*) dominated by flowering *Dombeya rotundifolia*.

16. Lower-Middle hillslope woodland (*Lpw-Lsw*) in a sheltered valley at the foor of Kisima Hill.

17. Kinondo Hill, north of Ibaya Camp, showing the effect of successive bush fires which have almost cleared the important rain-retaining woodland vegetation from its slopes.

18. George McGavin in full 'mist blow' at Pangaro (Chapter 22).

19. A sample of insects collected from tree canopies using an insecticide mist blower.

20. The tarantulas or baboon spiders are large and relatively frequent predators in Mkomazi. This is a female golden-brown theraphosid (*Pterinochilus sp.*), length 46 mm.

21. A flap-necked chameleon *(Chamaeleo dilepis)* spends the night resting on a euphorbia stem.

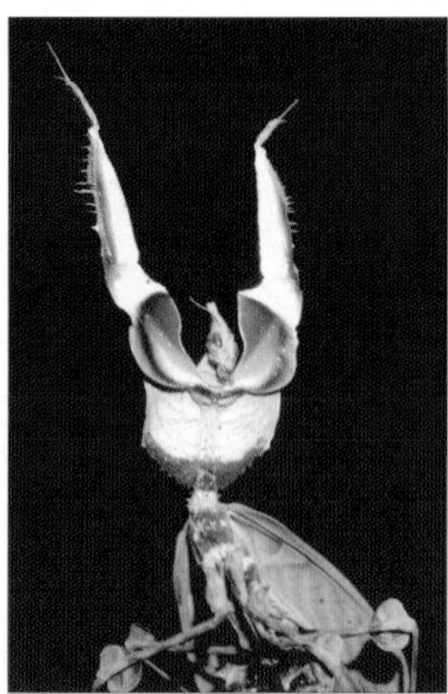

22. A myrmeleontid neuropteran hanging from grass stems. **23.** This 'science fiction' insect is the large mantid *Psuedocreobotra wahlbergi* (Hymenopodidae) whose brilliant lilac nymph is almost brighter than the adult. **24.** *Crematogaster* ants at the base of an *Acacia drepanolobium* pseudo-gall seeking food from the nectaries.

25. Just before the rains start, the fringes of the seasonally inundated grasslands are brightened by the brilliant pink blooms of *Ammocharis tinneana* (Amaryllidaceae).

26. The bright purple flowers and yellow throats of *Thunbergia holstii* (Acanthaceae), characteristic of the slopes of Vitewini and similar rocky hills.

27. The procumbent *Aristolochia bracteata* (Aristolochiaceae) is a prostate creeper in seasonally inundated grassland (*Sg*) and is rare throughout its range.

28. The sessile flowers of *Barleria spinisepala* (Acanthaceae) are thrust above the hard stony ground of the Ngurunga valley during the rains.

29. The globose inflorescences of *Acacia zanzibarica* (Mimosoideae).

30. The brilliant flowers of *Pentas parvifolia* (Rubiaceae) bring a blaze of colour to Vitewini ridge during the rains.

31. The seperate Kilimanjaro peaks of Kibo and Mawenzi rise above the clouds, viewed from Pangaro valley in Mkomazi.

32. A herd of buffalo leave the shrinking water supply at Dindira Dam at the end of the dry season. Observation Hill lies at the far end.

33. Overlooking Ndea from Vitewini ridge. The grassland (Sg) in the valley bottom has turned bright green at the beginning of the rains.

34. A giraffe (*Giraffa camelopardalis*) feeding on a low bush as effectively as it can stretch to a 5 m tree top. **35.** The male and female gerenuk (*Litocranius walleri*) in dense thicket. Almost exclusively tree foliage browsers, they are becoming increasingly rare in their restricted range between southern Somalia and northern Tanzania.

36. The diminutive Steinbuck (*Raphicerus campestris neumanni*) is particularly common in the scattered tree grasslands. Its numbers may well have increased due to increased burning of the denser woody habitats. **37.** Black-faced sandgrouse (*Pterocles decoratus*) on recently burnt grassland in the Mbono Valley.

38. The white-browed coucal (*Centropus superciliosus*), a common inhabitant of roadside vegetation.

39. The Eurasian roller (*Coracias garrulus*), a seasonal migrant sometimes arriving in large parties on its way south.

40. Ibaya Camp, base of the Mkomazi Ecological Research Programme, was renovated and extended with the support of Friends of Conservation to provide facilities for up to 12 visiting scientists.

41. Hamish Robertson and Omari Mohamed sort ants on the slope north of Ibaya. **42.** The December 1995 Mkomazi field team. Malcolm Coe (front left) is trying to keep order. The group includes Graham Stone (top centre), Alex Flemming (middle left) and botanists Emmanuel Mboya (middle left) and Raphael Abdallah (middle right).

CHAPTER 11

The spiders (Arachnida) of Mkomazi

Tony Russell-Smith

Introduction

Arachnids are an important but generally poorly studied group of arthropods which play a significant role in the regulation of insect and other invertebrate populations in most ecosystems. Although considerable effort has been invested in recording spider diversity in temperate habitats, only recently have studies on species diversity in tropical ecosystems been undertaken. While most studies have been on spider diversity in tropical forests (reviewed by Russell-Smith & Stork 1994), much less is known about the composition of the arachnid communities of savanna ecosystems. In Africa, most previous work on the inventory of savanna arachnids has either been undertaken for purposes other than biodiversity assessment (e.g. Blandin *et al.* 1981, Russell-Smith 1981) or has been conducted over a limited time span or in a limited area (Russell-Smith *et al.* 1981, Van der Merwe *et al.* 1996). Most previous studies have also used a very restricted range of sampling methods which arc likely to have provided a biased sample of the arachnid fauna as a whole.

The aims of the studies described here were two-fold. The first was to provide a baseline inventory of the diversity of arachnids other than mites in representative habitats throughout the reserve so that alterations in diversity as a result of future changes in land management can be monitored. The second was to study the short and medium term effects of uncontrolled burning on arachnid diversity in two specific vegetation types, hillside *Acacia-Commiphora* bushland and mixed grassland on hillside foot-slopes.

Methods

The importance of using as wide a range of sampling techniques as possible in invertebrate diversity surveys has been stressed by Hammond (1990) who convincingly shows that only a combination of different techniques adequately samples the extraordinary beetle diversity of tropical rain forest canopies. Although savannas are possibly less complex structurally than rain forests, the diversity of life-styles

Table 11.1 Estimated sampling effort and total number of habitats sampled using different techniques in Mkomazi Game Reserve, Tanzania, 1993–97.

sampling method	sampling effort	no. of habitats sampled
pitfall traps	12,370 trap/days	12
sweep net samples	30 samples of 10 x 20 sweeps	13
tree canopy fogging	183 trees, 29 species	n/a
hand collection	31 samples of 1–2 hours	16

and micro-habitats of savanna invertebrates is such that similar considerations apply. For the purpose of biodiversity survey work, it is convenient to sample the fauna of different layers or strata of each habitat separately. Conventionaly, four strata are recognised: the soil layer, the ground layer (species active on the soil surface), the field layer (species active in the grass or herb layer) and the shrub/tree layer. In the baseline survey we routinely used four different methods of sampling for arachnids; pitfall traps (ground layer), sweep-net samples (field layer), canopy fogging of trees and hand collecting (all strata). In addition, spiders were identified from Winkler bag samples (soil layer), Malaise traps (ground and field layers) and hand-sorted litter samples (ground layer). Table 11.1 provides an estimate of the sampling effort and number of habitats in the reserve sampled by each of the four principal techniques.

Although this represents a major sampling effort, probably the largest so far attempted for any single area in Africa, it is biased in two ways. Significantly more effort was spent on sampling with pitfall traps and canopy fogging than on any other technique and the sampling was heavily concentrated in the western and central parts of the reserve. Further sampling with other techniques, particularly in the dry eastern part of the reserve, would undoubtedly extend the species list considerably.

The composition of the Mkomazi arachnid fauna

Sorting and identification of the samples from Mkomazi is an on-going process. While the majority of material from pitfall traps, sweep net samples and hand collecting have now been sorted to morphotypes, sorting material from tree canopies has only just started. Identification and, in some families, description of the spiders has been initiated but will take several years as it is dependent on the availability of taxonomists in several different countries. Despite this, we now have a sufficiently large sample of the spider fauna to provide a reasonable assessment of the diversity and composition of the arachnid fauna as a whole. Table 11.2 shows the composition of the spider fauna as a whole. Other arachnid orders were poorly represented in Mkomazi. There were seven species of scorpion, an esti-

Table 11.2 Total numbers of families, genera and species of spiders recorded from Mkomazi Game Reserve. The order of families follows Platnick (1991).

family	genera	species	% of total	family	genera	species	% of total
Cyrtaucheniidae	1	6		Linyphiidae	9	17	3.3%
Idiopidae	1	8		Tetragnathidae	5	6	
Dipluridae	1	1		Araneidae	20	38	7.5%
Atypidae	1	1		Lycosidae	11	23	4.5%
Migidae	1	1		Pisauridae	9	17	3.3%
Barychelidae	2	2		Agelenidae	1	4	
Theraphosidae	3	4		Hahniidae	1	3	
Scytodidae	1	4		Dictynidae	2	3	
Sicariidae	1	1		Amaurobiidae	2	3	
Caponiidae	1	3		Oxyopidae	5	23	3.3%
Tetrablemmidae	2	2		Miturgidae	1	5	
Pholcidae	2	3		Clubionidae	4	5	
Telemidae	1	2		Liocranidae	4	6	
Ochyroceratidae	2	2		Corinnidae	10	15	3.0%
Segestriidae	1	2		Zodariidae	12	23	4.0%
Oonopidae	5	13	2.6%	Cithaeronidae	1	1	
Palpimanidae	5	9		Trochanteridae	1	1	
Mimetidae	2	4		Prodidomidae	2	4	
Eresidae	2	3		Gnaphosidae	22	51	12.0%
Oecobiidae	1	1		Ctenidae	2	7	
Hersiliidae	1	1		Selenopidae	2	2	
Uloboridae	2	4		Heteropodidae	2	8	
Cyatholipidae	1	1		Philodromidae	3	17	3.0%
Theridiidae	13	40	7.9%	Thomisidae	19	35	8.0%
Theridiosomatidae	1	1		Salticidae	35	70	13.0%
Anapidae	1	1					
Mysmenidae	1	1		totals	241	508	

mated ten species of solpugids (wind scorpions), five species of opilionid (harvestmen), one species of amblypygid (whip scorpions), an estimated six species of chelonethi (false scorpions) and one species of schizomid. Because these accounted for only 5% of all arachnids and are currently being studied by specialists, they are not considered further here. Their low diversity may be due partly to the fact that sampling techniques were not well adapted to these groups and also because most of them have maximal diversity in much drier habitats than those sampled in Mkomazi.

The most immediately evident feature of the spider fauna of the reserve is the very high family diversity. The 52 spider families recorded from Mkomazi represent half of all currently recognized spider families (total 105). The maximum number of families recorded from any other site in Africa is 33 from Ngome State Forest (Natal) by Van der Merwe and colleagues (1996). The estimate for the number of genera (Table 11.2) will undoubtedly be amended once poorly studied families have been revised and this estimate is considered a compromise between excessive 'splitting' and 'lumping'.

The species composition of the fauna is shown graphically in Figure 11.1. Among the 508 morphotypes recognised so far, the two most species rich families are the jumping spiders (Salticidae) with 70 species (13% of all species) and the ground spiders (Gnaphosidae) with 51 species (12%). The jumping spiders are the largest of all spider families and in Mkomazi were found in all vegetation strata, although most only occurred on the ground surface. The ground spiders, as the name suggests, are almost confined to the ground layer and are characteristic of semi-arid areas. Both jumping spiders and ground spiders actively hunt their prey but the former are largely diurnal and the latter largely nocturnal. Spiders belonging to the orb-web weavers (Araneidae) and comb-footed spiders (Theridiidae), represented by 38 and 40 species respectively, were abundant in the field, shrub and tree strata where they build a variety of different types of prey catching webs. The crab spiders (Thomisidae), with 35 species, are particularly abundant in the field layer but also occur on trees and, very occasionally, the ground surface. They

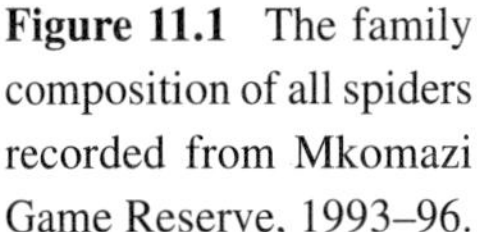
Figure 11.1 The family composition of all spiders recorded from Mkomazi Game Reserve, 1993–96.

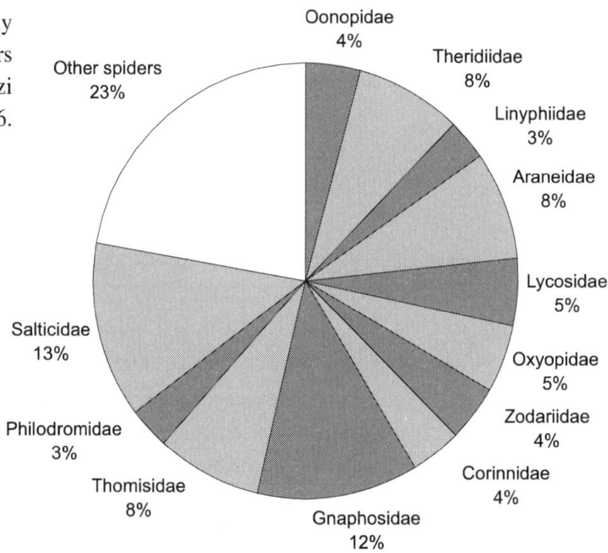

are diurnal species which have been described as 'lie in wait' predators, overpowering their prey with grasping front legs and a powerful venom. Wolf spiders (Lycosidae) and lynx spiders (Oxyopidae), both represented by 23 species, are long-legged active hunters, the former almost entirely on the ground and the latter mainly in the field layer. Wolf spiders are diverse in most areas studied but Mkomazi has an exceptionally high proportion of oxyopids in the fauna. Zodariidae (23 species) and Corinnidae (15 species) have no commonly used English names and are characteristic of semi-arid habitats in Africa where they are ground active nocturnal hunters, the former possibly specializing in ants and termites as prey. They are abundant in both grassland and bushland in Mkomazi, which has the largest recorded fauna of any African site. The majority of Corinnidae (formerly included in the sac spiders—Clubionidae) are apparent mimics of ants or, occasionally, velvet ants (Mutillidae: Hymenoptera).

The two other families with more than 10 species were the Oonopidae, minute hunting spiders (length < 2.5 mm) which capture collembola and other small insects in the litter layer, and money spiders (Linyphiidae) which spin small hammock webs in litter and grass.

Although no other family contributed more than 5% of all species recorded, many are of considerable interest. Mkomazi has a particularly rich fauna of mygalomorph spiders (the first seven families in Table 11.2). Nearly all of these relatively primitive spiders inhabit burrows from which they hunt at night. The most species rich families, Idiopidae and Cyrtaucheniidae both build retreats with trapdoors and males were common in pitfall traps at the outset of the rainy season in November 1994 and 1995. The Theraphosidae (so called 'bird-eating' spiders) included the largest species recorded from the reserve, a member of the genus *Pterinochilus* which measured 46 mm in length (Plate 20). The silk lined mouths of the burrows of this and other theraphosids are a conspicuous feature of the reserve in the early wet season. Many of the more interesting species are small and belong to families which are relatively poorly known in Africa. They include the species from the families Tetrablemmidae, Ochyroceratidae, Leptonetidae, Cyatholipidae, Theridiosomatidae and Anapidae. The last of these is represented by a single female of what is probably the smallest species from the reserve, total length just 0.8 mm!

Many of the earlier inventories of spider diversity in Africa suffered from inadequate or inappropriate sampling procedures. For example, the work of Blandin and colleagues in the humid savanna at Lamto (Côte d'Ivoire), one of the most extensive studies carried out in the region, relied primarily on collection of spiders from large cages of 4, 10 and 25 m^2 surface area (Blandin 1971). While this is an effective method for large species and has the advantage of providing estimates of absolute population densities, many small species will have been completely overlooked. This technique is also not well adapted for assessment of diversity because the results provide a 'snapshot' of the spider populations at a particular season and time of day. In particular, nocturnally active species are likely to be inadequately censused.

Recently however, a major survey of ground-active arachnids in Etosha National Park, Namibia has been carried out (Griffin, pers. comm.). The park is considerably larger than Mkomazi Game Reserve (220,000 km^2 compared to 3,250 km^2) and is also more arid, with rainfall ranging from 450 mm per annum in the west to 250 mm in the east. Pitfall trapping was undertaken in 12 different habitats using eight traps over 150 days in each site, giving a total trapping effort of 14,400 trap/days. This is about 15% greater than the total trapping effort to date in Mkomazi but for the purposes of estimating total spider species richness is approximately equivalent. Mkomazi is richer in both spider species (240 ground-active species versus 175) and spider families (52 versus 31) than Etosha National Park. This may be attributable to the lower rainfall in Etosha which is also not as well distributed as in Mkomazi (unimodal as opposed to bimodal rainfall distribution). However, historical and geographical factors may also play a role in maintaining the high diversity in Mkomazi. The proximity to the East Pare and Usambara Mountain massifs may contribute in this respect, as the Eastern Arc Mountains are known to be centres of diversity for some groups of spiders (Scharff 1992). The family composition of the spider fauna of the two areas also differs. In the more arid Etosha National Park, gnaphosids account for almost a quarter of all spider species whereas salticids only represent 3% of the species, as opposed to 19% in Mkomazi. As with overall diversity, it is not possible at this stage to say to what extent these differences are to be attributed to climatic differences between the two areas or to historical and biogeographical factors.

Geographical distribution of Mkomazi spiders

Relatively few spider families in Africa are sufficiently well known that the distribution of their species can be mapped with any reliability. However, for two of the larger families from Mkomazi, Salticidae and Araneidae, some estimate has been made of the distribution of species on the basis of data provided by Dr W. Weslowska (Salticidae) and Dr M. Grasshoff (Araneidae) (Table 11.3). In both families a relatively high proportion (43–63%) of all species have a poorly known distribution as they have been either recorded only from the type locality or from very few sites. Of the remainder, the araneids have roughly twice the proportion of species with a very wide distribution (40%) than do the salticids (20%). Amongst araneids, 17% appear to be restricted to east and central Africa and amongst salticids a similar proportion are restricted to east and southern Africa. What proportion of the poorly known species have relatively restricted distributions remains unclear. In both families a fairly high proportion (31–44%) appears to be currently undescribed species. If the average proportion of undescribed species for these two families (38%) were applied to all families from Mkomazi, the fauna would include 168 new species. However, given the high level of synonymy known to exist in African spiders, a more conservative proportion of 25% is probably more realistic, giving a total of 110 undescribed species. This is still a very high proportion compared to that for most northern European countries, where perhaps 5–10% of spider species remain to be described.

Spider diversity in contrasting habitats of Mkomazi

Ground layer spiders

Pitfall traps were operated in 10 habitats in the reserve, ranging from closed montane forest to open mixed grassland. The family composition of the spider communities varied with both habitat and season. The closed montane forest on Ibaya Hill had the most distinctive composition (Figure 11.2) in that Salticidae were completely absent from the ground layer and their place taken by Linyphiidae. The average proportion of Salticidae in woodland, bushland and grassland habitats was relatively constant. The proportion of both Zodariidae and Gnaphosidae increased twofold and threefold respectively between grassland and woodland and these families were clearly most abundant in shaded habitats. By contrast, there were double the number of wolf spider species (Lycosidae) in grassland habitats than in the closed montane forest (Figure 11.2). These active hunters are normally more speciose in open habitats than in wooded ones.

There were very clear seasonal effects on the composition of the spider communities in different habitats. For example, in two adjacent bushland communities near Dindira Dam, *Dichrostachys cinerea* and *Combretum* scrub, lycosids (principally a species of *Pardosa* which breeds during the rains) were abundant in both habitats in the wet month of April 1995. In the much drier period of January 1996, lycosids were less abundant and zodariids and gnaphosids, characteristic of dry environments, were dominant in both habitats.

Table 11.3 Geographical ranges of species of Salticidae and Araneidae from Mkomazi Game Reserve.

family	cosmo-politan	old world tropics	tropical Africa	east & central Africa	east & southern Africa	unknown	new species	total species
Araneidae	1	2	11	6	0	15	11 ?	35
	3%	6%	31%	17%	0	43%	31%	
Salticidae	0	0	11	0	10	35	25	56
	0	0	20%	0	18%	63%	45%	

Species richness was greatest in closed *Dichrostachys cinerea* scrub near Dindira Dam (38 species) (Figure 11.3) and was also high in well developed but open *Acacia senegal* woodland near Ndea (32 species) and in montane grassland on Ibaya Hill (33 species). Relatively low species richness was found in unburnt foot-slope grassland at Ibaya (22 species), seasonally waterlogged (vlei) grassland with scattered *Acacia drepanolobium* near Ndea (24 species) and in riverine scrub on the Umba River (24 species). As can be seen in Figures 11.3a and b, species diversity as measured by α was closely related to species richness and, across all ten habitats, there was a significant correlation between the two measures of diversity (least squares regression $r = 0.7508$, $p < 0.02$).

The family composition of ground-active spider communities in the ten different habitats studied in Mkomazi confirms results from other African semi-arid savanna ecosystems. Russell-Smith (1981) found that Gnaphosids and Zodariids were more abundant in Mopane woodland than in floodplain grassland in the Okavango Delta of Botswana. *Acacia-Commiphora* bushland sampled in the dry season in Kora National Reserve, Kenya, had a family composition similar to that in unburnt *Acacia-Commiphora* bushland at Ibaya sampled in August (Russell-Smith *et al.* 1987). In both cases Salticidae, Gnaphosidae and Zodariidae were the dominant families. However, this is the first published data that covers a large

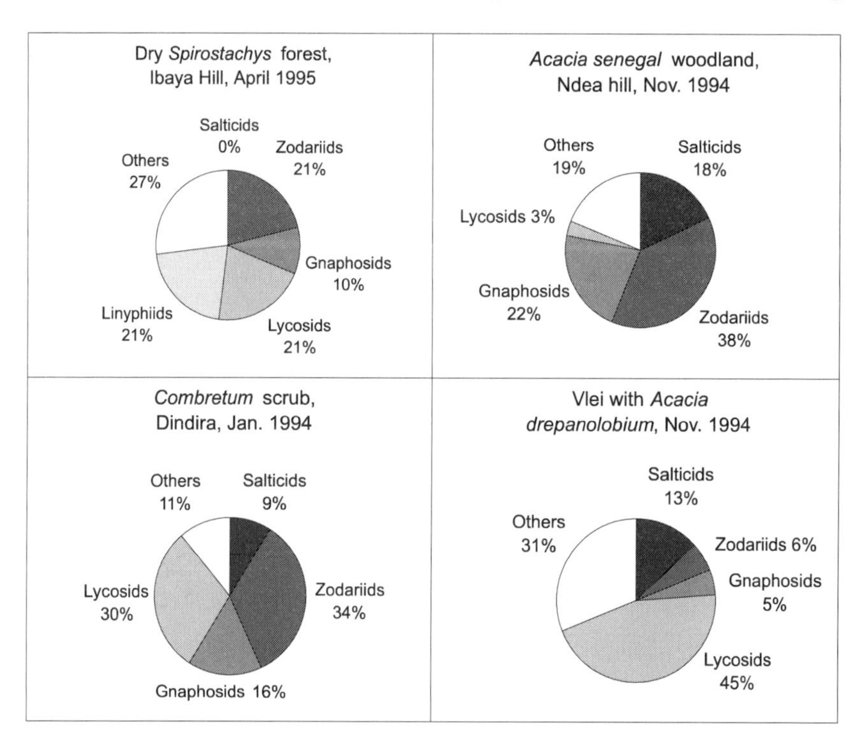

Figure 11.2 Composition of spider communities in montane forest, *Acacia* woodland, *Combretum* bushland and seasonally wet grassland, Mkomazi Game Reserve.

range of different savanna habitats and demonstrates a systematic variation in family composition between woodland, bushland and grassland habitats.

The factors controlling species richness and species diversity of spiders in the ground layer of the ten habitats studied are not at all clear. When averaged for each habitat type (forest, woodland, bushland or grassland) the range of values for these indices was quite small, 26 to 31 for species richness and 8.0 to 11.5 for species diversity, and they showed no particular trend across the different habitat types. Although the intensity of trapping was relatively high (180 trap-days in each site) it is possible that was insufficient to discriminate other than very large differences in species richness or diversity. This view is possibly supported by comparison of the diversity indices from these ten habitats with those obtained from monthly sampling of *Acacia-Commiphora* bushland and foot-slope grassland at Ibaya. All the values for species richness and diversity (α of the log series) for the ten separate habitats lay within the same range as those for the ten monthly samples from bushland and grassland at Ibaya. At Ibaya the maximum number of species recorded in any given month represented about 46% of all species ob-

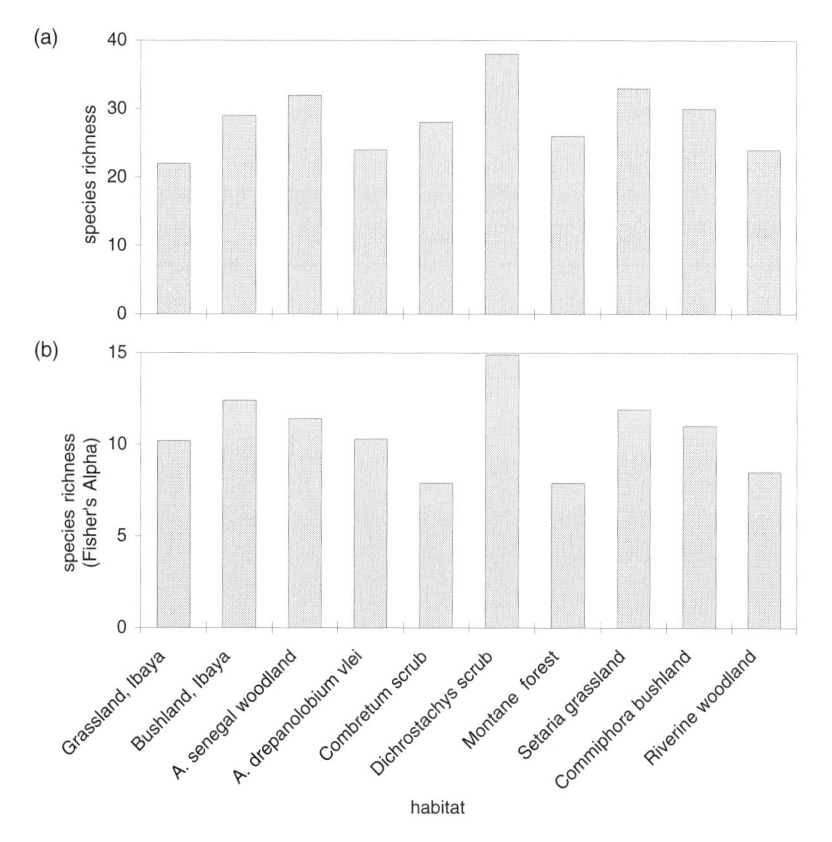

Figure 11.3 (a) Species richness and (b) species diversity of spiders in ten habitats. Mkomazi Game Reserve, November 1994 and April 1995.

tained over the whole ten month period. This suggests that the sampling in the ten individual habitats probably covered too short a time period (normally six days) to recover a sufficiently large proportion of total species richness to allow adequate representation of differences between habitats. However, from a conservation viewpoint, it is clear that woodland and some bushland habitats have particularly diverse spider communities and management should be aimed at conserving such components of the overall habitat mosaic of the reserve. Although the species richness of spiders from the ground layer of montane forest was not particularly high, many of the species were confined to this habitat in Mkomazi and protection of this type both from fire and illegal cutting, should be a conservation priority.

Field layer spiders

The dominant spider families in sweep net samples from 11 grassland sites in January 1996 were Thomisidae (mean 38%) and Araneidae (mean 22%) (Table 11.4). Other well represented families included Philodromidae (12%), Salticidae (12%) and Oxyopidae (5%). Proportions of all families were very variable between different grassland sites and there were no obvious correlations between the proportion of a particular family present and environmental variables except in the case of crab spiders (Thomisidae). Representatives of this family were always

Table 11.4 Composition of spider communities sampled by sweep netting in eleven grassland sites in Mkomazi Game Reserve, January 1996.

grassland type	location	Thomisids	Philodromids	Salticids	Araneids	Oxyopids	others
Footslope (unburnt)	Ibaya	21%	7%	16%	34%	1%	20%
Footslope (burnt)	Ibaya	55%	0%	10%	30%	0%	5%
A. senegal woodland	Ndea	28%	10%	10%	14%	7%	31%
Acacia/Commiphora	Ibaya/Simba	44%	3%	21%	21%	3%	9%
Vlei grassland	Ndea etc.	44%	9%	8%	22%	10%	7%
Themeda grassland	Ndea road	49%	21%	2%	21%	6%	1%
P. mezianum	Nyati plot	43%	22%	13%	17%	0%	4%
Grassland	Ubani track	46%	21%	27%	2%	0%	4%
Cynodon grassland	Kisima plot	48%	4%	26%	9%	9%	4%
Thick dead grass	Kikolo plot	11%	22%	16%	32%	11%	10%
Sparse dead grass	Cadaba plot	29%	11%	7%	36%	6%	89%
average		38%	12%	14%	22%	5%	17%

most abundant in open, unshaded grasslands and least abundant in shaded grassland in woodland or bush (Figure 11.4). The proportion of crab spiders in each habitat type showed a significant negative correlation with estimated percentage tree or bush cover ($r = -0.6642$, $p < 0.05$).

Species richness of grass layer spiders was highest in *Acacia senegal* woodland near Ndea hill and lowest in open grassland in the Nyati and Simba experimental plots (details of these plots are given in Chapter 7). Species richness was positively and significantly correlated with estimated percentage tree or bush cover for each site ($r = 0.7417$, $p < 0.01$).

Species diversity (α of the log series) was also greatest in *Acacia senegal* woodland but was lowest in *Cynodon* grassland below Kisima experimental plot and in *Themeda triandra* dominated grassland near Ndea Hill (Table 11.5). Species diversity was even more closely correlated with estimated tree or bush cover for each site than species richness ($r = 0.8711$, $p < 0.001$). It is not clear whether the effect of tree cover on species richness and diversity of grass layer spiders was a direct one due to the shading effect or was a result of other factors correlated with proportion of trees in the habitat.

There are virtually no published data on either the composition or the diversity of spider communities in the field layer of savanna habitats from Africa. Earlier

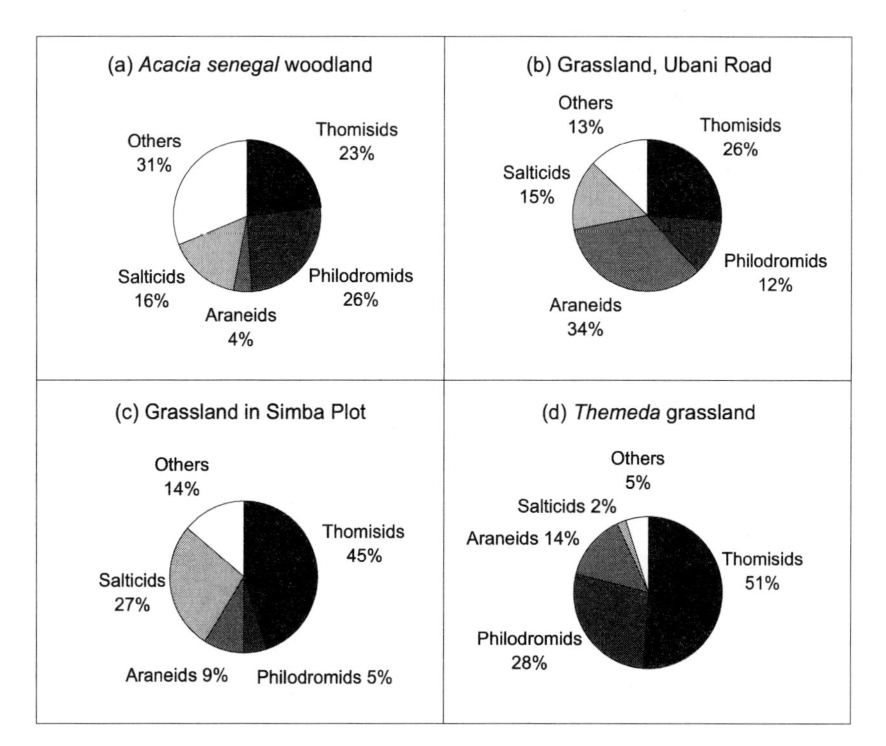

Figure 11.4 Family composition of spider communities in the field layer of shaded (A,B) and unshaded (C,D) grasslands, Mkomazi Game Reserve, January 1996.

Table 11.5 Diversity indices for grassland spiders swept from 11 sites, Mkomazi Game Reserve, January 1996.

	site[1]										
index	1	2	3	4	5	6	7	8	9	10	11
species number	35	10	9	21	10	16	9	9	16	26	13
Simpson's dominance	0.063	0.150	0.222	0.080	0.136	0.083	0.217	0.188	0.172	0.070	0.124
alpha of the log series	17.40	7.96	7.16	12.90	8.54	4.82	5.44	5.97	4.77	12.70	6.33
estimated tree cover (%)	60	0	40	40	10	0	0	10	0	50	0

[1] Key to sites:
1 *Acacia senegal* woodland
2 Burnt grassland, Ibaya
3 Cadaba plot sample 1
4 Cadaba plot sample 2
5 Kikolo plot
6 *Cynodon* below Kisima plot
7 Nyati plot
8 Simba plot
9 *Themeda* grassland
10 Ubani road
11 Vlei near Simba plot

work from the humid savannas of Lamto (Côte d'Ivoire) used large cages to sample spiders, the results from which do not discriminate between ground and field layer components of the fauna (Blandin 1971 & 1972, Blandin & Celerier 1981). The present studies have limitations due to the use of sweep nets which only sample a part of the grass layer (the proportion depending on the stage of growth of the grasses) and which present only a 'snapshot' of the community at one particular point in time. Despite this, the studies have demonstrated very clear differences between communities in different types of grassland, with Thomisids dominating open grasslands and other families more important in shaded grasslands within woodland or tall bushlands. The fact that both species richness and species diversity increase with the degree of tree cover within the grassland underscores the importance of wooded and bushed grasslands for the conservation of arachnid diversity.

The effects of burning on activity and diversity of spiders in bushland and grassland

Pitfall trapping at Ibaya was undertaken over 12 months from April 1995 until March 1996. Catches from pitfall traps are a product both of the numbers of spiders on the ground surface at any given time and of the level of activity of the spiders. Since the density of vegetation effects activity of spiders (and other soil surface arthropods) comparisons between different vegetation types, such as those made here between burnt and unburnt areas, do not necessarily reflect differences in spider numbers alone. The term 'activity' is therefore used for results from pitfall traps to distinguish them from more direct measures of abundance.

Activity of spiders in burnt and unburnt bushland and grassland

Figure 11.5 shows the activity of spiders in each month in these two habitats. In the hillside bushland active numbers in the burnt habitat were nearly double those in the unburnt in all months except December. By contrast, in the foot-slope grassland, although activity was somewhat higher in the burnt habitat in five out of nine months, differences in activity between the two habitats were very much smaller. In both the bushland and grassland, activity was greatest in the wetter months of April, May and November.

Figure 11.6 shows seasonal activity of the four most species rich spider families in these habitats. As with the spider community as a whole, most families were greatly more active in burnt than unburnt hillside bushland, while levels of activity in the burnt and unburnt grassland were more similar. The exception were the Lycosidae (wolf spiders) which were also more active in the burnt grassland than the unburnt grassland. In general, numbers of spiders trapped in the burnt hillside were approximately double those in the burnt grassland but the differences between numbers trapped in unburnt bushland and grassland were less consistent.

Although activity of all spiders in the bushland was higher in the burnt than in the unburnt site, individual species showed a range of different activity patterns (Figure 11.7). *Stenaelurillus* sp. n. A (Salticidae), was clearly much more active in the burnt than in the unburnt bushland. By contrast, the zodariid *Mallinella* sp. A was more active in unburnt than burnt bushland, although the differences were relatively small other than in May. *Stenaelurillus* sp. n. A was more active in burnt

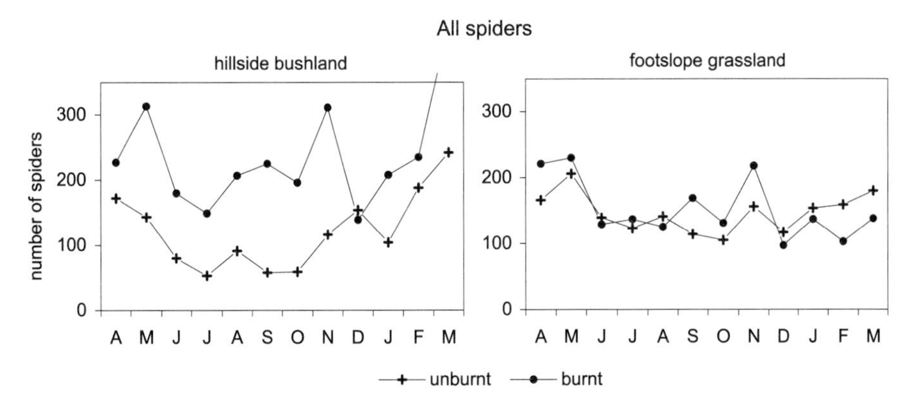

Figure 11.5 Spiders from pitfall traps in unburnt and burnt hillside bushland and unburnt and burnt grassland. Ibaya Camp, Mkomazi Reserve, 1995–96.

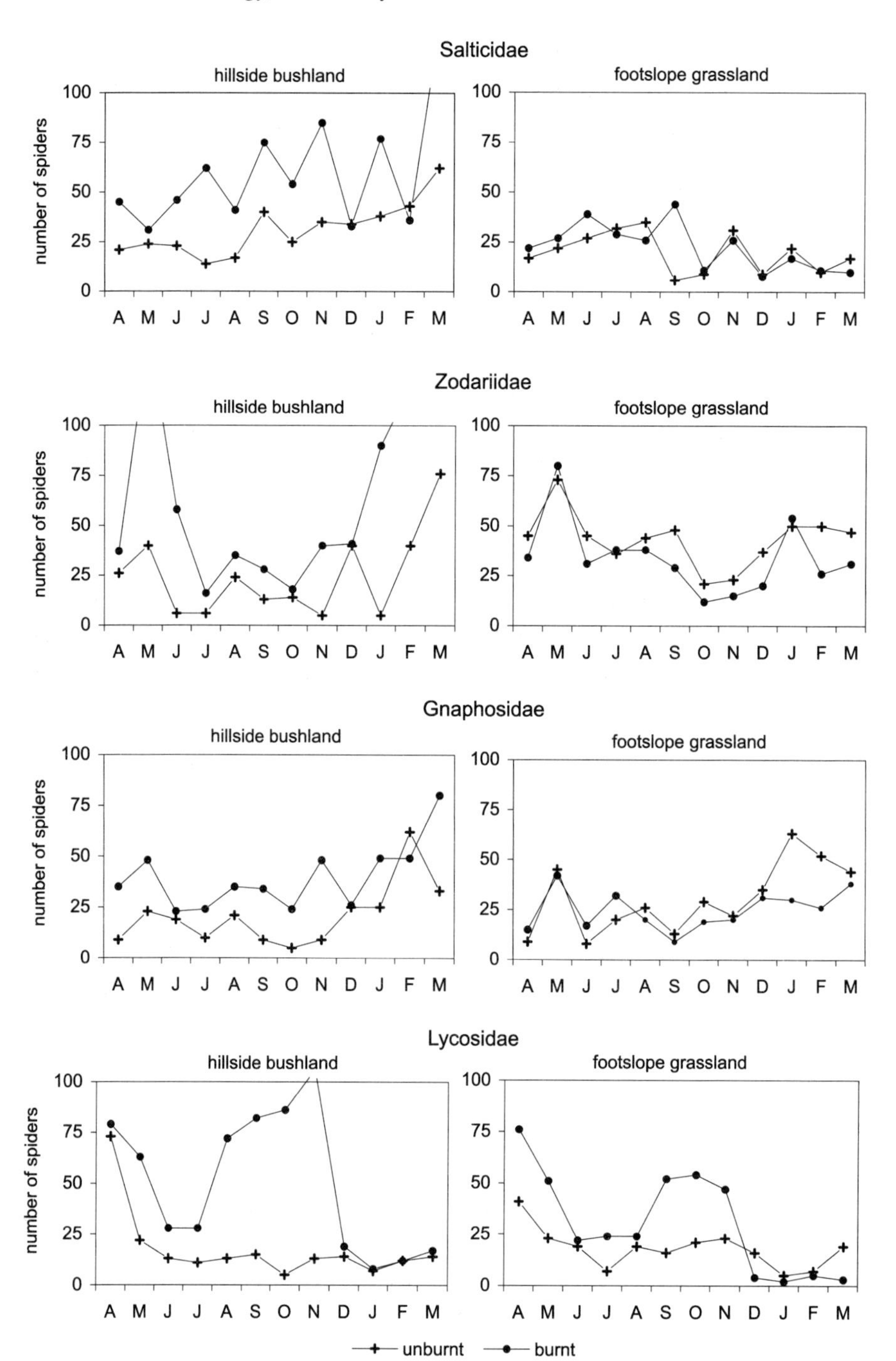

Figure 11.6 Abundance of spider families in pitfall traps in unburnt and burnt hillside bushland (left column) and footslope grassland (right column). Ibaya Camp, Mkomazi Game Reserve, 1995–96.

than unburnt grassland until September, after which activity was greater in the unburnt site until December. As in the bushland, activity of *Mallinella* sp. A was greater in unburnt than burnt areas but in this case substantially so through much of the sampling period.

The very much greater numbers of many species trapped in the burnt habitats were not expected. Burnt ground is hotter, drier and subject to greater extremes in temperature than corresponding better vegetated sites and, in principle, spiders in such areas would also be more susceptible to predation. To some extent, high temperatures and visual predators can be avoided by nocturnally active species. While most of the gnaphosid and perhaps zodariid species active in burnt areas are night hunters, salticid and most lycosid species are diurnally active.

Interpretation of spider activity data from pitfall trapping is notoriously difficult. Factors that influence trapping of spiders include species, sex, season and vegetation type (Southwood 1966). For example, peaks in activity of hunting species are often the result of males seeking out females in the breeding season. This is likely to have accounted for the peak of activity of the small zodariid, *Akyttara akagera* in May in the burnt hillside, where the sex ratio was four males to each female. Vegetation density can also influence numbers of a particular species trapped. For active ground hunters, a much larger proportion of the population is

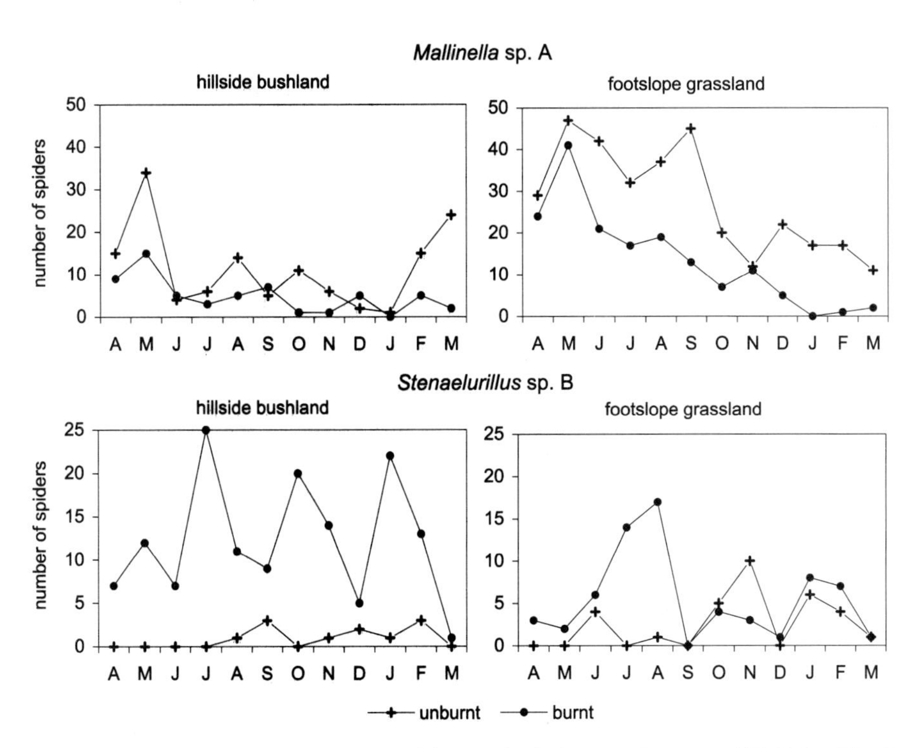

Figure 11.7 Activity of two common spider species in burnt and unburnt hillside bushland and grassland, Ibaya Camp, Mkomazi Reserve, 1995–96.

vulnerable to trapping in open ground than in densely vegetated areas. Although this may have influenced results in this study, it cannot be the only factor involved since species such as *Mallinella* sp. A were trapped in much greater numbers in unburnt than in burnt habitats. In addition, the fact that many species were trapped in large numbers in burnt areas suggests that bare ground plays an important role in their biology. Certainly species of Gnaphosidae and Zodariidae are particularly abundant in arid areas and it is possible that burnt areas provide the physical conditions they require in an otherwise more mesic environment in which other species are better able to compete.

Species richness and diversity of spiders in unburnt and burnt habitats

Species richness of spiders was slightly higher in unburnt than in burnt bushland in April, May and December but in the intervening months the burnt bushland site always had more species than the unburnt bushland (Figure 11.8). Species richness was greater in the wet months of April, May, November and December than in the intervening dry months in both unburnt and burnt bushland. During the dry months there were often significantly more species in burnt hillside, with a maximal difference of 42% more species in the burnt site in October 1995. In the

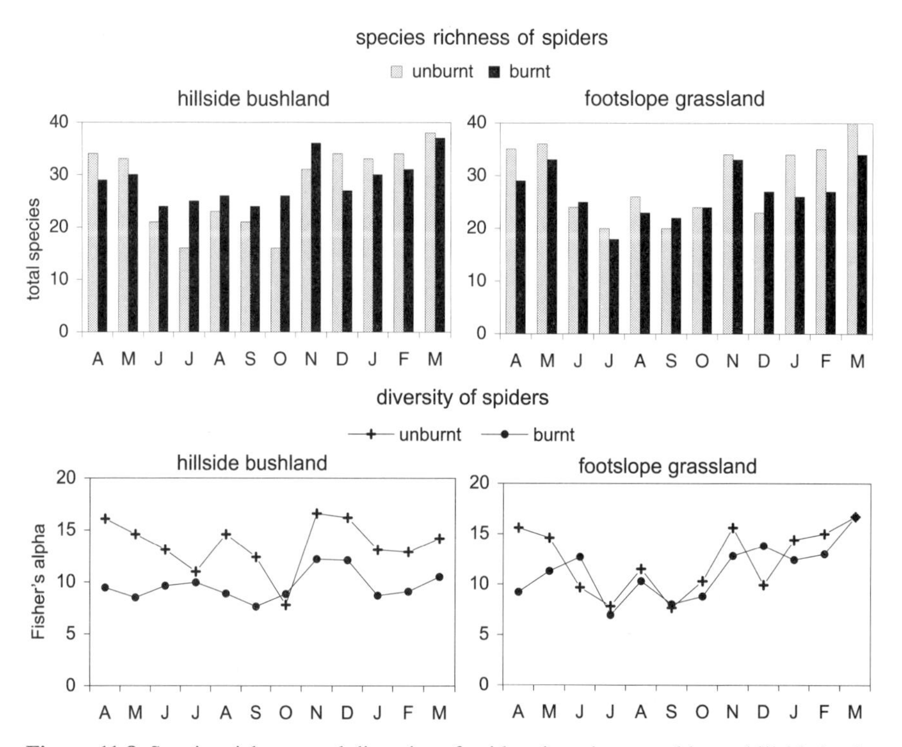

Figure 11.8 Species richness and diversity of spiders in unburnt and burnt hillside bushland and unburnt and burnt grassland. Ibaya Camp, Mkomazi Game Reserve, 1995–96.

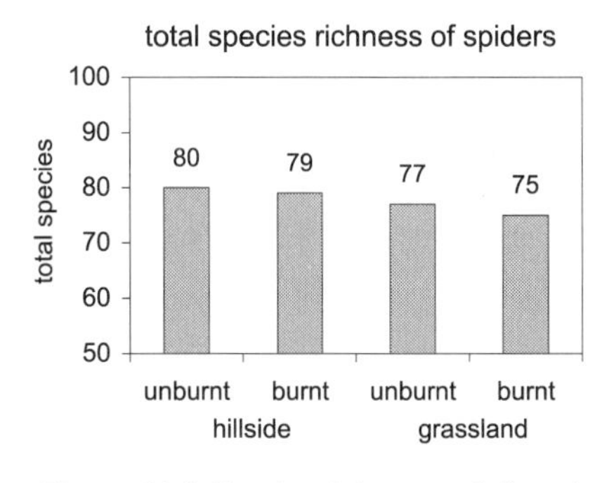

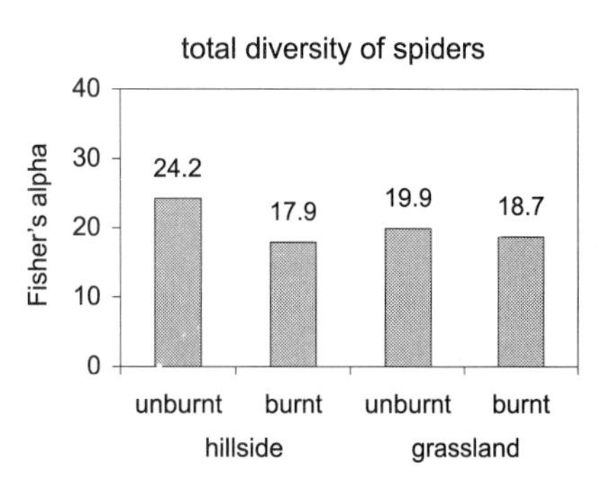

Figure 11.9 Species richness and diversity of spiders in unburnt and burnt hillside bushland and grassland over the whole period April to December 1995.

grassland by contrast, the unburnt site had more spider species in six out of nine months but the differences in numbers of species between burnt and unburnt were rarely very great, with a maximal difference of 18% more species in the unburnt grassland than burnt grassland in April 1995 (Figure 11.8).

Species diversity of spiders in bushland, as measured by α of the log series, was always considerably higher in the burnt area than in the unburnt except in July and October when it was virtually identical in burnt and unburnt bushland (Figure 11.8). The greatest difference was in the wet month of April when the index in the burnt area was double that in the unburnt area. Species diversity in the grassland was much higher in the unburnt site in April and May but was closely similar to that in the burnt site during the remainder of the study period. As in the bushland, diversity was highest during the wet months of the year.

Figure 11.9 shows species richness and diversity in the burnt and unburnt bushland and grassland sites over the 12 month period as a whole. Total species richness was hardly affected by burning, with just one or two species less in the burnt habitats. However, the index of species diversity was 26% lower in the burnt hillside bushland than in the unburnt. The index of diversity for the burnt grassland was only slightly lower than that for unburnt grassland.

Both species richness and diversity of spiders were clearly much more influenced by burning in the hillside bushland than in the grassland below. Although monthly species richness was higher in burnt bushland than unburnt in six out of 12 months, principally during the dry season, there was little difference between the two habitats over the whole period of study. Species diversity, however, was clearly reduced in the burnt habitat in most months and over the whole 12 month period was 26% lower in burnt than unburnt bushland. In the grassland differences in species richness or diversity between the burnt and unburnt sites were quite small and unlikely to be particularly significant.

At present, the reasons for the differences in the effect of burning between grassland and hillside bushland are a matter for speculation. The burnt hillside was quite badly eroded with a thin layer of soil between exposed stones. It is possible that the productivity of this habitat was much more severely affected by burning than the grassland below and that this was reflected in the diversity of predators in this environment. However, given the apparently elevated activity of certain species in the burnt bushland, it is more likely that habitat structure or physical constraints permitted only a limited number of species to compete in this relatively harsh environment. In the grassland, although burning left the soil surface exposed, rapid regrowth of a dense sward was evident following rains and this may have permitted re-invasion of a diverse fauna from adjacent unburnt grassland.

A synoptic study such as the present one has obvious limitations in studying the longer term effects of burning on invertebrate diversity. Without information on the frequency and intensity of previous fires in a given area, it is not possible to be sure that unburnt 'controls' have not themselves been subject to some level of burning. Only carefully controlled burning experiments, in which the previous fire history of both burnt and unburnt areas is well documented, could determine long term changes in diversity resulting from fire. Unfortunately, this was well beyond the resources of the present study. However, this study does suggest that greatest spider diversity would be maintained by a mosaic of burnt and unburnt patches. What we were unable to determine was the optimal frequency and intensity of burning or the optimal size of burnt areas in different habitats.

Acknowledgements

I would like to thank Tim Morgan and all the staff at Ibaya Camp for logistic support. Ramadani Makusi, Raphael Abdallah, Elias Kihumo, Daniel Mafunde, Paul Marenga and Ian Maxwell assisted with operation of the pitfall traps and initial sorting of spiders from the catch. Mark Ritchie, Hamish Robertson and Simon van Noort provided spider material from their trapping programmes and George McGavin carried out trapping in 1997 in areas which would otherwise have not been sampled. I acknowledge with thanks the taxonomic specialists who helped with identifications. They include Ansie Dipenaar (Thomisidae), Manfred Grasshoff (Araneidea), Norman Platnick (Gnaphosidae, Palpimanidae), Robert Raven (mygalomorphs) and Wanda Weselowska (Salticidae). This work was supported by the UK Darwin Initiative to which I make grateful acknowledgement.

References

Blandin, P. (1971) Recherches écologiques dans la savane de Lamto (Côte d'Ivoire): observations préliminaires sur le peuplement aranéologique. *La Terre et la Vie* 25: 218-229.

Blandin, P. (1972) Recherches écologiques sur les araignées de la savane de Lamto (Côte d'Ivoire): premières données sur les cycles des Thomisides de la strate herbacée. *Annales de l'Université d'Abidjan, ser. E* 5: 291-364.

Blandin, P. & Celerier, M.-L. (1981) Les araignées des savanes de Lamto (Côte d'Ivoire): Organisation des peuplements, bilans énergétiques, place dans l'écosystème. Fasc. 2. *Publications du Laboratoire de Zoologie. Ecole Normale Superieure* 21: 504-588.

Gillon, Y. (1983) The invertebrates of the grass layer. In: Bourlière, F. (ed.) *Tropical Savannas* (Ecosystems of the World: Vol. 13). Elsevier, Amsterdam, pp. 289-311.

Hammond, P.M. (1990) Insect abundance and diversity in the Dumoga-Bone National Park, N. Sulawesi, with special reference to the beetle fauna of lowland forest in the Toraut region. In: Knight, W.J. & Holloway, J.D. (eds.) *Insects and the Rain Forests of South East Asia (Wallacea).* Royal Entomological Society, London, pp. 197-254.

Russell-Smith, A. (1981) Seasonal activity and diversity of ground living spiders in two African savanna habitats. *Bulletin of the British Arachnological Society* 5: 145-154.

Russell-Smith, A., Ritchie, J.M. & Collins, N.M. (1987) The surface-active spider fauna of arid bushland in Kora Reserve, Kenya. *Bulletin of the British Arachnological. Society* 7: 171-174.

Russell-Smith, A. & Stork, N.E. (1994) Abundance and diversity of spiders from the canopy of tropical rainforests with particular reference to Sulawesi, Indonesia. *Journal of Tropical Ecology* 10: 545-558.

Scharff, N. (1992) The linyphiid fauna of eastern Africa (Araneae: Linyphiidae) – distribution patterns, diversity and endemism. *Biological Journal of the Linnean Society* 45: 117-154.

Southwood, T.E. (1966) *Ecological Methods.* Methuen, London.

Van der Merwe, M., Dippenaar-Schoeman, A.S. & Scholtz, C.H. (1996) Diversity of ground-living spiders at Ngome State Forest. Kwazulu/Natal: a comparative survey in indigenous forest and pine plantations. *African Journal of Ecology.* 34: 342-350.

Provisional checklist: Spiders of Mkomazi

Each name is followed by the collection method and habitat/location code. The codes are listed at the end of the checklist.

Class ARACHNIDA

Order ARANEAE

Family Atypidae
Calommata simoni. Pitfall B.

Family Cyrtauchenidae
Ancylotrypa sp. A. Pitfall A.
Ancylotrypa sp. B. Pitfall A,B.
Ancylotrypa sp. C. Hand A. Pitfall B.
Ancylotrypa sp. C. Hand A,B,H,J. Pitfall A,B,J.
Ancylotrypa sp. E. Pitfall B.
Ancylotrypa sp. F. Pitfall N.

Family Idiopidae
Ctenolophus sp. (imm.) Hand A.
Idiops sp. A. Pitfall B,N,O.
Idiops sp. B. Pitfall D.
Idiops sp. C. Pitfall N.
Idiops sp. D. Pitfall O.
Idiops sp. E. Pitfall O.
Idiops sp. F. Pitfall E.
Idiops sp. G. Pitfall B.

Family Migidae
Poecilomigas basileupi Benoit. Hand E.

Family Dipluridae
Thelechoris karschii. Hand K.
Thelechoris striatipes (Simon). Hand K.

Family Barychelidae
Cyphonisia sp. Pitfall A,B,R,S. Litter E. Hand M.
Pisenor sp. Pitfall A,B,E,M,R,S. Hand H.

Family Theraphosidae
Eucratoscelus sp. (imm.) Pitfall C.
Heterothele sp. ? Pitfall B,J.
Pterinochilus sp. Hand A,C,F,H. Pitfall A,Q.
Stromatopelma sp. ? (imm.) Pitfall B.

Family Sicariidae
Loxosceles sp. (imm.) Hand Q.

Family Scytodidae
Scytodes sp. A. Hand A. Pitfall B.
Scytodes sp. B. Litter E.
Scytodes sp. C. Litter E.
Scytodes sp. D. Hand V.

Family Caponiidae
Caponia cf. *natalensis* Simon. Hand K. Pitfall A,B,N,R.
Caponia sp. indet. B. Pitfall E.
Caponia sp. indet. C. Pitfall B.

Family Tetrablemmidae
Afroblemma cf. *thorelli* (Brignoli). Winkler Z (base of Kisima Hill).
Hexablemma cataphractum Berland. Hand E. Winkler E,Z.

Family Pholcidae
Smeringopus sp. A. Hand G.
Smeringopus sp. B. Hand X.
Pholcidae indet. sp. C. Litter E.

Family Telemidae
Telemidae indet. A. Pitfall E.
Telemidae indet. B. Pitfall A.

Family Ochyroceratidae
Ochyroceratidae indet. A. Pitfall A.
Ochyroceratidea indet. B. Pitfall O.

Family Dysderidae
Ariadna sp. A. Hand A,G.
Ariadna sp. B. Pitfall E. Litter E.

Family Oonopidae
Dysderina sp. A. Litter E.
Dysderina sp. B.
Gamasomorpha nr. *simoni* Berland. Pitfall A,N,O.
Gamasomorpha pusilla Berland. Litter E.
Gamasomorpha sp. A. Pitfall B.
Gamasomorpha sp. B. Pitfall B.
Gamasomorpha sp. C. Pitfall E.
Gamasomorpha ? sp. D.
Kijabe nr. *paradoxa* Berland.
Kijabe sp. A. Pitfall A,B.
Kijabe sp. ? B. Pitfall B.
Triaeris sp. Pitfall A,E. Litter E.
Oonops sp. Pitfall A.

Family Palpimanidae
Boagrius nr. *incisus* Tullgren. Hand A,B. Pitfall B.
Boagrius sp. B. Pitfall M.
Boagrius sp. C. Litter E,FF.
Diaphorocellus sp. Hand I (near pool).
Palpimanus sp. Hand K.
Scelidocteus sp. A. Hand B. Pitfall A.
Scelidocteus sp. B. Hand C.
Scelidocteus sp. C. Hand M.
Palpimanidae indet. Pitfall JJ.

Family Mimetidae
Mimetus sp. A. Pitfall D.
Mimetus sp. B. Hand M.
Mimetus sp. C. Hand E.
Mimetus sp. D. Malaise E.

Family Eresidae
Stegodyphus sp. Malaise A. Visual D.
Eresidae indet. A. Pitfall A.
Eresidae indet. B. Hand Q.

Family Oecobiidae
Oecobius amboseli Shear & Benoit. Hand H.

Family Hersiliidae
Hersilia alluadi Berland. Pitfall C,R.

Family Uloboridae
Miagrammopes sp. A. Swept A.
Miagrammopes sp. B. Hand Y.
Uloborus sp. A. Hand U.
Uloborus sp. B. Swept Y.

Family Cyatholipidae
Scharffia chinja Griswold. Hand E.

Family Theridiidae
Achaearanea sp. A. Hand V.
Achaearanea sp. B. Pitfall A.
Anelosimus sp. Swept Y.
Argyrodes sp. A. Swept C.
Argyrodes sp. B. Swept Y.
Chrysso sp. Hand M.
Coelosoma floridana Banks. Litter E.
Dipoena sp. A. Pitfall B,M.
Dipoena sp. B. Hand U.
Dipoena sp. C. Pitfall II.
Dipoena sp. D. Pitfall II.
Dipoena sp. E. Litter Forest at base of Kisima Hill.
Dipoena sp. ? Pitfall M.
Enoplognatha sp. ? Hand C.
Episinus sp. A. Swept C.
Episinus sp. B. Pitfall B.
Euryopis sp. A. Pitfall A.
Euryopis sp. B. on tree and Swept C.
Euryopis sp. C. Hand X.
Latrodectus geometricus C.L. Koch. Swept I.
Latrodectus sp. Swept C.
Phoroncidia sp. Hand C.
Steatoda sp. A. Pitfall A.
Steatoda sp. B. Pitfall B.
Steatoda sp. C. Pitfall B.
Steatoda sp. D. Pitfall A.
Theridion sp. A (*pallens* group). Hand G.
Theridion sp. B (*impressum* group). Swept C.
Theridion sp. C. Swept C.
Theridion sp. D (*pictum* group). Swept B.
Theridion sp. E (*pallens* group). Pitfall S.
Theridion sp. F. Pitfall B.
Theridion sp. G. Hand Y.
Theridion sp. ? Hand E (tree trunk).
Theridion sp. ? Swept C.
Theridion sp. ? Hand CC.
Theridula sp. Swept B.
Thwitesia sp. ? Swept C.
Theridiidae genus indet. Winkler A.
Theridiidae genus indet. Pitfall HH.

Family Theridiosomatidae
Theridiosomatidae indet. A. Pitfall B.

Family Anapidae
Anapidae indet. A. Pitfall A.

Family Mysmenidae
Mysmenidae indet. Pitfall B.

Family Linyphiidae
Callitrichia sp. n. Pitfall E. Litter E.
Ceratinopsis machadoi (Miller). Pitfall N.
Ceratinopsis benoiti ? Holm. Pitfall N.
Ceratinopsis sp. n. Hand I.
Chenideses sp. n. Winkler Z.
Locketidium cf. *couloni* Jocqué. Winkler GG.
Meioneta dentifera Locket. Pitfall B.
Meioneta prosectes Locket. Pitfall A,B,N. Swept D. Hand I.
Metaleptyphantes perexiguus (Simon & Fage). Pitfall M. Hand F.
Metaleptyphantes ovatus Scharff. Pitfall E. Litter E.
Nereine kibonotensis (Tullgren). Pitfall E. Litter E.
Pelecopsis sp. Pitfall N.
Pseudomicrocentria minutissima Miller. Pitfall II.
Tybaertiella kruegeri (Simon). Hand P.
Linyphiidae indet. A. Pitfall B.
Linyphiidae indet. B. Winkler S.
Linyphiidae indet. C. Pitfall II.

Family Tetragnathidae

Leucauge indet. sp. A. Litter E.
Leucauge indet. sp. B. Litter E.
Nephila senegalensis (Walckenaer). Hand D.
Nephilgenys cruentata (Fabricius). visual H,Q.
Pachyganatha leleupi Lawrence. Hand T. Pitfall B.
Tetragnatha boydi O.P-Cambr. Hand P.

Family Araneidae

'Araneus' cereolellus Strand. Swept C.
'Araneus' coccinella Pocock. Swept B,C,AA.
'Araneus' nr. *coccinella* Pocock. Swept B,D.
'Araneus' sp. s.l. Swept B.
'Araneus' sp. s.l. Hand M.
'Araneus' sp. s.l. Hand U.
Argiope suavissima Gerst. Swept BB. Pitfall N.
Argiope trifasciata (Forskal). Swept A,C,D,I.
Caerostris vinsoni Thorell. Hand M. Litter E.
Cyclosa insulana (Costa). Swept AA.
Cyclosa sp. B. Swept M.
Cyphalonotus larvatus Simon. Swept AA.
Cyrtophora citricola (Forskal). Hand D, Visual M.
Gasteracantha sanguinolenta. on tree C. Litter E.
Gasteracantha versicolor (Walck.) Hand E.
Gasteracanthinae indet. (imm.) Swept AA.
Hyposinga sp. Hand A. Swept B,C.
Isoxya stuhlmanni Bos. & Lenz. Swept Y.
Isoxya testudinaria Simon. Swept T.
Larinia trifida Tullgren. Pitfall B.
Larinia (Drexelia) sp. Swept B-D.
Larinia (Paralarinia) sp. Swept BB.
Neoscona moreli (Vinson). Swept AA.
Neoscona rufipalpis (Lucas). Hand M.
Neoscona subfusca (C.L. Koch). Swept M.
Neoscona theisiella. Hand E (tree trunk).
Pararneus cyrtoscapus (Pocock). Hand M.
Pycnacantha sp. Swept B,D.
Singa sp.? Hand V.
Araneidae indet. A. Pitfall A.
Araneidae indet. B. Hand V.
Araneidae indet. C. Hand L,U.
Araneidae indet. D. Swept V.

Family Lycosidae

Amblyothele sp. n. Pitfall A,B,N,O,S. Hand A.
Arctosa sp. A. Pitfall P,Q,Y.
Arctosa sp. B. Hand P,CC.
Evippa cf. *praelongipes* (O.P-Cambr.). Pitfall R,Y. Hand Y.
Hippasa sp. A. C,I,J.
Lycosinae indet. sp. A. Pitfall A-C,M,Y. Hand I,V,CC. Litter E.
Lycosinae nr. *'Pardosa lawrencei'*? Pitfall M,R,S. Hand EE.
Lycosinae indet. sp. E (W. African). Pitfall D,M,N,Q. Hand CC.
Lycosinae indet. sp. G (W. African). Hand P,CC.
Lycosinae indet. sp. H. B,J.
Lycosinae indet. sp. H. Pitfall vlei 1997.
Lycosa s.l. sp. A. Hand B. Pitfall O,S.
Lycosa s.l. sp. C. Hand A,J.
Lycosa s.l. sp. D1. Hand D.
Lycosa s.l. sp. D1 ? Hand P.
Pardosa karagonis ? Strand. Pitfall A,B,M, N,O,S. Hand B,I,L,CC.
Pardosa sp. B. Pitfall D. Hand T.
Trochosa sp. A. Hand E.
Trochosa sp. B. Pitfall N.
Trochosa sp. C. Pitfall M. Hand P,T.
Trochosa sp. C ? Pitfall B.
Wadicosa sp. ? Hand P.
Zenonina sp. Pitfall C,D (R imm.)
Lycosidae indet. Pitfall A.
Lycosidae indet. Hand M,T.
Lycosidae indet. Litter E. Pitfall R.
Lycosidae indet. Litter C,E,I,J 1994.

Family Pisauridae

Charminius nr. *camerunensis* Thorell. Pitfall M.
Cispius bidentatus Lessert. Malaise DD. Pitfall R.
Cispius nr. *variegatus* Simon. Hand M.
Cispius sp. B. Malaise B.
Eupresthenopsis armatus ? Strand. Hand B. Pitfall A.
Epresthenops sp. Hand Q.
Maypicius petrunkevitchi Lessert. Pitfall B.
Maypicius stuhlmanni ? (Bosenberg & Lenz). Pitfall A.
Perenethis simoni (Lessert). Pitfall A,M.
Perenethis sp. B. Pitfall C.
Rothus sp. ? Hand DD.
Tetragonophthalma sp. n. A. Hand A,B,J. Pitfall A,B,N,O,R.
Tetragonophthalma sp. B. Pitfall HH.
Thalassius spinosissimus (Karsch). Hand P.
Thalassius sp. indet. Pitfall E.
Thalassius sp. ? Swept C.
Pisauridae indet. A. Pitfall M.

Family Agelenidae

Mistaria sp. A. Hand I.
Mistaria sp. B. Pitfall M.
Mistaria sp. C. Pitfall E.
Mistaria sp. D. (imm.) Hand CC.

Family Hahniidae

Hahnia sp. A. Pitfall N,O.
Hahnia sp. B. Litter E.
Hahnia sp. C. Pitfall II.

Family Dictynidae
Dictyna sp. Hand A,B,H. Swept B.
Dictynidae indet. sp. B. Hand Q.
Dictynidae indet. sp. C. Pitfall B.

Family Amaurobiidae
Phyxelida bifoveata (Strand). Pitfall E,M.
Phyxelida nr. *tanganensis* (Simon & Fage). Pitfall E. Litter E.
Amaurobiidae indet. A. Pitfall B.

Family Oxyopidae
Hamataliwa sp. ? A. Pitfall A.
Hamataliwa sp. ? B. Hand T.
Hamataliwa sp. ? C. Hand T.
Hamataliwa sp. ? D. Swept C.
Hamataliwa sp. ? E. Swept Y.
Oxyopeidon sp. A. Pitfall A-D,G,N,O. Hand H,I.
Oxyopeidon sp. B. Hand H.
Oxyopeidon sp. ? C. Hand E.
Oxyopes sp. A. Pitfall A. Swept D.
Oxyopes sp. B. Pitfall B.
Oxyopes sp. C. Pitfall A,B,D,N. Swept C.
Oxyopes sp. D. Pitfall B,N,R,S.
Oxyopes sp. E. Pitfall A,B.
Oxyopes sp. F. Pitfall N,O.
Oxyopes sp. ? G. Pitfall B.
Oxyopes sp. H. Pitfall A.
Oxyopes sp. I. Swept C.
Oxyopes sp. J. Hand. Swept EE.
Oxyopes sp. K. Swept C.
Oxyopes sp. L. Hand CC.
Oxyopes sp. M. pan trap B.
Peucetia longipes Pocock. Pitfall N.
Oxyopidae Genus indet. Swept M.

Family Miturgidae
Cheiracanthium sp. A. Swept D.
Cheiracanthium sp. B. Litter E.
Cheiracanthium sp. C. Malaise trap C.
Cheiracanthium sp. D. Malaise trap M.
Cheiracanthium sp. E. Malaise trap M.

Family Clubionidae
Clubiona nr. *subtrivialis* Strand. Pitfall B.
Clubiona sp. B. Malaise trap C.
Clubionidae indet. sp. A. Litter E.
Clubionidae indet. sp. B. Pitfall A.
Clubionidae indet. sp. C. Pitfall B.

Family Liocranidae
Liocranidae indet. sp. A. Pitfall E. Litter E.
Liocranidae indet. sp. B. Pitfall B.
Liocranidae indet. sp. C. Hand I.
Liocranidae indet. sp. D. Hand Q.
Liocranidae indet. sp. E.
Andromma group ? Pitfall B.

Family Corinnidae
Brachyphaea proxima ? de Lessert. Hand V.
Brachyphaea hulli ? de Lessert. Pitfall A.
Copa benina ? Strand. Pitfall A,E,H,M,O.
Copa sp. B. Pitfall B,D,N.
Castianeriinae sp. A. Hand F.
Castianeriinae sp. B. Litter E.
Castianeriinae sp. C. Hand CC.
Castianeriinae sp. D. Pitfall O. Hand CC.
Corinninae indet. A. Hand D.
Corinninae indet. B. Pitfall E.
Corinninae indet. C. Pitfall A.
Graptartia sp. A. Pitfall B.
Graptartia sp. B. Hand CC.
Graptartia sp. C. Pitfall A.
Trachelinae sp. A. Hand DD.

Family Zodariidae
Akyttara akagera Jocqué. Pitfall A-D,M,O.
Chariobas sp. (imm.) Swept C.
Diores initialis Jocqué. Pitfall A-B,N,R. Hand H.
Diores nr. *strandi* Caporiacco. Pitfall A-C.
Diores murphyorum Jocqué. Pitfall D.
Heradida sp. n. Pitfall B,Y. Hand Q.
Mallinella sp. n. Pitfall A-E,M-O,Q,R,S. Hand A,B,M.
Palfuria sp. n. Pitfall B,C,O.
Ranops sp. n. A. Pitfall A-C,N.
Ranops ? sp. n. B. Pitfall B-D,N,O.
Systenoplacis sp. n. A. Pitfall A-B,J,Q. Hand A,M.
Systenoplacis sp.n. B. Pitfall E,M. Litter E.
Systenoplacis sp. n. C. Pitfall E,M.
Microdiores sp. n. Pitfall A.
Asceua sp. n. Pitfall E. Litter E. Hand M. Winkler FF.
Zodariidae indet. Pitfall R,S.
Zodariidae indet. Hand E.

Family Cithaeronidae
Cithaeron delimbata Strand. Hand Y.

Family Trochanteridae
Platyoides pusillus Pocock. Hand E (tree trunk).

Family Prodidomidae
Prodidomus sp. A. Pitfall A-C,N,O,Q,S.
Prodidomus sp. B. Pitfall A,E.
Prodidomidae indet. sp. A. Pitfall M,GG.
Prodidominae indet. sp. B. Pitfall N.

Family Gnaphosidae
Amusia cataracta ? Tucker. Pitfall C.
Aphantaulax sp. Swept B,C,D,I,V,Y,AA,BB. Hand P.
Asemesthes sp. A. Pitfall A-C,N,O,Q,R.

Asemesthes sp. B. Pitfall C,Y.
Berlandia plumalis (O.P-Cambr.). Pitfall C.
Camillina cordifera (Tullgren). Pitfall R,S,DD.
'Echemus' nr. *incinctus* Simon. Pitfall B,D.
Echemus group. sp. Pitfall A.
Haplodrassus sp. Pitfall A,B.
Leptodrassus sp. A. Pitfall A.
Leptodrassus sp. B. Pitfall Q,S.
Megamyrmaekion sp. Pitfall A,B.
Megamyrmekion group. sp. Hand E,M.
Micaria sp. Pitfall A,M,N.
Poecilochroa group. sp. Hand P.
Setaphis sp. Pitfall A,B,N,O,R,S. Hand I,J,V.
Trachyzelotes sp. Pitfall B.
Xerophaeus group. sp. Hand Q,U. Winkler GG.
Zelotes impexa A (Simon). Pitfall A,O,Q,R,S,Y.
Zelotes sp. n. C. Pitfall A,C,N,O,R,Y. Hand A,I,J,V.
Zelotes sp. n. E. Litter E. Hand I. Pitfall E, M. Winkler FF.
Zelotes sp. F. Litter E.
Zelotes sp. H. Pitfall S.
Zelotes sp. Pitfall S.
Zelotes sp. Pitfall A.
Zelotinae indet. sp. G. Pitfall R,S.
Zelotinae indet. sp. D. Pitfall A,C.
Zelotinae indet. Pitfall R.
Zelotinae indet. Pitfall R.
Zelotinae indet. Pitfall B.
Zelotinae indet. Hand Q.
Gnaphosidae indet. A. Pitfall A.
Gnaphosidae indet. B. Pitfall A-C,N,R,N, DD.
Gnaphosidae indet. C. Litter E.
Gnaphosidae indet. D. Hand H. Pitfall E.
Gnaphosidae indet. E. Pitfall E,M. Winkler GG.
Gnaphosidae indet. Pitfall B.
Gnaphosidae indet. Hand Q.
Gnaphosidae indet. Pitfall B.
Gnaphosinae indet. Hand V,CC.

Family Ctenidae
Ctenus sp. A. Hand Q,U.
Ctenus sp. B. Hand Q.
Cycloctenininae indet. A. Hand B. Pitfall A,B.
Cycloctenininae indet. B. Pitfall II.
Cycloctenininae indet. C. Pitfall A,II.
Cycloctenininae indet. D. Pitfall E. Hand M.
Cycloctenininae indet. E. Hand T.

Family Selenopidae
Selenops nr *radiata* Latreille. Hand H,K 1994.
Anyphops sp. n. Malaise E.

Family Heteropodidae
Sparassus sp. Pitfall B,M.
Sparassus sp. ? Hand Y.
Olios sensu lato sp. A. Pitfall N. Hand D,G, H.
Olios sensu lato sp. B. Pitfall C.
Olios sensu lato sp. C. Swept D.
Olios sensu lato sp. D. Pitfall A.
Olios sensu lato sp. E. Pitfall B.
Olios sensu lato sp. F. Hand H.

Family Philodromidae
Philodromus adjacens Walckenaer ? Pitfall C 1994.
Philodromus bigibba O. P-Cambr.
Philodromus brachycephalus Lawrence ? Swept C.
Philodromus mundus O.P-Cambr. ? Pitfall A-C,J,M,N,O. Swept C.
Philodromus partitus Lessert.
Philodromus cf punctisternis Caporiacco. Pitfall B 1993.
Philodromus sp. ? Pitfall A,B. Litter E. Hand H,M.
Philodromus sp. Litter E.
Philodromus sp. Hand M.
Suemus punctatus Lawrence.
Tibellus armatus Lessert. Pitfall N. Swept B,C,D,I. Hand H,I,J.
Tibellus flavipes Caporiacco. Swept D.
Tibellus hollidayi Lawrence. Swept T.
Tibellus minor Lessert. Pitfall N. Swept B,C,D,I. Hand H,I,J.
Tibellus seriopunctatus. Simon DD.

Family Thomisidae
Diaea puncta Karsch. Swept Y.
Heriaeuus buffoni (Audouin). Pitfall C.
Hewittia gracilis Lessert. Swept C.
Misumenops rubrodecorata Millot. Swept AA.
Monaesus austrinus Simon. Swept B-D,P,AA,DD,EE.
Monaesus fasiculigera Jezequel. Swept B-D,V,Y,AA,CC,EE.
Monaesus paradoxus (Lucas). Pitfall C,R.
Monaesus quadrituberculatus Lawrence. Swept B,C,U,AA.
Oxytate sp. Hand CC.
Ozyptila sp. ? Pitfall B,N,O.
Ozyptila sp. B. Hand C,P.
Parabomis lavanderi Kulczynski. Hand Q.
Paramystaria variabilis Lessert. Pitfall B.
Parasmodix quadrituberculatus Jezeq. Pitfall A,M.
Pheracydes cf *zebra* Lawrence.
Runcinia aethiops Simon. Hand M. Swept CC.
Runcinia affinis Simon. Swept CC.
Runcinia flavida (Simon). Swept A,B,C,I,V, Y,AA,CC,DD,EE.
Simorcus coronatus Simon. Hand H. Swept X.
Stiphropus intermedius Millot. Pitfall C.
Synaema langheldi Dahl. Swept C,AA.

Synaema mandibulare Dahl. Swept Y.
Synaema nigrotibiale Lessert. Swept B,C,I,J, V,Y,AA,BB,EE. Hand H,L.
Synaema viridsternis Jezequel. Swept Y.
Thomisops lesserti Millot. Pitfall A,B. Swept V,AA,BB,EE.
Thomisops sulcatus Simon. Litter E. Swept C,AA,CC.
Thomisus blandus Karsch. Swept A,B,C,J,V, Y,AA,EE. Litter E.
Thomisus citrinellus Simon. Swept T.
Thomisus kalaharicus Lawrence. Pitfallls O.
Thomisus spiculosus Pocock. Pitfallls O.
Thomisus sp. n. ? Swept C,AA.
Tmarus cameliformis Millot. Swept C.
Tmarus foliatus Lessert. Hand C.

Family Salticidae
Aelurillus lymphus Prochniewicz & Heciak. Pitfall A-D,N,O. Hand A.
Afrobeata sp. n. Pitfall Y.
Ant like genus. Juvenile. Hand M.
Asemonaea stella Wanless. Pitfall N. Swept C.
Bianor sp. Swept B.
Cembalea (Tularosa) plumosa (Lessert). Hand. Pitfall Q.
Cyrba bimaculata Simon. Pitfall B. Hand Q,CC.
Evarcha dotata (Hyllus) Peckham & Peckham. Litter E. Swept I. Pitfall R. Hand DD.
Evarcha sp. n. Swept. Pitfall A,B. Hand P.
Evarcha sp. n. B ? Pitfall S.
Evarcha sp. n. C. Pitfall B.
Festecula lawrencei Lessert. Swept A.
Habrocestum sp. n. A. Pitfall A,B,M,O,P,Q.
Habrocestum sp. n. B. Hand N,V,CC.
Harmochirus bianoriformis (Strand). Pitfall B.
Harmochirus sp. Hand P.
Hasarius roeweri Lessert. Winkler FF.
Heliophanus debilis Simon. Malaise C.
Heliophanus demonstrativus Weselowska. Hand Q.
Heliophanus undecimmaculatus Caporiacco. Swept Y.
Heliophanus sp. ? Pitfall D,S. Hand E.
Heliophanus sp. n. Pitfall A,B,C,S.
Hyllus argyrotoxus Simon. Pitfall A,B,C,E, N. Hand D,H,M,P,V,CC. Swept A,Y.
Hyllus sp. n. A. Pitfall S (19/4/95).
Hyllus sp. n. B. Hand Q.
Hyllus sp. n. C. Hand E (tree trunk).
Langaelurillus sp. n. A. Litter E. Pitfall M. Hand V.
Langaelurillus sp. n. B. Pitfall A,B,D,N,O. Swept B.
Langaelurillus sp. n. C. Pitfall R,S,Y.
Langaelurillus sp. n. D. Pitfall A,B.
Langona pecten Prochniewicz & Heciak. Pitfall Y,JJ.
Langona sp. n. Pitfall HH.
Leptorchestes sp. n. Hand B,D. Swept C. Pitfall A.
Marengo coriacea Simon. Pitfall A.
Menemerus congoensis Lessert. Hand Y. Pitfall HH.
Mexcala sp. (= *Cosmophasis*). Pitfall A.
Mogrus dillae Proszynski. Swept C.
Myrmarachne ichneumon (Simon). Hand H.
Myrmarachne kilifi Wanless. Pitfall E.
Myrmarachne uvira Wanless. Pitfall A.
Myrmarachne vanessae Wanless. Malaise CC.
Natta horizontalis Karsch. Hand C,E, FM,P,Q.
Neaetha sp. cf *oculata*. Swept A,C.
Pellenes simoni (O. P-Cambridge). Pitfall A,B,Y.
Pellenes sp. n. A. Pitfall C.
Pellenes sp. n. B. Hand Y.
Phintella ? sp. n. Hand P.
Phlegra bresnieri (Lucas). Pitfall B. Hand D.
Phlegra nuda Prochniewicz & Heciak. Hand E.
Phlegra sp. n. A. Pitfall A,B,N. Hand Q.
Phlegra sp. n. B. Pitfall B.
Phlegra sp. n. (cf. *chrysops*). Pitfall HH.
Piignus sp. n. ('*Mogrus*'). Pitfall A,B.
Pseudicius sp. n. Hand T (galls on *Acacia drepanolobium*).
Rhene sp. n. Hand M.
Schenkelia modesta Lessert. Hand E (tree trunk).
Stenaelurillus sp. n. A. Pitfall A,B,D,E,O, Q,Y. Hand P.
Stenaelurillus sp. n. B. Pitfall A-D,M,O, R,S.
Stenaelurillus sp. n. C. Pitfall A-D,N,R,S,Y. Hand E.
Stenaelurillus sp. n. D. Pitfall S (18/4/95).
Stenaelurillus sp. n. E. Pitfall Y.
Talavera sp. n. A. Pitfall A-C,N,O,II.
Talavera sp. n. B. Pitfall A,B.
Talavera sp. n. C. Swept B.
Euophryinae gen nov.
Thyene bucculenta (Gerstaecker). Hand H.
Thyene inflata (Gerstaecker). Hand A,CC. Swept B,C,I,Y,AA,EE.
Thyene semiargentata (Simon). Swept A,B,C,G,I,Y,AA,BB,DD,EE. Hand P,Q,V,CC.
Tusitala barbata Peckham & Peckham. Hand M,P,Q.
Tusitala sp. n. A. Hand Q.
T*usitala* sp. n. B. Hand P.
Unidentified sp. 2 *'Sitticus'* gen. nov. Litter E. Hand D,M.
Unidentified sp. 3 *'Pachyballus'*. Pitfall A,B. Swept B,C. Hand CC.

code	habitat and location
A	*Acacia-Commiphora* bushland on hillside behind Ibaya camp
B	Footslope grassland below Ibaya camp
C	Open *Acacia senegal* woodland, between Ndea and Mbula hills
D	Grassy vlei with scattered *Acacia drepanolobium*, between Ndea and Mbula
E	Dry *Spirostachys* forest, summit of Igire ridge, behind Ibaya camp
F	Litter in gully, hillside behind Ibaya camp
G	In termite mounds, around Ibaya camp
H	Hand collected, on walls & ground, Ibaya camp
I	Swept & hand-collected, grassland on vertisol, north of Simba plot
J	Grassland on vertisol, below Ibaya camp
K	Under bark of *Lannea* trees, Ibaya camp
L	Short grassland, ridge east of Maji Kununua hill
M	*Setaria* grassland, edge of Ibaya Hill forest
N	*Dichrostachys* scrub, Dindira Dam
O	*Combretum* bush, Dindira Dam
P	Edge of pool, stream 2 km NW of Ibaya camp
Q	Rocky gulley, 1 km NW of Ibaya camp (McGavin's Gulch)
R	*Commiphora* scrub, Umba river
S	Riverine bushland, Umba river
T	Vlei near Dindira Dam
U	Dense grass tussocks, Simba plot
V	Grass around rocky out crops, Simba plot
X	Thick dry grass beneath *Acacia* and *Commiphora* bushes, Kikolo plot
Y	Thick *Commiphora* woodland, Cadaba plot
Z	Montane forest, summit of Maji Kununua hill
AA	*Themeda trianda* grassland, *Commiphora* parkland below Ndea hill
BB	*Pennisetum mezianum* grassland, Nyati plot
CC	Grass tussocks, near rocky outcrop, Kisima plot
DD	In thick dead grass under *Acacia* trees, Kikolo plot
EE	Tall grass, by track 8.4 km from Ibaya on Ubani track
FF	Edge of montane forest, Kinondo hill
GG	Litter in *Combretum* thicket, base of Maji Kununua hill
HH	*Acacia reficiens*/*A. mellifera* woodland, Kisima Site
II	Vlei with *Acacia zanzibarica*, Ngurunga Site
JJ	Mixed *Acacia senegal*/*A. nilotica* woodland, Ubani plot

CHAPTER 12

Cicadas (Hemiptera, Homoptera: Cicadoidea) of Mkomazi

Martin H. Villet & Simon van Noort

Introduction

The cicadas are a group of insects belonging to the order Hemiptera which comprises a diverse range of insects including the bugs, leafhoppers, aphids, scale insects etc. The Hemiptera are characterised by specialised mouth parts that are adapted for piercing and sucking up sap from host plants or animal juices from their prey. All cicadas are plant feeders and can be extremely camouflaged when sitting and feeding on a branch, although the males give themselves away by their persistent, often high pitched calls. This conspicuous sound together with their large size has resulted in the familiarity of cicadas to many people. Cicadas do not usually adversely affect the health of their host plants and because of their substantial size they provide an important resource for insectivorous birds, predatory insects and spiders.

Our knowledge of Tanzanian cicadas compares favourably with neighbouring countries, but because cicadas have been poorly surveyed in Africa, current assessments of species richness are likely to be underestimates. The Tanzanian total of 31 species is likely to double with further comprehensive surveys. In this chapter we assess the richness of the local cicada fauna that is protected within Mkomazi Game Reserve, report on the ecology of the recorded species and discuss pertinent conservation and management practices to ensure the continued maintenance of existing cicada diversity in the reserve.

Ecology

The life cycles of very few African cicadas are known in any detail. After mating, a female may cut dozens of slits in vegetation with her ovipositor, and usually lays five to ten eggs in each slit. The eggs are off-white, elongated and usually pointed. Newly-hatched pronymphs are covered by their embryonic cutice which holds the limbs streamlined against the body. They wriggle from the egg slits and immediately moult into free-limbed nymphs that drop to the ground, dig down using their modified forelegs and find a root from which they can feed on xylem sap. They may construct a chamber in the soil beside the root, and some African species build small turrets above ground in waterlogged soils (Boulard 1965 & 1969). Cicada nymphs can account for a significant part of the herbivore biomass of some arid African ecosystems (Dean & Milton 1991 & 1992; Milton & Dean 1992), but this is probably unusual. However, a few North American species have been termed 'keystone species' and 'critical link species' because of their influence on host plants, predators, and the movement of water (Andersen 1994) in their environment.

Nymphs moult about four times over a period of one to (in the case of three North American species) 17 years, before emerging from the soil as adults. Adults may live for three or four weeks (Villet, pers. obs.).

The emergence of adults is seasonal, apparently being linked to the occurrence of good rains and warm periods in southern Africa (Villet & Capitao 1996), but this trend was not found in Australia (Coombs 1996). Precipitation in the Mkomazi Game Reserve is bimodal, with 'short rains' in November and December, and 'long rains' from March to May. Most of the Mkomazi cicadas were active during the 'long rains', and it is likely that synchrony with these rains is important in protecting adults of grassland species from the fires that have a major role in shaping the Mkomazi Game Reserve ecosystem. Fossorial living will certainly do the same for nymphs.

The principal activities of adult cicadas are feeding and seeking mates. Like their nymphs, adult cicadas are herbivores, feeding on sap from twigs and trunks. Many platypleurine species are specific in their use of certain plant hosts, and of certain parts of the plants (Pead 1910), as are some Neotropical cicadas (Johnson & Foster 1986).

Male cicadas produce loud, conspicuous calls (Villet 1987, Sanborn & Phillips 1995) from specialized abdominal organs as part of their courtship behaviour (Villet 1992). The mate-finding behaviour of the two Mkomazi cicada tribes is substantially different. Platypleurine cicadas are large-bodied and show a suite of adaptations for avoiding predators, which are primarily birds (Distant 1906a). They often have well camouflaged forewings, and tend to inhabit trees and bushes where the trunks or branches are thicker than their bodies, making them inconspicuous. Branch-inhabiting species commonly run around the other side of their perches when they see large predators approaching (Pead 1910). Perhaps as a further adaptation, males are sedentary, and will call from the same tree for several days if they are not disturbed by predators (Villet, pers. obs.). This habit is linked with the production of loud calls (Villet 1987) that tend to be a persistent whine (Distant 1906b, Villet 1988 & 1989). This call attracts females, which mate with the caller, and also attracts males in some species (Villet 1992), resulting in the formation of choruses. The females rove through the habitat, and may be found a few kilometres from the nearest males (Villet, pers. obs.).

On the other hand, the small-bodied, clear-winged tettigomyiine cicadas are very hard to follow with the eye as they flit rapidly and erratically between perches. They tend to perch on grasses and the outer twigs of bushes, and their main predators are robberflies (Asilidae) and web-spinning spiders (Aranaeidae). Males produce quieter calls than platypleurines (Villet 1987), often clicking, churring or croaking (Villet 1988). It is the males that move through the habitat, flying 1–20 m, and calling for perhaps a minute whenever they land. The role of the call in mating is not clear, since the males search out sedentary females that generally emit no readily perceivable signals (Villet, pers. obs.). These searching habits have facilitated the independent evolution of brachypterous females in several tettigomyiine genera, e.g. *Spoeryana*, *Stagira* and *Tettigomyia* (Boulard 1974). The most important ramification of this mating system is that females are unlikely to disperse very far, resulting in rather localized populations and higher rates of endemism in these cicadas.

Cicada diversity

The cicadas are a fairly diverse group, with over 1200 species catalogued worldwide, especially from the tropics and subtropics. Specialist revisions of African genera are currently being published (Heller 1980, Villet 1993, 1994a, b & 1997). Approximately 380 species have been recorded from Africa south of the Sahara. In the field cicadas' calls make their presence conspicuous, and because the calls are species-specific, censusing can be relatively easy, even without catching specimens (Villet & Capitao 1996). This is fortunate because they are wary beasts, and capturing them is generally not easy. A few species are sometimes drawn to light traps, but generally they must be stalked with a hand net. However, since initial censuses rely on captured material, most African countries are under sampled (Table 12.1), making faunistic generalizations harder to frame.

To assess cicada diversity, an inventory of African cicadas (Table 12.1) was compiled from catalogues (Dlabola 1962, Metcalf 1963, Duffels & van der Laan 1985), museum collections and personal records. The Platypleurini are used as a currency for assessment because they are often large and showy, which promotes their collection, and because the group occurs throughout the Afrotropical region and into eastern Asia. The South African cicada fauna has been especially well surveyed (Table 12.1), particularly in terms of platypleurines, and can therefore be used as a benchmark for assessments. Four issues can then be addressed: representativeness of the Mkomazi Game Reserve sample; likely biases in the Mkomazi Game Reserve sample; comparison of species richness with other African faunas; and species endemism.

Table 12.1 Faunistic diversity of regional cicada faunas in Africa.

	Cicadoidea						
	Cicadinae	Tibicininae					
	Platypleurini	Dazini	Parnisini	Taphurini	Tettigomyiini	Cicadettini	Prasiini
Tanzania	16	2		2	10		1
Mkomazi GR	6	1			4		
Kenya	10	2		1	7	1	1
Uganda	9			1	2		
South Africa	38		52	1	47	7	1
Grahamstown	9		4		6	2	
Mtunzini	7				2		
Mkuzi GR	5			1	1		
Dunstable Farm	9		1	1	2		
Thabazimbi	6			1			
Botswana	5						1
Zimbabwe	14	1	1	3	2		1
Kwe Kwe	6		1	2			1
Bulawayo	7		2	3			
Namibia	15		3	1			
Angola	23			7			1

Representativeness of the Mkomazi sample

There are about 38 platypleurine species in South Africa, while at any given site in the eastern half of the country there are generally about seven species (Table 12.1), or roughly 20% of the country's total. Tanzania has at least 16 species of Platypleurini, of which Mkomazi Game Reserve contains six species, or about 35% of the country's total (Table 12.1). Assuming that the South African pattern holds for all thorough samples, the Mkomazi Game Reserve sample is fairly representative of local species richness, and many more Tanzanian species remain to be recognized.

The Tanzanian Tibicininae are represented by 15 species, of which one species of Dazini and four congeneric species of Tettigomyiini occur in Mkomazi Game Reserve (Table 12.1). Kenya appears to have fewer species, but more tribes (Table 12.1) than Tanzania. These faunas are poor compared to that of South Africa, and it is almost trivial to deduce that further sampling in Mkomazi Game Reserve will return more tibicinine species and further tribes.

Sampling biases

Many African cicada species are specific to certain plants or habitats. Four major habitats were sampled in Mkomazi Game Reserve (Table 12.2), but due to logistic restrictions collecting effort was concentrated in the west of the reserve, particularly near Ibaya Camp. Additionally, seasonal coverage was limited, with specimens obtained over a period of only 36 collecting days. However, this bias was somewhat offset because the total effort was broken into two periods coinciding with both of the annual wet seasons. These are the principal periods when adult cicadas are active. Sampling of the cicada species was also subordinate to more intensive Hymenoptera collecting, and as such cicadas were only actively pursued when time allowed, and essentially were collected on an opportunistic basis. Hence, the species inventory presented here is likely to be an underestimate, but nevertheless already compares favourably with the knowledge of cicada faunas in areas of similar habitat elsewhere in southern and eastern Africa (Table 12.1). The drier central and eastern habitats remain largely unsampled for cicadas. Further collecting in these areas is required and may very well produce a different cicada fauna, typically representative of the drier habitats of the central region or riverine vegetation associated with the Umba River on the eastern reserve boundary.

Comparison of species richness with other African faunas

Except in South Africa, sampling in African savannas has not been systematic, and the faunas recorded from this area are unlikely to be representative, let alone comprehensive. Botswana is certainly unusually under-sampled, since three times as many platypleurine species (Table 12.1) have been reported from Zimbabwe and Namibia, neighbours that lie at the same latitude. Even in Zimbabwe, Namibia and Angola, the small numbers of tibicinine cicadas are certainly artefacts of sampling effort. Nevertheless, cicada species richness recorded from Tanzania compares well with these other countries, and Tanzania is therefore probably as well surveyed as most. However, a doubling of these faunal lists would not be unexpected.

Endemism

A general impression of the number of endemic (restricted to a specific region) cicada species can be gained from the catalogues of Metcalf (1963) and Duffels & van der Laan (1985), although there are many misleading records in the literature because of the taxonomic confusion present in taxa like *Platypleura*.

Ioba leopardina is one of the few non-endemic species being widespread in African savanna. Most platypleurine cicadas have a far more restricted geographical range. Each *Koma* and *Strumosella* species is known from fewer than three countries, and nearly all *Platypleura* species are endemic to a single country. The

Table 12.2 Cicada species diversity within Mkomazi Game Reserve habitats.

	montane forest	*Acacia/ Commiphora* bushland	open *Combretum* bushland	*Commiphora* woodland
Ioba leopardina			•	
Strumosella sp.			•	
Koma bombifrons	•		•	
Platypleura nigromarginata		•		
Platypleura sp. 1		•		
Platypleura sp. 2			•	
Orapa cf. *lateritia*	•			
Paectira ventricosa		•		
Paectira cf. *dispar*	•		•	•
Paectira cf. *ochracea*		•		
New genus	•			

type material of *Platypleura nigromarginata* is from Entebbe, Uganda, and Mkomazi Game Reserve is the only other known locality. The two undescribed *Platypleura* species from Mkomazi Game Reserve are not known from elsewhere. Tettigomyiines are even more localized in their distributions. While this pattern might be an artefact of sampling in some cases, the South African fauna (Villet 1993, 1994a & 1997) bears it out. The genus *Paectira* is known only from Kenya, Tanzania and Uganda, and individual *Paectira* species are very restricted in range (Heller 1980). This tendency to inhabit restricted geographical distributions seems to occur across the entire Cicadoidea.

It is not always clear whether endemism in cicadas is due to habitat preferences (e.g. *Pycna semiclara* which is largely restricted to forests), host-plant associations (e.g. *Platypleura maytenophila* which only feeds on *Maytenus heterophylla*) or edaphic (soil) constraints (e.g. *Azanicada zuluensis*, whose larvae are restricted to areas comprising sandy beach soils). The undescribed tettigomyiine genus and *Orapa* cf. *lateritia* come from montane forest on outliers of the Pare Mountains, and are probably restricted to that habitat within Mkomazi. Based on experience of other tibicinines (Villet, pers. obs.), it is unlikely that *Paectira* species are constrained by host plant distributions. The impact of cicada mating systems needs further exploration.

Distribution and ecology of cicadas within Mkomazi

Both cicada subfamilies (Cicadinae and Tibicininae) are widely distributed across habitat type and altitudinal variation within, at least, the western end of the reserve. Nine of the 11 cicada species recorded from Mkomazi are associated with bushland or woodland, either occurring on the trees (Cicadinae, all six species) or in the grassland habitat (Tibicininae, three species) (Table 12.2). The remaining two tibicinine species appear to be associated with montane forest.

The platypleurine, *Ioba leopardina*, the largest and most spectacular cicada occurring in the reserve with distinctive large, black paranotal lobes, was common on the lower slopes of the hill above Ibaya Camp in open *Combretum* bushland, feeding and calling on a tree species that unfortunately was not identified. Both sexes were often attracted to light at Ibaya Camp in November and December 1995. The species occurs from northeastern South Africa through Zimbabwe, Zambia and Malawi to Kenya. Adults are active from mid-September to early December, at least in South Africa and Zimbabwe, a pattern approximated in Mkomazi Game Reserve where adult emergence appears to be linked to the November/December 'short rains'. Adults were absent during the 'long rains' of late summer in April and May. Elsewhere in the species' distribution, *Ioba leopardina* has been recorded from several tree species, including *Sclerocarya birrea* (Maroela), *Julbernardia globiflora*, *Albizia amarah*, *Dichrostachys cinerea* and the exotic *Toona cilia* (Villet, pers. obs.). Males call throughout the day.

Many of the platypleurine cicada species are associated with *Acacia* species, a common and widespread host taxon within the reserve (Harris 1972, Coe & Windsor 1993), and hence the cicadas utilising these trees as feeding and calling sites are likely to be more widespread than indicated by current collecting data. The *Strumosella* sp. and *Platypleura* sp. 2 were collected calling and feeding on *Acacia senegal* in open *Combretum* bushland on Ibaya Hill. Very little has previously been documented regarding the biology of any *Strumosella* species. *Platypleura nigromarginata*, was erroneously synonymized with *Oxypleura quadraticollis*, a species from Botswana, Zimbabwe and South Africa. It does not belong to either *Platypleura* or *Oxypleura*, but its true generic affinities are not clear. This hitherto poorly known species was sampled between Ubani Plot and Ndea feeding on *Acacia senegal* and *A. etbaica* trees growing on the edge of an *mbuga*, and a similarly strong population was present near Ngurunga Pools feeding on unidentified *Acacia* species. Based on these habitat preferences *P. nigromarginata* is likely to be a common and widespread species within the reserve and possibly further afield, since the types are labelled "Entebbe, Uganda", indicating that the species may have a wider range in east Africa. A further platypleurine species, *Platypleura* sp. 1 was collected on the bottom slopes of Kisima Hill in very thick *Acacia-Commiphora-Combretum* bushland and if this is a habitat preference this species may have a more restricted distribution than its congeners. A further exception to the trend of widespread platypleurine distribution in the reserve may be *Koma bombifrons* whose distribution, based on limited data, appears to be altitude related, having only been collected on top of Kisima and Ibaya Hills. Hence this species may be restricted to the hill tops which are concentrated in the western half of the reserve.

A restricted habitat (determined by altitude) dependant distribution appears to be characteristic of two tibicinine species, *Orapa* cf. *lateritia* and the species representing the new tettigomyiine genus. Current data indicate that these two species are restricted to the threatened montane forest habitat occurring on top of a number of hills in the reserve, having only been collected on the peaks of Ibaya Hill and Maji Kununua. These species will probably also be present on other hill tops with montane forest, such as Igire Hill and Kinondo Hill. *Orapa* cf. *lateritia* feeds and calls on trees within the forest, a similar habit to that of a related forest-dwelling species from eastern Zimbabwe, *Orapa numa*. In contrast the new tettigomyiine genus is associated with the grass community that thrives in areas where a more open forest canopy allows establishment of such a habitat. Initially this floral community type may be favoured by the encroachment of fire and tree felling, but invariably the composition will be altered with continued degradation of the forest to the detriment of both these hill top endemics. However, in the case of the *Paectira* species more widespread intensive sampling is required to ascertain whether this species is restricted to hill top grassland.

Paectira species are east African endemics (Heller 1980), and in Mkomazi Game Reserve a precipitation dependant distribution pattern appears to be evident for adults of the genus. During the drier 'short rains' in November and December 1995 (the 'long rains' occur in March to May), *Paectira* species were common on the wetter hill tops with montane forest, but absent or rare (one specimen collected at Ngurunga Plot) from the lower lying savanna plains. Conversely during the wetter 'long rains' of early 1996 *Paectira* species were common and prevalent in grassland at lower altitudes. For example, *Paectira* cf. *dispar* was present in *Commiphora campestris* woodland and open disturbed *Combretum* bushland (burnt in November 1995) in this wetter period, whereas during the drier period of 'short rains' this species was absent in these habitats, but present on hill tops such as Ibaya Hill.

These seasonal distribution patterns suggest that the wetter hill tops may act as refugia for many species during relatively drier periods of adult emergence, a contention supported by species of Mecoptera, which follow a similar pattern (Chapter 15). Similarly *Paectira ventricosa*, which has a distinctive golden margin to the forewing, was absent in disturbed *Acacia-Commiphora* bushland at Ngurunga Plot (last burnt in August 1993) during December 1995 but present at this locality in April 1996. However, *Paectira* cf. *ochracea* which was common in April 1996 at Ngurunga Plot, was collected at this site in December 1995 (one specimen), indicating that suboptimal populations of this genus are able to survive in drier conditions. The Mandi Hills above Ngurunga plot and nearby Gulela Hills were not sampled, but it is possible these may act as a refuge for such species during drier periods, although the hills are devoid of montane forest and may not be any wetter than the surrounding areas. *Paectira ventricosa* has been collected elsewhere in Tanzania in April and May, at Kubulu, Neu Moshi and Tanga (Heller

1980), suggesting that the adults, unlike *Paectira hemaris*, may only emerge in the late summer during the 'long rains'. Species of *Paectira* were collected in Mkomazi by sweeping, which included sampling of both grasses and shrubs, or in Malaise traps, and hence specific plant associations were not recorded.

The restricted distributional information presented is an artifact of skewed collecting effort as outlined above under sampling biases. Many of the platypleurine species represented in the western end of the reserve are also likely to occur in the drier central and eastern habitats, and in seasons of good rainfall the *Paectira* species may also be present in the eastern half of the reserve. The latter could not be confirmed as no sampling of the grasslands in the eastern section of the reserve was carried out during the 'long rains'. Nevertheless, the distributional capabilities of *Paectira* species are poor (see under ecology for reasons) and this limitation in combination with a putative reliance on drier period refugia may exclude *Paectira* species from the flat, drier eastern end of the reserve.

Conservation and management

The *Paectira* species are probably the greatest conservation priority because of the impact of their mating system on rates of endemism in the tribe (see above).

Fire is not especially likely to affect cicadas directly because the adults are generally active in the wet months, and the nymphs live underground. This assumption is substantiated by the presence of *Paectira* species in recently rejuvenated grassland that had been burnt five months previously. Nymphs are probably able to survive the heat from a fast moving grassland fire protected in their subterranean habitat. Indirect effects of fires on cicadas include removal of host plant species and baking of soils. Woody host species, in particular, are susceptible to fire encroachment, both in forest and bushland habitats, and to tree felling for charcoal production, a practice which largely affects the montane forests. This latter habitat harbours at least two east African endemic cicada species, and may act as a dry period refuge for species that under favourable conditions enjoy a more widespread distribution. The presence of undisturbed montane refugia may be critical for the continued survival of many invertebrate species in Mkomazi and underlines the importance of conserving this habitat type within the reserve. Given the present susceptibility of this habitat to degradation in the reserve, pertinent conservation management of montane forest is a priority.

Mkomazi Game Reserve has a rich cicada fauna, protecting at least a third of the Tanzanian species. This is a similar local species richness to other surveyed African savanna habitats. The reserve harbours several endemic species and protects a healthy population of a species last reported in 1914. Four undescribed species were collected in montane forest and hillside bushland, one of which represents a new genus (Villet & van Noort, in prep.). This high local species richness underlines the value of Mkomazi Game Reserve as an important conservation area.

Acknowledgements

We would like to extend our thanks to all the staff at Ibaya Camp and to colleagues for logistical support and field assistance. The British Council provided logistic support in Dar es Salaam. Graham Stone critically read and improved the manuscript. This work was supported by grants awarded to SvN from the Commonwealth Science Council and the Foundation for Research Development.

References

Andersen, D.C. (1994). Are cicadas (*Diceroprocta apache*) both a "keystone" and a "critical-link" species in lower Colorado River riparian communities? *Southwestern Naturalist* 39: 26-33.

Boulard, M. (1965) Notes sur la biologie lavaire des cigales (Hom. Cicadidae). *Annales de la Société Entomologique de France* 1: 503-521.

Boulard, M. (1969) L'adaptation à la vie aquatique chez les larves de *Muansa clypealis* (Homopt. Cicadidae). *Compté Rendue du Academie de Science, Paris* 268: 2602-2604.

Boulard, M. (1974) *Speorryana llewelyni*, n.g., n.sp., une remarquable cigale d'Afrique orientale (Hom. Cicadoidea). *Annales de la Société Entomologique de France* 10: 729-744.

Coe, M. & Winser, N. (eds.) (1993) Introduction, Flora and Fauna Species lists and Bibliography for Mradi wa Utafiti, Mkomazi (Mkomazi Ecological Research Programme 1993-1996). Royal Geographical Society, London. 63 pp.

Coombs, M. (1996) Seasonality of cicadas (Hemiptera) on the northern tablelands of New South Wales. *Australian Entomologist* 23: 55-60.

Dean, W.R.J. & Milton, S.J. (1991) Emergence and oviposition of *Quintilia* cf. *conspersa* Karsch (Homoptera: Cicadidae) in the southern Karoo, South Africa. *Journal of the Entomological Society of Southern Africa* 54: 111-119.

Dean, W.R.J. & Milton, S.J. (1992) Emergence and density of *Quintillia* cf. *vitripennis* Karsch (Homoptera: Cicadidae) in the southern Karoo. *Journal of the Entomological Society of Southern Africa* 55: 71-75.

Distant, W.L. (1906a) *Insecta Transvaalensia: a contribution to a knowledge of the entomology of South Africa.* Volume 7, London.

Distant, W.L. (1906b) Some undescribed species of Cicadidae. *Annals and Magazine of Natural History* 17: 384-389.

Dlabola, J. (1962) Faunistics of Angolan cicadas from the Dundo Museum (Homoptera Cicadidae). *Publilcaçoes Culturais da Companhia de Diamantes de Angola* 54: 109-114.

Duffels, J.P. & van der Laan, P.A. (1985) Catalogue of the Cicadoidea (Homoptera, Auchenorhyncha) 1956-1980. *Series Entomologia* 33. Junk, The Hague.

Harris, L.D. (1970) *Some functional and structural attributes of a semi-arid East African ecosystem*. PhD. Thesis. Michigan State University.

Harris, L.D. (1972) *An ecological description of a semi-arid east African ecosystem.* Science Series No. 11, Colorado State Univ., Range Science Dept. 81 pp.

Heller, F.R. (1980) Revision der Gattung *Paectira* Karsch 1890 (Syn.: *Inyamana* Dist. 1905) (Homopt.: Cicadina). *Stuttgarter Beiträge zur Naturkunde* 339: 1-28.

Johnson, L.K. & Foster, R.B. (1986) Associations of large Homoptera (Fulgoridae and Cicadidae) and trees in a tropical forest. *Journal of the Kansas Entomological Society* 59: 415-422.

Metcalf, Z.P. (1963) General catalogue of the Homoptera. Volume VIII. Cicadoidea. North Carolina State College, North Carolina.

Milton, S.J. & Dean, W.R.J. (1992) An underground index of rangeland degradation: cicadas in arid southern Africa. *Oecologia* 91: 288-291.

Pead, C.H. (1910) Some Rhodesian Cicadidae and observations on economic entomology. *Proceedings of the Rhodesia Scientific Association* 10: 20-39.

Sanborn, A.F. & Phillips, P.K. (1995) Scaling of sound pressure level and body size in cicadas (Homoptera: Cicadidae; Tibicinidae). *Annals of the Entomological Society of America* 88: 479-484.

Villet, M.H. (1987) Sound pressure levels of some African cicadas (Homoptera: Cicadoidea). *Journal of the Entomological Society of Southern Africa* 50: 269-273.

Villet, M.H. (1988) Calling songs of some South African cicadas (Homoptera: Cicadidae). *South African Journal of Zoology* 23: 71-77.

Villet, M.H. (1989) Systematic status of *Platypleura stridula* L. and *Platypleura capensis* L. (Homoptera, Cicadidae). *South African Journal of Zoology* 24: 329-332.

Villet, M.H. (1992) Responses of free-living cicadas (Homoptera: Cicadidae) to broadcasts of cicada songs. *Journal of the Entomological Society of Southern Africa* 55: 93-97.

Villet, M.H. (1993) The cicada genus *Bavea* Distant 1905 (Homoptera Tibicinidae): redescription, distribution and phylogenetic affinities. *Tropical Zoology* 6: 435-440.

Villet, M.H. (1994a) The cicada genus *Stagea* n. gen. (Homoptera Tibicinidae): systematics. *Tropical Zoology* 7: 293-297.

Villet, M.H. (1994b) The cicada genus *Tugelana* Distant 1912 (Homoptera Cicadoidea): systematics and distribution. *Tropical Zoology* 7: 87-92.

Villet, M.H. (1997) The cicada genus *Stagira* Stål 1861 (Homoptera: Tibicinidae): systematic revision. *Tropical Zoology* 10: 347-392.

Villet, M.H. & Capitao, I.R. (1996) Cicadas (Homoptera: Cicadidae) as indicators of habitat and veld condition in valley bushveld in the Great Fish River Valley. *African Entomology* 4: 280-284.

Checklist: Cicadas of Mkomazi

Species determinations by M.H. Villet

Class: INSECTA

Order: HEMIPTERA

Suborder: HOMOPTERA

Superfamily: CICADOIDEA

Family: Cicadidae

Sub-family: Cicadinae

Tribe: Platypleurini
Ioba leopardina Distant 1881. Ibaya Camp.
Strumosella sp. (an undescribed species). Ibaya Hill.
Koma bombifrons (Karsch, 1890). Kisima Hill; Ibaya Hill.
Platypleura nigromarginata Ashton 1914. Between Junction 7 and Ubani plot; near Ngurunga Pools.
Platypleura sp. 1. (an undescribed species). Ascent of Kisima Hill.
Platypleura sp. 2. (an undescribed species). Ibaya Hill.

Sub-family: Tibicininae

Tribe: Dazini
Orapa cf. *lateritia* Jacobi 1910. Ibaya Hill; Maji Kununua.

Tribe: Tettigomyiini
Paectira ventricosa Melichar 1914. Ngurunga Plot.
Paectira cf. *ochracea* (Distant 1905). Ibaya Hill; Kikolo plot; between Dindira Dam and Viteweni Ridge.
Paectira cf. *dispar* Heller 1980. Ngurunga Plot.
Undescribed genus and species. Peak of Maji Kununua.

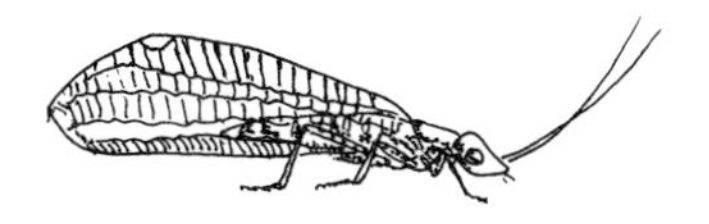

CHAPTER 13

Lacewings (Neuroptera) of Mkomazi

Mervyn W. Mansell & Simon van Noort

Introduction

Insects of the order Neuroptera are known as lacewings because of the delicate, filigreed appearance of their wings, imparted by the wing venation. The order includes many attractive species, including some of the largest insects with wingspans of 170 mm. The wings are held in a roof-like manner when at rest, and the patterns on the wings of many species have evolved to enhance the camouflage of these cryptic insects. The order is comprised entirely of beneficial species that prey upon other insects, including many pest species, while the adults of some families are pollinators of indigenous plants. Neuroptera are consequently of prime importance in agroecosystems as well as natural ecosystems.

Africa is richly endowed with lacewings, with at least 13 of the 17 families currently known from the Afrotropical Region. As in some other regions of the continent, the lacewing fauna of Tanzania is poorly known, although sporadic literature records indicate that most families are represented, and that species diversity is high. This survey of the Mkomazi Game Reserve has provided a valuable opportunity to explore the diversity of the Neuroptera of Tanzania and has revealed that a rich fauna is protected within the reserve.

Ecological role of lacewings

The larvae of all Neuroptera are specialist predators with elongated mouthparts that are adapted for piercing and sucking. These specifically modified mouthparts are formed from the maxillae and mandibles that fit closely together to enclose a canal, and are unique to the Neuroptera (Mansell 1986, 1992b). The adults of most families are also predacious, but with normal biting and chewing mouthparts, while some are nectar and pollen feeders.

The larvae of Neuroptera occupy a range of habitats ranging from arboreal (on trees or shrubs) to sandy places. Lacewing larval biology is varied. Some are parasitic on spider's egg cases and others live in the nests of termites where they share the termites food and are then called inquilines, still others prey upon freshwater sponges. Among the families recorded from Mkomazi Game Reserve, the arboreal Chrysopidae (green lacewings) are voracious predators of small insects and actively pursue their prey on plants. Owlfly (Ascalaphidae) larvae are sedentary ambush predators that lie in wait for their prey on branches or under rocks. The Berothidae and Rhachiberothidae are thought to be inquilines in the nests and galleries of termites (Isoptera), although little is known about the biology of the Afrotropical taxa. The Mantispidae (mantidflies) are parasites in the egg cases of spiders. The silky lacewings (Psychopsidae) live in crevices on tree trunks or in litter at the bases of trees, where they also ambush their prey. The well-represented family Myrmeleontidae (antlions) have mainly sand-dwelling larvae, and species in the genera *Hagenomyia*, *Cueta* and *Myrmeleon* construct the familiar conical pitfall traps in dry sand. The larger taxa such as *Palpares* and *Centroclisis* live freely in sand and ambush their prey from beneath the surface, while larvae of *Cymothales* live in detritus in tree-holes (Mansell 1987 & 1996a).

Neuroptera fulfil an important function as predators that control populations of other insects, particularly in savanna and arid ecosystems, where their biomass may be extremely high (Mansell, pers. obs.). They are ideal subjects for monitoring of biodiversity and habitat change, as they are very sensitive to such disturbance. They are also easy to collect and monitor (by means of light-traps) and, as the Afrotropical fauna is becoming increasingly well known, rapid identification of most species is possible. There also appears to be a very high diversity of these insects in Mkomazi Game Reserve that requires more intensive investigation.

Diversity

In Mkomazi Game Reserve, the recorded lacewings are represented by 37 species in 26 genera belonging to seven families (see checklist). About 120 species of Neuroptera are currently known from Tanzania, but the fauna is certainly far richer.

Sampling biases

As with the cicadas, sampling of lacewings in Mkomazi Game Reserve was secondary to more intensive collecting of Hymenoptera and, consequently, lacewings were collected on an opportunistic basis. Collecting effort was seasonally restricted, with only two short periods of 36 days in total available for sampling, although each of the biannual wet season peaks were covered. Collecting effort was concentrated in the western half of the reserve, and the species inventory presented here is certainly an underestimation of the diversity of lacewings in Mkomazi Game Reserve. Nonetheless, the annotated list is a significant indication that a very rich fauna of lacewings is protected within Mkomazi. The drier central and eastern habitats and the riverine vegetation associated with the Umba River on the eastern boundary of the reserve were essentially not sampled for lacewings. Further sampling in these relatively uncollected areas is also required and is predicted to produce many additional species.

Comparison of species richness with other African faunas

Only one previous survey, aimed specifically at Neuroptera, has been carried out in Tanzania (Hynd 1992), while Kolbe (1898) provided the first general report on the Neuroptera of eastern Africa. The studies of Tjeder (1957, 1959, 1960, 1961, 1966, 1967) also included Tanzanian material, while further species descriptions are scattered in literature cited below. Kenya has also never been specifically investigated for Neuroptera, although many taxa are common to both countries. All museum material has been collected on an *ad hoc* basis by specialists in other groups, but despite this, all but two (Dilaridae, Sisyridae) of the 13 families of Neuroptera known from the Afrotropical Region have been recorded from Tanzania and Kenya. About 300 species of Neuroptera can be predicted for Tanzania, and about 350 in Kenya because of the arid conditions that prevail in the northern regions, as certain families of Neuroptera (Myrmeleontidae and Nemopteridae) proliferate in warm, dry areas.

The southern African lacewing fauna, including those of Botswana, Lesotho, Namibia, South Africa, Swaziland, Zimbabwe and southern Mozambique, has been well surveyed, relative to other parts of the Afrotropical Region, and this collecting effort is reflected in numerous publications. Aspöck, U. & Aspöck, H. (1996) (Berothidae), Aspöck, U. & Mansell (1994), Aspöck, U. & Aspöck, H. (1997) (Rhachiberothidae), Handschin (1959 & 1960) (Mantispidae), Hölzel (1987, 1989, 1990a, 1991a & 1993) (Chrysopidae), Mansell (1980, 1981a, 1981b, 1985, 1987, 1988, 1990a, 1992 & 1996b) (Nemopteridae, Myrmeleontidae), Meinander (1983) (Coniopterygidae), Minter (1986) (Dilaridae), Oswald (1993a & 1994) (Psychopsidae), Smithers (1961) (Sisyridae), Tjeder (1957, 1959, 1960, 1961, 1966, 1967) (Coniopterygidae, Sisyridae, Osmylidae, Berothidae, Psychopsidae, Chrysopidae, Hemerobiidae, Nemopteridae), Tjeder & Hansson (1992) (Ascalaphidae), have all recently dealt with the southern African Neuroptera. Although this list is not exhaustive, these papers provide extensive references to further works and records of Afrotropical Neuroptera, including many pertinent to Tanzania. More general works that also provide extensive bibliographies include: Aspöck, U. (1996) (Berothidae), Brooks & Barnard (1990) (Chrysopidae), Hölzel (1990b, 1991b & 1992) (Chrysopidae), Mansell (1990b) (Myrmeleontidae), Mansell & Aspöck, H. (1990) (General), Meinander (1972) (Coniopterygidae), Monserrat (1990) (Hemerobiidae) and Oswald (1993b) (Hemerobiidae).

About 500 species of Neuroptera, including many endemic taxa, are currently known from southern Africa.

Comparatively little is known about the central and northern Afrotropical faunas of Neuroptera. Extensive surveys are required in Angola, Malawi, Mozambique,

Table 13.1 Known distribution of lacewing species collected in Mkomazi Game Reserve.

species	country records additional to Tanzania
Ascalaphus festivus	South Africa, Madagascar
Dicolpus primitivus	Kenya, Malawi, Zimbabwe
Disparomitis ?hovarthi	Kenya, South Africa
Dixonotus vansomereni	Kenya
Suhpalasca sp.	South Africa
Tmesibasis laceratus	South Africa, Mozambique
Tmesibasis ?alberti	Zambia
Italochrysa ?peringuey	South Africa, Zimbabwe
Italochrysa ?lyrata	South Africa, Zimbabwe
Chrysoperla congrua	Cameroon, Ethiopia, Lesotho, Madagascar, Mozambique, Namibia, Sierra Leone, Somalia, South Africa, Uganda, Zimbabwe
Chrysoperla pudica	Cape Verde Islands, Lesotho, Botswana, Gambia, Oman, Senegal, Somalia, S. Afr., St. Helena, Sudan, Zimbabwe
Chrysemosa jeanneli	Kenya, Namibia, South Africa
Dichochrysa nicolaina	Gambia, Mozambique, Senegal, Somalia, South Africa
Dichochrysa sjoestedti	Ethiopia, Gambia, Mozambique, Senegal, S. Afr., Uganda
Zygophlebius zebra	Kenya
Podallea vasseana	Kenya, Angola, Botswana, Comores, Ethiopia, Madagascar, Mozambique, Namibia, Nigeria, South Africa, Uganda, Zimbabwe
Mucroberotha copelandi	Kenya
Palpares papilionoides	Kenya
Palpares ?sparsus	Kenya, South Africa, Zimbabwe
Palpares torridus	Kenya, Zimbabwe
Palpares tristis	Kenya, Malawi, Mozambique, Zimbabwe
Centroclisis brachygaster	Kenya, Botswana, South Africa, Zimbabwe
Neuroleon guttatus	Namibia, South Africa, Zimbabwe
Gymnoleon elizabethae	Zimbabwe
Cueta mysteriosa	Kenya, Mozambique, South Africa
Cueta klugi	Kenya, Senegal, Sudan
Cueta secretus	Kenya
Hagenomyia lethifer	Kenya, Malawi, South Africa, Zimbabwe
Hagenomyia tristis	Kenya, Malawi, Mozambique, South Africa, Zaire, Zimbabwe - widespread in Africa and Madagascar
Banyutus lethalis	Kenya, Mozambique, South Africa, Zimbabwe
Cymothales bouvieri	Mozambique, Malawi, Madagascar, Namibia, Nigeria, South Africa, Zimbabwe
Myrmeleon quinquemaculatus	Kenya, Cameroon, Mozambique, South Africa, Zimbabwe
Creoleon nubifer	Botswana, South Africa, Zimbabwe

Uganda, Zaire, Zambia and countries to the north before accurate estimates of diversity can be made.

Endemism

The degree of endemism, reflected in the present Mkomazi Game Reserve material, appears to be low, as most of the species also occur in Kenya, as well as southwards to southern Africa. However, a number of species appear to be regional endemics, currently only known from Tanzania and Kenya (Table 13.1). These are: *Dixonotus vansomereni* (Ascalaphidae), *Zygophlebius zebra* (Psychopsidae), *Mucroberotha copelandi* (Rhachiberothidae), *Palpares papilionoides* and *Cueta secretus* (Myrmeleontidae).

Distribution and ecology of lacewings within Mkomazi

Fifteen of the lacewing species (represented by the Ascalaphidae, Myrmeleontidae and Chrysopidae) from Mkomazi were only collected by means of a UV light trap run at Ibaya Camp in open *Combretum* bushland, and it is difficult to predict their distribution within the reserve. However, a number of the species sampled using this method were also collected in other areas and habitats, such as *Acacia/ Commiphora/Combretum* bushland, *Commiphora* woodland and riverine and montane forest (Table 13.2), suggesting that these species, *Disparomitis ?hovarthi, Tmesibasis laceratus, T. ?alberti* (Ascalaphidae), *Dichochrysa nicolaina* (Chrysopidae), *Palpares tristis*, *Neuroleon guttatus* (Myrmeleontidae), and probably the others collected at UV light, are likely to be more widespread over the reserve than current data indicates. One of the Ascalaphid species (*Suhpalasca* sp.) collected at UV light may represent a new genus and species. The species of the widespread Afrotropical genus *Macronemurus* could not be identified, as this genus requires revision. Larval biology of the species attracted to light at Ibaya Camp varies from that of *Cueta mysteriosa* which construct pits in hard, exposed soil, often in rocky areas or along footpaths, to that of *Cymothales bouvieri* which inhabit holes in trees, where they live in accumulated detritus.

The highest percentage (62%) of the total species collected in Mkomazi were sampled in open *Combretum* bushland (Figure 13.1). However, this is purely indicative of unequal sampling effort. The UV light trap, which is an excellent means of collecting lacewings, was only deployed in this habitat. Additionally, although the UV light was orientated towards open *Combretum* bushland, it is feasible that the light trap was also attracting lacewings from *Acacia-Commiphora* bushland in the near vicinity as well as possibly from montane forest. An assessment of the number of species unique to each habitat indicates that montane and riverine forest harbour a high percentage (60% and 57% respectively) of species that are only found in those habitats (Figure 13.1). The high value (65% unique) for open *Comb-*

Table 13.2 Lacewing species diversity within Mkomazi Game Reserve habitats.

	montane forest	riverine bushland/ forest	*Acacia/ Commiphora* bushland	open *Combretum* bushland	*Commiphora* woodland
Ascalaphus festivus				•	
Dicolpus primitivus	•				
Disparomitis ?hovarthi		•		•	
Dixonotus vansomereni				•	
Suhpalasca sp.				•	
Tmesibasis laceratus			•	•	•
Tmesibasis ?alberti			•	•	
Italochrysa ?peringuey			•		
Italochrysa ?lyrata			•		
Chrysoperla congrua		•	•	•	•
Chrysoperla pudica			•		
Chrysemosa jeanneli				•	
Dichochrysa nicolaina			•	•	
Dichochrysa sjoestedti				•	
Mantispidae sp. 1					•
Mantispidae sp. 2	•				
Zygophlebius zebra		•			
Podallea vasseana			•		
Mucroberotha copelandi			•		•
Palpares papilionoides				•	
Palpares ?sparsus				•	
Palpares torridus				•	
Palpares tristis	•		•	•	
Centroclisis brachygaster		•			
Neuroleon guttatus			•	•	
Gymnoleon elizabethae				•	
Cueta mysteriosa				•	
Cueta klugi				•	
Cueta secretus				•	
Hagenomyia lethifer			•	•	
Hagenomyia tristis	•	•	•		
Macronemurus sp.				•	
Banyutus lethalis		•			
Banyutus sp.	•				
Cymothales bouvieri				•	
Myrmeleon quinquemaculatus		•			
Creoleon nubifer				•	

retum bushland is likely to be indicative of the unequal deployment of collecting methods. Many of the species collected in open *Combretum* bushland would probably also have been collected in other habitats if the UV light trap had been used more widely.

The occurrence of a number of species in both montane forest and bushland (Table 13.2) suggests that for many species there are no distribution restrictions imposed by altitude or habitat. However, *Dicolpus primitivus* (Ascalaphidae), *Banyutus* sp. (Myrmeleontidae), and Mantispidae sp. 2 were only collected in association with montane forest. The species of *Banyutus* may be new, as it is not represented in other collections examined by MWM. The Mantispidae are a poorly-known group in the Afrotropical Region, and in urgent need of revision. Many species are undescribed.

Two of the myrmeleontid species, *Hagenomyia tristis* and *Banyutus lethalis* are gregarious, congregating in groups of hundreds of specimens in grass patches bordering rivers, or pools of water in dry river courses. *Hagenomyia tristis* was recorded in these congregations along the Nakombo River at the base of the South Pare Mountains as well as in the river valley north-west of Ibaya Camp leading up Ibaya Hill. *Banyutus lethalis* was less abundant but present together with *H. tristis* in the valley north-west of Ibaya Camp. Superficially the adults of these two species look very similar, but the larvae of *H. tristis* construct pits, whereas the larvae of *B. lethalis* are free-living in sand. Together with the pyschopsid, *Zygophlebius zebra*, these two species are largely restricted to wetter riverine habitats within the reserve. Further species that were only collected in the river valley north-west of Ibaya Camp are: *Centroclisis brachygaster* whose large, black larvae do not construct pits but live freely in sand, often at the bases of tree trunks and are very

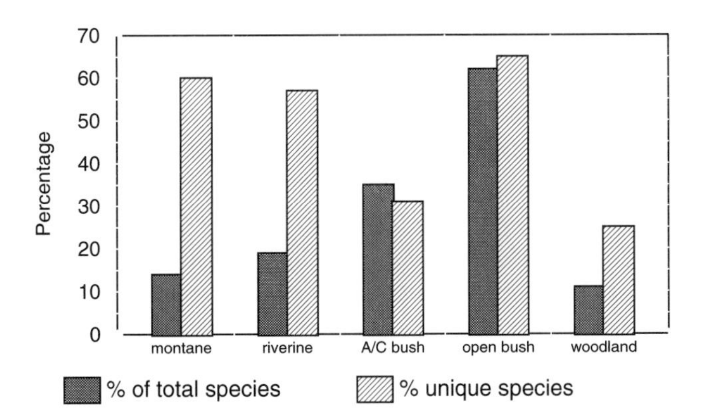

Figure 13.1 Contribution of habitat type to lacewing species richness in Mkomazi. Abbreviations: montane = montane forest, riverine = riverine forest/bushland, A/C bush = *Acacia-Commiphora* bushland, open bush = open *Combretum* bushland, woodland = *Commiphora* woodland.

slow-moving; and *Myrmeleon quinquemaculatus* the largest pit-building ant-lion currently known. Both species are likely to be more widespread. A probably undescribed species of *Distoleon* was collected in riverine forest along the Nakombo River at the base of the South Pare Mountains, but was not recorded from within the reserve.

The two *Italochrysa* species (Chrysopidae) were encountered congregating in a fig tree, *Ficus ingens* (Miq.) Miq., growing at the base of Kamakota Hill. This fig tree had a ripe crop of figs and it is possible that the chrysopids were attracted to and feeding on the sweet exudate of the figs. This may have attracted two other chrysopid species, *Dichochrysa nicolaina* and *Chrysoperla pudica*, and the rhachiberothid: *Mucroberotha copelandi* that were also collected in the vicinity of this tree. Two further specimens of *M. copelandi* were collected at Kikola Plot and the three specimens are paratypes of this recently described species (Aspöck & Aspöck, 1997). The holotype, which is the only other known specimen, was collected in the Kajiado District in southern Kenya, approximately 300 km north-east of Mkomazi. *Mucroberotha copelandi* may be localised in distribution, but is probably under-collected and may not be as rare as indicated by current records.

The chrysopid *Chrysoperla congrua* was common and widespread in the reserve, consistent with its widespread distribution across most of the Afrotropical region.

An indication of the known Afrotropical distribution of the Mkomazi Game Reserve species is provided in Table 13.1.

Conservation and management

Montane and riverine forest contributed a low percentage to overall lacewing species richness in Mkomazi, however, this is a function of the non-deployment of UV light as a collecting method in these habitats. The high percentage of species unique to these two habitats illustrates the presently small but critical role forests play in contributing to lacewing species richness. Further comprehensive collecting in these habitats will no doubt reinforce the importance of conserving forest within the reserve. Negative impacts on species richness will include habitat modification due to fire encroachment, both in forest and bushland habitats, or tree felling for charcoal production affecting the montane forests. Such fragmentation and habitat degradation could have a profoundly adverse effect on the diversity of the sensitive lacewings, as many are exclusively arboreal. These effects should be monitored as part of the overall assessment of habitat preservation.

Acknowledgements

We would like to extend our thanks to all the staff at Ibaya Camp and to numerous colleagues for logistical support and field assistance. The British Council provided logistic support in Dar es Salaam. Graham Stone critically read and improved the manuscript. This work was supported by grants awarded to SvN from the Commonwealth Science Council and the Foundation for Research Development.

References

Aspöck, U. (1996) The Berothidae of Africa: a review of present knowledge (Insecta: Neuroptera). In: Mansell, M.W. & Aspöck, A. (eds.) *Advances in Neuropterology. Proceedings of the Third International Symposium on Neuropterology, Berg en Dal, Kruger National Park, RSA, 1988*. Department of Agricultural Development, Pretoria. pp. 101-113.

Aspöck, U. & Aspöck, H. (1996) Revision des Genus Podallea Navás, 1936 (Neuroptera: Berothidae: Berothinae) *Mitteilungen der Münchener Entomologisches Gesellschaft* 86: 99-144.

Aspöck, U. & Aspöck, H. (1997) Studies on new and poorly-known Rhachiberothidae (Insecta: Neuroptera) from subsaharan Africa. *Annalen des Naturhistorischen Museums in Wien* 99: 1-20.

Aspöck, U. & Mansell, M.W. (1994) A revision of the family Rhachiberothidae Tjeder, 1959, stat.n. (Neuroptera). *Systematic Entomology* 19: 181-206.

Brooks, S.J. & Barnard, P.C. (1990) The green lacewings of the world: a generic review (Neuroptera: Chrysopidae). *Bulletin of the British Museum (Natural History). Entomology Series* 59: 117-286.

Handschin, E. (1959) Beiträge zu einer Revision der Mantispiden (Neuroptera). I Teil. Mantispiden des Musée Royal du Congo Belge, Tervuren. *Revue de Zoologie et de Botanique Africaines* 59: 185-227.

Handschin, E. (1960) Beiträge zu einer Revision der Mantispiden (Neuroptera). II Teil. Mantispiden des Musée Royal du Congo Belge, Tervuren. *Revue de Zoologie et de Botanique Africaines* 62: 181-245.

Hölzel, H. (1987) Descriptions of two new *Brinckochrysa* species from South Africa, with taxonomic notes on other African species of the genus (Neuropteroidea: Planipennia; Chrysopidae). *Journal of the Entomological Society of Southern Africa* 50: 261-268.

Hölzel, H. (1989) Chrysopidae (Neuroptera) der Afrotropischen Region: Genus *Chrysoperla* Steinmann. *Neuroptera International* 5: 165-180.

Hölzel, H. (1990a) *Crassochrysa*, a new genus of Chrysopinae from South Africa (Neuroptera: Chrysopidae). *Phytophylactica* 22: 285-288.

Hölzel, H. (1990b) The Chrysopidae of the Afrotropical Region (Insecta: Neuroptera). In: Mansell, M.W. & Aspöck, A. (eds.) *Advances in Neuropterology. Proceedings of the Third International Symposium on Neuropterology, Berg en Dal, Kruger National Park, RSA, 1988*. Department of Agricultural Development, Pretoria. pp. 17-26.

Hölzel, H. (1991a) Chrysopidae (Neuroptera) der Afrotropischen Region: Genus Glenochrysa Esben-Petersen. *Zeitschrift der Arbeitsgemeinschaft Österreichischer Entomologen* 43: 77-81.

Hölzel, H. (1991b) Beitrag zur Kenntnis der Chrysopidae von Somalia (Neuroptera, Chrysopidae). *Entomofauna, Zeitschrift fur Entomologie* 12: 49-70.

Hölzel, H. (1992) The African species of Ankylopterygini (Insecta: Neuroptera: Chrysopidae). In: Canard, M., Aspöck, H. & Mansell, M.W. (eds.) *Current Research in Neuropterology. Proceedings of the Fourth International Symposium on Neuropterology. Bagnères-de-Luchon, France, 1991.* Toulouse, France. pp. 183-188.

Hölzel, H. (1993) Neue Mallada-Spezies aus dem südlichen Afrika (Neuroptera: Chrysopidae). *Zeitschrift der Arbeitsgemeinschaft Österreichischer Entomologen* 45: 69-74.

Hynd, W.R.B. (1992) On some Neuroptera recently collected in Tanzania (Insecta). In: Canard, M., Aspöck, H. & Mansell, M.W. (eds.) *Current Research in Neuropterology. Proceedings of the Fourth International Symposium on Neuropterology. Bagnères-de-Luchon, France, 1991.* Toulouse, France. pp. 183-188.

Kolbe, H.J. (1898) Die Thierwelt Deutsch-Ost-Afrikas und der Nachbargebiete. Wirbellose Thiere 4. Netzflügler. In: Möbius, K. (ed.) *Deutsch-Ost-Afrika* 4: Dietrich Reimer, Berlin. pp. 1-42.

Mansell, M.W. (1980) The Crocinae of southern Africa (Neuroptera: Nemopteridae). 1. The genera *Laurhervasia* Navás and *Thysanocroce* Withycombe. *Journal of the Entomological Society of Southern Africa* 43: 341-365.

Mansell, M.W. (1981a) The Crocinae of southern Africa (Neuroptera: Nemopteridae). 2. The genus *Concroce* Tjeder. *Journal of the Entomological Society of Southern Africa* 44: 91-106.

Mansell, M.W. (1981b) The Crocinae of southern Africa (Neuroptera: Nemopteridae). 3. The genus *Tjederia* Mansell, with keys to the southern African Crocinae. *Journal of the Entomological Society of Southern Africa* 44: 245-257.

Mansell, M.W. (1985) The ant-lions of southern Africa (Neuroptera: (Myrmeleontidae). Introduction and genus *Bankisus* Navás. *Journal of the Entomological Society of Southern Africa* 48: 189-212.

Mansell, M.W. (1986) Southern African ant-lions. *Antenna* 10: 121-124.

Mansell, M.W. (1987) The ant-lions of southern Africa (Neuroptera: Myrmeleontidae): genus *Cymothales* Gerstaecker, including extralimital species. *Systematic Entomology* 12: 181-219.

Mansell, M.W. (1988) The Myrmeleontidae (Neuroptera) of southern Africa: genus *Tricholeon* Esben-Petersen. *Neuroptera International* 5: 45-55.

Mansell, M.W. (1990) The Myrmeleontidae of southern Africa: tribe Palparini. Introduction and description of *Pamares* gen. nov. with four new species. *Journal of the Entomological Society of Southern Africa* 53: 165-189.

Mansell, M.W. (1990) Biogeography and relationships of southern African Myrmeleontidae (Insecta: Neuroptera). In: Mansell, M.W. & Aspöck, A. (eds.) *Advances in Neuropterology. Proceedings of the Third International Symposium on Neuropterology, Berg en Dal, Kruger National Park, RSA, 1988*. Department of Agricultural Development, Pretoria. pp. 181-190.

Mansell, M.W. (1992a) The ant-lions of southern Africa: genus *Pamexis* Hagen (Neuroptera: Myrmeleontidae: Palparinae: Palparini). *Systematic Entomology* 17: 65-78.

Mansell, M.W. (1992b) Specialized diversity: the success story of the antlions. *Phoenix (Magazine of the Albany Museum, Grahamstown)* 5: 20-24.

Mansell, M.W. (1996a) Predation strategies and evolution in antlions (Insecta: Neuroptera: Myrmeleontidae). In: Canard, M., Aspöck, H. & Mansell, M.W. (eds.) *Pure and Applied Research in Neuropterology. Proceedings of the Fifth International Symposium on Neuropterology, Cairo, Egypt, 1995*. Toulouse, France. pp. 161-169.

Mansell, M.W. (1996b) The antlions of southern Africa (Neuroptera: Myrmeleontidae): genus *Palparellus* Návas, including extralimital species. *African Entomology* 4: 239-267.

Mansell, M. W. & Aspöck, H. (1990) Post-symposium neuropterological excursions. In: Mansell, M.W. & Aspöck, H. (eds.) *Advances in Neuropterology, Proceedings of the Third International Symposium on Neuropterology, Berg en Dal, Kruger National Park, RSA, 1988*. Department of Agricultural Development, Pretoria. pp. 287-298.

Meinander, M. (1972) A revision of the family Coniopterygidae (Planipennia). *Acta Zoologica Fennica* 136: 1-357.

Meinander, M. (1983) The Coniopterygidae (Neuroptera) of southern Africa and adjacent Indian Ocean Islands. *Annals of the Natal Museum* 25: 475-499.

Minter, L.R. (1986) The first record of Dilaridae (Neuroptera) from the Afrotropical Region *Journal of the Entomological Society of Southern Africa* 49: 87-94.

Monserrat, V.J. (1990) A systematic checklist of the Hemerobiidae of the world (Insecta: Neuroptera). In: Mansell, M.W. & Aspöck, A. (eds.) *Advances in Neuropterology. Proceedings of the Third International Symposium on Neuropterology, Berg en Dal, Kruger National Park, RSA, 1988*. Department of Agricultural Development, Pretoria. pp. 215-262.

Oswald, J.D. (1993a) Phylogeny, taxonomy, and biogeography of extant silky lacewings (Insecta: Neuroptera: Psychopsidae). *Memoirs of the American Entomological Society* 40: 1-65.

Oswald, J.D. (1993b) Revision and cladistic analysis of the world genera of the family Hemerobiidae (Insecta: Neuroptera). *Journal of the New York Entomological Society* 101: 143-299.

Oswald, J.D. (1994) Revision of the Africa silky lacewing genus *Zygophlebius* Navás (Neuroptera: Psychopsidae). *African Entomology* 2: 83-96.

Smithers, C.N. (1961) New records and a key to the genus *Sisyra* Burm. In Southern Africa (Neuropt.: Sisyridae). *Journal of the Entomological Society of Southern Africa* 24: 308-309.

Tjeder, B. (1957) Neuroptera-Planipennia. The Lace-wings of Southern Africa. 1. Introduction and Families Coniopterygidae, Sisyridae, and Osmylidae. In: Hanström, B., Brinck, P. & Rudebeck, G. (eds.) *South African Animal Life* 4: 95-188. Swedish Natural Science Research Council, Stockholm.

Tjeder, B. (1959) Neuroptera-Planipennia. The Lace-wings of Southern Africa. 2. Family Berothidae. In: Hanström, B., Brinck, P. & Rudebeck, G. (eds.) *South African Animal Life* 6: 256-314. Swedish Natural Science Research Council, Stockholm.

Tjeder, B. (1960) Neuroptera-Planipennia. The Lace-wings of Southern Africa. 3. Family Psychopsidae. In: Hanström, B., Brinck, P. & Rudebeck, G. (eds.) *South African Animal Life* 7: 164-209. Swedish Natural Science Research Council, Stockholm.

Tjeder, B. (1961) Neuroptera-Planipennia. The Lace-wings of Southern Africa. 4. Family Hemerobiidae. In: Hanström, B., Brinck, P. & Rudebeck, G. (eds.) *South African Animal Life* 8: 228-534. Swedish Natural Science Research Council, Stockholm.

Tjeder, B. (1966) Neuroptera-Planipennia. The Lace-wings of Southern Africa. 5. Family Chrysopidae. In: Hanström, B., Brinck, P. & Rudebeck, G. (eds.) *South African Animal Life* 12: 228-534. Swedish Natural Science Research Council, Stockholm.

Tjeder, B. (1967) Neuroptera-Planipennia. The Lace-wings of Southern Africa. 6. Family Nemopteridae. In: Hanström, B., Brinck, P. & Rudebeck, G. (eds.) *South African Animal Life* 13: 290-501. Swedish Natural Science Research Council, Stockholm.

Tjeder, B. & Hansson, C. (1992) The Ascalaphidae of the Afrotropical Region (Neuroptera). *Entomologica Scandinavica Supplement* 41: 1-237.

Checklist: Lacewings of Mkomazi

Species determinations by M.W. Mansell & U. Aspöck

Class INSECTA

Order NEUROPTERA

Family Ascalaphidae

Ascalaphus festivus (Rambur, 1842). Ibaya Camp.

Dicolpus primitivus (Van der Weele, 1909). Maji Kununua.

Disparomitis ?hovarthi van der Weele, 1909. Ibaya Camp, Kisiwani River.

Dixonotus vansomereni Kimmins, 1950. Ibaya Camp.

Suhpalasca sp. Ibaya Camp.

Tmesibasis laceratus (Hagen, 1853). Ibaya Camp, Kikolo Plot.

Tmesibasis ?alberti Navás, 1912. Ibaya Camp, Kisima Plot.

Family Chrysopidae

Italochrysa ?peringuey (Esben-Petersen, 1920). Kamakota Hill.

Italochrysa ?lyrata Tjeder, 1966. Kamakota Hill.

Chrysoperla congrua (Walker, 1853). Mzukune River, Ibaya Camp, Kikolo Plot, Pangaro Plot, near Zange Gate, Ubani Plot.

Chrysoperla pudica (Navás, 1914). Kamakota Hill.

Chrysemosa jeanneli (Navás, 1914). Ibaya Camp.

Dichochrysa nicolaina (Navás, 1929). Kamakota Hill, Ibaya Camp.

Dichochrysa sjoestedti (van der Weele, 1910). Ibaya Camp.

Family Mantispidae

Species 1. Kikolo Plot.

Species 2. Ibaya Hill.

Family Psychopsidae

Zygophlebius zebra (Brauer, 1889). Kisiwani River,Valley NW Ibaya Camp.

Family Berothidae

Podallea vasseana (Navás, 1910) Ibaya Camp, Ubani Plot.

Family Rhachiberothidae

Mucroberotha copelandi U. & H. Aspöck, 1997. Kamakota Hill, Kikolo Plot.

Family Myrmeleontidae

Palpares papilionoides (Klug, 1834). Ibaya Camp.

Palpares ?sparsus (McLachlan, 1867). Ibaya Camp.

Palpares torridus Navás, 1912. Ibaya Camp.

Palpares tristis Hagen, 1853. Ibaya Camp, Maji Kununua.

Centroclisis brachygaster (Rambur, 1842). Valley NW Ibaya Camp.

Neuroleon guttatus (Navás, 1914). Ibaya Camp.

Gymnoleon elizabethae Banks, 1911. Ibaya Camp.

Cueta mysteriosa (Gerstaecker, 1894). Ibaya Camp.

Cueta klugi Hölzel, 1982. Ibaya Camp.

Cueta secretus (Navás, 1914). Ibaya Camp.

Hagenomyia lethifer (Walker, 1853). Ibaya Camp.

Hagenomyia tristis (Walker, 1853). Valley NW Ibaya Camp, Nakombo River, Ibaya Camp, Ibaya Hill.

Macronemurus sp. Ibaya Camp.

Banyutus lethalis (Walker, 1853). Valley NW Ibaya Camp, Kisiwani River.

Banyutus sp. Ibaya Hill.

Cymothales bouvieri van der Weele, 1907. Ibaya Camp.

Myrmeleon quinquemaculatus Hagen, 1853. Valley NW Ibaya Camp.

Creoleon nubifer (Kolbe, 1897). Ibaya Camp.

CHAPTER 14

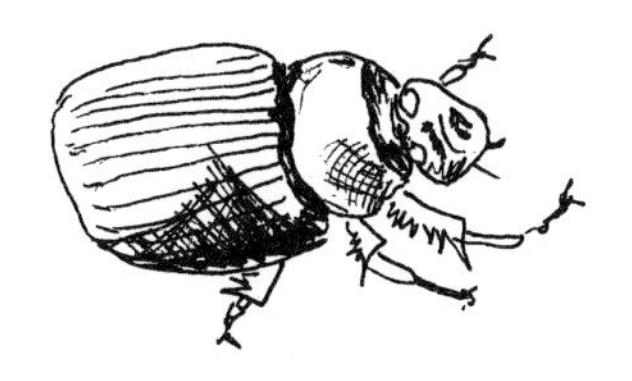

Beetles (Coleoptera) of Mkomazi

Jonathan G. Davies

Introduction

Studies of beetle diversity have been rare in savanna habitats, focusing more often on tropical rain forests, and in particular the tree canopies (see, for example, Stork *et al.* 1996). Yet even cursory investigation of ground-dwelling invertebrates in African savanna reveals an impressive diversity (and also biomass), which is orders of magnitude greater than that of vertebrates (Gillon 1983). Beetles represent a very significant proportion of this fauna. Indeed, Gillon describes the darkling beetles (Tenebrionidae), one of the three most abundant families in the savanna ground fauna along with the ground beetles (Carabidae) and dung beetles (Scarabaeidae), as one of the 'faunistic markers' of dry savannas.

The chief aim of the Mkomazi Ecological Research Programme was to inventory key elements of the biodiversity of Mkomazi, in order to facilitate conservation management of the reserve in the context of local human development. This requires selection and sampling of a representative sub-set of the plants and animals present. Of these, perhaps the most diverse group is the order Coleoptera, with its substantial variety of form, size and ecological function, ranging from fungivores and detritivores to herbivores, scavengers and predators. The ability to exploit such a wide range of food types, which has been one of the main factors in the success of beetles throughout the world, makes the group not only highly significant ecologically, but possibly more representative of overall biodiversity than most other groups of invertebrates.

There were four main objectives of this element of the programme:

1. to compile as comprehensive an inventory as possible of the Coleopteran fauna of Mkomazi;
2. to assess beetle diversity in a number of different habitat and tree types throughout the reserve;
3. to investigate the effects of savanna fires on beetle diversity and community composition; and
4. to study seasonal variations in both the diversity and composition of the beetle communities.

In order to fulfil the first and second objectives, sampling was carried out in as many different habitat types and tree species as time allowed, while repeat sampling of four areas situated close to Ibaya Camp was carried out throughout the year in order to achieve the third and fourth objectives.

Methods

Ground-dwelling beetles

Sampling protocol

Sampling of ground-dwelling beetles was carried out using pitfall trapping (a common sampling methodology in tropical savannas, owing to the particularly rich beetle faunas of these habitats). Each trap consisted of two plastic coffee cups (one inside the other) buried in the ground with the upper rim level with the surface of the substrate. The bottom cup serves as a lining for the pit, preventing it from collapsing, and allows for easy removal of the inner cup during servicing of the trap. Traps were quarter-filled with a water and detergent mixture and were emptied and refilled every 24 hours to avoid decomposition of the catch. At all trapping sites (see below) a total of 180 trap-days of sampling was completed, either by setting 30 traps over a period of six days (as done at the four Ibaya Camp sites) or by setting 60 traps and leaving them in place for three days (all other sites).

Study sites: sampling period 1 (11–29 November 1994)

IBAYA CAMP (burnt hillside, unburnt hillside, burnt grassland, unburnt grassland). Permanent grids of pitfall traps were established in four habitat types in the vicinity of the Camp at Ibaya to record the responses of the beetle fauna to various ecological parameters. In particular, changes in the diversity and community composition of beetles were monitored in relation to habitat type, savanna fire and seasonal influences.

Comparison was made: (i) between an area of hillside *Acacia-Commiphora* bush that had been burnt approximately three months previously, and an equivalent habitat where burning had not taken place (i.e. burnt and unburnt hillside, respectively), and (ii) between an area of footslope grassland that had been burnt three months previously, and a corresponding area that had last been burnt at least 18 months previously (burnt and unburnt grassland, respectively).

NDEA HILL (*Acacia senegal* parkland and *Acacia drepanolobium* vlei). Further trapping was carried out in two habitat types near Ndea hill, in the north-western corner of the reserve. The first was an area of *Acacia senegal* parkland, consisting of scattered mature *Acacia* trees and large patches of red, sandy soil. The area was very dry and bare, and may indeed have been burnt in the previous year, as suggested by the presence of numerous charred clumps of grass. The second area was a dried-up vlei (or floodplain) of black-cotton soil, dominated by metre-high grass and scattered bushes of whistling thorn *Acacia drepanolobium*. Neither area appeared to have received any of the rainfall experienced at Ibaya.

Study sites: sampling period 2 (1–22 April 1995)

IBAYA CAMP. Pitfall trapping was repeated in the four habitat types at Ibaya Camp, in order to monitor seasonal differences in faunal composition, and in particular to investigate the extent of the recovery of the beetle communities present in the two areas that had been burnt the previous year.

In order to increase the geographical coverage of the study, six further habitat types, at three different locations, were also sampled:

DINDIRA DAM (*Combretum* bush and *Dichrostachys* bush). These two habitats were situated at the base of the Vitewini ridge, to the north of Dindira Dam, in a distinct band of bush separating the grassland plain from the slopes of the ridge. The band was approximately 200 m wide, and apparently represented an ecological zone strongly influenced by rainwater run-off from the slope. Another ecological gradient appeared to act along the length of the band from one end to the other, a distance of perhaps 2 km, resulting in a gradual transition from almost pure *Combretum* at the western end to a similar 'monoculture' of *Dichrostachys* at the eastern end. Traps were set in each of these two distinct habitats.

IBAYA HILL (*Spirostachys* forest and *Setaria* grassland). Trapping was carried out on the top of Ibaya Hill, where there is a small relict patch of forest largely dominated by *Spirostachys* sp. On the very crown of the hill is also a small area (approximately 30 m^2) of *Setaria/Panicum* grassland, forming a quite distinct habitat type surrounded by the forest. The two sets of pitfall traps, despite being laid in these different habitats, were no more than 50 m apart.

UMBA RIVER (Riverine bush and *Commiphora* woodland). These final two habitats were sampled during a four-day expedition to the Umba River at the eastern end of the reserve. The intention had been to trap in 'riparian forest', but the vegetation present is better described as open riverine bush, consisting of *Grewia*, *Commiphora* and *Terminalia* bushes, the occasional *Acacia* tree, and patches of *Cenchris* grass smothered with a flowering *Epomoya* scrambler. In comparison with this site, an area of dry *Commiphora* woodland, approximately 3 km from the river and consisting of a patchwork of dense *Commiphora* thicket and bare areas of sandy soil, was also sampled. This eastern part of the reserve was notably hotter and drier than the area surrounding Ibaya Camp.

Study sites: sampling period 3 (May 1995 – March 1996)

In order to record seasonal variation in invertebrate community structure, all four Ibaya Camp habitat types were monitored throughout the year by monthly pitfall sampling, carried out by Ramadhani Makusi and Elias Kihumo from the Tropical Pesticides Research Institute. However, owing to the logistic constraints of processing all of the beetle material collected from 12 months of sampling, a full account of seasonal variation for this project was not possible. Instead, the collections made in August 1995 and January 1996 were used, in conjunction with those from the November and April samples from the earlier studies (above), to provide a relatively even distribution of sampling points throughout the year.

In summary, pitfall trapping was carried out: (i) in burnt and unburnt habitats at Ibaya (in order to investigate the effects of fire on invertebrate species diversity and composition); (ii) in the same Ibaya habitats every month for a whole year (May 1995 to April 1996) (to investigate seasonal differences in the fauna); and (iii) in a variety of different locations throughout the reserve (both to investigate the effect of habitat type and to contribute to an overall inventory for Mkomazi).

Arboreal beetles

Owing to the logistic constraints of processing all of the beetle material collected from the mist-blowing insecticide project (see McGavin, Chapter 22) the arboreal beetle study had to be restricted to the beetles from only a small sub-sample (17) of the total number of trees sampled. The beetles from this sample were used to investigate the diversity of, and faunal overlap between, different species of host plant. Sampling of these trees was carried out between 28 July and 5 August 1994. Their location and sampling details are provided in Table 14.1.

Processing of specimens

Beetles from both studies were pinned, mounted and labelled at the Natural History Museum in London, then sorted to family and species. A representative reference collection has been established at the Tropical Pesticides Research Institute in Arusha, while the remainder of the specimens are held at the Museum. Although the collections of ground-dwelling and arboreal species were compiled separately, any arboreal species already represented in the pitfall collections were given the same species number as previously allocated. A separate database was established for both studies, incorporating species' presence and abundance data for each habitat type (pitfall study) or each tree (arboreal study).

Data analysis

The diversity of the beetle communities investigated in this study has been described in terms of their species richness (that is, the number of species recorded

Table 14.1 Details of trees sampled for arboreal beetles by insecticide mist-blowing.

tree species	tree code	location	height (m)	collection area (m^2)
Lannea stuhlmanii	LSTL28	Ibaya Camp	6	4
Lannea stuhlmanii	LSTL29	Ibaya Camp	5	2
Terminalia sp.	TERM29	Ibaya Camp	10	2
Melia volkensii	MVOL29	Ibaya Camp	7	2
Albizia anthelmintica	AANT30	3.5 km N of Zange Gate	8	4
Acacia reficiens	AREF30	3.5 km N of Zange Gate	7	2
Acacia senegal	ASEN1A	1 km N of Zange Gate	8	2
Acacia senegal	ASEN1B	1 km N of Zange Gate	8	2
Acacia tortilis	ATORT1	1 km N of Zange Gate	10	2
Acacia drepanolobium	ADRP3A	Vitewini ridge	2	0.32
Acacia drepanolobium	ADRP3B	Vitewini ridge	2	0.32
Acacia drepanolobium	ADRP3C	Vitewini ridge	2	0.32
Acacia zanzibarica	AZAN4A	E of Ibaya Camp	3	0.64
Acacia zanzibarica	AZAN4B	E of Ibaya Camp	4	0.64
Acacia zanzibarica	AZAN4C	E of Ibaya Camp	7	0.64
Acacia zanzibarica	AZAN4D	E of Ibaya Camp	5	0.64
Acacia mellifera	AMELL5	3 km N of Zange Gate	4	2

from each habitat or tree) and the dominance (that is, the degree to which the fauna is dominated by a small number of species; calculated here as Simpson's dominance index, using the Biodiv software package). Most modern indices of diversity combine these two attributes in their calculation, but it is often easier to interpret the ecological implications of the results if species richness and dominance are examined separately. It is generally accepted that a community with high species richness and low dominance constitutes high 'biodiversity'.

Results and interpretation

The composition of the beetle fauna

A total of 675 species, belonging to 47 families, were collected by pitfall trapping, insecticide mist-blowing, and by hand collection. The pitfall sampling resulted in the collection of 5,338 individual beetles belonging to 421 species, while a total of 1,254 individuals, belonging to 167 species, were collected from the 17 trees sampled in the insecticide-spraying study.

The species composition of the different beetle families is shown in Table 14.2. The most species-rich families recorded in this study were the dung beetles

(Scarabaeidae, representing 13.9% of all species), leaf beetles (Chrysomelidae, 13.9%), ground beetles (Carabidae, 13.2%), weevils (Curculionidae, 9.6%), rove beetles (Staphylinidae, 6.5%) and darkling beetles (Tenebrionidae, 5.9%). With few similar studies providing comparative data on beetles, it is difficult to assess quite how rich the beetle fauna of Mkomazi is, relative to other areas of Africa, or how the family composition compares with that from other studies. However, the baseline inventory provided by this survey will serve as a standard against which future studies both in Mkomazi and elsewhere may be compared.

It is also, of course, impossible to know exactly what proportion of the total beetle fauna of Mkomazi has been collected during this study. Only a long-term, exhaustive survey of the reserve could provide an accurate overall figure. A species accumulation curve for the pitfall sampling to date (Figure 14.1) reveals that the number of species from the reserve continued to increase significantly as more habitats were surveyed during the study, with no evidence that a limit had been reached. Furthermore, even though approximately 90% of the species so far recorded were collected in one or other of the two wet seasons, sampling at different times of year is also likely to add considerably to the overall species list for Mkomazi. An accurate inventory of overall beetle species richness for Mkomazi would therefore require not only the detailed study of all unsurveyed habitat types, but also thorough sampling to be undertaken at different times of year. Large parts of the reserve were inadequately covered by this study, and these areas are likely to include habitat types not sampled to date.

The distribution of beetles across habitats in Mkomazi

Information regarding the distribution of beetle species in different parts of the reserve helps to identify the more diverse habitats and those areas which contain

Table 14.2 Species richness of the different beetle families (arranged in systematic order) from Mkomazi Game Reserve.

family	common name	no. of species
Carabidae	ground beetles	89
Cicindelidae	tiger beetles	8
Staphylinidae	rove beetles	44
Pselaphidae	short-winged mould beetles	5
Leiodidae	round fungus beetles	4
Scydmaenidae	antlike stone beetles	7
Scaphidiidae	shining fungus beetles	1
Histeridae	hister beetles	11

continued

Table 14.2 continued

family	common name	no. of species
Scarabaeidae	scarab beetles	94
Limnichidae	minute marsh-loving beetles	1
Buprestidae	jewel beetles	23
Elateridae	click beetles	12
Cantharidae	soldier beetles	2
Lycidae	net-winged beetles	2
Dermestidae	dermestid beetles	2
Anobiidae	death-watch beetles	8
Bostrichidae	branch-and-twig borers	3
Peltidae	bark-gnawing beetles	2
Trogositidae	bark-gnawing beetles	1
Cleridae	chequered beetles	2
Melyridae	soft-winged flower beetles	14
Nitidulidae	sap beetles	13
Cucujidae	flat bark beetles	1
Silvanidae	flat bark beetles	2
Phalacridae	shining mould beetles	11
Erotylidae	pleasing fungus beetles	1
Corylophidae	minute fungus beetles	7
Coccinellidae	ladybird beetles	17
Endomychidae	handsome fungus beetles	1
Lathridiidae	minute brown scavenger beetles	4
Biphyllidae	biphyllid beetles	1
Monommidae	monommid beetles	2
Colydiidae	cylindrical bark beetles	3
Tenebrionidae	darkling beetles	40
Lagriidae	long-jointed bark beetles	1
Alleculidae	comb-clawed beetles	4
Scraptiidae	false darkling beetles	2
Mordellidae	tumbling flower beetles	5
Meloidae	blister beetles	24
Anthicidae	antlike flower beetles	13
Aderidae	antlike leaf beetles	1
Cerambycidae	long-horned beetles	14
Chrysomelidae	leaf beetles	94
Bruchidae	seed weevils	5
Anthribidae	fungus weevils	1
Curculionidae	weevils	65
Scolytidae	bark-and-ambrosia beetles	8

rare or restricted species. Such information is provided both by the pitfall study of habitat types and by the insecticide fogging study of different tree types.

Habitat types

SPECIES RICHNESS AND DOMINANCE. Figure 14.2a shows the species richness of the ground-dwelling beetle faunas from ten different habitat types throughout Mkomazi. The clearest feature is the high species richness in the two Dindira habitats (*Combretum* bush and *Dichrostachys* bush, E and F respectively in the figure), the two Ibaya Hill habitats (*Spirostachys* forest and *Setaria* grassland, G and H), and the two Ibaya Camp habitats (hillside *Acacia-Commiphora* bush and footslope grassland, A and B). This suggests that these sites, all of which are situated in the west of the reserve, may contribute significantly to beetle diversity in Mkomazi. In contrast, the dry habitats around the Umba River (riverine bush and *Commiphora* woodland, I and J) and Ndea Hill (*Acacia senegal* woodland and *Acacia drepanolobium* vlei, C and D) produced much lower numbers of species.

In contrast to the low species richness in the two Ndea Hill sites, species dominance was relatively high (Figure 14.2b), indicating that a small number of species here were occurring in large numbers. In the dry and sandy *A. senegal* woodland the dominant species was a zophosine darkling beetle, or 'sand-swimmer'. It is likely that the ability of these beetles to bury themselves rapidly under the surface in avoidance of predators gives them a competitive advantage in the bare, sandy soil. This species was completely absent from the nearby vlei, but dominance was nevertheless also high here, with two blister beetles (Meloidae), *Coryna arussina* and *Ceroctis rufimembris*, particularly prevalent. The water-loving ground beetle *Styphromerus 4-maculatus* was also present in significant numbers, an indication

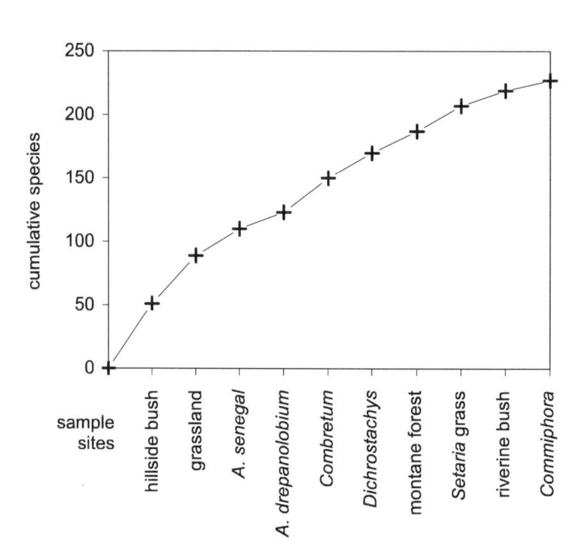

Figure 14.1 Cumulative numbers of beetle species from pitfall traps in Mkomazi Game Reserve.

that the vlei, despite being parched at the time of sampling, becomes waterlogged in the wet season. The beetle communities of the two Umba River habitats were also dominated by blister beetles and darkling beetles (including the same sand-swimmer as above), with very few species belonging to other beetle families.

COMMUNITY COMPOSITION. Although the wetter habitats had a greater diversity of beetles than the drier areas, a large proportion of their species seem to be relatively ubiquitous throughout Mkomazi. In contrast, despite yielding few species, the drier habitats were found to contain species not found elsewhere in the reserve, with the implication that they may contribute significantly to beta diversity (i.e. the overall species complement of the region). Thus, for example, although a total of 75 beetle species were collected in the two Dindira habitats in comparison with only 43 from the two *Acacia* habitats at Ndea, it was the latter, drier, site which had a

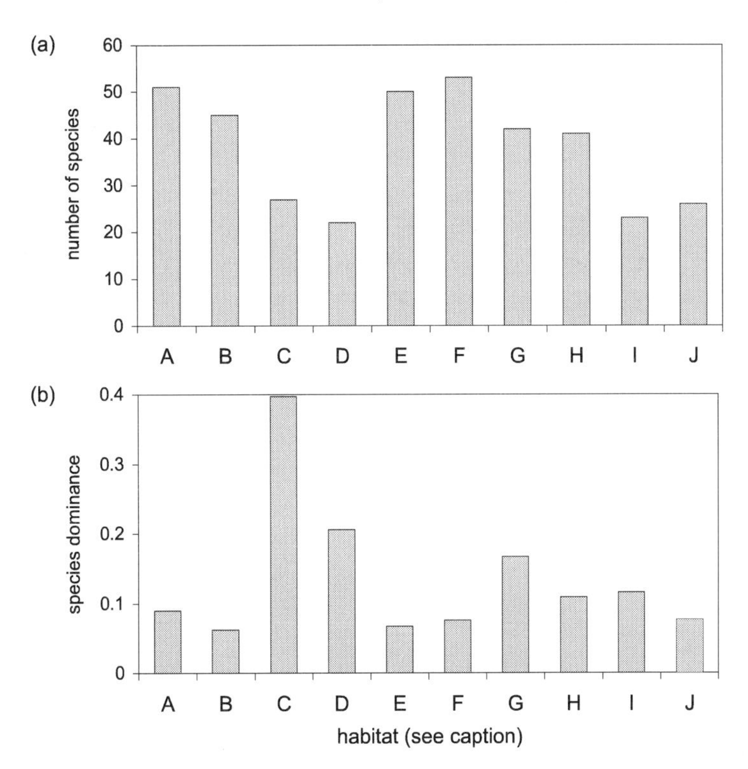

Figure 14.2 (a) Species richness and (b) dominance of beetle faunas of ten habitats in Mkomazi Game Reserve. Habitat codes:

A = hillside *Acacia-Commiphora* bush	B = footslope grassland	(Ibaya Camp)
C = *Acacia senegal* woodland	D = *Acacia drepanolobium* vlei	(Ndea Hill)
E = *Combretum* bush	F = *Dichrostachys* bush	(Dindira Dam)
G = *Spirostachys* forest	H = *Setaria* grassland	(Ibaya Hill)
I = Riverine bush	J = *Commiphora* woodland	(Umba River)

greater number of unique species: 22 of the species from Ndea (51% of the total) were only collected from this site while only 19 of the 75 species from Dindira (25% of the total) were found only here.

Beetle samples from the drier habitats around the reserve (including the Ibaya habitats during the dry season, and in particular the burnt sites) revealed a relatively high proportion of shared species, suggesting that similar communities of arid-land species may exist throughout Mkomazi when conditions become hot and dry. The most prominent of these 'xerophilic' specialists are the darkling beetles (particularly the sand swimmers) which are abundant in extensive areas of bare sandy soil, such as occur in burnt or arid habitats. Darkling beetles, therefore, would appear to be good indicators of restricted water availability in Mkomazi. They are most abundant and species-rich in the dry, eastern habitats of riverine bush (7 species, 23 individuals) and *Commiphora* woodland (8, 35) near the Umba River, and in the bare sandy soil of the *Acacia senegal* woodland (10, 190). Even on the top of Ibaya Hill, where the two habitats sampled are within 50 m of one

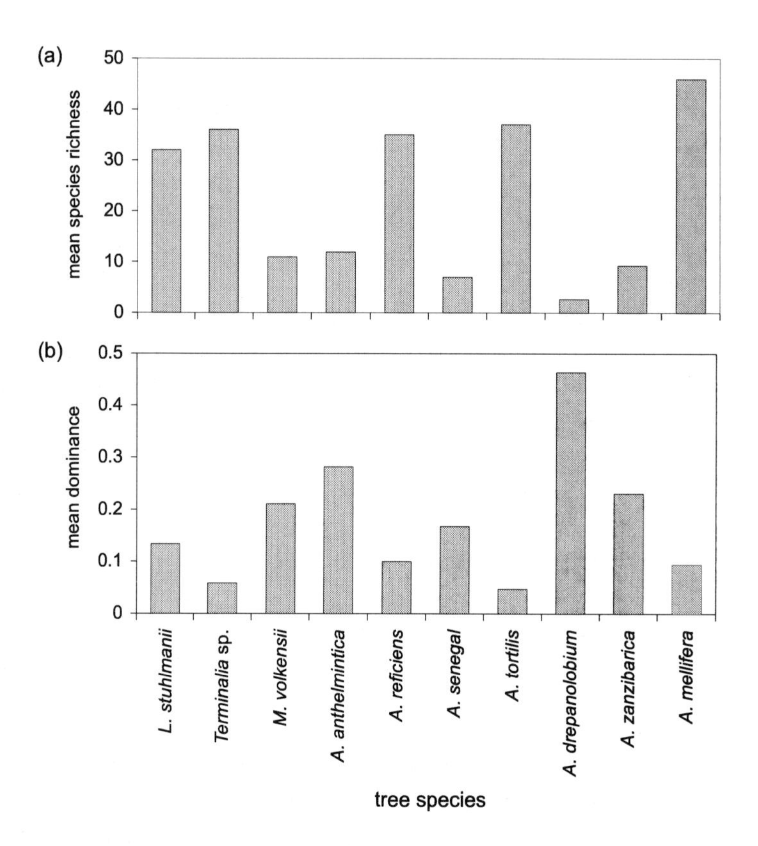

Figure 14.3 Species richness (a) and dominance (b) of beetles from the canopies of ten tree species in Mkomazi Game Reserve.

another, the open and drier *Setaria* grassland produced four species and 37 individuals, compared to none at all in the forest.

Analysis of community composition in the two Ibaya Hill habitats indicates that they support a very distinct and localised fauna. Of the 42 species collected in the forest, 31 (74%) were unique to this habitat, while in the *Setaria* grassland 41 species were collected, of which 17 (42%) were found only here. Clearly, the forest, and to a lesser extent the *Setaria* grassland, support highly distinctive as well as diverse beetle communities. This small patch of forest, rare in the Mkomazi landscape, is thus of particular importance for the biodiversity of the reserve.

Identifying further habitats which make a significant contribution to the biodiversity of Mkomazi, either in terms of the numbers of species they support or the distinctive nature (or rarity) of their faunas, is clearly important for the long-term management of the reserve. Further surveys, particularly in the rarer habitats of Mkomazi, are likely to produce many more beetle records for the reserve, and also help in any future identification of priority habitats for protection.

Tree type

Sampling of the 17 trees analysed in this study was carried out between 28 July and 5 August 1994. Details of the trees are given in Table 14.1.

SPECIES RICHNESS AND DOMINANCE. Species richness and dominance of beetles varied considerably between tree species (Figure 14.3). The greatest number of species (46) was collected from *Acacia mellifera*, which was a small tree with a limited canopy. Five other trees, the two *Lannea stuhlmannii*, *Terminalia* sp., *Acacia reficiens* and *Acacia tortilis*, which varied in height from five to ten metres, produced similar numbers of species (from 32 to 37), while each of the other trees had fewer than 15 species. The smaller number of species from the three *Acacia drepanolobium* bushes (which produced a total of only eight species and 11 individuals between them), and to a lesser extent the four *Acacia zanzibarica* trees, are probably related to the small size of the trees and the limited canopy area from which insects were fogged. However, *Melia volkensii*, *Albizia anthelmintica* and the two *Acacia senegal* also exhibited low species diversity despite their relatively large heights, canopy areas and collecting areas.

Unsurprisingly, therefore, analysis of the relationship between species richness and the various measures of tree and/or sample size revealed no significant correlations ($r^2 = 0.289$, $p>0.05$). This suggests that species richness was related more to the type of tree sampled than to the size or other physical features of the trees.

Species dominance also varied between tree species, with the species-poor trees generally having much higher dominance (Figure 14.3b). In *Albizia anthelmintica* for example, two beetle species were particularly common (an ant-like flower beetle with 30 individuals and a soft-winged flower beetle with 48 individuals), thus dominating the tree's beetle fauna. Similarly, dominance was high (and species richness low) in the three *Acacia drepanolobium* trees. This may be because only a limited number of species are able to share the bushes with the species of ant which guards this species from insect attack (see Stapley, Chapter 23). Conversely, none of the 37 species collected from the *Acacia tortilis* were represented by more than eight individuals, with the result that the dominance score was very low.

COMMUNITY COMPOSITION. Analysis of community composition reveals that trees from the same species shared more beetles than those from different species. The faunas of the four *Acacia zanzibarica* trees, for example, had a number of beetle species in common: two species (a leaf beetle and a weevil) were common to all four trees, while a further three were found on three of the trees (one minute fungus beetle, a sap beetle and another weevil). A similar pattern is revealed by the two *Lannea stuhlmannii* trees, which shared 16 (exactly 50%) of their species.

There was very little similarity between the faunas of different tree species. The one exception was the *Acacia mellifera* and the *Terminalia* sp. which, perhaps surprisingly for species from different genera, shared 14 species of beetles. However, 13 of these were also found in other trees around the reserve, suggesting that they were 'tourists' (that is, non-host specific). The fauna of *Acacia tortilis*, as well as being relatively rich in species, was also quite distinct from those found on the other trees in the study, suggesting that this species (similar to the *Spirostachys* forest in the habitat section above), might be of particular importance to the overall beta diversity of Mkomazi.

The low similarity between the faunas of the different tree species in this study suggests that the sampling of further trees would continue to augment the beetle species list for the reserve, particularly since the ten species sampled here represent only a small fraction of the total number of species recorded for the reserve. The arboreal beetles therefore clearly constitute a very significant proportion of the total beetle fauna of Mkomazi. The limited time available for analysing the beetle material from the mist-blowing study has greatly restricted the scope of the investigation. An assessment of the other taxa collected by mist-blowing, including an estimate of the total number of arboreal insect species in Mkomazi, is given elsewhere in this volume (see McGavin, Chapter 22).

The effect of burning on beetle diversity at Ibaya

Beetle species richness and abundance for burnt and unburnt habitats at Ibaya was recorded at four different times of year (January, April, August and November; Figure 14.4). None of the differences in species richness between burnt and unburnt habitats (Figures 14.4a and 14.4c) were found to be statistically significant, but there were significant differences between habitats in terms of the number of individuals caught. In the hillside *Acacia-Commiphora* bushland sites (Figure 14.4b), beetle abundance in November 1994 was significantly higher in the unburnt area (Chi-squared test: $\chi^2 = 26.5$, $p < 0.001$) while in April 1995 (Chi-squared test: $\chi^2 = 193.2$, $p < 0.001$) and January 1996 (Chi-squared test: $\chi^2 = 121.7$, $p < 0.001$) it was significantly higher in the burnt area. This suggests that although burning initially reduces beetle populations, in the longer term it results in significantly higher beetle abundance and activity. This may be due to the removal of the large biomass of dead grass (which is likely to physically impede beetle movement over the substrate), and/or through the rejuvenation of the site owing to the release of recycled nutrients to the soil.

By contrast, in the footslope grassland sites (Figure 14.4d) the only major difference between burnt and unburnt habitats was in April 1995, when beetle abundance was significantly higher in the unburnt patch (Chi-squared test: $\chi^2 = 9.22$, $p < 0.01$). This means that in April 1995 beetle abundance on the hillside was significantly higher in the burnt area while in the grassland it was higher in the unburnt area. This disparity perhaps relates to the recent history of burning events in the two sites. While there was no recent record of fire in the unburnt hillside, the 'unburnt' grassland had in fact experienced burning 18 months previously. The higher abundance recorded in the unburnt grassland relative to the unburnt hillside may, as above, be due to a lower biomass of dead grass and a higher nutrient status, resulting from the earlier fire.

Burning, therefore, can play a positive ecological role in the savanna by removing excess ground vegetation. However, the sparse and patchy vegetation left after fire will also temporarily reduce beetle numbers and diversity, owing to the increased vulnerability to predation. This may explain the lower diversity in the burnt habitats in November 1994 (only three months after the fire) relative to April 1995 (eight months after the fire). Optimal conditions are likely to exist some months after a burn, particularly if the subsequent wet season brings about a recovery in the vegetation. Regeneration of the grass layer is more than likely to have taken place in the burnt habitats between August 1995 and January 1996, following the November rains, which may explain the consistent increases in beetle species richness and abundance in the burnt habitats between these two dates (Figure 14.4). Indeed, the only habitat in which no significant increase in abundance was observed was the unburnt hillside, the one habitat that was already heavily overgrown, and in which further growth of the grass layer was unlikely to benefit the fauna.

The clearest and statistically most significant (see following section) difference in Figure 14.4 is between the high abundance and diversity in November 1994 and April 1995 (i.e. three and eight months after the fire, respectively) and the low values in the August 1995 and January 1996 samples (i.e. 12–17 months after the fire). Although this might be interpreted as an effect of time elapsed since fire, it is more likely to result from seasonal rainfall in November and April. This highlights the difficulties associated with attempting to isolate fire-induced popu-

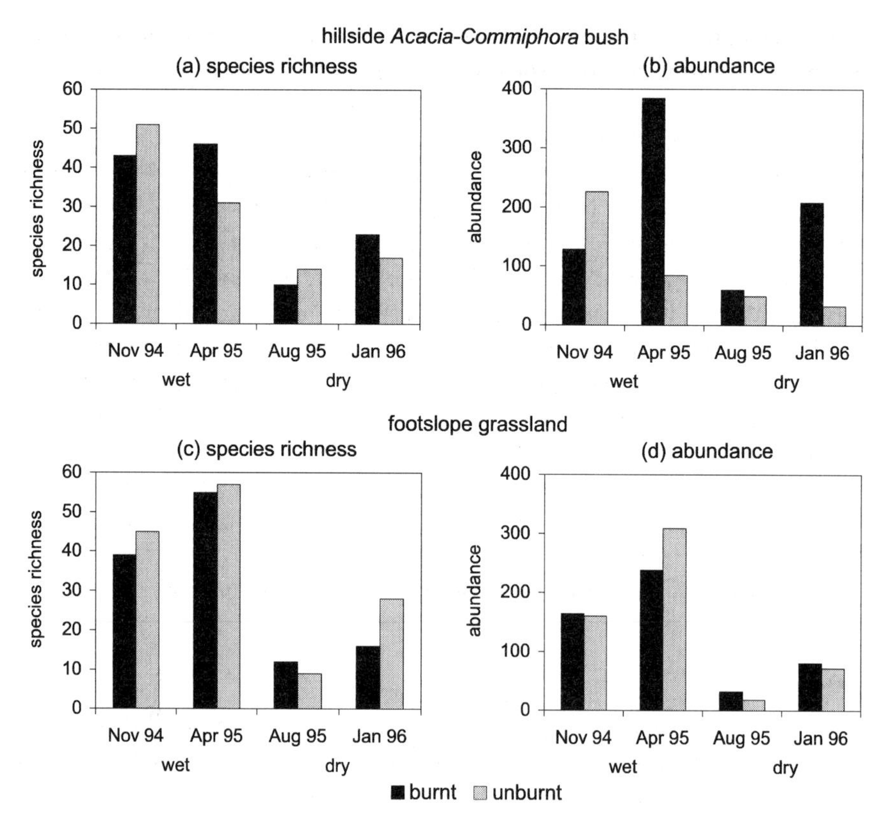

Figure 14.4 Effects of burning on species richness and abundance of ground-active beetles in hillside *Acacia-Commiphora* bushland (a and b) and footslope grassland (c and d) at Ibaya Camp, Mkomazi Game Reserve.

lation changes from background seasonal variation. In an attempt to control for this, sampling in this study was carried out not only in burnt and unburnt areas but also at four different times of year. However, in order to take proper account of seasonal variation it would probably be necessary to monitor population changes monthly for several years. Such shortcomings must always be borne in mind when investigating the impacts of fire on insect population dynamics.

Whelan (1995), in a review of studies on responses of ground-dwelling invertebrates to fire, found that results were equivocal; in some cases invertebrate abundance was reported to increase after fire, while in others there was a decrease. There are similar discrepancies in the recorded effects of fire on vegetational diversity, with, in one case, two authors in the same volume giving contradictory evidence; one suggests that protection from fire leads to an increase in plant diversity (Braithwaite 1996), while the other reports that the elimination of fire produces an accumulation of plant biomass, which reduces diversity (Bulla 1996). Potential causes of such contradictory results (for both plants and invertebrates) may be: (i) variations in fire intensity from study to study; (ii) differential responses by the various constituent taxa (i.e. some groups will suffer great losses while others benefit from the burn); and (iii) differences due to the use of a wide variety of sampling protocols (Whelan 1995).

The use of pitfall sampling for monitoring population dynamics following a fire has further, intrinsic, problems associated with it. Capture rates in pitfall traps are more an indication of beetle activity than true population size, and increased catches may simply reflect increased activity of a reduced population. There are three main reasons why such increases in activity are particularly likely after a fire: firstly, exposed and vulnerable invertebrates will increase active searching for holes to avoid predation; secondly, the distance and duration of foraging will need to increase in order to maintain adequate food input in an impoverished environment; and thirdly, the shift in complexity of the habitat from three-dimensional to two-dimensional (and the ensuing reduction in niche diversity) is likely to result in an increased pitfall catch of those species normally associated with the tree and shrub layer. Therefore, increases in the number of beetles caught directly after fire are as likely to be due to changes in behaviour as to real increases in population size.

Seasonal variations in the beetle fauna of Ibaya

As mentioned in the previous section, the year-long pitfall study at the Ibaya Camp sampling sites (Figure 14.4), showed that beetle species richness and abundance were significantly higher in the two wet season sampling periods (November and April) than in the two dry season sampling periods (January and August) for all four Ibaya habitats (Mann-Whitney tests: species richness: $U=6$, $p<0.05$; abundance: $U=19$, $p<0.05$). This suggests that a large proportion of the beetles of Mkomazi time adult emergence to coincide with the rains in order to profit from increased food availability.

Savanna invertebrates generally tend to be more abundant during the wet season, though different taxonomic groups respond in different ways. Gillon (1983), working in an area of relatively high annual rainfall (over 1,200 mm), identified four patterns of seasonal abundance exhibited by savanna arthropods: (i) peak abundance during the rainy season itself (e.g. cockroaches, grasshoppers, assassin bugs and ground beetles); (ii) an increase leading up to a peak at the beginning of the next dry season (e.g. bush crickets and lepidopteran caterpillars); (iii) peak abundance coinciding with the beginning of the annual grass fires (e.g. most homopteran bugs); and (iv) a multimodal pattern with two or more peaks (e.g. mantises and ruteline beetles). Some groups show no discernible variation in abundance at all (e.g. some spiders). The beetles in this study would appear to exhibit two peaks of abundance and diversity, associated with the two wet seasons of November and April, though a more long-term study would be required to investigate this more rigorously.

The evenness of the beetle populations also varied seasonally, with dominance significantly higher in the two dry season sampling periods than in the two wet seasons (Mann-Whitney test: $U = 19$, $p < 0.05$), a pattern that was particularly pronounced in the burnt hillside and burnt grassland. These habitats were dominated in the dry season by large numbers of the same species of sand swimmer (in both January and August), as well as by the blister beetle *Coryna chevrolati,* an ant-like flower beetle, and two hister beetles. This suggests that stressful conditions (i.e. those that are very hot and dry) favour a few species that are specifically adapted to them.

Most other savanna insects have evolved adaptive mechanisms to avoid the worst effects of the dry season. The most common of these 'escape strategies' is diapause, during which development is arrested. Many species thus spend periods of drought underground, burrowing into the relatively loose soil towards the end of the rainy season, with the hard soil of the dry season serving to protect them both from predators and from savanna fires. This explains the low numbers of beetle species and individuals caught during the dry season sampling periods.

Finally, seasonal variations of climate tend to be far less marked and much more predictable in wetter savannas than in the drier savannas (Gillon 1983). If a similar pattern was to exist between the wetter and drier parts of Mkomazi (from wetter in the west, to drier in the east), a study of seasonality in the eastern end of the reserve, similar to that carried out at Ibaya, may be expected to produce even greater differences in beetle species richness between in wet and dry seasons.

Discussion: implications for the management of Mkomazi

This study of the diversity and distribution of the beetle fauna of Mkomazi, and the ecological factors influencing it, provides information for the long-term management of the reserve. In particular, it identifies those areas of the MGR which are of particular significance for biodiversity, and which are thus most worthy of sympathetic management.

It also suggests that biodiversity conservation would be best served through the maintenance of a mosaic of different habitat types, each contributing unique species to the overall list for the reserve. In particular, areas such as the small patches of *Spirostachys* forest and *Setaria* grassland on the top of Ibaya Hill are of great importance owing to the richness and limited distribution of the beetle communities they contain. Nearly three-quarters of the beetle species collected in the forest were found nowhere else in the study, suggesting that if this restricted habitat were to be removed there would be a significant local extinction of species.

The climate gradient in the reserve, from relatively wet in the west to hot and dry in the east, also seems to have a marked influence on the distribution and diversity of beetles. Although the habitats in the west tend to support a greater diversity of beetle species, those in the east sustain important populations of arid-land specialists not found in the wetter areas, emphasising that those areas with low species richness are by no means lacking in value.

Two of the most fundamental ecological factors in the savanna, fire and rainfall, have a profound effect on beetle diversity, both directly and indirectly. The direct effects of both are usually catastrophic, with high beetle mortality often associated with burning and, less significantly, flooding events. Indirectly, though, the effects can often be more beneficial. Both fire and rainfall play key roles in plant productivity, the former through the removal of dead material and the recycling of nutrients, and the latter through the provision of plant-available moisture. The resultant higher plant productivity is strongly related to vegetational diversity (Bulla 1995), which is, in turn, likely to have a major influence on the productivity and diversity of both primary and secondary consumer insects. However, the correlation between productivity and diversity is not linear, with maximum plant species richness likely at intermediate levels of productivity. This is supported by the increased diversity of beetles recorded in areas around Ibaya which had burnt but in which the vegetation had recovered following subsequent rains (for example, the burnt hillside and burnt grassland in April, Figure 14.4).

Reductions in diversity seem to be associated with stress and limited resources at the lower end of productivity, and with competition at the higher end. At the lower end of productivity, this would seem to explain, firstly, the low levels of beetle diversity recorded in the recently-burnt habitats in November 1994 relative to the same areas five months later (suggesting that the vegetation in April 1995 had developed to a high, but not inhibiting level), and secondly, the very low figures for the dry season. At the higher end of productivity, the large accumulation of dead vegetation is likely to exclude some beetle species while favouring a small number of others, thus reducing diversity as well as equitability. This may explain the relatively low diversity, throughout all four seasons, recorded in the only heavily overgrown habitat (i.e. the unburnt hillside).

It therefore seems likely that controlled burning of patches of savanna, or at least a policy of non-intervention towards natural fires, may help to maintain the overall species richness of the reserve. In particular, early (low-intensity) dry season fires will remove dead grass without being too destructive. Such fires typically impact minimally on the canopies of trees, while higher intensity fires, resulting from much greater accumulation of dead plant biomass, can scorch high into the canopy and cause significant mortality in arboreal insects (Braithwaite 1996). It is important to stress that any increases in diversity after burning are likely to be transitory; according to Tainton & Mentis (1984), burnt tropical savannas may, in the long term, support only about 30% of the invertebrate fauna of unburnt savannas, as the moisture-loving arthropod groups tend to be excluded. This only serves to re-emphasise (a) the importance of maintaining a mosaic of habitats (including those that are both burnt and unburnt), and (b) the need for more research into the long term impacts of fire on savanna insect diversity.

As indicated by species accumulation curves, much of the beetle fauna of Mkomazi remains to be recorded. More extensive sampling, and particularly in locations not visited in the current study, will greatly enhance the overall assessment of the extent and distribution of beetle diversity throughout the reserve. When used in conjunction with information provided by the studies of other taxonomic groups, these results will facilitate the effective conservation management of the biodiversity of Mkomazi as a whole.

Acknowledgements
The author would like to thank Ramadhani Makusi and Elias Kihumo for their invaluable assistance with the field work, and Martin Brendell, Stuart Hine and Peter Hammond of the Natural History Museum, London, for their help with the identification of material. This work was supported by the UK Darwin Initiative.

References

Braithwaite, R.W. (1996) Biodiversity and fire in the savanna landscape. In: Solbrig, O.T., Medina, E. & Silva, J.F. (eds.) *Biodiversity and Savanna Ecosystem Processes*. Ecological Studies, Vol. 121, Springer-Verlag, Berlin.

Bulla, L. (1996) Relationships between biotic diversity and primary productivity in savanna grasslands. In: Solbrig, O.T., Medina, E. & Silva, J.F. (eds.) *Biodiversity and Savanna Ecosystem Processes*. Ecological Studies, Vol. 121, Springer-Verlag, Berlin.

Gillon, Y. (1983) The invertebrates of the grass layer. In: Bourlière, F. (ed.) *Tropical Savannas (Ecosystems of the World: Vol. 13)*. Elsevier, Amsterdam. pp. 289-311.

Menaut, J.C. (1983) The vegetation of African savannas. In: Bourlière, F. (ed.), *Tropical Savannas (Ecosystems of the World: Vol. 13)*. Elsevier, Amsterdam. pp. 109-150.

Samways, M.J. (1994) *Insect Conservation Biology*. Chapman and Hall, London.

Solbrig, O.T. (1996) Summary and Conclusions. In: Solbrig, O.T., Medina, E. & Silva, J.F. (eds.) *Biodiversity and Savanna Ecosystem Processes*. Ecological Studies, Vol. 121, Springer-Verlag, Berlin Heidelberg.

Stork, N.E., Adis, J. & Didham, R.K. (eds.) (1996). *Canopy Arthropods*. Chapman and Hall, London.

Tainton, N.M. & Mentis, M.T. (1984) Fire in grassland. In: P. de V. Booysen & N.M. Tainton (eds.) *Ecological effects of fire in South African ecosystems*. Ecological Studies Vol. 48, Springer-Verlag, Berlin.

Whelan, R.J. (1995). *The ecology of fire*. Cambridge University Press, Cambridge.

Checklist: Beetles of Mkomazi

In the limited time available, the following species list was made of those species collected by pitfall trapping. It includes only taxa identified to species or genus level. The full number of taxa recorded for each family is given in Table 14.2.

Carabidae
Chilanthia cavernosa
Chlaenius (coscinioderus ?)
C. fulvosignatus (v. *agraphus ?)*
C. obesus
C. (Stenodinodes) pachydinodoides
C. pauli
C. pulchellus (v. *fraternus* ?)
C. (pulchellus ?)
C. (subculcatus ?)
C. sellatus (ssp. *epistrophus* ?)
Clivina (montei ?)
Crepidogaster dollmani
C. (lateralis ?)
C. ornatus
Cylindera rectangularis
Cypholoba bihamata
C. cinereocincta
C. hamifera
C. perspicillaris
C. somereni
C. spatulata
C. tenuicollis
Dischiridium sp. 1
Disphaericus sp. 1
Disphaericus sp. 2
Dromica egregia v. *neumanni*
D. nobilitata
Eccoptopterus cupricollis
Eudema symei
Eudema sp. 1
Eudema sp. 2
Eudema sp. 3
Euryarthron sp. 1
Galeritiola procera
Graphipterus interlineatus ssp.1
G. interlineatus ssp. 2
Hipparidium sp. 1
Lasiocera egregia
Megacephala regalis
Paussus antinorii
Pheropsophus (jurinei ?)
Prothyma quadripustulata
Psecadius (Isotarsus) eustalactus
Pterostichus sp. ?
Rhopaloteres (nr. *compressicornis)*
Scarites aestuans
S. molossus
Siagona (brunnipes ?)
Styphromerus 4-maculatus
S. ludicrus
S. titshacki
Systolocranius elongatus
S. (sulcipennis ?)
Tefflus carinatus
Tetragonoderus scitulus
Thermophilum binotata
Triaenogenius sculpturatus

Drilidae
Selasia pulchra

Scarabaeidae
Allogymnopleurus umbrinus
Anachalcos convexus
Bolbocerus (princeps ?)
Catharsius sesostris
Cleptocaccobius biceps
C. viridicollis
Copris amyntor
C. harrisi
C. mesacanthus
Digitonthophagus gazella

Onthophagus aeruginosus
O. atricolor
O. (nr. *atricolor)*
O. bicavifrons
O. crucenotatus
O. (fimetarius ?)
O. leucopygus
O. (misellus ?)
O. quadrimaculatus (or *talpa* ?)
O. rufobasalis
O. sansibaricus
O. variegatus
Orphnus (zanzibaricus ?)
Scarabaeus gangeticus goryi
Sisyphus crispatus
S. seminulum
Sisyphus sp. 1
Sisyphus sp. 2
Sisyphus sp. 3
Trox (asperulatus ?)
Trox (baccatus ?)
Trox (niloticus ?)

Buprestidae
Acmaeodera ruficaudis
A. virgo

Tenebrionidae
Psammodes sp. 1

Meloidae
Ceroctis rufimembris
Coryna ambigua
C. apicicornis
C. arussina
C. chevrolati
C. parenthesis
Coryna sp. 1
Cylindrothorax strangulata
C. (sudanica ?)
Epicauta albovittata
Meloe proscarabaeus
Mylabris aperta
M. aperta v. *bioculata*
M. bihumerosa (=amplectens)
M. bipartita
M. praestans
M. hypolachna
Mylabris sp. 1
Zonitopsis (elongaticeps ?)

Cerambycidae
Amphistylus pauli

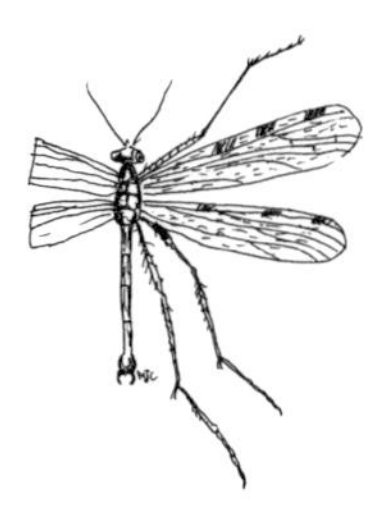

CHAPTER 15

Hangingflies (Mecoptera: Bittacidae) of Mkomazi

Jason G.H. Londt & Simon van Noort

Introduction

The Order Mecoptera is a very small group of insects related to the true flies (Diptera) and to the fleas (Siphonoptera). Two families, the Panorpidae and Bittacidae, include most of the approximately 500 known world species. The Panorpidae, commonly known as scorpion flies do not occur in Africa. The Bittacidae includes about 145 species spread throughout the temperate and tropical regions and is the only family present in Africa. Representatives of this family are called hangingflies because they have a habit of hanging from plants by their fore legs or sometimes their mid legs, freeing the other legs for prey capture. Adults are medium to large insects with an average wingspan of around 40 mm. Hangingflies are quite commonly encountered in their preferred habitats, however, adults may only be active for a few weeks every year and their emergence is easily missed.

Ecology

Very little is known about the ecology of African Bittacidae. Apart from a few examples of the eggs of about five South Africa species having been observed, the immature stages are otherwise completely unknown. Field collected imagoes frequent a wide range of vegetational situations ranging from the herbaceous ground-cover of forest floors to open grassland. The majority of species appear to prefer rank vegetation along the margins of forests or under large trees. Species found in more open situations appear to be more common in protected depressions or in rank vegetation bordering streams. A number of species are commonly attracted to lights at night and have been collected in good numbers during evenings of light rain.

Bittacidae are efficient predators of other insects. While no published study of their feeding preferences exists, the range of prey which has been observed is wide, suggesting catholic tastes. Despite their deceptively fragile appearance, hangingflies are capable of grappling with and overcoming fairly robust insects such as honey-bees. Their highly prehensile and raptorial tarsi very effectively overcome captured prey items, which are then swiftly killed by a bite coupled with the injection of powerful salivary secretions. The elongated rostrum may be an adaptation to protect the bittacid's eyes and slender antennae from being damaged by struggling prey. Under laboratory conditions adult hangingflies may capture more than one prey item at a time (one specimen has been observed hanging from a twig by one leg and in possession of five prey items, one restrained by each of the free tarsi). While hangingflies probably lie in wait for their prey, some observations suggest that they may select a perch close to the head of a flowering plant thereby making more possible the capture of insects which visit the flower, but these observations require confirmation.

The courtship behaviour of hangingflies is both fascinating and well documented. Males capture prey, attract a female by means of pheromones secreted from glands situated between the terminal abdominal terga, pass the prey item to the female, and copulate with her while she is busy feeding on the prey. Such observations have led to the suggestion that female hangingflies are incapable of capturing prey. This is, however, false: both sexes are capable of capturing prey.

Diversity

The African Mecoptera fauna consists of 53 species in three genera of Bittacidae (Londt 1994). *Afrobittacus* Londt is known only from west Africa (one species), *Anomalobittacus* Kimmins only from South Africa (one species), leaving the dominant genus, *Bittacus* Latreille, with its 51 species to which all east African species belong.

Five *Bittacus* species have been recorded for Mkomazi Game Reserve (see Checklist), including four of the nine species previously known to occur in Tanzania (according to the Natal Museum's computerised records). Previously recorded species were, *B. berlandi* Capra, *B. fumosus* Esben-Petersen, *B. leptocercus* Navás, *B. lineatus* Navás, *B. montanus* Weele, *B. moschinus* Navás, *B. sjostedti* Weele, *B. stanleyi* Byers and *B. weelei* Esben-Petersen. *Bittacus discors* Navás is here recorded for the first time from Tanzania. The known distributions of the Tanzanian species are summarised in Table 15.1.

Bittacus leptocercus Navás was described on material from Kinanga in southern Tanzania. The validity of this species, however, requires verification. Unfortunately the location of the type material is unknown.

Bittacus berlandi, B. discors, B. lineatus, B. sjostedti and probably *B. stanleyi* are species which appear to be regional endemics with limited ranges, while *B. moschinus*, *B. fumosus*, *B. montanus* and *B. weelei* are far more widespread, being found in at least some central, southern, and, in the case of *B. weelei*, west

Table 15.1 Known distribution of mecopteran *Bittacus* species recorded from Tanzania.

species	Mkomazi	also known from
berlandi	•	Kenya
discors	•	Kenya, Somalia
fumosus	•	Angola, Ethiopia, Kenya, Malawi, Zambia, Zimbabwe
leptocercus	-	-
lineatus	-	Kenya
montanus	-	Angola, Cameroon, Kenya, Malawi, Rwanda, Uganda, Zaire, Zambia, Zimbabwe
moschinus	•	Malawi, Zambia
sjostedti	-	Kenya
stanleyi	•	Congo, Malawi, Uganda
weelei	-	Angola, Ethiopia, Ghana, Kenya, Malawi, Mozambique, Nigeria, Sudan, Uganda, Zaire, Zambia, Zimbabwe

African countries (Table 15.1). Although *B. lineatus, B. montanus*, *B. sjostedti* and *B. weelei* have not been recorded from Mkomazi Game Reserve it is highly probable that with more intensive collecting they will be found there.

Sampling biases

As with the cicadas, sampling of Bittacidae was subordinate to more intensive Hymenoptera collecting, and as such hangingflies were only collected on an opportunistic basis. Furthermore, collecting effort was seasonally restricted with only two short periods, totalling 36 days, available for sampling, although each of the biannual wet season peaks were covered. Notwithstanding the seasonal restrictions in sampling, the known seasonal occurrence of the Tanzanian species indicates that for nine of the ten Tanzanian species all or part of the recorded adult seasonal

Table 15.2 Known seasonal incidence of Tanzanian species of Mecoptera.

species	J	F	M	A	M	J	J	A	S	O	N	D
berlandi	-	-	-	•	-	-	-	-	-	-	•	x
discors	-	-	-	•	•	-	-	-	-	-	-	-
fumosus	•	•	-	•	•	-	-	•	-	-	-	•
leptocercus	-	-	-	-	-	-	•	-	-	-	-	-
lineatus	-	-	-	-	-	-	-	-	-	-	•	-
montanus	•	•	•	•	-	-	-	-	-	-	•	•
moschinus	-	-	-	•	•	-	-	-	-	-	-	-
sjostedti	-	-	-	•	-	-	-	-	•	-	-	-
stanleyi	-	-	-	-	-	-	-	-	-	-	-	•
weelei	•	•	•	-	-	-	-	-	-	-	•	•

incidence fell within the sampled periods (Table 15.2). *Bittacus leptocercus* has only been recorded from July, a dry season period that was not sampled. Essentially only the western half of the reserve was surveyed, with most attention paid to areas within the vicinity of Ibaya Camp. Consequently, the species inventory presented here is likely to be an underestimate. The drier central and eastern habitats and the riverine vegetation associated with the Umba River on the eastern reserve boundary were not sampled for hangingflies. Further sampling in these relatively uncollected areas is required and may produce additional species.

Comparison of species richness with other African faunas

The African fauna has been treated by various specialists including Esben-Petersen (1921), Lestage (1929) and Byers (1971). Londt (1994) catalogued the African

Table 15.3 Occurrence of *Bittacus* species in Tanzania and bordering countries*.

species	Ken	Uga	Rwa	Bur	Zai	Zam	Mal	Moz	Tan
aequalis	•	-	-	-	-	-	-	-	-
africanus	-	-	-	-	•	-	-	-	-
berlandi	•	-	-	-	-	-	-	-	•
burgeoni	-	-	-	-	•	-	-	-	-
caprai	-	-	-	-	-	•	-	-	-
discors	•	-	-	-	-	-	-	-	•
elizabethae	-	-	-	-	•	-	-	-	-
erythrostigma	-	•	-	-	•	-	-	-	-
fumosus	•	-	-	-	-	•	•	-	•
lachlani	•	-	-	-	-	-	-	-	-
leptocercus	-	-	-	-	-	-	-	-	•
lineatus	•	-	-	-	-	-	-	-	•
livingstonei	-	-	-	-	-	-	•	-	-
montanus	•	•	•	-	•	•	•	-	•
moschinus	-	-	-	-	-	-	•	-	•
nebulosus	-	-	-	-	-	-	-	•	-
pobeguini	-	•	-	-	•	-	-	-	-
rossi	-	-	-	-	•	-	-	-	-
schoutedeni	-	-	-	-	•	-	-	-	-
sjostedti	•	-	-	-	-	-	-	-	•
stanleyi	-	•	-	-	•	-	•	-	•
tuxeni	-	-	-	-	-	-	•	-	-
weelei	•	•	-	-	•	•	•	•	•
zambezinus	-	?	-	-	-	-	•	•	-

* Kenya; Uganda; Rwanda; Burundi; Zaire; Zambia; Malawi; Mozambique; Tanzania

Mecoptera giving full bibliographic information as well as a table in which distributional information is summarised. This table provides the following data with respect to the number of species recorded for the faunas of west (9), east (17), central (16) and southern Africa (29). A direct comparison of the Tanzanian fauna with those countries bordering it is shown in Table 15.3.

Only the southern African fauna has been reasonably well surveyed, key references being: Wood (1933), Tjeder (1956) and Londt (1972). Although publications dealing with east African or Tanzanian species include a number of minor works there are a few key works worth special mention: van der Weele (1909), Capra (1939) and Byers (1971). Londt's 1981a & b papers dealing with the Malawian fauna may also be useful.

Distribution within Mkomazi

The Bittacidae are represented in most of the major habitat types (montane forest, *Commiphora* woodland, *Acacia-Commiphora* bushland, open *Combretum* bushland) within at least the western half of the reserve (Table 15.4).

Two species, *B. discors* and *B. moschinus* were collected in *Commiphora campestris* woodland, a habitat associated with red lateritic soils and reasonably prevalent in the centre of the reserve. *Bittacus stanleyi* was collected in montane forest on Ibaya Hill and Maji Kununua and is also likely to be present on Igire and Kinondo Hills. This species is probably restricted to montane forest within the reserve. Conversely, a widespread east African species, *B. fumosus*, was present in a wide variety of habitats (Table 15.4) including montane forest, bushland and woodland. The distribution of *B. fumosus* extends from the Ethiopian/Kenyan border south through Kenya, Tanzania and Malawi to the Zambian/Zimbabwe border (Londt 1976) and has been recorded from bushland and *Brachystegia* woodland from October to May (Londt 1976).

Bittacus fumosus commonly came to a light trap at Ibaya Camp during the 'long rains' in April–May 1996 and was present in a wide variety of habitats on the lower lying plains during this period. However, during the relatively drier period of 'short rains' in late November–December 1995, the species was only

Table 15.4 *Bittacus* species diversity across Mkomazi Game Reserve habitat types.

	montane forest	*Acacia-Commiphora* bushland	open *Combretum* bushland	*Commiphora* woodland
B. stanleyi	•			
B. fumosus	•	•	•	•
B. berlandi	•			
B. discors				•
B. moschinus				•

collected at altitude in montane forest and was not attracted to the light trap at Ibaya Camp, suggesting a seasonally determined, rainfall dependent distribution for the species. This montane refuge phenomenon is also exhibited by a number of tibicinine cicada species and is discussed further in the chapter dealing with cicadas of Mkomazi Game Reserve. During favourable periods *B. fumosus* is likely to be widespread within Mkomazi.

B. berlandi Capra was collected in a Malaise trap sited on the edge of a grassland patch, comprising *Setaria* and *Panicum* species, against the margin of montane forest and may be associated with the montane grassland habitat rather than with montane forest *per se*. This species will probably also be present on other hill tops with montane forest, such as Maji Kununua, Kinondo and Igire Hills.

The restricted distributional information presented for Mkomazi Game Reserve results from a biased collecting effort due to reasons outlined above. Some of the species represented in the western end of the reserve are also likely to occur in the drier central and eastern habitats, particularly *B. fumosus*. However, as with the tibicine cicadas, the possible reliance on higher altitude refuges during dry periods may restrict the occurrence of this species in the flat eastern end of the reserve. Further sampling in these relatively uncollected areas is required and may produce additional species.

Conservation and management

Habitat modification due to fire encroachment, both in forest and bushland habitats, or tree felling for charcoal production affecting the montane forests may adversely affect hangingfly species richness within the reserve. The presence of montane forest appears to play a critical contributory role to the high insect species richness conserved within Mkomazi. Besides contributing to the diversity of habitats, by acting as a refuge during drier periods montane forest enables many species to survive adverse conditions. These species are then able to subsequently re-expand their range from these core areas during favourable periods. Without these core areas, many species will probably become locally extinct within the reserve. Given the present susceptibility of montane forest to degradation in the reserve, pertinent conservation management of this habitat is a priority.

Mkomazi conserves at least half of the known Tanzanian species of hangingflies and it is likely that further collecting will establish that most of the remaining species also occur within the reserve. The high species richness found within Mkomazi illustrates the valuable role this conservation area plays in preserving a representative portion of a physiographically diverse savanna.

Acknowledgements

We would like to extend our thanks to all the staff at Ibaya Camp and to colleagues for logistical support and field assistance. The British Council provided logistic support in Dar es Salaam. Graham Stone critically read and improved the manuscript. This work was supported by grants awarded to SvN from the Commonwealth Science Council and the Foundation for Research Development.

References

Byers, G.W. (1971) An illustrated, annotated catalogue of African Mecoptera. *The University of Kansas Science Bulletin* 49: 389-436.

Capra, F. (1939) Planipennia, Mecoptera. *Missione biologica nel paese dei Borana*, vol. 3, no. 2, pp. 157-178. Accademia d'Italia, Roma. Centro studi per l'Africa. Publ. no. 4, 1939-1940.

Esben-Petersen, P. (1921) Mecoptera. Monographic Revision. *Collections Zoologiques de Baron Edm. de Selys Longchamps. Catalogue systématique et descriptif.* Fasc. 5, pp. 1-172. Bruxelles.

Lestage, J.A. (1929) Notes critiques et zoogéographiques sur les Bittacus Africaines. *Reveu de zoologie et de botanique africaines* 18: 1-22.

Londt, J.G.H. (1972) The Mecoptera of Southern Africa. *Journal of the Entomological Society of southern Africa* 35: 313-343.

Londt, J.G.H. (1976) New descriptions of *Bittacus fumosus* Esben-Petersen, and *Bittacus zambezinus* Nava, with a new synonymy (Mecoptera: Bittacidae). *Journal of the Entomological Society of southern Africa* 39: 175-183.

Londt, J.G.H. (1981a) *Bittacus livingstonei*, a new species from Malawi (Mecoptera: Bittacidae). *Annals of the Natal Museum* 24: 621-624.

Londt, J.G.H. (1981b) The Scorpionflies of Malawi (Mecoptera: Bittacidae). *Nyala* 7: 129-134.

Londt, J.G.H. (1994) A catalogue of Afrotropical Mecoptera. *Annals of the Natal Museum* 35: 45-59.

Tjeder, B. (1956) Mecoptera. *South African Animal Life* 3: 344-390.

van der Weele, H.W. (1909) Neuroptera, Planipennia et Panorpata. In: Sjöstedt, Y. (ed.) *Wissenschaftliche Ergebnisse der schedischen Expedition nach dem Kilimandjaro, dem Meru und den umgebenden Massaisteppen Deutsch-Ostafrikas 1905-1906.* Vol. 2, pp. 11-22. Stockholm.

Wood, H.G. (1933) Notes on some South African Bittacidae (Mecoptera). *Annals and Magazine of Natural History* 12: 508-531.

Checklist: Hangingflies of Mkomazi

Species determinations by J.G.H. Londt.

Class INSECTA

Order MECOPTERA

Family Bittacidae
Bittacus berlandi Capra. Ibaya Hill.
Bittacus discors Navás. Kikolo Plot.
Bittacus fumosus Esben-Petersen. Maji Kununua, Kikolo Plot, Ibaya Camp, Ubani Plot.
Bittacus moschinus Navás. Kikolo Plot.
Bittacus stanleyi Byers. Maji Kununua, Ibaya Hill.

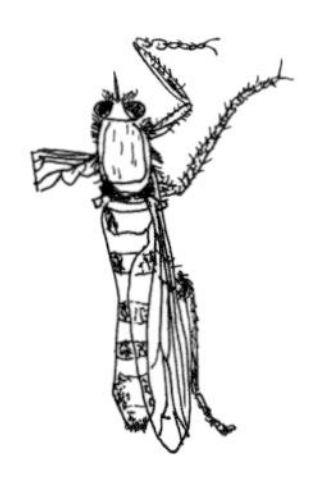

CHAPTER 16

Asilids (Diptera: Asilidae) of Mkomazi

Jason Londt, Malcolm Coe & Tony Russell-Smith

The asilids are a group of Diptera or two-winged flies which represent the largest family (Asilidae) in the sub-order Brachycera. Although they are commonly called robber flies, they do not in fact rob anything, other than the lives of the insects that they prey on, for they are adept and voracious predators. Their predatory life style is indicated by their long hairy legs with which they grasp their prey and the stout proboscis which they use to stab and digest their prey, which are almost entirely caught on the wing. Asilids lay their eggs in soil, sand, wood, leaf litter and living plant tissue where their exhibit a wide variety of feeding styles.

Holm & de Meillon (1986) estimate that there are probably at least 500 species and 70 genera in the southern African region, out of an estimated dipteran total of 16,000 species for the whole Afrotropical region (Barraclough & Londt 1985). Individual asilid species vary in size from 3 mm to a massive 40 mm, which enables them to tackle prey as large as the zylocopid or carpenter bees and a number of other large bees and insects, including small dung beetles (*Hoplistomerus nobilis*) and small grasshoppers (*Alcimus* spp.) (Barraclough & Londt 1985). In southern Africa species of the genus *Hyperechia* feed selectively on zylochopid bees (Skaife 1953). Interestingly Beeson (1941) informs us that the aptly named *Hyperechia xylocopiformis* preys on *Xylocopa tenuiscapa* in India.

Asilids were not systematically surveyed in Mkomazi and the list presented here is of species collected incidentally while sampling for other insects. The great majority of species were collected in pitfall traps in footslope grassland and *Acacia-Commiphora* bushland at Ibaya Camp (all those on the first page of the list). The fauna in the pitfall traps was dominated by *Neomochtherus* species which accounted for 44% of all asilids collected. A few individuals were collected in sweep net samples from the Ibaya grassland and *Acacia senegal* woodland between Nbeya and Mbula. Only seven specimens were taken in Malaise traps (specifically designed to sample flying insects) which suggests that the well developed eyes of these flies may allow them to avoid capture in this type of trap.

Asilid collections in Mkomazi

Species determinations by J.G.H. Londt.

Ibaya, burnt hillside

15	Nov 95	2M *Trichardis* sp.; 1F *Neolophonotus* sp,; 1M *Neomochtherus* sp.
18–19	Nov 94	1M *Scylaticus quadrifasciatus*
18	Nov 95	2M *Trichardis* sp.
17	Nov 95	1M1F *Neomochtherus* sp.
14	Oct 95	2M1F *Neomochtherus* sp.
13	Oct 95	2M *Neomochtherus* sp.
27–28	Nov 94	1M1F *Neomochtherus* sp.
15	Oct 95	1F *Neomochtherus* sp.
18	Nov 95	1M1F *Neomochtherus* sp.
16	Oct 95	1M1F *Neomochtherus* sp.
25–26	Nov 94	1M1F *Neomochtherus* sp.
16	Nov 95	1M1F *Neolophonotus* nr. *malawi*
7	Sept 95	1F *Neolophonotus* sp.
5	June 95	1M1F *Neolophonotus* nr. malawi; 1F *Promachus* sp.
8	June 96	1F *Promachus* sp.
7	June 96	1F *Promachus* sp.

Ibaya, burnt grassland

17–18	Nov 94	1F *Scylaticus* ? *quadrifasciatus*
15	Nov 95	1M *Nusa infumata*
16	Nov 95	1M *Nusa infumata*
19	Nov 95	1F *Pilophoneus krugeri*
4	Aug 95	1F *Neolophonotus* sp. nr. *chaineyi*
25–26	Nov 94	1F1? *Microstylum* sp.

Ibaya, unburnt hillside

20	Nov 95	1F *Trichardis* sp.; 2M *Neomochtherus* sp.
27–28	Nov 94	1M1F *Neolophonotus* sp.; 1F *Trichardis* sp.; 1M *Neomochtherus* sp.
17	Nov 95	1F *Trichardis* sp.
26–26	Nov 94	1M *Trichardis* sp.
16	Nov 95	1M *Neomochtherus* sp.
16–17	Sept 94	1M *Neomochtherus* sp.
18	Nov 95	1M1F *Neomochtherus* sp.
13	Oct 95	1M *Neomochtherus* sp.
11	June 96	1F *Neolophonotus* sp nr. *chaineyi*
17–18	Nov 94	1M1F *Microstylum* sp.

Ibaya Camp, Malaise trap

29 Jan 96 — 1F *Laxenecera albicincta*; 3M2F *Neomochtherus* sp.; 1F *Gonioscelis* sp.

Summit of Ibaya Hill, *Spirostachus* forest

28 Jan 96 — 1M *Neolophonotus porcellus*

Swept, *Acacia senegal* woodland between Mbula & Ndea

19 Nov 94 — 2F *Pegesimallus* sp.

Swept, pumphouse grassland, Ibaya Camp

18 Nov 94 — 1F *Pegesimallus* sp.

Swept, *Pennisetum* in *Acacia senegal* woodland 21.2 km S. Ibaya Camp

14 Jan 96 — 1M *Afroholopogon* sp.

SUMMARY OF ASILIDAE IN SAMPLE

Afroholopogon sp.	Genus needs revision. An interesting record.
Gonioscelis sp.	Genus needs revision. An interesting record.
Laxenecera albicincta	Common widespread species.
Microstylum sp.	Large genus needing revision.
Neolophonotus sp.	At least one possible new species; others difficult to identify as they are females.
Neolophonotus chaineyi	A valuable record and specimen.
Neolophonotus porcellus	A reasonably well-known species.
Neomochtherus sp.	Genus needs revision. There are at least two species here.
Nusa infumata	Common widespread species.
Pegesimallus sp.	Females not easy to identify, no males present in sample.
Pilophoneus krugeri	Known from few species. A valuable addition.
Promachus sp.	Large genus needing revision.
Seytancus quadrifasciatus	A nice record.
Trichardis sp.	Genus needs revision.

References

Barraclough, D.A. & Londt, J.G.H. (1985) Diptera. In: Scholtz, H. & Holm. E. (eds.) *Insects of Southern Africa*. Butterworths, Durban.

Holm, E. & de Meillon, E. (1986) *Struik Pocket Guide: Insects of Southern Africa*. Struik.

Beeson, C.F.C. (1941) *The ecology and control of forest insects of India and the neighbouring countries.* Government of India.

Skaife, S.H. (1953) *African Insect Life*. Longmans Green, London.

CHAPTER 17

Butterflies (Lepidoptera: Hesperioidea and Papilionoidea) of Mkomazi

Simon van Noort & Graham N. Stone

Introduction

The order Lepidoptera (butterflies and moths) includes approximately 150,000 species, and, as with the flies (order Diptera) and wasps, ants and bees (order Hymenoptera), is one of the largest groups of insects after the beetles (order Coleoptera). Adult butterflies and moths are characterised by a covering of overlapping scales, which are present on the two pairs of wings as well as the rest of the body. Usually the mouthparts are modified into a sucking tube called a proboscis. Butterflies and moths are some of the most conspicuous and easily recognisable insects, although many of the moths are extremely tiny, some with wingspans of only 3 mm. At the other end of the scale, Africa is also home to many large emperor moths and a swallowtail butterfly with one of the largest (23 cm) wingspans in the world. Of the world total of around 17,500 butterfly species, 3,607 occur in the Afrotropical region (Ackery *et al.* 1995). Many of these are widespread species, but there are also regions with a very rich local diversity usually including a relatively high percentage of butterfly species that are found nowhere else. Areas such as these urgently need to be afforded conservation status in order to protect as much of Africa's butterfly diversity as possible (Ackery *et al.* 1995).

Tanzania has an exceptionally rich butterfly fauna. Although the butterflies of Tanzania (Kielland 1990 & 1994 and Congdon & Collins 1998) along with those of Kenya (Larsen 1991, Collins & Larsen 1996) and southern Africa (Pringle *et al.* 1994) are the best documented in Africa, much work still needs to be done, particularly in Tanzania and Kenya. This study has allowed for a valuable preliminary assessment of the role that Mkomazi Game Reserve plays in the conservation of local butterfly diversity in east Africa.

Ecology

Adult butterflies use their proboscis to feed on sugar-rich liquid sources such as nectar, honeydew, tree sap and fluids from over-ripe fruit. Most will seek out nectar in flowers for their energy requirements, but some groups, such as members of the genus *Charaxes* and certain other nymphalids will feed on rotting fruit, tree sap or even carnivore or monkey droppings. During typical periods of hot and dry conditions in savanna areas many pierids, papilionids and some lycaenids will be found congregating besides water bodies where they feed on moisture from damp ground, particularly in areas where mammals have recently urinated.

In contrast to adult butterflies, the immature larval stages (caterpillars) have chewing mouthparts and most feed on plants, although some of the lycaenid caterpillars are predatory on other insects. Many caterpillar species will only feed on particular host-plant species or groups of related plant species. The majority of the food requirements of the whole life-cycle of a butterfly are acquired in the caterpillar stage. Caterpillars are eating machines, processing large quantities of plant material throughout their development and as such they are important primary consumers of vegetation. As such they play a major role in the recycling of nutrients. Their frazz (droppings) make nutrient resources re-available for uptake by plants. Caterpillars also form an important link in the food chain by converting plant material into a protein food source. Many predatory insects, spiders and vertebrates (birds, chameleons and lizards) feed on caterpillars and adult butterflies. Caterpillars are also utilised by a large number of parasitic insects (mostly wasps and some flies) for the development of their own larvae.

Butterflies have evolved some elaborate mechanisms for protection from attack. Many adult butterflies have a cryptically coloured or patterned underside to the wings, often resembling a dried leaf. When they are at rest with their wings folded, whether it be on the ground or on a branch they blend in with the surrounding environment, effectively camouflaging themselves from potential predators. The wing topside of some of these species is brightly coloured and may include eye spots. If disturbed these butterflies suddenly open their wings startling the would be predator. The Danainae and Acraeinae are distasteful to vertebrate predators and are brightly coloured to warn potential predators of their unpalatability. Vertebrate predators quickly learn to associate these colours with an unpleasant experience and avoid any further attempts to eat butterflies displaying such colours. Distasteful species obtain their poisonous properties from host plants and the assimilated poisons are retained through to the adult stage. A number of palatable butterflies have evolved a similar colour pattern to the distasteful species, and through this mimicry enjoy protection from predators as well. The phenomenon of mimicry is well represented in the Afrotropical region with three main warning colour combinations, black and orange; black and yellow; and black and white having evolved. Each colour combination has a suite of distasteful and palatable species.

Diversity

153 species of butterfly have been recorded from Mkomazi Game Reserve. The majority of these species were identified from collected specimens, but in a few cases were recorded through observation. The latter approach was only used for species that are unmistakable in the field.

Sampling biases

Except for the study on pollination of *Acacia* species (Stone *et al.*, Chapter 20), where all the butterfly species visiting *Acacia* flowers on the study trees were comprehensively documented, the sampling of butterflies in Mkomazi was done on an opportunistic basis. No attempt was made to sample all habitats and altitudes present in the reserve and those that were sampled did not receive exhaustive treatment. Sampling of butterflies was subordinate to more intensive collecting of other groups of insects, particularly the Hymenoptera. Most of the collecting was concentrated in the western end of Mkomazi leaving the majority of the reserve unsampled. Seasonal variation in species richness was not assessed, although collecting was carried out during the two annual peaks of butterfly emergence, corresponding with the bimodal rainfall pattern, the long rains of February to May, and the short rains of November and December. Consequently the checklist presented here is likely to be a gross underestimate of butterfly species richness in Mkomazi and must be assessed as a preliminary survey.

Comparison of species richness with other African faunas

The Tanzania butterfly species count of 1,387, comprising the 1,117 species documented (as recorded or probably occurring in Tanzania) in *The Butterflies of Tanzania* (Kielland 1990) and the subsequent 270 additional species recorded in Congdon & Collins (1998), is well over 853 species that occur in the whole southern African region (Pringle *et al.* 1994). This exceptionally high species richness also substantially exceeds the 901 species recorded from Kenya (Larsen 1991, Collins & Larsen 1996). The richness of Tanzania's butterfly fauna is supported by the comparatively high number of endemic species (species only found in Tanzania). Tanzania has 118 endemic species compared with 25 for Kenya and 34 for Uganda (Congdon & Collins 1998). Tanzania's high endemism is a function of a larger land area, greater diversity of habitat and the presence of the Eastern Arc Mountains, which are speciation hotspots (Congdon & Collins 1998).

Of the 1,387 Tanzanian species, 419 species could potentially occur in Mkomazi Game Reserve. This assessment is based on the distribution and habitat preferences of Tanzanian butterflies recorded in Kielland (1990) and Cordeiro (1995), i.e. species whose recorded distribution encompassed north-eastern Tanzania and whose habitat preferences are those that are found in Mkomazi were included in this potential list. Most of these are savanna species, but thirty-nine are montane forest species cited by these authors as occurring on the North Pare or South Pare Mountains bordering the reserve. These were included in the potential list as they may well be present in the isolated montane forests on outlying peaks of these mountains that fall within Mkomazi. A further 84 montane species recorded in the

Table 17.1 Taxonomic composition and comparative species richness of Mkomazi's butterfly fauna. The figures for Mkomazi include both recorded species and (potential species). The Tanzanian count was tallied from Kielland (1990) and Congdon & Collins (1998). The Kenyan figures are derived from Larsen (1991) and Collins & Larsen (1996) and the southern African totals are after Pringle *et al.* (1994).

	Mkomazi	Tanzania	Kenya	Southern Africa
HESPERIIDAE				
Coeliadinae	1 (4)	8	8	6
Pyrginae	5 (25)	65	54	45
Hesperiinae	5 (42)	167	94	75
PAPILIONIDAE	11 (15)	40	27	17
PIERIDAE				
Coliadinae	5 (8)	11	10	7
Pierinae	37 (52)	92	77	47
NYMPHALIDAE				
Satyrinae	7 (18)	90	52	82
Charaxinae	8 (27)	78	60	38
Apaturinae	0 (1)	1	1	1
Limenitinae	8 (25)	123	84	40
Nymphalinae	11 (24)	44	41	25
Argynninae	1 (3)	6	5	4
Acraeinae	16 (39)	109	84	50
Danainae	3 (8)	15	11	7
Libytheinae	1 (1)	1	1	1
LYCAENIDAE				
Theclinae	8 (45)	174	108	193
Lipteninae	6 (18)	159	60	31
Liphyrinae	0 (0)	11	4	3
Miletinae	1 (4)	20	5	31
Lycaeninae	0 (1)	2	1	2
Polyommatinae	19 (59)	166	113	148
RIODINIDAE	0 (0)	5	1	0
Total	153 (419)	1387	901	853

literature as occurring on the Usambara Mountains were not included in the potential list for Mkomazi because these mountains constitute a region of high species endemism (Rodgers & Homewood 1982) and are situated at a considerably greater distance from Mkomazi than are the Pare Mountains.

The total of 419 potential species for Mkomazi is almost a third of the total number of species recorded for Tanzania and is a reasonably high diversity for a savanna area of this size, a figure that can be attributed to the large geographical and associated habitat and floral diversity within the reserve. This is born out by the presence of different ecologically adapted assemblages of butterflies within Mkomazi (see below). The recorded butterfly species richness of 153 species for Mkomazi compares favourably with preliminary assessments of local species richness in Zambia. Terblanch & Henning (1993) recorded 160 species from the Mwinilunga region in north-western Zambia resulting from nine days of intensive collecting in an area comprising wetter Zambezian miombo woodland, sensu White (1983). However, comparison with local species richness in rainforest reinforces the relative low richness of savanna areas. For example, Kakum National Park in Ghana, comprising 350 km^2 of rainforest, an area one-tenth the size of Mkomazi, harbours 440 recorded species of an estimated 550–600 total species (Larsen 1995).

Biogeographical affinities

Mkomazi Game Reserve is situated within the Somalia-Masai regional centre of endemism which includes approximately 2,500 plant species of which around half are endemic (White 1983). Faunistically, Mkomazi lies in the southern region of the Somali Zone of open formations (Carcasson 1964, Larsen 1991), and includes elements of the Highland forest division, specifically the Tanzania-Nyasa zone, encompassing the highland forests of south-eastern Kenya, most of the mountains of Tanzania, mountains of Malawi, Mozambique and eastern Zimbabwe (Carcasson 1964).

At a regional level Mkomazi falls largely within zoogeographical subzone 4c as defined by Kielland (1990), which includes the East and West Usambara Mountains and the South Pare Mountains. The butterfly species of the South Pare Mountains have a higher affiliation with those of the West Usambara Mountains than with the North Pare Mountains, whose fauna is closer to that of the Northern Highlands, such as Mt. Kilimanjaro (Kielland 1990). The butterfly fauna of the South Pare Mountains and Usambaras also show some affinity with the Teita Hills in south-eastern Kenya (Kielland 1990). However, the outliers of the Pare Mountains around Ibaya Camp in the north-western part of the reserve can be considered as part of the North Pare system and consequently fall within subzone 6b of Kielland (1990), which encompasses Mt. Kilimanjaro, the North Pare Mountains and the mountains of the Lossogonoi Plateau. Since montane forest occurring within Mkomazi is restricted to these north-western outliers, the associated butterfly fauna would be expected to be representative of the faunal assemblage typical of the North Pare Mountains rather than that of the South Pare Mountains. This assertion could not be assessed due to the inadequate sampling of the montane butterfly community within the reserve. Nevertheless, a number of montane forest species, most of which had previously been recorded in the literature from both the North and South Pare Mountains, were collected in Mkomazi during the programme, justifying the inclusion of montane species recorded from these mountains in the potential list for Mkomazi.

The montane forest species recorded from Mkomazi included *Papilio phorcas nyikanus*, *Papilio echerioides wertheri*, *Mylothris sagala sagala*, *Belenois margaritacea intermedia*, *Acraea cerasa cerasa*, *Acraea quirina rosa*, *Acraea pharsalus pharsaloides*, *Acraea johnstoni johnstoni*, *Junonia tugela aurorina* and *Alaena nyassa major*. *Acraea cerasa* is very rare in Tanzania and has not previously been recorded from the Pare Mountains, although it is known from the East Usambara Mountains (Kielland 1990). Mkomazi is also the furthest north that *A. nyassa major* has been recorded, previously only having been recorded from the Usambara and Uluguru Mountains (Ackery *et al.* 1995). *Charaxes protoclea azota*, a species more typical of lowland forest (Henning 1989) but one that is found in forest up to 1,700 m (Kielland 1990), was also recorded on top of Ibaya Hill (1,400 m) in montane forest. However, another forest species, *Euphaedra neophron littoralis*, that occurs between 400 and 1,600 m (Kielland 1990) and is reasonably common just outside the reserve at the base of the South Pare Mountains, has not yet been collected from within the reserve and appears to be absent from the montane forest patches. *Euphaedra neophron* may still be recorded in the thicker woodland areas south of the Mandi and Gulela Hills or in the region of Kisiwani River. Two other forest species whose distribution is not restricted by altitude, *Acraea cabira* and *Celaenorrhinus galenus*, both of which are common and widespread African species, were recorded from montane forest in Mkomazi. Other species with a wider habitat association, typically present in woodland as well as forest, such as *Papilio constantinus*, *P. dardanus*, *Graphium leonidas*, *G. policenes*, *Eronia cleodora*, *Acraea esebria*, *A. natalica*, *Phalanta phalantha*, *Junonia terea*, *Pseudacrea boisduvali*, *Charaxes brutus*, *C. candiope* and *Libythea labdaca* were only encountered in association with montane forest within the reserve. Many of these species were locally abundant on Ibaya Hill and Maji Kununua at the forest margins where savanna elements are penetrating the forest due to fire encroachment and tree felling. These species are typically characteristic of forest margins and sub-climax lowland forest (Carcasson 1964).

Although a dozen or so montane species have already been recorded within Mkomazi there may be further species that are present on the South Pare Mountains, but whose range does not extend into the reserve. This contention is supported by *Amauris echeria*, *Neptis aurivillii*, and *Mylothris yulei* which were recorded during this programme on the South Pare Mountains but were not found within

Mkomazi, although this may equally be the result of under sampling within the reserve.

Although most butterfly species are capable of dispersing over large areas, isolated montane forests often act to reduce or restrict inter-population gene flow resulting in subsequent genetic divergence. One possible example of this phenomenon is *Belenois margaritaceae*, a species centered in the central Kenyan highlands (Larsen 1991) with four subspecifically defined populations, two of which are restricted to northern Tanzania. *Belenois margaritacea intermedia* was recorded on Maji Kununua, but the presence of this subspecies in Mkomazi is at odds with the known distribution of the subspecific populations of this taxon. The subspecies that would be expected to be present in Mkomazi is *Belenois margaritacea plutonica*, which is recorded from the North and South Pare and the Usambara Mountains (Kielland 1990). However, the series collected on Maji Kununua answers to *B. m. intermedia* as illustrated in Kielland (1990), a subspecies that is currently recorded from the Uzungwa Range (including Image Mountain) and the Nguru Mountains, the latter which are approximately 150 km further south than Maji Kununua. There seems to be no clear cut geographical division between the different subspecies of *Belenois margaritacea*, since specimens from one location on the North Pare Mountains are very close to subspecies *B. m. kenyensis* from the Teita Hills in Kenya, yet a population a few kilometres south in the North Pare Mountains answers to typical *B. m. plutonica* (Kielland 1990). As Kielland (1990) suggests, the variation within this taxon may only be the result of ecological factors, a view that seems to be supported by the presence of *B. m. intermedia* within the normal distribution of *B. m. plutonica* and indicates the futility of defining populations such as these at subspecific rank.

Species that are confined to montane forest in Tanzania occur at lower altitudes in southern Africa where latitude compensates for altitude e.g. *Acraea cerasa* and *Junonia tugela*, which are common at low altitudes in Kwazulu/Natal (South Africa), but only present at 1,400–1,600 m on top of Ibaya Hill and Maji Kununua in Mkomazi. This phenomenon is characteristic of other insect groups such as species in the three fly families: Diastatidae, Campichoetidae and Opomyzidae (Barraclough 1994).

The majority of species present in Mkomazi enjoy a widespread distribution that extends down to eastern South Africa and are broadly associated with the open biogeographical formations of Carcasson (1964). However, besides the montane forest butterfly assemblage, there is another unique butterfly community in the reserve. This assemblage is one that is typical of the arid scrub and dry grassland of the Somalia-Masai regional centre of floral endemism, *sensu* White (1983) and corresponding with the biogeographical Somali zone of Carcasson (1964) and Larsen (1991). Most of these species are associated with dry *Acacia* scrub and only just penetrate into northern Tanzania, enjoying a distribution centred in the dry horn of Africa that typically includes Ethiopia, Somalia, south-east

Sudan, Kenya, and for a few species, southern Arabia (Ackery *et al.* 1995). Arid-adapted species recorded from Mkomazi are: *Kedestes rogersi* (Hesperiidae), *Junonia limnoria*, *Neocoenyra duplex*, *Acraea chilo*, *Acraea pudorina* (Nymphalidae), *Colotis protomedia*, *Colotis halimede*, *Colotis vestalis*, *Colotis chrysonome* (Pieridae) and *Anthene opalina* (Lycaenidae). This is the first record of *Acraea (Acraea) pudorina* for Tanzania, confirming the predictions by both Larsen (1991) and Kielland (1990) that the species will be found in the arid country in northern Tanzania. Previously this species had only been reliably recorded from central and southern Kenya (Kielland 1990, Larsen 1991, Ackery *et al.* 1995).

Conservation and management

Montane forest and forest/savanna margins on Ibaya Hill and Maji Kununua contributed to 35% of the total species richness (Table 12.2) recorded for Mkomazi, with a further third of the remaining butterfly species recorded from elsewhere in the reserve also occurring on these hills (a total of 57% of all the butterflies collected in Mkomazi). This concentration of butterflies was obvious in the field with a noticeably higher species richness on these hill tops compared to the surrounding low lying areas. After the rains the open *Setaria/Panicum* grass glades fringed by montane forest on Ibaya Hill teemed with butterflies, even more so than the *Acacia-Commiphora* bushland. The combination and meeting of montane forest and savanna is probably the main contributory factor to this high local diversity. These regions of higher altitude comprise only a very small percentage of the reserve but proportionate to their area contribute the most to butterfly diversity in Mkomazi. As such, the geographically restricted montane habitats that are exceptionally prone to degradation require careful management and are a conservation priority.

Table 17.2 Contribution of habitat to butterfly species richness in Mkomazi Game Reserve.

family	montane forest	montane forest/ savanna margins	savanna
Hesperiidae	2	2	7
Papilionidae	3	7	1
Pieridae	3	8	31
Nymphalidae	6	17	32
Lycaenidae	2	4	28
Total	16 (10%)	38 (25%)	99 (65%)

Butterflies and *Acacia* pollination ecology

Acacias are a dominant feature of the Mkomazi vegetation (see Coe *et al.*, Chapter 7 and Stone *et al.*, Chapter 20), and butterflies are important flower visitors for a number of them. Mkomazi acacias are members of two subgenera within the genus *Acacia*—the subgenus *Acacia* and the subgenus *Aculeiferum*. The flowers of the former are usually regarded as nectarless, while flowers of the latter often secrete nectar (Stone *et al.*, Chapter 20). In practice, our work in Mkomazi has shown this distinction not to be clear-cut, and butterflies are a very sensitive indicator of nectar secretion. Among the Mkomazi acacias, the most abundant nectar secretors are *Acacia mellifera* and *A. senegal*. Both of these species have elongate 'spicate' inflorescences, and belong to the subgenus *Aculeiferum*. When in full flower, both species are visited by a huge diversity of insects, including many butterflies. Each floret on the inflorescences (which may contain up to 100 flowers) contains 1–2 microlitres of relatively dilute nectar (20–30% sucrose) when it opens in the morning, and flowering trees thus represent a very rich source of both water and sugar. The individual florets are c. 5 mm deep, and have a very narrow diameter, such that only flower visitors with long, fine mouthparts (such as butterflies) are able to reach the nectar. In Mkomazi, *A. senegal* was seen to be visited by 18 butterfly species (see Appendix in Stone *et al.*, Chapter 20), though this is certainly an underestimate of the total butterfly richness visiting this tree.

Contrary to the generalisation mentioned above, at least two species in the subgenus *Acacia* definitely do secrete nectar. *Acacia brevispica* and *A. zanzibarica* both have flowers in spherical 'capitate' inflorescences which secrete very small volumes of highly concentrated nectar, and are visited by a range of nectar feeding insects. The high sugar concentration (c. 70% sucrose) and small volumes (much less than 1 microlitre per flower) of the nectar in these two species means that large nectar foragers such as honey bees do not harvest nectar from the flowers. Both species are popular with butterflies, particularly small blues, coppers and hairstreaks in the family Lycaenidae, skippers in the Hesperiidae, and whites, sulphurs and orange-tips in the Pieridae. We have good data from Mkomazi for *A. zanzibarica*, which was visited by 16 butterfly species, one fewer than the more productive *A. senegal*. Butterfly visitors to *A. zanzibarica* included *Anthene opalina*, the Opal ciliate blue, which is regarded as a rare and local dry savanna species whose range extends northwards into the eastern Sahel (Larsen 1991). Adult butterflies do not harvest pollen, and flowers without nectar are usually ignored by butterflies. *Acacia tortilis* secretes only very tiny amounts of nectar, and though visited for nectar by a diversity of very small solitary bees (Stone *et al.*, Chapter 20) this species was exploited rather little by butterflies (only four species observed). This may reflect the availability of alternate, more productive nectar sources, and it is probable that this species would be exploited at least by small butterflies were other sources not in flower. Two Mkomazi acacias seem to secrete no nectar at all (*A. drepanolobium* and *A. nilotica*), and on each species we recorded very few flower visitation events by lycaenid blues (*Anthene otacilia* on *A. drepanolobium* and *Azanus ubaldus* on *A. nilotica*). These visits were almost certainly feeding 'mistakes' by butterflies which probably feed on these acacias as larvae.

Acacias are also important for many Lycaenid butterflies as foodplants, and adults and larvae are thus found even on non-flowering acacias (Larsen 1991). Our observations showed that most small lycaenids—particularly the cilate blues in the genus *Anthene*, zebra blues of the genus *Leptotes*, and babul blues of the genus *Azanus*—were extremely local in their activity, often spending the entire day on a single part of a single tree. Acacias are generally regarded as self-sterile, and so flower visitors which do not disperse between trees cannot be effective pollinators. Where butterflies do not disperse between trees but do take nectar, they should be regarded as nectar robbers rather than pollinators, and feeding them is a cost without benefit for the tree. The larger butterfly species recorded from *Acacia senegal* and *A. zanzibarica*, such as *Danaus chrysippus*, *Amauris ochlea* and *Catopsilia florella*, range widely, and would certainly be capable of effective pollen transfer between *Acacia* individuals.

Acknowledgements

We would like to extend our thanks to all the staff at Ibaya Camp and to numerous colleagues for logistical support and field assistance. The British Council provided logistic support in Dar es Salaam. Steve Collins (African Butterfly Research Institute, Nairobi) kindly commented on an earlier draft of the manuscript and assisted with identification of some of the more difficult taxa. Jonathan Ball very kindly made his own copies of some of the major literature on African butterflies available for reference. This work was supported by grants from the Commonwealth Science Council and the Foundation for Research Development (SvN), and the Darwin Initiative (GS).

References

Ackery, P.R., Smith, C.R. & Vane-Wright, R.I. (eds.) (1995) Carcasson's African Butterflies: an Annotated Catalogue of the Papilionidea and Hesperioidea of the Afrotropical Region. CSIRO, Melbourne.

Barraclough, D.A. (1994) First record of Diastatidae (Diptera: Schizophora) from east Africa, with descriptions of two new species. *African Entomology* 2: 111-116.

Carcasson, R.H. (1964) A preliminary survey of the zoogeography of African butterflies. *East African Wildlife Journal* 2: 122-157.

Collins, S.C. & Larsen, T.B. (1996) Butterflies new in Kenya since 1991, updating, and corrections. In: Larsen, T.B. *The Butterflies of Kenya and their Natural History,* pp. 491-500. Oxford University Press, Oxford.

Congdon, C. & Collins, S.C. (1998) *Kielland's Butterflies of Tanzania Supplement. A.B.R.I. – Lambillionea.* Nairobi and Tervuren.

Cordiero, N.J. (1995) Interesting distribution records of butterflies from northern Tanzania. *Metamorphosis* 6: 194-198.

D'Abrera, B. (1980) *Butterflies of the Afrotropical Region.* Lansdowne, Melborne.

Henning, S.F. (1989) *The Charaxinae Butterflies of Africa.* Aloe Books, Frandsen.

Kielland, J. (1990) *The Butterflies of Tanzania.* Hill House, Melbourne & London.

Kielland, J. (1994) Butterfly collecting in North-Western Tanzania II. *Metamorphosis* 5: 5-21.

Larsen, T.B. (1991) *The Butterflies of Kenya and their Natural History.* Oxford University Press, Oxford.

Larsen, T.B. (1995) Butterflies in Kakum National Park, Ghana. *Metamorphosis* 6: 138-145.

Pringle, E.L.L., Henning, G.A. & Ball, J.B. (eds.) (1994) *Pennington's Butterflies of southern Africa.* Second Edition. Struik Winchester, Cape Town.

Rodgers, W.A & Homewood, K.M. (1982) Species richness and endemism in the Usambara Mountain forests, Tanzania. *Biological Journal of the Linnean Society* 18: 197-242.

Terblanche, R.F. & Henning, G.A. (1993) List of butterfly species collected in Mwinilunga Region, North Western Zambia 18-26 Dec. 1991. *Metamorphosis* 4: 26-30.

White, F. (1983) *The Vegetation of Africa. A descriptive memoir to accompany the UNESCO/AETFAT/UNSO vegetation map of Africa.* UNESCO, Paris.

Checklist: Butterflies of Mkomazi

Nomenclature and hierarchical ordering follows Ackery *et al.* (1995). The 153 species recorded from Mkomazi are not indented. **Potential species not yet recorded from Mkomazi are included in the list, but are indented to distinguish them from the recorded species.** Of these potential species, montane forest species recorded in the literature from the North and South Pare Mountains are indicated as such. Species determinations by S. van Noort, G.N. Stone and S.C. Collins.

Class INSECTA

Order LEPIDOPTERA

Superfamily HESPERIOIDEA

Family Hesperiidae

Subfamily Coeliadinae

Coeliades anchises anchises (Gerstaecker). Ibaya Hill, widespread.

- *Coeliades forestan forestan* (Stoll)
- *Coeliades libeon* (Druce)
- *Coeliades pisistratus* (Fabricius)

Subfamily Pyrginae

Celaenorrhinus galenus (Fabricius). Maji Kununua.

- *Tagiades flesus* (Fabricius)
- *Eagris nottoana nottoana* (Wallengren)
- *Eagris sabadius astoria* Holland [N. Pares]
- *Eretis lugens* (Rogenhöfer)
- *Eretis melania* Mabille

Eretis umbra maculifera Mabille & Boullet. Kisima Hill.

- *Sarangesa lucidella* (Mabille)
- *Sarangesa maculata* (Mabille)

Sarangesa motozi (Wallengren). Maji Kununua.

Sarangesa phidyle (Walker). Kisima Plot, Mbono Valley.

- *Sarangesa seineri seineri* Strand
- *Caprona pillaana* Wallengren
- *Netrobalane canopus* (Trimen)
- *Abantis paradisea* (Butler)
- *Abantis venosa* Trimen

Spialia colotes transvaaliae (Trimen). Ibaya Camp, Kisima Plot, Maji Kununua, Kisima Hill.

- *Spialia confusa obscura* Higgins
- *Spialia depauperata depauperata* (Strand)
- *Spialia diomus diomus* (Hopffer)
- *Spialia dromus* (Plötz)
- *Spialia mafa higginsi* Evans [N. Pares]
- *Spialia spio* (Linnaeus)
- *Spialia zebra bifida* (Higgins)
- *Gomalia elma elma* (Trimen)

Subfamily Hesperiinae

- *Metisella medea medea* Evans [N. Pares]
- *Metisella midas midas* (Butler) [N. Pares]

Metisella orientalis orientalis (Aurivillius). Ibaya Hill, Maji Kununua.

- *Metisella quadrisignatus quadrisignatus* (Butler) [N. Pares]
- *Metisella willemi* (Wallengren)

Ampittia capenas capenas (Hewitson). Zange Gate.

- *Kedestes callicles* (Hewitson)
- *Kedestes mohozutza* (Wallengren)

Kedestes rogersi Druce. Kisima Plot.

- *Kedestes wallengrenii* (Trimen)
- *Gorgyra bibulus* Riley
- *Gorgyra johnstoni* (Butler)
- *Teniorhinus harona* (Westwood)
- *Teniorhinus herilus* (Hopffer) [S. Pares]
- *Pardaleodes incerta* (Snellen)
- *Parosmodes morantii morantii* (Trimen)
- *Acleros mackenii* (Trimen)
- *Acleros ploetzi* Mabille
- *Semalea arela* (Mabille)
- *Semalea pulvina* (Plötz)
- *Andronymus caesar philander* (Hopffer)
- *Andronymus neander neander* (Plötz)
- *Chondrolepis niveicornis niveicornis* (Plötz) [Pares]
- *Monza punctata punctata* (Aurivillius)
- *Fresna nyassae* (Hewitson)
- *Platylesches galesa* (Hewitson)
- *Platylesches moritili* (Wallengren)
- *Platylesches picanini* (Holland)
- *Zenonia zeno* (Trimen)
- *Pelopidas mathias* (Fabricius)
- *Pelopidas thrax inconspicua* (Bertolini)

Borbo borbonica borbonica (Boisduval). Ibaya Camp.

Borbo detecta (Trimen). Dindira Dam, Kisiwani River.

Borbo fallax (Gaede)
Borbo fatuellus fatuellus (Hopffer)
Borbo ferruginea ferruginea (Aurivillius) [Gonja lowland forest – S. Pares]
Borbo gemella (Mabille)
Borbo holtzi (Plötz)
Borbo lugens (Hopffer)
Gegenes hottentota (Latreille)
Gegenes niso brevicornis (Plötz)
Gegenes pumilio (Hoffmansegg)

Superfamily PAPILIONOIDEA

Family Papilionidae

Subfamily Papilioninae

Papilio (Princeps) constantinus constantinus Ward. Ibaya Hill.
Papilio (Princeps) dardanus tibullus Kirby. Ibaya Hill, Maji Kununua.
Papilio (Princeps) demodocus demodocus Esper. Widespread.
Papilio desmondi magdae Gifford [N. Pares]
Papilio (Princeps) echerioides wertheri Karsch. Ibaya Hill, Maji Kununua.
Papilio fuelleborni rydoni Kielland [S. Pares]
Papilio (Princeps) nireus lyaeus Doubleday. IbayaHill, Kisiwani River.
Papilio (Princeps) ophidecephalus ophidecephalus Oberthür. Ibaya Hill, Maji Kununua.
Papilio (Princeps) phorcas nyikanus Rothschild & Jordan. Ibaya Hill, Maji Kununua.
Graphium (Arisbe) angolanus angolanus (Goeze)
Graphium (Arisbe) antheus antheus (Cramer). Ibaya Hill, Kisiwane River.
Graphium (Arisbe) leonidas leonidas Fabricius. Ibaya Hill, Maji Kununua.
Graphium (Arisbe) philonoe philonoe (Ward). Ibaya Hill.
Graphium (Arisbe) policenes policenes (Cramer). Ibaya Hill.
Graphium (Arisbe) porthaon porthaon (Hewitson)

Family Pieridae

Subfamily Coliadinae

Catopsilia florella (Fabricius). Widespread.
Colias electo pseudohecate Berger. Widespread.
Eurema (Eurema) brigitta brigitta (Stoll). Widespread.
Eurema (Eurema) desjardinsii marshalli Butler. Widespread.
Eurema (Eurema) regularis (Butler)
Eurema (Terias) hapale (Mabille)
Eurema (Terias) hecabe solifera (Linnaeus). Widespread.
Eurema (Terias) senegalensis (Boisduval)

Subfamily Pierinae

Pinacopteryx eriphia melanarge (Butler). Widespread.
Nepheronia argia mhondana (Suffert)
Nepheronia buqueti buqueti (Boisduval). Zange Gate.
Nepheronia thalassina sinalata (Suffert)
Eronia cleodora dilatata Butler. Ibaya Hill, Maji Kununua.
Eronia leda (Boisduval). Widespread.
Colotis amata calais Cramer. Umba River, Simba Plot, Kisima Hill.
Colotis antevippe zera Lucas. Ibaya Hill, Simba Plot, widespread.
Colotis aurigineus Butler. Ibaya Hill.
Colotis auxo (Lucas). Simba Plot, widespread.
Colotis celimene celimene (Lucas). Ibaya Hill.
Colotis chrysonome (Klug). Ngurunga.
Colotis daira jacksoni (Sharpe). Ngurunga.
Colotis danae pseudacaste (Butler). Widespread.
Colotis dissociatus (Butler). Nyati Plot, Mbono valley, widespread.
Colotis eris eris Klug. Widespread.
Colotis euippe complexivus Butler. Ibaya Camp, Maji Kununua, widespread.
Colotis evagore antigone (Boisduval). Ibaya Camp, Simba Plot, Maji Kununua, widespread.
Colotis evenina casta (Gerstaeker). Dindira Dam, widespread.
Colotis halimede australis Talbot. Mbono Valley, widespread.
Colotis hetaera ankolensis Stoneham. Maji Kununua, Ibaya Hill, Kisima Plot.
Colotis hildebrandti (Staudinger). Kisima Plot, Ubani Plot, widespread.
Colotis ione (Godart). Kisima Plot, Ibaya Camp, Kisiwani River, widespread.
Colotis pallene (Hopffer)
Colotis phisadia rothschildi (Sharpe) [southern Kenya]
Colotis protomedia (Klug). Simba Plot, Ngurunga Plot.
Colotis regina (Trimen). Ibaya Hill.

Colotis venosus (Staudinger)
Colotis vesta catachrysops (Butler). Kisima Plot, widespread.
Colotis vestalis castalis (Staudinger). Kisima Plot.
Belenois aurota aurota (Fabricius). Dindira Dam, widespread.
Belenois creona severina (Stoll). Kisiwani River, widespread.
Belenois gidica gidica (Godart). Ibaya Camp. Ibaya Hill, Kisima Plot, widespread.
Belenois margaritacea plutonica (Joicey & Talbot) [N. & S. Pares]
Belenois margaritacea intermedia Kielland. Maji Kununua.
Belenois thysa thysa (Hopffer). Ibaya Hill, Kisiwani River.
Belenois zochalia agrippinides (Holland). Ibaya Hill.
Pontia distorta (Butler)
Pontia helice johnstoni (Crowley)
Dixeia doxo costata Talbot. Ibaya Hill.
Dixeia orbona vidua (Butler) [N. Pares]
Dixeia pigea Boisduval
Dixeia spilleri (Spiller)
Appias (Glutophrissa) epaphia contracta (Butler). Zange gate.
Appias (Glutophrissa) lasti lasti (Grose-Smith)
Appias (Glutophrissa) sabina phoebe Butler. Ibaya Hill, Kisiwani River.
Mylothris agathina agathina (Cramer). Zange gate.
Mylothris kilimensis kilimensis Kielland
Mylothris ruppellii tirikensis Neave
Mylothris sagala sagala Grose-Smith. Maji Kununua.
Mylothris yulei yulei Butler [N. & S. Pares]
Leptosia alcesta inalcesta Bernardi. Ibaya Hill, Maji Kununua.

Family Nymphalidae

Subfamily Acraeinae

Acraea (Acraea) acrita Hewitson
Acraea (Acraea) aganice montana (Butler). Ibaya Hill, Kisima Hill.
Acraea (Acraea) anemosa Hewitson. Kisima Hill.
Acraea (Acraea) braesia Godman
Acraea (Acraea) caecilia pudora Aurivillius
Acraea (Acraea) caldarena neluska Oberthür
Acraea (Acraea) chilo chilo Godman. Ibaya Camp.
Acraea (Acraea) cerasa cerasa Hewitson. Ibaya Hill, Maji Kununua.
Acraea (Acraea) egina egina (Cramer)
Acraea (Acraea) equatorialis anaemia Eltringham
Acraea (Acraea) insignis Distant
Acraea (Acraea) lygus Druce
Acraea (Acraea) natalica Boisduval. Ibaya Hill.
Acraea (Acraea) neobule neobule Doubleday. Ibaya Hill, Simba Plot, Mbula, nr. Kisima Hill, Zange Gate.
Acraea (Acraea) oncaea Hopffer.
Acraea (Acraea) petraea Boisduval
Acraea (Acraea) pudorella pudorella Aurivillius. Ibaya Hill.
Acraea (Acraea) pudorina Staudinger. Kisima Plot.
Acraea (Acraea) quadricolor leptis (Jordan)
Acraea (Acraea) quirina rosa Eltringham. Maji Kununua.
Acraea (Acraea) rabbaiae Ward
Acraea (Acraea) utengulensis Thurau
Acraea (Acraea) zetes acara Hewitson. Kisima Hill & Plot.
Acraea (Acraea) zonata Hewitson
Acraea (Actinote) acerata Hewitson
Acraea (Actinote) anacreon bomba Grose-Smith
Acraea (Actinote) aubyni Eltringham [Kenya, probably NE corner of Tanzania]
Acraea (Actinote) baxteri baxteri Sharpe [Pares]
Acraea (Actinote) cabira Hopffer. Ibaya hill.
Acraea (Actinote) encedana Pierre
Acraea (Actinote) encedon (Linnaeus). Ibaya Hill.
Acraea (Actinote) eponina (Cramer). Ibaya Hill, Kikolo Plot.
Acraea (Actinote) esebria Hewitson. Ibaya Hill.
Acraea (Actinote) johnstoni johnstoni Godman. Ibaya Hill, Maji Kununua.
Acraea (Actinote) perenna thesprio Oberthür [Pares]
Acraea (Actinote) pharsalus pharsaloides Holland. Ibaya Hill.
Acraea (Actinote) servona orientis Aurivillius
Acraea (Actinote) sotikensis Sharpe
Pardopsis punctatissima (Boisduval)

Subfamily Danainae

Danaus (Anosia) chrysippus aegyptius (Linnaeus). Widespread.

Tirumala formosa formosa (Godman)
Tirumala petiverana (Doubleday)
Amauris (Amauris) niavius dominicanus Trimen. Ibaya Hill, Maji Kununua.
Amauris (Amauris) tartarea damoclides Staudinger.
Amauris (Amaura) albimaculata hanningtoni Butler
Amauris (Amaura) echeria serica Talbot
Amauris (Amaura) ochlea ochlea (Boisduval). Widespread.

Subfamily Satyrinae

Gnophodes betsimena diversa (Butler). Ibaya Hill.
Melanitis leda helena (Westwood). Widespread in bushland and woodland.
Bicyclus anynana anynana (Butler)
Bicyclus ena (Hewitson)
Bicyclus safitza safitza (Westwood). Ibaya Hill.
Henotesia perspicua (Trimen). Ibaya Hill, Dindira Dam, Kisima Hill.
Henotesia simonsii (Butler)
Ypthima antennata antennata van Son
Ypthima asterope asterope (Klug). Kikolo Plot.
Ypthima granulosa Butler
Ypthima impura paupera Ungemach
Ypthima rhodesiana Carcasson
Ypthimomorpha itonia (Hewitson)
Physcaeneura jacksoni Carcasson
Physcaeneura leda (Gerstaecker). Ibaya Hill.
Coenyropsis carcassoni Kielland
Neocoenyra duplex Butler. Pangaro Plot.
Neocoenyra masaica Carcasson

Subfamily Argynninae

Lachnoptera ayresii Trimen
Phalanta eurytis eurytis (Doubleday)
Phalanta phalantha aethiopica (Rothschild & Jordan). Ibaya Hill, Maji Kununua.

Subfamily Nymphalinae

Hypolimnas anthedon wahlbergi (Wallengren)
Hypolimnas deceptor deceptor (Trimen)
Hypolimnas misippus (Linnaeus). Ibaya Hill, widespread.
Salamis anacardii nebulosa Trimen. Zange Gate.
Salamis parhassus (Drury)
Junonia actia (Distant)
Junonia antilope (Feisthamel). Ibaya Camp, Kisima Plot.
Junonia archesia (Cramer)
Junonia cuama Hewitson
Junonia hierta cebrene Trimen. Widespread.
Junonia limnoria taveta (Rogenhöfer). Ibaya Hill, Simba plot.
Junonia natalica natalica (Felder & Felder). Zange Gate, Mbono Valley.
Junonia octavia sesamus (Trimen)
Junonia oenone oenone (Linnaeus). Widespread.
Junonia orithya madagascariensis Guenée
Junonia pelarga (Fabricius)
Junonia sophia infracta Butler
Junonia terea elgiva Hewitson. Ibaya Hill, Maji Kununua.
Junonia tugela aurorina Butler. Ibaya Hill, Maji Kununua.
Catacroptera cloanthe cloanthe (Stoll). Mbono Valley.
Cynthia cardui (Linnaeus). Widespread.
Antanartia abyssinica jacksoni Howarth [probably on N. Pares]
Antanartia dimorphica dimorphica Howarth
Antanartia schaeneia dubia Howarth

Subfamily Limenitinae

Byblia anvatara acheloia (Wallengren)
Byblia ilithyia (Drury). Widespread.
Neptidopsis ophione nucleata Grünberg
Eurytela dryope angulata Aurivillius. Mbono Valley.
Eurytela hiarbas lita Rothschild & Jordon. Ibaya Hill.
Sallya boisduvali boisduvali (Wallengren)
Sallya moranti moranti (Trimen)
Sallya natalensis (Boisduval)
Cyrestis camillus sublineata Lathy. Ibaya Hill.
Neptis aurivillii aurivillii Schultze [Pares]
Neptis goochi Trimen
Neptis kiriakofi Overlaet
Neptis laeta Overlaet. Zange Gate.
Neptis nina Staudinger
Neptis saclava marpessa Hopffer. Zange Gate.
Neptis serena Overlaet
Neptis trigonophora trigonophora Butler
Harma theobene blassi (Weymer)
Pseudacraea boisduvali trimeni Butler. Ibaya Hill.
Pseudacraea eurytus conradti Oberthür
Pseudacraea lucretia expansa (Butler)
Euptera pluto kinungnana (Grose-Smith)

Euphaedra neophron violacea (Butler) [N. Pares]
Euphaedra neophron littoralis Talbot [S. Pares]
Hamanumida daedalus (Fabricius). Ibaya Camp, widespread.

Subfamily Charaxinae
Euxanthe (Euxanthe) wakefieldi (Ward)
Charaxes achaemenes achaemenes Felder & Felder
Charaxes acuminatus usambarensis van Someren [N. & S. Pares]
Charaxes aubyni aubyni van Someren & Jackson [N. & S. Pares]
Charaxes baumanni baumannii Rogenhöfer [N. & S. Pares]
Charaxes bohemani Felder & Felder
Charaxes brutus alcyone Stoneham. Ibaya Hill.
Charaxes candiope candiope (Godart). Ibaya Hill.
Charaxes castor flavifasciatus Butler
Charaxes cithaeron kennethi Poulton
Charaxes druceanus praestans Turlin [S. Pares]
Charaxes etesipe tavetensis Rothschild
Charaxes ethalion littoralis van Someren
Charaxes guderiana guderiana (Dewitz)
Charaxes hansali baringana Rothschild
Charaxes jahlusa kenyensis Joicey & Talbot. Kikola Plot.
Charaxes jasius saturnus Butler. Kamakota Hill, widespread.
Charaxes kirki kirki Butler. Kamakota Hill.
Charaxes lasti lasti Grose-Smith
Charaxes macclounii Butler
Charaxes pollux mira Ackery [N. & S. Pares]
Charaxes protoclea azota (Hewitson). Ibaya Hill.
Charaxes pythodoris nesaea Grose-Smith [S. Pares]
Charaxes varanes vologeses (Mabille). Ibaya Hill, Zange Gate, widespread.
Charaxes violetta melloni Fox
Charaxes xiphares maudei Joicey & Talbot [N. Pares]
Charaxes zoolina zoolina (Westwood). Ibaya Hill, Ibaya Camp, widespread.

Subfamily Apaturinae
Apaturopsis cleochares schultzei Schmidt [S. Pares]

Subfamily Libytheinae
Libythea labdaca laius Trimen. Ibaya Hill.

Family Lycaenidae

Subfamily Lipteninae
Alaena amazoula nyasana Hawker-Smith. Ibaya Hill.
Alaena caissa caissa Rebel & Rogenhofer
Alaena dodomaensis Kielland
Alaena nyassa major Oberthür. Ibaya Hill.
Alaena reticulata Butler
Pentila rogersi parapetreia Rebel
Pentila tropicalis mombasae (Grose-Smith & Kirby). Ibaya Hill, Maji Kununua.
Ornipholidotos peucetia peuceda (Grose-Smith)
Baliochila ?amanica Stempffer & Bennett. Kisima Plot.
Baliochila ?dubiosa Stempffer & Bennett. Ibaya Hill, Maji Kununua.
Baliochila fragilis Stempffer & Bennett
Baliochila hildegarda (Kirby)
Baliochila lipara Stempffer & Bennett
Baliochila minima (Hawker-Smith)
Baliochila ?pseudofragilis Kielland. Kisima Plot.
Baliochila stygia Stempffer & Bennett
Cnodontes vansomereni Stempffer & Bennett
Deloneura ochrascens littoralis Talbot. Kikola Plot.

Subfamily Miletinae
Spalgis lemolea Druce
Lachnocnema bibulus (Fabricius). Kikolo Plot.
Lachnocnema brimo Karsch.
Lachnocnema durbani Trimen

Subfamily Theclinae
Myrina dermaptera nyassae Talbot
Myrina silenus ficedula Trimen. Kamakota Hill.
Aphnaeus (Paraphnaeus) hutchinsonii Trimen
Spindasis apelles (Oberthür)
Spindasis ella (Hewitson)
Spindasis homeyeri (Dewitz)
Spindasis mozambica (Bertolini)
Spindasis tavetensis Lathy
Spindasis victoriae (Butler)
Chloroselas azurea Butler
Chloroselas esmeralda esmeralda Butler
Chloroselas overlaeti Stempffer
Chloroselas pseudozeritis tytleri Riley
Desmolycaena rogersi Riley
Axiocerses amanga (Westwood)
Axiocerses bambana Grose-Smith
Axiocerses harpax ugandana Clench. Ibaya

Hill , Mbula, nr. Kisima Hill, Kisiwani Gate.

Axiocerses styx Rebel
Aloeides conradsi talboti Tite & Dickson
Iolaus (Epamera) aemulus apatosa Talbot
Iolaus (Epamera) diametra diametra (Karsch)
Iolaus (Epamera) nasisii (Riley)
Iolaus (Epamera) sidus Trimen
Iolaus (Epamera) silanus silanus Grose-Smith
Iolaus (Epamera) tajoraca ertli Aurivillius
Iolaus (Aphniolaus) pallene (Wallengren)
Iolaus (Argiolaus) crawshayi littoralis Stempffer & Bennett
Iolaus (Argiolaus) lalos lalos (Druce)
Iolaus (Argiolaus) silarus Druce
Iolaus (Argiolaus) silas (Westwood)
Iolaus (Pseudiolaus) poultoni (Riley)
Iolaus (Stugeta) bowkeri mombasae (Butler)
Iolaus (Hemiolaus) caeculus littoralis (Stempffer)
Hypolycaena buxtoni rogersi Bethune-Baker

Hypolycaena pachalica Butler. Mbono valley, Cadaba plot, Mbula, Kisiwani gate.

Hypolycaena philippus philippus (Fabricius). Mbono valley.

Leptomyrina (Leptomyrina) hirundo (Wallengren). Ibaya Hill.

Gonatomyrina gorgias sobrina Talbot

Deudorix (Pilodeudorix) caerulea Druce. Kisima Plot.

Deudorix (Virachola) antalus (Hopffer). Maji Kununua.

Deudorix (Virachola) dinochares Grose-Smith. Kisima Plot.

Deudorix (Virachola) diocles Hewitson
Deudorix (Virachola) ecaudata Gifford
Deudorix (Virachola) livia (Klug)
Deudorix (Virachola) lorisona coffea Jackson

Subfamily Lycaeninae

Lycaena phlaeus abboti (Holland) [N. Pares]

Subfamily Polyommatinae

Anthene amarah amarah (Guérin-Méneville). Mbula, Kisiwani Gate, Zange Gate.

Anthene butleri stempfferi Storace [probably N & S. Pares]
Anthene contrastata mashuna (Stevenson)
Anthene crawsheyi crawsheyi (Butler)
Anthene definita definita (Butler)
Anthene hobleyi Neave [N. Pares]
Anthene indefinata (Bethune-Baker)
Anthene kersteni (Gerstaecker)

Anthene larydas (Cramer). Kisiwani River.

Anthene lasti (Grose-Smith & Kirby)

Anthene lemnos loa (Strand). Ibaya Hill.

Anthene ligures amanica (Strand)
Anthene liodes (Hewitson)
Anthene lunulata (Trimen)
Anthene minima (Trimen)

Anthene opalina Stempffer. Mbula, Kisiwani Gate.

Anthene otacilia otacilia (Trimen). Ubani Plot, Mbula, Kisiwani Gate.

Anthene princeps princeps (Butler)
Cupidopsis cissus (Godart)

Cupidopsis jobates jobates (Hopffer). Ibaya Hill.

Pseudonacaduba sichela sichela (Wallengren). Ibaya Camp, Ibaya Hill, Kisiwani River.

Lampides boeticus (Linnaeus). Ibaya Hill.

Uranothauma antinorii felthami (Stevenson)
Uranothauma cordatus (Sharpe) [N. & S. Pares]
Uranothauma falkensteini (Dewitz) [N. & S. Pares]
Uranothauma nubifer (Trimen) [N. Pares]
Uranothauma vansomereni Stempffer
Phlyaria heritsia intermedia Tite [S. Pares]

Cacyreus lingeus (Stoll). Kisima Hill.

Cacyreus palemon palemon (Stoll) [N. & S. Pares]
Cacyreus virilis Stempffer
Leptotes babaulti (Stempffer)
Leptotes brevidentatus (Tite)
Leptotes jeanneli (Stempffer)
Leptotes marginalis (Stempffer)

Leptotes pirithous (Linnaeus). Kisiwani River, Mbula.

Tuxentius calice gregorii (Butler)
Tuxentius margaritaceus (Sharpe) [N. Pares]
Tarucus grammicus (Grose-Smith & Kirby)
Zintha hintza hintza (Trimen)

Zizeeria knysna (Trimen). Mbula, nr. Kisima Hill, Zange Gate.

Zizina antanossa (Mabille)
Actizera lucida (Trimen)

Zizula hylax (Fabricius). Ubani Plot.

Azanus jesous (Guérin-Méneville). Ibaya Hill, Mbula, nr. Kisima Hill, Zange Gate.

Azanus mirza (Plötz). Ibaya Hill, Kisiwani River, Ubani Plot.
Azanus moriqua (Wallengren). Mbula, nr. Kisima Hill.
Azanus natalensis (Trimen)
Azanus ubaldus (Stoll). nr. Ubani Plot, Mbula, nr. Kisima Hill.
Eicochrysops hippocrates (Fabricius)
Eicochrysops messapus mahallakoaena (Wallengren)
Euchrysops barkeri (Trimen)
Euchrysops brunneus Bethune-Baker
Euchrysops malathana (Boisduval)
Euchrysops osiris osiris (Hopffer). Ibaya Hill.
Euchrysops subpallida Bethune-Baker. Nyati Plot.
Lepidochrysops lukenia van Someren [Pares]
Lepidochrysops neonegus neonegus (Bethune-Baker)
Freyeria trochylus (Freyer). Ubani Plot.

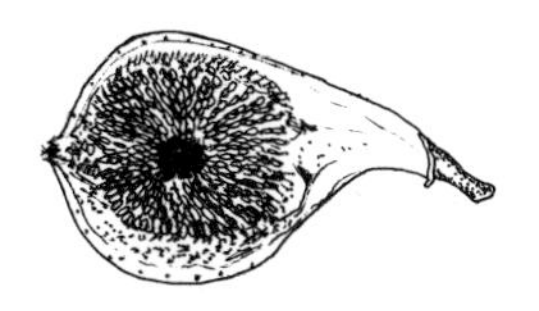

CHAPTER 18

Fig wasps (Hymenoptera: Chalcidoidea: Agaonidae) and fig trees (Moraceae: *Ficus*) of Mkomazi

Simon van Noort & Stephen G. Compton

Introduction

The order Hymenoptera (wasps, bees and ants) is second only to the beetles in terms of species richness and abundance, and includes a diverse range of morphological forms and biologies. The superfamily Chalcidoidea is a large and economically important group of wasps, whose representatives are mostly parasitoids of other insects. 'Fig wasp' is a broad term applied to wasps of the superfamily Chalcidoidea that solely breed in figs (*Ficus*, Moraceae), but excludes wasps from this superfamily that are parasitoids of moth, beetle and fly larvae that sometimes also breed in figs. From a taxonomic perspective the term 'fig wasp' encompasses representatives of three families: Agaonidae, Eurytomidae and Ormyridae. Of the latter two families only a small proportion of their species are associated with figs, whereas all of the species placed in the Agaonidae are fig wasps. Hence the majority of fig wasps belong to the Agaonidae, which currently includes six distinct subfamilies (Boucek 1988).

In this chapter we assesses species richness of both the fig trees and their associated fig wasps in Mkomazi Game Reserve. As with most African countries the fig wasp fauna of Tanzania is poorly known, and this survey has played a valuable role in furthering our knowledge of Tanzanian fig wasps.

Ecology

Fig wasps and their host fig tree species, are important components of tropical and subtropical ecosystems, from both an abundance and diversity perspective and as an integral part of the food chain. The fascinating relationship between pollinating fig wasps (Agaoninae) and their host fig trees is a classic example of an obligate mutualism, where neither partner can reproduce without the other (Galil 1977,

Janzen 1979). The interaction between figs and fig wasps is more complicated than first appears, because pollinating wasps are not the only fig wasps that utilise the fig flowers for propagation of their offspring. The fig also provides a suitable breeding site for a diverse array of non-pollinating fig wasps, which are either phytophagous (plant feeding), galling the ovules as do the pollinators, or parasitoids of the gall formers (Compton & van Noort 1992, Kerdelhué & Rasplus 1996a & 1996b).

The mutualism between pollinating fig wasps and fig trees is usually a one-to-one relationship (Ramirez 1970, Wiebes 1979, Wiebes & Compton 1990). Each fig tree species (approximately 750 worldwide) has a single pollinating fig wasp species and each wasp species is only associated with one fig tree species, although there are a few exceptions to this rule. Non-pollinating fig wasps are generally less specific with a number of species associated with more than a single host species, and often each fig tree species has two or three closely related non-pollinating fig wasp species breeding in its figs.

The developmental cycle of the fig comprises five distinct but inter-connecting stages with fig wasp larval development correlating strongly with fig development (Galil 1977). The cycle may encompass anything from 3 to 20 weeks (Bronstein 1992, Ware & Compton 1994). The fig is an urn-shaped receptacle containing hundreds of tiny flowers which line the inside walls of the central cavity and becomes receptive for pollination and oviposition early in the developmental cycle. Female pollinating wasps gain access to the inside of the fig through the ostiole (a tiny, narrow opening at the top end of the fig). The pollinating wasps are uniquely adapted to squeeze their way through the ostiole, having evolved a flattened head and body and many rows of backward pointing mandibular teeth situated on the underside of the head. Once inside the fig cavity, the female proceeds to unload pollen onto the stigmas and inserts her ovipositor down the style of the flower to oviposit within the ovule. The ovary swells up to form a gall and the wasp larvae feed on endosperm tissue in the galled ovary (Verkerke 1989). Although some non-pollinating wasp species also enter the fig for oviposition, and have then evolved similar physical adaptations to squeeze through the ostiole, most of the non-pollinators oviposit through the fig wall from the outside of the fig at various stages of fig development (Kerdelhué & Rasplus 1996a). These wasps often have extremely long ovipositors, the length of which is related to the wall thickness of their host fig. Fig size varies tremendously across species, and ranges from smaller than a marble to as large as a tennis ball. Towards the end of the fig developmental cycle, all the fig wasps breeding in a particular fig emerge from their galls within a short period of each other. Mating largely takes place within the confines of the fig before the males chew a hole through the fig wall to the exterior to allow the females to escape. Pollinator females actively load up pollen from the ripe anthers before emerging from the fig to search for young receptive figs to complete the cycle. Most of the figs within a crop on a fig tree are usually at the same stage of development, with the consequence that emerging female fig wasps need to find another fig tree to continue the reproductive cycle. This may require a long distance flight to locate a tree with receptive figs for oviposition and pollination. These tiny wasps, averaging between one and two millimetres in length, achieve this remarkable feat by homing in on gaseous chemicals, released by the figs when they are receptive for pollination (van Noort *et al.* 1989, Hossaert-McKey *et al.* 1994).

Once the female fig wasps have left the fig, it ripens and becomes attractive to fruit-eating birds, bats and monkeys. Because figs are produced throughout the year a continual supply of food is provided through periods when there is a seasonal dearth of other fruits. As such, fig trees are considered to be keystone species in many tropical and subtropical ecosystems (Terborgh 1986, Lambert & Marshall 1992), but see Gautier-Hion & Michaloud (1989) and Basset *et al.* (1997). To complete the reproductive cycle of the mutualism, fruit-eating vertebrates play an important role in the propagation of fig trees, acting as the dispersal agents of the seeds, which, at least in the case of birds, are positively affected by passage through the digestive tract, resulting in increased germination viability (Compton *et al.* 1996).

Fig trees

Regional richness

Of the 105 fig tree species that occur in the Afrotropical region (Berg & Wiebes 1992) an estimated 39 species are found in Tanzania (Berg & Hijman 1989). Ten of these species are distributionally restricted (endemic) to east Africa. Tanzania has a higher fig tree species richness than Kenya, but a lower richness than Uganda (Berg & Hijman 1989) (Table 18.1). The higher Ugandan species richness is attributable to the presence of ten species that are typical elements of the Guinea-Congolian forest region and whose distribution does not reach as far east as Kenya or Tanzania. 28 fig species occur in all three countries, five are shared between Tanzania and Kenya, four between Tanzania and Uganda and one between Uganda and Kenya (Berg & Hijman 1989).

Fig tree species richness is considerably higher in east Africa than southern Africa, where 32 species occur in the whole southern African subregion (defined as including Namibia, Botswana, Zimbabwe, Mozambique south of the Zambezi River, South Africa, Swaziland and Lesotho). 22 of these species are present in

Table 18.1 Fig tree species richness by country.

	Tanzania	Kenya	Uganda	South Africa
Ficus species	39	34	43	22

South Africa (Berg 1990). The lower southern African species richness can be ascribed to the temperate climate of large areas of southern Africa, making most of the region unsuitable for fig trees which prefer a tropical climate.

Species richness and distribution within Mkomazi Game Reserve

Nine fig tree species were recorded within Mkomazi Game Reserve (Table 18.2). It is highly probable that further species await discovery, and because of this there is value in assessing which of the remaining 30 Tanzanian species may potentially be present in the reserve.

Five of the Tanzanian fig species are restricted to rainforest in the north-west region effectively excluding them from Mkomazi Game Reserve. *Ficus capreifolia* Delile, *F. verruculosa* Warburg, and *F. trichopoda* Baker are associated with riverine or swamp conditions (Berg 1990, Berg & Wiebes 1992) and therefore unlikely to be found in the seasonally dry conditions in Mkomazi Game Reserve. However, the habitat along the Umba River, the only permanent water body in the reserve, has not been comprehensively surveyed for fig trees, with only a single limited visit to one locality within this area. Conceivably, these three riverine fig species may be present along the eastern boundary of the reserve. A further riverine species that is also found in ground-water forest (Berg & Hijman 1989), *F. vallis-choudae* Delile, is common just outside Mkomazi at the base of the South Pare Mountains where it is present in riverine forest. It is feasible that this species remains undetected within the reserve, although no suitable habitats were identified. *Ficus sur* Forsskål is usually associated with riverine conditions or moist forest but also occurs in woodland (Berg & Hijman 1989) and because it is a common and widespread species it is likely to be present in Mkomazi.

A further seven Tanzanian species are associated with forest (Berg & Hijman 1989), two of which, namely *F. exasperata* Vahl and *F. tremula acuta* (De Wild), were recorded on the South Pare Mountains during this programme. These two species together with *F. mucosa* Ficalho, *F. c. cyathistipula* Warb., *F. s. scassellatii* Pamp., *F. polita brevipedunculata* Berg and *F. chirindensis* Berg may be present in the limited montane forest patches within the reserve on hill tops such as Ibaya Hill and Maji Kununua. To date only *F. thonningii* Bl. has been recorded in montane forest in Mkomazi. The montane forest habitat within the reserve may be too degraded or limited in extent to support these forest endemics. *Ficus lingua depauperata* (Sim) Berg, *F. ottoniifolia ulugurensis* (Mildbr. & Burr.) Berg and *F. t. tremula* Warb. are associated with lowland dry forest or coastal bushland. These three species have been recorded by Hawthorne (1993) in the east African coastal forests, but are probably not present within Mkomazi.

Of the rock-splitters (species that often germinate and grow in cracks in rocks), *F. glumosa* Delile is a very common species within Mkomazi, occurring on many of the rocky ridges on the hills in the western end of the reserve and on isolated rock outcrops, such as Kamakota Hill in the central region. *Ficus ingens* (Miquel) Miquel is far less common, having only been recorded on Kamakota Hill, Kisima Hill and at Ngurunga Pools, but undoubtably occurs in other unsurveyed rocky areas as well. Further rock-splitters that were expected to be present, such as *F. cordata salicifolia* (Vahl) Berg, *F. abutilifolia* (Miquel) Miquel, *F. platyphylla* Delile, and *F. populifolia* Vahl, have not yet been recorded from Mkomazi.

A number of woodland species, *F. wakefieldii* Hutch., *F. n. natalensis* Hochst., *F. nigro-punctata* Mildbr. & Burr., *F. fischeri* Mildbr. & Burr., *F. amadiensis* De Wild, *F. faulkneriana* Berg, *F. usambarensis* Warburg and *F. ovata* Vahl may yet be recorded from the reserve. If they are present they are likely to be concentrated in the wetter western areas along perennial water courses, although this habitat in the vicinity of Ibaya Camp has been well surveyed for fig trees. *Ficus stuhlmannii* Warb., *F. s. sansibarica* Warb., *F. bubu* Warb. and *F. sycomorus* Linnaeus only occur along these seasonal river courses, such as the Mzukune River in Mbono Valley, or in wetter ravines on the slopes of hills such as the top of the valley north-west of Ibaya Camp leading up Ibaya Hill, where these species as well as *F. lutea* Vahl are present. *Ficus bussei* has only been recorded growing in rocky areas of the river course near Ngurunga Pools. The lower reaches of this river course, below the pools, have not yet been surveyed and promise to produce further records. The dryer central and eastern areas of Mkomazi appear to be unsuitable for these woodland species and apart from the concentrations of fig trees on the isolated rocky hills are relatively fig tree depauperate.

Fig wasps

Regional richness

On a world basis the Afrotropical fig wasps are probably the best documented, with systematics of two of the six subfamilies reasonably well known in the region: the Agaoninae, extensively studied by J.T. Wiebes, references in Berg & Wiebes (1992), and the Sycoecinae (van Noort 1993a, b, 1994a, b, c), although only an estimated 72% (Wiebes & Compton 1990) and 56% (van Noort 1994c) respectively of the total extant fauna of these two subfamilies is known. Additionally, the genus *Apocrypta* Coquerel (Sycoryctinae) has been revised on a world basis (Ulenberg 1985). Currently 230 fig wasp species have been described from the 105 Afrotropical species of *Ficus* (Moraceae) (Berg & Wiebes 1992, van Noort 1994c & 1998), a figure that probably represents about one third of the extant species, an estimation based on available undescribed material, host-specificity and extrapolation from host associations (van Noort & Rasplus 1997). Three of the remaining four subfamilies, the Epichrysomallinae (Rasplus, unpubl.), Sycophaginae (Rasplus & Kerdelhué, unpubl.) and the Otitesellinae (van Noort & Rasplus 1997, van Noort, unpubl.) are currently under revision.

Sampling biases

Fig crops are produced randomly throughout the year and individual trees produce crops at different times to each other, both essential traits to ensure the continued cycling of the mutualism. Because of this most of the fig trees that are located during field surveys either have no figs or have figs at the wrong stage of development for rearing of fig wasps. On average only one out of every 30 trees that is located during field work has a fig crop at the right stage of development. These sampling constraints in conjunction with the limited time spent in the reserve meant that fig wasps were not reared from three fig tree species (*F. lutea*, *F. thonningii* and *F. bussei*) recorded in Mkomazi. In addition, the single *F. ingens* tree that was sampled had a fig crop that had already released most of its wasps and hence produced an incomplete sample (Table 18.2).

Furthermore, not every fig wasp species associated with a particular fig tree species is reared from every sample of figs. There are two reasons for this. Firstly, not every fig crop borne by the tree has all the possible fig wasp species present. Some species normally associated with the tree may not have managed to located the fig crop, or alternatively some species may be absent from the local geographical area. Secondly, it is impossible to sample every fig in a particular crop and, because not all the fig wasp species associated with the fig crop will be breeding in every fig, some species which are rarer than others may be missed. To collect every fig wasp species associated with a fig tree species may require anything up to 23 samples from different trees at different times in a particular geographic area

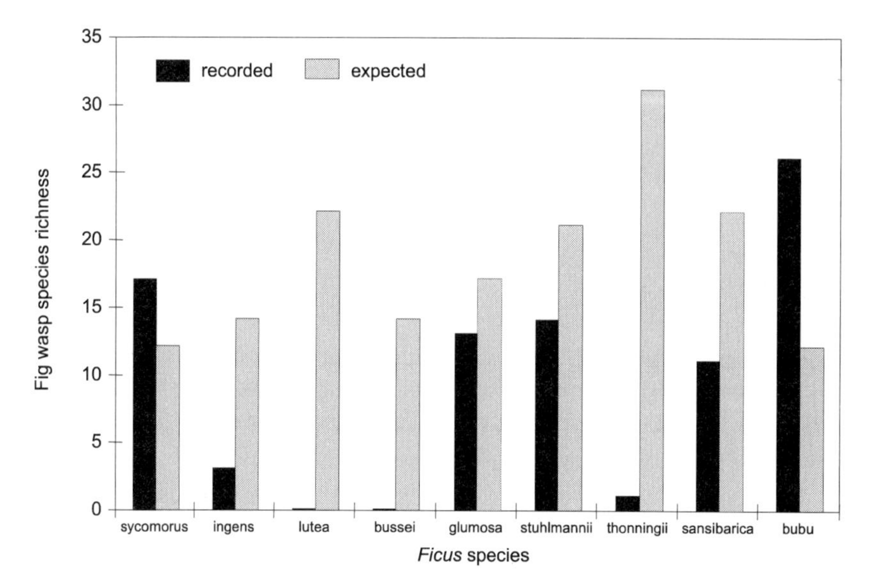

Figure 18.1 Recorded and expected species richness of the fig wasp assemblages associated with the nine fig tree species in Mkomazi Game Reserve.

(Compton and Hawkins 1992, Compton, *et al.* 1994, West *et al.* 1996). This is verified by an example from the New World where a single crop of *F. aurea* in Florida, USA produced four species of fig wasp, a sample of 12 crops over time from the same tree produced seven species, and a sample of 60 crops from 23 different trees produced a total of nine species (Bronstein & Hossaert-Mckey 1996), indicating that comprehensive sampling is required to collect the full compliment of the associated fig wasp assemblage. To be reasonably confident that all the associated fig wasp species have been collected from a fig tree species a species accumulation curve is plotted. This curve depicts the sequential acquisition of new fig wasp species associated with a fig tree species as the samples are collected. Once the curve starts levelling out to a plateau this indicates that the majority of species have been sampled. None of the fig tree species were sufficiently sampled in Mkomazi to attain a levelling off of the accumulation curves. One exception may be *F. bubu* where an excellent sample of figs was taken from a single tree with a large fig crop. This produced a record number of fig wasp species (25) reared from a single fig crop for *F. bubu*, and is likely to be close to the total number associated with this host species in the reserve. It is thus possible, but unusual, to sample the majority of fig wasps associated with a particular host through a single collection. This is verified by a single collection made from *F. thonningii* in Tanzania that produced 31 species of fig wasp (J.Y. Rasplus, pers. comm.), a total only achieved after 49 collections were made from this fig tree species in southern Africa.

There are thus two main reasons for the underestimation of fig wasp species richness in Mkomazi Game Reserve. Firstly, fig wasp species richness recorded from the host fig tree species in the reserve is currently an under representation, given the limited sampling effort, and secondly, it is likely that further species of *Ficus* still await to be recorded from Mkomazi Game Reserve each with its own host-specific fig wasp fauna.

Species richness within Mkomazi Game Reserve

Eighty-five species of fig wasp have been recorded from Mkomazi of which around three-quarters are undescribed (Tables 18.2 & 18.3). This is about half of the fig wasp species expected to be reared from the nine recorded fig tree species in Mkomazi (see under sampling biases for an explanation). However, because of the high host-specificity of fig wasps we can be reasonably confident that wasps previously recorded as being associated with the unsampled fig trees will probably also be present in Mkomazi Game Reserve (Figure 18.1). These previous records are from fig wasp collections made in other areas in eastern and southern Africa (Table 18.3). If these unrecorded fig wasps are taken into account the minimum total richness for the reserve is likely to be around 183 species. This total will still be an underestimate because it is probable that further fig tree species are present within the reserve. The unrecorded fig wasp species have been included in the

Table 18.2 Systematic composition of the fig wasp assemblages associated with each fig tree species in Mkomazi Game Reserve. Brackets enclose the expected species number (based on collections elsewhere in east and southern Africa). A dash (–) denotes instances where the subfamily or tribe is not associated with the host fig tree. * denotes where foundress female pollinating fig wasps were collected from immature figs.

Ficus sp.	no. of samples	Agaoninae	Epichryso–mallinae	Otitesellinae	Sycoecinae	Sycophaginae	Sycoryctinae	Eurytomidae	Ormyridae	total
F. sycomorus	5	2 (2)	1 (0)	–	–	4 (4)	7 (6)	3 (0)	0 (0)	17 (12)
F. ingens	1 (poor)	1 (1)	0 (1)	1 (2)	–	–	1 (4)	0 (6)	0 (0)	3 (14)
F. lutea	0	0 (1)	0 (6)	0 (2)	0 (1)	–	0 (5)	0 (7)	0 (0)	0 (22)
F. bussei	0	0 (1)	0 (3)	0 (0)	0 (1)	–	0 (6)	0 (3)	0 (0)	0 (14)
F. glumosa	7	1 (1)	2 (2)	2 (2)	2 (3)	–	3 (4)	2 (4)	1 (1)	13 (17)
F. stuhlmannii	1	1 (1)	3 (3)	2 (2)	2 (3)	–	6 (8)	0 (4)	0 (0)	14 (21)
F. thonningii	0*	1 (1)	0 (4)	0 (2)	0 (2)	–	0 (9)	0 (10)	0 (3)	1 (31)
F. sansibarica	3	1 (1)	1 (2)	1 (2)	1 (1)	–	6 (8)	1 (7)	0 (1)	11 (22)
F. bubu	2	1 (1)	4 (2)	2 (2)	1 (1)	–	9 (4)	9 (1)	0 (1)	26 (12)
total	19	8 (10)	11 (23)	8 (14)	6 (12)	4 (4)	32 (54)	15 (42)	1 (6)	85 (165)

checklist (but clearly marked as such), because this provides a more realistic interpretation of local fig wasp species richness.

18 more fig wasp species were recorded from Mkomazi fig trees than were expected based on fig wasp collections made in southern Africa (Table 18.2). These additional species were reared from *F. sycomorus* and *F. bubu* illustrating the higher species richness of the fig wasp assemblages associated with these two species in east Africa as compared to southern Africa (Figure 18.2). However, all families and subfamilies of fig wasps as well as the fig wasp assemblages from many of the fig trees are under-represented in Mkomazi Game Reserve (Table 18.2). This is simply because of the insufficient sampling that has been undertaken within reserve. Once the fig wasps and fig trees have been comprehensively surveyed a very different picture of actual species richness is likely to emerge.

Many of the undescribed fig wasp species collected in Mkomazi have already been collected from their hosts elsewhere in Africa and await taxonomic revisions, but a number of new, previously uncollected species of fig wasp were sampled. For example this was the first record of a third *Sycoscapter* species from *F. stuhlmannii*, and the first record of a third *Sycoryctes* species and a second *Watshamiella* species from *F. s. sansibarica*. The *Ficus bubu* collections produced 14 additional species to those previously recorded in southern Africa, although six of these have previously been recorded from Tanzania (Rasplus, pers. com.). Material collected from *F. ingens* contributed to the description of a new species of Otitesellinae: *Otitesella longicauda* van Noort (van Noort & Rasplus 1997).

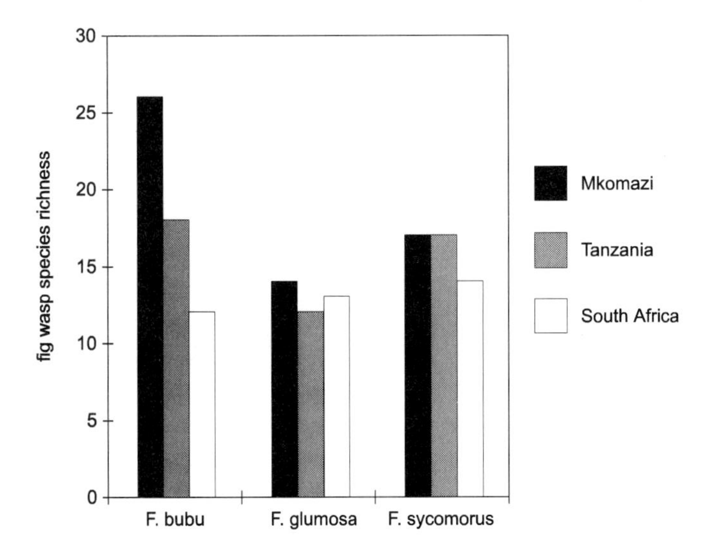

Figure 18.2 Comparative regional species richness of the three most common *Ficus* species in Mkomazi Game Reserve. The Tanznaian figure excludes Mkomazi and respresents collections made by J.Y. Rasplus and C. Kerdelhué.

The fig wasp assemblages associated with each fig species are listed in Table 18.3. For comparison fig wasps recorded from outside the reserve around Kisiwani and on the South Pare Mountains are listed in Table 18.4.

Comparative species richness and similarity

An assessment of fig tree species richness represented in Mkomazi Game Reserve can be achieved by comparing the number of recorded species with floral surveys carried out in other east African areas. A study of indigenous trees and shrubs of Bura, in the Tana River district (Kenya), a semi-arid region that is floristically similar to Mkomazi, identified four *Ficus* species (Gachathi *et al.* 1994). Two of the three identified species occur in Mkomazi and the remaining *F. capreifolia* could potentially be present along the Umba River. The *Ficus* species count in a similar habitat, but one-tenth the area of Mkomazi, Ol Ari Nyiro Ranch on the

F. sycomorus	F. ingens	F. lutea	F. bussei	F. glumosa
Ceratosolen arabicus	Platyscapa soraria	*Allotriozoon heterandromorphum	*Elisabethiella sp.	Elisabethiella glumosae
Ceratosolen galili	Otitesella longicauda	*Philocaenus silvestrii	*Philocaenus zambesiacus	Philocaenus warei
Sycoscapteridea sp. 1	*Otitesella rotunda	*Sycoryctes sp. 4	*Philotrypesis sp. 7	Crossogaster stigma
Sycoscapteridea sp. 2	Philotrypesis sp. 2	*Sycoryctes sp. 5	*Sycoryctes sp. 6	*Crossogaster quadrata
Sycoscapter sp. 1	*Sycoryctes sp. 2	*Sycoscapter sp. 3	*Sycoryctes sp. 7	Sycoryctes sp. 8
Watshamiella sp. 1	*Sycoryctes sp. 3	*Philotrypesis selenitica	*Sycoscapter sp. 4	Sycoryctes sp. 9
Watshamiella sp. 2	*Sycoscapter sp. 2	*Philotrypesis sp. 8	*Sycoscapteridea sp.	*Sycoscapter sp. 5
Apocrypta longitarsus	*Camarothorax sp. 1	*Otitesella sp. 1	*Watshamiella sp. 3	Philotrypesis sp. 5
Apocrypta sp. 1	*Sycophila sp. 4	*Otitesella sp. 2	*Camarothorax sp. 8	Otitesella sp. 3
Sycophaga sycomori	*Sycophila sp. 5	*Camarothorax sp. 2	*Camarothorax sp. 9	Otitesella sp. 4
Eukoebelea sycomori	*Sycophila sp. 6	*Camarothorax sp. 3	*Camarothorax sp. 10	Camarothorax sp. 11
Apocryptophagus gigas (Mayr)	*Sycophila sp. 7	*Camarothorax sp. 4	*Sycophila sp. 16	Camarothorax sp. 12
Apocryptophagus sp.1	*Sycophila sp. 8	*Camarothorax sp. 5	*Sycophila sp. 17	Sycophila sp. 19
Camarothorax sp. 22	*Acophila sp. 1	*Camarothorax sp. 6	*Sycophila sp. 18	Sycophila sp. 20
Sycophila sp. 1		*Camarothorax sp. 7		*Sycophila sp. 21
Sycophila sp. 2		*Sycophila sp. 9		*Sycophila sp. 22
Sycophila sp. 3		*Sycophila sp. 10		Ormyrus sp. 1
		*Sycophila sp. 11		
		*Sycophila sp. 12		
		*Sycophila sp. 13		
		*Sycophila sp. 14		
		*Sycophila sp. 15		

Table 18.3 *Ficus* species and their associated fig wasps in Mkomazi Game Reserve. * denotes fig wasp taxa not yet recorded from the reserve, but recorded from the *Ficus* species elsewhere in east and southern Africa. The majority of the fig wasps listed below are undescribed species.

Laikipia Plateau (Kenya), totalled six species (Muasya *et al.* 1994). Based on the floral diversity recorded, an estimated 10% of the Kenyan flora, Ol Ari Nyiro is touted as one of the most diverse non-forest areas in east Africa (Muasya *et al.* 1994). In a study of species richness and endemism of the Usambara Mountain forests, 12 *Ficus* species were identified (Rodgers & Homewood 1982), only three more than occur in Mkomazi from an area which constitutes one of the richest biological communities in Africa (Rodgers & Homewood 1982). Five of the Usambara species are shared with Mkomazi, but all five are fairly widespread savanna species that are not typically associated with montane forest. Hawthorne (1993) recorded 16 fig tree species occurring in the east African coastal forests. However, this survey encompassed a wide area and hence elevated species richness. Six of these species also occur in Mkomazi. From these comparisons Mkomazi appears to have a high fig tree species richness (Table 18.5), but all of these studies are likely to have underestimated species richness in their respective areas. A more

F. stuhlmannii	F. thonningii	F. sansibarica	F. bubu
Alfonsiella binghami	Elisabethiella stuckenbergi	Courtella armata	Courtella michaloudi
Philocaenus liodontus	*Philocaenus barbarus	Seres solweziensis	Seres wardi
*Philocaenus barbarus	*Crossogaster odorans	Otitesella sp. 8	Sycoryctes sp. A
Crossogaster odorans	*Philotrypesis parca	*Philosycus sp.	Sycoryctes sp. B
Otitesella sp. 5	*Philotrypesis sp. 1	Sycoryctes sp. 12.	Sycoryctes sp. C
Otitesella sp. 6	*Sycoscapter cornutus	Sycoryctes sp. 13	Sycoryctes sp. D
Sycoryctes sp. 10	*Sycoryctes remus	Sycoryctes sp. 14	Sycoryctes sp. E
Sycoryctes sp. 11	*Sycoryctes hirtus	*Sycoscapteridea sp. 4	Sycoryctes sp. F
*Sycoryctes sp. 21	*Sycoryctes sp. 1	*Sycoscapter sp. 9	Sycoscapter sp. 10
Sycoscapter sp. 6	*Watshamiella alata	Watshamiella sp. 6	Watshamiella sp. 8
Sycoscapter sp. 7	*Watshamiella sp. 4	Watshamiella sp. 7	Watshamiella sp. 9
Sycoscapter sp. 8	*Watshamiella sp. 5	Philotrypesis sp. 6	Otitesella sp. 9
Philotrypesis sp. 3	*Otitesella tsamvi	*Camarothorax sp. 16	Philosycus sp. 2
*Philotrypesis sp. 4	*Otitesella sp. 7	Camarothorax sp. 17	Camarothorax sp. 18
Camarothorax sp. 13	*Camarothorax brevimucro	*Sycophila sp. 27	Camarothorax sp. 19
Camarothorax sp. 14	*Camarothorax equicollis	*Sycophila sp. 28	Camarothorax sp. 20
Camarothorax sp. 15	*Camarothorax longimucro	*Sycophila sp. 29	Camarothorax sp. 21
*Sycophila sp. 23	*Sycotetra serricornis	*Sycophila sp. 30	Ficomila sp. 1
*Sycophila sp. 24	*Ficomila curtivena	*Sycophila sp. 31	Sycophila sp. 34
*Sycophila sp. 25	*Ficomila gambiensis	*Sycophila sp. 32	Sycophila sp. 35
*Sycophila sp. 26	*Eurytoma ficusgallae	Sycophila sp. 33	Sycophila sp. 36
	*Syceurytoma ficus	*Ormyrus sp. 2	Sycophila sp. 37
	*Sycophila flaviclava		Sycophila sp. 38
	*Sycophila kestraneura		Sycophila sp. 39
	*Sycophila modesta		Sycophila sp. 40
	*Sycophila naso		Sycophila sp. 41
	*Sycophila punctum		*Ormyrus sp. 3
	*Sycophila sessilis		
	*Ormyrus flavipes		
	*Ormyrus subconicus		
	*Ormyrus watshami		

useful comparison may be provided by examination of local species richness in southern Africa, where the presence of host fig trees and collections of fig wasps is better documented.

An assessment of local species richness in South Africa was achieved by demarcating a comparative region (in size and habitat) to that of Mkomazi Game Reserve. This region was centred around Mkuze Game Reserve (Kwazulu Natal) and encompassed four sixteenth degree squares contained between 27°–28°S and 32°–32°15'E, with an altitudinal variation of 80–700 m, and an area of 3,250 km^2. The vegetational types comprised Natal Lowveld Bushland, Lebombo Arid Mountain Bushland and Sweet Lowveld Bushland (Low & Rebelo 1996). Overall this is a similar habitat to that found within Mkomazi, although the species composition is disparate between the two locations. Twelve fig tree species have been recorded within this demarcated area, including five of the species recorded in Mkomazi. Ninety fig wasp species have been reared from seven of these host species. Based on collections from elsewhere in southern Africa, the remaining five fig tree species should produce at least a further 27 wasp species. The recorded fig wasp species richness from Mkuze is thus comparable with that from Mkomazi (Table 18.6). However, the Mkomazi count will undoubtably increase with further surveys in the region and *Ficus* species richness is expected to be higher than currently recorded, whereas the Mkuze region is well surveyed for fig trees. Once Mkomazi is completely surveyed the reserve will probably be shown to protect a higher species richness than is found in a comparable savanna area in South Africa.

Although Mkomazi Game Reserve lies within the Somalia-Masai regional centre of floral endemism (White 1983), the reserve is close to the transition point between this centre, the Zambezian regional centre of endemism and the Zanzibar-Inhambane regional mosaic, which extends down the coast into southern Mozambique (White 1983). These three systems have shared affinities. As a result many of the *Ficus* species occurring within Mkomazi would be expected to

Table 18.4 *Ficus* species and their associated fig wasps from areas adjacent to Mkomazi Game Reserve (Kisiwani and South Pare Mountains).

F. exasperata	*F. vallis-choudae*	*F. sycomorus*	*F. lutea*	*F. thonningii B*
Kradibia gestroi afrum	*Ceratosolen megacephalus*	*Ceratosolen arabicus*	*Allotriozoon heterandromorphum*	*Alfonsiella brongersmai*
Sycoryctes sp. 21	*Apocrypta robusta*	*Ceratosolen galili*	*Philocaenus silvestrii*	*Philocaenus medius*
Philotrypesis sp. 9	*Sycoscapteridea* sp.5	*Sycoscapteridea* sp.1	*Sycoryctes* sp. 4	*Philocaenus barbarus*
	Eukoebelea sp. 1	*Sycoscapteridea* sp.2		*Crossogaster vansomereni*
	Camarothorax sp.23	*Sycoscapter* sp. 1		*Sycoscapter* sp. 11
	Sycophila sp. 42	*Watshamiella* sp. 1		*Sycophila* sp. 43
	Sycophila sp. 43	*Apocrypta longitarsus*		*Sycophila* sp. 44
		Apocrypta sp. 1		
		Sycophaga sycomori		
		Apocryptophagus gigas		
		Apocryptophagus sp. 1		

be shared with the southern African subregional flora and in fact all nine of the *Ficus* species from Mkomazi enjoy a wide distribution that extends into southern Africa (Berg & Wiebes 1992). However, a number of these species are nearer the centre of their distribution in east Africa, whereas in southern Africa they are at the extreme of their range and relatively rare. This has important ramifications for species richness of the associated fig wasp assemblages, which become increasingly depauperate towards the periphery of their host species distribution (Compton *et al.* 1994). This is a trend exemplified by *Ficus bubu*, from which fig wasp fauna was previously only known from a few collections in South Africa and Tanzania, although the species is widespread, extending from eastern South Africa up to Kenya and across to the Ivory Coast, but supposedly rare or overlooked (Berg & Wiebes 1992). *Ficus bubu* is relatively common in Mkomazi in the vicinity of Ibaya Camp. This is a species that sometimes persists in disturbed areas (Berg & Hijman 1989) and this ability may explain the continued existence of this forest species in fire encroached areas in the Mbono Valley and in the valley north-west of Ibaya Camp. Prior to the Mkomazi collections, 12 species of fig wasp were recorded from *F. bubu* in southern Africa and 18 species (J.Y. Rasplus, pers. comm.) from this host in Tanzania. An exceptional 26 species were reared from this host species in Mkomazi Game Reserve (Figure 18.2). A record 25 of these species were collected from a single fig crop produced by a tree between Dindira Dam and Viteweni Ridge. *Ficus sycomorus* similarly produced three more species than the 13 species recorded from southern African samples (Compton & Hawkins 1992) and one less than the 17 species previously recorded from this host in Tanzania (J.Y. Rasplus, pers. comm.). By contrast *F. glumosa*, *F. stuhlmannii* and *F. s. sansibarica* produced less species than expected. This is likely to be the result of under sampling and is supported by collections made from *F. s. sansibarica* from just outside the reserve, where an additional 10 species over those collected from within the reserve were recorded. With further sampling these species will

F. thoninngii C	*F. tremula acuta*	*F. sansibarica*	
Elisabethiella socotrensis	*Courtella* sp. 1	*Courtella armata*	*Camarothorax* sp. 17
Otitesella sp. 7	*Philotrypesis* sp. 8	*Sycoryctes* sp. 12	*Ficomila* sp. 1
Sycoryctes sp. 22		*Sycoryctes* sp. 13	*Sycophila* sp. 27
		Sycoscapter sp. 9	*Sycophila* sp. 28
		Watshamiella sp. 6	*Sycophila* sp. 29
		Philotrypesis sp. 6	*Sycophila* sp. 30
		Otitesella sp. 8	*Sycophila* sp. 31
		Philosycus sp. 1	*Sycophila* sp. 32
		Camarothorax sp. 16	

undoubtably be recorded from within Mkomazi as well. Tanzanian data is available for three of the fig tree species that were not, or were poorly sampled in Mkomazi. Thirty-one, nine and eight species of fig wasp have previously been recorded from limited collections of *F. thonningii*, *F. bussei* and *F. ingens* respectively in Tanzania (J.Y. Rasplus, pers. comm.). A similar species richness can be expected for Mkomazi although based on southern African collections the richness is an underestimate for the latter two species and a figure of 14 species associated with *F. bussei* and *F. ingens* is probably more representative of real richness.

Table 18.5 Comparative local fig tree and fig wasp species richness. Numbers represent fig wasp species and * indicates fig tree presence but where no data is available on the associated fig wasps.

Ficus sp.	Mkomazi	Mkuze (S. Africa)	Bura (Kenya)	Ol Ari Nyiro (Kenya)	S. Pare Mtns.	Usambara Mtns.
F. exasperata					3	
F. capreifolia		*	*			
F. mucosa						*
F. vallis-choudae					7	*
F. sycomorus	16	11	*	*	10	*
F. sur		7				*
F. ingens	3	12		*		
F. c. salicifolia		*				
F. lutea	*				3	*
F. ?vasta				*		
F. abutilifolia		12				
F. glumosa	13	14		*		
F. stuhlmannii	14	14				
F. bussei	*		*			*
F. craterostoma		*				
F. n. natalensis				*		
F. n. leprieurii						*
F. usambarensis						*
F. burtt-davyi		*				
F. thonningii	*	20		*		*
F. "thonningii" B					7	
F. "thonningii" C					3	
F. scassellati						*
F. c. cyathistipula						*
F. tremula acuta					2	
F. bubu	26					
F. p. polita		*				
F. sansibarica	11				17	*
Ficus sp.			*			
total	9	12	4	6	8	12

Data from collections of the common species *F. glumosa* could be analysed in more detail because of the repetitive samples obtained in Mkomazi Game Reserve. An analysis of a comparative area in South Africa (a 45 km radius around Jozini in northern Kwazulu/Natal, which approximates the extent of the area in Mkomazi that was sampled for this fig tree species) illustrates that local species richness is comparable between the two areas: 13 fig wasp species were reared from *F. glumosa* in Mkomazi compared to 14 in the Jozini area. Previously 12 species had been recorded from three collections of *F. glumosa* in Tanzania (J.Y. Rasplus, pers. comm.). To evaluate whether the recorded fig wasp species richness from *F. glumosa* represented real species richness a species accumulation curve was produced for *F. glumosa* from each area. An accumulation curve plots the sequential addition of new species for each subsequent sample that is collected. When the curve starts levelling off, it indicates that the majority of the fig wasp species have been collected. As can be seen from Figure 18.3, only Jozini has been well sampled, whereas the Mkomazi curve is still rising. The data were further analysed using the program *EstimateS* (Version 4, R.K. Colwell, unpublished), which uses the incidence of species in each sample to estimate the number of uncollected species, and hence estimates the total fig wasp species richness associated with *F. glumosa* in each area. The resultant estimated local species richness for Mkomazi marginally exceeds that of Jozini by 15 to 14 species (Figure 18.3). The regional estimate for southern African fig wasp species richness associated with *F. glumosa* totals 20 species and if combined with the Mkomazi data the total increases to an estimated 24 fig wasp species associated with *F. glumosa* in east and southern Africa (Figure 18.3).

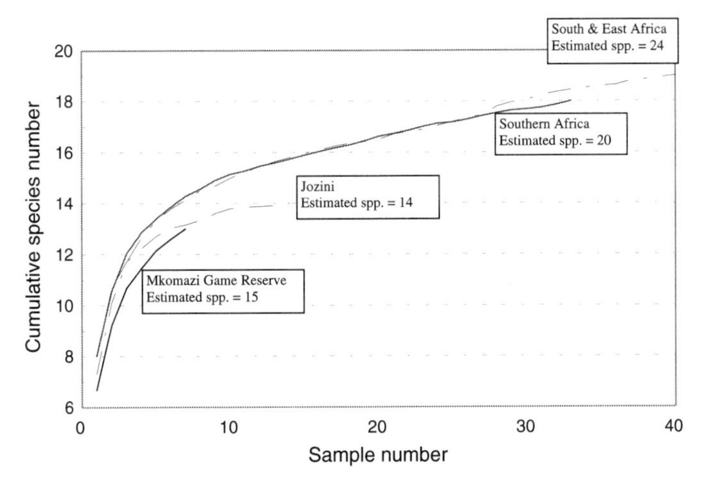

Figure 18.3 Species accumulation curves, including local and regional species richness estimates, for the fig wasp assemblage associated with *Ficus glumosa*.

Table 18.6 Comparative richness of fig trees and fig wasps between Mkomazi and a comparably demarcated area in South Africa, centred on Mkuze Game Reserve.

	Mkomazi	Mkuze
Fig tree species	9	12
Fig wasp species	85	90
Potential wasp species	183	117

The species accumulation curve from a particular host depends to a large extent on timing of samples and size of the fig crop. A large fig crop sampled at optimum development, i.e. just before the wasps emerge, will produce a high species count from a single sample. By contrast the same species count may only be achieved after 15 or 20 samples if the crop or sample sizes (due to non-optimum development) are small. These factors compound assessments of natural geographical or ecological variation of fig wasp species assemblage richness.

Conclusions

This study has shown that a typical savanna fig tree and fig wasp species richness is protected within Mkomazi Game Reserve. However, *Ficus bubu*, a fig tree that is generally assumed rare or overlooked in the Afrotropical region (Berg & Wiebes 1992) is locally abundant in Mkomazi and has a high associated fig wasp species richness. Although savanna fig tree species are well represented in the reserve, forest fig trees appear under-represented. The high fig tree and fig wasp species richness that is likely to be present on the mountains surrounding Mkomazi appears to be poorly represented in the isolated montane forests within the reserve. This is borne out by the four species of fig tree recorded on the South Pare Mountains during this programme that were absent within the reserve (Table 18.5).

Conservation and management

Fig trees are likely to be keystone species with many invertebrates and vertebrates depending on the presence of the resources provided by these trees. Since each species of fig tree is pollinated by its own species of fig wasp, an understanding of this complex obligate mutualism is critical for the future conservation and management of tropical ecosystems. The continued presence of many insects and vertebrates in Mkomazi therefore potentially hinges on the preservation of the fig trees within the reserve. Mature fig trees are probably not adversely affected by fire, given their habitat preference and resistance to burning by a fast moving grass fire. Nevertheless, too frequent burning will affect young trees, which even along the perennial watercourses are susceptible to fire destruction in their early years of growth. Although only one species, *F. thonningii*, has currently been recorded from the montane forests, it is probable that other forest species are present in these isolated patches and hence degradation of this habitat will adversely affect fig tree and fig wasp species richness within the reserve.

Acknowledgements

We would like to extend our thanks to all the staff at Ibaya Camp and to colleagues for logistical support and field assistance. The British Council provided logistic support in Dar es Salaam. Jean-Yves Rasplus (INRA) kindly allowed us to use unpublished data for comparative purposes. Graham Stone critically read and improved the manuscript. This work was supported by grants awarded to SvN from the Commonwealth Science Council and the Foundation for Research Development.

References

Basset, Y., Novotny, V. & Weiblen, G. (1997) *Ficus*: a resource for arthropods in the tropics, with particular reference to New Guinea. In: Watt, A.D., Stork, N.E. & Hunter, M.D. (eds.) *Forests and Insects*. Chapman & Hall, London. pp. 341-361.

Berg, C.C. (1990) Annotated check-list of the *Ficus* species of the African floristic region, with special reference and a key to the taxa of southern Africa. *Kirkia* 13: 253-291.

Berg, C.C. & Hijman, M.E.E. (1989) Chapter 11. *Ficus*. In: Polhill, R.M. (ed.) *Flora of Tropical East Africa*. A.A. Balkema, Rotterdam. pp. 43-86.

Berg, C.C. & Wiebes, J.T. (1992) African Fig Trees and Fig Wasps. *Koninklijke Nederlandse Akademie van Wetenschappen, Verhandelingen Afdeling Natuurkunde, Tweede Reeks* 89.

Boucek, Z. (1988) *Australasian Chalcidoidea (Hymenoptera). A Biosystematic Revision of Genera of Fourteen Families with a Reclassification of Species.* CAB International, Wallingford.

Bronstein, J.L. (1992) Seed predators as mutualists: ecology and evolution of the fig/pollinator interaction. In: Bernays, E. (ed.) *Insect-Plant Interactions Vol IV*. CRC Press, London.

Bronstein, J.L. & Hoassaert-Mckey, M. (1996) Variation in reproductive success within a subtropical fig/pollinator mutualism. *Journal of Biogeography* 23: 433-466.

Compton, S.G. & Hawkins, B.A. (1992) Determinants of species richness in southern African fig wasp assemblages. *Oecologia* 91: 68-74.

Compton, S.G., Rasplus, J.Y. & Ware, A.B. (1994) African fig wasp parasitoid communities. In: Hawkins, B.A. & Sheenan, W. (eds.) *Parasitoid Community Ecology*. Oxford University Press, Oxford. pp. 343-368.

Compton, S.G. & Van Noort, S. (1992) Southern African fig wasps (Hymenoptera: Chalcidoidea): resource utilization and host relationships. *Proceedings of the Koninklijke Nederlandse Akademie van Wetenschappen* 95: 423-435.

Compton, S.G., Craig, A.J.F.K. & Waters, I.W.R. (1996) Seed dispersal in an African fig tree: birds as high quantity, low quality dispersers? *Journal of Biogeography* 23: 553-564.

Gachathi, F.N., Johansson, S.G., & Alakoski-Johansson, G.M. (1994) A checklist of indigenous trees and shrubs of Bura, Tana River District, Kenya, with Malakote, Orma and Somali names. *Journal of East African Natural History* 83: 117-141.

Galil, J. (1977) Fig Biology. *Endeavour* 1: 52-56.

Gautier-Hion, A. & Michaloud, G. (1989) Are figs always keystone resources for tropical frugivorous vertebrates? A test in Gabon. *Ecology* 70: 1826-1833.

Janzen, D.H. (1979) How to be a fig. *Annual Review of Ecology and Systematics* 10: 13-51.

Hawthorne, W.D. (1993) East African Coastal Forest Botany. In: Lovett, J.C. & Wasser, S.K. (eds.) *Biogeography and ecology of the rain forests of eastern Africa*. Cambridge University Press, Cambridge.

Hossaert-Mckey, M., Gibernau, M. & Frey, J.E. (1994) Chemosensory attraction of fig wasps to substances produced by receptive figs. *Entomologia Experimentalis et Applicata* 70: 185-191.

Kerdelhué, C. & Rasplus, J.Y. (1996a) Non-pollinating Afrotropical fig wasps affect the fig-pollinator mutualism in *Ficus* within the subgenus *Sycomorus*. *Oikos* 75: 3-14.

Kerdelhué, C. & Rasplus, J.Y. (1996b) The evolution of dioecy among *Ficus* (Moraceae): an alternative hypothesis involving non-pollinating fig wasp pressure on the fig-pollinator mutualism. *Oikos* 77: 163-166.

Lambert, F.R. & Marshall, A.G. (1992) Keystone characteristics of bird-dispersed *Ficus* in a Malaysian lowland rain forest. *Journal of Ecology* 79: 793-809.

Low, A.B. & Rebelo, A.G. (eds.) (1996) *Vegetation of South Africa, Lesotho and Swaziland*. Department of Environmental Affairs, Pretoria.

Muasya, J.M., Young, T.P. & Okebiro, D.N. (1994) Vegetation map and plant checklist of Ol Ari Nyiro ranch and the Mukutan Gorge, Laikipia, Kenya. *Journal of East African Natural History*. 83: 143-197.

Ramirez, W.B. (1970) Host specificity of fig wasps (Agaonidae). *Evolution, N.Y.* 24: 680-691.

Rodgers, W.A & Homewood, K.M. (1982) Species richness and endemism in the Usambara Mountain forests, Tanzania. *Biological Journal of the Linnean Society* 18: 197-242.

Terborgh, J. (1986) Keystone plant resources in the tropical forest. In: Soulé, M.E. (ed.) *Conservation Biology*. Academic Press, New York. pp. 330-344.

Ulenberg, S.A. (1985) The systematics of the fig wasp parasites of the genus

Apocrypta Coquerel. *Verhandelingen der Koninklijke Nederlandse Akademie van Wetenschappen, Tweede Reeks* 83: 1-176.

van Noort, S. (1993a) Systematics of the sycoecine fig wasps (Agaonidae, Chalcidoidea, Hymenoptera), I (*Seres*). *Proceedings of the Koninklijke Nederlandse Akademie van Wetenschappen* 96: 233-251.

van Noort, S. (1993b) Systematics of the sycoecine fig wasps (Agaonidae, Chalcidoidea, Hymenoptera), II (*Sycoecus*). *Proceedings of the Koninklijke Nederlandse Akademie van Wetenschappen* 96: 449-475.

van Noort, S. (1994a) Systematics of the sycoecine fig wasps (Agaonidae, Chalcidoidea, Hymenoptera), III (*Crossogaster*). *Proceedings of the Koninklijke Nederlandse Akademie van Wetenschappen* 97: 83-122.

van Noort, S. (1994b) Systematics of the sycoecine fig wasps (Agaonidae, Chalcidoidea, Hymenoptera), IV (*Philocaenus*, in part). *Proceedings of the Koninklijke Nederlandse Akademie van Wetenschappen* 97: 311-339.

van Noort, S. (1994c) Systematics of the sycoecine fig wasps (Agaonidae, Chalcidoidea, Hymenoptera), V (*Philocaenus* concluded, generic key, checklist). *Proceedings of the Koninklijke Nederlandse Akademie van Wetenschappen* 97: 341-375.

van Noort, S. (1998) Afrotropical fig wasps and fig trees. http://www.mweb.co.za/ctlive/museums/sam/collect/life/ento/simon/figwasp.htm

van Noort, S. & Rasplus, J.Y. (1997) Revision of the otiteselline fig wasps (Hymenoptera, Chalcidoidea, Agaonidae), I: the *Otitesella digitata* species-group of the Afrotropical region, with a key to Afrotropical species of *Otitesella* Westwood. *African Entomology*, 5: 125-147.

van Noort, S., Ware, A.B. & Compton, S.G. (1989) Pollinator-specific volatile attractants released from the figs of *Ficus burtt-davyi*. *South African Journal of Science* 85: 323-324.

Verkerke, W. (1989) Structure and function of the fig. *Experientia* 45: 612-621.

Ware, A.B. & Compton, S.G. (1994) Responses of fig wasps to host plant volatile cues. *Journal of Chemical Ecology* 20: 785-802.

White, F. (1983) *The vegetation of Africa. A descriptive memoir to accompany the UNESCO/AETFAT/UNSO vegetation map of Africa*. UNESCO, Paris.

Wiebes, J.T. (1979) Coevolution of figs and their insect pollinators. *Annual Review of Ecology and Systematics* 10: 1-12.

Wiebes, J.T. & Compton, S.G. (1990) Agaonidae (Hymenoptera Chalcidoidea) and *Ficus* (Moraceae): fig wasps and their figs, VI (Africa concluded). *Proceedings of the Koninklijke Nederlandse Akademie van Wetenschappen* (C) 93: 203-222.

West, S.A., Herre, E.A., Windsor, D.M., & Green, P.R.S. (1996) The ecology and evolution of the New World non-pollinating fig wasp communities. *Journal of Biogeography* 23: 447-458.

Checklist: Fig wasps of Mkomazi

85 species recorded; 183 potential species. Potential species are those which have not yet been recorded in Mkomazi but are thought likely to occur there due to the presence of their host fig trees. **Potential species are included in the list, but are indented to distinguish them from the recorded species.** Species determinations by S. van Noort.

Class INSECTA

Order HYMENOPTERA

Superfamily CHALCIDOIDEA

Family Agaonidae

Subfamily Agaoninae

- *Ceratosolen arabicus* Mayr (ex *Ficus sycomorus*)
- *Ceratosolen galili* Wiebes (ex *F. sycomorus*)
- *Platyscapa soraria* Wiebes (ex *F. ingens*)
    - *Allotriozoon heterandromorphum* Grandi (*ex* F. lutea*)*
- *Elisabethiella glumosae* Wiebes (ex *F. glumosa*)
- *Elisabethiella stuckenbergi* Grandi (ex *F. thonningii*)
    - *Elisabethiella* sp. (ex *F. bussei*)
- *Alfonsiella binghami* Wiebes (ex *F. stuhlmannii*)
- *Courtella armata* (Wiebes) (ex *F. sansibarica sansibarica*)
- *Courtella michaloudi* (Wiebes) (ex *F. bubu*)

Subfamily Sycoryctinae

Tribe Apocryptini

- *Apocrypta longitarsus* (Mayr) (ex *F. sycomorus*)
- *Apocrypta* sp. 1 (ex *F. sycomorus*)

Tribe Philotrypesini

    - *Philotrypesis selenitica* Grandi (ex *F. lutea*)
    - *Philotrypesis parca* Wiebes (ex *F. thonningii*)
    - *Philotrypesis* sp. 1 (ex *F. thonningii*)
- *Philotrypesis* sp. 2 (ex *F. ingens*)
- *Philotrypesis* sp. 3 (ex *F. stuhlmannii*)
    - *Philotrypesis* sp. 4 (ex *F. stuhlmannii*)
- *Philotrypesis* sp. 5 (ex *F. glumosa*)
- *Philotrypesis* sp. 6 (ex *F. s. sansibarica*)
    - *Philotrypesis* sp. 7 (ex *F. bussei*)
    - *Philotrypesis* sp. 8 (ex *F. lutea*)

Tribe Sycoryctini

    - *Sycoryctes remus* Wiebes (ex *F. thonningii*)
    - *Sycoryctes hirtus* Wiebes (ex *F. thonningii*)
    - *Sycoryctes* sp. 1 (ex *F. thonningii*)
    - *Sycoryctes* sp. 2 (ex *F. ingens*)
    - *Sycoryctes* sp. 3 (ex *F. ingens*)
    - *Sycoryctes* sp. 4 (ex *F. lutea*)
    - *Sycoryctes* sp. 5 (ex *F. lutea*)
    - *Sycoryctes* sp. 6 (ex *F. bussei*)
    - *Sycoryctes* sp. 7 (ex *F. bussei*)
- *Sycoryctes* sp. 8 (ex *F. glumosa*)
- *Sycoryctes* sp. 9 (ex *F. glumosa*)
- *Sycoryctes* sp. 10 (ex *F. stuhlmannii*)
- *Sycoryctes* sp. 11 (ex *F. stuhlmannii*)
- *Sycoryctes* sp. 12 (ex *F. sansibarica sansibarica*)
- *Sycoryctes* sp. 13 (ex *F. sansibarica sansibarica*)
- *Sycoryctes* sp. 14 (ex *F. sansibarica sansibarica*)
- *Sycoryctes* sp. 15 (ex *F. bubu*)
- *Sycoryctes* sp. 16 (ex *F. bubu*)
- *Sycoryctes* sp. 17 (ex *F. bubu*)
- *Sycoryctes* sp. 18 (ex *F. bubu*)
- *Sycoryctes* sp. 19 (ex *F. bubu*)
- *Sycoryctes* sp. 20 (ex *F. bubu*)
    - *Sycoryctes* sp. 21 (ex *F. stuhlmannii*)
    - *Sycoscapter cornutus* Wiebes (ex *F. thonningii*)
- *Sycoscapter* sp. 1 (ex *F. sycomorus*)
    - *Sycoscapter* sp. 2 (ex *F. ingens*)
    - *Sycoscapter* sp. 3 (ex *F. lutea*)
    - *Sycoscapter* sp. 4 (ex *F. bussei*)
    - *Sycoscapter* sp. 5 (ex *F. glumosa*)
- *Sycoscapter* sp. 6 (ex *F. stuhlmannii*)
- *Sycoscapter* sp. 7 (ex *F. stuhlmannii*)
- *Sycoscapter* sp. 8 (ex *F. stuhlmannii*)
    - *Sycoscapter* sp. 9 (ex *F. sansibarica sansibarica*)
- *Sycoscapter* sp. 10 (ex *F. bubu*)
- *Sycoscapteridea* sp. 1 (ex *F. sycomorus*)
- *Sycoscapteridea* sp. 2 (ex *F. sycomorus*)
    - *Sycoscapteridea* sp. 3 (ex *F. bussei*)

Sycoscapteridea sp. 4 (ex *F. sansibarica sansibarica*)
Watshamiella alata Wiebes (ex *F. thonningii*)
Watshamiella sp. 1 (ex *F. sycomorus*)
Watshamiella sp. 2 (ex *F. sycomorus*)
Watshamiella sp. 3 (ex *F. bussei*)
Watshamiella sp. 4 (ex *F. thonningii*)
Watshamiella sp. 5 (ex *F. thonningii*)
Watshamiella sp. 6 (ex *F. sansibarica sansibarica*)
Watshamiella sp. 7 (ex *F. sansibarica sansibarica*)
Watshamiella sp. 8 (ex *F. bubu*)
Watshamiella sp. 9 (ex *F. bubu*)

Subfamily Otitesellinae

Otitesella longicauda van Noort (ex *F. ingens*)
Otitesella rotunda van Noort (ex *F. ingens*)
Otitesella tsamvi Wiebes (ex *F. thonningii*)
Otitesella sp. 1 (ex *F. lutea*)
Otitesella sp. 2 (ex *F. lutea*)
Otitesella sp. 3 (ex *F. glumosa*)
Otitesella sp. 4 (ex *F. glumosa*)
Otitesella sp. 5 (ex *F. stuhlmannii*)
Otitesella sp. 6 (ex *F. stuhlmannii*)
Otitesella sp. 7 (ex *F. thonningii*)
Otitesella sp. 8 (ex *F. sansibarica sansibarica*)
Otitesella sp. 9 (ex *F. bubu*)
Philosycus sp. 1 (ex *F. sansibarica sansibarica*)
Philosycus sp. 2 (ex *F. bubu*)

Subfamily Sycoecinae

Philocaenus silvestrii (Grandi) (ex *F. lutea*)
Philocaenus zambesiacus van Noort (ex *F. bussei*)
Philocaenus warei van Noort (ex *F. glumosa*)
Philocaenus liodontus (Wiebes) (ex *F. stuhlmannii*)
Philocaenus barbarus Grandi (ex *F. thonningii* & *F. stuhlmannii*)
Crossogaster stigma van Noort (ex *F. glumosa*)
Crossogaster quadrata van Noort (ex *F. glumosa*)
Crossogaster odorans Wiebes (ex *F. thonningii* & *F. stuhlmannii*)
Seres solweziensis van Noort (ex *F. sansibarica sansibarica*)
Seres wardi van Noort (ex *F. bubu*)

Subfamily Sycophaginae

Sycophaga sycomori (Linnaeus) (ex *F. sycomorus*)
Eukoebelea sycomori Wiebes (ex *F. sycomorus*)
Apocryptophagus gigas (Mayr) (ex *F. sycomorus*)
Apocryptophagus sp. 1 (ex *F. sycomorus*)

Subfamily Epichrysomallinae

Camarothorax brevimucro Boucek (ex *F. thonningii*)
Camarothorax equicollis Boucek (ex *F. thonningii*)
Camarothorax longimucro Boucek (ex *F. thonningii*)
Camarothorax sp. 1 (ex *F. ingens*)
Camarothorax sp. 2 (ex *F. lutea*)
Camarothorax sp. 3 (ex *F. lutea*)
Camarothorax sp. 4 (ex *F. lutea*)
Camarothorax sp. 5 (ex *F. lutea*)
Camarothorax sp. 6 (ex *F. lutea*)
Camarothorax sp. 7 (ex *F. lutea*)
Camarothorax sp. 8 (ex *F. bussei*)
Camarothorax sp. 9 (ex *F. bussei*)
Camarothorax sp. 10 (ex *F. bussei*)
Camarothorax sp. 11 (ex *F. glumosa*)
Camarothorax sp. 12 (ex *F. glumosa*)
Camarothorax sp. 13 (ex *F. stuhlmannii*)
Camarothorax sp. 14 (ex *F. stuhlmannii*)
Camarothorax sp. 15 (ex *F. stuhlmannii*)
Camarothorax sp. 16 (ex *F. sansibarica sansibarica*)
Camarothorax sp. 17 (ex *F. sansibarica sansibarica*)
Camarothorax sp. 18 (ex *F. bubu*)
Camarothorax sp. 19 (ex *F. bubu*)
Camarothorax sp. 20 (ex *F. bubu*)
Camarothorax sp. 21 (ex *F. bubu*)
Camarothorax sp. 22. (ex *F. sycomorus*)
Sycotetra serricornis Boucek (ex *F. thonningii*)

Family Eurytomidae

Acophila sp. 1 (ex *F. ingens*)
Ficomila curtivena Boucek (ex *F. thonningii*)
Ficomila gambiensis Boucek (ex *F. thonningii*)
Ficomila sp. 1 (ex *F. bubu*)
Eurytoma ficusgallae Boucek (ex *F. thonningii*)
Syceurytoma ficus Boucek (ex *F. thonningii*)
Sycophila flaviclava Boucek (ex *F. thonningii*)
Sycophila kestraneura (Masi) (ex *F. thonningii*)

Sycophila modesta Boucek (ex *F. thonningii*)
Sycophila naso Boucek (ex *F. thonningii*)
Sycophila punctum Boucek (ex *F. thonningii*)
Sycophila sessilis Boucek (ex *F. thonningii*)
Sycophila sp. 1 (ex *F. sycomorus*)
Sycophila sp. 2 (ex *F. sycomorus*)
Sycophila sp. 3 (ex *F. sycomorus*)
Sycophila sp. 4 (ex *F. ingens*)
Sycophila sp. 5 (ex *F. ingens*)
Sycophila sp. 6 (ex *F. ingens*)
Sycophila sp. 7 (ex *F. ingens*)
Sycophila sp. 8 (ex *F. ingens*)
Sycophila sp. 9 (ex *F. lutea*)
Sycophila sp. 10 (ex *F. lutea*)
Sycophila sp. 11 (ex *F. lutea*)
Sycophila sp. 12 (ex *F. lutea*)
Sycophila sp. 13 (ex *F. lutea*)
Sycophila sp. 14 (ex *F. lutea*)
Sycophila sp. 15 (ex *F. lutea*)
Sycophila sp. 16 (ex *F. bussei*)
Sycophila sp. 17 (ex *F. bussei*)
Sycophila sp. 18 (ex *F. bussei*)
Sycophila sp. 19 (ex *F. glumosa*)
Sycophila sp. 20 (ex *F. glumosa*)
Sycophila sp. 21 (ex *F. glumosa*)
Sycophila sp. 22 (ex *F. glumosa*)
Sycophila sp. 23 (ex *F. stuhlmannii*)
Sycophila sp. 24 (ex *F. stuhlmannii*)
Sycophila sp. 25 (ex *F. stuhlmannii*)
Sycophila sp. 26 (ex *F. stuhlmannii*)
Sycophila sp. 27 (ex *F. sansibarica sansibarica*)
Sycophila sp. 28 (ex *F. sansibarica sansibarica*)
Sycophila sp. 29 (ex *F. sansibarica sansibarica*)
Sycophila sp. 30 (ex *F. sansibarica sansibarica*)
Sycophila sp. 31 (ex *F. sansibarica sansibarica*)
Sycophila sp. 32 (ex *F. sansibarica sansibarica*)
Sycophila sp. 33 (ex *F. sansibarica sansibarica*)
Sycophila sp. 34 (ex *F. bubu*)
Sycophila sp. 35 (ex *F. bubu*)
Sycophila sp. 36 (ex *F. bubu*)
Sycophila sp. 37 (ex *F. bubu*)
Sycophila sp. 38 (ex *F. bubu*)
Sycophila sp. 39 (ex *F. bubu*)
Sycophila sp. 40 (ex *F. bubu*)
Sycophila sp. 41 (ex *F. bubu*)

Family Ormyridae

Ormyrus flavipes Boucek (ex *F. thonningii*)
Ormyrus subconicus Boucek (ex *F. thonningii*)
Ormyrus watshami Boucek (ex *F. thonningii*)
Ormyrus sp. 1 (ex *F. glumosa*)
Ormyrus sp. 2 (ex *F. sansibarica sansibarica*)
Ormyrus sp. 3 (ex *F. bubu*)

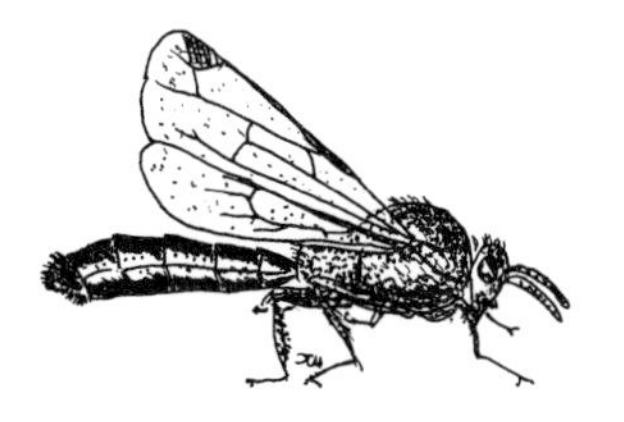

CHAPTER 19

Ants (Hymenoptera: Formicidae) of Mkomazi

Hamish G. Robertson

Introduction

Ants are a conspicuous and important component in the structure and functioning of terrestrial ecosystems (Hölldobler & Wilson 1990). The abundance of many animals is strongly influenced by the persistent predatory pressure of ants. Ants also influence plant survival through seed predation and protection of plant sucking bugs. They are also important in soil turnover although their effects are dwarfed by the much greater amounts of soil brought to the ground surface by termites (White 1983). Besides their ecological importance, ants are of great value as biological indicators not only because they are so abundant and found in a wide range of ecological niches but also because adult workers occur all year round so that short duration surveys can provide an adequate sample of the total diversity.

Termites are frequently confused with ants as both groups are social with reproductive and worker castes and with the capability of building up nests that in some species can persist for decades. However, there are obvious differences between them. Whereas termites are the sister group to the cockroaches and mantids (Thorne & Carpenter 1992), ants evolved from wasps (Baroni Urbani *et al.* 1992). Termites therefore have a hemimetabolous life cycle where the immatures become steadily more adult-like with each moult whereas in ants the immatures are grub-like and pass through a pupal stage to become adults. In both ants and termites, winged reproductives are released on a dispersal flight but whereas in ants the males mate and die, in termites the male ('king') joins the female ('queen') in starting a nest and they mate periodically through their lives (Watson & Gay 1991). In termites, workers can be of either sex whereas in ants all workers are female. Queen ants share with the rest of the Hymenoptera the ability to control the sex of their offspring, fertilised eggs usually producing females and unfertilised eggs producing males. Lastly, many of the higher termites (subfamily Termitinae) form mounds of soil whereas mound building by African ants is rare (no mound builders in Mkomazi Game Reserve).

The results of this study are based on two trips undertaken to Mkomazi Game Reserve in November–December 1995 and May 1996, totalling five weeks. A major goal of the project was to survey the ant species found in leaf litter in the hilltop forests of Mkomazi as well as in forests outside the reserve, including Gonja Forest Reserve, the riverine forest above Kisiwani and the South Pare and West Usambara forests. The results of this survey are used to assess the similarity between these forest zones and the implications this has for their conservation.

During the remaining time on these expeditions, I searched for and collected ants nests by breaking open dead wood, sifting soil, and digging up nests. A huge quantity of ant material was also obtained from the pitfall trapping survey conducted by A. Russell-Smith and co-workers and from sweeping, malaise trapping and light trapping by S. van Noort.

Species diversity

A total of 232 ant species have so far been recorded from Mkomazi, 165 of them only recorded in grassland/woodland, 32 of them only in hilltop forest and 35 were found in both vegetation types (Figure 19.1). Mkomazi now stands as one of the best collected sites for ants in Africa. In Africa south of the Sahara (Afrotropical region) 1,686 species of ants have been recorded and 9,538 species in the world (Bolton 1995).

Of the 78 species recorded in genera that have been revised to modern standards by Bolton (1973, 1974, 1975, 1976, 1980, 1981a, 1981b, 1982, 1983 & 1987), Bolton & Belshaw (1993) and Brown (1975, *Platythyrea* only, 1978), 28 species (31%) did not key out and most of these are probably undescribed species. The proportion of undescribed species in the remaining genera is probably greater than 31% because they have not been recently revised. One can therefore conclude that probably at least a third (77) of the 232 ant species recorded in Mkomazi Game Reserve are undescribed.

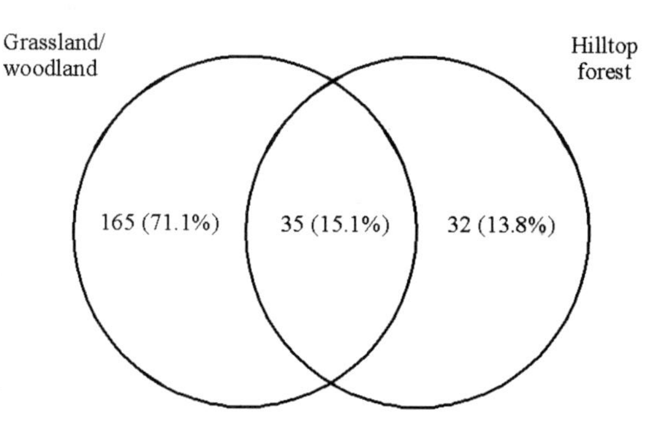

Figure 19.1 Number and percentage of ant species in Mkomazi, recorded from grassland/woodland only, from Hilltop forest only or recorded from both vegetation types.

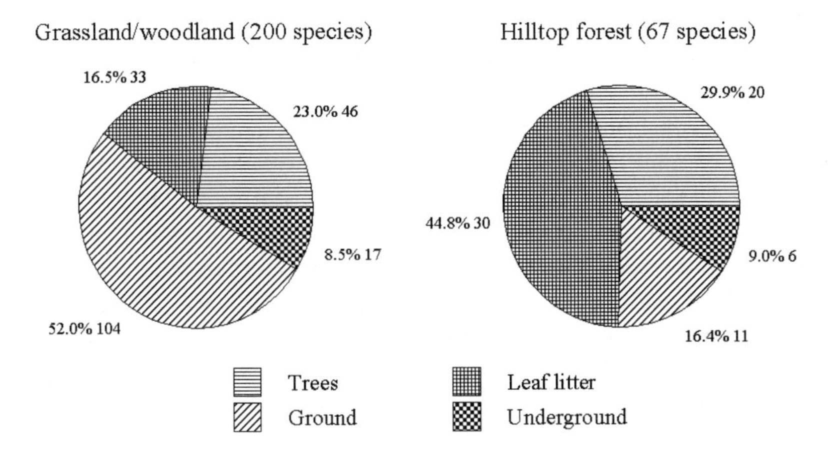

Figure 19.2 Number and percentage of ant species belonging to each of the four major nesting and foraging zones, analysed for each of the two major vegetation types in Mkomazi.

Ants can be divided up into four main groups based on where they nest and forage (Figure 19.2). The arboreal species nest mainly in cavities of dead wood on trees and bushes and spend most of their time foraging on vegetation. They fall mainly in the genera *Crematogaster*, *Camponotus*, *Polyrhachis*, *Tetraponera*, *Cataulacus*, *Monomorium*, *Tapinoma*, *Lepisiota*, *Leptothorax* and *Plagiolepis*. Many of the ground nesting ants also forage in trees (*Polyrhachis schistacea* is a good example) and likewise tree nesting ants often come to the ground—in fact some of the tree nesting species were only ever recorded from ground pitfall traps. Overall, one quarter of all the recorded ant species were classified as arboreal and the proportion of arboreal ants in grassland/woodland and hilltop forest was quite similar (Figure 19.2).

The leaf litter species spend most of their time foraging in the leaf litter and nest in the ground below the leaf litter or between rotten leaves or in cavities of rotten wood in and on the leaf litter. Important genera in this group include the dacetine ants (*Strumigenys*, *Smithistruma*, *Serrastruma*, *Glamyromyrmex*, *Microdaceton*), *Oligomyrmex*, *Calyptomyrmex*, *Tetramorium*, *Monomorium*, *Discothyrea* and *Hypoponera*. Some of the leaf litter species have become specialised predators of leaf litter fauna. For instance, species of *Strumigenys* have elongate mandibles that can open at an angle of 180 and snap shut round springtails (Collembola) (Dejean 1986). Distinguishing between surface and leaf litter species in forest can be quite arbitrary depending on whether one considers the ant species to forage in or on the leaf litter. About 45% of ant species in the hilltop forests are predominantly leaf litter inhabitants and the leaf litter in woodland also supports its own ant fauna which is of a similar diversity, but a lower proportion, to the forest leaf litter ants (33 versus 30 species—Figure 19.2).

The ground dwelling ant species usually nest underground but forage on the ground surface. Important genera include *Camponotus, Tetramorium, Pheidole, Monomorium, Aenictus, Anochetus, Anoplolepis, Lepisiota, Plagiolepis, Polyrhachis, Tapinoma, Technomyrmex, Leptogenys, Pachycondyla, Platythyrea* and *Plectroctena.* The majority of ants species in the grassland/woodland were in this category (Figure 19.2), recorded mainly through pitfall trapping.

Lastly, there are the subterranean species which both nest and forage underground and are either blind or have very reduced eyes. About the only time that they come to the surface is to release reproductives for their nuptial flight. Genera in this group include: *Acropyga, Aenictus, Amblyopone, Anillomyrma, Anochetus, Calyptomyrmex, Carebara, Cerapachys, Dorylus, Hypoponera, Leptanilla, Pachycondyla, Solenopsis* and *Tetramorium.* The proportion of subterranean species in grassland/woodland and hilltop forest was quite similar—about 9% in each case (Figure 19.2). Most of the subterranean species were obtained by digging holes in the ground so as to look for nest cells with brood and to search for workers in sifted soil. Excavating these holes was most rewarding and yielded two specimens of *Anillomyrma*, a new genus for Africa, as well as yielding many undescribed species.

Ecological interactions with plants

Plants often benefit from the presence of ants because the ants can protect the plant from herbivores (Beattie 1985, Huxley 1991). In some cases the plants have evolved mechanisms for increasing their attractiveness to ants for instance by producing extrafloral nectaries and by developing domatia in which the ants can nest. The most obvious example of domatia at Mkomazi Game Reserve are the swollen thorns produced by some species of *Acacia* and the interaction between these plants and ants is covered by Stapley (Chapter 23).

However, ants can also be detrimental to plant fitness. A close relationship often exists between plant-sucking homopteran bugs and ants where the bugs attract ants by producing a sugar-rich solution called honeydew and through their constant attendance around the bugs, the ants scare away predators and parasitoids thus encouraging proliferation of the bugs to the detriment of the plant (Hölldobler & Wilson 1990).

Specialised seed predators can also reduce plant fitness and at Mkomazi Game Reserve these species included *Messor galla, Monomorium abyssinicum, Tetramorium rothschildi, Pheidole* sp. 1 and possibly *Pheidole* sp. 6. *Messor galla, Monomorium abyssinicum* and *T. rothschildi* are all Sahel species and, in Mkomazi Game Reserve, *Messor galla* was only found on red laterite soils. It is puzzling that there is no *Messor* species inhabiting the grasslands on non-laterite soils such as those near Ibaya Camp.

Predation

The majority of ants are generalist predators and exert a persistent predatory pressure on other animal life. Probably the most noticeable predatory ant at Mkomazi Game Reserve is the driver ant *Dorylus molestus* which has brown eyeless workers of variable size and which does not have a single persistent nest site but moves nest sites in response to factors such as prey availability. When moving from one nest site to another, they form a column and when foraging a column is formed from the nest and terminates in a fan-like front of raiding workers (Gotwald 1995). Walking accidentally into one of these columns or fronts can be a painful experience as the workers bite readily with their sharp, sickle-shaped mandibles. A second species of driver ant recorded from Mkomazi Game Reserve was *Dorylus helvolus* but this was far less common than *D. molestus* and is mainly subterranean so does not form the conspicuous columns of the latter species.

Species of the genus *Aenictus* (subfamily Aenictinae) also have an army ant lifestyle and at Mkomazi Game Reserve I observed small yellow *Aenictus eugenii* workers in columns. According to Gotwald & Cunningham-van Someren (1976), this species raids the nests of *Pheidole* ants. *Cerapachys* species are similar to *Aenictus* in that they are nomadic and prey on *Pheidole* (Brown 1975). However, they belong to an entirely different subfamily (Cerapachyinae) and they are black thick-set ants that live in small colonies and do not form conspicuous columns.

Another column-raiding species found at Mkomazi Game Reserve is *Pachycondyla analis* (formerly known as *Megaponera foetens*). This is a large black ponerine ant with two sizes of workers (Crewe, Peeters & Villet 1984) and goes out on raiding parties for termites of the subfamily Macrotermitinae (Longhurst, Johnson & Wood 1978, Longhurst & Howse 1979, Longhurst, Baker & Howse 1979). A scout worker, upon finding foraging termites beneath soil sheets, returns to the nest laying a chemical trail from her poison gland. She recruits a column of workers from the nest and they return to the site, break open the soil sheeting protecting the termites, pick up a number of termites in their mandibles and return to the nest.

There are other ants species which are also specialist predators on termites. *Pachycondyla berthoudi* is found predominantly on the plains on red laterite soils and preys mainly on harvester termites *Hodotermes mossambicus*. According to Arnold (1915), *Pachycondyla sennaarensis* feeds on termites although he has also found seed stores in their nests. At Mkomazi Game Reserve, *P. sennaarensis* was only found along the Umba River. *Carebara vidua* is a subterranean species that has minute workers with potent stings (pers. obs.) that prey upon termites (Lepage & Darlington 1984). In contrast to the workers, the reproductive queens and males of *C. vidua* are massive and after the first heavy summer rains, the workers open up an entrance to the surface and release their reproductives on their nuptial flight

(Robertson & Villet 1989a, Robertson 1995). For a long time it was thought that the virgin queen took a few of the tiny workers with her on the nuptial flight to help her found a new nest but this has since been shown to be untrue (Lepage & Darlington 1984, Robertson & Villet 1989b).

Other specialist predators include *Leptogenys* species which feed mainly on isopods, *Plectroctena* species which feed predominantly on millipedes, and *Platythyrea arnoldi* which is a beetle predator and has beetle carcasses round its nest entrance (Arnold 1915).

Affinities of the grassland/woodland biota

African wooded grasslands and woodlands are found mainly in the Zambezian (southern Africa), Sudanian (west Africa–Sahel) and Somalia–Masai (east Africa) regional centres of endemism (White 1983). By using available distributional information for the 78 grassland/woodland species that fall within genera that have been recently revised (see above under *Species diversity*), one can assess the affinities of the Mkomazi Game Reserve fauna with the rest of the African grassland/

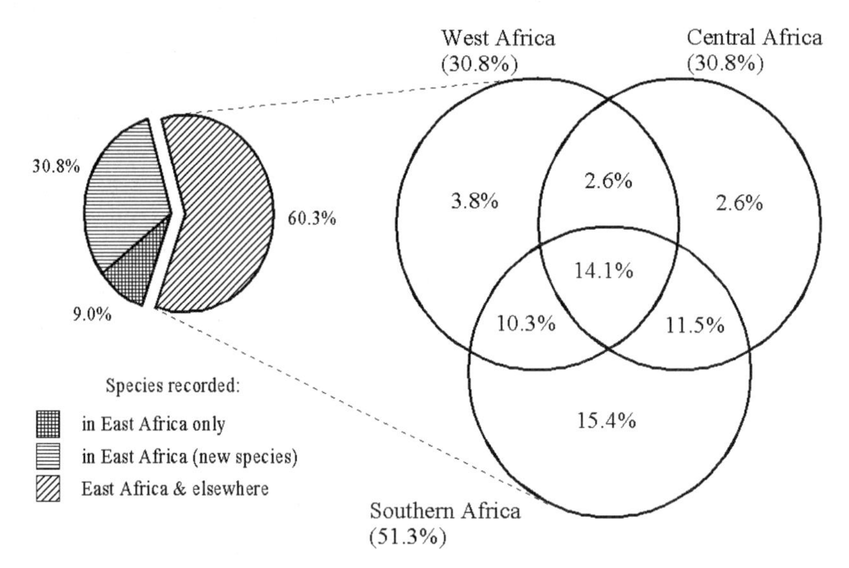

Figure 19.3 The affinities of the grassland/woodland ants in Mkomazi were analysed for the 78 species belonging to genera that have been recently revised, for which adequate distributional data exists. The left hand diagram partitions the species between those recorded only from east Africa and those with wider distributions than just east Africa. The right hand diagram partitions the 60.3% of ant species found outside east Africa by the three main areas in the Afrotropical region in which they could possibly occur. For instance, 14.1% of species recorded in Mkomazi have also been recorded from west, central and southern Africa.

woodlands (Figure 19.3). As the available distributional information is incomplete, the percentages in Figure 19.3 for the 60.3% of species found outside east Africa will increase with better sampling.

Whereas 51.3% of grassland/woodland species are shared with southern Africa (i.e. the Zambezian regional centre of endemism), only 30.8% are shared with west Africa (Sudanian) and with central Africa (Figure 19.3) and most of the shared species in the latter two regions are widespread and found in southern Africa as well. Only 3.8% of the species are shared exclusively with west Africa whereas 15.4% are shared exclusively with southern Africa. The species from Mkomazi Game Reserve that are found in the Sahel but not southern Africa include *Messor galla*, *Tetramorium rothschildi* and *Monomorium abyssinicum* and there are presumably a few others in the genera that were not included in the analysis.

Affinities of the forest biota

In Mkomazi Game Reserve there are two main forest categories, namely hilltop forest and closed woodland. The closed woodland is normally found in valleys where conditions are moist and where burning frequencies are low. Just outside the reserve, in Gonja Forest Reserve and along the Nakombo River, is lowland forest that contains a number of species not easily found within the reserve. Sampling of leaf litter using Winkler bags in these three main forest types shows that 21.5% of species were unique to the hilltop forest whereas only 6.3% were unique to the lowland forests. 83% of the lowland leaf litter fauna was shared with the other two forest types. The 24% of species unique to woodland is high (Figure 19.4) but this is only because many of these species are surface foragers rather than true leaf litter inhabitants. If all species that were also collected using other methods (e.g. pitfall trapping) are excluded, then one finds that 11% of leaf litter species are unique to woodland, also 11% unique to lowland forest and 37% are

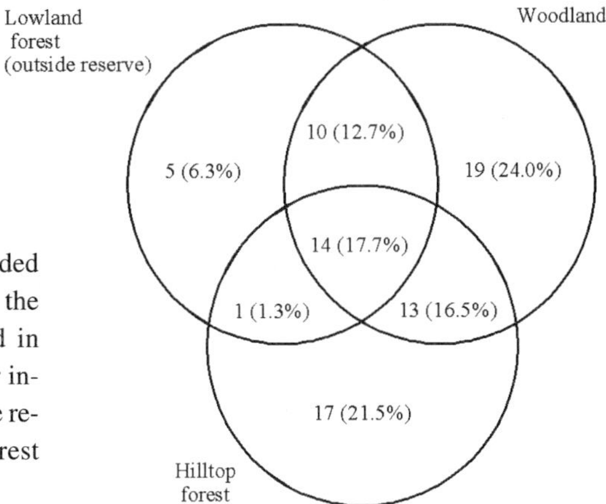

Figure 19.4 Ant species recorded from leaf litter partitioned by the three main forest types found in and just outside Mkomazi. For instance, 10 (12.7%) species were recorded from both lowland forest and woodland.

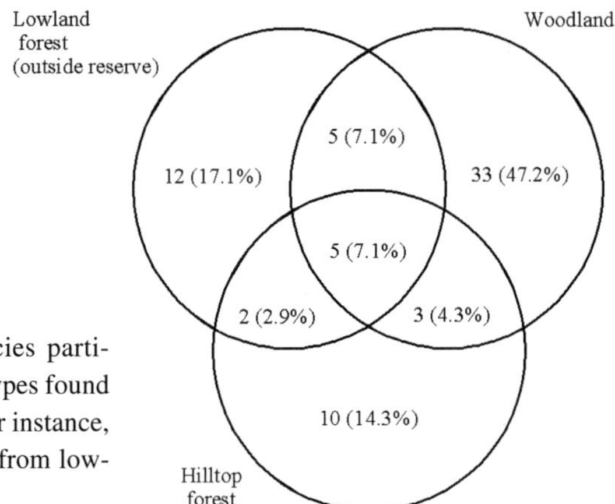

Figure 19.5 Arboreal ant species partitioned by the three main forest types found in and just outside Mkomazi. For instance, 12 (17.1%) were only recorded from lowland forest.

unique to hilltop forest. Increased sampling might change these figures to some extent, but it seems fairly clear that there is a unique element in the leaf litter ant fauna of each of the three forest types but that the hilltop forests in particular have a distinct and relatively diverse fauna.

Leaf litter species found repeatedly in only one forest type and not the others included for closed woodland: *Agraulomyrmex* sp. 1; in lowland forest *Paratrechina* sp. 1; and in hilltop forest: *Glamyromyrmex thuvidus* and *Discothyrea* sp. 1. The dacetine species *G. thuvidus* provides some hint of the affinities of the Mkomazi Hilltop forests to other forests in east Africa. It was not recorded in the Pares or Usambaras but has been collected in Kenya from Embu, Kirimiri Forest, west of Runyenje at 1,550 m (Bolton 1983). The hilltop forests at Mkomazi Game Reserve fall close to the latter altitude (they range from 1,400 to 1,600 m) and so perhaps altitude is a major determinant of the distribution of this species although it was not recorded in the South Pare forests which are at 1,598 m.

A similar comparison of lowland forest, woodland and hilltop forest was made using the arboreal ants that were collected mainly by breaking open dead branches (Figure 19.5). The lowland forest had relatively more unique arboreal than leaf litter species many of them being widespread species within tropical Africa such as *Polyrhachis militaris, Oecophylla longinoda* and *Simopone grandis* that seem to like warm moist conditions.

Of the 45 ants species extracted in Winkler bags from Mkomazi hilltop forests, 31% were shared with the South Pare and West Usambara forests. The South Pares were the least intensively sampled of the three forest types which would partly account for only 24 leaf litter species being collected there. However, this forest is also very disturbed with grassland/woodland species such as *Myrmicaria natalensis* found along its paths. Only 15 leaf litter ant species were recorded in the West Usambaras, this low diversity mainly being because these forests are at a high altitude (1,850 m) which reduces ant diversity—at one site which happened to be the most pristine site of all, we could not find one ant by hand collecting over a period of about three quarters of an hour although the leaf litter sample from this site did yield a few species.

Management implications

The grassland/woodland that covers most of Mkomazi Game Reserve has an ant fauna that has close affinities with the large expanses of grassland/woodland elsewhere in east Africa and southern Africa (Figure 19.3). Although the intensive sampling of this vegetation type resulted in many new species of ants being found, in all likelihood most of these species probably have quite wide distributions and are not endemic to the Mkomazi Game Reserve vicinity. The varied topography, soil types and burning regimes in Mkomazi Game Reserve have resulted in a wide range of habitats within grassland/woodland ranging from almost pure grassland through to dense woodland in valleys. This great diversity of habitats would explain the high diversity of 200 ant species recorded in grassland/woodland and it is this great habitat diversity which is one of the strengths of the reserve. From a management point of view, it is important that a mosaic of burning regimes be maintained so helping to maintain habitat diversity.

The hilltop forests have a distinctive fauna that does not overlap all that closely with the South Pares and West Usambaras although it does seem to have affinities with other forest patches at similar altitudes elsewhere in east Africa. Habitat availability for these species is much more limited than for grassland/woodland species and it is clearly a high priority that these forests are adequately conserved. It is unfortunate that they are right on the border of the reserve and open to uncontrolled exploitation from outside. It is also important that they protected from fires as these eat away the forest borders.

Unfortunately Mkomazi Game Reserve does not have any lowland forest within its borders as this would contribute greatly to increasing the range of habitats that it conserves and would also increase its total biodiversity. The limited collecting done in the lowland forests resulted in 19 species not recorded in the far more intensively sampled reserve so that incorporation of lowland forest in the reserve would have amounted to a minimum increase 8% in its total biodiversity.

Acknowledgements

I am most grateful to Simon van Noort, Tony Russell-Smith, Daniel Mafunde and Omari Mohamed for helping me with sampling of the ants in Mkomazi Game Reserve, to Tandi Russell, funded by the Royal Geographical Society (with IBG), Dawn Larsen for sorting samples, and to Barry Bolton (Natural History Museum London) for identifying the dacetine ants.

References

Arnold, G. (1915) A monograph of the Formicidae of South Africa. Part I. Ponerinae, Dorylinae. *Annals of the South African Museum* 14: 1-159.

Baroni Urbani, C., Bolton, B. & Ward, P.S. (1992) The internal phylogeny of ants (Hymenoptera: Formicidae). *Systematic Entomology* 17: 301-329.

Beattie, A.J. (1985) The Evolutionary Ecology of Ant-plant Mutualisms. Cambridge University Press, Cambridge.

Bolton, B. (1973) The ant genus *Polyrhachis* F. Smith in the Ethiopian region. *Bulletin of the British Museum (Natural History) (Entomology)* 28: 283-369.

Bolton, B. (1974) A revision of the Palaeotropical arboreal ant genus *Cataulacus* F. Smith. *Bulletin of the British Museum (Natural History) (Entomology)* 30: 1-105.

Bolton, B. (1975) A revision of the ant genus *Leptogenys* Roger in the Ethiopian region, with a review of the Malagasy species. *Bulletin of the British Museum (Natural History) (Entomology)* 31: 235-305.

Bolton, B. (1976) The ant tribe Tetramoriini. Constituent genera, review of smaller genera and revision of *Triglyphothrix* Forel. *Bulletin of the British Museum (Natural History) (Entomology)* 34: 281-379.

Bolton, B. (1980) The ant tribe Tetramoriini. The genus *Tetramorium* Mayr in the Ethiopian zoogeographical region. *Bulletin of the British Museum (Natural History) (Entomology)* 40: 193-384.

Bolton, B. (1981a) A revision of the ant genera *Meranoplus* F. Smith, *Dicroaspis* Emery and *Calyptomyrmex* Emery in the Ethiopian zoogeographical region. *Bulletin of the British Museum (Natural History) (Entomology)* 42: 43-81.

Bolton, B. (1981b) A revision of six minor genera of Myrmicinae in the Ethiopian zoogeographical region. *Bulletin of the British Museum (Natural History) (Entomology)* 43: 245-307.

Bolton, B. (1982) Afrotropical species of the myrmicine ant genera *Cardiocondyla, Leptothorax, Melissotarsus, Messor* and *Cataulacus*. *Bulletin of the British Museum (Natural History) (Entomology)* 45: 307-370.

Bolton, B. (1983) The Afrotropical dacetine ants. *Bulletin of the British Museum (Natural History) (Entomology)* 46: 267-416.

Bolton, B. (1987) A review of the *Solenopsis* genus-group and revision of Afrotropical *Monomorium* Mayr. *Bulletin of the British Museum (Natural History) (Entomology)* 54: 263-452.

Bolton, B. & Belshaw, R. (1993) Taxonomy and biology of the supposedly lestobiotic ant genus *Paedalgus*. *Systematic Entomology* 18: 181-189.

Brown, W.L., Jr. (1975) Contributions toward a reclassification of the Formicidae. 5. Ponerinae, tribes Platythyreini, Cerapachyini, Cylindromyrmecini, Acanthostichini, and Aenictogitini. *Search Agriculture (Ithaca, New York)* 5(1): 1-115.

Brown, W.L., Jr. (1978) Contributions toward a reclassification of the Formicidae. Part VI. Ponerinae, tribe Ponerini, subtribe Odontomachiti. Section B. Genus *Anochetus* and bibliography. *Studia Entomologica* 20: 549-652.

Crewe, R.M., Peeters, C.P. & Villet, M. (1984) Frequency distribution of worker sizes in *Megaponera foetens* (Fabricius). *South African Journal of Zoology* 19: 247-248.

Dejean, A. (1986) Étude du comportement de prédation dans le genre *Strumigenys* (Formicidae – Myrmicinae). *Insectes Sociaux* 33: 388-405.

Gotwald, W.H., Jr. & Cunningham-van Someren, G.R. (1976) Taxonomic and behavioral notes on the African ant, *Aenictus eugenii* Emery, with a description of the queen (Hymenoptera: Formicidae). *Journal of the New York Entomological Society* 84: 182-188.

Gotwald, W.H., Jr. (1995) *Army Ants: the Biology of Social Predation*. Cornell University Press, Ithaca, New York.

Hölldobler, B. & Wilson, E.O. (1990) *The Ants*. Harvard University Press, Cambridge, Massachusetts.

Huxley, C.R. (1991) Ants and plants: a diversity of interactions. In: Huxley, C.R. & Cutler, D.F. (eds.) *Ant-Plant Interactions*. Oxford University Press, Oxford. pp. 1-11.

Lepage, M.G. & Darlington, J.P. (1984) Observations on the ant *Carebara vidua* F. Smith preying on termites in Kenya. *Journal of Natural History* 18: 293-302.

Longhurst, C., Baker, R. & Howse, P.E. (1979) Termite predation by *Megaponera foetens* (Fab.) (Hymenoptera: Formicidae). Coordination of raids by glandular secretions. *Journal of Chemical Ecology* 5: 703-719.

Longhurst, C. & Howse, P.E. (1979) Foraging, recruitment and emigration in *Megaponera foetens* (Fab.) (Hymenoptera: Formicidae) from the Nigerian Guinea savanna. *Insectes Sociaux* 26: 204-215.

Longhurst, C., Johnson, R.A. & Wood, T.G. (1978) Predation by *Megaponera foetens* (Fabr.) (Hymenoptera: Formicidae) on termites in the Nigerian southern Guinea savanna. *Oecologia* 32: 101-107.

Robertson, H.G. (1995) Sperm transfer in the ant *Carebara vidua* F. Smith (Hymenoptera: Formicidae). *Insectes Sociaux* 42: 411-418.

Robertson, H.G. & Villet, M. (1989a) Mating behaviour in three species of myrmicine ants (Hymenoptera: Formicidae). *Journal of Natural History* 23: 767-773.

Robertson, H.G. & Villet, M. (1989b) Colony foundation in the ant *Carebara vidua*: the dispelling of a myth. *South African Journal of Science* 85: 121-122.

Thorne, B.L. & Carpenter, J.M. (1992) Phylogeny of the Dictyoptera. *Systematic Entomology* 17: 253-268.

Watson, J.A.L. & Gay, F.J. (1991) Isoptera (termites). In: *The Insects of Australia. A textbook for students and research workers*. Volume 1. Second Edition. Melbourne University Press, Melbourne. pp. 330-347.

White, F. (1983) *The Vegetation of Africa*. UNESCO, Paris.

Checklist: Ants of Mkomazi

Class INSECTA

Order HYMENOPTERA

Family FORMICIDAE

Subfamily Aenictinae

Aenictus eugenii Emery. Grassland/woodland; nomadic, forms columns.
Aenictus sp. 01. Grassland/woodland; subterranean.
Aenictus sp. 02. Grassland/woodland; male caught at light.

Subfamily Cerapachyinae

Cerapachys ?kenyensis Consani. Grassland/woodland gorge nr Ibaya; nomadic.
Cerapachys ?lamborni Crawley. Grassland/woodland; nomadic, forms columns.
Cerapachys wroughtoni Forel. Closed woodland & montane forest; nomadic.

Subfamily Dolichoderinae

Tapinoma sp. 01. Grassland/woodland; ground forager, nests in soil.
Tapinoma sp. 03. Grassland/woodland; nests in dead branches on trees.
Tapinoma sp. 04. Closed woodland; arboreal.
Tapinoma sp. 05. Grassland & open woodland; ground forager, nests in soil.
Tapinoma sp. 06. Woodland & forest; nests in dead branches on trees.
Tapinoma sp. 08. Grassland/woodland; single specimen ex pitfall trap.
Technomyrmex sp. 01. Grassland/woodland; ground forager.
Technomyrmex sp. 02. Grassland/woodland; nests found in soil under rocks.
Technomyrmex sp. 03. Grassland/woodland; nest found in soil under rock.
Technomyrmex sp. 04. Hilltop forest fringe, woodland in gorge; collected on plants and on ground.

Subfamily Dorylinae

Dorylus helvolus (Linnaeus). Forest fringe, open woodland; nomadic, subterranean.
Dorylus molestus (Gerstäcker). Woodland & forest; nomadic, column raider (‘*siafu*’).

Subfamily Formicinae

Acropyga sp. 01. Hilltop forest; subterranean.
Agraulomyrmex sp. 01. Closed woodland; in leaf litter.
Anoplolepis sp. 01. Grassland/woodland; ground forager; nests in soil.
Anoplolepis sp. 02. Open woodland with red soils; ground forager; nests in soil.
Anoplolepis sp. 03. Open woodland; nest in soil beneath leaf litter.
Camponotus acvapimensis Mayr. Open woodland; nests in dead branches on trees.
Camponotus braunsi Mayr. Woodland; nests in dead branches on trees.
Camponotus chrysurus Gerstäcker. Moist woodland (e.g. with *Ficus*); nests in dead branches on trees.
Camponotus cinctellus (Gerstäcker). Grassland & open woodland; ground forager, nests in soil.
Camponotus epinotalis Santschi. Hilltop forest; nests in dead branches on trees.
Camponotus etiolipes Bolton. Grassland & open woodland; ground forager, nests in soil.
Camponotus eugeniae Forel. Open woodland on rocky slopes; nest in soil under rock.
Camponotus sericeus (Fabricius). Grassland & open woodland; ground forager, nests in soil with turret (made of grass) over entrance.
Camponotus sp. 01 (*maculatus*-group). Grassland & open woodland; ground forager, nest in soil with 2 mm wide circular entrance.
Camponotus sp. 02. Hilltop forest; nests in dead branches on trees.
Camponotus sp. 04. Grassland & open woodland; ground forager, nest in soil with 2.5 mm diameter round entrance hole.
Camponotus sp. 05 (nr *guttatus* Emery). Grassland & open woodland; ground forager, nest in soil.
Camponotus sp. 06 (nr *debellator* Santschi). Grassland & open woodland with bare red clay soils; ground forager, nest in soil, nest entrance surmounted by flanged turret (wider at rim than at base).
Camponotus sp. 07. Grassland & open woodland; ground forager, nest in soil, entrance surmounted by turret of soil.
Camponotus sp. 08. Hilltop forest; nests in dead branches on trees.
Camponotus sp. 09. Woodland; probably nests in dead branches on trees.
Camponotus sp. 11. Hilltop & lowland forest; nest in dead branch on ground.
Camponotus sp. 14. Hilltop forest; nests in dead branches on trees.

Camponotus sp. 15. Grassland & open woodland; ground forager, presumably nests in soil.

Camponotus sp. 16. Single pitfall record from *Dichrostachys* scrub; ground forager, probably nests in soil.

Camponotus sp. 17. Recorded only from pitfalls in *Combretum* & *Dichrostachys* scrub; ground forager, probably nests in soil.

Camponotus sp. 18 (*foraminosus*-group). Single record from *Commiphora* woodland; probably nests in dead branches on trees.

Camponotus sp. 19 (*foraminosus*-group). Single record from open woodland; probably nests in dead branches on trees.

Camponotus sp. 20 (*foraminosus*-group). Open woodland; probably nests in dead branches on trees.

Camponotus sp. 21 (*rufoglaucus*-group). Open woodland; ground forager.

Camponotus sp. 22 (*rufoglaucus*-group). Open woodland on red laterite soil; ground forager.

Camponotus sp. 23 (near *petersii* Emery). Grassland & open woodland; ground forager, nests in soil.

Camponotus sp. 24 (near *maculatus* (Fabricius)). Open woodland; ground forager, nests in soil, sometimes under rocks.

Camponotus sp. 25 (?*vestitus* (F. Smith)). Grassland & open woodland; ground & plant forager, nests in soil.

Lepisiota foreli (Arnold). Single record from open *Commiphora* woodland on red laterite soil; Nest in soil.

Lepisiota santschii (Arnold). Hilltop & lowland forest; ground & plant foragers; nest in rotten branch on leaf litter.

Lepisiota spinosior (Forel). Grassland & open woodland; ground & plant forager; nests in soil.

Lepisiota sp. 01. Moist woodland; tree forager.

Lepisiota sp. 03. *Combretum* scrub; ground forager.

Lepisiota sp. 04. Hilltop forest; ground & plant forager.

Lepisiota sp. 05. Grassland & open woodland; ground forager.

Lepisiota sp. 06. Grassland & open woodland; ground & plant forager, nests in soil.

Lepisiota sp. 07. Grassland & open woodland; ground forager.

Plagiolepis sp. 01. Woodland & forest; forages on ground & on plants, nests in dead branches mainly on trees.

Plagiolepis sp. 02. Grassland & open woodland; ground & plant forager; nests in soil.

Plagiolepis sp. 03. Hilltop forest; nests in dead wood on trees and in leaf litter.

Plagiolepis sp. 04. Woodland & forest; ground & plant forager; nests in dead wood on ground.

Plagiolepis sp. 05. Woodland; ground & plant forager.

Polyrhachis cubaensis Mayr. Open woodland; nests in cavities of dead branches on trees.

Polyrhachis gagates F. Smith. Open woodland; ground forager.

Polyrhachis schistacea (Gerstäcker). Open woodland; ground & plant forager; nest in soil, entrance surmounted by woven dome of grass fragments with a number of large round entrance holes.

Polyrhachis schlueteri Forel. Hilltop & lowland forest; nests in dead branches on trees.

Polyrhachis viscosa F. Smith. Open woodland, especially on red laterite soils; ground forager, nests in soil.

Subfamily Leptanillinae

Leptanilla sp. 01. Woodland; subterranean, only males collected.

Leptanilla sp. 02. Woodland; subterranean, only males collected.

Leptanilla sp. 03. Woodland; subterranean, only males collected.

Subfamily Myrmicinae

Anillomyrma sp. 01. Open woodland; subterranean, single record from sifted soil.

Atopomyrmex mocquerysi André. Moist woodland; nests in dead branches on trees.

Calyptomyrmex piripilis Santschi. Moist woodland, lowland forest; nests in soil beneath leaf litter.

Calyptomyrmex sp. 01. Woodland; in leaf litter and sifted soil.

Cardiocondyla emeryi Forel. Open woodland; ground & herbacious plant forager.

Cardiocondyla sp. 01. Open woodland; ground & plant forager.

Carebara vidua F. Smith. Open woodland; subterranean termite predators, 1 queen recorded from pitfall.

Cataulacus intrudens (F. Smith). Open woodland; nests in dead branches on trees.

Cataulacus jeanneli Santschi. Lowland forest & woodland; nests in dead branches on trees.

Cataulacus wissmanni Forel. Hilltop & low–land forest; nests in dead branches on trees.

Crematogaster prelli Forel. Open woodland; nests in swollen thorns of *Acacia drepanolobium*.

Crematogaster sp. 01. Forest & woodland; nests in dead branches on trees and in *Acacia zanzibarica* spines.

Crematogaster sp. 02. Grassland & open woodland; ground & plant foragers.

Crematogaster sp. 03. Woodland & lowland forest; nests in dead branches on trees.
Crematogaster sp. 04. Woodland & hilltop forest; nests in dead branches on trees.
Crematogaster sp. 05. Open woodland; nests in swollen thorns of *Acacia zanzibarica*.
Crematogaster sp. 06. Open woodland; nests in dead branches on trees.
Crematogaster sp. 07. Open woodland; nests in dead branches on trees.
Crematogaster sp. 08. Open woodland; nest in carton structures attached to underside of branches.
Crematogaster sp. 09. Hilltop forest; nests in dead branches on trees.
Crematogaster sp. 12. Woodland; nests in dead branches on trees.
Crematogaster sp. 14. Open woodland; ground forager, nests in soil.
Crematogaster sp. 15. Hilltop forest; one record from leaf litter and one from pitfall.
Crematogaster sp. 17. Hilltop forest; single record from leaf litter.
Crematogaster sp. 18. Hilltop forest; single record from leaf litter.
Crematogaster sp. 19. Closed woodland; single record from leaf litter.
Crematogaster sp. 20. Open woodland; single record from sweep.
Crematogaster sp. 21. Open woodland; single record from sweep.
Crematogaster sp. 22. Open woodland; a sweep and a pitfall record.
Crematogaster sp. 23. Open woodland; single record from sweep.
Crematogaster sp. 24. Open woodland; single pitfall record.
Decamorium decem Forel. Grassland, open woodland, lowland forest; ground foragers.
Glamyromyrmex thuvidus Bolton. Hilltop forest; in leaf litter.
Leptothorax angulatus Mayr. Open woodland; nests in dead branches in trees.
Leptothorax humerosus Emery. Open woodland; probably mainly arboreal.
Meranoplus glaber Arnold. Open woodland; ground forager.
Meranoplus inermis Emery. Woodland; ground forager, nests in soil.
Meranoplus magrettii André. Grassland & open woodland; ground forager, nests in soil.
Messor galla (Mayr). Open *Commiphora* woodland on red laterite soils; seed harvesting ants, nest in soil with large crater of soil (up to 40 cm diameter) round entrance.
Microdaceton exornatum Santschi. Hilltop forest; single record from soil beneath leaf litter.
Monomorium abyssinicum Forel. Open woodland on red laterite soil; ground forager and seed harvester.
Monomorium bicolor Emery. Grassland & open woodland; ground forager, nests in soil.
Monomorium exiguum Forel. Woodland & lowland forest; forages in leaf litter and on trees, nests in dead wood on trees.
Monomorium hanneli Forel. Hilltop forest & closed woodland; nest in soil beneath leaf litter.
Monomorium junodi Forel. Open woodland; single record from pitfall.
Monomorium modestum Santschi. Hilltop & lowland forest, closed woodland; nest in soil beneath leaf litter.
Monomorium oscaris Forel. *Dichrostachys* scrub; single record from pitfall.
Monomorium osiridis Santschi. Open woodland; ground forager, nests in soil.
Monomorium robustior Forel. Open woodland; ground & plant forager, nests in soil.
Monomorium vaguum Santschi. Grassland & open woodland; ground & plant forager, nests in soil.
Monomorium sp. 01 (near *opacior* Forel). grassland & open woodland; ground forager, nest in soil.
Monomorium sp. 02 (*monomorium*-group). Closed woodland, hilltop forest, lowland forest; nests in dead wood on trees.
Monomorium sp. 03 (near *symmotu* Bolton). Hilltop forest & closed woodland; nests in dead branches on trees.
Monomorium sp. 04 (*monomorium*-group). Open woodland; ground & plant forager, nests in soil.
Monomorium sp. 05 (near *torvicte* Bolton). Open woodland; ground & plant forager, nests in soil.
Monomorium sp. 06 (*monomorium*-group). Grassland; single record from sweep.
Myrmicaria natalensis (Mayr). Woodland & hilltop forest; nest in soil with entrance often round base of vegetation or base of small tree trunk.
Ocymyrmex nitidulus Emery. Grassland & open woodland with bare ground; ground forager; nest in soil.
Oligomyrmex sp. 01. Woodland, hilltop & lowland forest; nest in soil beneath leaf litter.
Oligomyrmex sp. 02. Woodland, hilltop & lowland forest; nest in soil beneath leaf litter.
Oligomyrmex sp. 03. Woodland, hilltop & lowland forest; nest in soil beneath leaf litter.
Oligomyrmex sp. 04. Closed woodland, hilltop forest; in leaf litter & soil beneath leaf litter.
Oligomyrmex sp. 05. Closed woodland; in leaf

litter.

Paedalgus sp. 01. Closed woodland; in leaf litter.

Pheidole sp. 01. Open woodland; seed harvester, nests in soil.

Pheidole sp. 02. Open woodland; ground forager, nests in soil.

Pheidole sp. 03. Woodland & forest; ground & plant forager, nest in soil or in rotten wood on ground.

Pheidole sp. 04. Woodland, hilltop forest; ground & plant forager, nest in soil.

Pheidole sp. 05. Woodland, lowland forest; ground & litter forager, nest in soil.

Pheidole sp. 06. Open woodland; ground forager, nest in soil.

Pheidole sp. 07. Mainly hilltop & lowland forest but also localised in woodland; nests in soil beneath leaf litter.

Pheidole sp. 08. Hilltop forest; nest in soil beneath rock.

Pheidole sp. 09. Grassland & open woodland; nest in soil, entrance surmounted by small fragile turret of soil.

Pheidole sp. 10. Open woodland on red laterite soil; ground foragers.

Serrastruma lujae (Forel). Hilltop forest; ex leaf litter, nests found under bark of rotten logs on ground.

Serrastruma simoni Emery. Closed woodland & lowland forest; in leaf litter.

Smithistruma mandibularis (Szabó). Lowland forest & closed woodland; in leaf litter.

Smithistruma marginata (Santschi). Closed woodland, hilltop & lowland forest; in leaf litter.

Smithistruma truncatidens Brown. Hilltop forest; in leaf litter.

Solenopsis sp. 01. Open woodland; subterranean.

Strumigenys "percrypta" (name not available here). Hilltop forest; in leaf litter.

Strumigenys arnoldi Forel. Moist woodland; in leaf litter.

Strumigenys dromoshaula Bolton. Hilltop forest; in leaf litter.

Strumigenys faurei Arnold. Closed woodland & lowland forest; in leaf litter.

Strumigenys omalyx Bolton. Closed woodland, lowland & hilltop forest; in leaf litter.

Strumigenys sp. 07. Hilltop forest on Maji Kununua; in leaf litter.

Tetramorium agile Arnold. Open woodland; ground foragers.

Tetramorium argenteopilosum Arnold. Open woodland; ground foragers.

Tetramorium eminii (Forel). Open woodland; ground foragers, nest in soil.

Tetramorium ericae Arnold. Grassland & open woodland; ground foragers.

Tetramorium furtiva (Arnold). Woodland & forest; in leaf litter, nest in soil beneath leaf litter.

Tetramorium gazense Arnold. Grassland & open woodland; ground forager.

Tetramorium humbloti Forel. Woodland; nests in dead wood on ground or in soil beneath leaf litter.

Tetramorium inezulae (Forel). Woodland; ground forager, nest in soil.

Tetramorium longicorne Forel. Open woodland with bare ground; ground forager, nest in soil with wide entrance (up to 3 cm diameter).

Tetramorium phasias Forel. Grassland/woodland, lowland forest; ground forager, nest in soil or in dead wood on ground.

Tetramorium rothschildi Forel. Grassland & open woodland; seed harvester, nest in soil.

Tetramorium sericeiventre Emery. Grassland & open woodland; ground forager, nest in soil.

Tetramorium subcoecum Forel. Open woodland; subterranean.

Tetramorium weitzeckeri Emery. Hilltop forest; ground forager.

Tetramorium sp. 01 (near *pauper* Forel). Grassland, woodland & hilltop forest; largely subterranean, nests in soil.

Tetramorium sp. 02 (near *tersum* Santschi). Open woodland; ground forager, nest in soil.

Tetramorium sp. 03 (*dumezi*-group?). Hilltop forest; nests in hollow pithy centres of herbacious plants.

Tetramorium sp. 04 (*oculatum*-complex). Woodland; nest in soil under rock.

Tetramorium sp. 05 (*squaminode*-group). Grassland; ground forager.

Tetramorium sp. 06 (*bicarinatum*-group). Woodland; single record from pitfall.

Tetramorium sp. 07 (*simillimum*-group). Grassland, woodland & hilltop forest; ground foragers & in leaf litter.

Tetramorium sp. 11 (*simillimum*-group). Closed woodland; single record, from leaf litter.

Tetramorium sp. 12 (*sericeiventre*-group). Open woodland on red laterite soil; single record from pitfall.

Tetramorium sp. 13 (*simillimum*-group). Open woodland on red laterite soil; single record from pitfall.

Tetramorium sp. 14 (near *edouardi* Forel). Open woodland; ground & plant forager.

Tetramorium sp. 15 (near *squaminode* Santschi). Open woodland on slope; ground forager.

Tetramorium sp. 16 (near *agile* Arnold). Grassland & open woodland; ground forager.

Subfamily Ponerinae

Amblyopone sp. 01. Open woodland at valley base; subterranean.

Amblyopone sp. 02. Hilltop forest; subterranean, in soil below leaf litter.

Amblyopone sp. 03. Moist woodland; in leaf litter at base of large fig tree.

Anochetus katonae Forel. Closed woodland, hilltop & lowland forest; nest in soil beneath leaf litter.

Anochetus obscuratus Santschi. Hilltop forest on Kinondo hill; in leaf litter and found in rotten wood on ground.

Anochetus traegaordhi Meyer. Woodland & lowland forest; nest in soil beneath leaf litter.

Anochetus sp. 02 (near *pubescens* Brown). Open woodland; lone worker in subterranean cavity at depth of 40 cm.

Anochetus sp. 03 (*grandidieri*-group). Open woodland; in leaf litter & sifted soil.

Discothyrea sp. 01. Hilltop forest; in leaf litter.

Discothyrea sp. 02. Closed woodland, hilltop forest & lowland forest; in leaf litter.

Discothyrea sp. 03. Woodland; in leaf litter.

Discothyrea sp. 04. Hilltop forest on Maji Kununua; in leaf litter.

Discothyrea sp. 06. Hilltop forest; in leaf litter.

Hypoponera sp. 01. Closed woodland & hilltop forest; in leaf litter & soil beneath leaf litter.

Hypoponera sp. 02. Moist woodland; in soil beneath leaf litter.

Hypoponera sp. 03. Moist woodland & hilltop forest; in leaf litter & beneath bark of rotten trunk on ground.

Hypoponera sp. 04. Hilltop forest on Maji Kununua; one record from soil beneath leaf litter.

Hypoponera sp. 05. Closed woodland & lowland forest; nest between rotten leaves in leaf litter.

Hypoponera sp. 06. Closed woodland & hilltop forest; in leaf litter.

Hypoponera sp. 07. Woodland & hilltop forest; in leaf litter & soil beneath leaf litter.

Leptogenys pavesi Emery. Open woodland; nests in preformed cavities (e.g. made by termites) in soil.

Leptogenys stuhlmanni Mayr. Woodland; ground forager feed on isopods.

Leptogenys sp. 01. Open woodland on Mbuga soil; single record from pitfall.

Leptogenys sp. 02 (near *piroskae* Forel). Hilltop forest on Kinondo; single record from leaf litter.

Pachycondyla ambigua André. Open woodland; ground & maybe subterranean foragers.

Pachycondyla analis (Latreille). Grassland & open woodland; black major & minor workers form a column & go on raids for termites; nest in soil.

Pachycondyla berthoudi (Forel). Open woodland; Hodotermes termite predators, nest in soil.

Pachycondyla caffraria (F. Smith). Closed woodland & lowland forest; nest in soil beneath leaf litter.

Pachycondyla crassa (Emery). Grassland & woodland; ground forager; nest in soil, often under rock.

Pachycondyla sennaarensis Mayr. Open woodland and bare river banks on Umba River; nest in soil.

Pachycondyla suspecta (Santschi). Open woodland; subterranean.

Pachycondyla tarsata (Fabricius). Localised in grassland & open woodland; ground forager, nests in soil.

Pachycondyla sp. 01 (*crassa*-group). Hilltop forest, woodland & grassland; nest in soil, sometimes under rock.

Pachycondyla sp. 02 (near *sculpturata* Santschi). Woodland & lowland forest; nests in soil beneath leaf litter.

Phrynoponera sveni Forel. *Dichrostachys* scrub; single record from pitfall.

Platythyrea arnoldi Forel. *Acacia* woodland on red laterite soil; single record nest in soil with old beetle elytra round entrance.

Platythyrea cribrinodis (Gerstäcker). Widespread in grassland & woodland; nests in soil.

Platythyrea modesta Emery. Woodland & lowland forest; nests in dead branches on trees.

Platythyrea schultzei Forel. Grassland & open woodland; ground forager, nests in soil.

Plectroctena ?mandibularis F. Smith. Hilltop forest & closed woodland; millipede predator, nest in soil under log.

Plectroctena ?strigosa Emery. Grassland & open woodland; millipede predator, nests in soil, bleached millipede rings round entrance.

Subfamily Pseudomyrmecinae

Tetraponera sp. 01. Open woodland; nests in dead branches on trees.

Tetraponera sp. 02. Woodland & forest; nests in dead branches on trees.

Tetraponera sp. 03. Open woodland; nests in dead branches on trees.

Tetraponera sp. 04. Open woodland; single record of queen from yellow pan trap.

CHAPTER 20

The pollination ecology of Mkomazi *Acacia* species

Graham N. Stone, Pat Willmer, J. Alexandra Rowe, Bruno Nyundo & Raphael Abdallah

Introduction

Acacia trees are one of the dominant features of the Mkomazi environment, particularly so on the lower slopes of hills and in valley bottoms (see Coe *et al.*, Chapter 7, on vegetation and habitats). Acacias provide food and living environments for a huge diversity of animals, including large mammals, birds (see Cotgreave, Chapter 29) and, particularly, insects (see Willmer, Stone & Mafunde, Chapter 21; McGavin, Chapter 22; and Stapley, Chapter 23). This chapter concentrates on the ecology of *Acacia* flowering, a part of the biology of African acacias which has received very little attention. Unlike some plants, *Acacia* trees cannot fertilise their flowers with their own pollen, and so transfer between trees of a given species is essential. To achieve such transfer, during flowering acacias interact with a specific set of animals which fulfil the role of pollinators. At the same time, flowers are exposed to a specific set of herbivores which specialise in flower-eating. These two groups of animals, almost all of which are insects, have a crucial impact on how much seed the *Acacia* trees set, and so on the survival of the *Acacia* populations. In addition to these general considerations, acacias are interesting scientifically for two further reasons.

First, many of the Mkomazi species flower more or less synchronously ('coflower') after the short and/or long rains each year. This raises the possibility that *Acacia* species may compete for pollination. This can happen in two ways (Rathcke 1988).

(a) Some acacias may attract a higher proportion of the available community of insect pollinators than others (for example by having larger or more abundant flowers), and so deprive the other species of adequate pollen transport.

(b) Even if there are enough pollinators in general, if the same pollinators visit a range of different *Acacia* species then the pollen they carry will be a many-species mixture. Transfer of pollen between *Acacia* species has no benefit whatsoever, since successful production of seed set requires pollen of the same species; in fact, exchange of pollen between species may have real negative consequences. Specifically, pollen from one *Acacia* species can block the stigma (the receptive female part on which pollen must be deposited for successful fertilisation) on the flowers of another, and so prevent even the right pollen from pollinating the flower. Given that *Acacia* species do coexist and flower together, what mechanism prevents such negative interactions between species? This aspect of *Acacia* pollination is considered below. Those wishing for more in depth analyses should consult two resulting scientific papers: Stone, Willmer & Nee (1996) and Stone, Willmer & Rowe (1998).

Second, as described by Hocking (1970), Janzen (1974) and Stapley (this volume), some *Acacia* species have mutually beneficial associations with ants in the genus *Chromatogaster*. The ants live in hollow 'pseudogalls' (actually the swollen bases of thorns) formed by the tree, and effectively deter herbivores of all sizes from browsing on the foliage. An aspect of this interaction which hitherto has not been appreciated is that over-enthusiastic ant-guards may cause problems during pollination if they chase pollinators away! How is this potentially disastrous aspect of the acacia-ant symbiosis avoided? This aspect of *Acacia* pollination is discussed in the following paper by Willmer, Stone & Mafunde (Chapter 21) and in Willmer & Stone (1997).

Structuring of pollination in Mkomazi *Acacia* communities.

A lot of work in pollination biology has considered how plants might avoid competition for pollination. Plants may differ in the pollinators they recruit, and so have independent avenues of pollen transfer (Pleasants 1983, Armbruster & Herzig 1984, Rathcke 1988). East African *Acacia* flowers, however, are visited by a wide diversity of flower visitors, at least some of which are shared by more than one *Acacia* species (Tybirk 1989 & 1993, Stone, Willmer & Nee 1996). Where pollinators are shared, plant species may use effectively different populations of them by being separated in either space or time. Time here is measured on a seasonal timescale: for example, one species may flower in June and other in July, thus avoiding cross-species pollen contamination. However, in Mkomazi (as elsewhere in savanna ecosystems) acacias commonly grow in mixed-species assemblages and several species flower together in space and time after seasonal rainfall (Ross 1981; see Table 20.1 for Mkomazi acacias); neither of the space or seasonal time options are thus available.

One further option is that plants sharing pollinators may actually use different *parts* of the pollinator's body to deposit and collect their pollen (e.g. Armbruster *et al.* 1994). This solution is usually accompanied by the evolutionary divergence among the plant species sharing the pollinator(s) in some crucial aspect of floral structure associated with pollen transfer and retrieval. For example, divergence in

Table 20.1 Seasonal flowering patterns of Mkomazi *Acacia* species. An asterisk (*) indicates mass flowering, and a cross (+) slight, scattered flowering.

Acacia species	Jan	Feb	Mar	Apr	May	Jun	Jul	Aug	Sep	Oct	Nov	Dec
A. brevispica	+				*	*	+					+
A. bussei									+	*	*	+
A. etbaica								+	*	*		
A. reficiens	+							+	*	*	+	+
A. thomasii							*	*	+		+	
A. drepanolobium	*	+			+	+	+					*
A. nilotica	*	+				+	+					*
A. senegal	*	+				+	+				+	*
A. tortilis	*	+					+	+				*
A. zanzibarica	*					+	+				+	*

the length or positioning of pollen-bearing anthers and pollen-receiving stigma is a known example in a group of Australian plants (Armbruster *et al.* 1994). However, *Acacia* flowers are fundamentally all very similar in structure. The flowers are individually small (5–10 mm long, about 1 mm across, depending on the species), but are collected together into inflorescences containing from 10–20 to 500 flowers. The inflorescences are either spherical ('globose') or bottle-brush-like ('spicate'). The pollen-bearing anthers form a layer all over the surface of the inflorescence, and typically pollinators walking over the surface of the inflorescence are coated from below with pollen. With such a floral structure there is little potential for between-species difference in pollen placement on the body of a pollinator.

One remaining alternative is for plant species to diverge in their timing of pollen release (dehiscence). If each coflowering species releases its pollen only for a discrete period each day, then pollinators could be shared even through a limited flowering season. Daily partitioning of pollen release would avoid both of the mechanisms of competition for pollination mentioned earlier. Daily temporal partitioning of pollinator behaviour would result in most of the pollinators being concentrated on a single species at any given time, so minimising competition for actual pollinator visits. Second, because many pollinators (bees in particular) remove pollen from their bodies at regular intervals (Gilbert 1981, Roubik 1989, Stone 1994), temporal partitioning of their activity will result in pollinators carrying predominantly one type of pollen at any one time, so reducing interspecific pollen transfer.

Very few studies have demonstrated this sort of temporal partitioning on a daily scale (Armbruster & Herzig 1984, Armbruster 1985). Coflowering acacias in Mkomazi seemed a possible candidate for such a solution to the problems of competition, in part because alternative solutions appear unavailable to them. We set

Figure 20.1 A map of the study area.

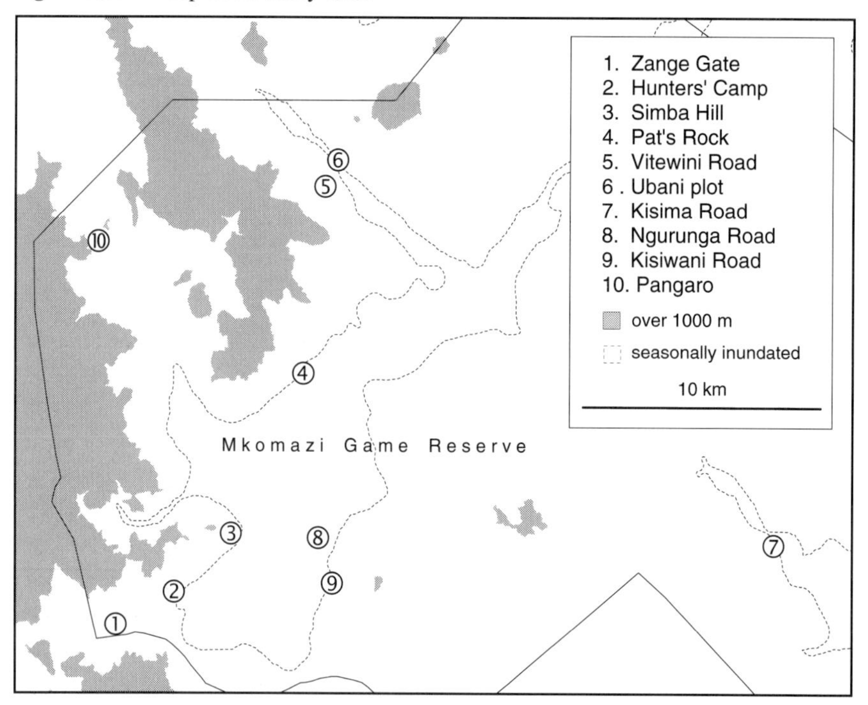

out to determine to what extent Mkomazi acacias overlap in their pollinator assemblages, and to what extent acacias differ in their daily timing of pollen release. If times of release vary, it is possible to ask whether the variation is compatible with the predictions of competition between species. Competition theory predicts that timings of release for competing species should be as evenly spaced through

Table 20.2 Acacia species composition at the chosen study sites ('+' = present).

Acacia species	Zange Gate	Hunters' Camp	Simba Hill	Pat's Rock	Vitewini Road	Ubani Plot
A. brevispica			+			
A. bussei					+	+
A. drepanolobium						
A. etbaica					+	+
A. nilotica			+			+
A. reficiens				+	+	+
A. senegal	+	+	+		+	+
A. thomasii			+			
A. tortilis	+	+	+			
A. zanzibarica				+	+	+

the day as possible (or 'overdispersed'), so minimising overlap in pollinator use between *Acacia* species. The theoretical basis of this prediction and the associated statistical methods are not described in detail here, but are presented in Stone, Willmer & Nee (1996) and Stone, Willmer & Rowe (1998).

Methods

Ten *Acacia* species, representing the two *Acacia* subgenera present in Africa (Ross 1981), were studied in an area 20 km by 50 km at the western end of the reserve (Figure 20.1). *Acacia brevispica, A. drepanolobium, A. etbaica, A. nilotica, A. reficiens, A. tortilis* and *A. zanzibarica* are members of the subgenus *Acacia,* while *A. bussei, A. senegal* and *A. thomasii* are members of the subgenus *Aculeiferum* (Ross 1981). We selected 10 sites (Figure 20.1) with different *Acacia* assemblages, listed in Table 20.2. Flowering times and associated pollinators for many plants are highly variable in space and time (Kephart 1983, Ashman & Stanton 1991) and to determine how characteristic particular timings of pollen release or pollinating insects are of each *Acacia* species we studied separate populations of each *Acacia* species that were as widely separated as logistically possible (10–30 km; Figure 20.1). The sampling effort devoted to analysis of pollen release and pollinator activity in each species is summarised in Table 20.3. To obtain detailed data showing how much variation there is between individuals of a single species between days, sites and years, we concentrated sampling effort on *A. zanzibarica* and *A. senegal.* Both of these species grow widely in the study area in a range of *Acacia* species assemblages (Table 20.2). Data allowing comparison between years for the same individual trees were also obtained for *Acacia nilotica, A. drepanolobium* and *A. senegal.* Data allowing comparison between spring and autumn rains within the same year were obtained for *A. nilotica.*

Table 20.2 continued

Acacia species	Kisima Road	Ngurunga Road	Kisiwani Road	Pangaro
A. brevispica				
A. bussei	+	+		+
A. drepanolobium				+
A. etbaica				+
A. nilotica				+
A. reficiens	+	+		+
A. senegal	+			+
A. thomasii				
A. tortilis				+
A. zanzibarica	+	+	+	

Table 20.3 Summary of the sampling effort (for both pollen release and flower visitation) associated with each *Acacia* species at each site.

Acacia species	site	pollen release (in tree days)	flower visitation (trees watched)	date
A. brevispica	Kisima Hill	1	0	3 Jan 97
	Simba Hill	1	1	4 Sep 96
	Simba Hill	part days	0	6–10 Jan 96
A. bussei	Ngurunga	2	2	22 Dec 96
	Ngurunga	2	2	23 Dec 96
A. drepanolobium	Ubani	2	2	9 Dec 95
	Ubani	4	4	12 Jan 97
A. nilotica	Pangaro	2	0	13 Jun 96
	Ubani	0	2	8 Dec 95
	Ubani	2	2	9 Dec 95
	Ubani	0	1	19 Dec 95
	Ubani	3	3	12 Jan 97
	Pangaro	2	0	13 Jun 96
A. senegal	Kisima Road	3	3	4 Jan 97
	Ubani	1	1	19 Dec 95
	Ubani	8	2	13 Jan 97
	Vitiwini Road	2	2	29 Dec 96
	Zange Gate	1	1	9 Jan 97
	Zange Gate	1	1	10 Jan 97
A. thomasii	Simba Hill	2	0	29 Jul 96
	Simba Hill	2	0	4 Aug 96
A. tortilis	Hunter's Camp	2	0	8 Jan 97
	Zange Gate	4	4	9 Jan 97
	Zange Gate	3	3	10 Jan 97
A. zanzibarica	Kisima Road	3	0	4 Jan 97
	Kisiwani Road	8	2	6 Jan 97
	Ngurunga	2	0	22 Dec 96
	Ngurunga	3	3	28 Dec 96
	Pat's Rock	2	0	18 Nov 95
	Pat's Rock	2	0	20 Nov 95
	Pat's Rock	2	2	21 Nov 95
	Pat's Rock	0	2	22 Nov 95
	Pat's Rock	2	2	24 Nov 95

Determining patterns of pollen and nectar availability through time

Flowers of *Acacia* species in the subgenus *Aculeiferum* (such as *A. senegal*) are usually in the form of 'bottle-brush' inflorescences and produce nectar, while those of species in the subgenus *Acacia* (such as *A. nilotica*) are usually in the form of spherical inflorescences and lack nectar. An exception to the latter generalisation is *A. zanzibarica,* which does in fact produce very tiny amounts of highly concentrated nectar (Stone, Willmer & Rowe 1998). The flowers of both subgenera contain both male and female parts, and last for only a single day. Individual inflorescences last for a single day for the study species in the subgenus *Acacia*, while species in the subgenus *Aculeiferum* have inflorescences in which discrete sets of flowers along the inflorescence axis open over 2–3 days.

Acacia pollen is presented in the form of compound masses of 16 or 32 pollen grains, called 'polyads' (Knox & Kenrick 1982). Eight polyads are present in each anther, and there are as many as 100 anthers per flower in the *Acacia* species studied here. Given that there are up to 120 flowers per inflorescence (depending on the species), the total number of pollen grains available to pollinators on each inflorescence is huge—of the order of three million per flower! Pollen release (dehiscence) in each species was evaluated by examining the relative abundance of pollen available on the surface of inflorescences sampled at intervals through the day. For each *Acacia* species, at each sample time, two or three inflorescences on each tree were chosen randomly and removed from the tree without touching the inflorescence surface. Each sampled inflorescence was rolled lightly across the adhesive side of a piece of clear adhesive tape. The tape was then placed sticky-side down over a microscope slide and examined with a light microscope. The progress of dehiscence over time was recorded by scoring the ratio of anthers to polyads collected on the tape. Prior to dehiscence, only unopened anthers were collected. Once anthers began to dehisce, polyads were also collected: the numbers increased as dehiscence took place, and then decreased as they were removed by pollinators. For each inflorescence, the ratio of anthers to polyads was recorded for 3–5 randomly chosen microscope fields and the mean calculated. For each *Acacia* and time interval, the mean ratio was calculated across the sampled inflorescences. Standardising the range in pollen to anther ratios among trees is necessary in order for each tree to contribute equally to means calculated across trees, days or sites. Pollen to anther ratios were therefore constrained to vary between 0 and 1 for each individual tree and day by dividing them by the maximum value recorded at any time interval for that tree and day. Differences among sites in patterns of pollen availability through time were analysed with two-way ANOVA of the standardised pollen/anther ratios. All proportion data were arcsine transformed prior to analysis (Sokal & Rohlf 1981).

Timing of pollen release in flowers is sensitive to a variety of environmental cues, particularly relative humidity (e.g. Corbet *et al.* 1988). To allow us to control for microclimatic variation in our analyses, we recorded temperature and relative humidity every hour during days on which pollen availability was scored. In summary, microclimatic variation explained some variation between sites within a species, but did not significantly affect any of the patterns described here. Microclimate is discussed in more detail in Stone, Willmer & Rowe (1998).

Table 20.4 Visitation by different taxa to *Acacia* species. Data are the percentage of total visits contributed by a particular taxon. * indicates less than 0.1% of total visitation per inflorescence over the entire activity period. † indicates a value between 0.0 and 1.0%.

site / date	Calliphorid flies	Syrphid flies	Megachilid bees	other solitary bees	honey-bees	wasps	Lepidoptera	total visits per inflorescence	total visits observed	% collecting nectar
a. Variation between sites and years for *Acacia senegal*.										
Zange	1.3	*	12.4	37.2	44.4	2.1	*	86.3	604	86.3
Ubani 95–96	2.5	3.6	9.6	0.0	59.9	3.8	20.6	23.9	357	90.0
Ubani 96–97	*	*	7.9	3.7	86.4	*	1.5	46.9	352	92.1
Vitiwini	4.9	0.5	4.0	18.4	0.0	24.3	45.7	141.7	992	90.6
Kisima	9.7	*	15.8	58.0	0.0	6.1	10.1	29.3	278	74.5
b. Variation between days at a given site for *A. zanzibarica* at Pat's Rock.										
21 Nov 95	33.8	2.8	10.3	34.8	0	0	18.3	4.5	73	
22 Nov 95	27.3	6.5	10	12.25	0	0	44.5	1.6	96	
24 Nov 95	11.2	3.1	14.6	39.8	0	20	10.4	3.6	131	
c. Variation between sites for *Acacia zanzibarica*. Mean values are used for the Pat's Rock site.										
Pat's Rock	24.5	4.2	24.9	15.5	0.0	7.4	23.4	3.25	100 ± 14	
Kisiwani	22.7	*	21.7	34.0	0.0	21.7	*	10.9	185	
Ngurunga	88.6	11.3	0.0	0.0	0.0	0.0	*	28.7	273	
d. Variation between days at a given site for *A. nilotica* at Ubani.										
8 Dec 95	2.5	0.0	97.6	0.0	0	0	0	16.1	370	
9 Dec 95	11.9	0.2	86.5	1.3	0	0	0	14.4	446	
18 Dec 95	22.8	29.1	47.4	0.0	0	0	0	5.1	127	
e. Comparisons between *Acacia* species in % of all visits contributed by particular visitor taxa (± 1 standard error).										
A. drepanol.	†	0.0	98.0	0.0	1.3	†	†	6.4	930	
A. nilotica	12.0 ± 5.0	10.0 ± 0.8	76.0 ± 12.0	2.0 ± 1.0	0.0	0.0	0.0	11.9 ± 3.3	314 ± 78	
A. senegal	4.1 ± 1.9	0.3 ± 0.1	10.1 ± 2.2	29.7 ± 10.2	33.1 ± 17.9	8.2 ± 4.8	14.5 ± 9.3	76.0 ± 21.6	557 ± 139	
A. tortilis	46.0	†	8.3	34.1	11.1	†	†	46.9	468	
A. zanzibarica	40.7 ± 20.3	5.5 ± 2.7	17.7 ± 7.6	17.3 ± 7.25	0.0	12.0 ± 4.8	6.7 ± 5.2	15.8 ± 5.3	186 ± 41	

Temporal patterns of flower visitation and comparisons of visitor assemblages across *Acacia* species

Detailed analyses of flower-visitor behaviour were completed during the autumn rains in 1995–96 and 1996–97 (Table 20.4). Pollinator visitation was quantified in the same way for each *Acacia* species by watching the same set of 20 selected inflorescences for 30 minutes every 1–1.5 hours through the day from before the onset of foraging until after it ceased. In our analysis, visitation is quantified in terms of the number of inflorescence visits made by each named insect group. Differences in patterns of activity of foragers through time were analysed using Kolmogorov-Smirnov tests (Sokal & Rohlf 1981). Specimens of flower visitors were captured and identified by Chris O'Toole at the Hope Department of Entomology, Oxford University (bees), Donald Baker at the Natural History Museum, London (bees) and by John Deeming at the National Museum of Wales, Cardiff (flies). Similarities in the pollinator insect communities between pairs of *Acacia* species were quantified using the proportional similarity statistic (PS; Schoener 1970, Kephart 1983, Horvitz & Schemske 1990), which ranges from 1 (maximum similarity) to 0 (no overlap whatsoever).

Results

The distribution of *Acacia* flowering patterns in seasonal time

Months during which flowering by component *Acacia* species was observed are summarised in Table 20.1. The highest diversity of flowering *Acacia* species in this habitat was present in December and January, following the autumn rains, when up to eight species flowered concurrently. In contrast to the majority of *Acacia* species, *Acacia thomasii* and *A. etbaica* flowered only during the dry season, at a time when few individuals of other species were in flower. These species occupy seasonal flowering niches in which competition for pollinators with other *Acacia* species is either much reduced or avoided altogether.

Do coflowering *Acacia* species share pollinators?

A full list of all of the insect flower visitors observed at each *Acacia* species is presented in the appendix at the end of this chapter.

Acacia senegal is the only species in the sampled community to produce abundant nectar. Flowers contained their maximum volume of nectar (a mean of 1.35 ± 0.1 µl per floret at $19.8 \pm 2.1\%$ sucrose by mass) at the onset of dehiscence, and the available volume per flower fell rapidly (as a result of evaporative loss of water and flower visitation by insects) to reach barely detectable levels by noon. Nectar foragers consistently made up a high proportion of all visits to *A. senegal* (74–92%, Table 20.4a), including large, common and highly mobile insects capable of moving between trees, and so of potential value as pollinators. Honeybees

(*Apis mellifera*) were important visitors at two sites, and collected nectar only from *A. senegal*. Other visitors contributing significant visitation were butterflies, solitary leaf-cutter bees (family Megachilidae), spider wasps and two species of sunbirds (*Nectarinia hunteri* and *N. venusta*).

Acacia tortilis and *A. zanzibarica* flowers both produced minute quantities of highly concentrated nectar (<0.1 μl of >70% sucrose per floret) in comparison to *A. senegal*. Both species received substantial visitation from nectar foraging insects (Table 20.4b, c), including the same leaf-cutter bees seen at *Acacia senegal*. Over a third of visits to *A. tortilis* were by small halictid solitary bees foraging for nectar (Table 20.4e). The nectar-foragers recorded at both *A. tortilis* and *A. zanzibarica* were mainly small insects, without the larger nectar-feeding insects and sunbirds recorded from *A. senegal*. In contrast to *A. senegal*, both *A. tortilis* and *A. zanzibarica* received c. 40% of their total daily visitation from a wide diversity of pollen-feeding flies, particularly hoverflies (family Syrphidae) of the genus *Eristalinus* and calliphorid flies of the genus *Rhyncomya* (Table 20.4b, c, e). *Acacia tortilis* received significant visitation from honeybees (*Apis mellifera*), which in marked contrast to their behaviour at *A. senegal* at the same location only collected pollen at *A. tortilis*.

Acacia nilotica and *A. drepanolobium* produced no nectar, and almost no nectar foraging butterflies, wasps or honeybees were observed at flowers of either species (Table 20.4e). By far the most important visitors were solitary leaf-cutter bees of the same species observed at *A. senegal*, *A. tortilis* and *A. zanzibarica*. While *A. drepanolobium* received almost all of its visitation from a single megachilid bee species, *A. nilotica* at the same site were also visited by the same calliphorid flies and hoverflies observed at *A. tortilis* and *A. zanzibarica*.

Similarity in patterns of visitation within and among *Acacia* species can be quantified with the proportional similarity (PS) statistic. All *Acacia* species showed intraspecific variation in their flower visitor assemblages. For all species, variation among sites (Table 20.5a, b) was greater than variation at a given site among days or years (Table 20.5c, d). Despite such intraspecific variation, differences in visitor taxa result in a gradient of proportional similarity values across *Acacia* species (Table 20.5e). *Acacia senegal* is most similar to *A. tortilis,* then *A. zanzibarica, A. nilotica* and *A. drepanolobium*. This gradient correlates with differences among species in nectar production and is repeated for PS values for all of the pairwise comparisons among *Acacia* species.

Flower visitor assemblages in this *Acacia* community are thus characterised both by partial separation and by overlap. Substantial differences in visitor guilds across *Acacia* species correspond to variation in the nectar available to foragers. *A. senegal* has a highly diverse assemblage of nectar-feeding insects, including large and mobile species, and the vast majority of visits by insects to this *Acacia* were for nectar. Conversely, flower visits to *A. drepanolobium* and *A. nilotica* were almost entirely for pollen, and there were very few visits by nectar-feeding

Table 20.5 Proportional similarities in levels of visitation by pollinator taxa for Mkomazi *Acacia* species, calculated from data in Table 20.4 using the formula given in Schoener (1970).

a. *Acacia senegal*, across four sites in the autumn rains, 1996–97

	Ubani	Vitewini Road	Zange
Kisima Road	0.13	0.44	0.54
Ubani		0.10	0.57
Vitiwini Road			0.26

b. *Acacia zanzibarica*, across three sites in the autumn rains, 1996–97

	Ngurunga	Pat's Rock
Kisiwani Road	0.23	0.67
Ngurunga		0.29

c. *Acacia zanzibarica*, across days for 21, 22 and 24 November 1995

	22 Nov 95	24 Nov 95
21 Nov 1995	0.70	0.70
22 Nov 1995		0.47

d. *Acacia nilotica*, across days for 8, 9 and 18 December 1995

	9 Dec 95	18 Dec 95
8 Dec 1995	0.89	0.50
9 Dec 1995		0.60

e. Across *Acacia* species

	A. tortilis	*A. zanzibarica*	*A. nilotica*	*A. drepanolobium*
A. senegal	0.53	0.47	0.16	0.11
A. tortilis		0.67	0.22	0.10
A. zanzibarica			0.37	0.18
A. nilotica				0.81

insects. Acacia flower visitor assemblages nevertheless overlap. Megachilid leaf-cutter bees in the genera *Creightonella, Chalicodoma* and *Megachile* visited all of the coflowering *Acacia* species (Table 20.4a). Syrphid hoverflies and calliphorid flies visited *A. nilotica, A. tortilis* and *A. zanzibarica*, and honeybees were important foragers (though for differing resources) at *A. senegal* and *A. tortilis*.

Do *Acacia* species release their pollen at different times of day?

This question really needs to be divided into two. First, does each *Acacia* species have a characteristic time when pollen is released each day? Second, are the times utilised by a set of acacias flowering together spread out through the day?

The answer to the first question is generally yes. As shown in Figure 20.2, individuals of a given *Acacia* species flowering in a given site on a given day tend to release their pollen at a similar time. For all but *A. drepanolobium* the peaks of pollen release shown by individuals of a given species flowering during the autumn rains are significantly clustered together (*Var* statistic, V, after Williams 1995: *A. nilotica* V=0.177, p<0.01; *A. tortilis* V=0.09, p<0.05; *A. zanzibarica*

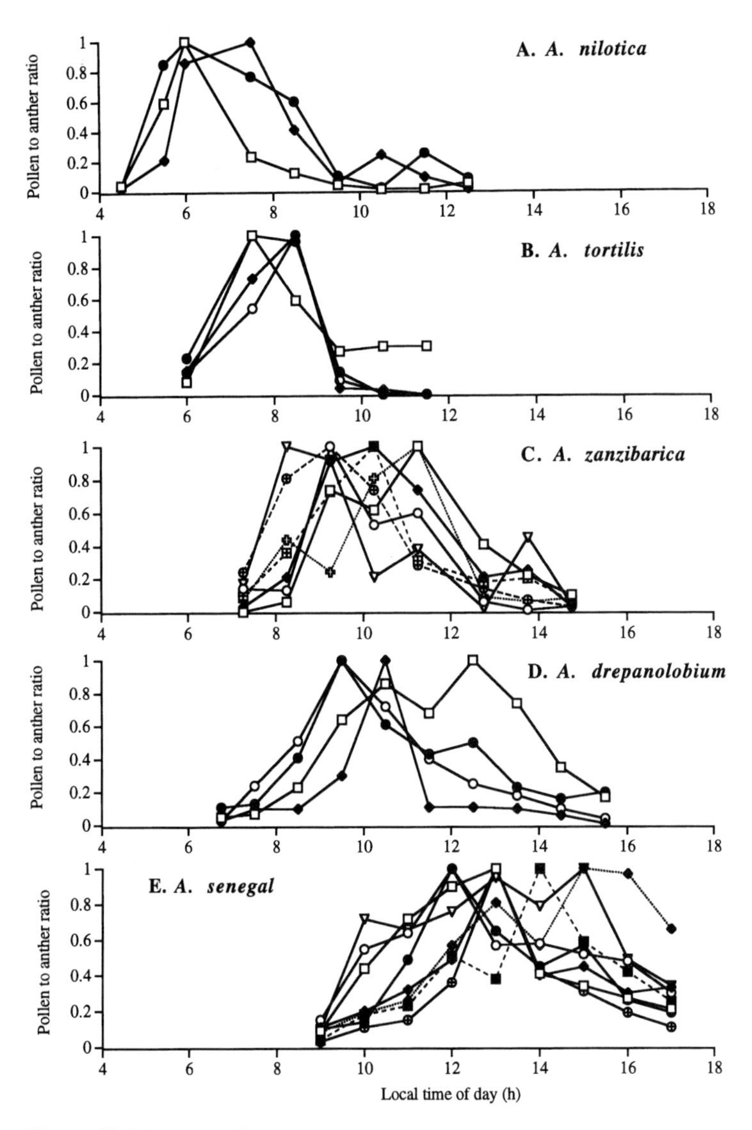

Figure 20.2 Patterns of pollen availability through time for individual trees of *A. nilotica, A. tortilis, A. zanzibarica, A. drepanolobium* and *A. senegal*. Each symbol type represents an individual tree.

V=0.035, p<0.025; *A. senegal* V=0.028, p<0.025). Timing of pollen release is not only common to individuals at a single site on a single day, but is also generally consistent across sites, and even across years. This is shown well by data for *A. zanzibarica* at different sites in the 1996–97 season (Figure 20.3a). Timing of pollen release in *A. nilotica* was recorded for the same trees in the autumn rains of 1995–96 and 1996–97 at the Ubani site, and for different trees at the Pangaro site following the 1996 Easter rains. In all three cases, flower opening and the onset of anther dehiscence took place at the same time (Figure 20.3b). Similar similarity between years was seen in *A. drepanolobium* and *A. senegal*. Species flowering together after the autumn rains can thus be said to have genuinely species-characteristic daily times of pollen release (summarised in Table 20.6a). *A. nilotica* and *A. tortilis* both opened their flowers well before dawn, and released their pollen at or shortly after sunrise. *A. drepanolobium* and *A. zanzibarica* flowers opened from

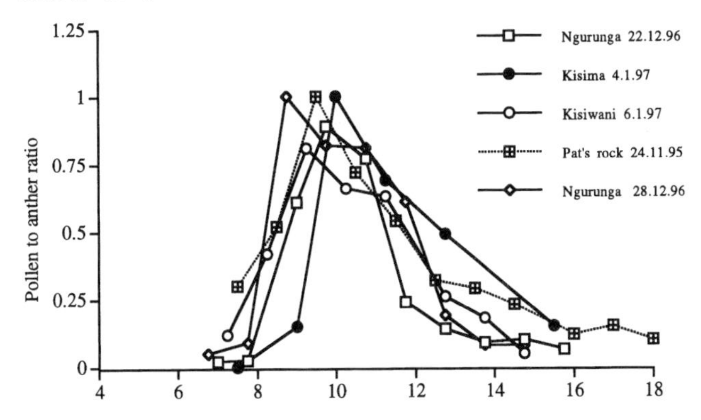

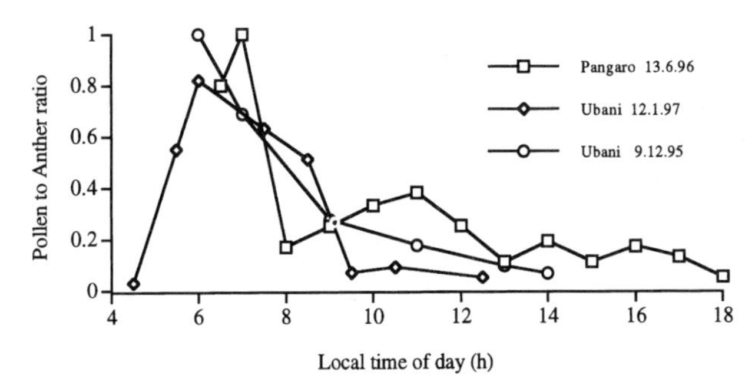

Figure 20.3 (a) Intraspecific variation in patterns of pollen release over time between sites in *A. zanzibarica*. (b) Intraspecific variation in patterns of pollen release over time between years and between flowering bouts in a single year in *A. nilotica*. Data points are means over individual trees at each site.

07.00 h and began to release their pollen from 08.00–09.00 h. *A. bussei* released its pollen at a similar time to these two, but differed in that its flowers opened well before dawn and delayed anther dehiscence for 2–3 hours. *A. senegal* flowers opened in mid to late morning, and anther dehiscence followed immediately, giving peak availability during the middle of the day. Flowers of both *A. brevispica* and *A. reficiens* only began to open from 11.00 h, and *A. brevispica* showed peak pollen availability in the afternoon. Although data on pollen release could not be collected for *A. reficiens*, the timing of flower opening means that peak pollen availability in this species must also occur in the afternoon.

Table 20.6 (a) Daily patterns of floral reward availability in the sampled *Acacia* species. An asterisk indicates that relevant data have not been collected. Values given for each variable indicate the ranges within species observed across days, sites and years. Relative humidity at dehiscence represents the mean value over the period between onset of pollen release and maximum availability.

Acacia species	nectar	flowers begin to open (h)	dehiscence begins (h)	peak pollen availability (h)	relative humidity at dehiscence (%)
Subgenus *Acacia*					
A. nilotica	none	2–3.00	4–5.00	6–7.00	90–95
A. tortilis	trace	4–5.00	6–7.00	7.50–9.00	80–90
A. drepanolobium	none	6–7.00	7–8.00	10–12.00	60–70
A. zanzibarica	trace	7–8.00	8–9.00	10–11.00	55–65
A. brevispica	none	10–11.00	10–11.00	15–16.00	40–50
A. reficiens	*	10–11.00	*	*	*
Subgenus *Aculeiferum*					
A. bussei	present	4–5.00	7–8.00	10–11.00	65–85
A. senegal	present	9–10.00	9–10.00	11–13.00	50–60
A. thomasii	*	throughout	throughout	no obvious peak	

(b) Values and significance of Williams' V for given *Acacia* assemblages (an asterisk indicates that a given *Acacia* species is included in the series of peaks tested), assuming that peaks of pollen release can be distributed anywhere between dawn and dusk. This assumption is justified with the assumption that despite observation before dawn and after dark very few nocturnal visitors to *Acacia* flowers were observed.

Acacia species							
nilotica	*tortilis*	*zanzibarica*	*bussei*	*senegal*	*brevispica*	V	p value
*	*	*	*	*	*	0.00408	<0.025
*	*	*		*	*	0.00435	<0.025
*		*	*	*	*	0.00528	<0.025
*		*		*	*	0.00047	<0.001

Are the peaks of different species spaced through the day in a manner consistent with community structuring among them? Tests for structuring between *Acacia* species in timing of pollen release compare the distribution of peaks observed for species that grow together with patterns of spacing predicted if timings of pollen release for all species were random (Williams 1995). Regularity of spacing, as for clustering, is expressed using a statistic called *Var,* V. For regularly spaced peaks, V is higher than expected for a random set of release times. For a range of *Acacia* assemblages encountered in Mkomazi, peaks of pollen release across species are indeed spaced in such a way that overlap among *Acacia* species is very low (Table 20.6b). A fuller analysis of spacing of peaks is given in Stone, Willmer & Rowe (1988).

Interestingly, a quite different pattern was seen in *A. thomasii*, which flowers alone in the dry season: the four *A. thomasii* trees sampled showed far more variation in patterns of pollen availability than any of the co-flowering species (compare Figure 20.2 with Figure 20.4).

Is the activity of shared flower visitors structured by daily patterns of pollen availability?

Differences in the timing of pollen release by co-flowering acacias can only help to reduce pollen flow between species if pollinators which are shared by *Acacia* species actually respond to those differences, i.e. is there a sequence of pollinator activity across *Acacia* species which tracks the sequence of pollen release? The answer without doubt is that there is. Daily sequences are described for the three most important groups of shared pollinators in turn.

Megachilid leaf-cutter bees

Megachilid bees show daily structuring of pollen-collecting activity that follows the sequence of pollen release in the *Acacia* species (for example, see Figure 20.5a).

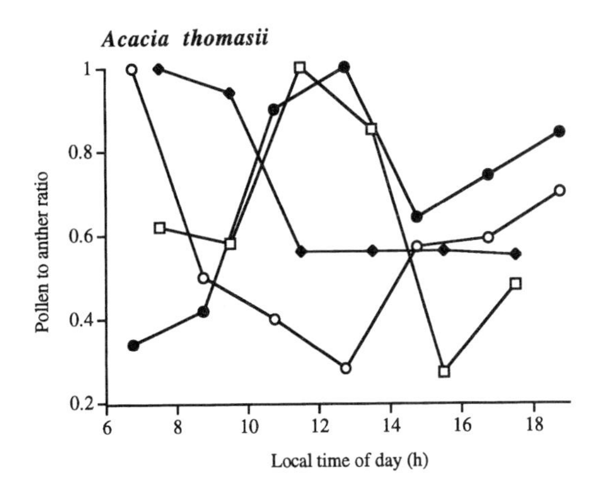

Figure 20.4 Patterns of pollen availability through time for four individual *A. thomasii* trees (identified by different symbols).

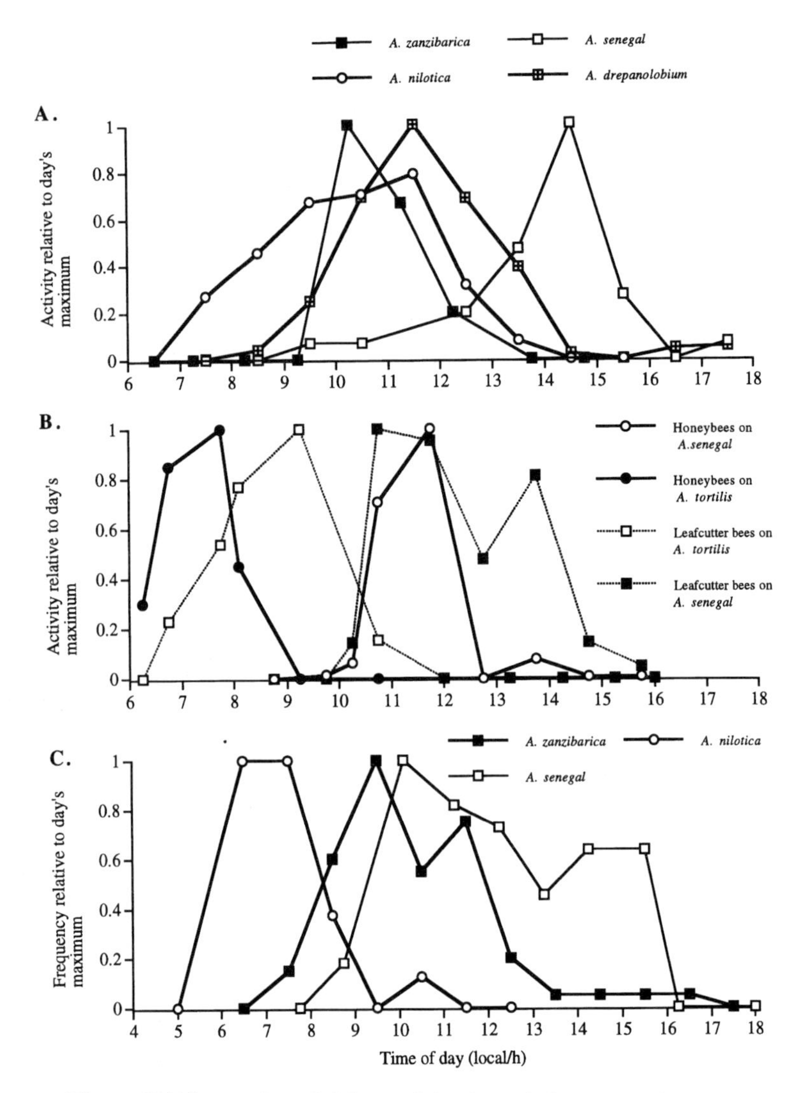

Figure 20.5 Structuring of visitor activity through time across *Acacia* species for (a) megachilid bees at Ubani on 19 December 1995, (b) honeybees and megachilid bees at Zange on 10 January 1997, and (c) calliphorid flies at Ubani on 19 December 1995.

Although there was overlap of bee activity among *Acacia* species, peak activity periods by the bees at each *Acacia* species are significantly separated in time (Kolmogorov-Smirnov test, $p<0.001$ for all six pairs). Activity patterns of these bees at *A. tortilis* and *A. senegal* at Zange showed no overlap in time at all (Figure 20.5b).

Honeybees

At Zange honeybees visited *A. tortilis* for pollen and *A. senegal* for nectar. The distributions of activity at these two floral sources showed no overlap in time (Figure 20.5b), and paralleled the timing of pollen release by the two *Acacia* species.

Calliphorid flies

Activity by calliphorid flies showed significant differences among *Acacia* species at the Ubani site (Kolmogorov-Smirnov test, $p<0.001$ for all species pairs, Figure 20.5c), and again tracked the sequence of dehiscence.

Discussion

Four mechanisms contribute to structuring of flowering of *Acacia* species growing in Mkomazi: spatial separation, seasonal separation of flowering in time, partial division of flower visitor assemblages, and daily patterning of pollinator behaviour through timing of pollen release. Separation of *Acacia* species by habitat type is dealt with in more detail elsewhere (Coe *et al.*, Chapter 7). Partial structuring of flowering through separation of flowering season is apparent in this community; while a majority of species flower either predominantly or at least partially during the autumn rains, *A. brevispica* flowers predominantly after the Easter rains, and *A. bussei, A. etbaica, A. reficiens* and *A. thomasii* flower during the dry season. These species thus do not interact with the diversity of species flowering during the autumn rains. *Acacia thomasii* achieves almost complete seasonal separation on rocky hill slopes in Mkomazi, with no other *Acacia* species flowering significantly in the same area at the same time. Those *Acacia* species flowering at the same time do have partially separated groups of pollinators, but there is substantial overlap as well, requiring an additional mechanism if pollen flow between species (and associated negative impacts) are to be avoided. The most exciting new discovery arising from this work is the demonstration that separation of *Acacia* species is achieved by daily partitioning of pollen release—and associated partitioning of activity by shared pollinators. An additional piece of evidence supports the conclusion that release of pollen at a specific time of day may be a response to competition with other coflowering acacias. *A. thomasii*, which flowers alone during the dry season, does not show the highly synchronised pollen release across individuals demonstrated by co-flowering species characteristic of the autumn rains. Although our small sample size precludes firm conclusions, one interpretation of this finding is that in the absence of competing coflowering acacias, *A. thomasii* has not needed to evolve a narrow daily window of pollen release, and so is able to exploit a far broader daily temporal niche for pollen release.

Wider significance of diurnal resource partitioning in flowering acacias

Separation in seasonal time, space or pollinator assemblages, and exploitation of separate regions on the bodies of shared pollinators are the four principal resource axes that have been discussed in the context of competition for pollination (Pleasants 1983, Waser 1983, Rathcke 1988, Ollerton & Lack 1992, Armbruster *et al.* 1994). Coflowering *Acacia* species overlap substantially in all four, and by metrics applied to seasonal flowering phenology, should compete substantially (Pleasants 1980). Daily structuring of pollen release and shared pollinator activity, as described here and elsewhere (Armbruster & Herzig 1984, Armbruster 1985) provides two cautionary notes applicable to many existing studies of flowering phenology (timing). First, the importance of a particular flower visitor in the pollination of a given *Acacia* species is very dependent on the daily time of observation. If all *Acacia* species in this study were watched only between 11.00 and 12.00 h each day, the daily structuring of shared pollinator activity would not be revealed, and some important visitation in the system (almost all the visitation to *Acacia nilotica*, for example) would be missed entirely. Not only would differences among *Acacia* species in pollinator assemblages appear greater than they are, but the temporal structuring that may reduce competition within the system would not be detected. Second, estimates of the severity of competitive interactions among plant taxa based only on the degree of pollinator sharing and overlap in flowering seasons may need to be re-evaluated.

The demonstration of fine-scale daily partitioning raises a number of interesting questions. Just as species may diverge in seasonal flowering phenology (Rathcke & Lacey 1985), so we might expect to find populations of a single species in different competitive environments to diverge in their daily timing of pollen release. The possibility of significant intraspecific variation on a large geographic scale does exist for several of the *Acacia* species studied here. *Acacia nilotica*, *A. senegal* and *A. tortilis* are widely distributed through Africa to Arabia and (for the first two species) India (Ross 1981), and form part of a wide diversity of *Acacia* assemblages with different pollinator guilds (Tybirk 1993). Comparison of data for Mkomazi with data collected in 1998 and 1999 for a site in central Kenya shows that individual *Acacia* species can indeed show substantial geographical variation in timing of pollen release.

Lastly, there has been considerable discussion of constraints that may act on flowering phenology (Kochmer & Handel 1986, Wright & Calderon 1995), and other constraints may act on daily timing of pollen release. Are there any phylogenetic constraints acting on daily timing of pollen release in acacia? What characteristics of species (floral or otherwise) are associated with release of pollen at a particular time of day? Why do some *Acacia* species release their pollen around dawn, and others in the afternoon? Activity of bees and other insects can be strongly affected by daily fluctuations in temperature and humidity (Willmer

1983, Herrera 1990, Stone 1994), and this may place constraints on the pollinators available to acacias flowering at particular times of day. Is there a best time to release your pollen?

Acknowledgements

We would like to thank Linsey Stapley, George McGavin, Tim Morgan, Ian Cooksey, Will Dixon, Nigel Raine and Daniel Mafunde for their help in Tanzania, Malcolm Coe, Nigel Winser and Venetia Simonds for their support in Britain, and the Department of Wildlife of the Government of the Republic of Tanzania. We are particularly thankful to Nick McWilliam and Mike Packer for the map in Figure 20.1, and to Donald Baker, Chris O'Toole and John Deeming for their taxonomic expertise. GNS was supported by a Varley-Gradwell fellowship in field entomology and the Apgar Prize from Magdalen College, Oxford. Both GNS and PGW are supported by a grant from NERC (ref GR9/03553). This work was supported by the Royal Geographical Society (with IBG) and the Darwin Initiative, and forms part of the Mkomazi Ecological Research Programme.

References

Armbruster, W.S. (1985) Patterns of character divergence and the evolution of reproductive ecotypes of *Dalechampia scandens* (Euphorbiaceae). *Evolution* 39: 733-752.

Armbruster, W.S., Edwards, M.E. & Debevec, E.M. (1994) Floral character displacement generates assemblage structure of Western Australian triggerplants (*Stylidium*). *Ecology* 75: 315-329.

Armbruster, W.S. & Herzig, A.L. (1984) Partitioning and sharing of pollinators by four sympatric species of *Dalechampia* (Euphorbiaceae) in Panama. *Annals of the Missouri Botanical Garden* 71: 1-16.

Ashman, T.-L. & Stanton, M. (1991) Seasonal variation in pollination dynamics of sexually dimorphic *Sidalcea oregana* ssp. *spicata* (Malvaceae). *Ecology* 72: 993-1003.

Corbet, S.A., Chapman, H. & Saville, N. (1988) Vibratory pollen collection and flower form: bumble bees on *Actinidia, Symphytum, Borago* and *Polygonatum*. *Functional Ecology* 2: 147-155.

Gilbert, F.S. (1981) Foraging ecology of hoverflies: morphology of the mouthparts in relation to feeding on nectar and pollen in some common urban species. *Ecological Entom*ology 6: 245-262.

Herrera, C.M. (1990) Daily patterns of pollinator activity, differential pollinating effectiveness, and floral resource availability, in a summer-flowering Mediterranean shrub. *Oikos* 58: 277-288.

Hocking, B. (1970) Insect associations with the swollen thorn *Acacia* species. *Transactions of the Royal Entomological Society of London* 122: 211-255.

Horvitz, C.C. & Schemske, D.W. (1990) Spatiotemporal variation in insect mutualists of a neotropical herb. *Ecology* 71: 1085-1097.

Janzen, D.H. (1974) Swollen-thorn *Acacia* species of Central America. *Smithsonian Contributions to Botany* 13.

Kephart, S.R. (1983) The partitioning of pollinators among three species of *Asclepias*. *Ecology* 64: 120-133.

Knox, R.B. & Kenrick, J. (1982) Polyad function in relation to the breeding system of *Acacia*. In: Mulcahy, D. & Ottavianopp, E. (eds.) *Pollen Biology*. North Holland Press, Amsterdam, Holland. pp. 411-418.

Kochmer, J.P & Handel, S.N. (1986) Constraints and competition in the evolution of flowering phenology. *Ecological Monographs* 56: 303-325.

Ollerton, J. & Lack, A.J. (1992) Flowering phenology: an example of relaxation of natural selection? *Trends in Ecology and Evolution* 7: 274-276.

Pleasants, J.M. (1980) Competition for bumblebee pollinators in Rocky Mountain plant communities. *Ecology* 61: 1446-1459.

Pleasants, J.M. (1983) Structure of plant and pollinator communities. In: Jones, C.E. & Little, R.J. (eds.) *Handbook of Experimental Pollination Biology*. Van Nostrand Reinhold, USA. pp. 375-393.

Rathcke, B. (1988) Flowering phenologies in a shrub community: competition and constraints. *Journal of Ecology* 76: 975-994.

Rathcke, B. & Lacey, E.P. (1985) Phenological patterns of terrestrial plants. *Annual Review of Ecology and Systematics* 16: 179-214.

Ross, J.H. (1981) An analysis of the African *Acacia* species: their distribution, possible origins and relationships. *Bothalia* 13: 389-413.

Roubik, D.W. (1989) Ecology and natural history of tropical bees. Cambridge University Press, Cambridge, Massachusetts, USA.

Schoener, T.W. (1970) Non-synchronous spatial overlap of lizards in patchy habitats. *Ecology* 51: 408-418.

Sokal, R.R. & Rohlf, F.J. (1981) *Biometry*. W.H. Freeman and Co., New York, USA.

Stone, G.N (1994) Activity patterns of females of the solitary bee *Anthophora plumipes* in relations to temperature, nectar supplies and body size. *Ecological Entomology* 19: 177-189.

Stone, G.N., Willmer, P.G. & Nee, S. (1996) Daily partitioning of pollinators in an African *Acacia* community. *Proceedings of the Royal Society of London B* 263: 1389-1393.

Stone, G.N., Willmer, P.G. & Rowe, J.A. (1998) Partitioning of pollinators during flowering in an African *Acacia* community. *Ecology* 79: 2808-2827.

Tybirk, K. (1989) Flowering, pollination and seed production of *Acacia nilotica*. *Nordic Journal of Botany* 9: 375-381.

Tybirk, K. (1993) Pollination, breeding system and seed abortion in some African *Acacia* species. *Botanic Journal of the Linnean Society* 112: 107-137.

Waser, N.M. (1983) Competition for pollination and floral character differences among sympatric plant species: a review of the evidence. In: Jones, C.E. & Little, R.J. (eds.) *Handbook of Experimental Pollination Biology*. Van Nostrand Reinhold, USA. pp. 277-293.

Williams, M.R. (1995) Critical values of a statistic to detect competitive displacement. *Ecology* 76: 646-647.

Willmer, P.G. (1983) Thermal constraints on activity patterns in nectar-feeding insects. *Ecological Entomology* 8: 455-469.

Willmer, P.G. & Stone, G.N. (1997) Ant deterrence in *Acacia* flowers: how aggressive ant-guards assist seed-set. *Nature* 388: 165-167.

Wright, S.J. & Calderon, O. (1995) Phylogenetic patterns among tropical flowering phenologies. *Journal of Ecology* 83: 937-948.

Appendix

Insect species recorded from Mkomazi acacias, by *Acacia* species (brev. = *A. brevispica*, drep. = *A. drepanolobium*, nilot. = *A. nilotica*, seneg. = *A. senegal*, tort. = *A. tortilis*, zanz. = *A. zanzibarica*).

	Acacia species					
	brev.	drep.	nilot.	seneg.	tort.	zanz.
DIPTERA (flies)						
Asilidae (Robber flies)						
Dasypogonina						
Oligopogon sp.				*		
Bombyliidae						
Anthracinae						
Exhyalanthrax abruptus (Loew)	*					
Exhyalanthrax sp. 1						*
Exoprosopa sp. 2						*
Ligyra enderleini Paramonov		*				
Petrolosia sp.	*					
Bombyliinae						
Bombylisoma sp. 1						*
Bombylisoma sp. 2			*			*
Syrphidae (Hoverflies)						
Syrphinae						
Asarkina africana Bezzi				*		
Allobaccha sapphirina Wiedemann					*	*
Paragus barbonicus Masquart						*
Paragus capriconi Stuckenberg						*
Milesiinae						
Eristalinus tabanoides Jaennicke			*	*	*	*
Eristalinus flaveolus Bigot			*	*	*	*

cont.

Appendix, continued

	Acacia species					
	brev.	drep.	nilot.	seneg.	tort.	zanz.
Eristalinus plurivittatus Macquart				*		
Mesembrius tarsatus Bigot	*					
Phytomia incisa Wiedemann		*			*	
Phytomia natalensis Macquart						*
Calliphoridae						
Rhiniinae						
Chrysomya regalis (Rabineau-Desvoidy)					*	
Cosmina sp. nr. *undulata* Malloch						*
Isomyia pubera Villeneuve						*
Neomyia albigena (Stein)						*
Neomyia macrops (Curran)					*	*
Physiphora sp. nov.						*
Rhyncomya cassatis (Walker)			*			
Rhyncomya dasyops Bezzi	*				*	
Rhyncomya forcipata Villeneuve	*			*	*	*
Rhyncomya hassei Zumpt			*			*
Rhyncomya sp. nov. nr. *io* Peris						*
Rhyncomya peraequa Villeneuve					*	
Rhyncomya pruinosa Villeneuve				*	*	*
Rhyncomya sp. nr. *soyauxi* Karsch					*	*
Rhyncomya trispina Villeneuve					*	*
Rhyncomya zumpti Peris					*	*
Rhyncomya sp. indet.			*		*	*
Tricyclea semicinerea Bezzi				*		
Zumba rhinoides Peris					*	*
Conopidae						
Conops sp.						*
Muscidae						
Muscinae						
Musca xanthomelas Wiedemann					*	
Musca lusoria Wiedemann					*	*
Coenosiinae						
Haematobia minuta (Bezzi)				*		
Mycetophilidae						
Mycetophilinae						
Phronia sp.				*		
Sarcophagidae						
Miltogramminae						
Hoplocephalopsis sp. nov.				*		
Pterella sp.				*		
Tabanidae						
Sp. 1						*
Sp. 2						*

cont.

Appendix, continued

	Acacia species					
	brev.	drep.	nilot.	seneg.	tort.	zanz.
Tachinidae						
Goniinae						
Aplomya sp.					*	*
Dolichocolon sp.						*
Exorista sp.					*	
Sp. 1					*	*
Sp. 2					*	
Tachininae						
Leskiini gen nr. *Subfischeria*					*	
Tephritidae						
Tephritinae						
Goniurellia sp.				*		
Paramacronychiinae sp.						*
HYMENOPTERA						
Apoidea (Bees)						
Anthophoridae						
Amegilla caligata (Gerst.)				*		
Amegilla africana (Fr.)				*		
Xylocopa somalica Magretti				*	*	
Apidae						
Apis mellifera L.			*	*		
Halictidae						
Curvinomia somalica (Erichson)						*
Lipotriches sp.						*
Nomia capensis Friese						*
Nomia sp. aff. *chrysogona* Cockerell					*	
Nomia lamellicornis Friese				*		*
Nomia patellifera Westwood						*
Nomia sp aff. *pulchella* Friese				*		*
Nomia tridentata Smith						*
NomiaI sp. innom.						*
Megachilidae						
Creightonella discolor (Smith)	*		*	*	*	*
Chalicodoma mossambicum (Grib.)	*		*	*	*	*
Heriades sp.			*			*
Megachile sp. aff. *microxanthops*						*
Megachile rufohirtula Cockerell						*
LEPIDOPTERA (butterflies and moths)						
Lycaenidae						
Polyommatinae						
Anthene amarah Guér.					*	*
Anthene opalina Stempffer						*
Anthene otacilia Trimen		*				*

cont.

Appendix, continued

	Acacia species					
	brev.	drep.	nilot.	seneg.	tort.	zanz.
Azanus jesous Guér.				*		*
Azanus moriqua Wallengr.				*		*
Azanus ubaldus Cramer			*	*		
Leptotes pirithous L.				*		
Ziseeria knysna Trimen				*		*
Theclinae						
Axiocerses harpax Clench				*		*
Hypolycaena pachalica Butler						*
Nymphalidae						
Acraeinae						
Acraea neobule Doubleday				*		
Danainae						
Danaus chrysippus L.			*	*	*	
Amauris ochlea Bois.			*	*	*	
Nymphalinae						
Byblia ilithyia Drury				*		*
Hypolimnas misippus L.				*	*	*
Junonia oenone L.				*		*
Junonia hierta Trimen				*		*
Pieridae						
Coliadinae						
Catopsilia florella Fab.				*		*
Colotis danae Fab.				*		*
Colotis eucharis Klug			*			
Colotis halimede Talbot				*		*
Hesperiidae						
Coeliades anchises Gerstücker				*		*

CHAPTER 21

Ants, pollinators and acacias in Mkomazi

Pat Willmer, Graham N. Stone & Daniel Mafunde

Introduction

Most green plants interact with animals in a variety of ways; particular animals may act as pollinators, or seed dispersers (which have advantages for the plant), but most commonly they act as herbivores, in which case the effect is obviously deleterious for the plant. Many plants therefore have anti-herbivore defences, either physical (spines, stinging hairs, etc.) or chemical (gums or resins, distasteful chemicals that deter feeding, or toxins). Such defences may be relatively ineffective against large vertebrate grazers but highly deterrent to the much more numerous and potentially more serious insect herbivores. This can result in very specialist interactions between particular insects and particular plants, so that only those few species that have mechanisms of avoiding the defence or of overcoming its effects will feed on a given plant species.

Acacia trees are a spectacular example of anti-herbivore defences; in most of the tropical and sub-tropical areas of the world they are a dominant part of the vegetation, and may offer almost the only green leaf in the dry seasons, so that they are bound to be grazed upon. Their spectacular large spines and prickles deter many antelope, and even larger grazers such as rhino and giraffe treat them with care. But they also have very small scale spines to impede the activities of herbivorous insects, and most of them contain chemical defences in the leaves as well. Thus they rarely get seriously defoliated.

While the main insect herbivores come from amongst the Coleoptera, Lepidoptera and Orthoptera (essentially the beetles, caterpillars and grasshoppers and their kin), many plants also have an association with ants, and a series of rather peculiar interactions occur, often of a mutually beneficial character, so that the phenomenon is called 'ant-plant mutualism'. There may be several key reasons for these special associations. In particular, ants are usually tiny and most of them cannot fly. They are therefore too small and slow moving to be good pollen carriers, yet they are highly attracted to the sugary nectar contents of flowers. Some plants use them instead as seed dispersers, but this only works for very small seeds. Ants are also highly social and potentially aggressive, so that when one worker is disturbed or excited many other workers can be recruited by the use of pheromones and an organised attack can be launched. So, how can plants best make use of the attributes of ants? Most obviously, they can use them as guards, and exploit their communal aggression in complex interactions with the plant's potential herbivores.

Ants as plant guards

A whole range of plants have 'extrafloral nectaries' (EFNs), giving out sugary secretions well away from any flower, often in the leaf petioles or blades or just on stems (Koptur 1992). Bentley (1977) provided an early review, stressing the great range of plant families and genera that have EFNs, and also giving the evidence that these nectaries really have a purpose in attracting and using ants as protection against herbivores. The ants stay on the plant because of the sugary reward, and without getting into the flowers; and they also use the plant as a source of insect prey so they will attack the plant's herbivores. This is a very non-specific interaction, and more or less any plant will do for any ant.

But there are also much more specific and much more interesting relationships where 'coevolution' between particular ants and specific plants appears to be occurring (Janzen 1966, Huxley 1986, Huxley & Cutler 1991). This phenomenon of ants as plant guards is best known in the genus *Acacia*, which is a large genus (>700 species) of tropical and savanna trees and shrubs worldwide, particularly in Central America, Africa and Australia (see Stapley, Chapter 23). In the first two continents, some of the spiny acacias—maybe about 10% of the species—have some of their thorns greatly expanded and these expanded 'pseudogalls' make homes for small colonies of ants. These are not true galls, because the ants do not actually induce swelling, they just move in through a pore once the swelling has happened, hollowing out the inside into a home for a complex mini-colony with a queen, nursery areas and living quarters. The ants concerned come particularly from the genera *Crematogaster* (Africa) and *Pseudomyrmex* (New World). The swollen-thorn acacias that participate in this are known as 'myrmecophytic' plants, and they are very obvious in the savanna ecosytems of east Africa, including those at Mkomazi.

The ants depend on the plant for shelter; but they also rely on it for food, provided by extrafloral nectaries on the stems or petioles (outside the flowers; Knox *et al.* 1986), and/or in a few cases by special proteinaceous food bodies at the tips of leaflets, called Beltian bodies, that are chewed by the ants. In return the plant gets protection, as the ants will aggressively attack any herbivore that tries to eat the plant leaves, not only other insects but even mammals such as giraffes, gazelles and even rhinos. Ants may also deter vegetation such as lianas and vines that may try to grow over their tree, by nipping them off. They tend to destroy all other seedlings growing within a basal ring around the tree's trunk, perhaps up to

1–2 m diameter. This cleared space may also help the *Acacia* tree to resist fire, a major factor in habitats such as Mkomazi.

The deterrence to herbivores by this alternative mode of biotic defence by ant-guards is considerable. In a Mexican species, Janzen showed that shoots with ants had 2.7% occupancy with herbivore insects, whereas non-ant-attended shoots had 38.5% herbivore occupancy in daylight. When an ant detects an intruding herbivore it not only bites at it itself but liberates alarm pheromone that brings in many other ants who also attack. At any one time 25% of the ants may be out of their thorns patrolling their host plants, and up to 50% (nearly all the old workers) may come out in response to attack and to the presence of the alarm pheromone. Both partners (the ants and the tree) clearly benefit.

The swollen thorn acacias are especially those that are in leaf all year round, since they may provide the only green leaf left at the end of a savanna dry season and are therefore under heavy herbivore pressure. *Acacia* species that have ant-guards often invest less heavily in chemical defences; for example, swollen-thorn acacias that are unoccupied by ants are severely defoliated (Janzen 1966), and they taste much blander to the human tongue than the normal bitter *Acacia* leaf. The trees are making some saving on other defences by 'paying' the ants instead. Thus biotic defence by ants might be cheaper than either chemical or physical defence.

Ant-guards and pollinators in Mkomazi

The work that follows is described fully in Willmer & Stone (1997), and details of pollinator activity in Mkomazi acacias are given in Stone, Willmer & Rowe (1998).

At least 16 species of *Acacia* occur in Mkomazi (see Coe *et al.*, Chapter 5 and Stone *et al.*, Chapter 20), of which at least two have pseudogalls housing ant-guards (see Stapley, Chapter 23). While the general phenomenon has been well known for many years, and has attracted extensive scientific study in terms of anti-herbivore defence, one particular aspect remains unexplored: what effects do the patrolling ants have on pollinators? There is a clear conflict of interest here, since the tree needs insects to cross-pollinate its flowers and ensure its reproductive success, but employs ants to deter all incoming insects and increase its vegetative success. How do pollinators get 'through' the defence and do their job? Do the ants behave differently towards these potentially beneficial kinds of visitor? Of course the tree could just open its flowers at times when the ants were not about; but if there are such times, herbivory ought to become a real problem again, and the trade-offs between defence and pollination success would be quite complex.

Acacia zanzibarica is a classic ant-guarded tree where resident ants—usually a species of *Crematogaster*—spend all their time patrollling around the plant, and will attack any other insect they find. The first part of our study therefore involved time spent just watching the flowers very carefully, and scoring all visitors of all kinds to a particular group of flowers throughout the daylight hours. We were also measuring pollen availability and depletion through time, nectar concentrations and volumes on offer, and flower visitor behaviours.

The flowers on this species (see Plate 29) are golden yellow, presented in spherical inflorescences of many small florets in a manner characteristic of the subgenus *Acacia*. The anthers protrude and present their pollen right at the surface where it is accessible to any visitor. They open from the bud stage around 05.00 hours, and start to dehisce (present pollen) around 08.00–09.00 h. They are fresh and pristine through their first day of life, with the pollen gradually being lost; but they persist on the tree for at least one more day as slightly tattered yellow balls, having little or no pollen but with stigmas now protruding beyond the withered anthers. After day two the whole structure starts to wither and fall off, leaving some small proportion of the individual flowers with setting fruit.

The daily pattern in ant activity is very clear. Figure 21.1 shows that ants patrol their twigs from the very early morning, but start to decline in numbers from 09.00–10.00 h, persisting at low numbers till around 13.00–14.00 h, then increasing again to be abundant through most of the afternoon. But we noticed that in the morning they are mainly patrolling the old (day two) flowers and the unopened buds, while in the afternoon they switch to patrolling the young (day one) flowers as well. This is reflected in the length of visits to each kind of flower through the day, shown in Figure 21.2. In the morning, visits to young flowers when they do occur are extremely short, and ants tend to appear agitated, with rapid antennal tapping and mandible activity, and often with the abdomen cocked upwards. In the afternoon their visits to young flowers are much longer and their behaviour is more normal.

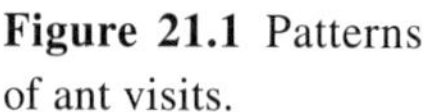

Figure 21.1 Patterns of ant visits.

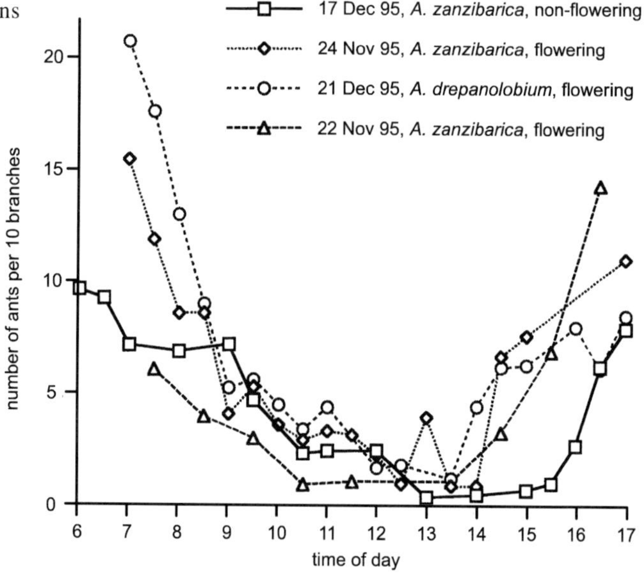

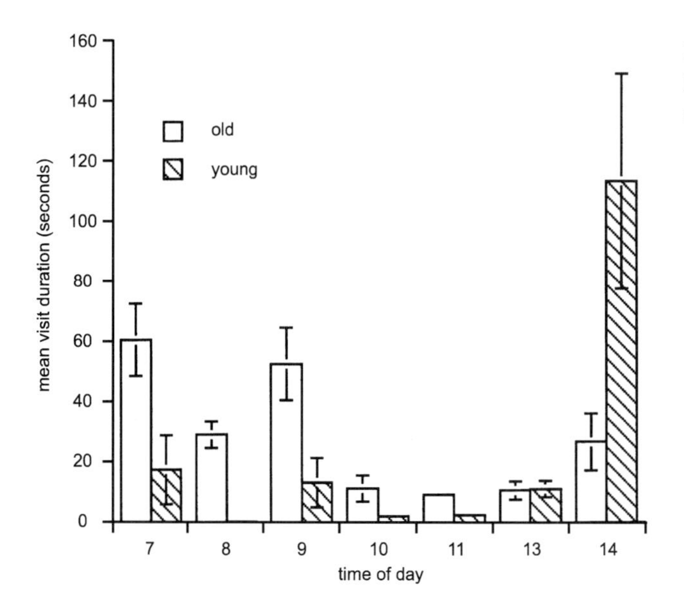

Figure 21.2 Duration of ant visits.

So when do flower visitors and potential pollinators come to the trees? Flies (such as calliphorids and syrphids—see Stone *et al.*, Chapter 20, for graphs of daily activity patterns), and some small beetles, come in rather early in the day, and make long visits, mostly pollen-gathering; but these are not moving between flowers much, and are not likely to be good pollinators. If found by ants before about 10.00 h they do get nipped at and fought off. Butterflies have a peak through the hotter hours, purely taking nectar. But above all bees (solitary bees, from the families Megachilidae and Anthophoridae) are important, arriving on the tree from the time of first dehiscence and with a big peak in activity at 11.00–14.00 h (see Stone *et al.*, Chapter 20). They are by far the most efficient visitors, making neat fast somersaults all over the surface of the inflorescence, raking in the superficial pollen to their collecting apparatus (either on their legs or on their abdomens) and then moving on to the next inflorescence.

Bees and butterflies, because of the timing of their visits, almost never meet an ant. In other words, there are brief times each day when ants are not active, and these correspond to times of dehiscence and times of visitation of flowers. However, it may be that this is all just fortuitous. Ants may go away in the midday hours just because it is too hot for them to patrol, and that gives the pollinators a chance. But this hypothesis is unlikely to be true for several reasons. Firstly, on purely theoretical grounds if it is too hot for a small walking ant, it ought to be much too hot for a larger flying insect. Secondly, at the times of ant absence we used thermocouples to show that it is often much hotter inside the pseudogalls occupied by the ants than it is out on the branches. And thirdly we recorded very much the same pattern of ant behaviour on different days of varying weather patterns, including days when it was cool and overcast.

This leaves the other obvious possibility, that the young flowers actively deter ants from patrolling them or even being anywhere near them, so that during the early hours of flowering the ants retreat to their pseudogalls temporarily. We began to wonder whether the flowers released a chemical that the ants did not like; specifically, whether the freshly dehiscing flowers had a volatile ant-deterrent, possibly even within the pollen itself. This would also explain some other observations; in particular the curious ant behaviour on young flowers, and why the ants did not leave the inflorescences on a day when it had rained in the first half of the morning, suggesting that the rain may have washed away the chemical cue.

Our hypothesis, then, was that the *Acacia* tree manipulates its own ants using chemical signals, so that they briefly allow visitors in to the flowers for a couple of hours, and then can re-invade the branches and defend the newly pollinated flowers from further disturbance, protecting them right through the seed-set phase. This allowed us to make and to test two key predictions: a) Something on or in the dehiscing flowers is anathema to ants. It could be purely physical cues, but this seems unlikely. If it is chemical, and volatile, then it might be transferable, i.e. wiping an unattractive fertile dehiscing flower onto the surfaces of an older one might make the latter unattractive to ants too; b) Ants should actually aid pollination; they keep the very young flowers pristine and unchewed, then they allow pollinators in at peak flower fertility, before going back and keeping off any further chewers. The abundance of ants should therefore be positively related to seed set on any particular branch.

To test the first prediction, on one particular day at around 07.00 h we set up and labelled the various treatment categories that we needed:

(a) Untouched young (day 1) flowers;
(b) Untouched old (day 2) flowers;
(c) Young flowers wiped onto old flowers (using flowers from the same tree, only handled with tweezers, to avoid spurious chemical effects);
(d) Old flowers wiped onto old flowers (as a control for purely handling effects).

Then we watched the treated flowers very carefully over the next few crucial hours, recording all visits. Figure 21.3 shows these results. Essentially, old flowers (b) get more visitors, and wiping them in itself (d) does not make them any more unattractive. Wiping them with a young flower (c) makes them nearly as unattractive as a normal young flower (a) and they are largely avoided. This fits with the idea that some chemical signal has been transferred from the surface of young flowers; either a liquid or solid exudate from the flowers that contains a volatile chemical, or perhaps even the pollen itself, which in the genus *Acacia* occurs as rather large 'polyads' of accreted pollen grains. We think the hypothesis that the pollen itself bears the deterrent is most likely, because then depletion of the pollen by visitors would directly trigger the release of the ants from repellency, and ant activity could be directly tied to pollinator activity.

To test out second prediction we used a group of 11 trees, growing close to-

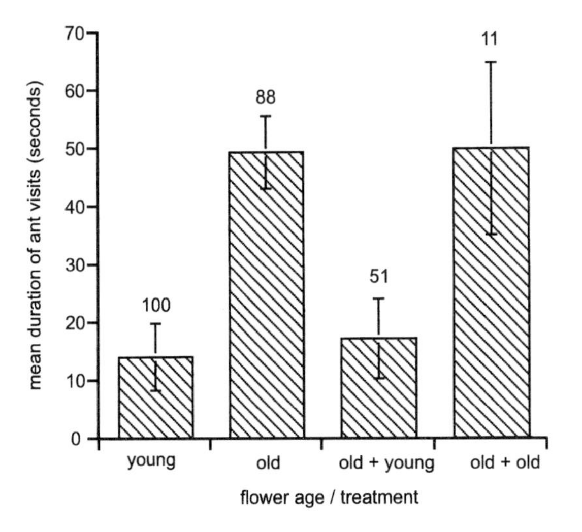

Figure 21.3 Comparison of ant visits for different treatments (numbers above bars show sample sizes).

gether to reduce other variables such as water stress or soil type, and selected and tagged equal lengths of five branches on each tree; then we scored ant activity every hour on each branch. We also recorded the numbers and sizes of pseudogalls on each branch, the numbers of flowers of each age, the height and aspect of the branch, and the tree size and estimated age, to try and rule out other possible causes of variation in seed set. At the end of the day we scored fruit set at each position on the tagged lengths of the branches, giving a range 0–11 seeds per inflorescence. Figure 21.4 shows our results. The mean ant score per branch is significantly and positively correlated with the mean fruit set per inflorescence. No other factor measured is a significant predictor of fruit set. From this we conclude that on these ant-guarded plants the presence of ants does indeed assist fruit set.

This means the *Acacia* ant-guards are not 'potential problems' for pollinators, and not even neutral in terms of pollination; while still being effective as anti-

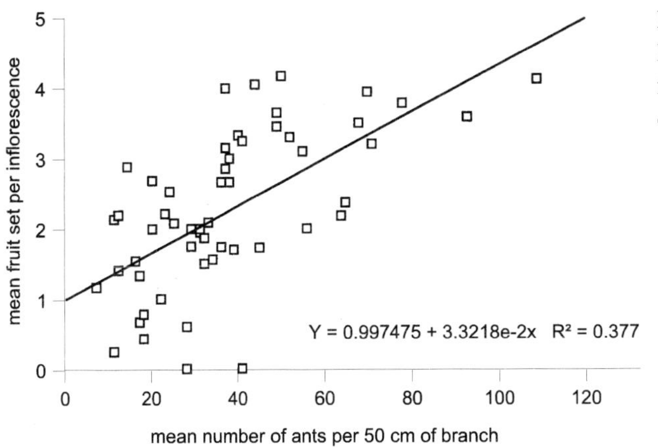

Figure 21.4 Relationship between fruit set and number of ants.

herbivore systems (i.e. increasing the plant's vegetative growth), they are potentially beneficial to the plant's reproductive efforts too, by putting their protective mantle over the pollinated flowers. The whole mutualism is more sophisticated than suggested in existing literature, and provides a complex and very neat series of benefits for all the participants.

Further implications of this work

This is the first demonstration of interactions between guarding insects and pollinators, and a rare documented case of ant-repellency in flowers (Feinsinger & Swarm 1978). Our work at Mkomazi is therefore leading us into new fields of research. We are hoping to extend our field studies of the ant-acacia-pollinator interactions to other east African acacias, and to carry out parallel experiments on ant-acacias in west Africa and in Mexico. This will provide an indication of the generality of ant-repellent floral volatiles in ant-*Acacia* interactions. By comparison of ant responses to flowers from ant-associated species and species without ants we will be able to discover whether ant repellency is a general feature of *Acacia* flowers during pollen release, or whether the pattern we have found is specific to ant-guarded species and resident ants. We can explore the possible effects of age of tree, as ant communities may change with age (Young *et al.* 1997).

We are also pursuing the idea of ant repellency in flowers more generally, since there are obvious good reasons why almost any plant may wish to exclude ants from its flowers. We know that a few plants that flower at ground level do have ant-repellent nectar (Prys-Jones & Willmer 1992), and we have preliminary evidence that some temperate flowers from the UK also have pollen with deterrent properties (Willmer, unpublished). We are hoping to be able to identify and localise the chemical(s) responsible for this ant deterrence, in *A. zanzibarica* and in other systems. We can explore how far the chemicals are a deterrent to some or to all ants, or are specific to resident ant-guards, and whether they are (as we suspect from our observations) mimicking a natural ant alarm pheromone. All of this would enhance our understanding of ant-pollen interactions, and open up new research areas on floral repellency and ant-pollinator competition, exciting and largely unexpored fields for the future.

References

Bentley, B.L. (1977) Extrafloral nectaries and protection by pugnacious bodyguards. *Annual Review of Ecology and Systematics* 8: 407-427.

Hocking, B. (1970) Insect associations with the swollen-thorn acacias. *Transactions of the Royal Entomological Society of London* 122: 211-255.

Huxley, C.R. (1986) Evolution of benevolent ant-plant relationships. In: Juniper

B.E. & Southwood T.R.E. (eds.) *Insects and the Plant Surface*. Edward Arnold. pp. 257-282.

Huxley, C.R. & Cutler, D.F. (1991) *Ant-Plant Interactions.* Oxford Scientific Publishing.

Janzen, D.H. (1966) Coevolution of mutualism between ants and Acacia in Central America. *Evolution* 20: 249-275.

Knox, R.B., Marginson, R., Kenrick, J. & Beattie, A.J. (1986) The role of extrafloral nectaries in Acacia. In: Juniper B.E. & Southwood T.R.E. (eds.) *Insects and the Plant Surface*. Edward Arnold. pp. 295-307.

Koptur, S. (1992) Extrafloral nectaries mediate interactions between insects and plants. In: Bernays, E. (ed.) *Insect-Plant Interactions*. CRC Press, Boca Raton. pp. 81-129.

Feinsinger, P. & Swarm, L.A. (1978) How common are ant-repellent nectars? *Biotropica* 10: 238-239.

Prys-Jones, O.E. & Willmer, P.G. (1992) The biology of alkaline nectar in the Purple Toothwort (*Lathraea clandestina*): ground level defence. *Biological Journal of the Linnean Society* 45: 373-388.

Stone, G.N., Willmer, P.G. & Rowe, J.A. (1998) Partitioning of pollinators during flowering in an African *Acacia* community. *Ecology* 79: 2808-2827.

Willmer, P.G. & Stone, G.N. (1997) Ant deterrence in Acacia flowers: how aggressive ant-guards assist seed-set. *Nature* 388: 165-167.

Young, T.P., Stubblefield, C.H. & Isbell, L.A. (1997). Ants on swollen thorn Acacias: species coexistence in a simple system. *Oecologia* 109: 98-107.

CHAPTER 22

Arthropod diversity and the tree flora of Mkomazi

George C. McGavin

Background

During the last 15–20 years, a number of studies have provided a more complete understanding of insect communities in tropical forest tree canopies (Erwin 1982 & 1983, Moran & Southwood 1982, Adis *et al.* 1984, Stork 1987a & 1987b, Morse *et al.* 1988, Basset & Kitching 1991, Stork 1991, Basset 1996). A review of the history of these and other major studies of arthropod diversity in tropical forest canopies is given by Erwin (1995). Despite the fact that savannas cover over 40% of the surface area of the tropics (Cole 1986, Solbrig 1996), the insect communities of tropical savanna tree canopies are still virtually unknown and there are practically no estimates of insect diversity in savannas in general (Lewinsohn & Price 1996). With the exception of two studies, that of West (1986) who sampled arthropods from *Acacia* and *Commiphora* canopies in Kora National Reserve, Kenya and Wagner (1997) who studied beetles sampled from the canopies of *Lannea fulva*, *Teclea nobilis* and *Carapa grandiflora* in forests in Rwanda and east Zaire, there have been no large scale studies conducted in the African continent.

The pressures of increasing human population and associated large scale habitat destruction (e.g. fire and over grazing) are reducing the time available for an assessment of the insect fauna of savannas. Ecological inventories are an essential tool for environmental management and the assessment of habitats for conservation (Campbell 1993). As insects have a much greater impact on terrestrial habitats than all other animal groups put together and are a major component of diversity they must form a very large part of any such effort. Studying arthropod communities in savanna habitats can be difficult because of a lack of clear habitat boundaries (Lewinsohn & Price 1996) but savanna trees provide an excellent setting for insect community research, because they can be considered a discrete ecological unit (Southwood & Kennedy 1983). Trees also have great niche diversification because of structural complexity (Lawton 1978, Lawton & Price 1979, Lawton 1986), are a stable resource (Southwood 1978) and their inhabitants are all more or less trophically interlinked (Moran & Southwood 1982). As Moran & Southwood (1982) conclude, "in almost all respects it is easier to sample arboreal communities more completely, more widely and more accurately, [...], than it is to sample other very complex communities". This is especially true for savanna tree canopies where there might be no overlap with neighbouring trees.

Methods

The Mkomazi Game Reserve (MGR) was visited on five separate occasions to collect arthropods from tree canopies using an insecticidal knockdown technique. Samples were taken during the following five periods:

Trip 1: 28 July 1994–14 August 1994
Trip 2: 4 April 1995–21 April 1995
Trip 3: 29 December 1995–18 January 1996
Trip 4: 28 March 1996–3 April 1996
Trip 5: 2 January 1997–13 January 1997

Sampling was carried out using a Hurricane Minor petrol-driven mist blower (Cooper-Pegler Ltd.) fitted with an ultra low volume delivery nozzle and charged with undiluted Pybuthrin 216 (Roussel Uclaf). The advantage of this method over other mass-collection techniques such as fogging is the degree to which the mist can be directed accurately into the canopy from ground level. Pybuthrin 216, a pyrethroid formulation synergised with piperonyl butoxide, is ideal for use in the field since it gives rapid knockdown with non-persistency. Sampling was carried out whenever possible in dry, still conditions. Experience showed that best time of day was between dawn and 10:00 hours after which time winds, largely katabatic, made sampling impossible. Foliage above 10 m high was not sampled in any case.

Arthropods knocked down were collected on purpose-built, square, funnel-shaped trays (each 1 m^2). The trays, made from rip-stop nylon balloon fabric, braced with 16 gauge aluminium tube, were tied to wooden stakes by strips of inner tube (see Plate 18). For small trees whose foliage was too low or spiny to permit the use of suspended nylon trays, plastic washing up bowls (each 0.16 m^2) were used instead.

Trees whose canopies were isolated from surrounding trees were selected for sampling and the calculation of canopy cover area assumed a near-circular canopy. Collecting trays or bowls were placed below the tree in positions likely to maximise samples (branches which were bare or did not appear healthy were avoided). Canopy foliage above the collecting trays was sprayed for a minimum of 15 seconds in three five-second bursts from different directions. When trees were in full leaf, the spray time was increased to a maximum of 30 seconds in three bursts to ensure thorough penetration through the canopy. For most trees one hour was used as the standard drop time period as most material was knocked down in the first 15 minutes and nothing was collected after 45 minutes. For some trees, such as those less than 3 m high or with a small or low canopy, the sample time was reduced to

30 minutes. Again the vast majority of the total catch was obtained during the first quarter. During the sampling period GPS coordinates and measurements of tree height, canopy cover and tree girth were recorded. The height of trees over 2.5 m was estimated, by visual comparison against measured heights. Canopy depth was not estimated. Other factors such as the presence or absence of flowers and seed pods were also recorded. At the end of the sampling period all material and fallen foliage was gently brushed into the collecting jars and pooled. Catches were examined, separated from debris and plant material and transferred to 70% alcohol (see Plate 19).

Arthropods were initially sorted to class or subclass. Insects were sorted to the level of order, family and then morphotyped into Recognisable Taxonomic Units (RTUs). The classification followed is that given by CSIRO (1991) Larvae have been excluded from the analyses presented here to avoid counting two developmental stages as different RTUs. Five individuals of each RTU were measured to the nearest 0.01 mm and the median was then taken to calculate the dry biomass per RTU using the formula given in Moran & Southwood (1982).

While the analysis of the samples continues, reference material will be kept in alcohol. At the end of the study, material will be dry-mounted as appropriate and a voucher collection will sent to Tropical Pesticides Research Institute in Arusha. Material representing any new taxon described will be deposited in Oxford, London and Arusha. It is likely that somewhere between 20–40% of the RTUs sampled will be undescribed.

The spiders and beetles sampled during the first field trip have been analysed separately (see Russell-Smith, Chapter 11 and Davies, Chapter 14) as have ants (see Roberston, Chapter 19).

Trees sampled

Quantitative samples of arthropods have been obtained from the canopies of 266 trees representing 30 species (Table 22.1). The total canopy area sampled for all tree species is a little over 485 m^2. In total, an estimated 500,000 specimens have been collected and preserved. Full details of the samples are given in the Appendix at the end of the chapter. This work represents the biggest single study of savanna tree canopies ever undertaken.

Composition of the canopy fauna

As the tree flora is dominated by species of *Acacia* and *Commiphora*, these were sampled extensively and have been studied in detail first. So far, six species of *Acacia* (Table 22.2a) have been analysed by Oliver Krüger (Krüger 1997, Krüger & McGavin 1997, 1998a & b) and three species of *Commiphora* (Table 22.2b) by Jennifer Slyker (Slyker 1998).

Table 22.1 Tree canopy sampling in Mkomazi Game Reserve.

tree species	number sampled	area sampled (m^2)
Acacia ancistroclada	1	4.96
Acacia brevispica	3	2.56
Acacia bussei	5	10.00
Acacia drepanolobium	15	8.12
Acacia etbaica	1	3.00
Acacia mellifera	13	22.35
Acacia nilotica	28	53.60
Acacia reficiens	21	45.56
Acacia senegal	37	80.76
Acacia thomasii	1	0.64
Acacia tortilis	14	25.20
Acacia zanzibarica	33	45.08
Albizia anthelmintica	1	4.00
Apodytes dimidiata	2	2.00
Boswellia sp.	2	2.00
Combretum molle	17	20.00
Commiphora africana	3	4.00
Commiphora campestris	30	83.00
Commiphora holtziana	3	6.00
Craibia brevicaudata	1	2.00
Craibia brownii	1	4.00
Dichrostachys cinerea	8	6.40
Drypetes parvifolia	3	3.00
Grewia sp.	2	2.56
Heywoodia lucens	9	16.00
Lannea schweinfurthii	6	17.28
Lannea triphylla	1	1.28
Melia volkensii	1	2.00
Ochna holstii	2	4.00
Terminalia brownii	2	4.00
totals		
30 species	266	485.35

Table 22.2a Summary statistics of the *Acacia* trees analysed to date.

species	no. sampled	area sampled (m^2)	RTUs			individuals		
			mean	min.	max.	mean	min.	max.
A. etbaica	1	3.00	58	–	–	652	–	–
A. mellifera	4	6.28	62	35	79	386	210	592
A. nilotica	12	23.72	98	65	188	1,445	400	2,998
A. reficiens	4	8.28	86	65	103	1,099	814	1,485
A. senegal	8	16.00	107	78	145	1,786	1,497	2,716
A. tortilis	2	4.00	105	84	125	872	439	1,304

Table 22.2b Summary statistics of the *Commiphora* trees analysed to date.

species	no. sampled	area sampled (m^2)	RTUs			individuals		
			mean	min.	max.	mean	min.	max.
A. etbaica	30	83	37.0	11	75	459	22	2,606
A. senegal	3	4	32.3	18	41	360	267	427
A. tortilis	3	6	26.7	21	111	166	111	497

Material sampled between 30 December 1995 and 18 January 1996 from 31 trees belonging to six species of *Acacia* has been analysed. The detailed results are given in Krüger & McGavin 1997, 1998a & b. A total of 41,099 insect specimens, belonging to 14 orders, 133 families and 492 RTUs were recognised from the 61.28 m^2 area sampled. The average insect density was 666 per m^2 (if larvae are included this figure rises to 800–900 per m^2) and the mean number of RTUs per tree was 93. A total of 20,535 specimens belonging to 12 orders, at least 80 families and 605 RTUs have been recognised from the three species and 93.0 m^2 of *Commiphora* sampled.

One of the problems inherent in survey work based on a small number of sampling occasions is knowing what proportion of the total fauna is represented in the collections. For the six species of *Acacia* analysed to date, a graph of cumulative area/RTU relationship for the entire fauna sampled shows that, according to the two true richness estimators, between 77 and 100% of the insect fauna is represented in the samples. Similar percentages were found for Hemiptera, Coleoptera and Hymenoptera (between 80 and 100%) with the exception of Diptera, where the one estimator indicated that only 53% of the total was represented (Figure 22.1). For *Commiphora*, analysis indicates that while Orthoptera are probably well represented, taxa such as the Hemiptera, have not yet been fully sampled.

For the *Acacia*s analysed, the RTU richness was highest for the Hemiptera, followed by the Coleoptera, Hymenoptera and Diptera (Figure 22.2a). The most diverse families were the planthoppers (Cicadellidae), leaf beetles (Chrysomelidae), plant bugs (Miridae) and ants (Formicidae) with 7.9, 5.5, 4.1 and 2.9% of the total RTU richness respectively.

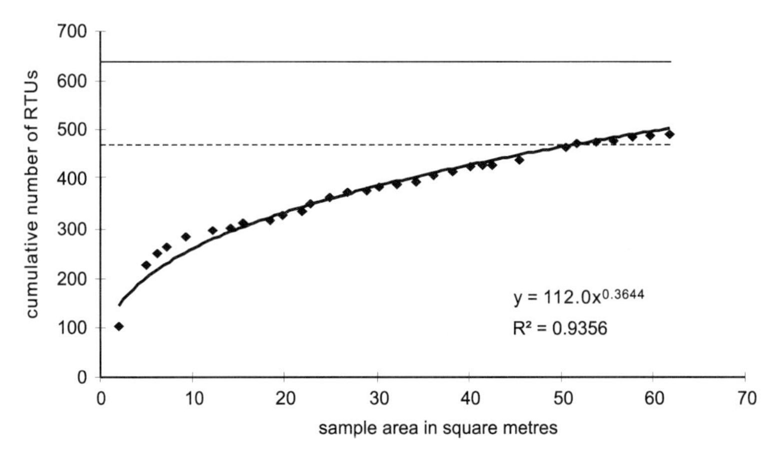

Figure 22.1 Cumulative area / RTU curve for the entire insect fauna sampled from six species of *Acacia*. The two horizontal lines are asymptotes of two true richness estimators, the dashed line from a negative exponential regression (Holdridge *et al.* 1971) and the continuous line from a non-parametric approach (Chao 1984).

For the *Commiphora* species analysed, the RTU richness was highest for the Coleoptera, Hymenoptera and Diptera and Hemiptera, accounting for 29, 27, 17 and 16% of the total RTU richness respectively (Figure 22.2b).

In terms of abundance, thrips (Thysanoptera) and ants (Formicidae) become much more important than Coleoptera but the Hemiptera had the highest abundance in *Acacia* canopies (Figure 22.3a). The families with the highest abundances were the common thrips (Thripidae) ants (Formicidae), planthoppers (Cicadellidae)

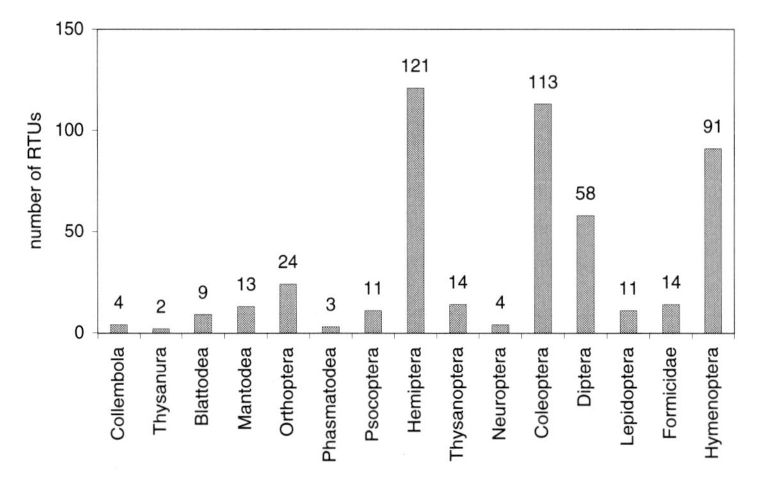

Figure 22.2a Distribution of RTU richness among the 14 orders recorded from six species of *Acacia*. Ants (Formicidae) are separated from the rest of the Hymenoptera as their biologies are diverse and they have a significant impact of the composition of insect communities (Krüger & McGavin 1998b).

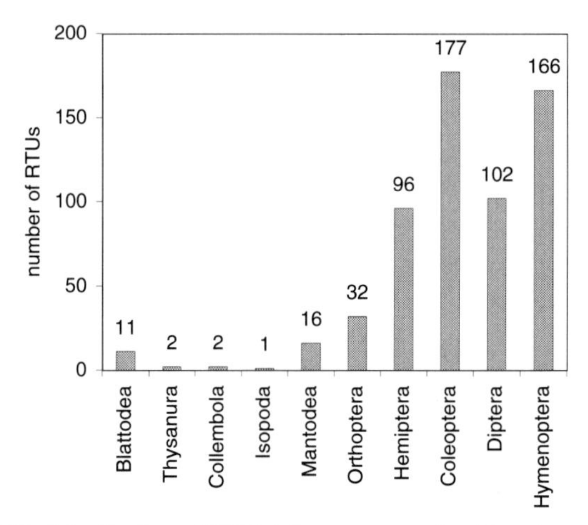

Figure 22.2b Distribution of RTU richness among the 10 major taxa so far analysed from three species of *Commiphora*. Ants are included in the Hymenoptera is this analysis.

and plant bugs (Miridae) with 19.2, 18.2, 12.3 and 12.1% of the total abundance respectively. In *Commiphora* canopies, ants sap-sucking bugs and woodlice were the most numerically abundant taxa comprising 30, 24 and 15% of the total respectively (Figure 22.3b).

The distribution of biomass among the taxa has been examined for the *Acacia* samples (Figure 22.4). Ants constitute the highest fraction of the total biomass followed by larvae and praying mantids (Mantodea). Although the Mantodea had a small share of the RTU richness and abundance they were among the biggest insects collected.

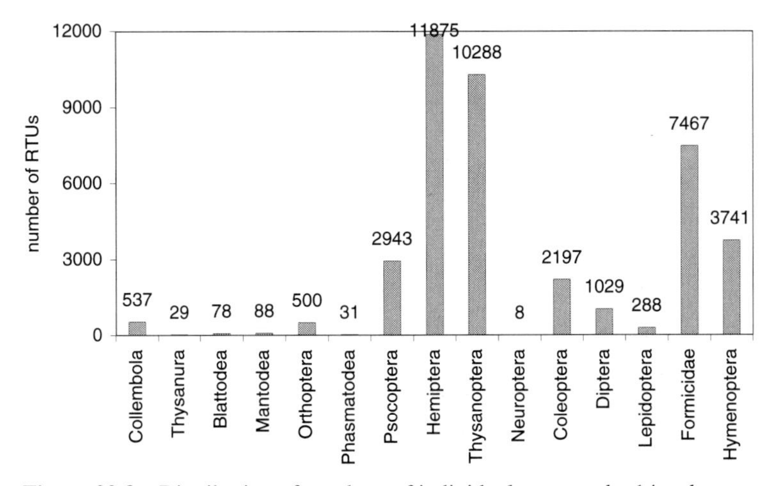

Figure 22.3a Distribution of numbers of individuals among the 14 orders recorded from six species of *Acacia*.

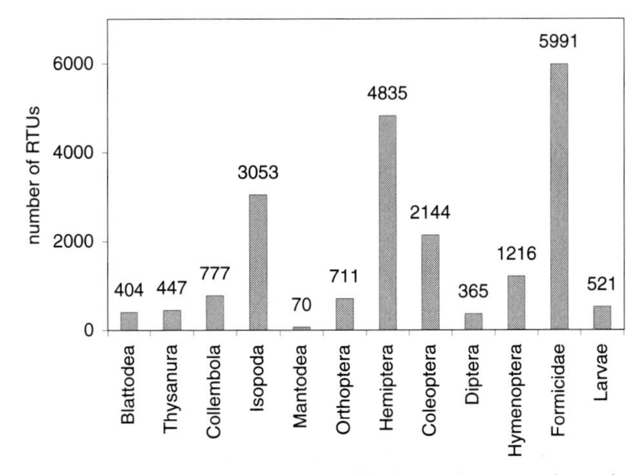

Figure 22.3b Distribution of numbers of individuals among the major groups so far analysed from three species of *Commiphora*.

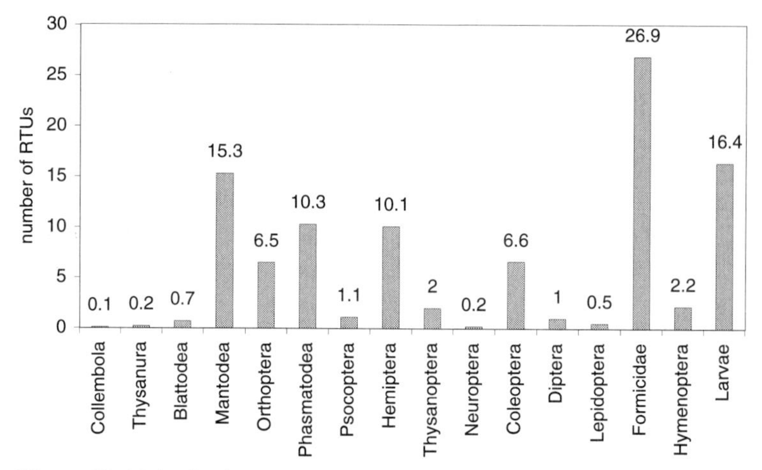

Figure 22.4 Distribution of percentage biomass share among the 14 orders recorded from six species of *Acacia*.

There is a significant relationship between body weight and RTU richness such that the entire fauna and four major orders (Hemiptera, Coleoptera, Diptera and Hymenoptera) sampled from *Acacia* species show bell-shaped curves with maxima of RTU richness and abundance at intermediate body weights. These findings support recent evidence that the peak in species richness might occur at intermediate body weights. Data are not yet available for the *Commiphora* canopy samples.

Species abundance distributions

Rank/abundance plots are a useful way of presenting data dealing with the relative abundances of a number of species. A rank / abundance graph for the entire *Acacia* fauna is shown in Figure 22.5 ($R^2 = 0.984$). For the four main orders (Hemiptera, Coleoptera, Diptera and Hymenoptera) the results were similar. *Commiphora* communities analysed so far show a very similar pattern ($R^2 = 0.973$) with a few species (ants and woodlice) being very abundant and a great number of species being much less abundant. The Coleoptera and Hemiptera are dominated by a few species. In the Coleoptera a single species of flea beetle (Halticinae) was the most abundant RTU. In the Hemiptera the most species-rich families were the plant-hoppers (Cicadellidae), plants bugs (Miridae) and jumping plant lice (Pysllidae).

For the entire insect fauna from *Acacia* species, the observed abundance distribution is best described by a log normal model ($R^2 = 0.908$). For the entire fauna from *Commiphora* a log series model provided a better fit to the observed abundance distribution than a log normal model ($R^2 = 0.83$). The data from *Commiphora* must be interpreted with caution at present as two large orders, the Hymenoptera and Diptera have not yet been incorporated.

That a log normal distribution fits the most complete data obtained best makes

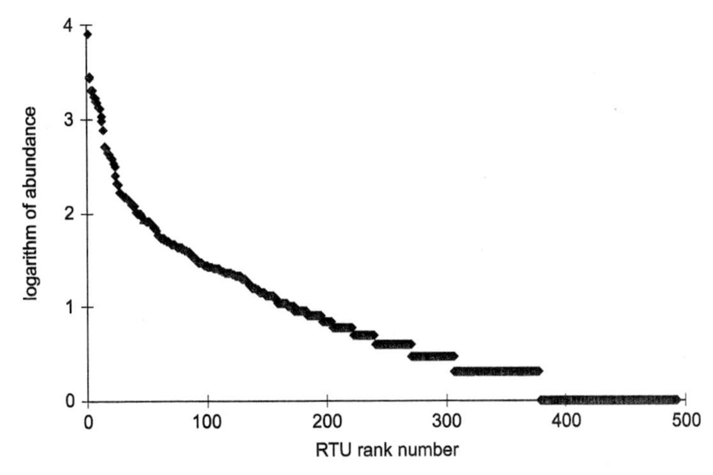

Figure 22.5 Rank/abundance plot for the entire *Acacia* fauna. RTUs are ranked from the most abundant (rank no. = 1) to the least abundant (rank no. = 490). The y-axis is shown on a log scale for convenience.

this system similar to most others studied and suggest that these canopies contain mature and varied insect communities. *Acacia* and *Commiphora* canopies are likely to represent a complex system where the interaction of a number of factors at random creates the observed log-normal distribution.

While species richness in Mkomazi tree canopies was lower compared to rain forest canopies, comparison with other studies shows that insect density is higher that in some rain forest or temperate tree canopies (Table 22.3).

Table 22.3 Comparison of the Mkomazi data to data from other studies in tropical and temperate regions. The 4th and 5th columns give the insect density per m^2 and the individual to species (=RTU) ratio*. ‡ The figures here include data for the Diptera and Hymenoptera.

source	country	habitat	ind. / m^2	ind. : species*
this study (*Acacia* spp.)	Tanzania	savanna	666.3	41,099 : 492
this study (*Commiphora* spp.)	Tanzania	savanna	203.8	20,535 : 605‡
West (1986)	Kenya	savanna	51.1	–
Adis *et al.* (1984)	Brazil	rain forest	32–161	–
Basset (1991b)	Australia	subtropical forest	19–46	–
Southwood *et al.* (1982)	South Africa	subtropical forest	76	–
Southwood *et al.* (1982)	Britain	forest	389	–
Watanabe & Ruaysoongnern (1989)	Thailand	rain forest	123–256	–
Stork (1991)	Borneo	rain forest	51–218	–
Basset (1991a)	Australia	subtropical forest	–	51,600 : 759

Guild structure

For the six species of *Acacia*, insects were assigned to nine major ecological guilds (*sensu* Moran & Southwood 1982) and the distribution of species diversity and biomass was analysed. The data cover a 1,000,000-fold body weight range. Between tree variation in the proportion of RTUs and biomass in most guilds is large

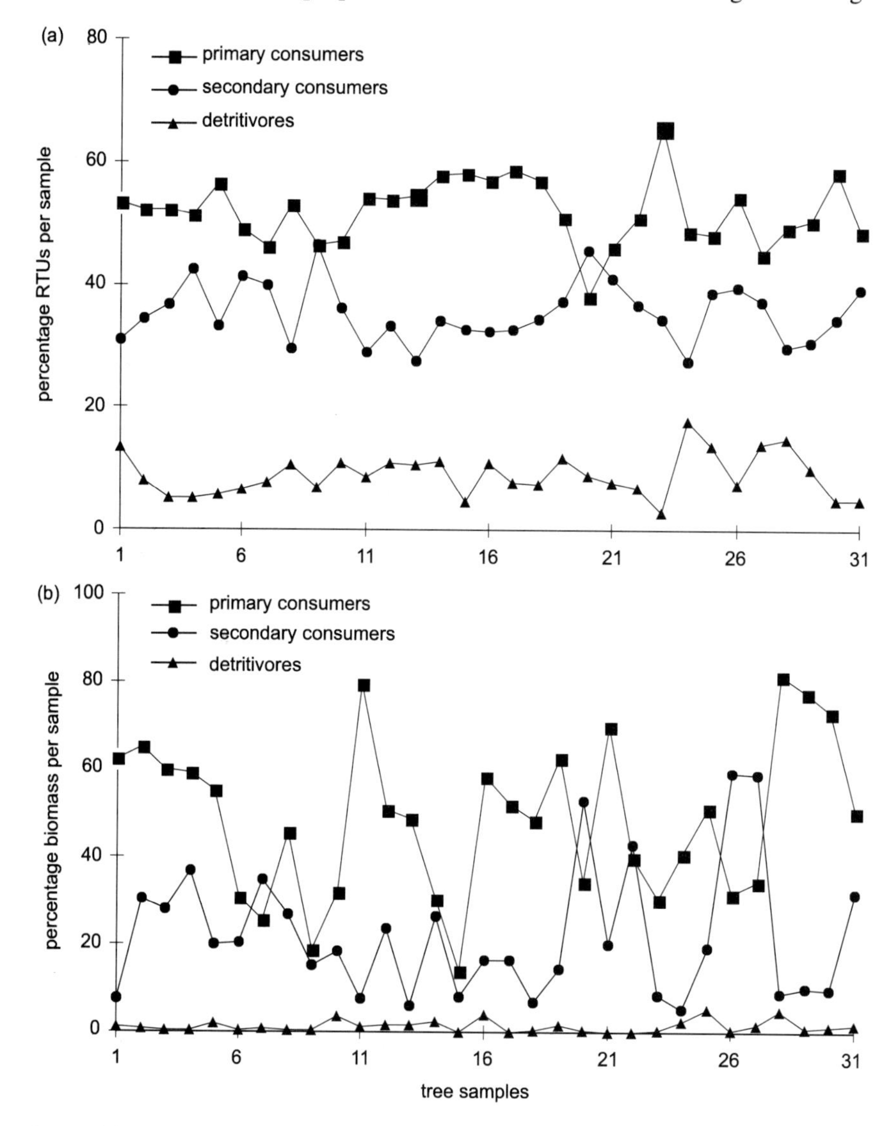

Figure 22.6 (a) Between-tree variation in the percentage RTU of three trophic levels. Primary consumers = phytophagous chewers (*pc*), phytophagous sapsuckers (*ps*), phytophagous nectarivores (*pn*), epiphyte grazers (*e*); secondary consumers = predators (*p*), parasitoids (*pa*); detritivores = scavengers (*s*). (b) Between-tree variation in the percentage biomass for the trophic levels.

and covers a 2.5-fold range. RTU diversity was highest in the phytophagous sapsucker guild (*ps*: mainly Hemiptera and some Diptera), followed by the parasitoid (*pa*: Hymenoptera) and phytophagous chewer (*pc*: Orthoptera, Phasmatodea and some Coleoptera) guilds. Biomass share was highest in the ant guild, followed by the *pc* and the *ps* guilds. This structure is consistent with the expectations of energy flow in communities, since the biomass share of predators (*p*), parasitoids and scavengers (*s*) was much lower than that of herbivores.

To extract general patterns of variation between trees, guilds were grouped into primary consumers, secondary consumers and detritivores and, while the level of between tree variation decreases greatly in the RTU composition (Figure 22.6a), it is still apparent for the biomass shares (Figure 22.6b).

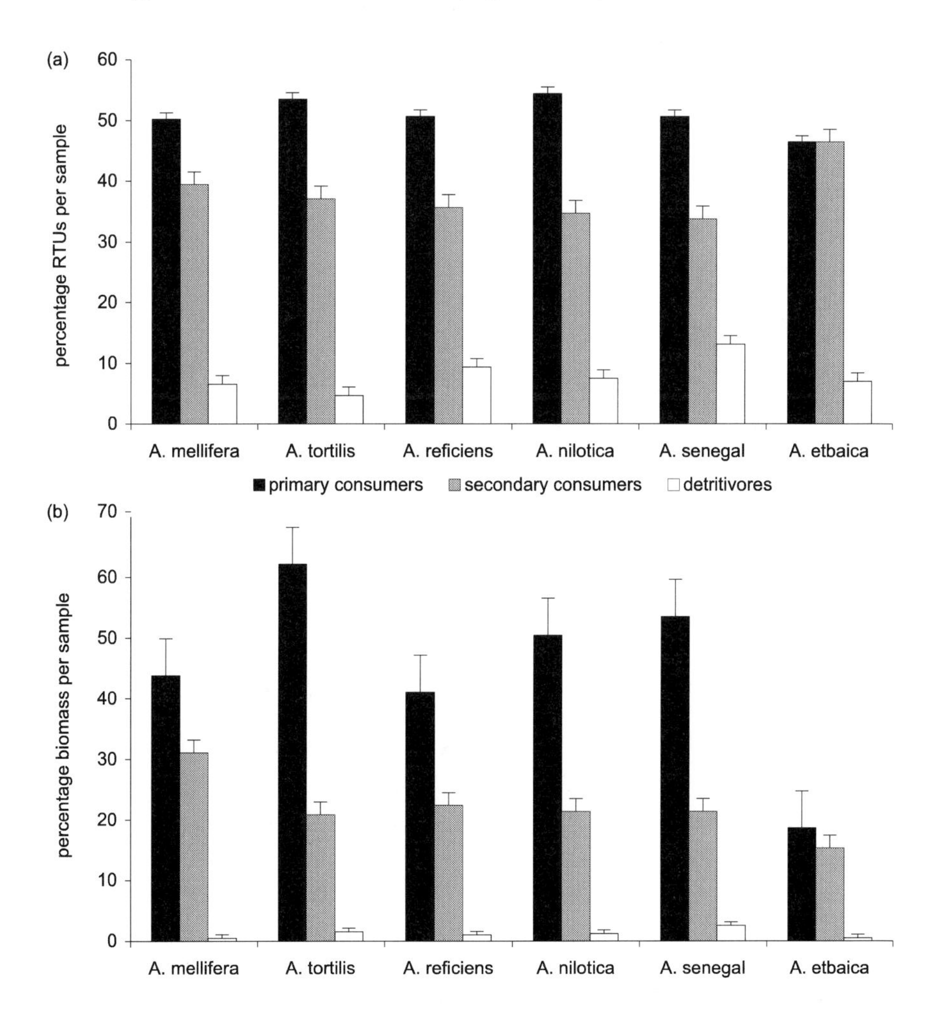

Figure 22.7 (a) Mean percentage RTUs in the three trophic levels for each of the six *Acacia* species. (b) Mean percentage biomass in the three trophic levels for the six *Acacia* species.

A comparison between the six *Acacia* species with regard to RTU composition reveals no significant difference for primary and secondary consumers (Figure 22.7a, ANOVA, $F_{5,25} = 1.022$, $P = 0.426$ and $F_{5,25} = 2.381$, $P = 0.067$ respectively), while there is a significant difference for the detritivores (ANOVA, $F_{5,25} = 6.994$, $P < 0.0001$). This consistency also holds true for the biomass percentage of primary and secondary consumers and detritivores (Figure 22.7b, ANOVA, $F_{5,25} = 1.136$, $P = 0.367$, $F_{5,25} = 0.291$, $P = 0.913$ and $F_{5,25} = 2.536$, $P = 0.055$ respectively).

The percentage biomass of ants was correlated positively with the diversity share of sapsuckers and negatively with the diversity share of tourists. Negative correlations were found with the biomass share of predators, phytophagous chewers and parasitoids. Diversity and abundance share was much higher in egg and coccoid parasitoids compared to larval parasitoids, probably due to predation by ants on larval parasitoids. Their low diversity supports the hypothesis of a decline towards the equator in ichneumonid diversity (Krüger & McGavin 1998b).

Discussion

The results presented here are a first step in the analysis of the arthropods collected from Mkomazi tree canopies. Completion of the study may take many years and specialists in specific taxa will need to be involved to be sure of accurate identifications in certain taxa. Despite the huge volume of material collected, the total sample is very small indeed relative to the extent of the MGR in particular and east African savannas in general. The timing of fieldwork was constrained by a number of factors and it is almost certain that samples were not taken at times of peak arthropod abundance. Nevertheless, This work represents the biggest single study of savanna tree canopies ever undertaken.

Quantitative arthropod samples have been obtained from the canopies of 266 trees representing 30 species). The total canopy area sampled for all tree species is a little over 480 m^2. Details of the samples are given in the Appendix.

Only 67 of the 266 tree samples taken, comprising a little over 30% of the total area sampled and 60,000 specimens has been sorted, analysed and identified to the level of RTU. The six species of *Acacia* and three species of *Commiphora* analysed to date have yielded 492 and 605 RTUs respectively. The degree of overlap in species composition between these two communities is less than 10%. Species richness and abundance is very much greater than previously recorded in the only similar savanna study of six species of *Acacia,* five species of *Commiphora* and one species of *Lannea* carried out in the Kora National Reserve, Kenya (West 1986). In West's study, conducted from early November 1983 to late January 1984, 6,742 specimens, assigned to 496 morphotypes (=RTUs), were collected from 49 tree samples each of 4 m^2 (a total of 196 m^2). The mean density of arthropods in the Kora study was 34.4/m^2 as compared with a mean density of 666/m^2 (*Acacia* excluding larvae) and 149/m^2 (*Commiphora*) in the present study. The huge difference in the relative numbers of specimens collected must be largely due to the to the superior sampling efficiency of a motorised mistblower over a manually-operated backpack sprayer.

Acknowledgements

The author would like to thank Oliver Krüger and Jennifer Slyker for their immense contribution to specimen sorting and data analysis. M. Atkinson made the collecting trays and Thunder and Colt provided the balloon nylon. Raphael Abdallah, Jon Davies, Elias Kihumo, Daniel Mafunde, Ramadani Makusi and Ian Maxwell gave much field assistance. We acknowledge with thanks the taxonomic specialists who helped with identifications. They include Jon Davies (Coleoptera), John Ismay (Diptera), Brian Levy (Coleoptera), John Noyes (Hymenoptera), Chris O'Toole (Hymenoptera) and Mark Robinson (Coleoptera). The final "Asante sana" goes to Tim Morgan and all the staff at Ibaya Camp.

References

Adis, J., Lubin, Y.D. & Montgomery, G.G. (1984) Arthropods from the canopy of inundated and terra firme forests near Manaus, Brazil, with critical consideration of the pyrethrum-fogging technique. *Studies of Neotropical Fauna and Environment* 19: 223-236.

Basset, Y. (1991a) The seasonality of arboreal arthropods foraging within an Australian rainforest tree. *Ecological Entomology* 16: 265-278.

Basset, Y. (1991b) The taxonomic composition of the arthropod fauna associated with an Australian rainforest tree. *Australian Journal of Zoology* 39: 171-190.

Basset, Y. (1996) Local communities of arboreal herbivores in Papua New Guinea: predictors of insect variables. *Ecology* 77: 1906-1919.

Basset, Y. & Kitching, R.L. (1991) Species number, species abundance and body length of arboreal arthropods associated with an Australian rainforest tree. *Ecological Entomology* 16: 391-402.

Campbell, D.G. (1993) Scale and patterns of community structure in Amazonian forests. In: Edwards, P.J., May, R.M. & Webb, N.R. (eds.) *Large scale ecology and conservation biology: the 35th symposium of the British Ecological Society with the society for Conservation Biology.* Blackwell, Oxford.

Chao, A. (1984) Non-parametric estimation of the number of classes in a population. *Scandinavian Journal of Statistics* 11: 265-270.

Cole, M.M. (1986) *The savannas. Biogeography and geobotany.* Academic Press, London.

CSIRO (1991). *The insects of Australia: a textbook for students and research workers.* 2nd edition. Melbourne University Press, Melbourne.

Erwin, T.L. (1982) Tropical forests: their richness in Coleoptera and other arthropod species. *Coleopterists Bulletin* 36: 74-75.

Erwin, T.L. (1983) Tropical forest canopies, the last biotic frontier. *Bulletin of the Entomological Society of America* 29: 14-19.

Erwin, T.L. (1995) Measuring arthropod biodiversity in the tropical forest canopy. In: Lowman, M.D. & Nadkarni, N.M. (eds.) *Forest canopies*. Academic Press, London. pp. 109-127.

Holdridge, L.A., Grenke, W.C., Hatheway, W.H., Liang, T. & Tosi, J.A. (1971) *Forest environments in tropical life zones*. Pergamon Press, Oxford.

Krüger, O. (1997) Community structure and biodiversity of the insect fauna of *Acacia* species in Mkomazi Game Reserve, north-east Tanzania. MSc Thesis, University of Oxford.

Krüger, O. & McGavin, G.C. (1997) The insect fauna of *Acacia* species in Mkomazi Game Reserve, north-east Tanzania. *Ecological Entomology* 22: 440-444.

Krüger, O. & McGavin, G.C. (1998a) Insect diversity of *Acacia* canopies in Mkomazi Game Reserve, north-east Tanzania. *Ecography* 21: 261-268.

Krüger, O. & McGavin, G.C. (1998b) The influence of ants on the guild structure of *Acacia* insect communities in Mkomazi Game Reserve, north-east Tanzania. *African Journal of Ecology* 36(3): 213-220.

Lawton, J.H. (1978) Host-plant influences on insect diversity: the effects of space and time. In: Mound, L.A. & Waloff, N. (eds.) *Diversity of insect faunas*. Blackwell, Oxford. pp. 105-215.

Lawton, J.H. (1986) Surface availability and insect community structure: the effects of architecture and fractal dimension of plants. In: Juniper, B.E. & Southwood, T.R.E. (eds.) *Insects and the plant surface*. Edward Arnold, London. pp. 317-322.

Lawton, J.H. & Price, P.W. (1979) Species richness of parasites on hosts: Agromyzid flies on the British Umbelliferae. *Journal of Animal Ecology* 48: 619-637.

Lewinsohn, T.M. & Price, P.W. (1996) Diversity of herbivorous insects and ecosystem processes. In: Solbrig, O.T., Medina E. & Silva P. (eds.) *Biodiversity and savanna ecosystem processes*. Springer-Verlag, Berlin. pp. 143-157.

Moran, V.C. & Southwood, T.R.E. (1982) The guild composition of arthropod communities in trees. *Journal of Animal Ecology* 51: 289-306.

Morse, D.R., Stork, N.E. & Lawton, J.H. (1988) Species number, species abundance and body length relationships of arboreal beetles in Bornean lowland rain forest trees. *Ecological Entomology* 13: 25-37.

Slyker, J. (1998) Community structure and biodiversity of the insect fauna of *Commiphora* species in Mkomazi Game Resereve, Tanzania. MSc Thesis, University of Oxford.

Solbrig, O.T. (1996) The diversity of the savanna ecosystem. In: Solbrig, O.T., Medina E. & Silva P. (eds.) *Biodiversity and savanna ecosystem processes*. Springer-Verlag, Berlin. pp. 1-27.

Southwood, T.R.E. (1978) *Ecological methods*. Chapman and Hall, London.

Southwood, T.R.E. & Kennedy, C.E.J. (1983) Trees as islands. *Oikos* 41: 359-371.

Stork, N.E. (1987a) Guild structure of arthropods from Bornean rain forest trees. *Ecological Entomology* 12: 69-80.

Stork, N.E. (1987b) Arthropod faunal similarity of Bornean rain forest trees. *Ecological Entomology* 12: 219-226.

Stork, N.E. (1991) The composition of the arthropod fauna of Bornean lowland rain forest trees. *Journal of Tropical Ecology* 7: 161-180.

Wagner, T. (1997) The beetle fauna of different tree species in forests of Rwanda and East Zaire. In: Stork, N.E., Adis, J. & Didham, R.K. (eds.) *Canopy Arthropods*. Chapman and Hall, London. pp. 169-183.

Watanabe, H. & Ruaysoongnern, S. (1989) Estimation of arboreal arthropod density in a dry evergreen forest in northeastern Thailand. *Journal of Tropical Ecology* 5: 151-158.

West, C. (1986) Insect communities in tree canopies. In: Coe, M. (ed.) *Kora: an ecological inventory of Kora National Reserve*. Royal Geographical Society, London. pp. 209-222.

Appendix: Summary data of all trees

trip/ tree	tree species	sample time	sample date	UTM east, north	tree height (m)	canopy circ. (m)	canopy area (m²)	sample area (m²)
1/1	*Lannea schweinfurthii*	08.45	28/7/94	367010, 9561496	6	35	97.5	4
1/2	*Lannea schweinfurthii*	08.30	29/7/94	367010, 9561496	5	24	45.8	2
1/3	*Terminalia* sp.	08.30	29/7/94	367010, 9561496	10	25	49.7	2
1/4	*Melia volkensii*	08.30	29/7/94	367010, 9561496	8	24	45.8	2
1/5	*Acacia tortilis*	11.00	29/7/94	369157, 9553733	13	75	447.6	4
1/6	*Albizia anthelmintica*	07.15	30/7/94	368769, 9553414	8	30	71.6	4
1/7	*Acacia reficiens*	07.45	30/7/94	368907, 9553494	7	24	45.8	2
1/8	*Acacia senegal*	12.30	1/8/94	366347, 9551601	8	32	81.5	2
1/9	*Acacia senegal*	12.40	1/8/94	366347, 9551601	8	26	53.8	2
1/10	*Acacia tortilis*	12.40	1/8/94	366347, 9551601	10	30	71.6	2
1/11	*Acacia drepanolobium*	13.30	3/8/94	375036, 9570639	2	8	5.1	0.32
1/12	*Acacia drepanolobium*	13.50	3/8/94	375036, 9570639	2	4	1.3	0.32
1/13	*Acacia drepanolobium*	13.55	3/8/94	375036, 9570639	1.7	5	2	0.32
1/14	*Acacia drepanolobium*	14.00	3/8/94	375036, 9570639	1.6	4.4	1.5	0.32
1/15	*Acacia zanzibarica*	10.30	4/8/94	371354, 9559866	3	11	9.6	0.64
1/16	*Acacia zanzibarica*	11.30	4/8/94	371354, 9559866	3.5	14	15.6	0.64
1/17	*Acacia zanzibarica*	12.30	4/8/94	371808, 9560232	7	18	25.8	0.64
1/18	*Acacia zanzibarica*	13.40	4/8/94	371076, 9559833	5	17	23	0.64
1/19	*Acacia drepanolobium*	14.20	4/8/94	369911, 9559499	1.5	2	0.3	0.32
1/20	*Acacia mellifera*	06.50	5/8/94	368499, 9552809	4	20	31.8	2
1/21	*Acacia mellifera*	07.10	5/8/94	368344, 9552776	5	20	31.8	2
1/22	*Acacia drepanolobium*	08.35	7/8/94	374981, 9570783	1.5	6	2.9	0.32
1/23	*Acacia drepanolobium*	08.35	7/8/94	374981, 9570783	2	5	2	0.32
1/24	*Acacia drepanolobium*	08.35	7/8/94	374981, 9570783	1.5	5	2	0.32
1/25	*Acacia senegal*	10.50	7/8/94	375427, 9568661	3	4	1.3	2
1/26	*Acacia mellifera*	06.50	10/8/94	366869, 9551723	2.2	12	11.5	0.96
1/27	*Acacia bussei*	09.15	10/8/94	368022, 9552322	3	14	15.6	2
1/28	*Acacia bussei*	09.15	10/8/94	368022, 9552322	3	18	25.8	2
1/29	*Acacia tortilis*	07.20	12/8/94	391555, 9545721	2.2	10	8	0.96
1/30	*Acacia ancistroclada*	10.45	12/8/94	391091, 9544051	10	41	133.8	4.96
1/31	*Acacia tortilis*	11.30	13/8/94	375162, 9555169	2	3.5	1	0.64
1/32	*Acacia tortilis*	09.15	14/8/94	366810, 9564494	3	7	3.9	0.96
1/33	*Acacia nilotica*	10.30	14/8/94	367151, 9563221	5	11	9.6	0.96
1/34	*Acacia tortilis*	12.00	14/8/94	367151, 9563221	2	8	5.1	2
1/35	*Acacia nilotica*	12.30	14/8/94	367151, 9563221	2	13	13.5	0.96
2/1	*Acacia senegal*	09.30	4/4/95	368934, 9566706	2	7	3.9	0.32
2/2	*Acacia tortilis*	09.30	4/4/95	368934, 9566706	2	8.5	5.8	0.64
2/3	*Acacia senegal*	10.35	4/4/95	370577, 9567703	2.5	4	1.3	0.32
2/4	*Acacia thomasii*	10.35	4/4/95	370577, 9567703	4	8	5.1	0.64
2/5	*Acacia brevispica*	17.30	4/4/95	366862, 9563884	3.5	8	5.1	0.64
2/6	*Acacia brevispica*	17.30	4/4/95	366862, 9563884	2.5	12.5	12.4	0.64
2/7	*Commiphora campestris*	10.20	5/4/95	366214, 9551667	10	26	53.8	4
2/8	*Dichrostachys cinerea*	12.40	5/4/95	368510, 9552798	6	70	389.9	1.28
2/9	*Grewia* sp.	16.30	5/4/95	369264, 9553440	3	10	8	1.28
2/10	*Lannea schweinfurthii*	17.20	5/4/95	370705, 9555355	3.5	10	8	1.28
2/11	*Lannea schweinfurthii*	09.30	6/4/95	367010, 9561496	6	35	97.5	4
2/12	*Acacia brevispica*	11.25	6/4/95	367060, 9563624	3	10	8	1.28
2/13	*Commiphora campestris*	11.55	7/4/95	365881, 9551467	6	25	49.7	4
2/14	*Commiphora campestris*	11.55	7/4/95	365881, 9551467	5	22	38.5	4
2/15	*Grewia* sp.	11.55	7/4/95	365881, 9551467	3	23	21.5	1.28
2/16	*Commiphora campestris*	16.00	7/4/95	365881, 9551467	8	33	86.7	4

cont.

trip/ tree	tree species	sample time	sample date	UTM east, north	tree height (m)	canopy circ. (m)	canopy area (m^2)	sample area (m^2)
2/17	*Commiphora campestris*	16.00	7/4/95	365881, 9551467	8	35	97.5	4
2/18	*Combretum molle*	09.50	8/4/95	367845, 9567478	5	3	0.72	1
2/19	*Combretum molle*	09.50	8/4/95	367845, 9567478	3.5	3	0.72	1
2/20	*Combretum molle*	09.50	8/4/95	367845, 9567478	4.5	3	0.72	1
2/21	*Combretum molle*	09.50	8/4/95	367845, 9567478	3.5	3	0.72	1
2/22	*Combretum molle*	09.50	8/4/95	367845, 9567478	4	4	1.3	1
2/23	*Combretum molle*	09.50	8/4/95	367845, 9567478	3.5	5	2	1
2/24	*Combretum molle*	09.55	9/4/95	367690, 9567478	5	5	2	1
2/25	*Combretum molle*	09.55	9/4/95	367690, 9567478	5	6	2.9	1
2/26	*Combretum molle*	09.55	9/4/95	367690, 9567478	6	12	11.5	1
2/27	*Combretum molle*	09.55	9/4/95	367690, 9567478	5	9	6.5	1
2/28	*Combretum molle*	09.55	9/4/95	367690, 9567478	4.5	8	5.1	1
2/29	*Combretum molle*	09.55	9/4/95	367690, 9567478	5	7	3.9	1
2/30	*Combretum molle*	09.55	9/4/95	367690, 9567478	5	5	2	1
2/31	*Combretum molle*	09.55	9/4/95	367690, 9567478	4.5	7	3.9	1
2/32	*Dichrostachys cinerea*	14.45	9/4/95	368845, 9567015	1.5	3	0.7	0.64
2/33	*Dichrostachys cinerea*	14.45	9/4/95	368845, 9567015	1.5	3	0.7	0.64
2/34	*Dichrostachys cinerea*	10.00	10/4/95	368845, 9567015	1.5	3	0.7	0.64
2/35	*Dichrostachys cinerea*	10.00	10/4/95	368845, 9567015	1.5	3	0.7	0.64
2/36	*Dichrostachys cinerea*	10.00	10/4/95	368845, 9567015	1.5	3	0.7	0.64
2/37	*Dichrostachys cinerea*	10.00	10/4/95	368845, 9567015	1.5	3	0.7	0.64
2/38	*Commiphora campestris*	09.05	12/4/95	373638, 9562158	7	39	121	4
2/39	*Commiphora campestris*	09.05	12/4/95	373638, 9562158	9	44	154	4
2/40	*Acacia zanzibarica*	09.15	12/4/95	373638, 9562158	3.5	14	15.6	1.282
2/41	*Acacia zanzibarica*	09.15	12/4/95	373638, 9562158	3.5	9	6.5	0.64
2/42	*Acacia zanzibarica*	09.50	12/4/95	373638, 9562158	3.5	8	5.1	0.64
2/43	*Acacia zanzibarica*	10.55	12/4/95	373859, 9562313	4	19	28.7	1.28
2/44	*Commiphora campestris*	16.25	12/04/95	371420, 9560353	6	37	108.9	4
2/45	*Commiphora campestris*	16.25	12/4/95	371420, 9560353	6	39	121	4
2/46	*Acacia zanzibarica*	10.00	13/4/95	371242, 9559855	4	18	25.8	1
2/47	*Acacia zanzibarica*	10.00	13/4/95	371242, 9559855	3	12	11.5	1
2/48	*Acacia zanzibarica*	10.00	13/4/95	371242, 9559855	3	10	8	1
2/49	*Acacia zanzibarica*	10.00	13/4/95	371242, 9559855	5	16	20.4	1
2/50	*Acacia zanzibarica*	10.00	13/4/95	371242, 9559855	4	10	8	1.28
2/51	*Acacia senegal*	16.45	13/4/95	375651, 9567821	4	15	17.9	1
2/52	*Acacia tortilis*	16.45	13/4/95	375651, 9567821	6	36	103.1	2
2/53	*Acacia senegal*	16.45	13/4/95	375651, 9567821	5	31	76.5	2
2/54	*Acacia senegal*	16.45	13/4/95	375651, 9567821	4	24	45.8	1
2/55	*Acacia senegal*	16.45	13/4/95	375651, 9567821	5	25	49.7	2
2/56	*Acacia senegal*	16.45	13/4/95	375651, 9567821	3	15	17.9	1.28
2/57	*Acacia tortilis*	10.15	15/4/95	375506, 9567843	5	12	11.5	2
2/58	*Acacia tortilis*	10.15	15/4/95	375506, 9567843	4.5	20	31.8	2
2/59	*Acacia tortilis*	10.15	15/4/95	375506, 9567843	6	15	17.9	2
2/60	*Acacia senegal*	10.15	15/4/95	375506, 9567843	3	14	15.6	1.28
2/61	*Acacia senegal*	10.15	15/4/95	375506, 9567843	4	20	31.8	2
2/62	*Commiphora campestris*	10.05	18/4/95	449157, 9503228	4	25	49.7	2
2/63	*Commiphora africana*	10.05	18/4/95	449157, 9503228	3.5	11	9.6	1
2/64	*Commiphora campestris*	10.05	18/4/95	449157, 9503228	5	22	38.5	2
2/65	*Commiphora campestris*	10.05	18/4/95	449157, 9503228	4	16	20.4	1
2/66	*Commiphora campestris*	10.05	18/4/95	449157, 9503228	5	32	81.5	2
2/67	*Commiphora campestris*	09.00	19/4/95	450942, 9506567	3	12	11.5	1
2/68	*Boswellia* sp.	09.00	19/4/95	450942, 9506567	4	24	45.8	1
2/69	*Boswellia* sp.	09.00	19/4/95	450942, 9506567	3	15	17.9	1
2/70	*Commiphora africana*	09.00	19/4/95	450942, 9506567	4	12	11.5	1

cont.

trip/ tree	tree species	sample time	sample date	UTM east, north	tree height (m)	canopy circ. (m)	canopy area (m²)	sample area (m²)
2/71	*Commiphora campestris*	09.00	19/4/95	450942, 9506567	4	25	49.7	2
2/72	*Commiphora africana*	09.00	19/4/95	450942, 9506567	5	30	71.6	2
2/73	*Commiphora holtziana*	11.55	21/4/95	392020, 9547302	9	25	49.7	2
2/74	*Commiphora holtziana*	11.55	21/4/95	392020, 9547302	6	15	17.9	2
2/75	*Commiphora holtziana*	11.55	21/4/95	392020, 9547302	6.5	13	13.5	2
2/76	*Acacia tortilis*	11.55	21/4/95	392020, 9547302	6	37	108.9	2
3/1	*Dichrostachys cinerea*	08.25	29/2/96	368345, 9567181	2			1.28
3/2	*Acacia reficiens*	07.00	30/12/95	375236, 9570694	3	20	31.9	2
3/3	*Acacia nilotica*	07.00	30/12/95	375236, 9570694	4	22	38.5	3
3/4	*Acacia nilotica*	07.00	30/12/95	375236, 9570694	3	13	13.5	1.28
3/5	*Acacia nilotica*	07.00	30/12/95	375236, 9570694	4	18	25.8	1
3/6	*Acacia nilotica*	07.00	30/12/95	375236, 9570694	3.5	25	49.8	2
3/7	*Acacia reficiens*	08.50	30/12/96	375281, 9570672	3.5	22	38.5	3
3/8	*Acacia reficiens*	08.50	30/12/96	375281, 9570672	3.5	13	13.5	2
3/9	*Acacia reficiens*	08.50	30/12/96	375281, 9570672	4	13	13.5	1.282
3/10	*Acacia etbaica*	08.50	30/12/96	375281, 9570672	5	20	31.8	3
3/11	*Acacia nilotica*	06.55	31/12/95	375203, 9570827	4	14	15.6	1.28
3/12	*Acacia nilotica*	06.55	31/12/95	375203, 9570827	3	17	23	2
3/13	*Acacia senegal*	06.55	31/12/95	375203, 9570827	5	15	17.9	1
3/14	*Acacia nilotica*	06.55	31/12/95	375203, 9570827	5	25	49.7	2
3/15	*Acacia senegal*	06.55	31/12/95	375203, 9570827	3	19	28.7	2
3/16	*Acacia nilotica*	08.50	31/12/95	375447, 9571159	2.5	12	11.5	2
3/17	*Acacia nilotica*	08.50	31/12/95	375447, 9571159	2	7	3.9	1.28
3/18	*Acacia nilotica*	08.50	31/12/95	375447, 9571159	4	18	25.8	2
3/19	*Acacia nilotica*	08.50	31/12/95	375447, 9571159	5	16	20.4	2
3/20	*Acacia senegal*	08.50	31/12/95	375447, 9571159	4	14	15.6	2
3/21	*Acacia mellifera*	08.05	2/1/96	391037, 9543277	4	18	25.8	2
3/22	*Acacia mellifera*	08.05	2/1/96	391037, 9543277	4	17	23	2
3/23	*Commiphora campestris*	08.05	2/1/96	391037, 9543277	3	13	13.5	2
3/24	*Acacia mellifera*	08.05	2/1/96	391037, 9543277	2	12	11.5	1.28
3/25	*Commiphora campestris*	08.05	2/1/96	391037, 9543277	3	13	13.5	1
3/26	*Acacia mellifera*	08.05	2/1/96	391037, 9543277	4	15	17.9	1
3/27	*Commiphora campestris*	09.45	2/1/96	391059, 9543211	3.5	16	20.4	2
3/28	*Commiphora campestris*	09.45	2/1/96	391059, 9543211	3	13	13.5	1
3/29	*Commiphora campestris*	09.45	2/1/96	391059, 9543211	5	22	38.5	3
3/30	*Acacia bussei*	09.45	2/1/96	391059, 9543211	5	20	31.8	2
3/31	*Acacia zanzibarica*	07.25	4/1/96	375225, 9570639	7	19	28.7	2
3/32	*Acacia drepanolobium*	07.25	4/1/96	375225, 9570639	1.5	5	2	1
3/33	*Acacia senegal*	07.25	4/1/96	375225, 9570639	5	23	42.1	3
3/34	*Acacia senegal*	09.00	4/1/96	374949, 9569202	8	37	108.9	5
3/35	*Acacia nilotica*	09.00	4/1/96	374949, 9569202	2	8	5.1	1.28
3/36	*Commiphora campestris*	06.55	6/1/96	373693, 9562103	5	34	92	4
3/37	*Commiphora campestris*	06.55	6/1/96	373693, 9562103	6	40	127.3	4
3/38	*Acacia zanzibarica*	10.45	7/1/96	373683, 9561937	3	10	8	1
3/39	*Acacia zanzibarica*	10.45	7/1/96	373683, 9561937	3	10	8	2
3/40	*Acacia zanzibarica*	10.45	7/1/96	373683, 9561937	3	10	8	1
3/41	*Heywoodia lucens*	08.55	10/1/96	363292, 9562499	10	21	35.1	2
3/42	*Heywoodia lucens*	08.55	10/1/96	363292, 9562499	9	35	97.5	2
3/43	*Apodytes dimidiata*	08.55	10/1/96	363292, 9562499	10	30	71.6	1
3/44	*Apodytes dimidiata*	08.55	10/1/96	363292, 9562499	5	24	45.8	1
3/45	*Craibia brevicaudata*	08.55	10/1/96	363292, 9562499	6	24	45.8	2
3/46	*Heywoodia lucens*	10.40	10/1/96	363292, 9562499	5	10	8	1
3/47	*Drypetes parvifolia*	10.40	10/1/96	363292, 9562499	4	8	5.1	1
3/48	*Drypetes parvifolia*	10.40	10/1/96	363292, 9562499	5	10	8	1
3/49	*Drypetes parvifolia*	10.40	10/1/96	363292, 9562499	4	6	2.9	1

cont.

trip/ tree	tree species	sample time	sample date	UTM east, north	tree height (m)	canopy circ. (m)	canopy area (m^2)	sample area (m^2)
3/50	*Heywoodia lucens*	10.40	10/1/96	363292, 9562499	10	23	42.1	2
3/51	*Heywoodia lucens*	10.40	10/1/96	363292, 9562499	3	9	6.5	1
3/52	*Heywoodia lucens*	09.15	12/1/96	363292, 9562499	12	18	25.8	2
3/53	*Heywoodia lucens*	09.15	12/1/96	363292, 9562499	12	15	17.9	2
3/54	*Heywoodia lucens*	09.15	12/1/96	363292, 9562499	10	19	28.7	2
3/55	*Heywoodia lucens*	09.15	12/1/96	363292, 9562499	15	23	42.1	2
3/56	*Ochna holstii*	11.45	12/1/96	363292, 9562499	5	10	8	2
3/57	*Ochna holstii*	11.45	12/1/96	363292, 9562499	5	16	20.4	2
3/58	*Craibia brownii*	11.45	12/1/96	363292, 9562499	12	33	86.7	4
3/59	*Acacia senegal*	08.30	14/1/96	377966, 9569927	3	16	20.4	2
3/60	*Acacia senegal*	08.30	14/1/96	377966, 9569927	4	20	31.8	2
3/61	*Acacia senegal*	08.30	14/1/96	377966, 9569927	6	24	45.8	2
3/62	*Commiphora campestris*	10.10	14/1/96	375411, 9565725	6	37	108.9	4
3/63	*Combretum molle*	08.15	15/1/96	367679, 9567500	4	10	8	2
3/64	*Combretum molle*	08.15	15/1/96	367679, 9567500	5	10	8	2
3/65	*Combretum molle*	08.15	15/1/96	367679, 9567500	4	12	11.5	2
3/66	*Lannea schweinfurthii*	08.15	16/1/96	367010, 9561496	5	20	31.8	2
3/67	*Lannea schweinfurthii*	08.15	16/1/96	367010, 9561496	7	36	103.1	4
3/68	*Terminalia brownii*	08.15	16/1/96	367010, 9561496	10	25	49.8	2
3/69	*Acacia tortilis*	07.30	18/1/96	367303, 9552432	4	14	15.6	2
3/70	*Acacia bussei*	07.30	18/1/96	367303, 9552432	5	18	25.8	2
3/71	*Acacia tortilis*	07.30	18/1/96	367303, 9552432	5	17	23	2
3/72	*Acacia bussei*	07.30	18/1/96	367303, 9552432	4	25	49.7	2
4/1	*Acacia senegal*	08.00	28/3/96	364690, 9568281	4	18	25.8	2
4/2	*Acacia senegal*	08.00	28/3/96	364690, 9568281	4.5	12	11.5	1
4/3	*Acacia senegal*	08.00	28/3/96	364690, 9568281	5	13	13.5	2
4/4	*Acacia senegal*	08.00	28/3/96	364690, 9568281	4	19	28.7	2
4/5	*Acacia senegal*	09.55	28/3/96	364879, 9568104	5	26	53.8	7
4/6	*Commiphora campestris*	11.00	2/4/96	391500, 9545400	5	30	71.6	3
4/7	*Commiphora campestris*	11.00	2/4/96	391500, 9545400	6	25	49.7	3
4/8	*Commiphora campestris*	11.00	2/4/96	391500, 9545400	5	20	31.8	2
4/9	*Lannea triphylla*	11.00	2/4/96	391500, 9545400	1.5	10	8	1.28
4/10	*Commiphora campestris*	08.15	3/4/96	376403, 9570254	5	26	53.8	2
4/11	*Commiphora campestris*	08.15	3/4/96	376403, 9570254	5	18	25.8	2
4/12	*Commiphora campestris*	08.15	3/4/96	376403, 9570254	4	25	49.7	2
4/13	*Commiphora campestris*	08.15	3/4/96	376403, 9570254	5	30	71.6	2
4/14	*Acacia senegal*	08.15	3/4/96	376403, 9570254	2.5	10	8	1.28
4/15	*Acacia senegal*	10.00	3/4/96	375369, 9570617	6	26	53.8	2
4/16	*Acacia nilotica*	10.00	3/4/96	375369, 9570617	6	27	58	2
4/17	*Acacia nilotica*	10.00	3/4/96	375369, 9570617	5	26	53.8	2
4/18	*Acacia senegal*	10.00	3/4/96	375369, 9570617	5	20	31.8	2
4/19	*Acacia mellifera*	10.00	3/4/96	375369, 9570617	1.5	12	11.5	1.28
5/1	*Acacia reficiens*	08.00	2/1/97	393436, 9555353	4	18	25.8	2
5/2	*Acacia mellifera*	08.00	2/1/97	393436, 9555353	2.5	10	8	1.28
5/3	*Acacia reficiens*	08.00	2/1/97	393436, 9555353	3	20	31.8	2
5/4	*Acacia reficiens*	08.00	2/1/97	393436, 9555353	4	14	15.6	2
5/5	*Acacia reficiens*	08.00	2/1/97	393436, 9555353	3	14	15.6	2
5/6	*Acacia reficiens*	09.30	2/1/97	393436, 9555353	6	32	81.5	4
5/7	*Acacia reficiens*	08.00	2/1/97	393436, 9555353	4	20	31.8	4
5/8	*Acacia mellifera*	08.00	2/1/97	393436, 9555353	3	13	13.5	1.28
5/9	*Acacia reficiens*	08.10	3/1/97	393524, 9555288	2.5	16	20.4	2
5/10	*Acacia zanzibarica*	08.10	3/1/97	393524, 9555288	3	17	23	2
5/11	*Acacia mellifera*	08.10	3/1/97	393524, 9555288	2.5	18	25.8	1.28
5/12	*Acacia mellifera*	08.10	3/1/97	393524, 9555288	4	20	31.8	2
5/13	*Acacia reficiens*	08.10	3/1/97	393524, 9555288	3	18	25.8	2

cont.

trip/ tree	tree species	sample time	sample date	UTM east, north	tree height (m)	canopy circ. (m)	canopy area (m²)	sample area (m²)
5/14	*Acacia senegal*	09.20	3/1/97	393524, 9555288	6	24	45.8	5
5/15	*Acacia senegal*	09.20	3/1/97	393524, 9555288	5	18	25.8	3
5/16	*Acacia zanzibarica*	09.20	3/1/97	393524, 9555288	2	8	5.1	1.28
5/17	*Acacia zanzibarica*	08.30	4/1/97	393296, 9555510	3	10	8	1.28
5/18	*Acacia reficiens*	08.30	4/1/97	393296, 9555510	3	13	13.5	2
5/19	*Acacia reficiens*	08.30	4/1/97	393296, 9555510	2	11	9.6	2
5/20	*Acacia reficiens*	08.30	4/1/97	393296, 9555510	4	16	20.4	2
5/21	*Acacia senegal*	08.30	4/1/97	393296, 9555510	4	16	20.4	2
5/22	*Acacia reficiens*	10.20	4/1/97	393436, 9555353	4	16	20.4	2
5/23	*Acacia reficiens*	10.20	4/1/97	393436, 9555353	2	10	8	1.28
5/24	*Acacia reficiens*	10.20	4/1/97	393436, 9555353	3	20	31.8	2
5/25	*Acacia mellifera*	10.20	4/1/97	393436, 9555353	4	20	31.8	4
5/26	*Acacia zanzibarica*	07.20	6/1/97	374778, 9553640	5	14	15.6	2
5/27	*Acacia zanzibarica*	07.20	6/1/97	374778, 9553640	2	7	3.9	1.28
5/28	*Acacia zanzibarica*	07.20	6/1/97	374778, 9553640	5	12	11.5	2
5/29	*Acacia reficiens*	07.20	6/1/97	374778, 9553640	3	14	15.6	2
5/30	*Acacia zanzibarica*	07.20	6/1/97	374778, 9553640	5	20	31.8	2
5/31	*Acacia zanzibarica*	09.45	6/1/97	374778, 9553640	7	13	13.5	4
5/32	*Acacia zanzibarica*	09.45	6/1/97	374778, 9553640	6	10	8	2
5/33	*Acacia zanzibarica*	09.45	6/1/97	374778, 9553640	4	10	8	2
5/34	*Acacia zanzibarica*	08.15	7/1/97	373989, 9555958	2	7	3.9	1.28
5/35	*Acacia zanzibarica*	08.15	7/1/97	373989, 9555958	3	7	3.9	1
5/36	*Acacia zanzibarica*	08.15	7/1/97	373989, 9555958	2.5	8	5.1	1
5/37	*Acacia zanzibarica*	08.15	7/1/97	373989, 9555958	2	10	8	1
5/38	*Acacia reficiens*	08.15	7/1/97	373989, 9555958	3	13	14.8	2
5/39	*Acacia nilotica*	08.10	8/1/97	375304, 9571083	4	20	31.8	4
5/40	*Acacia senegal*	08.10	8/1/97	375304, 9571083	2	7	3.9	1.28
5/41	*Acacia nilotica*	08.10	8/1/97	375304, 9571083	3.5	16	20.4	2
5/42	*Acacia nilotica*	08.10	8/1/97	375304, 9571083	3	14	15.6	2
5/43	*Acacia nilotica*	07.45	9/1/97	375201, 9570990	4	20	31.8	2
5/44	*Acacia zanzibarica*	07.45	9/1/97	375201, 9570990	3.5	12	11.5	2
5/45	*Acacia zanzibarica*	07.45	9/1/97	375201, 9570990	2.5	10	8	1.28
5/46	*Acacia senegal*	07.45	9/1/97	375201, 9570990	4	19	28.7	2
5/47	*Acacia senegal*	07.45	9/1/97	375201, 9570990	4	20	31.8	2
5/48	*Acacia drepanolobium*	09.15	9/1/97	375014, 9570881	1	4	1.3	0.64
5/49	*Acacia drepanolobium*	09.15	9/1/97	375014, 9570881	2	6	2.9	0.64
5/50	*Acacia senegal*	08.15	11/1/97	375330, 9571120	5	27	58	4
5/51	*Acacia nilotica*	08.15	11/1/97	375330, 9571120	4	18	25.8	2
5/52	*Acacia senegal*	08.15	11/1/97	375330, 9571120	4	20	31.8	2
5/53	*Acacia nilotica*	08.15	11/1/97	375330, 9571120	3	10	8	1.28
5/54	*Acacia senegal*	07.05	12/1/97	375378, 9571003	4	23	42.1	4
5/55	*Acacia nilotica*	07.05	12/1/97	375378, 9571003	4.5	14	15.6	2
5/56	*Acacia nilotica*	07.05	12/1/97	375378, 9571003	2	10	8	1.28
5/57	*Acacia nilotica*	07.05	12/1/97	375378, 9571003	4	17	23	2
5/58	*Acacia drepanolobium*	08.40	12/1/97	374884, 9571017	1	4	1.3	0.64
5/59	*Acacia drepanolobium*	08.40	12/1/97	374884, 9571017	2	3	0.7	1
5/60	*Acacia drepanolobium*	08.40	12/1/97	374884, 9571017	1.5	3.5	1	0.64
5/61	*Acacia drepanolobium*	08.40	12/1/97	374884, 9571017	1.5	3	0.7	1
5/62	*Acacia nilotica*	08.00	13/1/97	375103, 9570896	4	12	11.5	2
5/63	*Acacia nilotica*	08.00	13/1/97	375103, 9570896	5	16	20.4	3
5/64	*Acacia nilotica*	08.00	13/1/97	375103, 9570896	6	12	11.5	3

CHAPTER 23

The interaction of *Crematogaster* ants and *Acacia* trees in Mkomazi

Linsey Stapley

Introduction

The relationship between ants and plants provides examples of a diverse range of interactions, some of which appear to be coevolved. Ants can act as:

(a) defoliators, for example leaf cutter ants of the genus *Atta* (Formicidae; Myrmicinae), which harvest leaves on which they grow the fungus they eat (Cherret *et al.* 1989);

(b) as seed dispersers (myrmecochory) where ants harvest seeds containing nutrient-rich bodies (elaiosomes), thus dispersing the seeds away from the parent plant (Anderson 1991), and

(c) as nutrient providers (myrmecotrophy), for example the epiphyte *Dischidia major* (Asclepiadaceae) absorbs carbon dioxide and nitrogen waste produced by *Philidris* ants (Dolichoderinae) living on the plant (Treseder *et al.* 1995).

Perhaps the most specialised ant-plant relationship is that found between ants and a group of plants known as myrmecophytes. Myrmecophytes have evolved preformed cavities (domatia) which are inhabited by species of ants that are often specific to one plant species. The plants provide the ants with food and shelter and, in return, the ants are thought to defend the plant against herbivorous mammals and insects.

Myrmecophytic plants are found in a range of families with perhaps the best examples coming from the Leguminosae, including the acacias studied in this work. Some myrmecophytes show a much looser relationship between the ants and the plant than is shown between many acacias and their resident ant species. In such loose relationships, the ants may occupy the plants opportunistically as is the case with the myrmecophyte *Clerodendrum fallax* (Jolivert 1985). It may be that the relationship has not stabilised evolutionarily and that eventually a true myrmecophytic plant will evolve from what is, at present, an imperfect and loose association.

The relationship between *Acacia cornigera* and its resident ant species *Pseudomyrmex ferruginea* (Pseudomyrmicinae) in South America is a well-known and well-documented example of an *Acacia* myrmecophyte. It was studied extensively in a series of papers by Janzen (1966, 1967a & 1967b) who showed that this species of *Acacia* provides food and shelter for the ants in return for being defended by them against herbivores, fire and encroaching vegetation. The costs and benefits to the parties involved in this interaction are very obvious. This chapter is concerned with east African examples of the relationship. These are intriguing systems to study because the costs and benefits to each party are not very clear. Although some work has been done on these (Hocking 1970, Willmer & Stone 1997, Willmer *et al.* this volume), there are many questions left to be answered.

Acacias in Mkomazi Game Reserve

Myrmecophytes from the genus *Acacia* are characterised by having stipular thorns that have become swollen and inflated at the base into so-called pseudogalls. There are examples of other acacias which also have these pseudogalls but which do not have a symbiotic relationship with ants (see list below). Unlike the true galls on many other plants e.g. poplar trees and oaks, formation of these *Acacia* pseudogalls is not induced by another organism, but happens whether ants are present or not (Monod & Schmitt 1968, Hocking 1970). In the case of *A. drepanolobium*, the pseudogalls appear within three months of germination (pers. obs.).

There are six species of east African acacias that possess pseudogalls (Coe & Beentje 1991):

- *Acacia bussei* Sjöstedt
- *Acacia drepanolobium* Sjöstedt *
- *Acacia elatior* Brenan subsp. *elatior*
- *Acacia horrida* (L.) Willd. subsp. *benadirensis* (Chiov.) Hillcoat & Brenan
- *Acacia seyal* Del. var. *fistula* (Schweinf.) Oliv. *
- *Acacia zanzibarica* (S. Moore) Taub. var. *zanzibarica* *

Of these, three are known to have ant-inhabitants, shown above by '*'. Mkomazi Game Reserve (MGR) is an ideal place to work, because there is an abundance of *A. drepanolobium* and *A. zanzibarica*.

Acacia drepanolobium (the whistling thorn)

In MGR *Acacia drepanolobium* is a small (usually less than two metres in height) many-branched tree, although in undisturbed areas it can reach seven metres high. It inhabits areas of seasonally-inundated black cotton soil and is especially abundant on the *mbugas* at Dindira and Ndea. The plants flower at the onset of and during the wet season (see Stone *et al.*, Chapter 20 and Coe *et al.*, Chapter 8), with flowers and fruit being present at the same time on any one tree. This species typically bears an impressive array of long, sharp white thorns up to 7.6 cm in length with many swollen at the base into pseudogalls (see Plate 24). On a tree

1.35 m high, a total of 211 pseudogalls were counted. The pseudogalls are round and globular in shape, the largest here measuring 2.86 cm across. On the trees in MGR these pseudogalls can reach volumes of up to 5.1 cm^3 although in other regions they can get much larger (pers. obs.). When they are young, the pseudogalls are hollow, green and soft with the inside covered in pithy material and a petiole leaf attached to the midpoint of the gall face. As they age, the petiole leaf is lost which leaves a scar behind. The pseudogalls gradually darken and harden with age, becoming red/brown and eventually black or even white when very old.

Acacia zanzibarica

Acacia zanzibarica can reach 12 m high in undisturbed areas, although in MGR it is more commonly a many branched tree less than three metres in height. It favours well-drained soil on the edge of black cotton areas and is especially common along the road leading to Mbula water-hole from Ibaya camp. As with *A. drepanolobium*, the trees generally flower at the onset of and during the wet season, although some may flower earlier, and fruits and flowers are found on the same tree at the same time. It is also a very spiny plant with some thorns reaching up to 5.1 cm in length and again some have swollen at the base into pseudogalls. Unlike the pseudogalls of *A. drepanolobium*, which are globular in shape, these are bi-lobed, reaching up to 4.74 cm in width and 5.6 cm^3 in volume (pers. obs.). The two halves are separated by a septum, effectively sectioning off the inside of the pseudogall. When young, they are soft and green in colour and as they age, they lose the petiole leaf and become harder and darker, eventually turning black/grey.

Both plants have single glands on the upper surface of the leaf petiole ('petiole glands', although they may be referred to as extra-floral nectaries) which are also found in species of *Acacia* lacking pseudogalls (*A. drepanolobium* also has glands on the leaf rachis at the base of the leaflet pairs). It is thought that these are similar in function to the extra-floral nectaries found on other plant species such as *Catalpa speciosa* trees (Stephenson 1982) and vetches (Koptur 1979). The extra-floral nectaries in species like these secrete a sugary solution (nectar) that may or may not contain other nutrients, for example proteins or amino-acids. The secretions attract worker ants (amongst other insects) which collect the nectar and take it back to the colony where it is fed to the brood. The presence of ants is a benefit to the plant because the ants are found to defend the plant from attack from herbivorous insects (Stephenson 1982). The petiole glands on ant-acacias also attract ants, but what the reward from the gland is and whether the ants do defend the plant has yet to be fully resolved. Some species of acacia-ant are known to chew the petiole glands (Coe, Stanton, pers. comm.).

The east African ant-acacias differ from the South American ant-acacias in that they lack Beltian bodies (protein rich bodies on the leaflet tips) which provide the resident ants with protein. Where then, do the ant inhabitants of these east African ant-acacias get their protein? This is an important question because protein is a necessary part of any insect's diet. This suggests that the benefits these acacia-ants receive from these acacias are not as straightforward nor as obvious as those seen in the South American examples. Further studies are requiredto ascertain the sources of dietary protein, which the ants exploit.

The acacia-ants

The ants on both species of *Acacia* have been identified as belonging to the genus *Crematogaster* (Myrmicinae). The ants found on *A. drepanolobium* in MGR have been identified as *C. nigriceps* Emery and are characterised by their red/brown abdomen and black head and thorax (Ward, pers. comm.). Those on *A. zanzibarica* are *C. sjöstedti* Mayr and are characteristically black all over (Ward, pers. comm.). In other areas, up to four species of ant can be found inhabiting *A. drepanolobium* (Young *et al.* 1997).

By collecting all the ants on a single tree and carrying out introduction experiments between trees, I have found that there is only a single colony of ants on any one *Acacia drepanolobium* tree (although this is different in other areas, Stanton, pers. comm.). On *A. zanzibarica*, a colony may inhabit more than one tree if the trees are less than three metres apart. The colonies are monogynous (having only one queen) and are established when a mated queen moves into a new gall on an unoccupied tree by chewing a hole in the softest part of the gall and settling in to rear her first brood. Once the first brood has reached adulthood, they begin to forage on the outside of the plant, visits to the petiole glands being especially common. These small colonies rarely attack and have to be severely provoked before they will do so. As the colony increases in size, the reluctance to attack decreases and the workers will readily attack the source of a disturbance.

The ants have a characteristic 'cocktail' alarm reaction. On detection of a disturbance, the workers already present on the outside of the tree begin to run around with their abdomens cocked up over their heads. More workers come out of the galls and the tree can seem to be covered in ants. They are most numerous on fresh growth, the area of the plant favoured most by herbivores, particularly mammals. The ants administer a sharp nip and are adept at finding the most tender areas to attack. The ants' reaction to invading insects is to surround the intruder and either throw it off the plant or immobilise it and carry it into a gall. If such a gall is opened up, it is found that the intruder has been consumed by the occupying ants.

Not all ants emerge from the galls to attack. If galls are collected during a disturbance, ants are still found inside them, looking after the brood. I made measurements of the heads of these 'Defender' and 'Domestic' ants and found that Defender ants have significantly larger heads than the Domestic ants (Figure 23.1), a difference which is more obvious for ants from *A. drepanolobium* trees than from *A. zanzibarica* trees. I therefore concluded that there are physical worker castes in colonies of *Crematogaster nigriceps* and that this physical difference is correlated with an intra-colonial division of defensive behaviour (Stapley, 1999).

Winged reproductives (alates) are most common at the start of and during the wet season (generally November to January and March to May). There is much competition expected between newly mated queens for unoccupied *Acacia* seedlings (seedlings can produce pseudogalls within three months of germination). In MGR, seedlings are not very common. Trees (especially *A. drepanolobium*) get burnt on a regular basis and only a few reach maturity and produce seeds. Most of the surrounding ant-acacias are already occupied and it is unlikely that a new queen would be able to exert enough influence over an established colony to be able to oust the already established queen.

The ants move into pseudogalls when the latter are still young and green and soft enough for the ants to chew an entrance-hole. The ants clean the inside of pith, making a smooth inner surface. Once the pseudogalls have become harder, and therefore less vulnerable to attack, eggs are deposited in the pseudogalls by

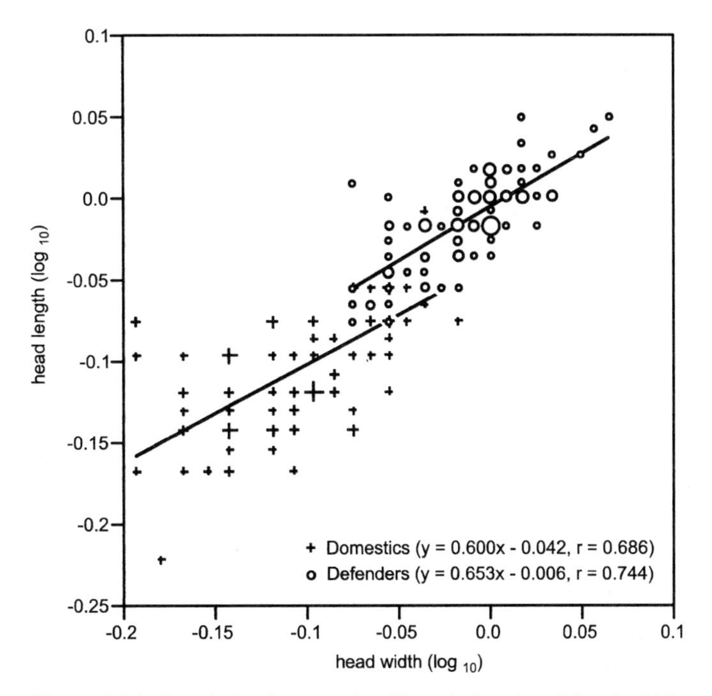

Figure 23.1 Correlation between head length ($\log_{10}$) and head width ($\log_{10}$) for *Acacia drepanolobium* Defender and Domestic ants. The size of each point indicates number of ant heads represented by that point (range 1–7 in each case).

workers. The eggs remain in the pseudogall until the young ants emerge from the pupae. Ants have never been seen to be carrying young larvae or pupae around on the branches of the trees. Neither is the queen ever seen out of her pseudogall (it is unlikely that she will be able to fit through the hole).

By collecting and examining each pseudogall on a tree, I have shown that there is a general pattern of pseudogall use especially on *A. drepanolobium*: new pseudogalls (on the outside, more exposed areas of the plant) are generally inhabited by workers only; those further towards the trunk (in the more protected areas of the plant) by workers and young, and pseudogalls right next to the trunk of the tree (i.e. the older pseudogalls on the plant) are inhabited just by workers or are abandoned and taken over by caterpillars or spiders. The older pseudogalls are often more brittle than those towards the outside of the tree, making them more prone to damage. By abandoning these or avoiding putting brood in them, the colony effectively ensures that the brood is concentrated in the least accessible and vulnerable pseudogalls on the tree.

There is no obvious distinguishing feature of the pseudogall in which the queen is found. Finding the queen entails the collection and dissection of almost all the pseudogalls on a tree. Scale insects are sometimes found in *A. drepanolobium* pseudogalls that are inhabited by other species of *Crematogaster* ant (Stanton, Young, pers. comm.) although this was not the case with *C. nigriceps*.

On *A. zanzibarica*, the pattern of pseudogall use follows a similar, if less obvious, pattern to that described above. A slightly different arrangement exists here in that the colony may extend within the trunk of the tree and down towards the ground and the roots. It was not possible to locate a queen from an *A. zanzibarica* tree despite extensive pseudogall collection. She is probably housed within the main body of the trunk or underground.

The ants inhabit the pseudogalls on *A. zanzibarica* in a characteristic manner. The pseudogalls are bi-lobed (as previously described) and there is usually one entrance hole in only one side of the pseudogall. When this is the case, the brood is found in the half of the pseudogall opposite to the side with the entrance hole. There is usually a passage-way through the septum dividing either side of the pseudogall. This arrangement allows the worker ants to guard the brood efficiently as there is only one possible way in for any prospective predator.

On both species of tree, the entrance holes to the pseudogalls are guarded all the time. A ring of worker ants surround the hole from the inside and it is possible to see their antennae protruding from the hole, especially when a worker on the outside attempts to gain access to that pseudogall. I marked individual ants with fluorescent powder and found that the ants appear to have specific pseudogalls that they occupy and return to after a disturbance. Out of 60 ants marked and followed for five minutes, 49 ants returned to the same pseudogalls they had emerged from.

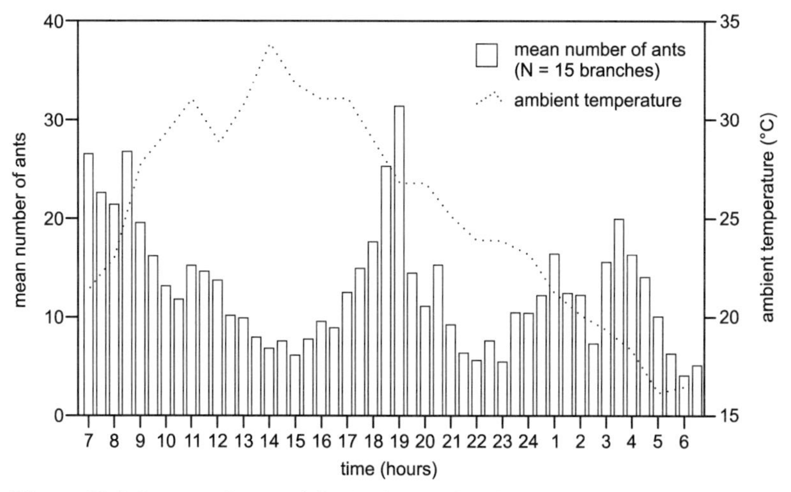

Figure 23.2 Pattern of ant activity (columns) in relation to ambient temperature (dotted line) on new branches of *Acacia drepanolobium*.

Acacia-ant foraging patterns

I carried out 24-hour ant counts on branches on individual trees. I did this by marking 30 cm stretches on 15 branches and then every 30 minutes, counting the number of ants present within those stretches. This showed that the ants, whilst being active all day, are more so in the early morning and evening (Figure 23.2; see also Willmer & Stone 1997 and Willmer *et al.*, Chapter 21). Activity appears to be correlated with both time and ambient temperature although this awaits statistical analysis.

Ants are seen patrolling the plant surface, feeding from petiole glands and grooming themselves. I have observed them several times when they appear to be doing nothing, not an uncommon occurrence within an ant colony as workers can spend over 70% of their total time standing still, Cole (1986). In the case of *A. drepanolobium*, few observations have been made to suggest that these ants are active foragers on the ground. Over a period of five days, I set up a series of pitfall traps around these plants but failed to trap any acacia-ants.

A. zanzibarica ants, however, are more commonly found foraging on the ground. A series of pitfall traps around these trees yielded some samples of acacia-ants. Close examination of foraging ants failed to show what they were foraging for. If it is the case that petiole glands do not provide protein or amino acids in sufficient amounts, then it may be that the ants are foraging to meet their needs for proteinaceous material.

I found that the ants on both species of tree would readily collect proteinaceous material, such as pieces of cheese or other insects that I placed on the thorns. Their show of interest in these offerings suggests that protein may be a limiting resource for them or else they are simply keeping the plant free of alien material. The fact that both the cheese and the insects were taken into the pseudogalls rather than being thrown from the plant or ignored, suggests that the former may be the case. Their reaction to pieces of tape, which were used to mark branches, was different from their reaction to proteinaceous objects. I observed them chewing at the tape, in some cases causing it to become detached from the branch but I did not observe ants carrying pieces of the tape to the pseudogalls. When pieces of paper were stuck onto thorns, the ants showed interest at first, feeling it with their antennae and holding it in their jaws. After a few minutes the paper was ignored and the ants did not come back to it.

Effects of ants and thorns on herbivore feeding behaviour

The armature of long, sharp thorns is a characteristic of ant-acacias and some other *Acacia* species. In the ant-acacias a number of these thorns are swollen at the base into pseudogalls. The effects of unswollen thorns in defending the plant against mammalian browsers remains uncertain (Cole 1986, Potter & Kimmerer 1988, reviewed Myers & Bazely 1991). In *A. drepanolobium* for example, intense browsing pressure from mammals induces an increase in both the density and number of longer thorns produced (Young 1987, Milewski *et al.* 1991), suggesting that thorns are important in defending the plant. In another *Acacia* species (*A. tortilis* (Forssk.) Hayne), however, an increase in the density of thorns protects the twigs but not necessarily the leaves of the plant from being browsed by goats (Gowda 1997). Cooper & Owen-Smith (1986) showed that thorns are effective in as much as they slow down the feeding rate of generalised browsing mammals (goat, lesser kudu, impala), but that the mammals compensate for this by feeding for a longer period of time. Thorns affect larger browsers less than small browsers. For example, Foster & Dagg (1972) and Pellew (1984) showed that the ingestion rate of giraffes (*Giraffa camelopardalis*) is not affected by thorns of *A. drepanolobium*. Therefore, although thorns appear to be an inducible response to mammalian browsing pressure, their actual effectiveness as an anti-herbivore defence, against specialised herbivores at least, is not very clear.

Whilst the effectiveness of symbiotic ants (Madden & Young 1992) and thorns (Young 1987) on their own as a means of defence of *A. drepanolobium* has been considered, the interaction of ants and thorns as a defensive measure has been ignored. The fact that the trees invest energy and resources in thorns as well as ants suggests that there is an important interaction between them which provides a more effective defence than either on their own. I carried out two experiments to look at the effectiveness of the thorns and ants, both on their own and together, as defences against browsing mammalian herbivores (Stapley, 1998). Goats were used as representative mammalian herbivores.

In Experiment one, the goat was allowed to feed for a certain amount of time, from a series of branches and trees with different physical defences present. For example, some branches lacked ants and thorns, others had just ants on them, some had just thorns present, and the remainder had both ants and thorns present (the natural defence condition). Measurements were made of the mean amount eaten from each branch, the mean number of bites taken, the average bite size and the reaction to the defences present. In the Experiment two, the goat was taken to *A. drepanolobium* trees where it was allowed to feed from trees with ants and thorns present, and from trees with just thorns present. This was designed to test the effectiveness of a whole tree's worth of ants rather than those from just one branch. This is an important consideration because when a tree is disturbed, ants from all over the tree will react, not just those from one or two branches. Measurements were made of how long the animal fed for, the approximate percentage of vegetation consumed and its reaction to the defences on the tree.

Statistical analysis of the results showed that the type of defences present did affect the length of time the goat spent browsing and how much vegetation it ate. Figure 23.3 shows the results from the Experiment one and Table 23.1 shows the results from the second experiment.

From the results, it was concluded that the type of defence present on a branch or tree did affect the amount of vegetation eaten (Figure 23.3(i)) and the amount of time the goat spent browsing (Table 23.1). The effect of the ants was more obvious when a whole tree's worth was considered. The ants would swarm on to the goat's head and neck and move to soft areas of skin, for example around the ears, eyes and up the nose. Once in a soft area they would take a bite and not let go. The goat had to stop feeding so that it could dislodge the ants that had climbed onto its body.

The ants congregate on thorn tips, which the goat brushes against, picking the ants up on its coat. Without the thorns, the ants cannot gain access to the goat as easily. However, thorns on their own are a better defence than just ants on their own. This is because the feeding rate of the goat is reduced. The goat has to negotiate the thorns in order to get at the vegetation, which not only reduces the number

Table 23.1 The mean number of bites taken and the median amount of vegetation eaten (%) from trees inhabited by ants and from trees that were lacking ants. The sample size is five in each case. Results were analysed using Student's t-test and Mann-Whitney U-test. Significant results are shown with a *.

day	number of bites taken (in five minutes)			amount of vegetation eaten (% / five minutes)		
	ants	no ants	t-test	ants	no ants	U-test
1	11.2 ± 2.0	80.8 ± 3.5	-17.09 p<0.0001*	5.0	70.0	15.0 p<0.011*
2	11.2 ± 2.6	83.6 ± 6.7	-10.07 p<0.0002*	30.0	75.0	15.0 p<0.011*
3	11.0 ± 3.9	78.4 ± 6.9	-8.52 p<0.0001*	5.0	75.0	15.0 p<0.011*
4	11.0 ± 4.1	100.0 ± 8.4	+9.54 p<0.0002*	5.0	85.0	15.0 p<0.011*

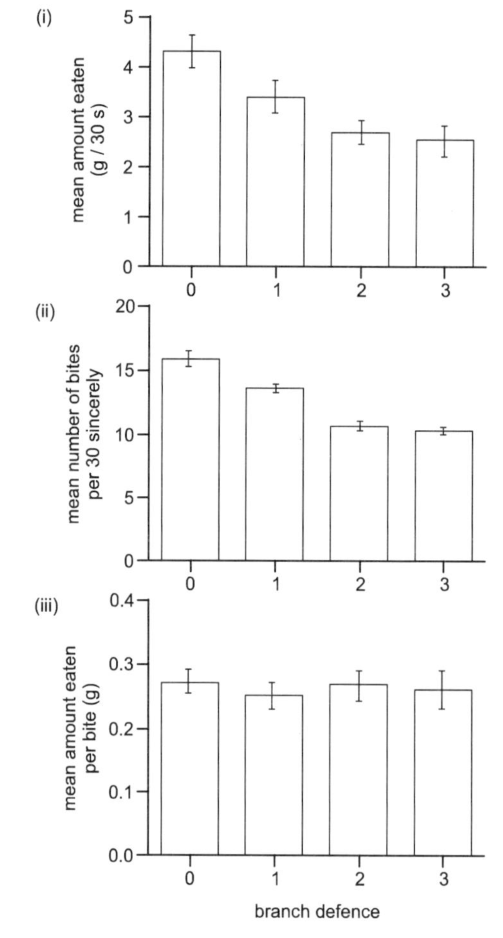

Figure 23.3 (i) Mean amount of vegetation eaten (g) in 30 seconds; (ii) mean number of bites taken in 30 seconds; (iii) mean bite size (g).
Branch defence:
0 = no ants, no thorns
1 = ants only
2 = thorns only
3 = ants and thorns
Error bars show standard error of the mean.

of bites taken in the allotted time but also the amount of vegetation consumed. If thorns are not present, then the goat is able to quickly strip all the leaves from a branch. The ants cannot get on to the goat as easily if thorns are not present.

Ants act as a deterrent to browsing on a particular tree but it is found that thorns do not. If the goat had already been attacked by ants from a particular tree, then it would refuse to return to that tree to browse. It would however, move on to another ant-inhabited tree and feed until the ants present there attacked it. It may be that given enough exposure to ant-inhabited *Acacia drepanolobium* the goat will eventually ignore the trees and not feed from them at all. The goat did not appear to acknowledge the presence of thorns and did not stop feeding because of them.

Therefore, one of the benefits that *A. drepanolobium* gains from its ant inhabitants is defence against browsing mammals, as shown by the reduced browsing pressure experienced on ant-inhabited trees in this investigation. The ants significantly add to the effectiveness of thorns as a defence measure, the combination of these being more effective than either on their own.

Evidence for the evolution of an ant-*Acacia* relationship

Whilst there has been a lot of work done on the ecology of various ant-plant interactions, little work has focused on the reasons why such relationships may have evolved. The main hypothesis for the evolution of the ant-*Acacia* and other myrmecophytic relationships is that the ants act as defences against herbivorous, browsing mammals (Brown 1960). The 'Browsing Mammal Theory' is based on observations made of acacias in Australia and the New and Old Worlds (Belt 1874, Brown 1960).

Brown (1960) notes that most of the 400 Australian *Acacia* species (suggested to now be 800 species: Ross 1981) lack myrmecophytic adaptations (petiole glands, pseudogalls) as well as well-developed thorns. On other continents, however, acacias are predominantly thorny and many have myrmecophytic adaptations (13 known ant-acacias in the New World (Seigler & Ebinger 1995), three known ant-acacias in Africa (Hocking 1970) as well as others with pseudogalls and/or extra-floral nectaries).

Brown suggests that the selection pressures which led to the evolution of thorns and myrmecophytic adaptations amongst the New and Old World acacias are no longer present in modern-day Australia. Whilst large, tree-browsing herbivores are still relatively abundant in the former areas (acacias provide one of the main food sources for mammals such as giraffe, gazelle and antelope) the modern-day endemic, large mammal fauna of Australia does not include similar large herbivores. It may be that the selective pressure of mammalian browsing did once exist in Australia but due to the decline in the large mammal fauna, the presence of thorns and myrmecophytic adaptations now has little adaptive value. It is interesting to note that many of the smaller, shrubby plants and grasses do have thorns on them and that these are browsed by smaller mammals (rodents, kangaroos and other marsupials, for example: Brown 1960).

These ecological differences are also highlighted in the taxonomy of the *Acacia* genus (Ross 1981). The majority of Australian *Acacia* species are from the sub-genus *Heterophyllum* Vassal, a sub-genus that has very few examples in the New or Old Worlds. In contrast, the majority of the *Acacia* species in the latter two areas are classified in the sub-genera *Acacia* and *Aculeiferum*, with ant-acacias occurring in the former. There are only ten or so species from the subgenus *Acacia* known in Australia, none of these have pseudogalls and often the thorns are present only in young plants. There is only one known species from the sub-genus *Aculeiferum* in Australia. There is some dispute as to which of the three sub-genera is the more ancestral (Robbertse (1974), Guinet and Vassal (1978)) and which sub-genus first appeared in Australia (Ross 1981). However there is little doubt that the separation (and hence almost complete isolation) of Australia from the rest of Gondwanaland (the timing of which is not agreed upon but thought to be in the mid-Cretaceous, Raven & Axelrod 1974) has led to the evolution of unique and significantly different Australian *Acacia* species compared to the species of the New and Old Worlds. The lack of thorns and myrmecophytic adaptations are just two of these significant differences (Ross 1981).

Conclusions

The work carried out in Mkomazi has been of a mainly experimental and manipulative nature, attempting to answer a series of questions about these relationships. Answers to some of the questions have already been reported above and come as a direct result of the more structured pieces of work.

There are still many questions relating to this relationship that remain to be answered. It is hoped that the analyses of the data collected in MGR will provide some of those answers. The most important points to be considered include the analysis of the chemical defences of the acacias and the structure and function of the petiole glands. Examination of the range of other insects found on the trees will also provide further information and insight in to this complicated and unusual relationship.

Acknowlegdements

I would like to thank the following for their help in the field and also here in Cambridge: Dr William Foster, Dr George McGavin, Dr Malcolm Coe, Mr Tim Morgan, Mr Nick McWilliam, Dr Graham Stone, Mr Daniel Mafunde, Mr Hamisi Ayubu, Mr Maneno Myinga, Mr Emmanuel Mboya, Mr Raphael Abdallah, Dr Phil Ward, Professor Maureen Stanton; all the support staff at the Royal Geographical Society (with IBG), especially those at FOC Ibaya camp; and the Department of Zoology, University of Cambridge and the BBSRC for funding my work.

References

Anderson, A.N. (1991) Seed harvesting by ants in Australia. In: Huxley, C.R. & Cutler, D.F. (eds.) *Ant-Plant Interactions.* Oxford Science Publications. pp. 493-503.

Belt, T. (1874) *The Naturalist in Nicaragua.* London.

Brown, W.L. (1960) Ants, *Acacias* and browsing mammals. *Ecology* 41: 587-592.

Cherret, J.M., Powell, R.J. & Stradling, D.J. (1989) The mutualism between leaf-cutting ants and their fungus. In: Wilding, N., Collins, N.M., Hammond, P.M. & Webber, J.F. (eds.) *Insect-fungus interactions.* Academic Press, London. pp. 93-120.

Coe, M.J. & Beentje, H. (1991) *A Field Guide to the Acacias of Kenya.* Oxford University Press.

Cole, B.J. (1986) The social behaviour of *Leptothorax allardycei*: time budgets and the evolution of worker reproduction. *Behaviour, Ecology and Sociobiology* 18: 165-173.

Cole, M.M. (1986) *The Savannas: Biogeography and Geobotany*. Academic Press, London.

Cooper, S.M. & Owen-Smith, N. (1986) Effects of plant spinescence on large mammalian herbivores. *Oecologia* 68: 446-455.

Foster, J.B. & Dagg, A.I. (1972) Notes on the biology of the giraffe. *East African Wildlife Journal* 10: 1-16.

Gowda, J.H. (1997) Spines of *Acacia tortilis*: What do they defend and how? *Oikos* 77: 279-284.

Guinet, P.H. & Vassal, J. (1978) Hypotheses on the differentiation of the major groups in the genus *Acacia* (Leguminosae). *Kew Bulletin* 32: 509-527.

Hocking, B. (1970) Insect associations with the swollen thorn Acacias. *Transactions of the Royal Entomological Society of London* 122: 211-255.

Janzen, D. (1966) Co-evolution of mutualisms between ants and Acacias in Central America. *Evolution* 20: 249-275.

Janzen, D. (1967a) Interaction of the Bull's-Horn Acacia (*Acacia cornigera* L.) with an ant inhabitant (*Pseudomyrmex ferruginea* F. Smith) in Eastern Mexico. *University of Kansas Science Bulletin* 47: 315-558.

Janzen, D. (1967b) Fire, Vegetation Structure and the Ant X *Acacia* interaction in Central America. *Ecology* 48: 26-35.

Jolivert, P. (1985) Un mrymécophyte hors de son pays d'origine: *Clerodendrum fallax* Lindley, 1844 (Verbenaceae). *Bulletin de la Société Linneienne* Lyon 54: 122-128.

Koptur, S. (1979) Facultative mutualism between weedy vetches bearing extrafloral nectaries and weedy ants in California. *American Journal of Botany* 66: 1016-1020.

Madden, D. & Young, T.P. (1992) Symbiotic ants as an alternative defence against giraffe herbivory in spinescent *Acacia drepanolobium*. *Oecologia* 91: 235-238.

Milewski, A.V., Young, T.P. & Madden, D. (1991). Thorns as induced defences: experimental evidence. *Oecologia* 86: 70-75.

Monod, T. & Schmitt, C. (1968) Contribution à l'étude des pseudo-galls formicaires chez quelques *Acacias* africains. *Bulletin de l'Institute Fondamentale d'Afrique Noire Series A* 30: 1302-1333.

Myers, J.H. & Bazely, D. (1991) Thorns, spines, prickles and hairs: are they stimulated by herbivory and do they deter herbivores? In: Tallamy, D.W. & Raupp, M.J. (eds.) *Phytochemical Induction by Herbivores*. Wiley. New York.

Pellew, R.A.P. (1984) Giraffe and okapi. In: Macdonald, D. (ed.) *Encyclopaedia of Mammals*. Facts on File, New York.

Potter, D.A., Kimmerer, T.W. (1988) Do holly leaves really deter herbivory? *Oecologia* 75: 216-221.

Raven P.H. & Axelrod, D.I. (1974) Angiosperm biogeography and past continental movements. *Annals of the Missouri Botanical Garden* 61: 539-673.

Robbertse, P.J. (1974) The genus *Acacia* in South Africa, II, with special reference to the morphology of the flower and inflorescence. *Phytomorphology* 24: 1-15.

Ross, J.H. (1981) An analysis of the African *Acacia* species: their distribution, possible origins and relationships. *Bothalia* 13: 389-413.

Seigler, D.S. & Ebinger, J.E. (1995) Taxonomic revision of the ant-Acacias (Fabaceae, Mimosoideae, *Acacia*, series Gummiferae) of the New World. *Annals of the Missouri Botanical Garden* 82: 117-138.

Stapley, L. (1998) The interaction of thorns and symbiotic ants as an effective defense mechanism of swollen-thorn acacias. *Oecologia* 115: 401-405.

Stapley, L. (1999) Physical worker casts in colonies of an acacia-ant (*Crematogaster nigriceps*) correlated with an intra-colonial division of defensive behaviour. *Insectes Sociaux* 46: 146-149.

Stephenson, A.G. (1982) The role of extra-floral nectaries of *Catalpa speciosa* in limiting herbivory and increasing fruit production. *Ecology* 63: 663-669.

Treseder, K.K., Davidson, D.W. & Ehleringer, J.R. (1995) Absorption of ant-provided carbon dioxide and nitrogen by a tropical epiphyte. *Nature* 375: 137-139.

Willmer, P.G. & Stone, G.N. (1997) How aggressive ant-guards assist seed-set in *Acacia* flowers. *Nature* 388: 165-167.

Young, T.P. (1987) Increased thorn length in *Acacia drepanolobium*—an induced response to browsing. *Oecologia* 71: 436-438.

Young, T.P., Stubblefield, C.H. & Isbell, L.A. (1997) Ants on swollen-thorn *Acacias*: Species co-existence in a simple system. *Oecologia* 109: 98-107.

CHAPTER 24

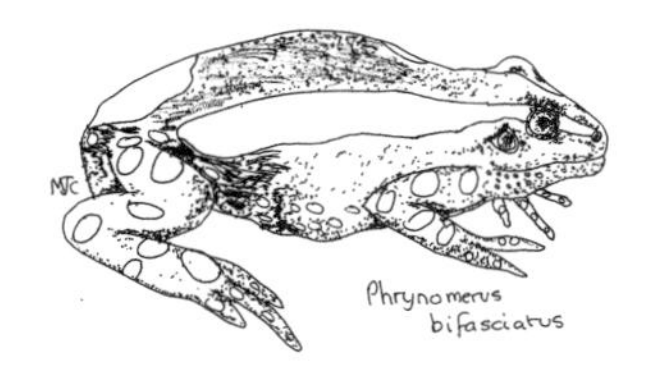

Amphibians of Mkomazi

Michael I. Cherry, Melissa J. Stander & John C. Poynton

Introduction

This study provides a checklist of 14 species of amphibians from the Mkomazi Game Reserve, as part of a wider study of the biodiversity of this reserve. The amphibian fauna is characteristic of a moderately arid savanna. The reserve contains outlying hills of the Pare Mountains, but despite extensive searching, no representatives of the Afromontane fauna were found within its boundaries, and as far as amphibians are concerned, the conservation value of the Mkomazi Game Reserve appears to be confined primarily to savanna forms. A more detailed study of the transition from a fairly arid savanna to an Afromontane fauna is merited.

Methods

Comprehensive sampling of the anuran fauna of the Mkomazi Game Reserve was conducted in November–December 1995 by MIC, and again in April–May 1996 by MJS. Specimens were collected by hand and using pitfall traps. Netting in permanent waterbodies was not employed, and the absence of members of the genus *Xenopus* from this checklist can almost certainly be attributed to this. Specimens were fixed in 70% alcohol. Localities for specimens are listed in the appendix.

Results

Order: Anura

Family: Microhylidae

Phrynomantis bifasciatus (Smith). Recorded from Ubani *mbuga*. Widely distributed in savanna areas from north-eastern South Africa, Botswana and Swaziland, northwards to Kenya.

Family: Hemisotidae

Hemisus marmoratum marmoratum (Peters). Recorded from Simba plot, Nyati plot, Dindira Dam, Mbula waterhole and Ibaya Ravine. This species is found in savanna areas from north-eastern South Africa, Botswana and Swaziland, northwards to southern Somalia.

Family: Ranidae

Pyxicephalus edulis Peters. Recorded on the road between Mbula waterhole and Ibaya, and from Ubani *mbuga*. Found in savanna areas from north-eastern South Africa, Botswana, and Swaziland northwards through east Africa to Nigeria.

Hildebrandtia ornata (Peters). Recorded from Ubani *mbuga*, Kavateta Dam, Dindira Dam and Mbula waterhole. Found in savanna from north-eastern South Africa and northern Namibia northwards to Kenya.

Ptychadena anchietae (Bocage). Recorded from Maji Kininua (at 1,540 m), Kavateta Dam, Ibaya Hill, a buffalo wallow near Mwasumbi plot (3°59.24'S, 37°48.64'E), Kisima plot, Ngurunga Dam and Dindira Dam. The most widespread grass frog in the reserve, and very abundant after rain. Found in savanna areas from north-eastern South Africa and Botswana northwards to Eritrea.

Ptychadena mossambica (Peters). Recorded from Ubani *mbuga*, Dindira Dam and Mbula waterhole. Found in savanna areas from north-eastern South Africa and Botswana northwards to southern Somalia.

Ptychadena schillukorum (Werner). Recorded from Dindira Dam, Kavateta Dam and Ubani *mbuga*. Found from Mozambique northwards to the Sudan.

Phrynobatrachus acridoides (Cope). Recorded from Dindira Dam, Ibaya Ravine and near the Mzukune River. Found from north-eastern South Africa northwards through east Africa to west Africa.

Phrynobatrachus natalensis (Smith). Recorded from Ibaya Ravine. This species is very widespread, being found in savanna areas south of the Sahara.

Family: Rhacophoridae

Chiromantis petersii Boulenger. Recorded 1.73 km from Ubani plot (3°53.78'S, 37°54.74'E), Dindira Dam, Mbula waterhole, and a large waterhole at 3°55.09'S, 37° 49.64'E. Found from northern Tanzania to northern Somalia and Ethiopia.

Hyperolius pusillus (Cope). Recorded from Dindira Dam. This species is found from eastern South Africa to southern Somalia.

Kassina somalica Scortecci. Recorded from Mbula waterhole and Ibaya camp. This species occurs only from northern Tanzania to southern Somalia.

Family: Bufonidae

Bufo xeros Tandy *et al.* Recorded from the Umba River, and the plain between Kisiwani and Ibaya camp. This species is found in savanna areas from Tanzania westwards to west Africa.

Bufo gutturalis Power. Recorded from Ibaya Hill and Ibaya Ravine. The species is widely distributed from eastern South Africa northwards to Kenya.

Discussion

The amphibian fauna is characteristic of fairly dry savanna, as shown by the presence of *Kassina somalica, Chiromantis petersii* and *Bufo xeros*, approaching here their southern limits from the Horn of Africa (Poynton 1990). As the amphibian fauna of Tanzania is still in need of comprehensive taxonomic analysis, the above identifications are tentative. Particular difficulty is presented by some of the *Ptychadena* specimens which do not allow ready separation into *P. mossambica* and *P. schillukorum*: some degree of hybridization is indicated.

The reserve contains outlying hills of the Pare Mountains, but despite extensive searching, none of the amphibian taxa distinctive of rain forests (Schiøtz 1981), or even representatives characteristic of the Afromontane region (White 1983), were found. As far as amphibians are concerned, the conservation value of the Mkomazi Game Reserve is confined primarily to savanna forms, although the Afromontane species *Callulina kreffti* Neiden and *Rana angolensis chapini* Noble were collected in the Pare Mountains, close to the boundaries of the reserve. This transition from a fairly arid savanna to an Afromontane fauna awaits a more detailed study, of which this collection should form an important part. *Kassina somalica, Chiromantis petersii* and *Bufo xeros* are probably the most important species conserved in the reserve, as their distributions are the most restricted.

Acknowledgments

We gratefully acknowledge financial support from the Commonwealth Science Council, which enabled our participation in the Mkomazi Ecological Research Programme.

References

Bocage, J.V.B. (1867) Batriciens nouveaux de l'Afrique occidentale (Loanda et Benguella). *Proceedings of the Zoological Society of London* 35: 843-846.

Boulenger, G.A. (1882) *Catologue of the Batrachia Salientia Ecaudata in the collection of the British Museum (2nd Edition).* British Museum (Natural History), London.

Cope, E.D. (1862) On the limits and relations of the Raniformes. *Proceedings of the Academy of Natural Sciences of Philadelphia* 1864: 181-183.

Cope, E.D. (1867) On the families of the raniform Anura. *Journal of the Academy of Natural Sciences of Philadelphia (2)* 6(4): 189-206.

Lanza, B. (1981) A check-list of the Somali amphibians. *Monitore Zoologico Italiano N.S., Suppl.* 15(10): 151-186.

Peters, W. (1866) Ueber neue amphibien des zoologisches Museum zu Berlin. *Mber. Akad. wiss. Berl.* 1866: 86-94.

Peters, W. (1878) Über die von Hrn. J. M. Hildebrandt während sienen letzten ostaafrikanischen Reise gesammelten Säugethiere und Amphibien. *Monatsb. Akad. wiss. Berlin* 1878: 194-209.

Peters, W. (1882) *Naturwissenschaftliche Reise nach Mossambique. Zoologie. III. Amphibien.* G. Reimer, Berlin.

Power, H.F. (1926) Note on the occurrence of hybrid Anura at Lobatsi, Bechuanaland Protectorate. *Proceedings of the Zoological Society of London* 1926: 777-778.

Poynton, J.C. (1990) Composition and subtraction patterns of the east African lowland amphibian fauna. In *Vertebrates in the Tropics*, Museum Alexander Koenig, Bonn. pp. 285-296.

Schiøtz, A. (1981) The Amphibia in the forested basement hills of Tanzania: a biogeographical indicator group. *African Journal of Ecology* 19: 205-207.

Smith, A. (1849) *Illustrations of the Zoology of South Africa. Vol. 3. Reptilia.* Smith, Elder & Co., London.

Tandy, M., Tandy, J., Keith, R. & Duff-Mackay, A. (1976) A new species of *Bufo* (Anura: Bufonidae) from Africa's dry savannas. *Pearce-Sellards Ser. Texas Mem. Mus.* 24: 1-20.

Werner, F. (1907) Ergebnisse der mit Subvention aus der Erbschaft Treitl unternommenen zoologischen Forschungsreise Dr Franz Werner nach dem ägyptischen Sudan und Nord-Uganda. XII. Die Reptilien und Amphibien. *Sber. Akad. Wiss. Wien.* 116: 1823-1926.

White, F. (1983) *The Vegetation of Africa. A descriptive memoir to accompany the UNESCO-AETFAT-UNSO Vegetation Map of Africa.* UNESCO, Paris.

Checklist: Amphibians of Mkomazi

Class AMPHIBIA

Order ANURA

Family Microhylidae
Phrynomantis bifasciatus (Smith) – Banded rubber frog

Family Hemisotidae
Hemisus marmoratum marmoratum (Peters) – Mottled shovel-nosed frog

Family Ranidae
Pyxicephalus edulis Peters – African bullfrog
Hildebrandtia ornata (Peters) – Ornate frog
Ptychadena anchietae (Bocage) – Plain grass frog
Ptychadena mossambica (Peters) – Broad-banded grass frog
Ptychadena schillukorum (Werner)
Phrynobatrachus acridoides (Cope) – East African puddle frog
Phrynobatrachus natalensis (Smith) – Snoring puddle frog

Family Rhacophoridae
Chiromantis petersii Boulenger
Hyperolius pusillus (Cope) – Water-lily frog
Kassina somalica Scortecci

Family Bufonidae
Bufo xeros Tandy *et al.*
Bufo gutturalis Power – Guttural toad

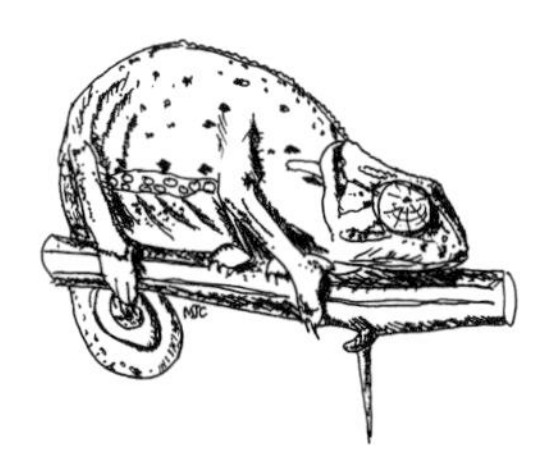

CHAPTER 25

Reptiles of Mkomazi

Alexander F. Flemming & Michael F. Bates

Introduction

Tanzania has a rich reptile fauna of over 270 species, but little is known of their biology, distribution and zoogeographical affinities. Broadley & Howell (1991) published a comprehensive list with synoptic keys of reptiles in Tanzania; the only previous checklist available for this country was that of Loveridge (1957). Broadley & Howell (1991) attempted to analyse the zoogeographical affinities of the Tanzanian reptiles and assigned the taxa to the floristic regions defined by White (1983). They proposed a number of regions, of which the Somalia-Masai and Afromontane ones are relevant to the present study. According to Broadley and Howell (1991), the Somalia-Masai extends into northern Tanzania in the form of extensive *Acacia-Commiphora* deciduous bushland. The Afromontane element is represented by the Usambara and Pare Mountains, which lie adjacent to the boundary of the Mkomazi Game Reserve (MGR).

The aim of the present study is to provide a check list of the reptiles collected in the MGR and to comment on the zoogeographical affinities of its fauna.

Materials and methods

Comprehensive sampling of the reptile fauna in the MGR was conducted during November–December 1995 and in April–May 1996 by one of us (AFF). Specimens were collected by noose or pit-fall traps. The occasional specimen was collected by the staff at the Ibaya Research Centre, or by visiting scientists. Specimens were fixed in 10% formalin and preserved in 70% ethanol, and are housed in the collection of the National Museum, Bloemfontein. Specimens were identified (by MFB) using the synoptic keys of Broadley & Howell (1991). Dr D. Broadley (Natural History Museum of Zimbabwe) kindly identified a few of the problem specimens.

Species accounts

Scientific and common names (where available) are given, together with the National Museum catalogue numbers. Exact localities are given in the Appendix.

Order: Testudines
Family: Pelomedusidae
Pelomedusa subrufa subrufa (Lacepède) – Helmeted Terrapin
MATERIAL: One specimen: NMB R7946, near Kisima. Adult, carapace length 103.4 mm, plastron length 89.2 mm, found in a pool on the road. Carapace dull brown, plastron black with a few scattered cream patches, especially mid-ventrally.
REMARKS: Some authors (e.g. Boycott & Bourquin 1988) regard *P. subrufa* as monotypic, but the status of the three subspecies (*subrufa*, *olivacea* [Schweigger], *nigra* [Gray]) listed by Iverson (1992) remains unresolved.

Family: Testudinidae
Geochelone pardalis (Bell) – Leopard Tortoise
MATERIAL: One specimen: NMB R7947, Simba Hill Plot. Juvenile with carapace length 54.0 mm, plastron length 47.7 mm; found near mud puddle in grassland.
REMARKS: Broadley (1989a) and Broadley & Howell (1991) recognize *G. p. babcocki* (Loveridge) which has a widespread range in southern and eastern Africa, including Tanzania. The Simba Hill specimen essentially agrees with Broadley's (1989a) description of *G. p. babcocki*, but pending a more thorough review we follow Boycott & Bourquin (1988) and Branch (1988) in treating the latter as a synonym of *G. pardalis*.

Kinixys spekii (Gray) – Speke's Hinged Tortoise
MATERIAL: Photographic record: Ibaya Camp.
REMARKS: *Kinixys spekii* Gray, long treated as a subspecies of *K. belliana*, was revived as a full species by Broadley (1989b & 1993). Broadley (1989b) and Broadley & Howell (1991) recognized *Kinixys belliana zombensis* Hewitt as a southern and eastern subspecies. However, after examining additional material Broadley (1992 & 1993) relegated *K. b. zombensis* to the synonomy of *K. b. belliana*, but continued to recognise *K. b. nogueyi* (Lataste) of west Africa. In Tanzania *K. spekii* occurs in the north and west with a few *K. b. belliana* localities in the east; the Ibaya record is situated near recorded localities for both species (Broadley 1992).

Malacochersus tornieri (Lindholm) – Pancake Tortoise
MATERIAL: Photographic record: Kamakota Hill.
REMARKS: *M. tornieri* inhabits rocky hills in central Kenya and northern and central Tanzania (Broadley 1989c). The Kamakota record is situated about 250 km SE of the nearest recorded locality (see map in Broadley 1989c).

Order: Squamata
Suborder: Sauria
Family: Gekkonidae

Lygodactylus gravis (Pasteur) – Usambara Dwarf Gecko

MATERIAL: One specimen: NMB R7939, Maji Kununua Mountain. Female measuring 37.9 mm SVL found under rotten log; six transverse mid-dorsal scale rows per caudal verticel; seven lamellae under 4th toe (right foot).

REMARKS: Maji Kununua represents a small north-westerly range extension as this species was previously known only from the Usambara Mountains (Broadley & Howell 1991).

Lygodactylus manni (Loveridge) – Mann's Dwarf Gecko

MATERIAL: One specimen: NMB R7307, about 10 km E of Kivingo. Female (30.1 + 26.8 = 56.9 mm) found in a tree; dense black network on throat.

REMARKS: This species is restricted to northern Tanzania and the Lake Victoria region of south-western Kenya (Broadley & Howell 1991). Pasteur (1964) referred *L. picturatus ukerewensis* Loveridge to the synonomy of *L. manni*, but as female *L. p. ukerewensis* display merely a fine grey stipple on the throat, unlike the largely black throat of the Kivingo female, two taxa may be involved (D.G. Broadley, pers. comm., May 1998).

Lygodactylus luteopicturatus luteopicturatus (Pasteur)—Yellow-headed Dwarf Gecko

MATERIAL: Two specimens: NMB R7308, 3 km E of Kisiwani; NMB R7909, river at Zange Gate. Two males, one collected in a tree, the other in a shrub, with nine and seven preanal pores, respectively.

REMARKS: Found only in eastern Tanzania and south-eastern Kenya; the other subspecies, *L. l. zanzibaritis* Pasteur, is endemic to Zanzibar Island (Broadley & Howell 1991).

Cnemaspis africana (Werner) – Usambara Forest Gecko

MATERIAL: Eleven specimens: NMB R7894–7896, Maji Kununua Mountain; NMB R7309-7312, top of Maji Kununua Mountain; NMB R7890, Ibaya Hill; NMB R7891-7893, South Pare Mountains. The seven Maji Kununua specimens were found on trees and rotting logs; the largest (NMB R7309, female) measures 38 mm SVL; six are males with 8–10 pre-cloacal pores. The male (NMB R7891), found on a tree trunk, was the largest of the three South Pare specimens, measuring 48.4 mm SVL with 12 pre-cloacal pores; the two females were in rocky outcrops. The Ibaya Hill specimen (NMB R7890), collected in April 1996 in the grass of a forest clearing, is a hatchling measuring 17.2 + 17.1 = 34.3 mm.

REMARKS: Found only in central Kenya and eastern Tanzania (Broadley & Howell 1991).

Hemidactylus brookii angulatus (Hallowell) – Angulate Tropical Gecko
MATERIAL: One specimen: NMB R7897, Ibaya Campsite B. Male (49.1 mm SVL) with 30 preanofemoral pores found on a building.

Hemidactylus mabouia (Moreau de Jonnès) – Tropical House Gecko
MATERIAL: One specimen: NMB R7898, river at Zange Gate. Male (53.3 mm SVL) with 35 preanofemoral pores found in a tree.
REMARKS: Broadley & Howell (1991) give a preanofemoral pore range of 22–40 for *H. mabouia*. The pore count of 35 for the Zange Gate male compares to counts of 26–28 and 29–33 in recent material (44–66 mm SVL) from Mocambique (NMB R6507-6508, 12 km E of Moatize) and Malawi (NMB R6530, 15 km N of Club Makololo, Mangochi district; NMB R6519, 6522, Kachula village, Lake Chilwa district; NMB R6499-6500, Nkhotakota village) respectively.

Hemidactylus platycephalus (Peters) – Baobab Gecko
MATERIAL: 21specimens: NMB R7313–7315, 3 km E of Kisiwani; NMB R7316–7317, Ibaya Hill; NMB R7318–7319, 100 m SW of Ibaya Camp; NMB R7320–7321, 1 km W of Ngurunga Dam; NMB R7322, Ngurunga Dam; NMB R7323, Kisima Mountain; NMB R7324, half-way between Mabata and Kamakota Hill; NMB R7325, 3 km S of Kisiwani (half-way up South Pare Mountain); NMB R7899, between Vitewini Ridge and Ndea *mbuga*; NMB R7900–7902, near Ibaya Camp; NMB R7903, river at Vitewini; NMB R7904, Ubani; NMB R7905–7906, Lake Chala. Specimens were collected on and under the bark of trees (including baobabs), in rock crevices and on rocky outcrops. The largest male (NMB R7317) measures 83.6 + 80.4 = 164.0 mm; largest female, gravid with two eggs (NMB R7899), 83.7 + 53.8r (tail regenerated) = 137.5 mm. Preanofemoral pores in males (65.0–83.6 mm SVL) 40–48 (mean = 44.1, SD = 2.32, n = 14).
REMARKS: In their key to the genus *Hemidactylus* Broadley & Howell (1991) report 45-57 preanofemoral pores for *H. platycephalus* and 22–40 for *H. mabouia*. Pore counts in recent material (77–89 mm SVL) from Malawi were 44 (NMB R6494, Luloma Peninsula) and 50–54 (NMB R6533, 15 km N of Club Makokolo, Mangochi district; NMB R6504, Karonga; NMB R6509, Nkhata Bay; NMB R6514–6515, Otter Point, Cape McClear). Pore counts for the Mkomazi area are therefore generally low for *H. platycephalus* and mostly intermediate between the two species. In their key for *Hemidactylus* Broadley & Howell (1991) state that the dorsal granules and tubercles of *H. platycephalus* are smooth. However, in the Mkomazi specimens, several of these are weakly but distictly ribbed, somewhat similar to *H. mabouia*.

Hemidactylus squamulatus squamulatus (Tornier) – Squamate Tropical Gecko
MATERIAL: Two specimens: NMB R7326, 100 m SW of Ibaya Camp; NMB R7907, Ibaya Campsite B. Two males: NMB R7326 (37.9 mm SVL) with 14 precloacal

pores, collected from a rock crevice; NMB R7907 (39.0 mm SVL) with 12 pores, found on building.

Hemidactylus tanganicus (Loveridge) – Tanzanian Tropical Gecko
MATERIAL: Two specimens: NMB R7908, Ibaya Campsite B; NMB R7327, Pangaro Plot, about 5 km N of Njiro Gate. Male (NMB R7327) 78.1 mm SVL with 42 preanofemoral pores found under bark of burnt tree; female (NMB R7908) 63.2 + 42.1r (tail regenerated) = 105.3 mm on tree.
REMARKS: Previously known only from the holotype collected at Dutumi near Kisaki in Tanzania (see Broadley & Howell 1991). The two specimens are largely in agreement with Loveridge's (1929) description and plate, but differ in that the free edges of lateral ventral scales are smooth and not "markedly serrate", and there is no "more or less regular but interrupted line of enlarged round tubercles" bordering the belly laterally. Loveridge (1929) noted that a male from Mbunyi in Kenya, previously assigned by him (Loveridge 1920) to *H. ruspolli* Boulenger, may be referable to *H. tanganicus*. No details on number of preanofemoral pores were provided, and NMB R7327 is thus the first confirmed male *H. tanganicus*. Its high preanofemoral pore count (42) distinguishes this species from both *H. ruspolii* Boulenger (28–34 pores) and *H. barodanus* Boulenger (6–11, eight in type), considered closely related by Loveridge (1920, 1929 & 1947). A more detailed description of the Mkomazi specimens is in preparation (Bates & Flemming).

Family: Agamidae
Agama agama usambarae (Barbour & Loveridge) – Usambara Rock Agama
MATERIAL: Fifteen specimens: NMB R7911, Ibaya; NMB R7333–7335, 100 m SW of Ibaya Camp; NMB R7328–7330, Ibaya Hill; NMB R7331–7332, half-way up Ibaya Hill; NMB R7340, Kamakota Hill; NMB R7339, half-way between Kamakota Hill and Mabata Hill; NMB R7336, Ngurunga Dam; NMB R7910, South Pare Mountains; NMB R7338, half-way up South Pare Mountains, about 3 km S of Kisiwani; NMB R7337, Zange Gate. Specimens were collected on rocky outcrops and in rock crevices; the Zange Gate specimen was found among buildings. Largest males: NMB R7910 (117 + 129 = 246 mm), NMB R7339 (113 + 211 = 324 mm); largest female: NMB R7333 (94 + 160 = 254 mm). The six males have 11–14 large preanal pores, often with a parallel row of smaller pores (generation glands?) anterior thereto, numbering 12–14. Venter blue and throat pink, but underparts uniform cream in NMB R7337.
REMARKS: According to Broadley & Howell (1991) this subspecies is endemic to the Usambara Mountains of north-eastern Tanzania.

Agama sp.
MATERIAL: One specimen: NMB R8039, Kikolo plot. Female (68 + 77 = 145 mm) found in savanna. Dorsum tan with four pairs of paravertebral dark blotches be-

tween axilla and groin; dorsals large, more-or-less uniform, mucronate, strongly keeled and spinose posteriorly, much larger than the mucronate, weakly but distinctly keeled ventrals; 57 scales around midbody; ear opening more than half diameter of eye; 3rd toe longer than 4th, lamellae under 3rd toe 19 left, 18 right, under 4th toe 17 left, 18 right; no vertebral, caudal or nape crest.
REMARKS: This specimen could not be identified to species level (see key in Broadley & Howell 1991).

Family: Chamaeleonidae
Chamaeleo dilepis dilepis (Leach) – Common Flap-necked Chameleon
MATERIAL: NMB R7343, Ibaya Camp; NMB R7341, top of Ibaya Hill; NMB R7342, Zange Gate. NMB R7341 is an adult (132.5 mm SVL) found on a shrub in the forest; the other two specimens are juveniles, NMB R7343 found on a tree and NMB R7342 on a shrub.

Chamaeleo bitaeniatus (Fischer) – Side Striped Chameleon
MATERIAL: Specimen collected in grassland near northern side of Vitewini Ridge (Coe, pers comm).

Rhampholeon sp. (near *platyceps* Günther) – Pigmy Chameleon
MATERIAL: Two specimens: NMB R7913-7914, South Pare Mountains. NMB R7913 (43.2 + 22.9 = 66.1 mm) in leaves of banana tree; NMB R7914 (43.7 mm) on tree trunk. Small flexible rostral appendage on snout; no tufts of scales on the chin, or spines on the palms; distinct pits in axilla and groin.
REMARKS: Using Broadley & Howell's (1991) diagnostic key the two specimens key out to "*Rhampholeon* sp."; this was in reference to a specimen from Bondwa in the Uluguru Mountains.

Family: Scincidae
Melanoseps loveridgei (Brygoo & Roux-Estéve) – Loveridge's Limbless Skink
MATERIAL: One specimen: NMB R7361, 500 m SW of Ibaya Camp. One specimen (40.0 + 11.8 = 51.8 mm, tail length 29.5% SVL) collected under leaf litter in valley; 139 scales in a row between mental and vent; 18 scales around midbody.
REMARKS: This record represents a considerable range extension as the species was previously known only from southern Tanzania and north-eastern Zambia (Broadley & Howell 1991). *M. ater longicauda* Tornier (known from only two specimens), treated as a full species by Broadley & Howell (1991), is the only other *Melanoseps* occuring in north-eastern Tanzania, but it is easily distinguished from *M. loveridgei* by its lower mental to vent count (less than 130) and long tail (more than half SVL).

Mabuya brevicollis (Wiegmann) – Short-necked Skink
MATERIAL: Seven specimens: NMB R7344, Ibaya Camp; NMB R7917, near Ibaya

Camp; NMB R7345–7347, 100 m SW of Ibaya Camp; NMB R7918, waterhole near Dindira Dam; NMB R7348, locality unknown. Collected from crevices in rocky outcrops, in an open sandy area (NMB R7344) and at a waterhole (NMB R7918). Largest specimen (NMB R7917) 138.6 + 209 = 347.6 mm; three juveniles (NMB R7345–7347, 56.5–59.9 mm SVL) have black dorsa with numerous pale spots, very different to the striped colour pattern of adults; 31–33 scale rows around midbody.

Mabuya maculilabris maculilabris (Gray) – Speckle-lipped Skink
MATERIAL: One specimen: NMB R7919, Lake Chala. Adult (91.8 mm SVL) found amongst rocks at the edge of the lake.

Mabuya planifrons (Peters)
MATERIAL: One specimen: NMB R7349, Ibaya Camp. Adult male (85.7 mm SVL) found in tree; 29 scales rows round midbody.
REMARKS: Broadley (1997) lists *M. binotata* (Bocage) as a subspecies of *M. planifrons* without explanation.

Mabuya quinquetaeniata margaritifer (Peters) – South-eastern Rainbow Skink
MATERIAL: One specimen: NMB R7350, half-way up South Pare Mountain, about 3 km S of Kisiwani. Adult male (78 + 156 = 234 mm) found in rock crevice; 37 scale rows around midbody.
REMARKS: According to Broadley *et al.* (1998) "This form is sympatric with *Mabuya quinquetaeniata* (Lichtenstein) in southeastern Kenya and is to be reinstated as a full species (Broadley & Bauer, in prep.)."

Mabuya striata striata (Peters) – Common Striped Skink
MATERIAL: Five specimens: NMB R7922, Ibaya; NMB R7356, 100 m SW of Ibaya Camp; NMB R7920-7921, Ibaya Hill; NMB R7357, Pangaro Plot, about 5 km N of Njiro Gate. Collected under bark, on dead tree, amongst loose stones in rocky outcrop, and on building. Largest specimen (NMB R7922) 96.0 mm SVL.

Mabuya varia (Peters) – Variable Skink
MATERIAL: Seven specimens: NMB R7354–7355, 100 m SW of Ibaya Camp; NMB R7351, about 5 km E of Ibaya Camp; NMB R7352–7353, Ibaya Hill; NMB R7923, Ngurunga Dam; NMB R7924, Ubani. Collected in rocky outcrops and crevices, under stones in an open area (NMB R7351), and on a tree trunk (NMB R7924). Largest specimens: NMB R7354 (59.3 mm SVL), NMB R7923 (48.2 + 71.6 = 119.8 mm). Four specimens have, in additional to pale lateral and dorsolateral stripes, vague indications of a pale vertebral stripe.
REMARKS: Although Broadley (1966) analysed and treated *M. v. nyikae* Loveridge from the Nyika Plateau in Malawi as a synonym of *M. varia*, Broadley & Howell

(1991) list "*M. v. varia*" without explanation regarding subspecific status. We follow recent authors (e.g. Branch 1988) in treating *M. varia* as a monotypic species.

Lygosoma afrum (Peters) – Peters' Writhing Skink
MATERIAL: One specimen: NMB R7359, 500 m SW of Ibaya Camp. Male (92 + 75 = 167 mm) from a rock crevice in the valley; dorsum mostly uniform brown, but lateral band on body with white-centred scales; 28 midbody scale rows, supranasals discrete.
REMARKS: Similar to *L. s. sundevallii*, but distinguished therefrom by its larger size (80-140 mm SVL) and (usually) speckled dorsum (Broadley & Howell 1991).

Lygosoma sp. (near *somalicum* [Parker]) – Somali Writhing Skink
MATERIAL: Three specimens: NMB R7358, 7916, Ibaya Camp; NMB R7915, Ibaya Hill. Collected in open sandy areas and under a loose stump (NMB R7915). Largest specimen: (NMB R7358) 74.5 mm SVL.
REMARKS: The Ibaya specimens have the colour pattern and long fifth toe of *L. somalicus*, known only from north-western Somalia. D.G. Broadley (pers. comm.) compared the three Ibaya lizards with two *L. sundevallii* (A. Smith) from Kimana (N of Mt Kilimanjaro) in the collection of the Natural History Museum of Zimbabwe (NMZB-UM 6535, 6536). He reports that the Kimana specimens lack dark head markings (present in the Ibaya specimens), a condition typical of *L. sundevallii* throughout its extensive range; also, the Ibaya specimens have more strongly depressed and pointed snouts. *L. somalicum* was described as a subspecies of *Riopa modesta* but elevated to full species status by Lanza (1990); as *Sepacontias modestus* Gunther (= *L. s. modestum*) was placed in the synonomy of *L. s. sundevallii* by Broadley & Howell (1991), *L. sundevallii* reverts to binomials.

Panaspis wahlbergii (A. Smith) – Savanna Snake-eyed Skink
MATERIAL: One specimen: NMB R7360, about 10 km E of Kivingo. Specimen (36.4 mm SVL) found under leaf litter; 24 midbody scale rows.

Family: Lacertidae

Adolphus jacksoni (Boulenger) – Jackson's Forest Lizard
MATERIAL: One specimen: NMB R7362, Kisiwani. Adult (81.0 mm SVL) found amongst small reddish shrubs; six longitudinal rows of ventrals; 18 femoral pores, nine on each thigh.

Nucras boulengeri (Neumann) – Boulenger's Scrub Lizard
MATERIAL: Five specimens: NMB R7363, Ibaya Camp; NMB R7938, Mbulu Dam; NMB R7936, Mwasumbi Plot; NMB R7935, 7937, Simba Hill Plot. NMB R7363 was collected in an open sandy area. Largest specimens: 58.4 mm SVL (NMB R7935); 50.5 + 93.4 = 143.9 mm (NMB R7938). NMB R7363 (52.8 mm SVL) is a male with 26 femoral pores (14 left, 12 right).

Heliobolus spekii spekii (Günther) – Southern Speke's Sand Lizard
MATERIAL: Seven specimens: NMB R7931, Ibaya Plot; NMB R7364, Ibaya Hill; NMB R7365, about 10 km E of Kivingo; NMB R7932, Mbulu Dam; NMB R7929, Mwasumbi Plot; NMB R7928, Simba Hill Plot; NMB R7930, Vitewini Ridge Plot. Found in open sandy areas (NMB R7364–7365), at the base of a grassy hillside (NMB R7928) and at a large waterhole (NMB R7930). Largest specimen: 59 + 95 = 154 mm (NMB R7364). Two males have 28 femoral pores each—NMB R7364 (15 left, 13 right), NMB R7365 (14 on either side).
REMARKS: This species is restricted to Kenya (south of the Tana River) southwards to central Tanzania (Broadley & Howell 1991).

Latastia longicauda revoili (Vaillant) – Southern Long-tailed Lizard
MATERIAL: Two specimens: NMB R7933, Ngurunga Plot; NMB R7934, Mwasumbi Plot. NMB R7933 found in sandy area. NMB R7934 (76.2 mm SVL) is the larger of the two specimens. Dorsals granular and small, moderately keeled, 66–67 across midbody; ventrals in six longitudinal rows; group of small irregular plates in anterior part of pectoral region; nostril bordered by three nasals and first supralabial.
REMARKS: In their key Broadley & Howell (1991) separate the two Tanzanian *Latastia* on the basis of, *inter alia*, unkeeled dorsals in *L. l. revoili* versus usually stongly keeled dorsals in *L. johnstonii*. The two Mkomazi specimens are somewhat ambiguous in this regard as their dorsals are moderately keeled.

Family: Gerrhosauridae
Gerrhosaurus flavigularis (Wiegmann) – Yellow-throated Plated Lizard
MATERIAL: Four specimens: NMB R7927, waterhole near Dindira Dam; NMB R7925, Ibaya Hill; NMB R7367, Kisima Mountain; NMB R7926, South Pare Mountains. Found amongst rocks or in clearings in grassy areas, including a forest clearing (NMB R7925). Largest specimens: 112 mm SVL (NMB R7926), 83 + 213 = 296 mm (NMB R7367).

Gerrhosaurus major major (Duméril) – Southern Tawny Plated Lizard
MATERIAL: One specimen: NMB R7366, 100 m SW of Ibaya Camp. Large adult (222 + 253 = 475 mm) found in a rock crevice. Ventrals in 10 distinct longitudinal rows with an extra row of very small scales on either side; supraciliaries 5 left, 4 right; 25 femoral pores (11 left, 14 right). Another specimen collected at Ibaya Camp from a hole in a tree (Coe, pers comm).

Family: Varanidae
Varanus niloticus niloticus (Linnaeus) – Nile Monitor
MATERIAL: Photographic record: Ngurunga Dam.
REMARKS: The nominate subspecies is replaced by *V. n. ornatus* Daudin in the evergreen forests of west Africa (Broadley & Howell 1991).

Varanus albigularis albigularis (Daudin) – Southern Savanna Monitor
MATERIAL: Specimen from Mbono Valley (Coe, pers comm), near Observation Hill.
REMARKS: Broadley & Howell (1991: 19) recognised *V. albigularis* as a distinct species but abandoned the recognition of subspecies because of "...great variation in both scale counts and colour pattern in *V. albigularis*, which is not correlated with distribution of the various described subspecies [*angolensis* Schmidt, *ionidesi* Laurent, *microstictus* Boettger]...". However, Broadley (1991) later examined material from northern Zambia which he assigned to *V. a. angolensis* mainly on account of a darker throat and vertebral band of enlarged scales from occiput to base of tail.

Suborder: Serpentes
Family: Leptotyphlopidae
Leptotyphlops scutifrons merkeri (Werner) – Merker's Worm Snake
MATERIAL: NMB R7370, 10 km W of Kisiwani. Found under a stone.

Family: Typhlopidae
Rhinotyphlops schlegelii mucroso (Peters) – Eastern Schlegel's Blind Snake
MATERIAL: Specimen collected at Ibaya Camp (Coe, pers comm).
REMARKS: Listed under the genus *Typhlops* by Broadley & Howell (1991), but referable to *Rhinotyphlops* (see Wallach 1994).

Rhinotyphlops unitaeniatus (Peters) – Stripe-backed Blind Snake
MATERIAL: One specimen: NMB R7941, Maji Kununua Mountain. Found in a rotten stump. Headless, remaining body black with a broad cream-yellow vertebral stripe extending to middle of tail; vent near tip of tail, 26–28 scales round middle of tail; about 32 subcaudals; 24 midbody scale rows.
REMARKS: Listed under the genus *Typhlops* by Broadley & Howell (1991), who omitted this species from their key, but referable to *Rhinotyphlops*, as discussed by Wallach (1994).

Family: Boidae
Python sebae natalensis (A. Smith) – Southern African Python
MATERIAL: Photographic record: Ibaya Camp.
REMARKS: The two subspecies of *P. sebae* (Gmelin), namely *sebae* and *natalensis*, are parapatric in northern Tanzania and Kenya, where apparent intergrades have been recorded; both subspecies have been recorded from the vicinity of Mkomazi Game Reserve (Broadley 1984).

Family: Viperidae
Bitis arietans arietans (Merrem) – Puffadder
MATERIAL: One specimen: NMB R7945, Ngurunga Plot. Subadult found in a sandy area; 136 ventrals; 19 pairs of subcaudals; 33 midbody scale rows.

Family: Elapidae

Naja pallida (Boulenger) – Red Spitting Cobra

MATERIAL: One specimen: NMB R7376, 1 km E of Kisiwani. Male (742 + 163 = 905 mm) found dead on the road. Dorsum pink-brown, venter pink-white; black band encircles neck, including ventrals 3–11; 3rd supralabial on either side enters orbit; 205 ventrals; 70+ subcaudals (tip of tail missing); 25 midbody scale rows.

REMARKS: Broadley (1968) noted that the black band on the throat encircles the neck only in juveniles. *N. nigricollis pallida* Boulenger and *N. n. katiensis* Angel, treated as subspecies of *N. mossambica* Peters by Broadley (1968), are now treated as full species (Broadley & Howell 1991; Spawls & Branch 1995).

Family: Colubridae

Lamprophis fuliginosus (Boie) – Common House Snake

MATERIAL: One specimen: NMB R7371, 100 m SW of Ibaya Camp. Female (ovaries examined) measuring 349 + 48 = 397 mm found under loose stone. Dorsum dark brown; narrow cream stripes from lower part of nostril though eye to gape of mouth, and from snout over eye to back of head, also a few small cream patches on either side of neck; two postoculars on either side, 4th and 5th enter orbit; 229 ventrals; 52 pairs of subcaudals; 33 midbody scale rows.

REMARKS: In a preliminary report on the taxonomy of *Lamprophis* Hughes (1997) considers *L. capensis* (Dumeril & Bibron) a valid species; he plots localities for *L. fuliginosus*, *L. maculatus* (Parker) and *L. capensis* in the vicinity of the Kenya-Tanzania border.

Dromophis lineatus (Duméril & Bibron) – Lined Olympic Snake

MATERIAL: One specimen: NMB R7943, Ibaya Camp. Adult (691 + 248 = 939 mm) found at camp; 184 ventrals; 116 subcaudals; 19 midbody scale rows; nostril in single nasal on right side, between two nasals on left.

REMARKS: In Tanzania this species has been recorded only in the west (Broadley & Howell 1991).

Psammophis sibilans sibilans (Linnaeus, 1758) – Hissing Sand Snake

MATERIAL: Specimens collected at Mwsumbi Plot (Coe, pers comm).

Psammophis phillipsii (Hallowell) – Olive Grass Snake

MATERIAL: Two specimens: NMB R7372, 8038, Ibaya Camp. NMB R7372: adult (738 + 323 = 1061 mm) found at camp, 168 ventrals, 97 pairs of subcaudals, 17 midbody scale rows; NMB R8038: adult (150 cm total length), only head and neck retained, 17 scale rows around body anteriorly, eight supralabials, 4th and 5th entering orbit, anterior four pairs of infralabials in contact with anterior sublinguals.

Aparallactus jacksoni jacksoni (Günther) – Kilimanjaro Centipede-eater

MATERIAL: One specimen: NMB R7942, Ibaya Camp. Adult (173 + 41 = 214 mm)

found in sandy clearing; 140 ventrals; 44 subcaudals; 15 midbody scale rows; 1st pair of infralabials narrowly separated behind mental; two postoculars on either side, in contact with anterior temporal; six supralabials on either side, 3rd and 4th entering orbit.

Aparallactus lunulatus (Peters) – Reticulated Centipede-eater
MATERIAL: One specimen: NMB R7374, Ibaya Camp. Adult with 132 ventrals, 52+ subcaudals (tip of tail missing); 15 midbody scale rows, 13 rows posteriorly; Dorsum grey-black; venter grey; anal plate and subcaudals single; 6 supralabials, 3rd and 4th entering orbit.
REMARKS: Broadley & Howell (1991: 28) rejected *A. l. scortecci* Parker as a northern subspecies because of "great variation in local populations"; *A. lunulatus* thus reverts to binomials.

Philothamnus semivariegatus (A. Smith) – Variegated Bush Snake
MATERIAL: One specimen: NMB R7373, 100 m SW of Ibaya Camp. Adult (528 + 274 = 802 mm) found in a shrub among rocky outcrops; 198 ventrals; 145 pairs of subcaudals; 15 midbody scale rows, 11 posteriorly; three supralabials enter orbit on either side of head.

Dispholidus typus typus (A. Smith) – Boomslang
MATERIAL: Sight record (AFF): near Kisima Hill. One specimen observed in a small tree.

Thelotornis capensis mossambicanus (Bocage) – Mozambique Vine Snake
MATERIAL: One specimen: NMB R7944, Ibaya Camp. Adult female (ovaries examined) measuring 541 + 339 = 880 mm collected in a vegetable garden. Top of head uniform 'grey', temporal region with dark speckling, black lateral markings on neck, labials with black speckling; 158 ventrals; 145 pairs of subcaudals; 19 midbody scale rows; two loreals on either side of head; supralabials 8 (left side: 1st and 2nd largely fused; small scale budded off on lower anterior part of 5th supralabial, not counted), 4th and 5th entering orbit; infralabials 10 left, nine right.
REMARKS: According to Broadley (1979) *T. capensis mossambicanus* from Somalia, Kenya and the Usambara Mountains are similar to *T. kirtlandii* (Hallowell) in that many have the top of the head and temporal region uniform green, while the rostral and nasals also recurve to the upper surface of the head in some specimens; ventral counts (156–166 in males, 159–172 in females) are intermediate between the two species, the number of loreals vary (1–2), but infralabial counts (9–13, usually 10–12) are high as in *T. c. mossambicanus*. These populations were referred to as "*T. c. mossambicanus* with *T. kirtlandii* features". Rasmussen (1997) recorded a male from Kwamgumi Forest Reserve near Mkomazi Game Reserve which had 161 ventrals, (only!) 136 pairs of subcaudals, one loreal on the left and

two on the right, and 13 infralabials. He did not comment on the patterning on the head, but referred his specimen to the above category. The Ibaya Camp specimen is referable to Broadley's (1979) categorisation of "*T. c. mossambicanus* with *T. kirtlandii* features" on account of its intermediate number of ventrals (158) and low infralabial counts (9–10; in *T. kirtlandii*: 7–11, mostly 8–10).

Dasypeltis scabra (Linnaeus) – Common Egg-eater
MATERIAL: One specimen: NMB R7375, Ibaya Camp. Subadult (211 + 38 = 249 mm) found at camp. Dorsum grey-brown with 92 squarish, somewhat transversely enlarged, dark vertebral blotches from nape to base of tail; blotches separated by a transverse row of mostly white scales; a narrow dark band extends from (or just below) the middle of each blotch but does not reach the belly; underparts immaculate cream-white; 211 ventrals; 65 subcaudals; 25 midbody scale rows; two postoculars on either side.
REMARKS: The Ibaya Camp specimen has as many as 92 pattern cycles from nape to base of tail (85–107 in *D. atra* Sternfield, 42-79 in *D. scabra*), but it is referable to *D. scabra* on account of its clearly patterned dorsum (not black, brown, or brown and faintly patterned) and possession of two postoculars on either side (see Broadley & Howell 1991; Hughes 1997). With regard to colour pattern, Hughes (1997: 69) writes "...I am inclined to place more emphasis on the absence of pattern in defining *atra*."

Discussion

The Mkomazi Game Reserve and its immediate vicinity has a rich diversity of reptile life. The 52 species recorded comprise one terrapin, three tortoises, 32 lizards and 16 snakes. The area is particularly rich in scincids (10 species), gekkonids (nine) and colubrids (10). Several specimens recorded represent extensive new distribution records, including the first Tanzanian record of *Lygosoma somalicum*; while material of poorly known taxa are also documented, including only the second and third known specimens (and first male) of *Hemidactylus tanganicus*.

In their survey of the Kora National Reserve in Kenya, considered an Acacia-Commiphora bushland area (as is the Mkomazi Game Reserve), Cheptumo *et al.* (1986) recorded the Nile crocodile, 19 lizard, 15 snake and three chelonian species. In the present study, 31 lizard, 14 snake and four chelonian species were recorded. As lizards, in particular, were systematically collected from rocky outcrops, trees and grassland areas, the recorded diversity is probably a fair representation of what should occur in the area. Fifty-eight per cent of the lizard species collected are considered distinctive of the Somalia-Masai vegetation region (see Broadley & Howell 1991). Snakes were difficult to collect and infrequently encountered—the number of species recorded is probably an under-representation (see Broadley & Howell 1991). The chelonian species recorded are about 80% of those expected to be found here.

Savanna vegetation in the MGR is greatly influenced along its central and north-western limits by the Usambara and Pare Mountains, which lie adjacent to its southern boundary (Coe & Stone 1995). These mountains fall within the Afromontane vegetation type as defined by White (1983). One would thus expect the presence of a number of the reptiles considered as typical of this vegetation type. However, less than 1% of the reptiles collected are considered typical of the Afromontane region (Broadley & Howell 1991).

The present study reflects only a part of the rich reptile fauna expected to occur in the MGR, and there is much scope for future documentation of particularly the snakes. Careful surveying of the forest elements occurring in the MGR may also prove that a larger component of the Afromontane reptile fauna is present than currently anticipated.

Acknowledgements

We are grateful to the Council of Directors of the National Museum, Bloemfontein, for providing financial support and allowing AFF to visit the MGR on two occasions. Participation was funded by a grant from the Commonwealth Science Council. We are also grateful for support received from the Royal Geographical Society (with IBG). The assistance of Mr E. Mohapi (National Museum, Bloemfontein) in the preparation and sorting of material is acknowledged. Ms M. Stander (University of Stellenbosch) assisted with pit-fall collecting of reptiles in the reserve.

References

Bates, M.F. (1996) New reptile distribution records for the Free State province of South Africa. *Navors. nas. Mus., Bloemfontein* 12(1): 1-47.

Boycott, R.C. & Bourquin, O. (1988) The South African Tortoise Book – A Guide to South African Tortoises, Terrapins and Turtles. Southern Book Publishers, Johannesburg.

Branch, W.R. (1988) Field Guide to the Snakes and other Reptiles of Southern Africa. Struik, Cape Town.

Broadley, D.G. (1966) *The herpetology of south-east Africa*. PhD thesis, University of Natal, Pietermaritzburg.

Broadley, D.G. (1968) A review of the African cobras of the genus *Naja* (Serpentes: Elapinae). *Arnoldia (Rhodesia)* 3(29): 1-14.

Broadley, D.G. (1979) Problems presented by geographical variation in the African vine snakes, genus *Thelotornis. South African Journal of Zoology* 14: 125-131.

Broadley, D.G. (1984) A review of geographical variation in the African python, *Python sebae* (Gmelin). *British J. Herpetol.* 6: 359-367.

Broadley, D.G. (1989a) *Geochelone pardalis*. In: Swingland, I.R. & Klemens, M.W. (eds.) *The Conservation Biology of Tortoises*. Occ. Pap. IUCN Species Survival Commission (SSC), No. 5. pp. 43-46.

Broadley, D.G. (1989b) *Kinixys belliana*. In: Swingland, I.R. & Klemens, M.W. (eds.) *The Conservation Biology of Tortoises*. Occ. Pap. IUCN Species Survival Commission (SSC), No. 5. pp. 49-55.

Broadley, D.G. (1989c) *Malacochersus tornieri*. In: Swingland, I.R. & Klemens, M.W. (eds.) *The Conservation Biology of Tortoises*. Occ. Pap. IUCN Species Survival Commission (SSC), No. 5. pp. 62-64.

Broadley, D.G. (1991) The herpetofauna of northern Mwinilunga District, northwestern Zambia. *Arnoldia Zimbabwe* 9(37): 519-538.

Broadley, D.G. (1993) A review of the Southern African species of *Kinixys* Bell (Reptilia: Testudinidae). *Ann. Transvaal Mus.* 36(6): 41-52.

Broadley, D.G. (1997) Family Scincidae. pp. 66-75. In: J.H. van Wyk (ed.) *Proceedings of the FitzSimons Commemorative Symposium – South African Lizards: 50 years of Progress and Third H.A.A. Symposium on African Herpetology*. Herpetological Association of Africa, Stellenbosch.

Broadley, D.G. & Howell, K.M. (1991) A checklist of the reptiles of Tanzania, with synoptic keys. *Syntarsus* 1: 1-70.

Broadley, D.G., Chidavaenzi, R.L., Rasmussen, G.S.A. & Broadley, S. (1998) The herpetology of the Dande Communal Lands, Guruve District, Zimbabwe. *African Herp News* No. 27: 3-12.

Cheptumo, M, Madsen, I., Duff-Mackay, A., Hebrard, J. Rotich, D. & Lohman, J. (1986) A survey of the reptiles and amphibians of Kora National Reserve. In: Coe, M. & Collins, N.M. (eds.) *Kora: An Ecological Inventory of the Kora National Reserve, Kenya.* Royal Geographical Society, London. pp. 235-239.

Coe, M. & Stone, G. (1995) *A preliminary report on the field research of the Mkomazi Research Programme: Progress Report, July (1995)* Royal Geographical Society (with IBG), London.

Hughes, B. (1997) *Dasypeltis scabra* and *Lamprophis fuliginosus* – two pan-African snakes in the Horn of Africa: a tribute to Don Broadley. *Afr. J. Herpetol.* 46(2): 68-77.

Iverson, J.B. (1992) *A Revised Checklist with Distribution Maps of the Turtles of the World.* Privately Printed, Richmond, Indiana.

Loveridge, A. (1920) Notes on East African Lizards collected 1915-1919, with Description of a new Genus and Species of Skink and new subspecies of Gecko. *Proc. Zool. Soc., London* 131-167.

Loveridge, A. (1929) East African reptiles and amphibians in the United States National Museum. *U.S. National Mus. Bull.* 151: 1-135.

Loveridge, A. (1947) Revision of the African lizards of the family Gekkonidae. *Bull. Mus. comp. Zool, Harvard* 98(1): 1-469.

Loveridge, A. (1957) Check list of the reptiles and amphibians of East Africa (Uganda, Kenya; Tanganyika; Zanzibar). *Bull. Mus. comp. Zool. Harv.* 117: 151-362.

Pasteur, G. (1964) Recherches sur l'Evolution des Lygodactyles, lezards Afro-Malagaches actuels. *Trav. Inst. scient. cherif.* Ser. zool., No. 29: 1-132.

Rasmussen, J.B. (1997) Tanzanian records for vine snakes of the genus *Thelotornis*, with special reference to the Udzungwa Mountains. *African Journal of Herpetology* 46(2): 137-142.

Spawls, S. & Branch, B. (1995) *The Dangerous Snakes of Africa*. Southern Book Publishers, Halfway House.

Wallach, V. (1994) The status of the Indian endemic *Typhlops acutus* (Dumeril & Bibron) and the identity of *Typhlops psittacus* Werner (Reptilia, Serpentes, Typhlopidae). *Bull. Inst. roy. Sci. nat. Belg. Biol.* 64: 209-229.

Welch, K.R.G. (1982) *Herpetology of Africa: A Checklist and Bibliography of the Orders Amphisbaenia, Sauria and Serpentes*. R.E. Kriegler, Malabar, Florida.

White, F. (1983) *The vegetation of Africa. A descriptive memoir to accompany the UNESCO-AETFAT-UNSO Vegetation Map of Africa*. UNESCO, Paris.

Checklist: Reptiles of Mkomazi

Class REPTILIA

Order TESTUDINES

Family Pelomedusidae
Pelomedusa subrufa subrufa (Lacepède) – Helmeted Terrapin

Family Testudinidae
Geochelone pardalis (Bell) – Leopard Tortoise
Kinixys spekii (Gray) – Speke's Hinged Tortoise
Malacochersus tornieri (Lindholm) – Pancake Tortoise

Order SQUAMATA

Suborder SAURIA

Family Gekkonidae
Lygodactylus gravis (Pasteur) – Usambara Dwarf Gecko
Lygodactylus manni (Loveridge) – Mann's Dwarf Gecko
Lygodactylus luteopicturatus luteopicturatus (Pasteur) – Yellow-headed Dwarf Gecko
Cnemaspis africana (Werner) – Usambara Forest Gecko
Hemidactylus brookii angulatus (Hallowell) – Angulate Tropical Gecko
Hemidactylus mabouia (Moreau de Jonnès) – Tropical House Gecko
Hemidactylus platycephalus (Peters) – Baobab Gecko
Hemidactylus squamulatus squamulatus (Tornier) – Squamate Tropical Gecko
Hemidactylus tanganicus (Loveridge) – Tanzanian Tropical Gecko

Family Agamidae
Agama agama usambarae (Barbour & Loveridge) – Usambara Rock Agama
Agama sp.

Family Chamaeleonidae
Chamaeleo dilepis dilepis (Leach) – Common Flap-necked Chameleon
Chamaeleo bitaeniatus (Fischer) – Side Striped Chameleon
Rhampholeon sp. (near *platyceps* Günther) – Pigmy Chameleon

Family Scincidae
Melanoseps loveridgei (Brygoo & Roux-Estéve) – Loveridge's Limbless Skink
Mabuya brevicollis (Wiegmann) – Short-necked Skink
Mabuya maculilabris maculilabris (Gray) – Speckle-lipped Skink
Mabuya planifrons (Peters)
Mabuya quinquetaeniata margaritifer (Peters) – South-eastern Rainbow Skink
Mabuya striata striata (Peters) – Common Striped Skink
Mabuya varia (Peters) – Variable Skink

Lygosoma afrum (Peters) – Peters' Writhing Skink
Lygosoma sp. (near *somalicum* [Parker]) – Somali Writhing Skink
Panaspis wahlbergii (A.Smith) – Savanna Snake-eyed Skink

Family Lacertidae
Adolphus jacksoni (Boulenger) – Jackson's Forest Lizard
Nucras boulengeri (Neumann) – Boulenger's Scrub Lizard
Heliobolus spekii spekii (Günther) – Southern Speke's Sand Lizard
Latastia longicauda revoili (Vaillant) – Southern Long-tailed Lizard

Family Gerrhosauridae
Gerrhosaurus flavigularis (Wiegmann) – Yellow-throated Plated Lizard
Gerrhosaurus major major (Duméril) – Southern Tawny Plated Lizard

Family Varanidae
Varanus niloticus niloticus (Linnaeus) – Nile Monitor
Varanus albigularis albigularis (Daudin) – Southern Savanna Monitor

Suborder SERPENTES

Family Leptotyphlopidae
Leptotyphlops scutifrons merkeri (Werner) – Merker's Worm Snake

Family Typhlopidae
Rhinotyphlops schlegelii mucroso (Peters) – Eastern Schlegel's Blind Snake
Rhinotyphlops unitaeniatus (Peters) – Stripe-backed Blind Snake

Family Boidae
Python sebae natalensis (A. Smith) – Southern African Python

Family Viperidae
Bitis arietans arietans (Merrem) – Puffadder

Family Elapidae
Naja pallida (Boulenger) – Red Spitting Cobra

Family Colubridae
Lamprophis fuliginosus (Boie) – Common House Snake
Dromophis lineatus (Duméril & Bibron) – Lined Olympic Snake
Psammophis sibilans sibilans (Linnaeus, 1758) – Hissing Sand Snake
Psammophis phillipsii (Hallowell) – Olive Grass Snake
Aparallactus jacksoni jacksoni (Günther) – Kilimanjaro Centipede-eater
Aparallactus lunulatus (Peters) – Reticulated Centipede-eater
Philothamnus semivariegatus (A. Smith) – Variegated Bush Snake
Dispholidus typus typus (A. Smith) – Boomslang
Thelotornis capensis mossambicanus (Bocage) – Mozambique Vine Snake
Dasypeltis scabra (Linnaeus) – Common Egg-eater

CHAPTER 26

Birds of Mkomazi

Peter C. Lack, with appendix by Neil E. Baker & Elizabeth M. Baker

Introduction

The Mkomazi Ecological Research Proramme collected information on the birds of Mkomazi Game Reserve from mid-1993 to 1997. Several people have contributed records but the main sources include: first, a visit of three and a half weeks in July–August 1993 by myself accompanied by Agustino Peter as driver and helper and for some of the time by Diane Ridgley. During this time all records of all birds seen were kept and a series of censuses was undertaken in the savanna habitats of the north-western third of the reserve. Second, an eight week visit by six students from St Peter's College, Oxford in July–August 1993. This group also carried out some census work in the savanna habitats of the northern part of the reserve and made a two-day visit to the Umba River in the south. Third, Neil and Liz Baker, often accompanied by others, have made several trips for periods of up to two weeks into the northern half of the reserve during 1994–96. All records were kept although specific censuses were not made, and they carried out some mist-netting in several areas (see Appendix, this chapter). Many of these visits were as part of the Important Bird Areas project of BirdLife International. Mkomazi Game Reserve is one of these. Finally, a two and a half week visit by myself in December 1995–January 1996 during which a repeat series of censuses was undertaken to those of 1993. In addition records have been received from several other observers contributing to the Tanzania Bird Atlas—see Checklist, next chapter.

Previously Larry Harris worked on mammals in the reserve in the 1960s during which time he kept note of bird species he saw (a list is given in Harris 1972). In addition the general area of 'Mkomazi' was visited many times in the 1930s by Reg Moreau, who was stationed as an agriculturalist at the Amani Station in the Usambaras. It is not certain exactly which area he was referring to but it is likely that he was mainly, if not entirely, referring to the area around Mkomazi Village and by the Pangani River, in the gap between the South Pare and Usambara Mountains rather than the present day reserve. Certainly it is this village area where David Lack did his study of the Black-winged Red Bishop *Euplectes hordeaceus* (Lack 1935) while he was staying with Moreau.

Bird work in the reserve 1993–97

The reserve has been visited by several ornithologists in this period and all have kept at least species lists of what they have seen. In particular visits to water bodies have proved quite productive and a fair number of the records of individual species result from what can loosely be described as casual observations. A full species list with brief notes of their occurrence is included in Chapter 27.

Census work

I carried out some census work in both wet (December 1995–January 1996) and dry (July–August 1993) seasons. In each case censuses were conducted along reserve roads. Ten-minute stops were made every kilometre (or sometimes 2 km) during which time all birds seen and/or heard were recorded. All such censuses were carried out in the mornings between dawn and about midday.
Each census point was allocated into one of 12 major habitat types. These were:

GRASSLAND (G) No bushes or trees present, and with the ground usually covered with thick grass. In the dry season all attributed to this habitat type were in unburnt areas.

BUSHED GRASSLAND (BG) No trees, bushes less than 10% canopy cover, grass cover variable from very thick and up to about 0.5 m to, in the dry season, burnt and therefore almost bare ground.

WOODED AND BUSHED GRASSLAND TYPE A (WBGa) Trees dominated by *Acacia tortilis* of less than 10% canopy cover, bushes less than 10% canopy cover and of variable height, grass as above. This type was WBG in the areas along the road from the reserve boundary towards Ngurunga Dam and near Zange Gate.

WOODED AND BUSHED GRASSLAND TYPE B (WBGb) Trees of various, and often mixed, species, bushes and grass as above. (This was all other areas of WBG.)

BUSHLAND (B) No trees, bushes of more than 10% canopy cover, grass usually present but in a variable amount. (In general fairly restricted in extent.)

WOODED BUSHLAND (WB) As bushland but with some trees present. (Also rather limited in extent.)

ACACIA WOODLAND TYPE A (AcWa) Trees dominated by *Acacia*, typically *A. tortilis*, of more than 10% canopy cover, bushes very dense and up to 3 m high, grass amounts variable. This type was restricted to the area along the Zange Gate to Kisiwani road which was adjacent to border of reserve and Forest Reserve.

ACACIA WOODLAND TYPE B (AcWb) Similar to above but elsewhere in the reserve.

NGURUNGA THICKET (NgT) Very dense trees and bushes (mixed species) close to the watercourse leading to the Ngurunga Dam.

COMMIPHORA WOODLAND TYPE A (CoWa) Woodland as above but dominated, often nearly exclusively, by *Commiphora* species. This type was the area from

Njiro Gate to about 6 km towards Kisima Camp which contained generally richer vegetation than elsewhere and with the trees up to 12–15 m in places.

COMMIPHORA WOODLAND TYPE B (CoWb) Other woodland dominated by *Commiphora* species, mostly in very dry areas and with trees often only up to 5 m and sometimes with a near closed canopy at about 3–4 m.

FOREST (F) A small area of groundwater/riverine forest near Kisiwani village with trees up to 20 m and impenetrable bush layer on the reserve boundary.

In the dry season a total of 326 ten-minute counts was made with the number in each habitat ranging from seven to 55. In the wet season 252 counts were made with 5 to 48 in each habitat. Although there was no significant difference between the seasons in the proportion of each habitat type visited ($\chi^2 = 18.3$, df 11, $p<0.10$), Bushed Grassland and Commiphora Woodland Type A were visited proportionately more in the dry season, and Wooded Bushland more in the wet.

For some of the analyses presented here only the 'common' bird species are considered and with raptors and aerial species (swifts and swallows) excluded. 'Common' species are defined as those which were recorded at a rate of more than one per ten hours censusing. In practice this means that at least six were seen in censuses in the dry season and at least five in the wet season, and applied to 83 species in the dry and 112 in the wet. These species have been divided into six ecological categories based on the main foods they eat: Insects (and other arthropods), Fruits, Seeds, Insects + Fruits, Insects + Seeds and Insects + Nectar. Species were also classified as a Resident, a Visitor from elsewhere in Africa or a Palaearctic Migrant. Both sets of categories were based largely on the categories defined by my earlier studies in Tsavo East National Park in the 1970s (Lack 1985).

Results

In the following only English names of birds are used. For scientific names see the Checklist, next Chapter.

The species list

The full list of species recorded in the reserve by the various observers is in Chapter 27, with very brief notes of their occurrence. The total now stands at 402 with a few others claimed and considered possibles.

Among these there are six species not previously recorded in Tanzania: Violet Wood Hoopoe, Friedmann's Lark, Somali Long-billed Crombec, Yellow-vented Eremomela, Three-streaked Tchagra (Lack 1994) and Shelley's Starling.

Records of several other species are towards the southernmost part of their range, for example Pygmy Batis, Pringle's Puffback and Rufous Bush Chat. There are also several records of migrants from the Palaearctic during the northern sum-

Table 26.1 The numbers of six main ecological groups of birds seen per hour's censusing ('x' indicates less than 0.5) in the 12 main habitat types in Mkomazi Game Reserve in (a) the dry season July–August 1993 and (b) the wet season December 1995–January 1996. Only species with at least six (dry season) or five (wet season) records in all censuses are included in the totals for each ecological group; the total for each season includes all species. The last line for each season is the number of 10 minute censuses carried out in that habitat.

season/	habitat (abbreviations as in text)												
ecological group	G	BG	WBGa	WBGb	B	WB	AcWa	AcWb	NgT	CoWa	CoWb	F	total
(a) dry season													
Insects only	10	40	36	36	12	39	15	32	18	13	23	20	27
Insects + Fruit	0	4	8	9	5	6	12	8	9	7	7	24	8
Insects + Seeds	4	12	12	5	18	17	5	9	19	11	7	2	10
Insects + Nectar	0	x	5	2	2	5	11	1	3	9	3	5	4
Fruits only	0	1	2	4	0	4	7	6	5	5	4	11	4
Seeds only	9	3	17	4	2	8	10	3	21	5	4	11	6
total	23	59	80	60	40	78	60	59	75	50	47	73	58
no. of 10 min. counts	7	50	28	55	16	17	24	22	9	44	46	8	326
(b) wet season													
Insects only	12	50	43	50	15	36	48	38	22	23	32	7	37
Insects + Fruit	0	5	8	6	2	8	10	6	6	9	6	2	6
Insects + Seeds	4	11	19	33	10	21	17	28	24	4	9	2	18
Insects + Nectar	0	1	1	1	x	3	4	2	3	7	7	5	3
Fruits only	0	4	8	5	2	4	6	5	4	4	8	0	5
Seeds only	1	25	23	30	4	19	19	29	12	5	5	5	18
total	23	105	108	135	37	98	116	117	79	56	72	27	95
no. of 10 min. counts	5	25	18	48	11	25	23	14	14	19	42	8	252

mer of which the Eurasian Swallow seen on 18 July 1993 must rank as the most exceptional.

Census work

Table 26.1 shows the numbers of birds per ten hours censusing in different ecological categories recorded in each season. Inspection of the table shows that, in the dry season, the total numbers seen did not vary very much or very consistently between habitats. However it must be remembered that these are not actual density figures, and indeed the counts in more open habitats were often over a larger area than those in the thicker ones. The proportions of the different ecological types of bird, however, show a few more interesting patterns.

The pure insectivores dominated all habitat types, and especially so in the more open wooded habitats (BG, WBGa, WBGb), except for the Ngurunga Thicket where seed eaters became more prominent.

All frugivores were commonest in Forest and then in Acacia Woodland type A. The reason for the latter is likely to be at least partly due to the location of this habitat type as the sites censused were all along the edge of the reserve and fairly close to Forest areas.

Seed eaters were most common in Ngurunga Thicket, then Wooded and Bushed Grassland Type A. This was probably due to the presence of Ngurunga Dam within a few kilometres. Seed eaters are dependent on surface water and need to drink every day, unlike most other birds.

Finally nectarivores were especially common in Acacia Woodland type A and Commiphora Woodland Type A. Both these habitats contained a higher proportion of mature trees than most other habitats, and some of the trees, especially *Albizia*, were flowering at the time of my visit.

Table 26.1 also shows similar figures for the wet season. It is clear that total numbers are considerably increased, especially in the more open habitats. This is largely, but not solely, due to the presence of visitors, both from elsewhere in Africa and from the Palaearctic. The only habitats showing smaller numbers were Bushland which was very poorly sampled in the wet season and Forest in which relatively few birds of any kind were actually seen and therefore scored for these censuses.

As in the dry season insectivores dominated all habitats, this time without exception, and again were especially so in the more open wooded habitats. For this group the increase is primarily migrants from the Palaearctic. All 14 'common' species which were such migrants were pure insectivores.

In the wet season the seed eaters were also much more prominent in all habitats except Grassland, Commiphora Woodland (both types) and Forest. The latter three were the habitats with the least grass (and hence seeds) present, but why they were so inconspicuous in Grassland is unclear. The increase applied both to the pure granivores and those eating a mixed diet, and seemed to be largely due to the large number of species visiting from elsewhere in Africa. Of 22 species classed as such visitors, 13 were at least partially granivorous and some of them occurred in quite large flocks, for example some of the weavers and waxbills. Two of these, Red-billed Quelea and Chestnut Weaver, were seen in both seasons but both were much commoner in the wet season (see below for further comparisons).

Frugivores were again widespread and numbers were very similar to those in the dry season. The majority of these species are residents, although my studies in Tsavo indicated that some of the Insect + Fruit eaters switched foods on a seasonal basis, rather than eating both all the year.

Finally the nectarivores at this season showed a strong preference for *Commiphora* Woodland of both types and appeared to be much more concentrated than

Table 26.2 Some 'resident' species showing large differences in abundance between wet and dry seasons in Mkomazi. Figures are the numbers seen per ten hours' censusing ('x' indicates less than one).

	season	
species	dry	wet
commoner in the dry season		
Northern/Red-faced Crombecs[1]	10	4
Rosy-patch Bush-Shrike	5	x
White-crested Helmet-Shrike	11	4
Golden-breasted Starling	6	2
Hildebrandt's Starling	41	15
Amethyst Sunbird	9	x
Green-winged Pytilia	8	2
commoner in the wet season		
Helmeted Guineafowl	0	9
Blue-naped Coly	3	25
Flappet Lark	5	8
Cisticola species (small)[2]	0	5
Fischer's Starling	15	30
Purple-banded Sunbird	3	6
Red-billed Buffalo-Weaver	29	48

[1] The two species are combined here. In the wet season they were distinguished, but in the dry season Red-faced was not recorded and it is now not certain that all were in fact Northern.

[2] The small Cisticolas are notoriously difficult to identify. As will be noted from the Checklist (Chapter 27) the Desert Cisticola was certainly present as almost certainly was Zitting, but there may also have been one or more others included in this total.

in the dry season. Presumably their nectar supply was much reduced in habitats other than Commiphora Woodland. Certainly *Albizia* was not flowering and this had been major attraction in the dry season.

In the dry season four species classed as Visitors came into the 'common' category but all were considerably increased in numbers in the wet season. In addition a further 18 Visitor and 14 Palaearctic Migrant 'common' species were not seen at all in the dry season. There were some differences among those classed as Residents as well. Some of the larger differences are listed in Table 26.2.

Number of species

Table 26.3 shows the number of species in each habitat calculated in three ways. By definition the longer that is spent in a habitat the more species are likely to be seen. So, the total figures (the third pair of columns) are not the ideal comparison, and this is likely to be part of the reason why Wooded and Bushed Grassland Type B has the highest total number of species in both seasons. More directly comparable figures are those in the other pairs of columns. These show that Acacia Woodland type A, Wooded Bushland and Wooded and Bushed Grassland of both types were

Table 26.3 The number of species recorded in the different habitat types.

	'common' species seen more than 1 per hour[1]		species seen and heard, average per hour[2]		all species during census[3]	
habitat	dry	wet	dry	wet	dry	wet
Grassland	7	9	16	18	20	20
Bushed Grassland	15	23	27	40	89	99
Wood Bush Grass A	23	31	41	46	82	85
Wood Bush Grass B	23	30	30	46	96	133
Bushland	10	17	26	32	46	50
Wooded Bushland	28	25	39	53	73	109
Acacia Wood A	23	40	43	58	87	98
Acacia Wood B	20	28	34	43	79	71
Ngurunga Thicket	22	17	33	44	45	73
Commiphora wood A	20	23	31	41	85	73
Commiphora Wood B	14	23	36	44	86	107
Forest	16	8	23	33	39	41

[1] The number of 'common' species seen at a rate of more than 1 per hour in that habitat.

[2] The average number of species noted (seen and heard) in one hour of censusing in that habitat.

[3] The total number of species seen in that habitat during censuses including those heard only and those flying over.

the most diverse habitats in both seasons. It seems that *Acacia* trees, which dominate both Acacia Woodland and much of the Wooded and Bushed Grassland, are a very important habitat component and perhaps especially for visiting species.

It will also be noticed that Bushed Grassland and Bushland supported fewer species than any habitat containing trees, presumably due to the lesser diversity of the habitat structure, and that Grassland and Forest held very low numbers al-

Table 26.4 A comparison of the numbers of species in different groups in Tsavo East National Park and Mkomazi Game Reserve.

(a) The total number of species recorded.

group[1]	total	Tsavo only	Mkomazi only
Waterbirds	90	38	3
Birds of prey	58	9	5
Large land birds	30	8	3
Aerial feeders	21	3	2
Non-passerines	93	12	4
Passerines	207	30	20
total	499	100	37

[1] 'Waterbirds' are all ducks, waders *etc.*; 'Large land birds' are such as bustards, plovers *etc.*; 'Aerial feeders' are swifts and swallows; 'Non-passerines' are pigeons, Coraciiformes and to woodpeckers in the taxonomic order; and 'Passerines' are the order Passeriformes excepting the swallows.

(b) The numbers of 'common'[2] species in each area.

group	in both	Tsavo only	Mkomazi only
Game birds	6	2	1
Pigeons, Parrots *etc.*	10	-	-
Cuckoos	3	-	1
Coraciiformes	10	5	3
Piciformes	6	-	2
Larks, Pipits	4	3	1
Shrikes, Bulbuls	7	3	4
Thrushes *etc.*	12	7	11
Tits, Sunbirds	3	-	5
Starlings	5	4	1
Finches, Weavers	10	6	6
total	76	30	35

[2] 'Common' species are defined as those which were seen at the rate of more than one per ten hours' censusing in the respective areas.

though it should be noted that the actual species recorded in these, especially Forest, were often different than those in other habitats.

Comparison with Tsavo East National Park

This study in Mkomazi allows some direct comparisons with my earlier studies in Tsavo East National Park (Lack 1980 & 1985). Table 26.4a compares the total number of species recorded divided into different groups between the two areas. It must be remembered that, firstly, Tsavo East is three to four times the size and, secondly, that the records cover a longer time period. It is clear that the major difference lies among waterbirds where there were far more in Tsavo presumably due to the presence of a permanent river, the Galana, and a fairly large permanent lake, Aruba Dam. More interesting is that Mkomazi has 37 species not recorded in Tsavo and 20 of these were passerines and especially warblers Sylviidae.

Table 26.4b only considers the 'common' species. This is a more direct comparison and shows that even though the two areas are only 100 to 150 km apart there is a considerable difference in the components of their common land avifauna. Mkomazi seems especially well represented by thrushes and warblers and by sunbirds. All of these are tree dwelling species and Tsavo's predominance of Coraciiformes, starlings, larks and pipits perhaps reflects the more open ground prevalent there, these groups being more dependent on ground feeding than others.

Some other, arguably more interesting, comparisons concern the habitat preferences of individual species. Table 26.5 gives some examples of major differences between the habitats occupied by species which are classed as common in both areas. Clearly there are some substantial differences. Unfortunately too little is known about the general ecology of most of these species to know which is the preferred habitat. It must also be noted that there were quite a few species which were about equally common in the two areas and which showed more or less the same habitat preferences, for example Black-throated Barbet (common) and Bare-eyed Thrush (fairly common but elusive) occurring almost exclusively in *Commiphora* woodland.

Mkomazi in a biogeographic context

In a biogeographic sense the Mkomazi Game Reserve is the southernmost extension of the Somali semi-arid belt which extends through much of eastern and south-eastern Kenya, including the Tsavo National Parks, although not the coastal strip, and into Somalia. With the western, southern and eastern boundaries of the reserve largely being mountains (Usambaras and North and South Pares), Mkomazi forms very much an isolated patch of semi-arid savanna habitat cut off from the rest of Tanzania. Table 26.6 shows that the avifauna reflects this. The majority of all species and the common species which occur in Mkomazi are widespread. In

Table 26.5 Some examples of species occurring in different habitat types in Tsavo East National Park and Mkomazi Game Reserve.

species	main habitats in Tsavo[1]	main habitats in Mkomazi[1]	differences[2]
Yellow-necked Spurfowl	WB,R	BG,WBG,B	commoner, more widespread, more open
Crested Bustard	WBG,B	CW,NT	thicker
Black-faced Sandgrouse	BG,WBG,B	CW	thicker, much more restricted
Orange-bellied Parrot	CW,WB	WB,WBG	more open, more widespread
White-faced Go-Away-Bird	CW	AW,CW,WBG	more open, more widespread
Diederik Cuckoo	WBG	WB,AW	thicker
Black-and-White Cuckoo	CW,WB,WBG	AW	thicker
White-browed Coucal	R	BG,WBG,AW	much more open
Speckled Mousebird	R	AW,F,B	more open
Lilac-breasted Roller	R	WBG,BG,AW,WB	more open, more widespread
Red-billed Hornbill	CW,R,WB	BG,WBG,WB,AW,CW	more widespread, more open
Spot-flanked Barbet	R	AW,F	more widespread
Pink-breasted Lark	B,BG,WBG,WB	CW,WBG	thicker, more restricted
Pangani Longclaw	G,BG	BG,WBG,AW,CW,WB	more widespread, moved into thicker in wet
Bare-eyed Thrush	CW	CW,AW,WB	more widespread
Upcher's Warbler	WBG,B,BG,CW	CW,NT,F	thicker
Willow Warbler	CW,WB,WBG	CW,NT,F	thicker
Tiny Cisticola	CW	WBG,AW,CW	more widespread
Ashy Cisticola	G,BG	BG,WBG,WB,G	thicker, more widespread
Grey Flycatcher	CW,WBG	CW	more restricted
Black-headed Batis	R	AW,CW	more widespread
Amethyst Sunbird	R	WBG,AW,CW	more widespread
Golden-breasted Starling	CW,WB,B,WBG	NT,CW	thicker, more restricted

[1] Habitat abbreviations for Mkomazi as in text except types A and B have been combined where appropriate. Habitat abbreviations for Tsavo are the same except that R = Riverine Forest.

[2] These compare Mkomazi to Tsavo. For example, 'thicker' will mean that there is a habitat preference for thicker habitats in Mkomazi.

addition the table shows clearly that there are several species which are very restricted in their total range, and more importantly it shows that there are far more species which occur predominantly to the north than to the south. The range of most of these northern species is primarily the Somali semi-arid zone.

For this reason if for no other Mkomazi is unique in Tanzania, and the group of northern species includes all six of the species not previously recorded in the country. Consequently none of them can be considered totally unexpected. All have been recorded not far across the border in southeastern Kenya, in some cases quite commonly, and the fact that they have not previously been recorded is at least in part due to the fact that ornithologists have not previously spent much time in Mkomazi Game Reserve.

Table 26.6 The biogeographic context of Mkomazi, showing the number of species in various range categories.

range[1]	common species[2]	all species[3]
widespread	100	319
restricted	9	24
northern	18	55
southern	4	4

[1] Ranges refer to overall ranges defined as follows:
'widespread' – over much of Kenya and Tanzania
'restricted' – confined to a fairly small area in the vicinity of Mkomazi
'northern' – mainly or entirely to the north of Mkomazi
'southern' – mainly or entirely to the south of Mkomazi.

[2] 'Common species' are those which were scored as 'common' in either season in Mkomazi (see text for definition).

[3] 'All species' include the above and all others in the Checklist.

Discussion

The patterns of occurrence of the different ecological groups of birds between the various habitats and between the seasons show no great surprises from what might be expected although there are some interesting points, some referred to briefly above. Savannas in Africa tend to be dominated by insectivorous and granivorous birds although there are very few detailed studies for which precise figures are available. Indeed one of the few studies is my own from Tsavo East National Park (Lack 1980, 1985 & 1987). Frugivores always tend to be commoner in forest areas as they are in Mkomazi and these habitats, extending into the thicker parts of woodland, tend also to be more even in their numbers between seasons, leaving the more open habitats to be the predominant ones where visiting species are most common. The visitors are almost all insectivores or granivores or both, although in the case of Mkomazi I suspect that more data may show that some of the nectarivores are also migrants to some degree. Palaearctic migrants are almost exclusively insectivorous and I have argued elsewhere that this is probably more because of the situation in the Palaearctic than in their wintering grounds (Pearson & Lack 1992).

It was noted that some of the species classed as residents showed some differences in numbers between the two seasons, sometimes considerable. Some of these differences could be due to changes in conspicuousness at the two seasons. In the dry season there were no leaves on most of the woody vegetation which made many birds inherently more conspicuous, and in the wet season many species are breeding (no direct evidence of this from Mkomazi but it was certainly true in Tsavo, Lack 1980) and therefore advertising themselves more. I doubt that this accounts for all of the changes and I think it is likely that several of these species are subject to movements at least on a local scale.

In this study, African birds were classed as residents by default with a classification as visitor only made if my earlier studies in Tsavo East National Park indicated a significant change in numbers through the year (Lack 1980 & 1985). The classification is certainly conservative. It would not for example show as a migrant a species which moved between habitats at different seasons, for which there was some evidence in Tsavo, or a species of which only a small proportion moved. Both these could have profound effects on a species' biology and potential survival especially in such a potentially harsh environment. Only more detailed data and data collected all the year round will enable definitive conclusions on this. It is certainly apparent that movements and migration are arguably the normal situation in African savannas, especially the more open ones, with true residency, in the sense of individuals being completely sedentary, being fairly rare.

The differences between Mkomazi and Tsavo can be mainly attributed to the amount of rainfall, and the consequent growth of the vegetation, and perhaps related to this the proximity of forest areas. As noted above Mkomazi seemed to be especially well represented by thrushes, warblers and sunbirds and Tsavo by larks, starlings and Coraciiformes. Many of the former group are bush and tree dwellers which take insects from leaves, and the sunbirds are more or less dependent on nectar sources especially flowering trees and shrubs. However there are several more grass warblers as well, such as those in the genus *Cisticola,* where five were 'common' in Mkomazi and only two were in Tsavo. Those species well represented in Tsavo are primarily ground feeders and most of them feeding mainly from patches of bare ground.

Mkomazi's rainfall is higher than Tsavo's and this creates a more luxuriant growth which is seen both in the woody vegetation and in the grass layer. The woody vegetation in Mkomazi was both more varied and in general thicker with much richer woodland and even extending into forest at times. Woody vegetation in Tsavo was relatively limited and it was only in some of the drier parts of Mkomazi where the vegetation in the two areas was similar. This was particularly in the southern part of the reserve rather than the area where most of the Mkomazi work has been done. Among the birds in this part there seemed to be larger numbers of species such as Taita Fiscal and Superb Starling which were among the commonest birds in Tsavo and yet were largely replaced in the northern parts of Mkomazi by Long-tailed Fiscal and Hildebrandt's Starling respectively. How important to the avifauna of Mkomazi is the vicinity of the forest reserves is unclear at present although several species were only seen along the southwestern edge of the reserve adjacent to these areas.

The differences in the grass layer are also very important to the differences in the avifauna. Firstly, although there may well be more food if there is a lot of grass, it is no good to a bird trying to eat the insects or seeds which may be there if this food cannot be reached. This serves to re-emphasise a fact too often forgotten that food abundance is a very different thing to food availability and it is the latter which is critical to the use which birds and other animals can put an area. The numbers of the different groups clearly show that such food was easier to get in Tsavo than in Mkomazi. Secondly many areas of Mkomazi are burnt in the dry season, a feature not present during my studies in Tsavo. The relationship between birds and fires would be well worth studying in itself but it was obvious superficially that birds were using burnt areas quite extensively and that the species involved were such as starlings and Coraciiformes and not such as cisticolas.

The other really obvious difference between the two areas which appears to be very important for the bird life is the presence in Mkomazi of acacias. This is both in areas of woodland and in more open habitats where the dominant woody species are *Acacia* of various species. Birds seem to be more common and more obvious in acacias than in most other woody species. It appears that there are more insects in acacias than other species (see McGavin, Chapter 22), and acacias are usually larger, have more twigs and stay in leaf for longer. In terms of bird numbers it was noted above that *Acacia* habitats were by some margin the most diverse of Mkomazi's habitats for birds. More work is again needed.

Finally it must be noted that one habitat in Mkomazi remains essentially unsampled for birds and it is one which could be a very important one for increasing the number of species recorded and for assessing the importance of the area in relation to others. This is the forest patches on top of several of the hills. The total areas involved are small but they have the potential of harbouring many fairly rare and interesting species. All the major mountain areas in the vicinity, the Usambaras and the Pare Mountains and, further afield, such as the Taita Hills and Kasigau, hold a very different avifauna to the surrounding savannas and the few brief visits to this habitat in Mkomazi suggest that forest species and not savanna species are the predominant. These other areas also hold a remarkable number of rare and often endangered and endemic species or subspecies and it is quite possible that some of the patches within Mkomazi could harbour a few of these. This remains for detailed study in the future.

Conclusions

Mkomazi Game Reserve has a high diversity of birds at all seasons. This is due mainly to the wide range of habitat types present, of which it appears that acacias are the most important. All habitats, but again especially *Acacia* dominated ones, attract large numbers of migrants from the Palaearctic and visitors from elsewhere in Africa. Indeed it appears that migrancy to some degree should be considered the norm in this and other such savanna areas, a fact which has very important implications for the conservation of these birds.

Mkomazi Game Reserve is unique within Tanzania due in particular to the range of species whose range is predominantly to the north.

Finally, there is a great deal more to find out!

Acknowledgements

I am grateful to the Royal Geographical Society (with IBG) for inviting me to participate in the programme in Mkomazi and for providing the facilities to enable me to carry out the studies. Special thanks to Tim Morgan, Hugh Watson and staff at Ibaya for making life pleasant there. I am also grateful to the Royal Society for the generous grant which enabled me to go to Tanzania in 1993.

References

Harris, L.D. (1972) An ecological description of a semi-arid East African ecosystem. *Range Science Department, Science Series* 11: 1-89. Colorado State University.

Lack, D. (1935) Territory and polygamy in a bishop-bird, *Euplectes hordeacea hordeacea* (Linn.). *Ibis* 1935: 817-836.

Lack, P.C. (1980) *The habitats and feeding stations of the birds in Tsavo National Park, Kenya*. Unpublished D.Phil. thesis, University of Oxford.

Lack, P.C. (1985) The ecology of the land-birds of Tsavo East National Park, Kenya. *Scopus* 9: 2-23, 57-96.

Lack, P.C. (1987) The structure and seasonal dynamics of the bird community of Tsavo East National Park, Kenya. *Ostrich* 58: 9-23.

Lack, P.C. (1994) Three-streaked Tchagra *Tchagra jamesi*: a new record for Tanzania. *Scopus* 17: 140-141.

Lack, P.C., Leuthold, W. & Smeenk, C. (1980) Check-list of the birds of Tsavo East National Park, Kenya. *Journal of the East Africa Natural History Society and National Museum* 170: 1-25.

Pearson, D.J. & Lack, P.C. (1992) Migration patterns and habitat use by passerine and near passerine migrant birds in eastern Africa. *Ibis* 134, suppl. 1: 89-98.

Appendix: Bird ringing in Mkomazi

Neil E. Baker and Elizabeth M. Baker

Between the 25 December 1994 and 28 March 1996 a total of 368 birds of 76 species was ringed at four sites in the western half of the reserve (Table 26.8). Full mensural data were collected and stored on the database of the Tanzania Bird Atlas Project. As all species were caught using mistnets, only those species which move through the habitat below 3 m were trapped. It is safe to assume these birds were utilising the habitat and not merely passing through. Much of the ringing took place opportunistically, undertaken by interested individuals when time and weather conditions allowed. During December 1995 a more concentrated effort was made at two of the wooded botanical plots: Ubani plot near Kavateta Dam and Kisima plot to the south of Kisima Camp (for a botanical assessment of these two plots refer to Coe *et al.*, Chapter 7).

Although this was primarily a training session for the Tanzania Important Bird Areas Project, it was hoped in some way to be able to quantify the importance of these vegetation types for Palaearctic migrants. However, there were only two qualified ringers supervising six trainees. The large numbers of birds moving through the habitat at the time proved to be too many to process and nets had to be furled much earlier than envisaged. Very few of the Palaearctic migrants were carrying fat reserves indicating that they had arrived during the early hours of the same day after an overnight flight and came down in Mkomazi in preference to habitat further north. Palaearctic birds accounted for 20% of the species total but 30% of the numbers of birds caught (Table 26.7).

All the Palaearctic birds are insectivorous species whereas only 28 (nearly half) of the Afrotropical species are classed as insectivores. Clearly the carrying capacity of the habitat for Afrotropical insectivores is dependent on food supply during the dry season and they are unable to fully exploit the super abundance available at the onset of the short rains. This excess food is heavily exploited by the more mobile Palaearctic insect eaters and, as noted above, this mobility has potentially important implications for the conservation of these and other species. Similarly

Table 26.7 Numbers of individuals and species ringed in western Mkomazi Game Reserve, 1994–96.

	species	individuals
Palaearctic birds	15	111
Afrotropical birds	61	257
total	76	368

Table 26.8 Numbers of birds ringed in Mkomazi Game Reserve 1994–96. For scientific names, see the Checklist, Chapter 27.

Species	No.	Species	No.
Crested Francolin	2	Rattling Cisticola	2
Emerald-spotted Wood Dove	4	Ashy Cisticola	1
Black-and-White Cuckoo	1	Grey Wren-Warbler	7
Eurasian Cuckoo	1	Grey-backed Camaroptera	7
White-browed Coucal	1	Northern Crombec	4
African Scops Owl	1	Abyssinian White-eye	1
Pearl-spotted Owlet	1	Chin-spot Batis	2
Donaldson-Smith's Nightjar	1	Northern White-crowned Shrike	5
Little Bee-eater	3	Red-tailed Shrike	3
Abyssinian Scimitarbill	1	Long-tailed Fiscal	1
Red-billed Hornbill	3	Brubru	2
Von der Decken's Hornbill	1	Three-streaked Tchagra	2
Red-fronted Tinkerbird	1	Grey-headed Bush-Shrike	1
Black-throated Barbet	7	Black Cuckoo Shrike	2
d'Arnaud's Barbet	14	Common Drongo	3
Greater Honeyguide	1	Black-headed Oriole	1
Lesser Honeyguide	3	Eurasian Golden Oriole	1
Nubian Woodpecker	1	Hildebrandt's Starling	8
Zanzibar Sombre Greenbul	1	Superb Starling	9
Northern Brownbul	9	Golden-breasted Starling	2
Yellow-bellied Greenbul	1	Fischer's Starling	9
Common Bulbul	11	Eastern Violet-backed Sunbird	1
Sprosser	12	Collared Sunbird	1
White-browed Scrub Robin	7	Hunter's Sunbird	4
Irania	10	Purple-banded Sunbird	7
Common Rock Thrush	3	Parrot-billed Sparrow	2
Spotted Flycatcher	8	Yellow-spotted Petronia	13
African Grey Flycatcher	2	White-headed Buffalo-Weaver	2
Marsh Warbler	7	Red-billed Buffalo-Weaver	1
Olivaceous Warbler	4	Vitelline Masked Weaver	12
Upcher's Warbler	3	Chestnut Weaver	5
Olive-Tree Warbler	4	Red-headed Weaver	1
Barred Warbler	2	Green-winged Pytilia	12
Common Whitethroat	47	Red-cheeked Cordon-bleu	5
Garden Warbler	3	Blue-capped Cordon-bleu	6
Willow Warbler	25	Purple Grenadier	2
Tiny Cisticola	2	Somali Golden-breasted Bunting	3
Croaking Cisticola	1		

CHAPTER 27

Checklist: Birds of Mkomazi

Peter C. Lack, Neil E. Baker & Elizabeth M. Baker

The following lists, with a brief note of status, all those species claimed to have been seen in the Mkomazi Game Reserve by various observers and which are considered to be correct by the authors. Some other species have been claimed by others at times. Some of these were clearly errors, most of which have been agreed to be so by the observer concerned when questioned, and others are not considered sufficiently authenticated: for example a single record of a species rather out of range and for which a similar species is present fairly commonly and which was not seen by the observer concerned. Where possible the observer has been contacted to confirm or otherwise. Not quite all have yet been fully accepted by the East Africa Natural History Society Rarities Committee. A fully annotated checklist, detailing individual records where appropriate and more details of habitat preferences and migratory status will be published elsewhere in due course.

Records from the following observers are included (with approximate dates they were in the reserve where known): the authors PCL (two visits July–August 1993 and December 1995–January 1996), NEB and EMB (several visits between 1994 and early 1997); several associates of the latter including Zul Bhatia (two visits December 1995 and April 1997) and Stan Davies (several visits 1994–96); Larry Harris (for a few years in the 1960s); Clide Carter (in part of 1979); Alan Tye (at intervals while he was working at Amani in the early 1990s); John and Annabel Chapple (January 1997); Kim Ellis and David Bygott; Kath Shurduff and Dave Houghton (July 1995); Trevor Charlton and Wesley Krause (a waterfowl count in January 1995); Don Turner and Miles Coverdale (two overnight stays in February 1994); and Paul Strecker, Angus Jackson, Ryan Mellor, Caroline Peciuch, Joanne Taylor and Ben Underwood from St Peter's College, University of Oxford (July–August 1993).

Species are described as Resident; Visitor (from elsewhere in Africa) if there are sufficient records to warrant the distinction being known; or Migrant (from the Palaearctic). For species known to be visitors or migrants, months are given for which records are extant.

Nomenclature and order follows Zimmerman, D.A., Turner, D.A. & Pearson, D.J. (1996) *Birds of Kenya and Northern Tanzania*. Helm, London.

Class AVES

Ostrich: Struthionidae

Common Ostrich *Struthio camelus*, often seen especially in grassland

Grebes: Podicipedidae

Little Grebe *Tachybaptus ruficollis*, regular at water when levels are suitable

Pelicans: Pelecanidae

Great White Pelican *Pelecanus onocrotalus*, 2 records of birds flying over

Pink-backed Pelican *Pelecanus rufescens*, a few records, mostly of birds flying over

Cormorants: Phalacrocoracidae

Great Cormorant *Phalacrocorax carbo*, not seen since 1960s

Herons: Ardeidae

Black-crowned Night Heron *Nycticorax nycticorax*, not seen since 1960s

Cattle Egret *Bubulcus ibis*, regular at Dindira Dam and a few elsewhere

Little Egret *Egretta garzetta*, 2 records (Dindira, Ibaya)

Squacco Heron *Ardeola ralloides*, not seen since 1960s

Green-backed Heron *Butorides striatus*, 1 record (Umba River)

Yellow-billed Egret *Mesophoyx intermedia*, 2 records (Dindira)

Great Egret *Casmerodius albus*, not seen since 1960s

Grey Heron *Ardea cinerea*, regular records at water sources

Purple Heron *Ardea purpurea*, not seen since 1960s

Black-headed Heron *Ardea melanocephala*, regular records at water sources

Hamerkop: Scopidae

Hamerkop *Scopus umbretta*, resident in small numbers near water

Storks: Ciconiidae

White Stork *Ciconia ciconia*, uncommon migrant in flocks Dec-Jan

Black Stork *Ciconia nigra*, a few records

Abdim's Stork *Ciconia abdimii*, regular visitor Dec-Jan

Woolly-necked Stork *Ciconia episcopus*, uncommon mainly at water sources

Saddle-billed Stork *Ephippiorhynchus senegalensis*, regular records

Marabou Stork *Leptoptilos crumeniferus*, regular records

African Open-billed Stork *Anastomus lamelligerus*, several records of flocks by water especially Dindira

Yellow-billed Stork *Mycteria ibis*, regular records at water

Ibises and Spoonbills: Threskiornithidae

Sacred Ibis *Threskiornis aethiopicus*, regular records

Hadada Ibis *Bostrychia hagedash*, 2 records (Umba River, Ibaya)

Glossy Ibis *Plegadis falcinellus*, 1 record (Umba River)

African Spoonbill *Platalea alba*, a few records by water

Flamingos: Phoenicopteridae

Lesser Flamingo *Phoeniconaias minor*, 1 record (immature at Dindira)

Ducks: Anatidae

White-faced Whistling Duck *Dendrocygna viduata*, 1 record

Spur-winged Goose *Plectopterus gambensis*, flocks up to 15 at Dindira, scarce elsewhere

Egyptian Goose *Alopochen aegyptiacus*, resident at Dindira when water present, scarce elsewhere

Knob-billed Duck *Sarkidiornis melanotos*, not seen since 1960s

Cape Teal *Anas capensis*, 1 record (Dindira)

Garganey *Anas querquedula*, 1 record (Dindira)

Red-billed Teal *Anas erythrorhyncha*, resident Dindira when water present, rare elsewhere

Hottentot Teal *Anas hottentota*, regular records

Southern Pochard *Netta erythrophthalma*, 2 records (Dindira)

Secretarybird: Sagittariidae

Secretary Bird *Sagittarius serpentarius*, common and widespread in more open areas

Hawks and allies: Accipitridae

Black-shouldered Kite *Elanus caeruleus*, uncommon in dry season, common in wet

Black Kite *Milvus migrans*, a few records

Lammergeier *Gypaetus barbatus*, 1 record

Egyptian Vulture *Neophron percnopterus*, 3 records (Dindira (2), Ndea)

Hooded Vulture *Necrosyrtes monachus*, uncommon resident

African White-backed Vulture *Gyps africanus*, common and widespread

Ruppell's Vulture *Gyps rueppelli*, widespread but uncommon

Lappet-faced Vulture *Torgos tracheliotus*, uncommon resident

White-headed Vulture *Trigonoceps occipitalis*,

uncommon resident but more widespread than Lappet-faced
Black-chested Harrier-Eagle *Circaetus pectoralis*, several records
Brown Snake Eagle *Circaetus cinereus*, fairly common resident
Bateleur *Terathopius ecaudatus*, common and widespread over all habitats
Harrier Hawk *Polyboroides radiatus*, uncommon resident
Pallid Harrier *Circus macrourus*, migrant, 1 definite record
Montagu's Harrier *Circus pygargus*, regular migrant Nov-Apr
(Ring-tailed Harrier *Circus spp.*, fairly common migrant, most not distinguished as to *macrourus* or *pygargus*; the proportion of males suggests latter is commoner)
African Marsh Harrier *Circus ranivorus*, regular records
Eurasian Marsh Harrier *Circus aeruginosus*, uncommon migrant
Gabar Goshawk *Melierax gabar*, common but elusive, black form also seen
Eastern Pale Chanting Goshawk *Melierax poliopterus*, common and widespread
African Goshawk *Accipiter tachiro*, 2 records (Njiro, Kisima)
Shikra *Accipiter badius*, uncommon resident in thicker habitats
Little Sparrowhawk *Accipiter minullus*, a few records
Grasshopper Buzzard *Butastur rufipennis*, common visitor Nov-Jan
Lizard Buzzard *Kaupifalco monogrammicus*, uncommon resident in forested areas
Steppe Buzzard *Buteo buteo vulpinus*, a few records Nov-Dec
Augur Buzzard *Buteo augur*, common resident around hills
Fish Eagle *Haliaetus vocifer*, uncommon
Palm-nut Vulture *Gypohierax angolensis*, a few records
Lesser Spotted Eagle *Aquila pomarina*, 2 records
Tawny Eagle *Aquila rapax*, common and widespread resident
Steppe Eagle *Aquila nipalensis*, common migrant Nov-Jan
Wahlberg's Eagle *Aquila wahlbergi*, common visitor July-Apr
Verreaux's Eagle *Aquila verreauxii*, regular records
African Hawk Eagle *Hieraaetus spilogaster*, scarce resident
Booted Eagle *Hieraaetus pennatus*, 1 record (Kisima)
Long-crested Eagle *Lophaetus occipitalis*, uncommon resident
Martial Eagle *Polemaetus bellicosus*, uncommon resident

Falcons: Falconidae

Pygmy Falcon *Polihierax semitorquatus*, common and widespread resident
Lanner Falcon *Falco biarmicus*, a few records
Peregrine Falcon *Falco peregrinus*, a few records
Eurasian Hobby *Falco subbuteo*, rare migrant
African Hobby *Falco cuvieri*, 1 record (Ndea)
Amur Falcon *Falco amurensis*, a few records in April
Eleonora's Falcon *Falco eleonorae*, 1 record (Dindira)
Sooty Falcon *Falco concolor*, 1 record
Lesser Kestrel *Falco naumanni* 1 recent record
Common Kestrel *Falco tinnunculus*, uncommon migrant Nov-Jan

Quail and Francolins: Phasianidae

Blue Quail *Coturnix adansonii*, 2 records (Ndea, Ibaya-Zange)
Harlequin Quail *Coturnix delegorguei*, very common visitor Dec-Mar
Shelley's Francolin *Francolinus shelleyi*, uncommon resident in open grassy areas
Crested Francolin *Francolinus sephaena*, very common and widespread resident
Hildebrandt's Francolin *Francolinus hildebrandti*, resident, heard commonly near hills
Yellow-necked Spurfowl *Francolinus leucoscepus*, very common and widespread resident

Guineafowl: Numididae

Crested Guineafowl *Guttera pucherani*, uncommon resident in hill forests
Vulturine Guineafowl *Acryllium vulturinum*, uncommon resident
Helmeted Guineafowl *Numida meleagris*, uncommon resident

Buttonquails: Turnicidae

Common Button-Quail *Turnix sylvatica*, fairly common resident

Crakes and allies: Rallidae

Buff-spotted Flufftail *Sarothrura elegans*, 1 record
African Crake *Crex egregia*, 1 record
Corncrake *Crex crex*, 2 records
Black Crake *Amaurornis flavirostris*, several records
Red-knobbed Coot *Fulica cristata*, 2 records

Finfoot: Heliornithidae

African Finfoot *Podica senegalensis*, 1 record (Umba River)

Bustards: Otididae

Kori Bustard *Ardeotis kori*, uncommonly seen all year

Crested Bustard *Eupodotis ruficrista*, common and widespread resident especially in thicker habitats

White-bellied Bustard *Eupodotis senegalensis*, fairly common and widespread resident

Black-bellied Bustard *Eupodotis melanogaster*, 1 record ('Superbowl')

Hartlaub's Bustard *Eupodotis hartlaubii*, uncommon widespread resident

Jacanas: Jacanidae

African Jacana *Actophilornis africanus*, 1 record

Painted-snipes: Rostratulidae

Greater Painted-snipe *Rostratula benghalensis*, 1 record

Avocets and Stilts: Recurvirostridae

Pied Avocet *Recurvirostra avosetta*, 1 record of 35 (Dindira)

Black-winged Stilt *Himantopus himantopus*, common resident at water

Thick-knees: Burhinidae

Spotted Thick-knee *Burhinus capensis*, uncommon and elusive resident

Coursers and Pratincoles: Glareolidae

Two-banded Courser *Rhinoptilus africanus*, several records

Heuglin's Courser *Rhinoptilus cinctus*, 3 recent records

Temminck's Courser *Cursorius temminckii*, fairly common visitor July-Jan and April

Plovers: Charadriidae

Blacksmith Plover *Vanellus armatus*, seen all year at water

Senegal Plover *Vanellus lugubris*, a few records

Crowned Plover *Vanellus coronatus*, fairly common resident in open areas

Kittlitz's Plover *Charadrius pecuarius*, a few records

Three-banded Plover *Charadrius tricollaris*, seen all year at water

Sandpipers and allies: Scolopacidae

Little Stint *Calidris minuta*, common at water in northern winter

Curlew Sandpiper *Calidris ferruginea*, only records are of failed breeders in July-Aug

Ruff *Philomachus pugnax*, a few records

Marsh Sandpiper *Tringa stagnatilis*, regular migrant Dec-Jan

Greenshank *Tringa nebularia*, fairly common migrant in northern winter, a few in summer

Green Sandpiper *Tringa ochropus*, fairly common migrant Aug-Jan

Wood Sandpiper *Tringa glareola*, fairly common migrant July-Apr

Common Sandpiper *Actitis hypoleucos*, common migrant, a few in summer

Gulls and Terns: Laridae

Gull-billed Tern *Sterna nilotica*, 1 record

Whiskered Tern *Chlidonias hybridus*, 1 record

Sandgrouse: Pteroclidae

Black-faced Sandgrouse *Pterocles decoratus*, uncommon resident in thicker *Commiphora* habitats

Pigeons and Doves: Columbidae

African Green Pigeon *Treron calva*, uncommon resident in large trees

Tambourine Dove *Turtur tympanistria*, 2 records in forest

Emerald-spotted Wood Dove *Turtur chalcospilos*, very common resident in all thicker habitats

Namaqua Dove *Oena capensis*, uncommon in dry, common in wet seasons

Red-eyed Dove *Streptopelia semitorquata*, common resident in thicker habitats

Ring-necked Dove *Streptopelia capicola*, very common and widespread resident

Laughing Dove *Streptopelia senegalensis*, resident at Ibaya and near water, much commoner and more widespread in wet season

Parrots: Psittacidae

African Orange-bellied Parrot *Poicephalus rufiventris*, very common resident wherever there are trees

Turacos: Musophagidae

White-bellied Go-Away Bird *Corythaixoides leucogaster*, very common resident wherever there are trees

Cuckoos and Coucals: Cuculidae

Black-and-White Cuckoo *Oxylophus jacobinus*, fairly common migrant Nov-Apr

Great Spotted Cuckoo *Clamator glandarius*, uncommon migrant

Black Cuckoo *Cuculus clamosus*, uncommon visitor into wooded areas

Red-chested Cuckoo *Cuculus solitarius*, uncommon visitor

Eurasian Cuckoo *Cuculus canorus*, scarce migrant

African Cuckoo *Cuculus gularis*, 1 definite record

Asian Lesser Cuckoo *Cuculus poliocephalus*, 2 records

African Emerald Cuckoo *Chrysococcyx cupreus*, 1 record

Klaas's Cuckoo *Chrysococcyx klaas*, 1 record in dry, several in wet seasons

Diederik Cuckoo *Chrysococcyx caprius*, common visitor Dec-Apr

Yellowbill *Ceuthmochares aereus*, 1 record (Umba River)

White-browed Coucal *Centropus superciliosus*, fairly common and widespread resident

Black Coucal *Centropus grillii*, uncommon visitor Dec-Feb

Typical Owls: Strigidae

African Scops Owl *Otus senegalensis*, a few records Oct-Dec and April

White-faced Scops Owl *Otus leucotis*, 2 records

Spotted Eagle-Owl *Bubo africanus*, scarce resident

Verreaux's Eagle-Owl *Bubo lacteus*, scarce resident

Pearl-spotted Owlet *Glaucidium perlatum*, fairly common and widespread resident

African Wood Owl *Strix woodfordii*, 1 record

Nightjars: Caprimulgidae

Donaldson-Smith's Nightjar *Caprimulgus donaldsoni*, fairly common resident

Freckled Nightjar *Caprimulgus tristigma*, not seen since 1960s

Plain Nightjar *Caprimulgus inornatus*, a few records

European Nightjar *Caprimulgus europaeus*, no definite record since 1960s, but 1 possible recent record

Gabon Nightjar *Caprimulgus fossii*, not seen since 1960s

Slender-tailed Nightjar *Caprimulgus clarus*, common visitor Dec-Jan

Swifts: Apodidae

Scarce Swift *Schoutedenapus myoptilus*, 1 record of 10 (Ibaya-Zange)

African Palm Swift *Cypsiurus parvus*, uncommon but widespread resident

Common Swift *Apus apus*, common migrant Dec-Jan

Nyanza Swift *Apus niansae*, several records

Mottled Swift *Apus aequatorialis*, several records

White-rumped Swift *Apus caffer*, common resident near hills

Little Swift *Apus affinis*, common and widespread resident

Alpine Swift *Apus melba*, 1 record (Ndea)

Mousebirds: Coliidae

Speckled Mousebird *Colius striatus*, uncommon resident in thick habitats

White-headed Mousebird *Colius leucocephalus*, uncommon resident

Blue-naped Mousebird *Urocolius macrourus*, common and widespread resident

Trogons: Trogonidae

Narina Trogon *Apaloderma narina*, 2 records (Umba River, Ibaya)

Kingfishers: Alcedinidae

Grey-headed Kingfisher *Halcyon leucocephala*, uncommon visitor

Brown-hooded Kingfisher *Halcyon albiventris*, scarce resident

Striped Kingfisher *Halcyon chelicuti*, fairly common and widespread resident

Malachite Kingfisher *Alcedo cristata*, 3 records

Giant Kingfisher *Megaceryle maxima*, 1 record (Umba River)

Pied Kingfisher *Ceryle rudis*, 1 record (Umba River)

Bee-eaters: Meropidae

Eurasian Bee-eater *Merops apiaster*, common migrant Oct-Apr

Madagascar Bee-eater *Merops superciliosus*, common visitor Sept-Mar

Blue-cheeked Bee-eater *Merops persicus*, common migrant Nov-Apr

Carmine Bee-eater *Merops nubicus*, several records

White-throated Bee-eater *Merops albicollis*, a few records

Little Bee-eater *Merops pusillus*, very common and widespread resident

Rollers: Coraciidae

Eurasian Roller *Coracias garrulus*, very common migrant Nov-Apr

Lilac-breasted Roller *Coracias caudata*, very common and widespread resident

Rufous-crowned Roller *Coracias naevia*, uncommon resident

Broad-billed Roller *Eurystomus glaucurus*, a few records

Hoopoe: Upupidae

Hoopoe *Upupa epops*, fairly common resident in more open habitats

Woodhoopoes: Phoeniculidae

Green Wood Hoopoe *Phoeniculus purpureus*, uncommon resident in wooded habitats

Violet Wood Hoopoe *Phoeniculus damarensis granti*, uncommon resident in wooded habitats

Common Scimitarbill *Rhinopomastus cyanomelas*, uncommon resident

Abyssinian Scimitarbill *Rhinopomastus minor*, fairly common resident in wooded habitats

Hornbills: Bucerotidae

Southern Ground Hornbill *Bucorvus leadbeateri*, local resident

Red-billed Hornbill *Tockus erythrorhynchus*, very common and widespread resident

Eastern Yellow-billed Hornbill *Tockus flavirostris*, uncommon resident in open woodland

Von der Decken's Hornbill *Tockus deckeni*, very common and widespread resident

Crowned Hornbill *Tockus alboterminatus*, uncommon resident in Acacia woodland and forest

African Grey Hornbill *Tockus nasutus*, uncommon but widespread resident

Trumpeter Hornbill *Bycanistes bucinator*, common resident in forest

Barbets: Lybiidae

Red-fronted Tinkerbird *Pogniulus pusillus*, common resident, rarely seen

Red-fronted Barbet *Tricholaema diademata*, 1 record

Spot-flanked Barbet *Tricholaema lacrymosa*, fairly common resident in Acacia woodland and forest

Black-throated Barbet *Tricholaema melanocephala*, common resident in *Commiphora*

White-headed Barbet *Lybius leucocephalus*, scarce resident

Brown-breasted Barbet *Lybius melanopterus*, scarce resident in forest

Red-and-Yellow Barbet *Trachyphonus erythrocephalus*, uncommon but widespread resident

d'Arnaud's Barbet *Trachyphonus darnaudii*, very common and widespread resident,

Honeyguides: Indicatoridae

Greater Honeyguide *Indicator indicator*, fairly common resident

Lesser Honeyguide *Indicator minor*, fairly common resident

Woodpeckers: Picidae

Nubian Woodpecker *Campethera nubicus*, common resident wherever there are trees

Cardinal Woodpecker *Dendropicos fuscescens*, uncommon widespread resident wherever there are trees

Bearded Woodpecker *Dendropicos namaquus*, scarce resident

Larks: Alaudidae

Singing Bush Lark *Mirafra cantillans*, 3 records

Friedmann's Lark *Mirafra pulpa*, uncommon visitor Dec, Jan, Mar, Sept

Red-winged Bush Lark *Mirafra hypermetra*, uncommon resident in thick grass

Flappet Lark *Mirafra rufocinnamomea*, very common resident

Fawn-coloured Lark *Mirafra africanoides*, a few records

Pink-breasted Lark *Mirafra poecilosterna*, uncommon resident in drier areas

Fischer's Sparrow-Lark *Eremopterix leucopareia*, uncommon resident in more open areas

Chestnut-backed Sparrow-Lark *Eremopterix leucotis*, a few records

Pipits, Wagtails and Longclaws: Motacillidae

African Pied Wagtail *Motacilla aguimp*, common resident in villages, scarce in reserve

Yellow Wagtail *Motacilla flava*, uncommon migrant

Grassland (Richard's) Pipit *Anthus cinnamomeus*, a few records

Tree Pipit *Anthus trivialis*, 2 records

Golden Pipit *Tmetothylacus tenellus*, fairly common visitor Dec-Apr, scarce outside this period

Pangani Longclaw *Macronyx aurantiigula*, common resident in grassy areas

Swallows and Martins: Hirundinidae

Banded Martin *Riparia cincta*, fairly common resident

Sand Martin *Riparia riparia*, common migrant Oct-Jan

Wire-tailed Swallow *Hirundo smithii*, common in surrounding villages, a few in reserve

Barn Swallow *Hirundo rustica*, very common migrant Sept-Apr, also July and August

Red-rumped Swallow *Hirundo daurica*, fairly common and widespread resident

Mosque Swallow *Hirundo senegalensis*, 1 recent record

Lesser Striped Swallow *Hirundo abyssinica*, uncommon but widespread resident

African Rock Martin *Hirundo fuligula*, uncommon but widespread especially around hills

Common House Martin *Delichon urbica*, a few records

Black Rough-wing *Psalidoprocne holomelas,* 3 records

Bulbuls: Pycnonotidae

Zanzibar Sombre Greenbul *Andropadus importunus*, local resident in thick vegetation

Northern Brownbul *Phyllastrephus strepitans*, uncommon but widespread in thick vegetation

Yellow-bellied Greenbul *Chlorocichla flaviventris*, common resident in forest areas

Common Bulbul *Pycnonotus barbatus*, very common resident especially near base of hills

Eastern Nicator *Nicator gularis*, a few records

Babblers and Chatterers: Timaliidae

Rufous Chatterer *Turdoides rubiginosus*, uncommon resident

Scaly Chatterer *Turdoides aylmeri*, 2 records (Kisima area)

Thrushes and Chats: Turdidae

Red-capped Robin-Chat *Cossypha natalensis*, a few records

White-browed Robin-Chat *Cossypha heuglini*, scarce resident

Spotted Morning Thrush *Cichladusa guttata*, uncommon but widespread resident (rarely seen)

Sprosser *Luscinia luscinia*, uncommon migrant to thick undergrowth

White-browed (White-winged) Scrub Robin *Cercotrichas leucophrys (vulpina)*, fairly common resident in wooded habitats

Rufous Bush Chat *Cercotrichas galactotes*, uncommon migrant

Irania *Irania gutturalis*, uncommon migrant

Whinchat *Saxicola rubetra*, 2 records

Northern Wheatear *Oenanthe oenanthe*, common migrant Oct-Jan

Pied Wheatear *Oenanthe pleschanka*, common migrant Oct-Feb

Isabelline Wheatear *Oenanthe isabellina*, uncommon migrant Nov-Jan

Capped Wheatear *Oenanthe pileata*, a few records

Red-tailed Chat *Cercomela familiaris*, 1 record

Cliff Chat *Thamnolaea cinnamomeiventris*, a resident pair at Ibaya

Common Rock Thrush *Monticola saxatilis*, common migrant Nov-Apr

Bare-eyed Thrush *Turdus tephronotus*, uncommon but widespread resident (rarely seen)

Flycatchers: Muscicapidae

Spotted Flycatcher *Muscicapa striata*, very common migrant Oct-Apr

Ashy Flycatcher *Muscicapa caerulescens*, scarce resident in forests

Southern Black Flycatcher *Melaenornis pammelaina*, scarce resident

African Grey Flycatcher *Bradornis microrhynchus*, uncommon resident in woodland

Pale Flycatcher *Bradornis pallidus*, scarce resident

Warblers: Sylviidae

Great Reed Warbler *Acrocephalus arundinaceus* or Basra Reed Warbler *Acrocephalus griseldis*, 1 record (at the time these forms were subspecies so it is uncertain to which the record referred)

Marsh Warbler *Acrocephalus palustris*, uncommon migrant Dec-Jan

Olivaceous Warbler *Hippolais pallida*, very common migrant Nov-Feb

Upcher's Warbler *Hippolais languida*, fairly common migrant Dec-Feb

Olive-Tree Warbler *Hippolais olivetorum*, a few records

Barred Warbler *Sylvia nisoria*, uncommon migrant Dec-Apr

Common Whitethroat *Sylvia communis*, very common migrant Dec-Feb

Garden Warbler *Sylvia borin*, uncommon migrant Dec-Apr

Blackcap *Sylvia atricapilla*, 1 record (Kisima)

Willow Warbler *Phylloscopus trochilus*, very common migrant Oct-Apr

Broad-tailed Warbler *Schoenicola brevirostris*, 1 record

Red-faced Cisticola *Cisticola erythrops*, 1 record

Rock Cisticola *Cisticola aberrans*, 1 record

Tiny Cisticola *Cisticola nanus*, common and widespread resident

Winding Cisticola *Cisticola galactotes*, uncommon resident in thick grass especially watercourses

Croaking Cisticola *Cisticola natalensis* 1 record

Rattling Cisticola *Cisticola chiniana*, common and widespread resident

Ashy Cisticola *Cisticola cinereolus*, very common and widespread resident

Siffling Cisticola *Cisticola brachypterus*, a few records

Zitting Cisticola *Cisticola juncidis*, fairly common resident (probably)

Desert Cisticola *Cisticola aridulus*, fairly common resident in grassland

(In grasslands there are at least two species of small Cisticola. One is Desert *C. aridulus*, another is probably Zitting *C. juncidis* and there may or may not be others.)

Tawny-flanked Prinia *Prinia subflava*, uncommon resident in thick bushes

Grey Wren-Warbler *Calamonastes simplex*, very common and widespread resident

Grey-backed Camaroptera *Camaroptera brachyura*, uncommon but widespread resident in thick bushes

Yellow-breasted Apalis *Apalis flavida*, 4 records in dry season, 1 in wet

Red-fronted Warbler *Spiloptila rufifrons*, 2 records

Northern Crombec *Sylvietta brachyura*, common resident

Red-faced Crombec *Sylvietta whytii*, fairly common resident

Somali Long-billed Crombec *Sylvietta isabellina*, 3 records

Yellow-bellied Eremomela *Eremomela icteropygialis*, fairly common resident

Yellow-vented Eremomela *Eremomela flaviventris*, 1 record

White-eyes: Zosteropidae

Abyssinian White-eye *Zosterops abyssinicus*, common and widespread resident

Tits: Paridae

Northern Grey Tit *Parus thruppi*, a few records

White-bellied Tit *Parus albiventris*, 3 records

Penduline Tits: Remizidae

Mouse-coloured Penduline Tit *Remiz musculus*, uncommon resident usually in Acacias

Monarch Flycatchers: Monarchidae

African Paradise Flycatcher *Terpsiphone viridis*, uncommon resident in forest areas

Batises: Platysteiridae

Chin-spot Batis *Batis molitor*, uncommon resident

Black-headed Batis *Batis minor*, uncommon resident in woodland and forest

Pygmy Batis *Batis perkeo*, 1 record

Helmet-shrikes: Prionopidae

White-crested Helmet-Shrike *Prionops plumatus*, fairly common and widespread resident

Retz's Helmet Shrike *Prionops retzii*, uncommon resident in forests

Northern White-crowned Shrike *Eurocephalus rueppelli*, very common and widespread resident

Shrikes: Laniidae

Red-backed Shrike *Lanius collurio*, fairly common migrant Dec and Feb-Apr

Red-tailed Shrike *Lanius isabellinus*, very common migrant Nov-Mar

Lesser Grey Shrike *Lanius minor*, common migrant Apr

Long-tailed Fiscal *Lanius cabanisi*, very common and widespread resident

Taita Fiscal *Lanius dorsalis*, uncommon resident in drier areas

Bush-shrikes: Malaconotidae

Brubru *Nilaus afer*, fairly common resident

Black-crowned Tchagra *Tchagra senegala*, fairly common and widespread resident

Brown-crowned Tchagra *Tchagra australis*, fairly common and widespread resident

Three-streaked Tchagra *Tchagra jamesi*, fairly common resident in drier areas

Sulphur-breasted Bush-Shrike *Malacanotus sulfureopectus*, fairly common resident in thicker vegetation

Four-coloured Bush-Shrike *Malaconotus quadricolor*, 1 record

Grey-headed Bush-Shrike *Malacanotus blanchoti*, uncommon resident especially near hills

Rosy-patch Bush-Shrike *Rhodophoneus cruentus*, common and widespread resident

Tropical Boubou *Laniarius aethiopicus*, fairly common resident in forest areas

Slate-coloured Boubou *Laniarius funebris*, common resident in thick bushes but rarely seen

Black-backed Puffback *Dryoscopus cubla*, fairly common resident in forests

Pringle's Puffback *Dryoscopus pringlii*, 4 records

Cuckoo-shrikes: Campephagidae

Black Cuckoo Shrike *Campephaga flava*, uncommon resident in wooded areas

Drongos: Dicruridae

Common Drongo *Dicrurus adsimilis*, very common resident

Orioles: Oriolidae

Black-headed Oriole *Oriolus larvatus*, uncommon but widespread resident

African Golden Oriole *Oriolus auratus*, a few records

Eurasian Golden Oriole *Oriolus oriolus*, common migrant Nov-Apr

Crows and Ravens: Corvidae

Pied Crow *Corvus albus*, uncommon resident especially in villages

White-naped Raven *Corvus albicollis*, a few records

Starlings and Oxpeckers: Sturnidae

Red-winged Starling *Onychognathus morio*, local resident

Blue-eared Starling *Lamprotornis chalybaeus*, uncommon resident

Hildebrandt's Starling *Lamprotornis hildebrandti*, very common resident

Shelley's Starling *Lamprotornis shelleyi*, 1 record

Superb Starling *Lamprotornis superbus*, common resident

Golden-breasted Starling *Cosmopsarus regius*, uncommon but widespread resident

Violet-backed Starling *Cinnyricinclus leucogaster*, a few records Jan-Apr

Fischer's Starling *Spreo fischeri*, common resident

Wattled Starling *Creatophora cinerea*, fairly common all year

Red-billed Oxpecker *Buphagus erythrorhynchus*, scarce resident

Yellow-billed Oxpecker *Buphagus africanus*, uncommon resident

Sunbirds: Nectariniidae

Eastern Violet-backed Sunbird *Anthreptes orientalis*, common and widespread resident

Collared Sunbird *Anthreptes collaris*, common resident in forest, scarce elsewhere

Amethyst Sunbird *Nectarinia amethystina*, uncommon but widespread resident

Hunter's Sunbird *Nectarinia hunteri*, common and widespread resident

Variable Sunbird *Nectarinia venusta*, common resident around base of hills

Mariqua Sunbird *Nectarinia mariquensis*, fairly common resident in woodland

Purple-banded Sunbird *Nectarinia bifasciata*, uncommon resident restricted to *Commiphora*

Black-bellied Sunbird *Nectarinia nectarinioides*, uncommon resident

Beautiful Sunbird *Nectarinia pulchella*, scarce resident

Sparrows: Passeridae

Grey-headed Sparrow *Passer griseus griseus*, a few records

Parrot-billed Sparrow *Passer griseus gongonensis*, uncommon resident

House Sparrow *Passer domesticus*, 1 record (Ibaya), another in Same

Chestnut Sparrow *Passer eminibey*, 1 record (Kisima)

Yellow-spotted Petronia *Petronia pyrgita*, common and widespread resident

Weavers: Ploceidae

White-headed Buffalo-Weaver *Dinemellia dinemelli*, fairly common resident

Red-billed Buffalo-Weaver *Bubalornis niger*, very common resident

White-browed Sparrow-Weaver *Plocepasser mahali*, common resident in villages, few elsewhere

Grosbeak Weaver *Amblyospiza albifrons*, 1 record

Black-necked Weaver *Ploceus nigricollis*, uncommon but widespread resident in thicker habitats

African Golden Weaver *Ploceus subaureus*, a few records

Vitelline Masked Weaver *Ploceus velatus*, very common visitor Oct-Jan

Masked Weaver *Ploceus intermedius*, a few records

Black-headed Weaver *Ploceus cucullatus*, common locally Oct-Mar

(Note that there are some 'yellow weavers' present all year which may well be *P. velatus*, *P. intermedius, P. cucullatus* or some other species)

Chestnut Weaver *Ploceus rubiginosus*, common all year

Red-headed Weaver *Anaplectes rubriceps*, uncommon but widespread resident

Red-billed Quelea *Quelea quelea*, common and widespread all year

Cardinal Quelea *Quelea cardinalis*, 2 records

Fire-fronted Bishop *Euplectes diadematus*, 2 records

Black-winged Red Bishop *Euplectes hordeaceus*, 1 record (Kisima)

Zanzibar Red Bishop *Euplectes nigroventris*, several records

Yellow Bishop *Euplectes capensis*, a few records

White-winged Widowbird *Euplectes albonotatus*, common visitor Nov-Mar

Waxbills and Whydahs: Estrildidae

Green-winged Pytilia *Pytilia melba*, common and widespread resident

Peters's Twinspot *Hypargos niveoguttatus*, uncommon resident in forests

Red-billed Firefinch *Lagonosticta senegala*, a few records Oct-Feb

Jameson's Firefinch *Lagonosticta rhodopareia*, uncommon resident

Red-cheeked Cordon-Bleu *Uraeginthus bengalus*, fairly common and widespread resident

Blue-capped Cordon-Bleu *Uraeginthus cyanocephalus*, a few records

Purple Grenadier *Uraeginthus ianthinogaster*,

uncommon resident in woodland
Crimson-rumped Waxbill *Estrilda rhodopyga*, uncommon, records all year
Waxbill *Estrilda astrild*, 2 records
Black-faced Waxbill *Estrilda erythronotus (delamerei)*, a few records
Quail-Finch *Ortygospiza atricollis*, 1 record (Maore)
African Silverbill *Lonchura cantans,* uncommon in flocks all year
Grey-headed Silverbill *Lonchura griseicapilla*, uncommon in flocks all year
Bronze Mannikin *Lonchura cucullata*, uncommon visitor
Rufous-backed Mannikin *Lonchura bicolor nigriceps*, a few records
Cut-throat Finch *Amadina fasciata*, uncommon visitor
Indigobird species *Vidua sp.*, 3 records
Pin-tailed Whydah *Vidua macroura*, common visitor Dec-Apr
Straw-tailed Whydah *Vidua fischeri*, not seen since 1960s
Paradise Whydah *Vidua paradisaea*, uncommon all year

Seedeaters and Canaries: Fringillidae
Yellow-rumped Seedeater *Serinus reichenowi,* uncommon all year especially near water
Yellow-fronted Canary *Serinus mozambicus*, 2 records

Buntings: Emberizidae
Cinnamon-breasted Rock Bunting *Emberiza tahapisi*, local resident
Somali Golden-breasted Bunting *Emberiza poliopleura*, uncommon resident

CHAPTER 28

Is Mkomazi a potential reservoir of avian crop pests?

James J. Matee

Introduction

In July 1995, Tropical Pesticides Research Institute (TPRI) was invited to a workshop organised by the Mkomazi Ecological Research Programme at Same, near the Mkomazi Game Reserve. At this workshop, TPRI was invited to prepare a proposal for its contribution to the programme in terms of research. TPRI was represented by a botanist, an entomologist, a rodentologist and an ornithologist, and they indicated possible areas of their contributions that would enrich the programme. Further meetings were held between TPRI and the programme staff and more areas of research were incorporated subject to availability of funds. Such areas included GIS, environmental research and natural products. It was decided that the ornithological section could start its work immediately (March 1996) with an emphasis on the Red-billed Quelea *(Quelea quelea* Linnaeus) which was observed in large numbers around the Ibaya Research Centre.

The presence of birds within an area is largely dependent on such factors as suitable habitat, availability of food and water as well as shelter and security (e.g. absence of predators). Quelea are known to be gregarious and migratory. Their behaviour is triggered by the need for survival and reproduction. The presence of big flocks of bird pests like *Q. quelea* within the vicinity of cereal growing areas usually raises alarm to the farmers and the farming community.

Quelea are reported to cause substantial damage to cultivated cereal grains especially wheat, rice, sorghum and millets (Haylock 1959). He also reported that the birds caused plague in Dodoma, Tanzania, which resulted in a high death rate in 1881. High infestations of Quelea have been recorded in Tanzania since 1944 and continue to the present time.

The natural habitat of the Quelea is semi-arid *Acacia* country where it relies on wild grass seeds (*Setaria* spp., *Echinochloa* spp., *Sorghum* spp., *Pannicum* spp., *Eragrostis* spp., *Digitaria* spp., *Brachystegia* spp., *Cynodon* spp. etc.) and natural water holes, migrating locally in a seasonal search for sustenance.

The expansion and modernization in agriculture, overgrazing leaving vast areas void of grass cover, global environmental changes and inadequate rains to promote wild grass germination have all contributed to the change in Quelea feeding behaviour, from its natural food and habitat, to an increase in its pest potential. The population dynamics of Quelea have not been monitored and studied in the Mkomazi Game Reserve. This study was proposed to find whether there is any relationship between Quelea found in the reserve and the crop damage experienced in the surrounding agricultural settlements.

Methodology

Surveys were initially carried out along the reserve roads to establish the presence of Quelea and possible locations for roosts or colonies. For comparison, surveys were concurrently conducted along the main roads from Same to Mkomazi village (where the Same–Kisiwani–Kihurio road joins the Same–Hedaru–Mombo tarmac road) and the irrigated rice fields along these areas. The farms are along the southern border of the reserve.

Information recorded included number of birds observed, habitat and activity. Birds were mist-netted during the morning and afternoon feeding hours (08:00–10:00 hours and 15:00–18:00 h) and dissected (Ward 1971) to establish the type of food eaten, age, sex and gonad size. The study was carried out at least one week in each month from March to November 1996.

Results

Flocks of Quelea in breeding plumage (containing both adults and juveniles) were observed along the valleys east and west of Ibaya centre, Kavateta Dam and the flood plain/swamp to the east and south of Simba plot.

The monthly records of Quelea observed and counted within and outside the MGR between March and November 1996 are shown in Table 28.1. There were no Quelea recorded in Kisima, Kifukua, Maore, Kamakota and Umba within the reserve and Gonja Maore outside the reserve during the study. Most of the Quelea were recorded during March–June (Table 28.1) followed by decline and absence of the birds between July–November).

Quelea were observed in Ndungu, Kihurio and east Mkomazi throughout the survey period. No surveys were carried out in April and July. A flock of 5 million birds was recorded during the study at Kavateta Dam in May (Table 28.1). The birds, in breeding plumage, were seen at 10:00 hours drinking water and roosting on the *Acacia* trees (*A. mellifera* and *A. tortilis*) around the dam. Quelea were mist-netted along the valleys around Ibaya camp (Pangaro/Mzukune/Kinongo valley) and dissected for analysis as shown in Tables 28.2a–c. Samples 1–5 represent birds netted on 22/3/96, 23/3/96, 10/5/96, 11/5/96 and 16/5/96 respectively.

A total of 215 Quelea were mist-netted at Ibaya and 15 at Kisiwani village between March and May. The ratio of females to males was 1:2.7. Of the 215 birds netted and dissected, 181 (84%) were adults and 34 (16%) were juveniles, while 93 (43%) were in breeding plumage and 122 (57%) were out of breeding plumage (Table 28.2). Most of these birds were netted between March and May.

A flock of five million birds was observed drinking water and roosting in Acacia trees at Kavatela Dam in May. Observations in the following days revealed no birds around that area. Of the birds netted at Ibaya in the morning and evenings, Tables 28.2b–c show that more birds were caught during the morning hours. These birds were netted in their feeding grounds along the valleys. The habitat mainly consisted of *Pannicum maximum* and *Acacia* species. which provided the roosting/nesting sites.

At Kisiwani the birds were netted while coming into the roost, which was located close to the rice fields in *Acacia* species and tall *Pannicum maximum*. The birds here attempted nest building although no eggs were laid.

At Ndungu, the birds roosted in reeds *Typha latifolia* and wild sorghum grasses along the canals, while at Mkomazi village the quelea habitat consisted of *Acacia xanthophloea*, *Typha latifolia* and *Phragmites* species.

In August, all the major water bodies in the MGR (Dindira, Ngurunga, Mbula, Kavateta and Maore), the source of water for both birds and animals, had already dried out. Kauzeni Dam, situated north of Pangaro (16 km north of Ibaya), was visited in September and it contained enough water to suffice for domesticated animals around the area for the whole of the dry season. More than two million birds were observed drinking water and day roosting in tree species around the dam. The habitat consisted of *Acacia nilotica*, *A. mellifera*, *A. tortilis*, *A.*

Table 28.1 Population of Quelea recorded inside and outside MGR during different months in 1996. See text for sample dates and locations.

	month						
area/site	Mar	May	June	Aug	Sept	Oct	Nov
inside MGR							
Ibaya/Pangaro/Dindera	29,500	6,200	30	0	0	0	0
Kavateta	*	5×10^6	1,200	0	0	0	0
Njiro	12,000	500	0	0	0	0	0
outside MGR							
Kisiwani	56,000	500	0	0	0	0	0
Ndungu	*	300	200	400	*	3,700	200
Kihurio	*	*	200	200	*	*	*
Mkomazi Village	*	*	500	500	*	*	*

* No surveys were made in these areas during these months.

ancistroclada, *A. albida*, *A. busseii* and *A. brevispica*. Most of these birds were out of breeding plumage. Eight birds were netted and dissected at this site between 10:00 and 15:00 h but their crops were empty. Neither grass nor cereal seeds were seen.

Surveys carried out in November showed that no Quelea were observed during the morning peak drinking period (10:00–12:00 h).

Discussion

The flocks of Quelea observed in the MGR were in breeding plumage (Table 28.2a) which is an indicator of either the birds are in the transition (moulting) into breeding or out of breeding plumage.

Surveys made throughout the MGR could not locate a colony, which indicates that the birds bred somewhere outside the reserve and were in the process of local migrations. The presence of juveniles (16% of the population) netted also indicated the likelihood that the adults had bred somewhere and migrated leaving the juveniles behind. This is also supported by the fact that 55% of the birds netted were out of breeding plumage (Table 28.2a). Attempt of nest building was observed late in May in tall *Pannicum maximum* grasses along the Pangaro/Kinongo valley. The nests were half built with fresh nesting material.

Further observations in June indicated that the colony was abandoned. The presence of the Quelea along this valley was due to the availability of fresh grass seeds which form their main diet—*Pannicum maximum, Eleusine jaegeri, Eriochloa* and *Brachiaria* species.

The results of the analysis of the birds netted within the MGR during morning and evening feeding hours did not show the presence of cultivated cereal seeds in their crops (Table 28.2c). Of the 215 birds analysed at Ibaya, 73% had grass seeds or insects in their crops while 27% had empty crops. The absence of cereal seeds in their crops indicate that these were unlikely to be causing damage in the rice growing areas around the reserve. The results of the birds netted at Kisiwani rice fields and those dissected for analysis showed that although they were feeding in the rice fields, no rice seeds were found in the crop. The absence of rice seeds in the crops (Table 28.2c) can be explained by either the presence of bird scarers; or since the rice crop was in the dough (milky) stage, the juice was digested and absorbed faster in the stomach; or the presence of these birds in these fields was due to the abundant presence of wild grass-seeds (their preference) along the canals.

The result of the surveys made within MGR showed that Quelea were only seen around the plain between Dindira and Zange. No Quelea were observed at Kisima, Kifukua, Maore, Kamakota and Umba. This may be attributed to the presence or absence of the preferred grass seed species.

The explanation for the 5 million birds observed drinking water at Kavateta in

Table 28.2a Age classification and reproductive condition of Quelea netted in MGR according to cranial pneumatization and gonad size. See text for sample locations and dates.

		cranial pneumatization				gonad size			
sample	total	adult		juveniles		breeding plumage		non-breeding plumage	
number	netted	no.	%	no.	%	no.	%	no.	%
1	50	50	100	0	0	25	50	25	50
2	40	40	100	0	0	20	50	20	50
3	56	42	71	17	29	25	42	34	58
4	55	42	76	13	24	19	35	36	66
5	11	7	64	4	36	4	36	7	64
total/ average	215	181	84	34	16	93	43	122	55

Table 28.2b Sex differences of Quelea netted at Ibaya at different times of day.

		morning				afternoon			
sample	total	female		male		female		male	
number	netted	no.	%	no.	%	no.	%	no.	%
1	50	13	43	17	57	11	55	9	45
2	40	20	50	20	50	0	0	0	0
3	59	18	38	29	62	4	33	8	67
4	55	10	30	23	70	7	32	15	68
5	11	4	36	7	64	0	0.0	0	0
total/ average	215	65	40	97	60	22	41	32	59

Table 28.2c Stomach (crop) contents of Quelea netted at Ibaya in the morning and afternoon hours.

		morning				afternoon			
sample	total	grass seed/ insects		no food		grass seed/ insects		no food	
number	netted	no.	%	no.	%	no.	%	no.	%
1	50	6	20	24	80	0	0	20	100
2	40	1	3	39	98	0	0	0	0
3	59	14	30	33	90	12	100	0	0
4	55	14	42	19	58	13	59	9	41
5	11	8	73	3	27	0	0	0	0
total/ average	215	43	27	118	73	25	46	29	54

only one day could be that these birds were migrating and had just stopped for water. This is further supported by the fact that their droppings under the trees where they were perching were fresh and that observations made in subsequent days revealed no Quelea visiting the dam or its vicinity.

Results of surveys made along the settlements neighbouring the reserve indicated the presence of Quelea at Kisiwani, Ndungu, Kihurio and east Mkomazi. Although damage was reported to occur in these areas, more damage was witnessed at east Mkomazi which is located further away from the MGR. Most of the birds were seen feeding on the remains around the milling grounds at Ndungu and Kihurio. Ndungu settlement has a better-defined irrigated rice scheme where the crop is cultivated in rotation, thus allowing the presence of wild grass seeds along the canals and spillways.

The decline in Quelea within the MGR in June, and complete absence between August and November, might be attributed to the absence of water in all the dams. A flock of two million Quelea was seen in August drinking water in a dam (Kauzeni Dam) 16 km north of Ibaya. The dam had enough water until November. Birds netted at this site had neither grass nor cereal seeds in their crops. These birds were thought to be scavenging on the scarcely available grass seeds. During that time of the year, there are no cultivated cereals around the area and most of the wild grass seeds have already been grazed by cattle.

Conclusion

The presence of a short lived roost/colony at Ibaya and the absence of any cultivated cereal seeds in the birds netted within the reserve cannot be associated with the damage of rice crop in the surrounding settlements. Also, the presence of short lived colony/roost at Kisiwani and the absence of cultivated cereal seeds in the crops of the netted birds does not imply that this is the same population as the one at Ibaya.

The absence of cultivated cereal seeds in populations at Ibaya and Kisiwani does not rule out the possibility that birds may be causing damage to such crops. The earlier assumption was that the crop might have been damaged while in a dough/milky stage, whereupon no remains would have been seen in the birds.

Quelea can forage within a radius of 20 km from their colony/roost. The distance between the roost observed at Ibaya and that at Kisiwani is more than 20 km. Given the prevailing situation of food availability and energy expenditure while foraging, it is unlikely that those at Ibaya will forage as far as Kisiwani and *vice versa*. The habitat between Kisiwani and Ibaya—tall and dense *Acacia*, *Commiphora* and other tree species, with less preferred grass seeds—might deter the birds flying from Ibaya to Kisiwani and returning to their roost. No permanent or long lived roost/colony was located in Kisiwani, Ndungu, Kihurio or east Mkomazi.

Recommendations

Since no detailed studies were made on the presence and damage by Quelea in areas surrounding the reserve, there is a need for further research to monitor the movements and feeding of these birds and come up with reliable information on whether the birds causing damage to crops in these areas reside in the MGR or are local residents and not related to the reserve. The extent of damage needs to be quantified, including the type of cereals experiencing more damage and at which period/stage.

Acknowledgements

I would like to thank the Director of the Mkomazi Ecological Research Programme who permitted and funded the study and the Director, TPRI for his permission to conduct the study.

References

Anon. (1995). Mkomazi Research Programme. Progress Report. 76 pp.

Harris, L.D. (1965) *List of Recorded Bird Species for the Mkomazi Game Reserve*. Annual Progress Report to Ministry of Agriculture, Forests & Wildlife.

Haylock, J.W. (1959) Investigation on the habits of *Quelea quelea* and their control. Department of Agriculture, Nairobi, Kenya.

Roming, T. (1988) Behaviour of *Quelea quelea* in invasion of the species in Turkana, north-west Kenya. *Scopus* 2: 96.

Williams, J. G. and Arlott, N. (1980) *A Field Guide to the Birds of East Africa*. Harper Collins, London.

Ward, P. (1971) *Research into the control of grain-eating birds (*Quelea quelea*). Manual of Techniques used in Research on Quelea Birds*. UNDP/FAO.

CHAPTER 29

Hornbills in Mkomazi as a case study of resource partitioning

Peter Cotgreave

Introduction

Approximately 400 species of birds have been recorded in the Mkomazi Game Reserve, representing about 4% of the entire bird diversity of the world. Thus, the birds of Mkomazi make ideal subjects for studying the biological processes that allow exceptionally high levels of biodiversity to persist.

The hornbills (Bucerotidae) of Mkomazi are a particularly appropriate species of study for a number of reasons. First, they are relatively large, and thus easy to see and identify accurately. Second, there are seven recorded species of hornbills at Mkomazi, making a practical assemblage to study. Third, hornbills of the genus *Tockus* are typical representatives of the important African savanna habitat, and most species share ecological and behavioural characteristics (Kemp 1995). Lastly, the geographic distributions of the species suggest that similar species are not always capable of coexistence. For example, the widely distributed von der Decken's hornbill *Tockus deckeni* is not found in an area of northern Uganda and northwestern Kenya, where it is replaced by the closely-related Jackson's hornbill *Tockus jacksoni* (van Perlo 1995).

This short paper presents some preliminary results concerning the ways in which the hornbills of Mkomazi partition food resources, together with initial results that are consistent with the hypothesis that ecological competition causes the pattern. It focuses on the five *Tockus* hornbills present in the reserve, namely the African red-billed hornbill *Tockus erythrorhynchus*, the Eastern yellow-billed hornbill *Tockus flavirostris*, the von der Decken's hornbill *Tockus deckeni*, the African grey hornbill *Tockus nasutus* and the African crowned hornbill *Tockus alboterminatus*. These five species are similar in size, have similar nesting requirements, and have broadly similar diets (Kemp 1995), so that the mechanisms by which they manage to coexist are not obvious. The study was carried out at the height of the dry season (July–August 1996), so that ecological competition is expected to be as intense as at any time of the year.

Coexistence

To confirm that the species genuinely coexist, I drove or walked over as many of the tracks as possible throughout the reserve, recording every hornbill sighting on a grid of five-kilometre squares. The results are presented in Table 29.1, which shows that each *Tockus* species shares at least two thirds of its range with at least one other *Tockus* species and that, on average, each species shares more than 50% of its range with at least two others. Omitting the yellow-billed hornbill (which was recorded in only one out of 88 squares), the proportion of a species' range that is shared with at least two other congeners (members of the same genus) remains high at 42%.

Table 29.1 Results of *Tockus* hornbill survey in Mkomazi Game Reserve.

hornbill	number of 5km squares recorded (out of 88)	% of range shared with at least one other *Tockus* species	% of range shared with at least two other *Tockus* species
red-billed	55	78	31
von der Decken's	53	83	34
grey	24	96	71
yellow-billed	1	100	100
crowned	6	67	33
average		85	54

At a finer scale, a more detailed study was carried out along a 7.7 km transect of track near Ndea, concentrating on the three commonest species. The track was surrounded by open *Acacia-Commiphora* woodland, and at the time of the study the *Commiphora* were beginning to produce large quantities of fruit, which are an important part of the diet of *Tockus* hornbills. On 18 separate occasions and at different times of the day, the transect was driven at approximately 20 km per hour and the position of every hornbill was recorded to the nearest 100 m along the transect. Observations were made to a distance of approximately 125 m from the track, so that they were recorded in blocks of 100 m x 250 m. At this scale, each species shares an average of almost 40% of its range with both of the other common hornbills. The data on distribution, therefore, show that similar hornbill species really do coexist, and somehow each species manages to avoid being outcompeted by the others.

Niche partitioning

To examine how the species share food resources, 437 different individual birds were observed and for each one, I recorded the height above ground at which the birds were first seen. For each of the three species the mean and standard error of the height above ground were as follows: red-billed: 2.5 ± 0.2 m; grey: 3.6 ± 0.6 m; yellow-billed: 4.5 ± 0 m; von der Decken's: 5.0 ± 0.3 m; crowned: 12.0 ± 3.0 m. This suggests that hornbill species may reduce competition with each other by foraging at different heights in trees.

To investigate further the possibility of the species coexistence being mediated by ecological competition, more detailed investigations were made of the two commonest species along the 7.7 km track passing Ndea. This part of the study concentrated on the two commonest species and results were similar to those above. Red-billed hornbills fed mainly on the ground and in the lower branches (mean: 3.0 ± 0.2 m) and von der Decken's hornbills fed mainly at the tops of the trees (4.2 ± 0.4 m). However, on occasions when there were relatively few von der Decken's hornbills in the area, the red-billed hornbills expanded the range of heights at which they foraged into the upper reaches normally exploited by the von der Decken's (Figure 29.1).

This natural experiment strongly suggests that ecological competition is responsible for the partitioning of food resources between these species of hornbill. Moreover, when the lower branches were removed from trees along a 200 m stretch of the track, the number of red-billed hornbills feeding on the stretch fell relative to the rest of the track. In other words, deprived of their normal feeding areas, the red-bills moved away rather than moving upwards, presumably because they were prevented from moving upwards by competition with the other species.

Further evidence that ecological competition causes the stratification of feeding heights among species comes from a comparison of Mkomazi with Lake Mburo National Park in Uganda. Here, the grey hornbill is the only regularly occurring *Tockus* species in the *Acacia-Commiphora* woodland and the range of heights at which it occurs is wider than in Mkomazi ($F_{8,13} = 2.9$, $p = 0.04$), although the mean heights are the same ($t_{21} = 1.71$, $p = 0.10$).

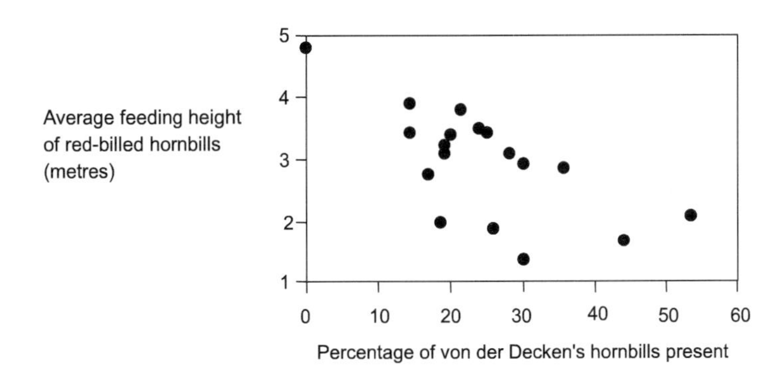

Figure 29.1 The tops of trees are normally dominated by von der Decken's hornbills, but on days when there are few von der Decken's hornbills in the area, red-billed hornbills are able to utilise the higher branches.

In summary, it appears that the coexistence of several species of *Tockus* hornbills in the Mkomazi Game Reserve is, at least in part, promoted by the partitioning of resources through ecological competition for food at different heights in *Acacia-Commiphora* woodland.

References

Kemp, A. (1995) *The Hornbills*. Oxford University Press.

Coe, M. & Stone, G. (eds.) (1995) *Mkomazi Research Programme 1993–97: Progress Report, July 1995*. Royal Geographical Society (with IBG), London.

van Perlo, B. (1995) *Birds of Eastern Africa*. HarperCollins, London.

CHAPTER 30

Small mammals of Mkomazi

Robert C. Morley

Introduction

While there have been studies of small mammals in east Africa (for review see Delany 1972) very little work has been conducted in Tanzania since Kingdon's research in the late 1960s (Kingdon 1974). More recent studies in Tanzania have concentrated on mountain environments (Demeter & Hutterer 1986, Shore & Garbett 1991) and no systematic study has been carried out on the small mammal fauna in Tanzania north of the Usambara and Pare Mountains. Recent studies (Alibhai & Key 1985 & 1986 and Duckworth *et al.* 1993) have shown that this region should contain a diverse and interesting fauna as far as its southern limit Mkomazi.

This study was carried out in response to the problems and priorities for small mammal research outlined by Schlitter (1978). Mkomazi could provide a valuable research site due to its location, importance as a conservation area and physiographic variation. Furthermore, there is a great need for increased small mammal research in the tropics, especially in Africa. Due to previous studies usually being opportunistic, with small data sets, and a general lack of faunal survey in Africa, there is geographic and taxonomic confusion over the status of small mammals.

Small mammals do not constitute a defined taxonomic group and for the purposes of this study comprised Insectivora and Rodentia as these were considered to be the mammals likely to enter the type of traps used.

The features of faunal structure relevant in a small mammal community are taxonomic composition; species diversity; relative abundance; and biomass and density (Hayward & Phillipson 1979). Of these only the lattermost was not included in this study.

The main aim and objective of the study was to produce a species list, information on species abundance and distribution between different habitat types within the reserve, especially for existing Mkomazi Ecological Research Programme plots. In addition the study was designed to show if the trap types used were equally successful in trapping and the density and rate at which animals were trapped for each of the selected study sites. The hypotheses to be considered were:

- The two trap types would show very similar success rates in both number of species and number of individuals captured.
- The number of individuals and species caught per site would be constant.
- That Mkomazi would contain a diverse small mammal fauna.

Materials and methods

Study sites

An extensive small mammal trapping programme was carried out on ten sites in the western sector of the Mkomazi Game Reserve. Coe (1995) broadly classifies the western sector of the reserve into four botanical types:

- open treeless grassland, seasonally-inundated.
- scattered tree grassland
- almost closed canopy *Acacia-Commiphora* wood land,
- *Spirostachys-Brachylaena* forests on hill summits.

Ten study sites were chosen as representative of narrower habitat types within Coe's broad classification. Seven sites were within existing Mkomazi Research Programme one hectare study plots (see Coe *et al.*, Chapter 5), one site was in an area which was to have had a plot positioned in it (Zange) and the remaining two plots were chosen on the recommendations of Kingdon for their dissimilarity to the other sites (Ibaya Amphitheatre, Ibaya Gully).

1. Mwasumbi plot 3° 59.23'S, 37° 48.66'E. Open treeless grassland, on heavily cracked black cotton soils which are seasonally-inundated. This fire-derived grassland savanna on an open plain has over 90% cover of *Urochloa* grass species, with for *Leucas* and *Dolichos* herb species and limited *Maerua* shrub.
2. Simba plot 4° 1.22'S, 37° 50.21'E. A low hillside of scattered tree grassland, with rocky substrate. *Aristida* grasses covered >80% with *Melhania* and *Hermannice* herbs dominant and *Dolichus* under shrubs.
3. Cadaba plot 3° 53.66'S, 37° 56.52'E. Thick *Acacia-Commiphora* woodland with over 80% canopy cover, and an understory of*Aristida* grasses, *Commelania* ground creeper, *Crassocephalum* and *Truefetta* herb species, and *Cordia* and *Grewia* shrubs. The soil was hard red quartz derived soil with gravel and large stones. Cadaba plot was situated on the extensive *Acacia-Commiphora* covered plain which spreads east across Mkomazi.
4. Ibaya Amphitheatre. No GPS fixes. Ibaya Amphitheatre was covered by hillside *Combretum* bush. Dominant vegetation types were *Aristidia* and *Panicum* grasses, *Truefetta* and *Plectranthus* herbs, *Cissus* creeper and *Croten, Vellozia* and chiefly *Combretum* shrubs giving almost total canopy cover. Situated on the east slope of a Ibaya Hill, the site included hill slope and ridge and large rock outcrops. The soil was rocky and the slope of the hill >45°. The plot ran across the slope face.

5. IBAYA GULLY. No GPS fixes. Ibaya Gully had the same species and vegetation s as Ibaya Amphitheatre, but with almost no grass cover and far less herb cover. It was a deep gully cut into the east face of Ibaya Hill and the plot ran from the top of the gully on a ridge, to the bottom of the hill, an undulating run-out zone.
6. PANGARO PLOT 3° 53.64'S, E 37° 46.67'E. Pangaro plot was scattered tree grassland on a firm red iron rich soil. Thick *Themeda* grass was dominant at lower levels, with shrub thickets of *Blepharispermum* and *Maerua* and *Acacia brevispica*. *Tephrosia villosa* and *Ipomea* herbs were also present in large numbers. Pangaro plot was almost flat, situated in a wide valley between hills in the reserve and outliers the Pare Mountains at the western boundary of the reserve.
7. NYATI PLOT 4° 0.40'S, 37° 49.21'E. Nyati plot was open treeless grassland very similar to Mwasumbi plot, also on black cotton seasonally-inundated soils with heavy cracking on an open plain. Seasonal burning led to an almost 100% cover of *Urochloa* grass species.
8. COE PLOT 3° 58.61'S, 37° 47.65'E. Coe plot was covered by tall, closed canopy *Spirostachys – Brachylaema* hill summit forest and is part of the Igere forest block. Coe plot runs from near the summit of Ibaya Hill down slope, facing south. The substrate is dark soil with high organic content and leaf litter.
9. UBANI PLOT 3° 53.50'S, 37° 53.80'E. Ubani plot was scattered tree grassland on hardpan red soil substrate. *Aristida* grass species dominate, although bare soil comprises about 15% of the area. *Truefetta* and *Crassocephalum* herbs covered under 2% of ground area and isolated *Grewia* shrubs and *Acacia* and *Commiphora* trees were present. Ubani plot sloped gently to the south southeast.
10. ZANGE 4° 3.32'S, 37° 47.75'E. Zange plot was scattered tree grassland on rich dark brown sandy soil, dominant species were *Aristida* grasses. *Dolichus* under shrubs and *Acacia tortilis* trees giving 20% canopy cover. Zange plot is on a flat area at the foot of Zange Hill.

Traps and trapping design

Trap type

The key factor in trap type selection was trap availability, ease of transport both to Tanzania and in the field, ease of maintenance and the time required to set and run traps. 'Longworth' (Penton Ltd.) and the 'Sherman' folding trap (H B Sherman Traps) have been used in previous African field studies and were considered the best traps under the prevailing circumstances being robust, tried and tested.

Trap numbers

40 Shermans and 30 Longworths were utilised plus extra traps as spares. More Longworth traps arrived in the field after Plot 4 had been sampled, allowing equal numbers of each trap type to be used.

Patterns of trapping points

The two most common trap patterns are trap lines and trap grids (Gurnell & Flowerdew 1994). Trap grids allow population size, density and home range studies and were used for the first two plots, Mwasumbi and Simba. However, when trap success was considered (see results) and allowance was made for the removal of specimens for preservation, it was decided that a reasonable volume of data of this type would be unattainable.

Trap lines were initiated for the rest of the study because this was an adaptive method for providing an index (Duckworth *et al.* 1993) and for covering a large number of habitats over a relatively short space of time (Barnett & Dutton 1995, Rose, Slade & Honaki 1977, Southern 1995). Trap lines were also suggested by Kingdon (pers. comm.) who used them effectively in his seminal studies of east African mammals (Kingdon 1974). At all sites traps were placed 5 m apart with trap configurations as shown in Table 30.1.

These configurations differ because the trapping programme had to be adaptive to trap availability and the practicality of trap placement: due to the nature of the trapping environment at Plot 4 (steep slopes and rock faces) Longworths were considered too difficult to carry, position, bait and re-bait. Usually one Sherman and one Longworth were placed together.

Trap lines were a maximum of 20 points so as to fit into the one hectare study plots of the research programme. When two lines were used, the lines were spaced 25 m apart running in the same direction.

Trapping period

Trapping was carried out for five days at each plot with no pre-baiting. The reasons for choosing this regime were as follows.

Table 30.1 Trap configurations on each plot.

plot	name	trap pattern	trap type
1 & 2	Mwasumbi, Simba	grid of 28 traps, 4 lines of 7 stations	1 Longworth and 1 Sherman at each station
3	Cadaba	2 lines of 15 stations	1 line of 1 Sherman and 1 Longworth, 1 line of 2 Shermans
4	Ibaya Amphitheatre	2 lines of 20 stations	2 Shermans at each station
5	Ibaya Gully	1 line of 20 stations	2 Shermans and 1 Longworth at each station
6–10	Pangaro, Nyati, Coe, Ubani, Zange	2 lines of 20 stations	1 Longworth and 1 Sherman at each station

- The period would be long enough to counter neophobic responses to the trap and short enough to prevent animals becoming 'trap happy' or immigrating to the food source.
- at least 200 trap nights could be conducted per site.
- 10 sites could be studied in the time period of field study.

Three sites were not trapped for this desired period. At site 5, Ibaya Gully and 8, Coe, the time and effort needed to service the traps was insufficiently rewarded by catch data obtained (see results) and at site 10 Zange the end of study period was reached before the fifth day. These three sites were each studied for three days.

Bait

A bait was formulated which it was hoped would appeal to a wide range of small mammals. Peanut butter is a standard attractant (Willan 1986) and millet and rice was used as a bulk filler. Animal fat was added to give the bait a higher calorific value, banana and tinned fish were added to broaden the potential catch. Initially maize and beans were also included as a filler but blocked the trapping mechanism in Shermans so use of these foodstuffs was curtailed.

Traps were baited in the early evening, usually between 17:00 and 18:00 hours, and left open for the night. Bait volume was about 1.5–2 cm^3 and uneaten bait was removed when traps were checked .

Checking traps

Traps were checked every morning usually by 09:00 h. Captures were processed and all traps were emptied of remaining bait and any faeces. Traps were then closed for the day and repositioned for evening baiting. Day time trapping was not conducted due to climate constraints (heat).

Processing of captures

Measurements

Captures were emptied into a clear plastic sack, pinned to the ground and had head, body and tail measurements taken. In hand, hind foot and ear length were measured. Sex was determined by applying gentle pressure to the genital region and by looking at status of nipples or presence of testes.

Animals were placed in a cloth bag and weighed to the nearest gram on a spring balance. All linear measurements were carried out to the nearest millimetre. Non-specimen animals were marked by fur clipping then released.

Killing animals

When animals were taken as specimens they were dispatched by breaking the neck between thumb and fore finger, which proved quick, caused minimal suffering and ensured good skin and skull integrity.

Preparation of specimens

Specimens were skinned and placed on card. Skulls were cleaned by exposing them to ants which stripped away the flesh. *Crocidurans, Suncus* and some small rodents were transported to UK as whole specimens in alcohol for identification.

Field identification

Animals were initially identified using a guide compiled by the author including photographs taken in the British Natural History Museum and notes from Kindon's *Mammals of East Africa* (Kingdon 1974) and Smither's text (Smithers 1983).

Full identification

Fuller identification, especially of the more difficult species, was conducted by the author at the Natural History Museum in London. Key texts were used for rodents (Delany 1975, Foster & Duff-Mackay 1966, Kingdon 1974, Meester 1971, Meester 1968 & Smithers 1983) and identification was carried out using measurements and specimens collected in the field, including looking at skull/teeth morphology.

Crocidura and *Suncus* species were handed to Paula Jenkins of the Natural History Museum for specialist identification and analysis. Richard Harbord also of the Natural History Museum checked the identifications made by the author.

Results

Taxonomic composition

In total, 252 individual animals were caught. 23 different species were captured from 13 different genera. Table 30.2 shows the species caught and the number of individuals of each species in each of the study plots. The total number of animals of each species by type can be seen in Table 30.3, which also shows the percentage of animals of that species caught by trap type.

Tables 30.2 and 30.3 show that 12 species dominated the data set, comprising 221 individuals. The remaining 11 species total only 31 captures, with the three least caught species each comprising less than 1% of the total catch.

Trap success

Table 30.3 shows that many species showed a marked trap type preference and this is supported by the statistical analysis of trap success in Tables 30.4a and 30.4b. There were significant differences between plots for trap success (χ^2 = 21.666, df = 9, p<0.01) and very significant differences between the success rate for the two different trap types ($\chi^2 = 6.635$, df = 2, $p<0.01$). The difference between success and failure was very significant for overall trap success and for trap type success. Table 30.5 shows trap success by trap per plot.

Summary of capture

Capture data is summarised in Table 30.6, and from this table it can be seen that no single plot contained more than ten species, the least diverse plot containing only two species. The number of species captured for each plot is shown as a percentage of the total number of species, with 43.5% of species present in the most diverse plot, and 8.7% of species present in the least diverse plot.

Table 30.2 Captures by species/plot.

	plot										
species	1	2	3	4	5	6	7	8	9	10	total
Aethomys chrysophilus	–	1	–	3	1	–	–	–	–	–	5
Acomys cahirinus	–	5	–	–	–	–	–	–	–	–	5
Acomys wilsoni	–	–	10	1	–	2	4	–	7	3	27
Acomys ignitus	–	–	3	1	–	–	–	–	1	–	5
Acomys spinosissimus	1	–	3	3	–	8	–	–	–	1	16
Acomys species	–	1	–	–	–	–	–	–	–	–	1
Arvicanthis niloticus	–	–	1	–	–	–	26	–	–	–	27
Grommomys dolichurus	–	1	–	–	–	–	–	–	–	–	1
Graphiurus murinus	–	1	–	2	1	1	–	–	–	–	5
Lemniscomys barbarus	–	–	5	–	–	3	–	–	2	–	10
Lenmiscomys griselda	–	–	–	2	–	3	1	–	4	–	10
Mus minutoides	2	1	5	–	–	1	1	1	–	3	14
Mus tenellus	–	–	1	–	–	–	–	–	–	–	1
Myomys fumatus	1	–	–	6	–	9	5	6	–	1	28
Tatera nigricauda	–	1	5	–	–	–	–	–	1	–	7
Tatera robusta	–	–	–	10	–	6	–	–	–	1	17
Saccostomus campestris	2	–	–	–	–	–	2	–	1	–	5
Crocidura hirta	5	–	–	–	–	–	26	–	–	–	31
Crocidura parvipes	–	–	–	–	–	7	–	–	4	–	11
Crocidura (=Afrosorex) voi	–	–	10	–	–	–	–	–	–	–	10
Crocidura species	–	–	–	–	–	–	2	–	–	–	2
Suncus lixus	–	–	–	–	–	–	–	2	–	–	2
Elephantulus rufesens	–	–	10	–	–	–	–	–	2	–	12
total individuals	11	11	52	28	2	40	68	9	22	9	252
total species	5	7	10	8	2	9	8	3	8	5	23
total genera	5	6	7	6	2	7	7	3	6	4	13

Table 30.3 Species (numbers and percentages) by trap type.

species	trap type				total
	Sherman		Longworth		
A. chrysophilus	4	80%	1	20%	5
A. wilsoni	20	74%	7	26%	27
A. cahirinus	3	60%	2	40%	5
A. spinossimus	13	81%	3	19%	16
A. ignitus	3	60%	2	40%	5
A. species	0		1	100%	1
A. niloticus	23	85%	4	15%	27
G. dolichurus	0		1	100%	1
G. murinus	4	80%	1	20%	5
L. barbarus	4	40%	6	60%	10
L. griselda	8	80%	2	20%	10
M. minutoides	7	50%	7	50%	14
M. tenellus	1	100%	0		1
Myomys fumatus	22	79%	6	21%	28
T. nigricauda	7	100%	0		7
T. robusta	17	100%	0		17
S. campestris	4	80%	1	20%	5
C. hirta	22	71%	9	29%	31
C. parvipes	1	9%	10	91%	11
C. afrosorex voi	8	80%	2	20%	10
C. sp.	0		2	100%	2
S. lixus	0		2	100%	2
E. rufesens	12	100%	0		12
total	183	73%	69	27%	252

Table 30.4a Trap success by plot.

plot	trap nights		
	success	failure	total
1	11	269	280
2	11	269	280
3	52	248	300
4	28	192	220
5	2	148	150
6	40	340	380
7	68	332	400
8	9	231	240
9	22	378	400
10	9	231	240
total	252	2,638	2,890

Table 30.4b Trap success by trap type.

trap type	trap nights		
	success	failure	total
Sherman	183	1,456	1,640
Longworth	69	1,182	1,250
total	252	2,638	2,890

Also shown in Table 30.6 are the numbers of animals caught per plot, and the percentage of the total volume of the catch that this represents. Finally trap success and trapping duration are shown. Trap success over all was 7.9% and varied widely between plots.

Data loss

Table 30.7 shows how data were lost through different processes. False releases were recorded when a trap was sprung without containing a capture. This was a common occurrence and it was felt that ants could be responsible for many of these false releases. Large numbers of ants were observed at all plots and often removed bait from traps.

At Plot 6, Ubani, more than 90 traps were disturbed over three nights by predator activity, and it would be impossible to calculate how many of these traps contained animals or would have captured. Traps were often moved several metres, and as far as 10 m from their bait station, and Longworths had the tunnel removed. The traps did not show claw or bite marks and this factor, when combined with observations of animal tracts and faeces, indicated to the author and ranger that the animal responsible was either a large mongoose, a serval cat or a civet. Non-target captures also reduced trapping effort, with slugs in Plot 8 being most problematic.

Escapes were not a significant problem, only four animals were lost this way without having all of their measurements taken. Recaptures were not included in the main data set because specimens were removed in large numbers. In total more than 25% of the total number of captures were as removed specimens.

Only one trap was lost to animal damage, a Sherman trap crushed while in position at Plot 7. The nature of the damage and the sightings of animals in the area indicated that zebra or giraffe were responsible.

Trap injuries and deaths

Some animals showed grazing of their muzzles from trying to force their way out of traps. The only serious injuries caused were damage to tail end by Sherman traps especially to the large *Tatera*. In these cases skin was lost from the tail or less frequently the bone was also severed. Only one animal was seen to die in the capture process. An adult *L. barbarus* died during handling and was included in the data set. She was found to be both old and pregnant. It is assumed she asphyxiated in the plastic bag used to process captures. No animals died in traps.

Other small mammals observed

Although not captured, the unstriped ground squirrel *Xerus rutilus* was often observed, especially at Plot 3 and Plot 9. Mole rat *Bathyergidae* mounds were observed near Plot 4 and in other areas in the reserve.

A single porcupine *Hystrix galeata* was observed near Plot 9 and quills were found in Plots 4 and 5, and near Plots 1, 3 and 7. Quills found elsewhere suggest that *H. galeata* is widespread and numerous in the reserve, as do the observations of the rangers.

Rock hyraxes *Procavidae* were observed in large numbers in Plots 4 and 5. According to Kingdon (pers. comm.), tree hyraxes *Dendrohyrax* were present in

Table 30.5 Trap success by trap type per plot, measured by the percentage of trap nights which were successful.

plot	Longworth (%)	Sherman (%)	combined (%)
1	1.4	6.4	3.9
2	3.6	4.3	3.9
3	13.0	19.5	16.2
4	-	12.7	12.7
5	2.5	0.9	1.7
6	7.8	13.2	10.5
7	9.5	24.5	17.0
8	3.3	4.2	3.7
9	3.5	7.5	5.5
10	1.7	5.8	3.7
total	5.1	9.9	7.9

Table 30.6 Summary of capture data by plot.

plot	species		individuals		trap success (%)	trapping duration (days)
	number	% of total	number	% of total		
1	5	21.7	11	4.7	3.9	5
2	7	30.4	11	4.7	3.9	5
3	10	43.5	52	20.6	16.2	5
4	8	34.8	28	11.1	12.7	5
5	2	8.7	2	0.8	1.7	3
6	9	39.1	40	15.9	10.5	5
7	8	34.8	68	27.0	17	5
8	3	13.0	9	3.6	3.7	3
9	8	34.8	22	8.7	5.5	5
10	5	21.7	9	3.6	3.7	3
total	23	n/a	252	100.6	7.9	44

Plot 8. *Lagomorpha* were observed on roads but were not identified by the author. Various species of Chiroptera were observed but considered outside the remit of this study.

Hypotheses testing

Further statistical tests for the data generated in this study would prove difficult and hard to validate. Different methods were used for different plots and so are not directly comparable, but the main aim was always to catch the most species possible, using the most suitable trapping regime to attain that goal.

Discussion

Taxonomic composition

17 rodent species from ten genera (including one *Acomys* species yet to be identified), five insectivore species from two genera, and one Macroscelidae species were captured, totalling 23 species and 13 genera, a more diverse group than recorded in recent literature. Bond *et al.* (1980) caught six rodent species and two shrew species in their study. More recent studies in Tanzania (Demeter & Hutteer 1986, Shore & Garbett 1991) report fewer species captured, although both these studies are of mountain fauna.

In more comparable research, Stephenson (1994) caught 12 species including one introduced species, and Duckworth *et al.* (1993) recorded 10 species caught or seen. Cheeseman & Delany (1979) undertook a 17-month study in Rwenzori Park, Uganda, and caught 12 species of rodent using a grid of Sherman traps. Their captures were mainly of five species and of the other five species analysed they caught 66 animals in the 17th month of study. The most frequently caught five species comprised the remaining 283 individuals captured.

The most similar trapping programme to this Mkomazi study was conducted by Alibhai & Key (1985) in the Kora National Reserve in Kenya. They caught 11 species and fewer individuals using more trap nights. When Delany reviewed the small mammals of tropical Africa (1972), of the 17 studies reviewed no study contained as many species captured, than this study of Mkomazi. Delany's review reports there being 26 genera of small mammal in Tanzania, with 59 species being present in Tanzania's 343,726 square miles of land.

The small mammal study conducted in Mkomazi caught 50% of the genera reported as present in Tanzania and 39% of the reported species.

It should be concluded that the Mkomazi Game Reserve contains a significant range of the Tanzanian small mammal fauna, and so an important range of tropical African small mammal species.

Trap success

The overall trap success rate was 7.9%. Sherman success was 9.9% and Longworth trap success was 5.14%. This overall success rate is better than reported by Alibhai & Key (1985 & 1986) and should be considered a high success rate (Coe and Kingdon, pers. comm.) for the type of study conducted. Furthermore, overall trap success would be improved if escapes and recaptures were included in the data set.

Further analysis of Cheeseman & Delany's (1979) trapping programme shows a total trap success rate for live captures of 34%. This however includes 3,126 recaptures.

49 individuals were caught over 10,125 trap nights giving a trap success rate for individuals of 3.4%. Over a comparable period of time, June–August 1973, their programme trapped less than 200 animals from eight species and June–August 1972 about 100 animals from eight species. When considering the success of other studies, it must be remembered that different methods of trapping were often used. The significant differences between trap types soon became obvious with Sherman traps catching more individuals of more species than Longworths. Longworth-only captures comprised only six individuals from four species, including one *Acomys* species yet to be identified, compared to Sherman only captures

Table 30.7 Other trap releases by plot.

plot	false releases[1]	non-targets[2]	escapes[3]	recaptures
1	15	2	1	0
2	10	0	0	1
3	2	0	1	11
4	11	2	0	7
5	0	0	0	0
6	–[4]	1	0	9
7[5]	29	1	1	31
8	15	14	0	0
9	17	0	1	7
10	5	2	0	3
total	94	22	4	69

[1] False releases were recorded when a trap was sprung without containing a capture.

[2] Non-target releases: two ant groups, two lizards, two beetles, 14 slugs, one cockroach and one millipede.

[3] Escapees lost in handling are not included in the main capture data.

[4] Plot 6: traps disturbed by predator activity, 96 disturbed/released traps.

[5] Plot 7: one trap crushed by zebra or giraffe.

comprising 37 individuals from four species. In total, 73% of captures were made in Sherman traps, and 27% of captures were made in Longworth traps.

Inter-plot difference

The difference between plots are marked but due to the differences between methods employed at different sites, statistical analysis was not used to describe these differences. Of the 12 most common species (10 or more individuals), four were found in more than half the study plots and eight of the most common species were found in three or fewer study plots. If these are used as an indicator of overall inter-plot difference, it can be seen that there were significant inter-plot differences in both species present and number of individuals, as can be seen in Tables 30.2 and 30.5.

Interplot differences for trap success also vary greatly and it was because some plots produced such low success that the trapping programme was terminated in these sites after three days. It can be seen (Table 30.4) that some plots showed higher trap success, with four plots showing better than average trap success and six plots showing less than average success.

While there appears to be a relationship between trap success /numbers caught, and species diversity for that plot, this relationship is not as strong as could be expected. The presence of just one animal of a given species will increase the diversity of that plot by 4.35%. So while Plot 2 has seven species present, six of these species are represented by one individual animal only. Plot 7 also has seven species present but most are present at a greater density than the animals in Plot 2. From analysis of the data for individual plots it can be argued that there are significant inter-plot differences for animals caught and for species caught. However, these inter-plot differences may not be as great as the intra-plot differences.

Data loss

When all factors which lead to traps being unavailable for animal capture are considered, a significant amount potential capture data must have been lost. In total, at least 281 trap nights were lost due to false release, the capture of non-targets, predator disturbance or the non-inclusion of recaptured animals, as summarised in Table 30.7.

Ants proved to be the most persistent problem and were abundant at all sites. Ants would remove bait, were the expected agents in the false release of many traps, and also attacked captured animals. Ants may have caused target animals to avoid the traps, but conversely they may have attracted insectivorous animals, especially *E. rufesens* and *G. murinus* which are usually only captured after long pre-bait periods (Alibhai & Key 1985). *Sacctomus campestris* also proved highly insectivorous when held for several days in captivity.

The level of trap disturbance by predators was perhaps unusual as Cheeseman & Delany (1979) comment that they had no traps disturbed, and this level of disturbance is not mentioned in the literature available. At Plot 6, one trap line was visited on the third night and both lines were visited on the fourth and fifth nights. In total a minimum of 96 traps were considered to be disturbed by predators, often moved and usually turned over. It was not possible to tell how many of the traps had contained captures but the first and second nights had been very successful. It was not apparent how the Sherman traps had been opened, but a lack of damage would suggest that a predator was able to open the spring loaded gates.

Such high level of trap disturbance was not expected, however more false captures were expected, especially of snakes and lizards.

Trap injuries and deaths

Trap injuries were not common and only one trap mortality occurred (see Results, Trap Injuries and Deaths).

Hypotheses testing

The data generated in this study did not lend itself well to many of the statistical methods known to the author, and only the statistical analysis proposed by the Department of Applied Statistics, University of Reading, was used. While the hypotheses proposed cannot be accepted or rejected through statistical analysis, except for the first hypothesis, the results do give an indication as to the validity of the other hypotheses.

Statistical analyses show that there was significant difference between trap type and success. Non-statistical analysis suggests that the second hypothesis—that number of individuals and species per site would remain constant—should be considered incorrect. The number of species caught should be considered to support the third hypothesis—that Mkomazi contains a diverse small mammal fauna.

Experimental improvements and recommendations

The problems encountered working alone in a difficult field environment and a lack of prior knowledge of the study area meant that the study had to be flexible.

For better experimental integrity and statistical analysis, a more constant, rigid trapping programme would need to be carried out. For an in-depth study more traps would be needed, with several people to service them. Important information on small mammal ecology could be gathered by using large grids and Sherman traps would be recommended as these are easier to carry, place and service, and have been shown to catch more animals and species.

Experimental design would improve greatly if capture-mark-recapture techniques were fully utilised, and if whenever possible fewer animals would be removed for specimens. Alternatively all animals caught could be removed and killed. As high numbers were caught, more people and equipment would be needed to process the captures.

Indicators of high small mammal population

Mkomazi Game Reserve contains a broad range of animals which predate on small mammals. Some of these predators were seen in high numbers. The author observed many raptors in and around study plots. In the reserve, white-tailed mongoose *Ichseumia albicauda,* genet *Genetta genetta,* civet *Viverra civetta* and serval *Felis serval* were all observed. Puff adders *Bitis arietans* were also observed, and one was found to have two *Tatera* in its stomach.

Location of the specimens taken in the field

All captures taken are now held in the Natural History Museum in London. Richard Harbord has checked the author's identifications and made some corrections in the genus *Acomys* due to changes in taxonomy. The identifications of *Crocidurans, Suncus* and *Acomys* are all now correct, these being the most difficult of the animals to identify. Other species are far less problematic so they should have been correctly identified by the author.

Concluding remarks

This study has shown that Mkomazi Game Reserve has a diverse small mammal fauna, and that these animals are often at high densities. The main aim of the programme has been achieved, a specimen list and distribution outlines have been produced. However the projected hypotheses have not been fully investigated and many of the results are unsuitable for statistical comparison. This lack of statistical analysis does not prohibit the understanding of important general themes.

Mkomazi Game Reserve is an important area for tropical small mammal fauna and contains an unusually high diversity of species, especially in its western end. The differences in vegetation and physiography leads to differences in taxonomic composition, and population density within habitat types.

Acknowledgements

I wish to thank the following: Dr Malcolm Coe, Programme Director, for his help and support both in the UK and in Tanzania; the Department of Wildlife Tanzania and Nigel Winser, Deputy Director of the Royal Geographical Society (with IBG), for the opportunity to work on the Mkomazi Ecological Research Program. For their help in Tanzania: Jonathan Kingdon, for all his advice on the small mammal fauna and methods; Raphael Abdallah and Emmanuel Mboya from the National Herbarium of Tanzania for botanical information; Tim Morgan, Field Director, for logistical support and especially to Daniel Mafunde and Max, my rangers and guides in the bush, for their constant help, protection, bush knowledge and good humour. Asante sana, kwa heri ya kuonana. On return to the UK, I thank Paula Jenkins and Richard Harbord for help in species identification pre and post field work, and Dr Alec Jones and Professor Richard M. Sibly for their advice and supervision. Finally I must thank my parents for funding and supporting me throughout my studies.

References

Alibhai, S.K. & Key, G. (1985) A preliminary investigation of small mammal biology in the Kora National Reserve, Kenya. *Journal of Tropical Ecology* 1: 321-327.

Alibhai, S.K. & Key, G. (1986) Biology of small mammals in the Kora Reserve. In: Coe, M.J. & Collins, N.M. (eds.) *Kora: an ecological invetory of the Kora National Reserve, Kenya*. Royal Geographical Society, London.

Barnett, A. & Dutton, J. (1995) *Expedition Field Techniques: Small Mammals (excluding bats)*. Expedition Advisory Centre, London.

Bond, W., Ferguson, M. & Forsyth, G. (1980) Small mammals and habitat structure along altitudinal gradients in the Southern Cape Mountains. *South African Journal of Zoology* 15 (1): 34-43.

Cheeseman & Delany (1979) The population dynamics of small rodents in a tropical African grassland. *Journal of Zoology* 188: 451-475.

Coe, M. (1995) Botanical studies in the Mkomazi Game Reserve, in *Mkomazi Research Program Progress Report July 1995*. Royal Geographical Society (with IBG), London.

Delany, M.J. (1972) The ecology of small rodents in tropical Africa. *Mammal Review* 2: 1-37.

Delany, M.J. (1975) *The Rodents of Uganda*. British Museum (National History) London.

Demeter, A. & Hulterer, R. (1986) Small Mammals from Mt. Meru and its environs (Northern Tanzania). *Cimbebasia* 8: 199-207.

Duckworth, J.W., Harrison, D.L. & Timmins, R. J. (1993) Notes on a selection of small mammals from the Ethiopian Rift Valley. *Mammalia* 57: 278-284.

Forster, J.B. & Duff-Mackay, A. (1966) Keys to the genera of Insectivora, Chiroptera and Rodentia of East Africa. *Journal of East Africa Natural History Society* 15 No. 3 (112): 189-204.

Gurnell, J. & Flowerdew, J.R. (1994) *Live Trapping Small Mammals: A Practical Guide*. 3rd edition. The Mammal Society, London.

Hayward, G.F. & Phillipson, J. (1979) Community structure and functional role of small mammals in ecosystems. In Stoddart, D.M. (ed.) *Ecology of Small Mammals*. Chapman and Hall, London.

Kingdon, J. (1974) *East African Mammals, Volume II Part B (Hares and Rodents)*. Academic Press, London.

Meester, J. (1968) *Preliminary Identification Manual for African Mammals 19: Rodentia*. Smithsonian Institution, Washington.

Meester, J. (1971) *Mammals of Africa*. Smithsonian Institution Press, Washington.

Rose, R.K, Slade, N.A. & Honacki, J.H. (1977) Live trap preference among grassland mammals *Acta Theriologica* 22: 296-307.

Schlitter, D.A. (1978) Problems and priorities of research on the taxonomy and ecology of African small mammals. *Bulletin Carnegie Museum of Natural History* 6: 211-214.

Shore, R.F. & Garbett, S.D. (1991) Notes on the small mammals of the Shira Plateau, Mt. Kilimanjaro. *Mammalia* 55: 601-607.

Smithers, R.H.N. (1983) *The Mammals of the Southern African Subregion*. University of Pretoria, Pretoria.

Southern, H.N. (1965) The trap-line index to small mammal populations *Journal of Zoology* 147: 216-238.

Stephenson, P.J. (1994) Seasonality effects on small mammal trap success in Madagascar. *Journal of Tropical Ecology* 10: 439-444.

Willan, K. (1986) Bait selection in Laminate-toothed rats and other Southern African small mammals. *Acta Theriologica* 31/26: 359-363.

CHAPTER 31

Large mammals of Mkomazi

S. Keith Eltringham, Ian A. Cooksey, William J.B. Dixon, Nigel E. Raine, Chris J. Sheldrick, Nicholas C. McWilliam & Michael J. Packer

Mkomazi Game Reserve has never been noted for large populations of large mammals but the diversity of species is as great as in any other east African protected area. The size criterion defining a large mammal must, ultimately, be arbitrary. In ecological terms, individual size correlates with individual contribution to the rate at which ecological processes occur. The threshold size criterion for a large mammal in Mkomazi was set at a weight of 3–5 kg or a height of about 50 cm, and so would include, for instance, baboon and dikdik. The threshold is ecologically and logistically justified for it includes individual 'large mammals' that are likely to play significant roles in specific ecological processes and means that the chances of observing individuals of the smaller species is reasonable across a range of habitats, ensuring that widely distributed species have a fair chance of being observed throughout their range.

Information on large mammal distribution and abundance patterns, and their seasonal variations, is important to understanding the significance of large mammals as an ecological driving force in Mkomazi. This information is also vital in evaluating various management questions, such as the conservation importance of Mkomazi for the persistence of large mammal species. For example, Mkomazi has the only recorded gerenuk population in a protected area in Tanzania, as well as nationally important populations of oryx and lesser kudu (TWCM 1991). Other management considerations include the tourist potential of Mkomazi for game viewing, the nature and extent of potential wildlife-human conflicts, possibilities for sustainable harvesting of bush meat by local communities and the re-introduction of formerly-present species, such as the black rhino.

Large mammal species in Mkomazi

A full list of species recently and currently present in the reserve is given in the checklist (Chapter 32). Megaherbivores, those weighing more than 1,000 kg, comprise elephant and giraffe. The equids are represented by plains zebra, and the pig family by warthog and bush-pig. There are no hippopotamus, probably because there is no natural source of permanent water in the reserve except for the Umba River in the far south-east. The eland and buffalo are the largest artiodactyls but there are many medium-sized ungulates such as waterbuck, lesser kudu, fringe-eared oryx, kongoni (Coke's hartebeest), impala, Grant's gazelle, gerenuk, bushbuck and Bohor reedbuck. Smaller large mammal species include bush duiker, klip-springer, steinbuck and dikdik.

The large carnivores are well represented in Mkomazi and include lion, leopard, cheetah and two hyaenas, the spotted and striped. Smaller species comprise wild dog (possibly extirpated from the reserve, although a group of 15 was sighted during a seven-day period in March 1997 near Kisima), black-backed jackal, bat-eared fox, aardwolf, ratel, serval, small-spotted genet, civet, five species of mongoose, zorilla, serval, caracal and wild cat. Some of these carnivores are very rare and populations of some may be threatened with extinction.

Other species have been recorded in the past. The wildebeest was common in the 1930s and the greater kudu occurred until the mid-1950s. Occasional sightings of the sable were reported up to the early 1950s. A group of 16 wildebeest was introduced to the reserve in 1966 but it does not seem to have become established. The black rhinoceros, which is the subject of a re-introduction programme, was present until relatively recently.

Most species that might be expected in the reserve from its geographical position are present with the exception of the hippopotamus. Information on the status of the large mammals has been derived from a number of sources. The only previous study in any detail was that made by Harris (1972) from 1964 to 1967. He established a series of line transects for estimating numbers supplemented by game counts around water holes and some aerial censuses. Harris extrapolated his data to provide estimates of the population size of some species and recorded a gradation in large mammal numbers from high densities in the north-west to low densities in the south-east. This trend reflects the rainfall pattern (see Chapter 2). Not all species were zoned in this way. Elephants, for example, were found throughout the reserve and oryx and zebra tended to be confined to the west-central and central areas. The distribution varies between water dependent and water-independent species and is influenced by vegetation, which is spatially heterogeneous. For several of the larger mammal species, seasonal movements between Tsavo West National Park in Kenya and Mkomazi mean that estimates of population size in the reserve show marked seasonal variation.

Following Harris' (1972) work, large mammal counts in Mkomazi were made from the air by the Kenya Range Management Unit (now the Department of Resource Surveys and Remote Sensing). These were sample, not total counts, and they were primarily carried out to count the elephants in the Tsavo ecosystem but some other species were included and the results of the most recent count, made in April 1994 (Inamdar 1994 & 1996), enables a comparison to be made with conditions some thirty years earlier. Ground surveys of large mammals conducted during the Mkomazi Ecological Research Programme (MERP) sought to determine current distributions and, where feasible, to estimate population sizes within the reserve. In addition, systematic observations made in the west of the reserve aimed to clarify seasonal variation in distribution and abundance.

Survey methods

The spatial and temporal patterns in distribution and abundance of large mammals in Mkomazi are very variable. Elucidating these patterns necessitates the use of a variety of survey methods, which would ideally sample all habitats several times a year to capture seasonal variation.

Ground-based and, to a far lesser extent, aerial surveys of large mammal populations were used in this study. For ground surveys, constraints of accessibility to different parts of the reserve and of resource availability mean that in practice the surveys are biased in various ways. A spatial bias exists in that the north-western part of the reserve is the most intensely surveyed. In addition, the 'sample' for the whole reserve is small and is associated almost entirely with the road network. A temporal bias exists as observations were not necessarily made at times of day when most species are likely to be active, and the observations were not seasonally representative, particularly outside the periodic survey sector. Finally, there are observation biases, caused by the varying visibility in different vegetation types and at different times of year.

Ground surveys

Ground-based surveys provide the majority of our presence records. Almost all surveys were conducted from vehicles using existing roads in Mkomazi. When an animal was sighted, the vehicle was stopped and the species, group size, location and date were recorded. In most cases, the location was read from a global positioning system (GPS), using the same co-ordinate system and projection parameters as the 1:50,000 maps of Tanzania. When animals were over 200 m from the vehicle, their distance was estimated by eye and a bearing was taken using a standard compass. These data were used within the Mkomazi geographical information system (GIS—see Chapter 4) to calculate individual or group locations. Survey routes were essentially confined to existing roads in Mkomazi, which introduced various biases, as explained above. Bearing in mind these sampling biases, three types of ground-based survey were employed.

Opportunistic records

Records of individuals or groups of large mammals (2,125 in total) were made along all routes used by MERP researchers during 1994 to 1997 (see Figure 31.1a). Observations were not systematic, being uncontrolled in terms of sampling effort and with respect to time of day, time of year and location (most were made in the western third of the reserve where most research occurred). Many of the observations were made during work to GIS map the roads, waterholes and other physical features of Mkomazi. All species of large mammals were recorded. Further observations were made during the course of other research activities, when there was a tendency to record only the less common species. The observations provide reliable, geo-referenced species presence data. Of records made without GPS location

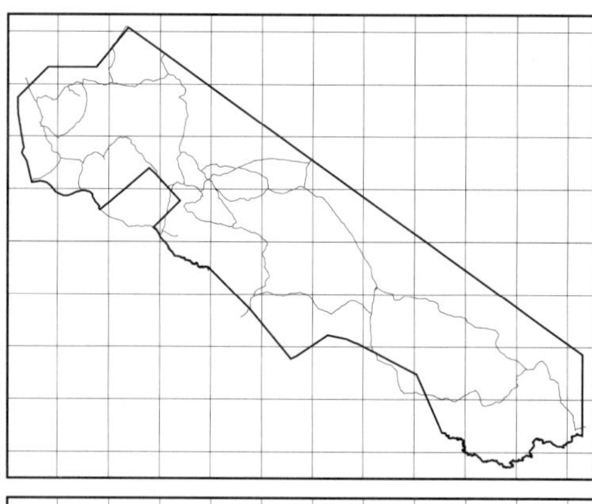

Figure 31.1 Routes of ground surveys of large mammals in Mkomazi, with 10 km grid squares. (a) All tracks in the reserve, along which most of the opportunistic sightings were made.

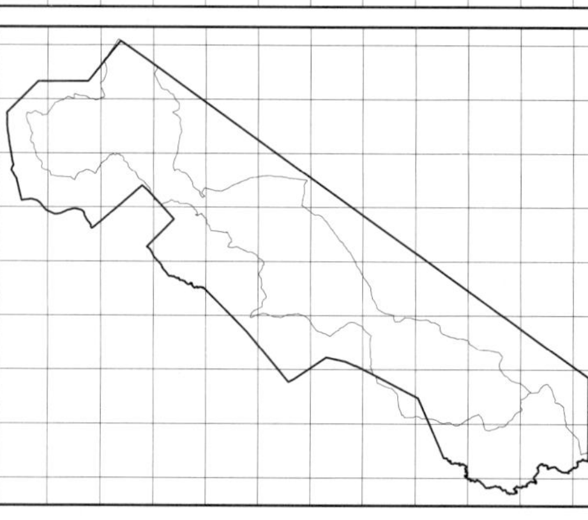

(b) The 1996 dry season survey route, surveyed four times.

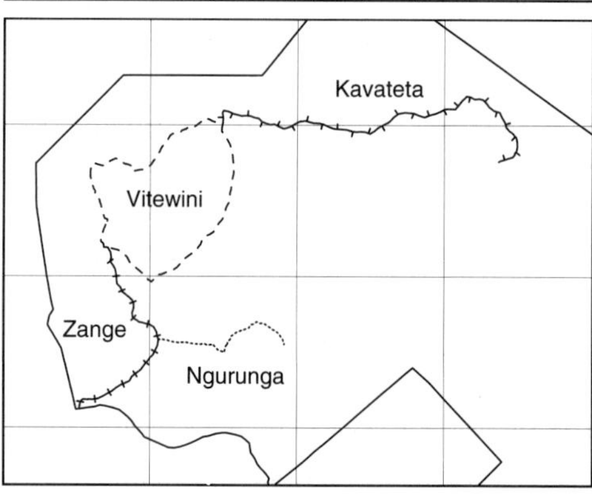

(c) The 1996 periodic survey routes in north-west Mkomazi.

data, only those which could be precisely located, using described positions in relation to Mkomazi GIS map features, were used for distribution mapping.

Dry season surveys

Systematic, dry season surveys were conducted along roads throughout the reserve (see Figure 31.1b). Four such surveys, each lasting three to four days in July to September 1996 (13–16 July, 27–30 July, 19–22 August and 31 August–2 September), provided a spatially extensive (the survey route was 393 km) although seasonally restricted, and very modest sample of the reserve. A total of 669 sightings of individuals or groups mammals were made. In addition, the duration of each survey meant that given locations were 'observed' at different times throughout the day. Diurnal variation in animal behaviour means that the probability of observing a given species changes through the day, and that surveying in late morning and early afternoon is the least 'efficient' for detecting individuals.

Apart from generating presence data for species, the dry season surveys were used to estimate numbers of individuals of certain species in Mkomazi. As already indicated, large mammal populations are generally bigger in the wet season. The dispersed nature of wet season populations and the difficulties of observing them in thicker vegetation resulted in inefficient sampling. The more open nature of vegetation in the dry season means that transects 'sample', on average, a greater area than they would in the wet season.

Transect widths were calculated from the average distances at which animals were sighted, and multiplied by the route length to give an approximate sample area. Estimates of total species abundance in the reserve were made by scaling up the sample area density to the area of the reserve, for species which were sighted ten or more times and when the distance of the animal from the road was greater than two metres.

This method of abundance estimation ignores the influence of spatial heterogeneity in habitat type on variation in species abundance as it assumes that species recorded during the survey have an equal chance of being observed in any part of the reserve. Taken together, the high degree of habitat heterogeneity in Mkomazi, the various biases associated with ground-based surveys (see above) and the fact that the method is sensitive to low numbers of sightings mean that these abundance estimates can only be treated as approximate indications.

At the same time as the dry season surveys were made, densities of large mammals were calculated from counts made around three of the waterholes that were surveyed by Harris (1972). This study was an attempt to detect changes or trends that may have occurred over the past three decades. The values for biomass units used in these calculations were the same as those used by Harris (1972). The calculated densities cannot be exactly compared because Harris classified individuals by age and sex whereas we did not distinguish between adults, instead using his biomass figure for unclassified adults. In addition, any comparison between Harris' and our observations must be made with great care because we could not take account of the possible differences in water availability (itself a strong influence on mammal densities) between observation periods.

Periodic surveys

The aim of these surveys was to represent temporal changes in species distribution across the western third of the reserve during a one year study, in 1996. The surveys were systematic and spatially intensive. They were carried out along four road routes in north-west Mkomazi (see Figure 31.1c), chosen to represent major habitats in north-west Mkomazi. Surveys were carried out twice each month during January to November 1996, although for practical reasons the number of surveys of each route in each month varied (see Table 31.1).

Surveys involved two or more observers inside a vehicle, and were standardised to occur as soon after dawn or before dusk as was possible, when animal activity was at its greatest. Presence data only were used in this study: a total of 875 sightings were made of individuals or groups.

Combining all of the ground-based surveys, a total of 3,542 geo-referenced sightings, comprising 24,033 individuals, was made (see Table 31.2).

Aerial surveys

In addition to the ground-based surveys, regular flights, for various purposes, have been made across the entire reserve by Tony Fitzjohn, resident at Kisima since 1989. The largely opportunistic sightings have been used to supplement data on species distributions and also to provide information on large mammal movements. Several, more systematic, aerial surveys of the reserve were made by SKE in July or August of 1994, 1995 and 1996.

Results

In analysing the results of surveys of the distribution and abundance of large mammals in Mkomazi, it is important to keep in mind that the reserve is part of the greater Tsavo ecosystem and is at its southern limit. Animals move widely over

Table 31.1 Number of surveys on regular routes in north-west Mkomazi in 1996.

route name & length	Jan	Feb	Mar	Apr	May	June	July	Aug	Sept	Oct	Nov
Kavateta (27.9 km)	1	3	2	1	1	0	1	2	1	2	1
Ngurunga (11.2 km)	0	4	2	1	1	1	1	2	1	3	1
Vitewini (37.6 km)	4	0	4	1	3	2	2	2	1	2	2
Zange (13.5 km)	0	4	4	2	0	1	2	2	1	2	0

Table 31.2 Summary of presence data for large mammal species in Mkomazi, gathered during ground-based surveys 1994–97.

species	sightings	individuals
dikdik	735	1,178
giraffe	477	2,761
kongoni	442	2,584
zebra	376	6,754
Grant's gazelle	264	1,278
impala	248	1,886
lesser kudu	156	263
eland	138	1,187
steinbuck	118	138
gerenuk	110	207
buffalo	102	3,376
warthog	76	204
elephant	71	1,291
reedbuck	43	73
oryx	40	387
lion	37	243
waterbuck	32	124
jackal	29	43
duiker	28	31
bushbuck	20	25
total	3,542	24,033

the whole area and their recorded presence or absence is sensitive to relatively small spatial changes in location, which may place them in Mkomazi or Tsavo. Movements in general are governed by rainfall (see Chapter 2) with Mkomazi acting as a wet season retreat for many of the animals because of the higher rainfall than that in the neighbouring Tsavo West National Park. The seasonal movements of some of the more important species are considered in *Species distributions* below.

Ground-based *versus* aerial survey methods

As indicated previously, aerial and ground-based surveys are subject to various significant biases. An idea of the variation in 'efficiency' of each of the survey methods used in this study can be gained from a comparison of observation data 'simultaneously' gathered by each method for the same area. A road transect was driven around Magunda at the same time as an aerial count of species was made. The results are given in Table 31.3.

Table 31.3 The numbers of animals observed during simultaneous aerial and ground counts of large mammals in part of Mkomazi on July 28, 1996.

species	aerial total	ground total	difference
buffalo	87	0	- 100%
eland	6	76	+ 92%
elephant	14	21	+ 33%
giraffe	46	26	- 43%
Grant's gazelle	5	11	+ 55%
kongoni	25	10	- 60%
waterbuck	3	3	0%
zebra	216	181	- 16%
lion	0	19	+ 100%

The results highlight ways in which each survey technique under-samples species presence and abundance. Aerial surveys are better at recording animals away from roads but are less efficient at detecting individuals or small groups of animals, especially of smaller large mammals. The aerial counters almost certainly flew over but failed to see a pride of eight lions with 11 cubs which was found by the ground team. Lions are notoriously difficult to detect from the air and are best surveyed from the ground. In addition, air-borne observers have less time in which to make repeated checks on numbers, and the survey usually takes far less time, reducing the relative opportunity to observe animals. In this case, the aerial survey was completed within 50 minutes while the ground counts took several hours so the surveys were not simultaneous. The difference in timing was probably responsible for the discrepancy between the aerial and ground totals for buffalo and eland. Most of the buffaloes seen from the air were in the hills south of Dindira and were moving towards thick country. It is unlikely that they could have been detected from the ground. The group of eland recorded on the ground count was certainly not present when the aircraft flew over the region where they had been seen. Where the animals were more widely distributed, the totals from the two methods agreed reasonably well.

Species abundance

Table 31.4 compares the estimates of the numbers of large mammals in Mkomazi Game Reserve made in the 1960s by Harris (1972) with those made in the 1990s by Inamdar and by this study in 1996. The 1996 estimates are conjectural and are based on the 1996 dry season systematic ground surveys, supplemented by Fitzjohn's aerial observations. The 1996 estimates are mainly of comparative value and probably do not represent the true totals. The 1994 aerial totals are also taken into account although aerial surveys are known to underestimate numbers, par-

Table 31.4 Minimum estimates of the numbers of large mammals based on aerial and ground counts made in Mkomazi Game Reserve between 1964 and 1967 by Harris (1972), in 1994 by Inamdar (1994) and in 1996 by the present authors. Not all species were counted on each occasion. Numbers in parentheses are standard errors.

	date and season		
species	1960s wet	1994 wet	1996 dry
buffalo	750	1,858 (1,569)	– [a]
eland	500	2,421 (1,279)	473 (1,313)
elephant	3,000	477 (304)	314 (149) [b]
gerenuk	250	17 (16)	933 (141)
giraffe	250	545 (76)	979 (84)
Grant's gazelle	–	–	306 (89)
impala	600	801 (348)	3,564 (2,470)
dikdik	–	–	55,978 (8,153)
kongoni	1,000	511 (200)	840 (229)
lesser kudu	250	426 (71)	5,739 (2,417)
oryx	400	102 (97)	– [c]
steinbuck	–	–	554 (339)
warthog	–	–	1,460 (704)
waterbuck	150	17 (16)	–
zebra	400	460 (178)	1,438 (741) [d]

[a] Buffalo were observed but numbers were too small for estimating population size
[b] 500 elephant were estimated by Harris (1972) to be present in 1960s dry season
[c] 100 oryx were estimated by Harris (1972) to be present in 1960s dry season
[d] 100 zebra were estimated by Harris (1972) to be present in 1960s dry season

ticularly of the smaller species. Any one-off count represents only a snapshot of the situation and needs to be treated with caution.

In view of the various techniques used and the seasonal differences when the counts were made, close comparisons are not justified but the results suggest that there has been little change except for a tendency towards an increase in numbers of the larger species (which may, however, be due to improved sampling techniques). The huge apparent increase in eland in 1994 is probably due to sampling error in the aerial survey total. Buffalo, giraffe and zebra have shown substantial increases and of the large mammals, only elephant has shown a marked decrease. Although not included in the analysis, the black rhino has also decreased, from several hundred to none. The declines in elephant and rhino are not surprising, given the known extensive poaching for ivory and horn in the intervening years.

The seasonal changes in numbers were investigated by Harris (1972) from counts made in three study areas surrounding semi-artificial water holes. Changes between wet and dry seasons are shown in Table 31.5. The biggest difference was recorded at Dindira, which is in the north-western corner of the reserve and which is the only one of the study areas to hold permanent water. Hence its attraction for wildlife in the dry season.

These counts were repeated in 1996 and the results are included in Table 31.5. Too much cannot be deduced from these comparisons because of the somewhat different techniques employed but they provide evidence of an increase in the numbers and biomass of large mammals in the northern sector of the reserve. Mbula is curious in that the biomass more than doubled although there was little increase in numbers. A similar trend is apparent in the 1960s when the biomass in the wet season was double that in the dry season although numbers remained the same. These observations suggest that the species composition of the large mammals around this waterhole is liable to fluctuate. Harris did not record eland or zebra at Mbula and the presence of these species in 1996 may explain the discrepancy. Alternatively it could be the sporadic appearance of elephants that is responsible. The Kavateta figures were influenced by the absence of giraffe and zebra in the 1960s and lower densities of impala and kongoni. These differences may be due to the presence of cattle in the 1960s and their absence in 1996.

Although the trends noted are not in themselves very convincing, they all point towards a possible increase in the numbers of most large mammals. Differences in densities between Harris' and our studies might, however, result from differences in environmental circumstances (such as water availability) at the times of observation, rather than from population trends. The increase at waterholes is largely attributable to two species, giraffe and zebra. These have relatively low standard errors in the road count estimates and it is very likely, therefore, that the perceived increases in these species are genuine. The decline in elephants is also likely to be real for the same reason.

Best estimates of herbivore abundance in Mkomazi are given in Table 31.6. These are based mainly on the 1996 dry season systematic ground surveys, whose totals are listed in Table 31.4, but include a subjective 'expert knowledge' element (SKE and Tony Fitzjohn).

Table 31.5 Numbers and biomass (kg) of large mammals per km^2 around Mkomazi waterholes in wet and dry seasons in the 1960s (from Harris 1972) compared with those recorded during the dry season in 1996. (Kavateta was named Mzara by Harris.)

	1960s				1996	
	dry		wet		dry	
waterhole	number	biomass	number	biomass	number	biomass
Dindira	23.7	12,705	7.7	2,082	31.6	17,329
Mbula	8.7	1,452	8.7	3,638	8.8	3,058
Kavateta	3.4	261	6.5	752	46.9	11,953

Table 31.6 Best estimates of the numbers of herbivores present in Mkomazi based on ground counts made in 1996, supplemented by incidental observations and subjective assessments.

species	number	comments
buffalo	2,000	wet season total
dikdik	100,000	subjective assessment
eland	500	
elephant	300 / 1,000	dry/wet season totals
gerenuk	1,000	imprecise total
giraffe	1,000	
Grant's gazelle	200	
impala	5,000	conservative estimate
kongoni	1,000	
lesser kudu	6,000	
steinbuck	600	conservative estimate
warthog	1,500	probably an overestimate
zebra	2,500	

Species distributions

Figures 31.2–31.13 show the distributions of species for which there are numerous geo-referenced sightings made in 1994–97 or which are important from a management point of view. In considering these maps, it is very important to keep in mind the biases associated with each survey technique. The maps do not necessarily represent the limits of species distributions in Mkomazi, although they may do so. Aerial observations, particularly of species movements, are incorporated in the following notes on elephant and buffalo. The letters in parentheses in the text refer to the seasonal movements shown on these maps.

Herbivores

BUFFALO. The seasonal movements of buffalo mirror those of the elephant to a large extent and the routes followed in the north-west are almost exactly the same. Some enter Mkomazi from the north-west (A) while a second wave (B) cross over from Tsavo West National Park and spreads out over the plain between Kavateta and Vitewini. In general, most of the buffalo occur either in the north-west or in the south-east of the reserve with few in between (Figure 31.2). They are quite common along the border with Kenya and three herds totalling some 350 to 500 criss-cross the border south-east of Kavuma Hill (C). Another 100 or so are resident along the Kenya border near Mabata. It is not possible to give an accurate

Figure 31.2
See text for explanation of movements

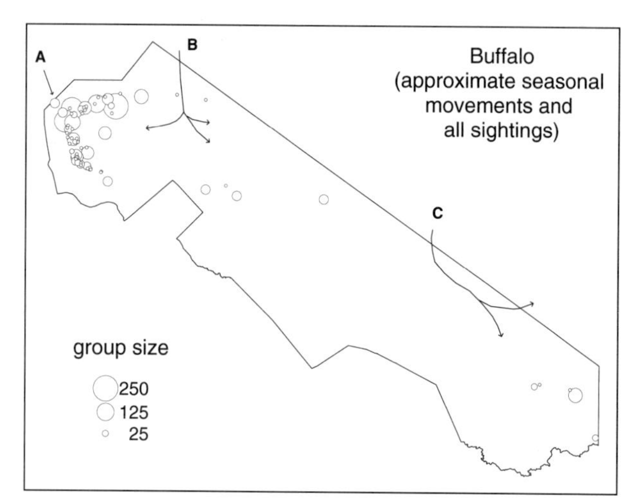

Figure 31.3

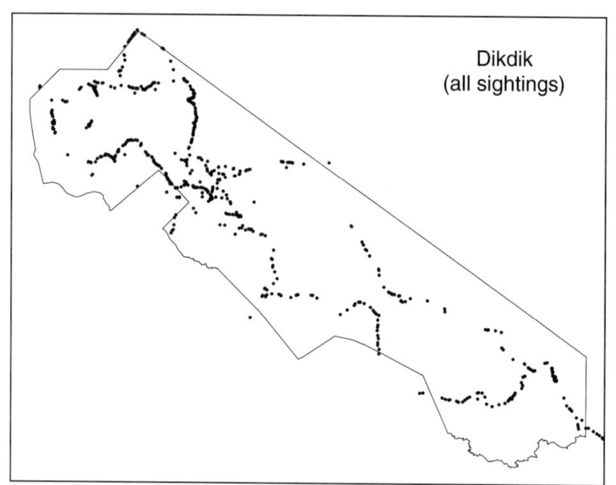

Figure 31.4

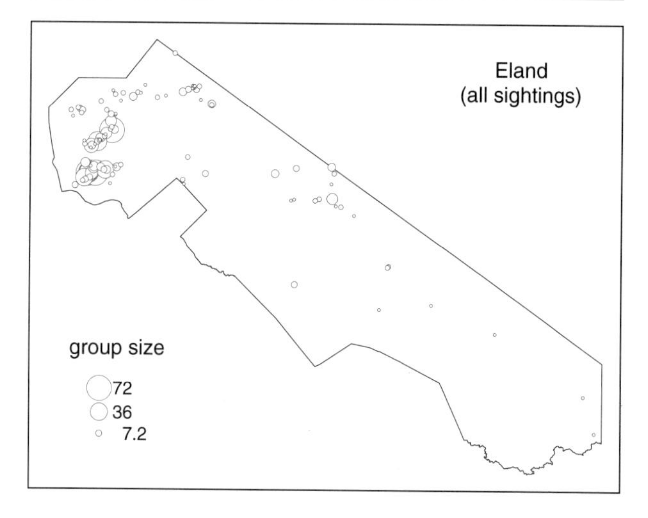

figure for total numbers in the reserve. A population of around 2,000 would seem to be reasonable for the wet season.

DIKDIK is the species most frequently seen from the roads in Mkomazi (Figure 31.3). It was almost never seen on road counts passing through the seasonal swamps or open vegetation.

ELAND are most likely to be seen in the central regions of the reserve, particularly around the Maore waterhole although they are found in most regions (Figure 31.4). Some 100 eland are known to enter the reserve from Kenya in the wet season between Maore and Kamakota. As far as total numbers are concerned, extrapolation from the ground surveys gives a figure of 473. General impressions, which admittedly are notoriously unreliable, tend to support this figure.

ELEPHANTS may be found anywhere within the reserve although their distribution is markedly clumped (Figure 31.5). In the north-west, elephants enter the reserve in the wet season and some (A) spread south-west to the Mbula and Gulela Hills and beyond. A second group (B) moves to the region between the Gulela and Mzara Hills. At the same time, 100 or more elephants (C) move out of the forested hills on either side of Dindira Dam and pass on to the plains between Zange and Ngurunga, where they mix with the Kenyan elephants. Smaller movements (D) across the border from Kenya to Kavateta occur if there is water in the dam. Similar small scale movements (E) across the border occur near Maore waterhole. A group of resident elephants occurs on the western side of the Mzara Hills but they may move out into Tsavo in the wet season (F). A more substantial wet season immigration takes place in the Mzara/Maore region (G). Some of these elephants pass to the west around Hafino Hill (H) and a few get as far as Kisiwani Village, where they may raid crops, and even fewer to Njiro Gate. Most pass between Kisima and Tussa Hills (I) to meet up with those that moved to the east of Hafino Hill. Mating is commonly observed in this region. Some migrate further to the south-east into the thick vegetation around Kamakota (J) where they mingle with elephants that have entered the reserve near Kavuma (K). Altogether some 400 or so elephants may be present in this area. Some 25 to 30 resident elephants occur north of Mabata in the far south-east of the reserve (L) but they may move a little way into Kenya from time to time or south to the Umba River. The total number of elephants present in the reserve during the wet season is around 1,000 but this falls to less than 100 in most dry seasons.

GERENUK are present throughout the reserve (Figure 31.6) but they are not easy to see and estimates of their numbers are probably too low. The 1996 dry season population size estimate of 933 suggests a healthy population commensurate with the size of the reserve.

Figure 31.5
See text for explanation of movements

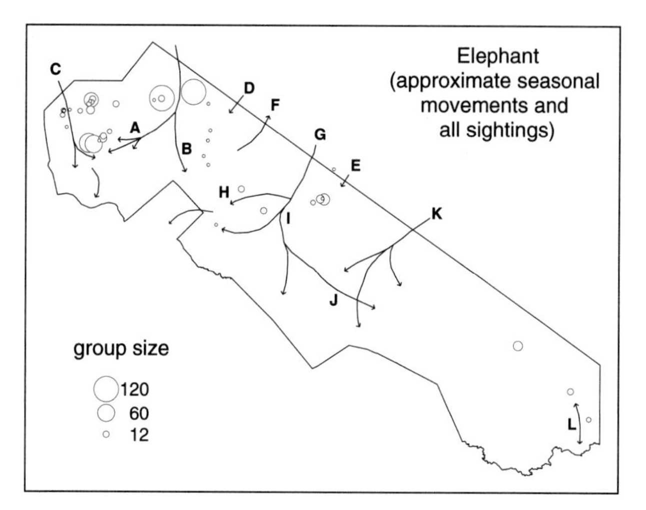

Figure 31.6

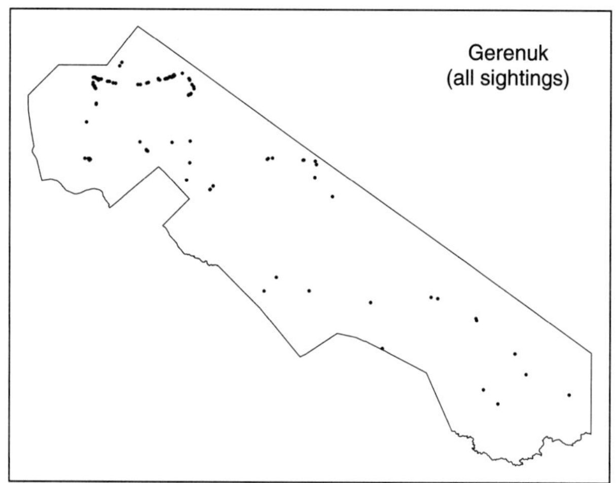

Figure 31.7

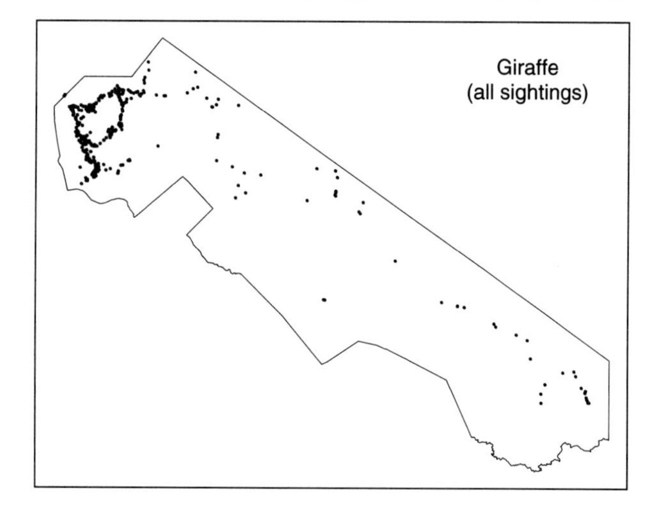

GIRAFFE were known to occur more or less throughout the reserve, although not uniformly so, and there are places where they are never seen, possibly because of heavy poaching in such regions. They are most numerous in the north-west but are present throughout the reserve (Figure 31.7), being comparatively rare in the central regions.

GRANT'S GAZELLE are restricted to the western half of the reserve (Figure 31.8), and while not present in large numbers are most numerous in the far west. The species was not recorded on the 1994 aerial survey, although it is not clear whether the species was excluded from the counts. The total from the 1996 ground counts was only 306. Even so this is likely to be too high as about half of the area included does not support gazelles and it is unlikely that there are more than a couple of hundred in the reserve altogether.

IMPALA is one of the commoner species and is generally widely distributed in the reserve although it is most abundant in the western half (Figure 31.9). Antelopes of this size and coloration are not easily seen from the air and, as with gerenuk, the 1994 aerial count of 801 is certainly too low. The 1996 dry season estimate of 3,564 suggests a population of several thousand.

KONGONI (Coke's hartebeest) are more or less distributed throughout the reserve although not uniformly so as there are regions of higher density, particularly near waterholes (Figure 31.10). Like other large ungulates, they show seasonal movements between Tsavo and Mkomazi and groups of 40–50 accompany eland and zebra on their passage through the Maore region.

LESSER KUDU are probably the most numerous of the larger antelopes in Mkomazi but because of their cryptic markings and the dense cover they inhabit, they are not easily seen. They are distributed widely throughout the reserve (Figure 31.11 but are rare in the Ibaya region due, no doubt, to the lack of suitably thick country there.

ORYX are widely distributed in Mkomazi (Figure 31.12), although the population size appears relatively low. A herd was regularly seen throughout the Mkomazi Ecological Research Programme at Kavateta.

ZEBRA are found most frequently in the western half of the reserve (Figure 31.13) but are present in the eastern half. The species is particularly numerous around Ibaya and in the vicinity of the Maore waterhole. Up to 400 zebra accompany eland and kongoni in the wet season movements from Kenya.

Figure 31.8

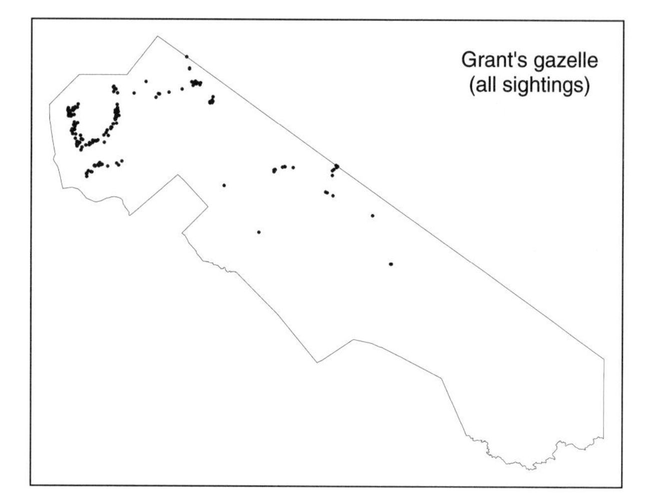

Figure 31.9

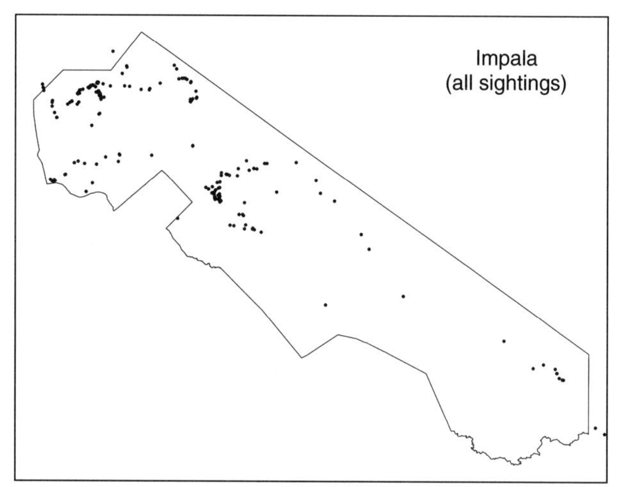

Figure 31.10

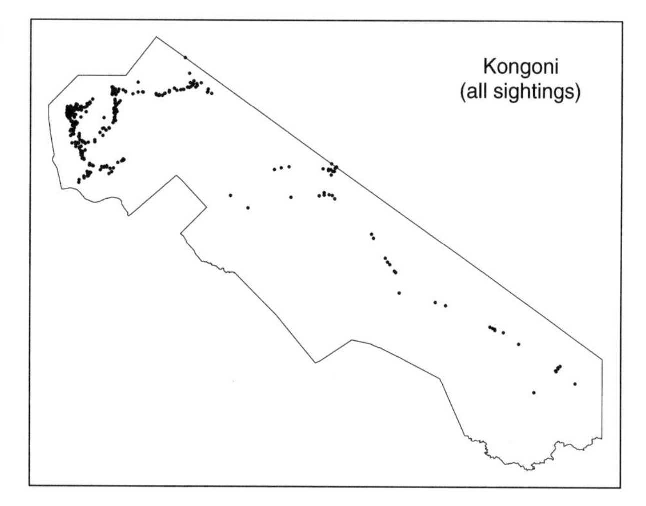

Carnivores

Data on the distribution of carnivores are limited, largely because relatively low population sizes mean that species are rarely encountered during surveys. The smaller carnivores are regularly seen and population sizes appear to be healthy. Compared with populations elsewhere in similar habitats, the spotted hyaena is very rare. The incidence of melanism in the serval seems to be high.

Domestic stock

Although cattle, sheep, goats and donkeys are not supposed to be present, large numbers of cattle, in particular, have been recorded within Mkomazi. This is perhaps not surprising given the problems of law enforcement, in part associated with the elongated shape of the reserve. The numbers involved run into thousands. Domestic stock were counted in the 1994 aerial survey (Inamdar 1994) and totals of 23,557 cattle (s.e. $\pm$ 12,530) and 4,739 sheep or goats (s.e. $\pm$ 2,356) were estimated although not all of these were within the reserve's borders. Cattle enter the reserve at Pangaro, where there is a dam just outside the reserve boundary, north of Ndea across to Kavateta, possibly including herds from Lake Jipe in Kenya, and on the southern boundary towards the Ngurunga region. Herds have been recorded at Kamakota in the centre of the reserve and in the south-eastern region.

Mammal species re-introductions

Following the judgement of the Tanzanian Wildlife Division in the late 1980s that Mkomazi was in a florally and faunally degraded state, a decision was made to rehabilitate the reserve. The Division formed the Mkomazi Project under the supervision of a project manager. One of the aims of the Project was to re-introduce species to the reserve that had recently been extirpated. Three species were initially selected for re-introduction: black rhinoceros, cheetah and wild dog. As cheetah re-established naturally in Mkomazi, attention was focused on the black rhino and wild dog. In view of the costs involved with re-introducing species, the UK-based George Adamson Wildlife Preservation Trust was invited to assist with the rehabilitation programme. The Trust provides funds for equipment and supports the activities of Tony Fitzjohn.

Black rhinoceros

There was a population of at least 150 rhinoceros in the reserve as recently as the mid-1960s but the species became extinct, largely due to poaching. A re-introduction programme obviously needs a source of animals but the general decline of the species throughout Africa to dangerously low levels was a problem. There

Figure 31.11

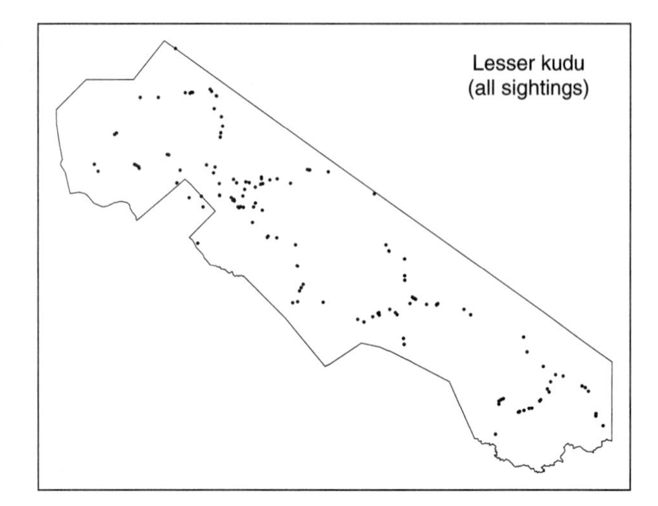

Figure 31.12

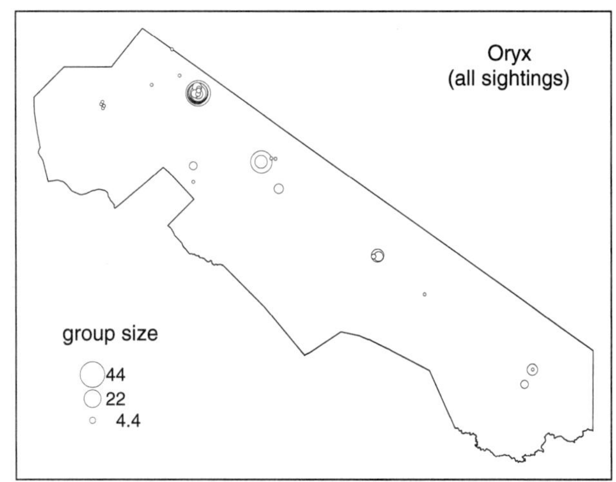

Figure 31.13

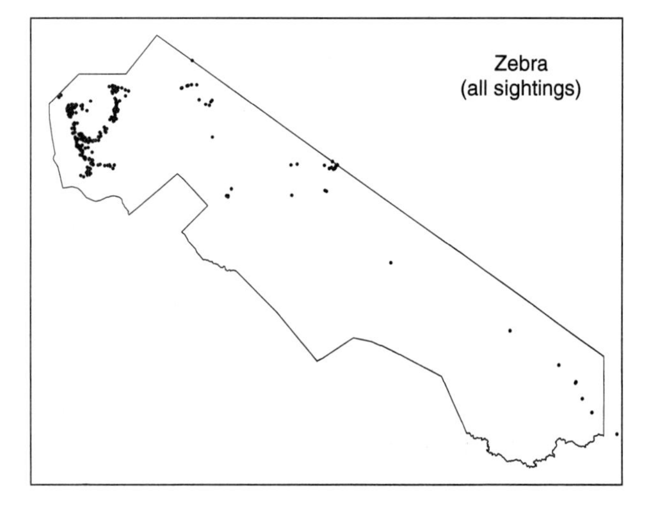

would be no justification for moving animals from where they were native unless they were at grave risk of being killed. It became known that authorities in South Africa wanted to dispose of a population of 35 black rhino in the Addo National Park because they were a subspecies that was not native to the area. The population was descended from a group of seven animals that had been shipped to South Africa in the early 1960s from the Tsavo ecosystem, and so belonged to the appropriate subspecies, *Diceros bicornis michaeli*, for re-introduction to Mkomazi.

Guidelines for the re-introduction of species have been prepared by the Species Survival Commission of IUCN (IUCN, 1987). These guidelines provided the framework for the Mkomazi rhino re-introduction project, which was assessed and approved by experts from the South African National Parks Board (Knight & Morkel, 1994). It was decided not to release the animals straight into the wild but to hold them for a number of years in a sanctuary enclosed by an electrified fence within the reserve. The fence was completed in late 1996 and covers an area of about 43 km^2. The first four rhinoceros arrived on 4 November 1997 and were kept in bomas for a few weeks to recover from the translocation and to acclimatise to their new circumstances, before being released into the sanctuary.

Wild dog

The wild dog is an endangered species and is still persecuted throughout its range in Africa. An attempt to re-establish the species in Mkomazi was considered to be an important contribution to the survival of the species. There are ethical problems, however, in taking specimens from the wild for re-introduction in former parts of the species' range, unless the population concerned is at immediate risk of being killed. This appeared to be the case with three groups that were located at Engassumet on the Maasai Steppe, about 100 km from Mkomazi. The dens were dug out in June 1995 and 25 pups (15 males and 10 females) were collected and brought to holding pens at Kisima, close to the rhino sanctuary, before threats to poison the dogs could be carried out. The parents of the captured pups survived long enough to produce further litters so the exercise did not seriously affect the status of the species in that area.

The captive dogs are breeding successfully, with the first young born in March 1997. The plan is to maintain a breeding stock in captivity and to release groups in a series of re-introductions as well as to provide source individuals for re-introductions elsewhere. Four males were sent to Kenya to provide 'fresh blood' for a re-introduction project. They were introduced to four wild-caught females with hunting experience and after some time, two males and all the females were released into Tsavo National Park. The pack eventually found its way back to Mkomazi where its progress is being monitored through radio-tracking.

Conclusion

The large mammal fauna of Mkomazi is diverse, resulting from both a high degree of habitat heterogeneity and the inclusion of the reserve in the greater Tsavo ecosystem. Although population sizes in general are not great, the reserve contains important populations of several herbivore and carnivore species, such as gerenuk, oryx, lesser kudu, leopard and cheetah. The seasonal movements of large mammals into and out of Mkomazi indicate that the reserve probably plays a key role in the population viability of several species, by providing important wet season resources.

Information on large mammal distribution and abundance is necessary for management planning purposes. The data generated by this study provide an essential baseline of information on species presence and status in Mkomazi. This information can be used to evaluate the potential for developing tourism within the reserve, to predict impacts of different kinds of utilisation of the reserve, as well as to provide the basis for monitoring impacts. The information would also be useful in planning the re-introductions of large mammal species to the reserve.

Acknowledgements

Our thanks to the members of MERP for many of the opportunistic large mammal observations and to the Mkomazi rangers who also participated in the systematic surveys.

References

Harris, L.D. (1972) *An ecological description of a semi-arid East African ecosystem.* Range Science Department Science Series No. 11, Colorado State University.

Inamdar, A. (1994) *Wildlife Census Mkomazi April 1994: an interim report on results of an aerial census of the Mkomazi Game Reserve.* Unpublished report, Worldwide Fund for Nature, Nairobi.

Inamdar, A. (1996) *The ecological consequences of elephant depletion.* Doctor of Philosophy thesis, University of Cambridge.

IUCN (1987) *The IUCN position statement on translocation of living organisms.* IUCN, Gland.

Knight, M. & Morkel, P. (1994) *Assessment of the proposed Mkomazi rhino sanctuary, Mkomazi Game Reserve, Tanzania.* Unpublished Report.

TWCM (1991) *Wildlife Census: Mkomazi 1991.* Tanzania Wildlife Conservation Monitoring, Arusha, Tanzania.

CHAPTER 32

Checklist: Mammals of Mkomazi

S. Keith Eltringham, Robert J. Morley, Jonathan Kingdon, Malcolm J. Coe & Nicholas C. McWilliam

The following list is based on: field records by the Mkomazi Ecological Research Programme in 1992–96; small mammal records made by Larry Harris in 1964–67 (marked LH) (Harris 1972) and collections made by Rob Morley (marked RM; see Chapter 30). Additional comments and sightings by Malcolm Coe (marked MJC), Keith Eltringham (KE) and Jonathan Kingdon (JK). Historical data collated by Nicholas McWilliam. Species thought not to occur wild any longer in Mkomazi are marked '*' and details are given at the end of this chapter. Nomenclature (Latin and common) follows Corbett & Hill (1991), Skinner & Smithers (1990) and Kingdon (1997).

MAMMALS (Class Mammalia)

PLACENTALS (Sub-class Eutheria)

PRIMATES (Order Primates)

Cheek-pouch monkeys (Cercopithecidae)

Yellow baboon *Papio cynocephalus.* Reported by JK and KE. JK has suggested mixing with olive baboon (*P. anubis*) in overlap of ranges.

Olive baboon *Papio anubis.* KE.

* Angola pied colobus *Colobus angolensis.* Reported possible sighting 19 June 1995 near Umba River. Formerly quite common.

Vervet monkey *Cercopithecus pygerythrus.*

Sykes' monkey *Cercopithecus mitis.*

Galagos (Galagonidae)

Small-eared galago *Otolemur garbetti.* Hills behind Ibaya camp (JK).

?Senegal galago *Galago senegalensis.* Expected in lower-lying woodlands (JK).

?Usambara galago *Galagoides orinus.* Possible recording from Ibaya Hill (JK).

BATS (Order Chiroptera)

Fruit bats (Sub-order Megachiroptera)

Pteropodidae

Rousette bat *Rousettus aegyptiacus.* LH, listed as *R. angolensis,* but not known from this region.

Epauletted fruit bat *Epomophorus sp.* LH. Probably *E. wahlbergi* but *Eidolon helvum* known from this area (MJC).

Insect bats (Sub-order Microchiroptera)

Slit-faced bats (Nycteridae)

Slit-faced bat *Nycteris thebaica.* LH, listed as 'large-eared hollow-faced bat'.

Leaf-nosed bats (Hipposiderinae)

Leaf-nosed bat *Hipposideros caffer.* LH.

Free-tailed bats (Molossidae)

Guano bat *Tadarida aegyptiaca*. LH, listed as *T. aegyptica.*, mastif bat.

INSECTIVORES (Order Insectivora)

Shrews (Soricidae)

White-toothed shrew *Crocidura hirta*. RM.

White-toothed shrew *Crocidura (Afrosorex) voi*. RM. New record for Tanzania. Not recognised at species or sub-generic level by Corbet & Hill (1991) or Kingdon (1997).

White-toothed shrew *Crocidura parvipes*. RM. Not recognised by Corbett & Hill (1991).

Crocidura sp. RM.

Musk shrew *Suncus lixus*. RM.

ELEPHANT SHREWS (Order Macroscelidea)

Lesser elephant shrew *Elephantulus rufescens*. RM.

Chequered elephant shrew *Rhynchocyon cirnei*. LH.

HARES (Order Lagomorpha)

Hares (Leporidae)

Scrub hare *Lepus saxatilis* (=*L. crawshayi*).

Cape hare *Lepus capensis*. LH.

RODENTS (Order Rodentia)

Squirrels (Sciuridae)

Unstriped ground squirrel *Xerus rutilus*.

Ochre bush squirrel *Paraxerus ochraceus*.

Dormice (Gliridae, syn. Myoxidae)

African dormouse *Graphiurus murinus*. LH, RM.

Blesmols (Bathyergidae)

Silky blesmol *Heliophobius argenteocinereus*.

Porcupines (Hystricidae)

Porcupine *Hystrix sp.* Reported by LH as *H. galeata*.

Murid rats & mice (Muridae)

Bush rat *Aethomys chrysophilus*. RM.

Spiny mouse *Acomys cahirinus*. RM, LH.

Spiny mouse *Acomys ignitus*. RM.

Spiny mouse *Acomys spinosissimus*. RM.

Spiny mouse *Acomys wilsoni*. RM, LH.

Acomys sp. RM.

Unstriped grass rat *Arvicanthis niloticus*. RM.

Gerbillus pusillus. LH, listed as Taita pygmy gerbil.

Narrow-footed woodland mouse *Grammomys dolichurus* RM.

Zebra mouse *Lemniscomys barbarus*. RM, LH.

Zebra mouse *Lemniscomys griselda*. RM, LH.

Multimammate rat *Mastomys natalensis*. LH, listed as 'shamba rat'.

Pygmy or common mouse *Mus minutoides*. RM.

Common mouse *Mus tenellus*. RM.

Meadow rat *Myomys fumatus*. RM.

Saccostomus campestris. RM.

Tatera nigricauda. RM.

Tatera gerbil *Tatera robusta*. RM & LH.

Taterillus osgoodi. RM. Not in Corbett & Hill (1991) or Kingdon (1997).

CARNIVORES (Order Carnivora)

Dogs (Canidae)

Side striped jackal *Canis adustus*. Reported by JK; not listed by KE.

Black backed jackal *Canis mesomelas*.

Bat-eared fox *Otocyon megalotis*.

*Wild dog *Lycaon pictus*. Captive population since 1992.

Mustelids (Mustelidae)

Zorilla *Ictonyx striatus*.

Ratel or honey badger *Mellivora capensis*.

Mongooses (Herpestidae)

Egyptian mongoose *Herpestes ichneumon*. Quite frequent & widespread (MJC).

Slender mongoose *Galerella (herpestes) sanguinea*. Sighting at 'klipspringer hill', north of Ibaya Camp (JK). Common (MJC).

Dwarf mongoose *Helogale parvula*.

Somali dwarf mongoose *Helogale hirtula*. Possible sighting (JK). Sighting on rock outcrop (MJC).

Banded mongoose *Mungos mungo*.

White-tailed mongoose *Ichneumia albicauda*.

? Bushy-tailed mongoose *Bdeogale crassicauda*. Expected, not seen (JK).

Hyaenids (Hyaenidae)

Striped hyena *Hyaena hyaena*.

Spotted hyena *Crocuta crocuta*.

Aardwolf *Proteles cristata*.

Genets & civets (Viverridae)

Common genet *Genetta genetta*.

Blotched genet *Genetta tigrina*. Sightings near (JK) and at (MJC) Ibaya camp.

African civet *Civettictis civetta*.

Cats (Felidae)

Wild cat *Felis sylvestris*.

Serval cat *Felis serval*.

Caracal *Felis caracal*.

Leopard *Panthera pardus*.

Lion *Panthera leo*.

Cheetah *Acinonyx jubatus*.

SCALY ANT-EATERS (Order Pholidota)

Pangolin (Manidae)

Ground pangolin *Phataginus (=Smutsia) temminckii*.

UNGULATES

AARDVARK (Order Tubulidentata)

Aardvark (Orycteropodidae)

Aardvark *Orycteropus afer*.

HYRAXES (Order Hyracoidea)

Hyraxes (Procavidae)

Rock hyraxes *Procavia sp.* Occasional.

Bush hyrax *Heterohyrax brucei*. Common throughout the MGR (MJC).

Tree hyrax *Dendrohyrax validus*. Signs on trees in Igire forest and skull at Pangaro (MJC).

PROBOSCIDS (Order Proboscidae)

Elephant (Elephantidae)

Elephant *Loxodonta africana*.

ODD-TOED UNGULATES (Order Perissodactyla)

Horses (Equidae)

Zebra *Equus burchelli*.

Rhinoceroses (Rhinocerotidae)

*Black (browse) rhinoceros *Diceros bicornis*. Captive re-introduction started Nov. 1997.

EVEN-TOED UNGULATES (Order Artiodactyla)

Pigs (Suidae)

Bush pig *Potamochoerus porcus*. Kingdon (1997) places in a separate species *P. larvatus*.

Common warthog *Phacochoerus africanus*.

Giraffe (Giraffidae)

Giraffe *Giraffa camelopardalis*.

Bovids, horned ungulates (Bovidae)

Bovines (Bovinae)

Buffalo *Syncerus caffer*.

Bushbuck *Tragelaphus scriptus*.

Lesser kudu *Tragelaphus imberbis*.

*Greater kudu *Tragelaphus strepsiceros*.

Eland *Taurotragus oryx*.

Bush duiker *Sylvicapra grimmia*.

Harvey's duiker *Cephalophus harveyi*. Sightings in Kisiwani road forest (JK) and Pangaro (MJC).

Suni *Neotragus moschatus*. Skull on Ibaya hill; sighting at Vitewini ridge 1996 (KE & JK).

Steinbuck *Raphicerus campestris*.

Klipspringer *Oreotragus oreotragus.*
Kirk's dikdik *Madoqua kirkii.*
Bohor reedbuck *Redunca redunca.*
Waterbuck *Kobus ellipsiprymnus.*
Grant's gazelle *Gazella granti.*
Gerenuk *Litocranius walleri.*
Impala *Aepyceros melampus.*
Kongoni *Alcelaphus buselaphus.*
*Wildebeest *Connochaetes taurinus.* Historically present, re-introduced, but now absent.
* Sable antelope *Hippotragus niger.*
Fringe-eared oryx *Oryx beisa.*

Large mammal species now extinct in Mkomazi

Historical records of species thought not to occur in Mkomazi Game Reserve today.

Black and White or Pied Colobus *Colobus angolensis*

Records of "Pangani black and white Colobus (*C. a. palliatus*)" from "Usambara Mountains ... middle Umba River" (Swynnerton & Hyman 1951); "the river bank of the Umba provides tall trees in which are Colobus and Sykes' monkeys" (Anstey 1958); "The colobus monkey (*Colobus angolensis*) was recorded by various game department personnel as late as 1957 (annual report, Game and Tsetse Division 1950, District Ranger's Report 1957). [This] species has [not] been recorded since 1957 and none presently exist in the area" (Harris 1970); one reported possible sighting on 19 June 1995 near Umba River Gate in river bank trees (Ferguson/Mafunde, pers. comm.).

Wild Dog *Lycaon pictus*

"...there are too many wild dogs" (Anstey 1956); "...hunting dog is present" (Harris 1970); A pack passed Kisima in 1997, the first wild dogs to be recorded in the reserve for decades.

Black or Browse Rhinoceros *Diceros bicornis*

Records of rhinoceros from "Northern Tanga, Pare, Umba Steppe" (Swynnerton & Hyman 1951); "These grasslands are the home of many rhinoceros ... The rhino have shown how much they like the open spaces now that it is safe to be there." (Anstey 1956); "Rhinoceros, lesser kudu, waterbuck and dikdik are not rare" (Harris 1970). Population estimates: in 1964–67: 45 (Harris 1970); in 1967: six (estimate based on a survey stratified to represent elephant habitat; Watson *et al.* 1969); in 1968: 150–250 (survey stratified specifically to represent rhino habitat; Goddard 1969); in wet season 1974: four (all in NW Mkomazi; Cobb 1979); in 1974: none observed (Cobb 1979). Re-introduction programme started in 1997.

Greater Kudu *Tragelaphus strepsiceros*

"The woodland game so far recorded includes: greater and lesser kudu…" (Anstey 1956); "Observations of the greater kudu were made by … the acting warden in official letter no. 451/8/46 of 17 October 1955. This, along with my finding of a greater kudu horn in the reserve, supports that species' former presence in the reserve" (Harris 1970).

Wildebeest or Brindled Gnu *Connochaetes taurinus*

"Wildebeest" marked in Mkomazi Gap, and between South Pare and West Usambara Mountains (Atlas of Tanganyika 1956); "Eastern white-bearded wildebeest were … common in Tanga Province and the area presently occupied by the reserve in the 1930s (Game and Tsetse Division annual report 1932; R. Bradstock pers. comm.). No wildebeest occurred in the area during the fifties and early sixties. Twenty were restocked in 1966… The introduced wildebeest population is growing after an initial decline…" (Harris 1970).

Sable Antelope *Hippotragus niger*

Records from Lake Jipe and Kisiwani. (Swynnerton & Hyman 1951); Anstey (1956): no mention; "Sable are still occasionally seen east of the reserve" (Harris 1970).

References

Anstey, D. (1956) Mkomazi Game Reserve. *Oryx* 3: 183-185.

Cobb, S. (1979) *The distribution and abundance of the large herbivore community of Tsavo National Park, Kenya.* D.Phil. Thesis, University of Oxford.

Corbett, G.B. & Hill, J.E. (1991) *A world List of Mammalian Species.* 3rd edition. Natural History Museum Publications, Oxford University Press.

Goddard, J. (1969) Aerial censuses of black rhinoceros using stratified random sampling. *East African Wildlife Journal* 7: 105-114.

Harris, L.D. (1972) *An ecological description of a semi-arid East African ecosystem.* Colorado State University Range Science Series 11.

Inamdar, A. (1994) *An interim report on results of an aerial census of the Mkomazi Game Reserve.* Unpublished report, Worldwide Fund for Nature, Nairobi.

Kingdon, J. (1997) *The Kingdon field guide to African mammals.* Academic Press, London.

Skinner, J.D. & Smithers, R.H.N. (1990) *The mammals of the Southern African Subregion* (New Edition). University of Pretoria.

Swynnerton, G.H. & Hayman, R.W. (1951) A checklist of the land mammals of the Tanganyika territory and the Zanzibar Protectorate. *Journal of the East African Natural History Society* 20: 274-392.

TWCM (1991) *Wildlife Census: Mkomazi 1991.* Tanzania Wildlife Conservation Monitoring, Arusha.

Watson, R.M., Parker, I., & Allan, T. (1969) A census of elephant and other large mammals in the Mkomazi region of N. Tanzanian and S. Kenya. *East African Wildlife Journal* 7: 11-26.

Human aspects

Human populations are an integral dimension of many ecosystems: their resource needs and patterns of utilisation shape important aspects of the ecology of an area. These needs and resource use patterns vary culturally: some groups are predominantly pastoralist, some cultivators, while others combine these livelihoods in a variety of ways.

The recently published *Wildlife Policy of Tanzania* recognises the importance of taking account of both national and local development needs in management planning and action for biodiversity conservation. The establishment of Wildlife Management Areas (WMAs) outside the 'core' protected area network is a key mechanism for actively involving local communities in the management of biological resources.

Whether or not the establishment of WMAs contributes successfully to conservation and development objectives will depend, in part, on whether local livelihoods, development needs and aspirations, and expectations of control over management of the areas are properly understood and taken account of during the planning process. This Section addresses issues of relevance to planning, and concerns resource utilisation in the context of local culture and livelihoods.

Several groups of people (Pare, Sambaa, Maasai, Kamba and Parakuyu) live around the margins of the Mkomazi/Umba Game Reserves, making a living in a variety of ways. The primary sources of sustenance and income are livestock husbandry, crop cultivation and wild resources exploitation.

In the chapter by Brockington and Homewood, patterns of utilisation of grazing resources within and around Mkomazi are studied from resource conservation and economic development perspectives. The chapter examines the impact of current conservation policy on livestock keeping and finds pastoralists have experienced costs that pose considerable dilemmas to attempts to reconcile conservation with development, or compensate people for lost resources.

Kiwasila and Homewood focus on the cultivator community, studying broad patterns of land use and wild resource exploitation. Agriculturalists dominate the populations around Mkomazi but their needs have been relatively neglected. This chapter demonstrates that there is a number of important resources other than pasture within Mkomazi that make important contributions to peoples livelihoods.

Together these chapters go some way towards addressing the need for more information about local livelihoods and development needs. The greater challenge is to work out ways of incorporating them into conservation plans acknowledging, perhaps, that humans have a role to play in shaping such dynamic ecosystems.

CHAPTER 33

Pastoralism around Mkomazi: the interaction of conservation and development

Dan Brockington & Katherine M. Homewood

Introduction

Protected areas do not exist in isolation. To understand the dynamics of the Mkomazi ecosystem, it is necessary to understand the way that people use this area and have used it in the past. Understanding current and future pressures on Mkomazi Game Reserve (MGR) means understanding the way needs and expectations of the reserve-adjacent population drive their use of natural resources. To evaluate the costs and benefits of MGR under different alternative styles of management it is necessary to grasp the impacts of current policies which prohibit resource use. In the long term, ecological sustainability depends on political and economic sustainability too. This chapter sets out the history of pastoralist involvement with Mkomazi and the way pastoralist populations have used and managed grazing resources in different areas in and around the reserve up to the time of their eviction. The chapter goes on to look at the impacts that eviction has had on pastoralist economy, and the ways these translate back into ecological pressures on MGR.

History

The history of land use in and around what is now MGR has been reconstructed through a detailed study of official archives, of contemporary accounts by missionaries and other visitors to the area, and through intensive oral history research consulting elders and eyewitnesses resident in the area for decades (Rogers *et al.* 1999, Brockington 1998, and see Table 33.1). The Mkomazi area has been used by Kamba, Maasai, Parakuyo, Pare and Sambaa herders of livestock for as long as records referring to Mkomazi can be traced. Lineages now present around Mkomazi can be found in the earliest records naming pastoralist families. Before gazettement that use appears to have taken the form primarily of wet season transhumance. Livestock herders entered the area that is now MGR to graze their herds during the wet season, when a flush of rainfed forage became available, and in the dry season withdrew to areas with permanent water (such as montane areas and the Umba and Ruvu river lines). Pastoralists resident around the borders and within the Lushoto part of MGR managed their grazing resources with systems of calf pastures (*milimbiko*) and dry season grazing reserves (*vitunga*). The archives and oral histories give evidence that these common property resource management systems entailed widely recognised and effective systems of negotiation (among councils of elders), rules of access, and enforcement of penalties for infringement. District and Game Reserve officials supported these systems. There was considerable co-operation and co-management between resident and reserve-adjacent pastoralists on the one hand, and Game Reserve and District authorities on the other over grazing management. Certainly pastoralists appealed to these authorities when that management was threatened (Table 33.1).

When the reserve was gazetted, limited use was permitted to some Parakuyo pastoralists in the east of the reserve. Legal and illegal use mounted in the 1960s. In dry times in 1969 and in the early 1970s, pastoralists sought and were granted access to the western part of MGR for grazing and watering livestock for the first time since its gazettement. From this time on ever greater numbers of pastoralists were attracted into the western part of MGR.

Historical and archival sources make it clear that some pastoralists resident around the Same part of MGR tried to appeal to the authorities to control this influx (Table 33.1). Government resolve to evict pastoralists intensified shortly afterwards. Eviction operations began in the mid-1980s and culminated in late 1988 when final orders were issued to revoke permission for pastoralists, who had been listed as legal residents, to stay in the reserve.

Livestock numbers in and around Mkomazi Game Reserve

Data on livestock numbers come from a variety of sources, each with its own problems. Putting together a clear picture of change is a complex task (Brockington 1998). Ground counts give potentially high quality data but are difficult and time consuming to carry out; surveys giving reported numbers may be subject to under-reporting, particularly where herd owners suspect subsequent taxation. Knowledgeable guestimates made by observers at the time have no formal basis, are at best based on unspecified extrapolations, and are open to many kinds of bias. Aerial counts may be divided into those carried out on a rigorous systematic sampling basis allowing estimates of variance, and those carried out in a less formal way whose validity is hard to assess. Either way, aerial censuses are not ideal for livestock estimates in this area given that herds are clumped (and variances are so high as to make estimates of little value).

The most reliable counts of livestock in and around Mkomazi up to the time of eviction are derived from ground counts and ground surveys. Aerial surveys performed before that time were carried out in an informal way, the results are not

Table 33.1 History of pastoralist use of Mkomazi Game Reserve (see Rogers *et al.* 1999). * indicates an oral history.

date	source	event / observation
1840s	Krapf (1860)	"Kwafi" (Wakwavi or Parakuyo) wilderness east and north of the Pare and Usambara Mountains
1891	Waller 1988, Smith 1894	Rinderpest clears herders from plains
early 1900s	Brockington 1998, 1999	Parakuyo pastoralists re-established north of Usambara mountains
1920s–1940s	TNA 723 vols. I–III TNA 11/S vols. II–III	Recovery from rinderpest: resurgence of pastoralism north of Pare and Usambara mountains; immigration from Maasai pastoralists
1951		Mkomazi Game Reserve gazetted
1952	TNA File 6/1	Several hundred resident pastoralists with 20,000–30,000 livestock in western half of MGR between Gonja and Same-Lushoto Districts border; 20,000 livestock at Katamboi where Same-Lushoto and Kenya-Tanzania borders meet
1952	TNA 723/111-3/5/54 *MM11/95 KK 10/5/96	List of Parakuyo pastoralists resident in Lushoto (eastern) part of Reserve drawn up and permitted to stay. Resident Parakuyo facilitate legal Maasai residence through intermarriage and adoption under Parakuyo names
1950s–1960s	Mangubuli (1992)	Dindira dam built in western (Same) part of MGR, Ngurunga and Kavateta dams in western (Same) MGR
1965	TNA G1/7-9/9/1965	Official concern on overstocking in Kalimawe
1968	Mussa Mwaimu TNA G1/7	New list of legal residents drawn up after three years of dispute
1969	KWLF DSG/F/40/1/77; wk23/28	Drought: pastoralists petition for access to western (Same) part of MGR with developed water points. Access granted
1976	KWLF 28/7/1977 KWLF 8/6/1977	Resident pastoralists complain of the uncontrollable influx. Areas round MGR borders seen as particularly badly affected as non-residents concentrate on formerly pastoralist-managed reserved grazing to allow daily access to MGR
1980	KWLF 20/2/1980	Meeting to establish *vitunga* grazing reserves, attended by District and Game Reserve authorities lending official support to resident pastoralist attempts to manage grazing resource
1982	*EK 16/4/96 ML/24/6/96	Further meeting attempting to establish *Mlimbiko* reserves, again attended by GR and District officials
1988	URT RG/C/MGR/77/91	Full scale eviction completed

TNA—Tanzania National Archive, KWLF—Kisiwani Ward Livestock File, URT—United Republic of Tanzania comparable with other aerial or ground datasets. The results from official ground censuses and surveys refer to livestock based in reserve-adjacent villages as well as those from bomas within MGR itself. It is clear that reserve-adjacent herds made frequent and regular use of reserve grazing. It is not possible to be precise about exact proportions of herds using MGR at any one time, nor about the exact proportion of their time spent within MGR. The figures therefore represent livestock using both MGR and also reserve-adjacent pastures. This in turn makes any calculation of stocking rates and grazing pressures dubious, so the discussion here focuses on trends in numbers, implying corresponding relative changes in grazing pressures for different areas.

As indicated above there are important contrasts between livestock population trends in the eastern and western parts of MGR in the decades leading up to eviction. Before 1970, there is a sizable and increasing population of cattle in and around the Lushoto part of MGR (see Table 33.2), but numbers in and around the Same part are small and poorly documented.

In 1970, following the granting of access to the western part of MGR to relieve drought hardship, there was a rapid influx from the Ruvu and areas west of Mkomazi. The first formal ground count in 1978 revealed the extent of the immigration. Cattle numbers in the west (estimated on the basis of circumstantial evidence to have been no more than 15,000) rose rapidly from 1970 to nearly 40,000 by 1978 and then remained around this level. This rapid increase in livestock provoked the occasional complaint from residents, referred to earlier.

No official figures exist for livestock numbers after 1984. Major evictions took place in 1988–89. The only data available after 1988 come from systematic aerial surveys (Huish 1993, Inamdar 1994). In 1988, there were wet season figures of 8,991 $\pm$ 5,259 cattle inside MGR itself rising to 17,625 $\pm$ 10,062 in the dry season of 1991. In 1994 wet season figures were 23,557 $\pm$ 12,530. They confirm that considerable numbers of cattle were still using the reserve but they give no information on reserve-adjacent cattle populations comparable to data for earlier periods. With standard errors over half the value of the estimates themselves, no trends can

Table 33.2 Cattle numbers in and around Mkomazi Game Reserve.

year	Lushoto	Same	total
1960	21,984	no data, probably not more than 15,000 *	21,984 + ?
1967	45,245	no data, probably not more than 15,000 *	45,245 + ?
1978	28,218	39,539	67,758
1984	48,233	39,977	88,210

* Harris 1970: intensive ground observations—negligible numbers in Same part of MGR. Parker & Archer (1970) map shows absence of stock from most of Same part of MGR and environs. No official record of or concern over pastoralism in Same part of MGR and environs cf. active issue in and around east MGR.

be inferred. There is little evidence for the general guestimate that 75–80% of cattle have left the area since eviction; however, livestock market data presented below indicate a dramatic fall in cattle numbers around Kisiwani and western MGR.

Impact of livestock population trends on vegetation in and around MGR

Conservation concerns expressed before and after eviction revolve round overgrazing, loss of ground cover, long term degradation of the productive capacity of the reserve, cutting and burning of forest areas and loss of wooded vegetation through fire impacts. Details of environmental trends await the results of current work using remote sensing (outlined in Packer *et al.*, Chapter 4). It was not possible in the course of this study to carry out systematic herd follows and plot the distribution of grazing movements across different areas of pasture around the reserve. This work was curtailed by the political problems of trying to investigate what turned out to be a continuing, if furtive and heavily penalised, high level of incursion and illegal grazing in Mkomazi. For the purpose of the present study we therefore made a very rough attempt to evaluate environmental trends and their relation to pastoralist presence by drawing on the preliminary work of Julie Cox (1994). This pilot study used Landsat images of the western half of MGR taken in 1975 and 1987. Despite the serious limitations of working by hand from a pair of photographs taken at different seasons with minimal opportunity for field checking and no possibility of ground truthing images from the past, the results are of heuristic interest for a number of reasons (Figure 33.1). First, the area and period covered are those corresponding to maximum potential pastoralist impact in the lead up to eviction. Second, though most changes suggested by Cox's analysis are

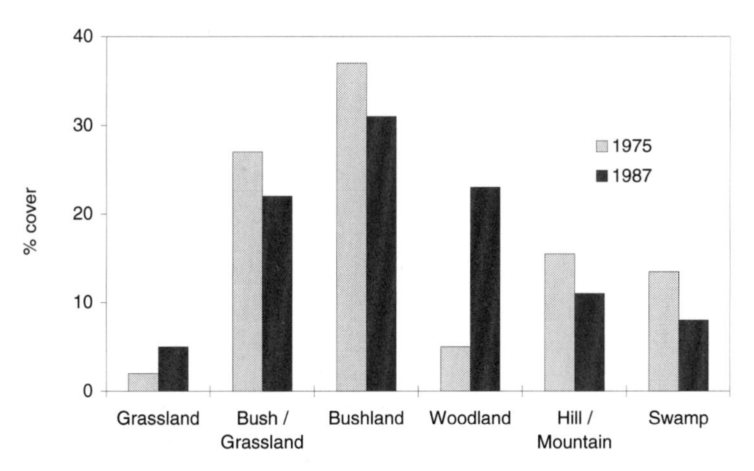

Figure 33.1 Changes in vegetation types in the western half of Mkomazi Game Reserve 1975–87.

minor and fall within the range of observational error, and all are very preliminary interpretations, they suggest a possible large increase in the area covered by heavily wooded vegetation (40–100% cover) at the expense of a range of other vegetation categories (<40% woody cover). Areas of open grassland show little change.

Overgrazing may be defined as grazing beyond some notional carrying capacity, thereby bringing about long term decline in range resources. There is a fierce debate over whether carrying capacity is a meaningful concept in variable arid and semi arid ecosystems which may be primarily driven by factors other than grazing (Sandford 1982, Caughley *et al.* 1987, Behnke & Scoones 1993, Homewood & Rodgers 1987). We are not aware of any evidence to suggest that grazing pressures have resulted in any long term degradation in Mkomazi. Attempts to study the long term effects of past heavy grazing by pastoralist livestock around Kifukua were abandoned by Mweka staff working alongside the Mkomazi Research programme when sites known to have been heavily overgrazed less than ten years previously revealed few differences from control sites (Kidegeshi & Masruli, pers. comm. 1996). There is evidence that vegetation in Mkomazi undergoes long term fluctuations between relatively open and relatively wooded phases. Similar fluctuations are well established in related rangeland areas (e.g. Serengeti-Mara: Dublin *et al.* 1990; Tsavo: Leuthold 1977 & 1996, Wijngaarden 1985). It seems increasingly clear that such changes are driven by the chaotic interplay of many factors, from rainfall through grazing pressures, through fire, to the physical damage inflicted by elephants.

The crude evidence that we currently have on changes in Mkomazi suggest that in the 1940s there was a period of relatively dense bush cover. This was followed by a period when burning carried out by Pare pastoralists in the 1950s, and perhaps the impact of elephant populations increasing within the protected area from 1951 on, may have combined to reduce woody cover and open up more grassland. Pastoralist burning would be expected to maintain and extend such grassland, though conversely heavy grazing might be expected to encourage bush encroachment in open grassland. Pastoralists aside, the crash in the elephant population during the 1970s would be expected to allow a new generation of woody vegetation to mature relatively undisturbed. If the changes Cox's preliminary analyses suggest for wooded areas 1975–87 are borne out, we would put forward as a hypothesis for investigation in the more detailed remote sensing analyses that these changes are consistent with areas of sparse, low bush or shrub grassland maturing to more thickly wooded areas (Figure 33.1). This is in turn would be more closely consistent with the changes expected to follow the 1970s crash in elephant populations, than with any impact of pastoralist burning and grazing.

Implications of eviction for people and livestock

This section summarises the findings of a two year field study of reserve-adjacent pastoralist communities. It looks at emigration after eviction, at the livestock economy pre and post-eviction, at herd dynamics, and at household economy. Besides official records (population census, livestock market records) the present study used multi-round survey (five repeats at two-month intervals) of a systematic sample of 32 reserve-adjacent households in Same and 20 in Lushoto District. The location of the main centres of pastoral settlement are shown in Figure 33.2. The study sites were at Kisiwani in Same District and Kisima, Mahambalawe and Mng'aro in Lushoto District.

Population trends: growth, eviction and emigration

In order to measure the diaspora of pastoralists in the wake of the eviction, people from the sample Parakuyo and Maasai households were asked where their siblings live and whether those siblings had been evicted from MGR. These data reveal the pattern of emigration following eviction, the area that has received evicted pastoralists, and also the distribution of the broader group of non- evicted pastoralists to whom reserve-adjacent pastoralists are related (Brockington 1998). There are two main caveats. Half siblings are likely to be under-reported. There is also a bias against sibling groups that fail to keep any members in the area at all. The poorest and smallest sibling groups, least able to draw on ties of kinship and reciprocity, drop out of sight altogether. The main points of interest to emerge are firstly that the great majority (84%) of the 416 siblings reported as currently living around the reserve had been evicted from MGR. Group and individual discussions with Pare, Kamba and Sambaa pastoralists not included in the detailed sibling survey suggest a similar result. Virtually all currently reserve-adjacent pastoralists have been excluded from reserve resource use since eviction. A significant minority of MGR's neighbours used to depend upon the reserve for their livelihoods.

Herd performance

Measures of cattle fertility, mortality and offtake provide useful indices of herd performance and reveal periods of stress affecting the animals. Fertility and mortality rates are calculated using a register of animals present in the herd, and relating numbers of births, deaths and transfers to the number of 'cow-years at risk' for that event (Brockington 1998). This can be done both for animals present during the period of the study itself and historically. Historical data suffer from the same bias against clusters of related individuals dropping out of sight that affects sibling studies. As with people, respondents are less likely to remember dead cattle that have left no live offspring. Mortality rates calculated from these data are *minimum estimate* mortality rates. By contrast the fertility rates are biased towards more fertile animals (barren cows leaving no offspring are less likely to appear in the register). Fertility rates are *maximum estimates*. These biases will act to obscure

Figure 33.2 Area surrounding Mkomazi Game Reserve. Circles indicate official villages and other settlements.

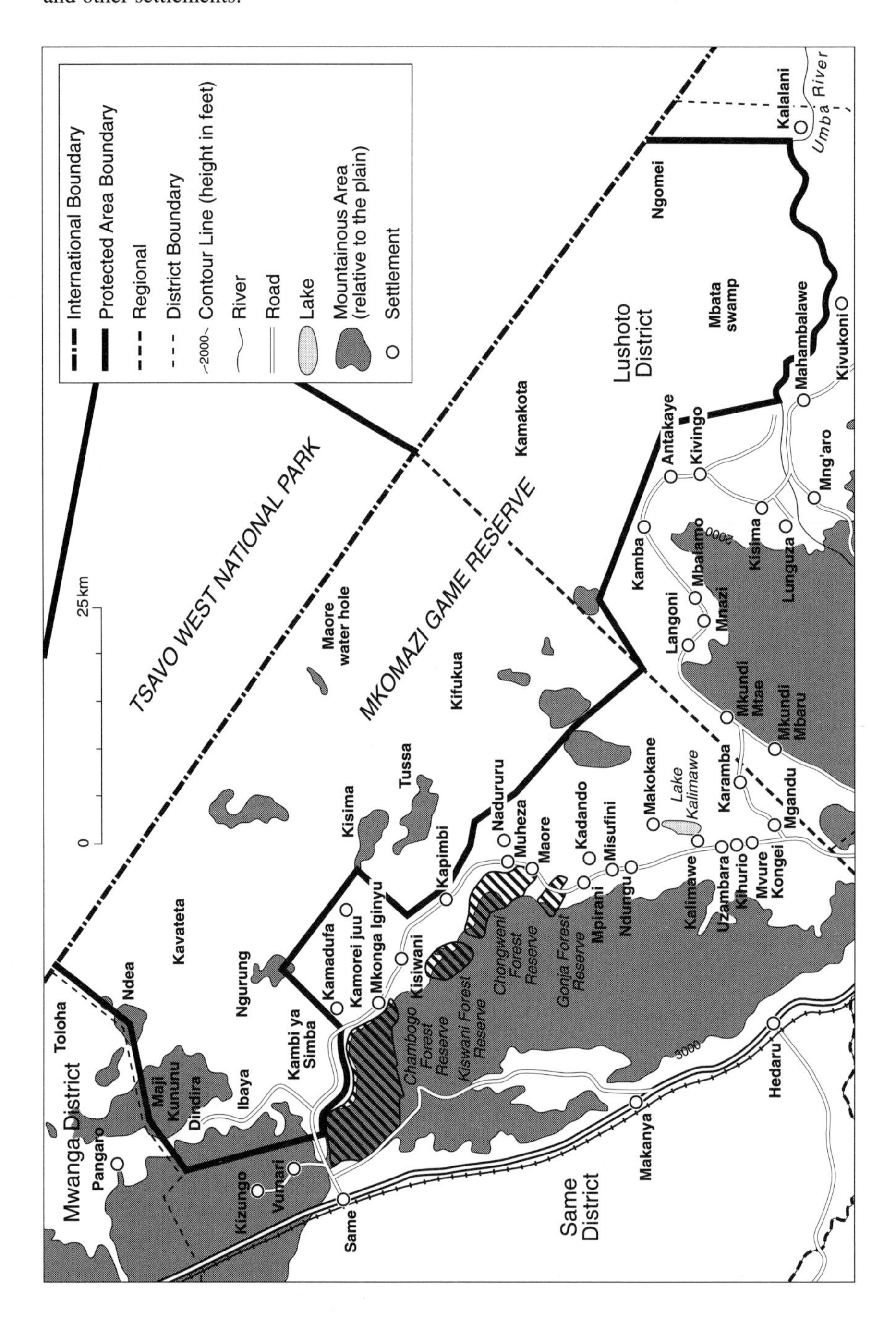

rather than create worsening trends consequent upon eviction. As the same problem affects data for other pastoralist populations, comparisons are legitimate.

Figure 33.3 plots fertility and mortality rates calculated on this basis for cattle in Same and Lushoto districts. The lowest recorded fertilities are associated with low rainfall years in Same (1991) and Lushoto (1992). Mkomazi fertility rates for non-drought years, particularly for Lushoto District, are overall somewhat lower than those recorded for other east African pastoralist herds in non-drought conditions (Figure 33.3). They are also particularly low for Same in 1988 and in Lushoto in 1989, possibly linked to the eviction. Mortality rates for adult cows are closely comparable to rates found elsewhere for non drought years. Unlike cow mortalities, non drought mortalities for calves of less than two years of age in Same are on average double those found in Lushoto and in other east African pastoralist areas (Figure 33.3). Cattle fertility in Mkomazi is rather lower and calf mortalities higher than expected for non-drought conditions in east African pastoralist systems. Interviews suggest eviction concentrated cattle into areas close to mountains where tick borne diseases are particularly prevalent, affecting fertility and calf mortality adversely through shortage of grazing and through transmission of infectious disease and parasites.

Livestock economy

In order to understand the ecological pressures facing MGR it is necessary to have a grasp of the economic and political stakes involved and the ways in which those translate into ecological pressures. Before it is possible to evaluate the long term sustainability of different alternative management policies in MGR it is useful to consider the scale and importance of the livestock economy, the extent to which it

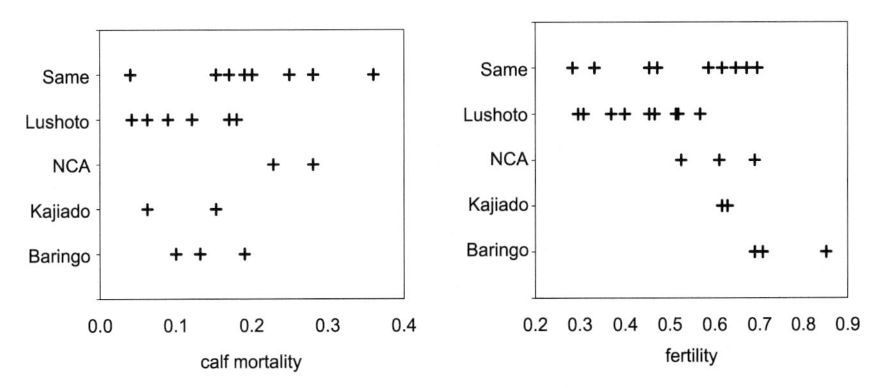

Figure 33.3 Calving rates and calf mortality rates for Mkomazi (Same and Lushoto District, 1988–94) compared to rates for other east African pastoralist groups in similar eco-climatic areas. NCA = Ngorongoro Conservation Area. Sources: NCA: Homewood, Rodgers & Arhem (1987); Kajiado: Bekure *et al.* (1991), Homewood (1992); Baringo: Homewood & Lewis (1987).

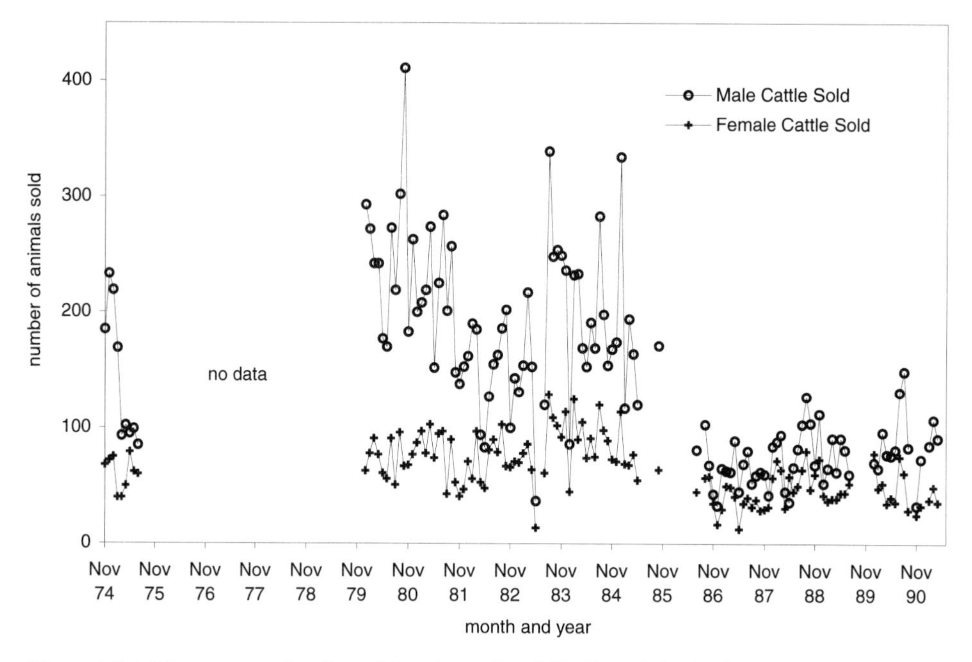

Figure 33.4 The number of male and female cattle sold in Same District livestock markets.

underpins local livelihoods, and the impact of eviction on that economy and those livelihoods.

Livestock sales provide an indication of patterns of livestock production. The livestock economy divides into official and unofficial sectors. The pattern of official sales can be traced through District records. These are an underestimate of the real volume of trade, because taxes on official transactions mean not all sales will be reported, and reported prices tend to be lower than actual price paid. In addition, any livestock-producing area near an international border is likely to encourage unofficial cross border trade. Animals are easily walked across the border, and price differences between countries can readily be exploited. In Ngorongoro, officials estimated that 70% of livestock transactions took place unofficially (Homewood & Rodgers 1991). As far as historical patterns are concerned official sources provide an indication of trends in Same District from 1974 onwards. The four markets in Same District (Same, Hedaru, Kisiwani and Makanya) occur four times a month. Separate records are available for each market except for the period 1986–June 1991 which gives only pooled data (Brockington 1998).

Figure 33.4 shows the numbers of cattle sold in Same markets 1974–95. Separate analyses show no long term rainfall trends, and, although there is a seasonal pattern and a clear link between dry years and sales, the main changes seen are not driven by drought. Briefly, sales are high 1976–86. There are gaps in the data 1986–91, covering the main period of conflict and eviction, but the data available

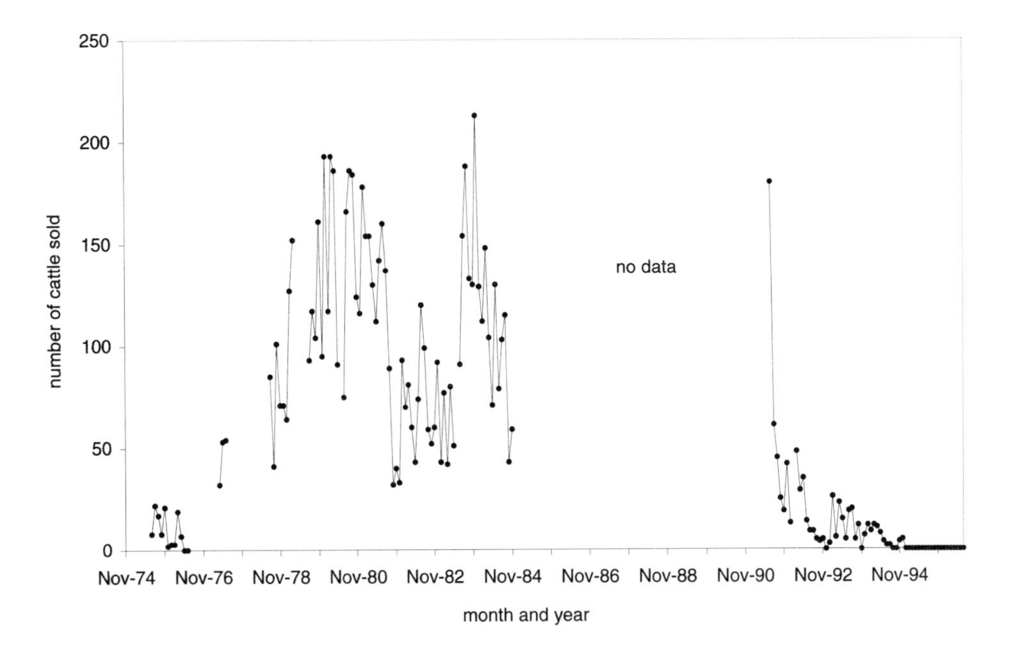

Figure 33.5 The number of cattle sold in Kisiwani Market 1974–94.

show sales fall to fluctuate around a much lower level from 1987 on, with a (still relatively low) peak in the drought years of 1990–92. Overall sales fell to half or less their earlier levels. A second important indication of changes in the livestock economy is given by the ratio of male to female cattle sales. Here data are only available for the period 1980–91. East African pastoralists needing to raise money tend to sell adult male steers and make every effort to retain female animals which are the productive nucleus of the herd (e.g. Dahl & Hjort 1976, Homewood & Rodgers 1991). In these groups it is a sign of real distress, or of a radical shift away from pastoralism, to sell young female animals. Same District data show a dramatic shift in the ratios of animals sold, from around 3 males to 1 female before 1985 to 2:1 or near unity after 1986 (Figure 33.4). Finally, and most telling, is the fate of Kisiwani livestock market (Figure 33.5). This opened in 1975 and rose to dominate official livestock sales for the District as cattle came into the western part of the reserve. Sales in early 1991 are extremely high, equal or greater to those of the entire District over the last 5 years. The peak is caused by the joint impact of drought and renewed clearance of pastoralists who had sought drought refuge grazing in MGR. These distress sales are the market's swansong. Thereafter it declines and eventually ceases to operate in the 1990s. The collapse of this livestock market, closest to MGR, reflects the cessation of any supply of cattle from the reserve.

It is difficult to translate these figures into meaningful monetary estimates of lost trade, because of seasonal and inter-annual variability as well as long term trends in prices. Inflation caused prices to rise four- or five-fold between the 1980s and 1990s, but terms of trade between cattle and sheep, and cattle and maize, remain steady despite those changes (Brockington 1998). A conservative average for official cattle sales in Same District 1994–95 was around 30 million shillings per quarter and the conservative estimate that official trade fell to a half of its pre-eviction levels would thus translate into a loss of 120 million shillings per year (c. £150,000 per year at the time). Bearing in mind that official figures are only the tip of the iceberg, with bush sales and cross border trade as or more important to the livelihoods of rural producers, this represents a serious economic collapse. Evidence from parallel work on reserve-adjacent farming communities shows that the wider economies of Same and Kisiwani suffered as traders, middlemen, butchers, transport contractors, stores, hotels and eating houses closed with the loss of trade (Kiwasila 1996). Any economic evaluation of conservation-related revenue generating and/or benefit-sharing initiatives proposed for MGR would be seriously incomplete without an understanding of the potential (and recent past performance) of the livestock economy in this area.

Household economy

During this study multi-round surveys gathered data on household production and consumption for an established sample of 52 pastoralist households. From this study it is possible to identify changes that have happened to households since eviction.

The importance of cultivation

Official records, pastoralist sources and information from reserve-adjacent farming communities all concur that prior to eviction Parakuyo and Maasai pastoralist families in the Mkomazi area did not engage in farming. It is therefore interesting to find that all the pastoralist households in the sample now engage in cultivation. Patterns of cultivation however differ among households and there are clear contrasts between Same and Lushoto District pastoralists. Briefly, the Lushoto District sample divide into families based in Mahambalawe and those based at Kisima. Mahambalawe households tend to have major livestock holdings (often based elsewhere) and also invested in land, sometimes in equipment such as irrigation pumps and hired labour to farm for them. By contrast the Parakuyo and Maasai based at Kisima had few livestock, and farmed using their own labour. Virtually no families reported yields that could begin to meet their requirements (Brockington 1998). Around Kisiwani in Same District there was little land available for pastoralists to farm, water for irrigation is controlled by established Pare communities, and yields to pastoralist cultivation were low.

The changing significance of women's income

As families' ability to rely upon livestock declined after eviction so they sought other means to sustain themselves. It was possible to establish how reliant families were on livestock by comparing income from, and expenditure on veterinary medicine for livestock as reported by the (generally male) household heads. Collecting such information is hard as people are particularly reluctant to disclose information on the income they earn. Although it is unlikely that the data on livestock sales and purchases give accurate absolute values, they show consistent rank orders among households on relative income and expenditure. Relatively high incomes from livestock sales correlate with relatively higher outgoings on medicine for livestock (Same: $r_s=0.74$, $n=36$, $p<0.001$; Lushoto $r_s=0.85$, $n=20$, $p<0.001$). If milk yields are also taken into account it is possible to rank families reliably according to their dependence on livestock (Brockington 1998), and to identify poorer families. Milk availability averages 0.42 and 0.16 kg/reference adult/day in Lushoto the wet and dry season respectively, and 0.41 and 0.26 kg/reference adult/day in Same. These correspond to very low contributions to recommended calorie intake levels compared to other pastoralist populations (13.6–5.2% in Lushoto; 12%–7.7% in Same; dry season values in NCA average 34%; mean annual values in Kajiado average 48%; reference adult recommended calorie intake 2,530 kcal/day: Homewood 1992). These mean values mask the fact that the poorer households had little or no milk intake at all. These were also unable to earn enough money from selling livestock or to grow enough to feed themselves, and had to find alternative sources of income. Money earned by women selling milk or traditional medicine was of vital importance in meeting day to day food needs. Women's income is always important for household diet and school expenses. Many Maasai women sell opportunistically or occasionally. However for women in these families opportunistic selling had become a daily necessity. The poorer the family, the more likely it became for food to be bought on a hand to mouth, daily basis in exchange for whatever products women could sell. Most of the women thus relied upon goods sold locally, but some traveled regularly and frequently to towns further afield in Kenya and Tanzania. In the poorest families with few livestock, wives and mothers lived an arduous life gathering wild plant medicines and spending long periods traveling all over northern Tanzania selling their products. During the study one died and several were injured in a road accident in the course of this activity. Wealthier families were more likely to have men periodically sell livestock and buy grain in bulk with the proceeds.

The patterns of men and women's income earning activities in Mkomazi, and of settlement near to or far from markets, is very much as expected in the context of sedentarising and declining pastoral economies (Sikana, Kerven & Behnke 1993). Wealthier families may sell milk, but only in the wet season when the surplus milk their herd produces is easily foregone. Wealthier families may also reserve more milk for their calves, deriving better growth rates and survival and profiting ultimately from the sale of male stock and the economies of scale in bulk buying (Sikana, Behnke & Kerven 1993). Poorer families move to locations where they can sell products or labour, while wealthier ones live in remote areas better for livestock production, or move the family to urban areas for trade and education while retaining the herd in remote areas conducive to livestock production. Poorer families may retain livestock but will sell the milk year round to benefit from terms of trade that typically allow them to get more calories in the form of grain in exchange. Whilst these patterns have been observed elsewhere, what distinguishes pastoralists at Mkomazi is the origin of this decline and impoverishment in their eviction and subsequent exclusion from the reserve. Those families most dependent on women's income were those who had lost large numbers of stock and started farming following eviction.

Conservation and development in Mkomazi: looking to the future

This chapter has outlined the way that pastoralist communities have used Mkomazi in the past, sketched the ecological outcomes both of use and of eviction, and documented the human costs of loss of access. The changes recorded give an insight into the depth and force of pastoralist opposition to current policies of eviction and exclusion. At the time of writing the government faces court action from a large number of pastoralists who allege that their constitutional rights have been denied them. Mustafa (1995) and Rogers *et al.* (1999) have described in detail how this struggle evolved.

The historical, social, legal, demographic, archival and economic data considered above suggest that there are pressing human needs around MGR. What might Mkomazi be like ten years from now? From a conservation viewpoint, the optimistic scenario might be that the international scientific presence will be maintained, that international conservation funds will continue to flow, that a (now reduced) Game Reserve staff will function effectively to limit illegal use of Reserve resources, and that revenue from tourism and hunting, though inevitably low, will be put to such good use that the reserve-adjacent communities will give the reserve their wholehearted support.

A more pessimistic view would see the scientific presence in MGR coming to an end, the decline of international conservation funds and a reduced game reserve staff unable to exercise effective control. With rising pressure to trespass on Reserve resources, increasingly punitive measures for those cases that are pursued would generate corruption and antagonism, while failing to stem resource degradation, possibly even eliciting vengeance targeting of valuable species as has happened elsewhere (Lindsay 1987, Western & Wright 1994).

Somewhere between these extremes there may be a workable compromise. If it is to be achieved it will be through a planning process that institutionalises participation by all stakeholders, and uses a forum for negotiation, a process of consultation, and a level of transparency that allows the emergence of solutions that all can own. If this is not done, management policies will not work in practice. Even if the losers do not have the power to get their own way, they have the power to undermine, block or destroy the aims of the 'winning' stakeholders (e.g. Lindsay 1987, Western & Wright 1994).

Given consultation, natural scientists could identify zones of importance—for example for biodiversity, rare species and migration corridors. Social scientists could help identify those resources that are of primary importance to reserve-adjacent communities, as well as the possibilities, and problems, of common property resource systems on which to base their management. Reserve-adjacent communities could benefit, not from negligible tourist or hunting revenue, but from the licensed and co-managed access to resources important to their livelihoods. MGR could be a model of the new Wildlife Management Areas envisaged in the recently published *Wildlife Policy of Tanzania* (MNRT 1998).

Unpredictable developments may drive the current impasse in unexpected and extreme directions. These include for example the controversial introduction of black rhino to a high security enclosure in MGR; the suggested shift to National Park status for MGR; or the remote possibility of a legal victory by pastoralists in the current court case challenging their eviction. All these present conceivable alternative futures for MGR. Whatever that future, it is unlikely to be workable in the long run unless the principles of participatory consultation, transparency, and joint ownership of the outcome are maintained.

Acknowledgements and disclaimer

This work was funded by the Overseas Development Administration. The authors also wish to acknowledge the support of the Royal Anthropological Institute's Emslie Horniman Fund, the Parkes Foundation, the Kathleen and Margery Elliot Trust, the Central Research Fund of London University and the Graduate School of University College London. The authors are grateful for the help and cooperation of the Institute of Resource Assessment, the University of Dar es Salaam, the Commission for Science and Technology of Tanzania and the Department of Wildlife. We are also indebted to the many members of local and District government and the countless other Tanzanians who live and work near to Mkomazi Game Reserve for their help, cooperation, welcome and warm hospitality.

The Department for International Development (DFID) is the British Government Department which supports programmes and projects to promote overseas development. It provides funding for economic and social research to inform development policy and practice. DFID funds supported this study and the preparation of the summary of findings. DFID distributes the report to bring the research to the attention of policy-makers and practitioners. However, the views and opinions expressed in the document do not reflect DFID's official policies or practices, but are those of the authors alone.

References

Bekure, S., de Leeuw, P.N., Grandin, B.E. & Neate, P.J.H. (1991) *Maasai Herding. An analysis of the livestock production system of Maasai pastoralists in eastern Kajiado District, Kenya.* International Livestock Centre for Africa, Addis Ababa.

Brockington, D. (1998) *Land loss and livelihoods: effects of eviction on pastoralists moved from Mkomazi Game Reserve, Tanzania.* PhD Thesis, University of London.

Brockington, D. (1999) *We just left it. An outline of the history of pastoralism in the borderland plains of northeast Tanzania 1800–1953.* Unpublished manuscript.

Caughley, G., Shepherd, N. & Short, J. (eds.) (1987) *Kangaroos: their ecology and management in the sheep rangelands of Australia.* Cambridge University Press, Cambridge.

Cox, J. (1994) *Remote sensing habitat survey: pilot study August–September 1993.* Mkomazi Ecological Research Programme report, Royal Geographical Society (with IBG), London.

Dublin, H., Sinclair, A. & McGlade, J. (1990) Elephants and fire as causes of multiple stable states in the Serengeti-Mara woodlands. *Journal of Animal Ecology* 59: 1147-64.

Homewood, K. (1992) Development and the ecology of Maasai food and nutrition. *Ecology of Food and Nutrition* 29: 61-80.

Homewood, K. (1994) Pastoralists, environment and development in East African rangelands In: Zaba, B. & Clarke, J. (eds.) *Environment and Population Change.* Ordina Editions. pp. 311-323.

Homewood, K., Kiwasila, H. & Brockington, D. (1997) *Conservation with Development? The case of Mkomazi, Tanzania.* Final research report to the Department for International Development, London.

Homewood, K. & Lewis, J. (1987) Impact of drought on pastoral livestock in Baringo Kenya 1983-5. *Journal of Applied Ecology* 24: 615-631.

Homewood, K. & Rodgers, W.A. (1987) Pastoralism, conservation and the overgrazing controversy. In: Anderson, D.M. & Grove, R. (eds.) *Conservation in Africa: People, policies and practice.* Cambridge University Press, Cambridge.

Homewood, K. & Rodgers, W.A. (1991) *Maasailand Ecology: Pastoralist Development and Wildlife Conservation in Ngorongoro, Tanzania.* Cambridge University Press, Cambridge.

Homewood, K., Rodgers, W.A. & Arhem, K. (1987) Ecology of Pastoralism in Ngorongoro Conservation Area, Tanzania. *Journal of Agricultural Science (Cambridge)* 108: 47-72.

Huish, S.A., Ole Kuwai, J. & Campbell, K.L.I. (1993) *Wildlife Census of Mkomazi 1991*. Tanzania Wildlife Conservation Monitoring. Unpublished report.

Inamdar, A. (1994) *Wildlife Census, Mkomazi April 1994*. Interim report on results of an aerial census of the Mkomazi Game Reserve. Unpublished Report.

Krapf, P. (1860) *Travels, Researches and Missionary Labours in East Africa*. London.

Leuthold W. (1977) Changes in tree populations in Tsavo East National Park, Kenya. *East African Wildlife Journal* 15: 61-69.

Leuthold W. (1996) Recovery of woody vegetation in Tsavo National Park, Kenya 1970–1994. *African Journal of Ecology* 34: 101-112.

Lindsay, W.K. (1987) Integrating Parks and pastoralists: some lessons from Amboseli. In: Anderson, D. & Grove, R. (eds.) *Conservation in Africa: People, Policies and Practice*. Cambridge University Press, Cambridge.

Mangubuli, M.J.J. (1992) Mkomazi Game Reserve—a Recovered Pearl. *Kakakuona* 4: 11-13.

MNRT (1998) *The Wildlife Policy of Tanzania*. Ministry of Natural Resources and Tourism, Dar es Salaam.

Mustaffa, K. (1995) *Evictions from Mkomazi Game Reserve*. IIED, London.

Rodgers, W.A. & Homewood, K. (1986) Cattle Dynamics in a Pastoralist Community in Ngorongoro, Tanzania during the 1982-3 drought. *Agricultural Systems* 22: 33-51.

Rogers, P.J., Brockington, D., Kiwasila, H. & Homewood, K. (1999) Environmental awareness and conflict genesis: people versus parks in Mkomazi Game Reserve. In: Granfelt, T. (ed.) *Managing the Globalized Environment—Local Strategies to Secure Livlihoods*. IT Publications, London.

Sandford, S. (1982) Pastoral strategies and desertification: opportunism and conservatism in drylands. In: Spooner, B. & Mann, D. (eds.) *Desertification and development*. Academic Press, London.

Sikana, P., Kerven, C. & Behnke, R. (1993) From subsistence to specialised commodity production: commercialisation nd pastoral dairying in Africa. *Pastoral Development Network* 34d: 1-46.

Smith, C.S. (1894) The Anglo-German boundary in East Equatorial Africa: Proceedings of the British Commission, 1892. *Geographical Journal* 4: 424-435.

Waller, R. (1988) Emutai: crisis and response in Maasailand. In: Spear, T. & Waller, R. (eds.) *Being Maasai—Ethnicity and Identity in East Africa*. James Currey, London.

Western, D. & Wright, R. M., (1994) *Natural Connections: Perspectives in Community-based Conservation*. Island Press, Washington, DC.

Wijngaarden W. van. (1985) *Elephants—trees—grass—grazers*. ITC Publication No. 4, Department of Natural Resources, Surveys and Rural Development of the Institute for Aerospace Survey and Earth Sciences (ITC) Enschede.

CHAPTER 34

Natural resource use by reserve-adjacent farming communities

Hildegarda L. Kiwasila & Katherine M. Homewood

Mkomazi's long term prospects for biodiversity conservation can only be meaningfully considered within a regional and community context. Together with a companion paper (Brockington & Homewood, Chapter 33), the present chapter summarises the results of a three year study on community use of natural resources in and around Mkomazi Game Reserve. The present paper deals only with more settled farming communities. It begins with a summary of census information available for the reserve-adjacent population and looks at the range of interest groups that population comprises. The paper then summarises broad patterns of natural resource management and use, and problems and priorities relating to natural resource use, identified in the course of Participatory Rural Appraisal (PRA) survey of reserve-adjacent villages. It goes on to look at detailed patterns of natural resource use as quantified by questionnaire and by multi-round household and market surveys over an 18 month period. Use of wild resources for food and fuelwood is quantified and analysed according to species, frequency of use, volumes and values sold, and importance to different socio-economic and age/sex classes of direct use and of sales.

Reserve-adjacent populations and interest groups

The number and distribution of people

Table 34.1 summarises the results of the 1978 and 1988 censuses showing changes in population recorded in reserve-adjacent villages and in the wider Mkomazi area. It shows the population and changes in population recorded in some 30 reserve-adjacent villages. The exact number of official villages is unclear because official records do not keep up with the growth, fission, fusion and decline of past settlements. However it is possible, for the purpose of monitoring general change, to aggregate results into meaningful statements of population change around the reserve. The total population of the wider area, including towns in the Ruvu valley area west of the Pare Mountains, grew from around 71,000 to around 89,000 between 1978 and 1988, an increase of 25% over ten years. The more immediately reserve-adjacent population grew at 13% between 1978 and 1988 (villages are shown in Figure 33.2, previous chapter). Population can be expected to have grown further since. The 1988 count of something over 48,000 is therefore a minimum estimate of the immediately reserve-adjacent population. The proximity of these people to the reserve boundary means that there is no buffer zone.

From the point of view of natural resource use and conservation-related issues, it is helpful to see the reserve-adjacent population as made up of a number of different interest groups. These range from groups representing large numbers of local residents whose primary concern is their subsistence livelihood, to groups representing locally small numbers of people but spearheading powerful and important national and international conservation concerns. These interest groups comprise cultivators, pastoralists, business entrepreneurs, district and local authorities, staff of the Ministry of Tourism Natural Resources and Environment,

Table 34.1 Population of villages around Mkomazi Game Reserve 1978–88. Areas immediately adjacent to the reserve are starred (*) and their populations are summed to give the 'MGR-adjacent total'. The basic spatial unit for collection of population data sometimes changed between censuses due to administrative changes. This table pools ward and village data into areas which remained more or less constant between censuses, despite administrative (and enumeration) area changes. Districts: S–Same, L–Lushoto, M–Mwanga. Data collected and collated by D. Brockington.

area	1978 district	1988 district	1978 popn.	1988 popn.	inter-censal change
Kwakoa	S	M	2,070	2,668	28%
Toloha*	S	M	830	1,573	89%
Kigonigoni	S	M	980	1,632	66%
Same Town	S	S	5,092	10,350	103%
Ruvu	S	S	3,734	4,122	10%
Vumari Kizungo*	S	S	1,975	3,949	100%
Kisiwani*	S	S	9,568	6,314	-34%
Maore*	S	S	10,480	9,703	-7%
Ndungu	S	S	10,368	10,430	1%
Bendera	S	S	6,238	11,421	83%
Mngaro*	L	L	1,724	4,053	135%
Mbaramo*	L	L	5,325	7,027	32%
Lunguza*	L	L	5,643	6,891	22%
Mnazi*	L	L	7,195	8,861	23%
Mkomazi: wider area total			71,222	88,994	25%
MGR-adjacent total*			42,740	48,371	13%

politicians, NGOs and foreign researchers. Each of these interest groups tends to hold a different perspective on natural resource use issues.

Different groups involved in Mkomazi

Cultivators

Cultivators, who make up the largest group numerically, are generally mixed farmers with livestock as well as irrigated and/or rainfed crops. The group includes many different tribes, Pare and Sambaa being the largest ethnic groups. The main concerns of this group are access to cultivable land (of which the reserve has little or none), and water for irrigation. Depending on the location, some farmers have in the past, or are currently, engaged in challenging reserve boundaries[1] over grazing and/or livestock watering facilities and/or farms which have been enclosed inside MGR boundaries. Other main uses this group report of reserve resources revolve around access for placing beehives, gathering fuelwood, wild foods and medicines; for mining gemstones, and for ritual use of long established sacred groves and other ceremonial sites located within MGR (Kiwasila 1996a, b).

In Same District, the majority of cultivators are Pare, the dominant ethnic group, in Lushoto the Sambaa predominate. Interviews with Pare and Sambaa born inside the area which later on became MGR made it possible to locate the sites where they once lived as well as former Parakuyo and Maasai settlements. Pare and Sambaa cultivated in MGR and grazed their livestock there, traded in *Zanthoxylum* (*Fagara*) *holtziana* leaves and used wild meat commonly obtained from Wandorobo hunters (Kiwasila 1996a, b, Werkgroup 1990). Agro-pastoral Pare report increased crop loss to wildlife since eviction (particularly buffaloes, elephants and birds). Villagers feel they subsidize MGR more than they benefit from it.

Pastoralists

Pastoralists include some Pare and Sambaa herdowners as well as large numbers of Maasai and Parakuyo. These are described in detail in Chapter 33. Put briefly, they have a strong interest in the grazing and water resources inside Mkomazi; those who were born and lived inside the reserve have emotional ties and customary tenure claims as well.

Entrepreneurs

Local entrepreneurs see an array of opportunities in and around Mkomazi, but these are contingent upon the reserve providing money earning opportunities and

[1] e.g. The Vumari court case over Igire area, described in the *Report of the Presidential Commission Inquiry into Land Matters* (URT 1993). Of the 11 villages where PRA was carried out in the present study, the majority reported concerns over the reserve's presence in terms of boundaries, grazing and/or other land shortage (Table 34.2).

being used in some locally profitable way. For Mkonga Iginyu and Kisiwani, entrepreneurs reported during PRA and follow up meetings that the most profitable income arose from the livestock economy. At its peak, this supported secondary services that collapsed in the wake of eviction (restaurants, beer bars, lodging, women tea room operators and food vendors, middlemen for grain and livestock, lorry transporters, etc.). Income for these is felt to be minimal following eviction and the return to low-level conservation-related enterprises such as wildlife tourism or hunting (briefly allowed, and highly profitable, but now considered incompatible with Game Reserve priorities). There are perennial illegal interests (poaching, mining of gemstones, tax-avoiding smuggling routes, timber extraction) which entrepreneurs still exploit.

Local government

District authorities complain that in the past they found themselves by-passed by other agencies on issues pertaining to Mkomazi (pers. comm. Mrs Tumbo, District Commissioner, 1994). District officials are responsible for resolving any crises that arise over MGR, including knock-on effects in areas that for example receive evicted people (Letters Ref. Ku/I/LO2 dated 4/3/91 from Kisiwani Ward Secretary to MGR Manager and Ref. GD/MKGR/J.1/81 dated 21/2/91 from MGR Manager to Kisiwani Ward Secretary). They have thus had to deal with problems instigated by other agencies, whether local, national or foreign, but have no budget of their own to facilitate community outreach activities and to foster good relations between conservation and community development initiatives. District authorities need to be involved in planning, decision making, facilitating, protecting and monitoring activities in the area.

National Government

The Ministry of Tourism, Natural Resources and Environment represent an extremely powerful and important group with full jurisdiction over MGR. Their main concern is to protect MGR from encroachment or illegal use (according to the Wildlife Conservation Act of 1974). The Ministry sees any concession as opening the way to potentially uncontrollable pressures (Kiwasila 1994). In this view de-gazetting would be no solution to shortages of land or grazing. The Ministry is interested in developing community-based conservation approaches outside the reserve, but has no funds to do this (I. Swai, former Reserve Manager, pers. comm.). The Ministry policy of dedicating 25% of protected area revenue to reserve-adjacent communities is hampered by the lack of revenue (Kauzeni & Kiwasila 1994). This is particularly the case in Mkomazi, and activities that could foster realistic and sustainable conservation with development are consequently constrained.

Politicians and NGOs

These are divided into those that support conservation and those which seek joint land use in MGR, and some that may shift between these positions for different audiences as a matter of political expediency, financial interest or employment (Kiwasila 1994).

International Organisations

There are several international organisations with interests in Mkomazi Game Reserve, their mandates to act deriving from invitations by the other groups to work in the area.

The George Adamson Wildlife Trusts have embarked on a large scale rehabilitation programme for the reserve involving road building, boundary clearance, and the introduction of endangered species into the area to restore the ecosystem to its full potential. Sometime after it began operations, the Trust supported an Outreach Programme. Although there has been no official assessment of this that we are aware of, it appears to have run into difficulties. These have been resolved by the government asking the outreach volunteers to leave Tanzania in early 1997.

The World Wide Fund for Nature is committed to spending nearly $95,000 on extension activities from the 'Beeswax fund'. This was set up by the Belgian government from the proceeds of smuggled ivory seized in Belgium and labelled as Beeswax. The money is earmarked for elephant conservation in Tanzania and, since 1989 Mkomazi was detailed to benefit in some way. How the money will be spent, and in cooperation with which partners, had not been decided in the last financial year (1996–97).

PRA: general patterns, problems and priorities in natural resource use

To assess the general attitudes and needs of some of these different groups a survey was made, using PRA methods (see Appendix 1 at the end of this Chapter) in 11 reserve-adjacent villages between May–October 1995. The survey identified key issues relating to natural resource use in general, and to relations with Mkomazi Game Reserve in particular. These visits involved liaising with local government authorities and interviewing them on their own, followed by group meetings organised with a wide cross section of villagers and individual meetings with key informant interviews. The results of the PRA survey are set out in greater detail in a village profile report (Kiwasila 1996b) and summarised in Table 34.2 and below.

The different villages had very different backgrounds, from long-established Pare settlements (Kisiwani and Mkonga Iginyu, Gonja Maore, Vumari) to villages arising from resettlement of pastoralists evicted from MGR (Mheza), and villages formed on areas originally expropriated by colonial settlers, where plantation land has been redistributed to smallholders (Mpirani). Recent histories also differ greatly, with some villages having experienced greater or lesser impacts of donor schemes such as the Traditional Irrigation Project (Maore) and the Japanese Rice scheme (Ndungu) as well as the presence of MGR, associated evictions, and the Kisiwani-based Mkomazi Outreach Programme. Finally, transport, communications and government provision of health and education all vary greatly between villages. Most villages have a range of ethnic groups living cooperatively side by side, but in some villages the pressure on land and water resources is sparking ethnic conflicts, exacerbated by conservation restrictions and by the nature of local development interventions (Mkundi Mbaru, Kisiwani, Mkonga Iginyu).

All villages saw their primary economic occupation as farming, usually mixed farming with livestock as well as irrigated and rainfed crops. All villages also have small businesses (shops, transport contractors, maize mill owners, petty traders). In addition to this some villages had people depending on employment with sisal plantation, rice project or other agricultural and/or development schemes. In a number of villages some people focused on beehives and honey production (this could be lucrative, with a single owner producing 600–1,000 litres of honey per

Table 34.2 The village PRA profiles. Other PRA group discussion meetings were held in Nadururu and Kiwanja sub-villages. There were also three non-PRA investigative meetings.

village	ward, district	problems
Vumari	Njoro, Same MGR 0–3 km	Land dispute; lack of grazing/farmland; wildlife damage; no market, veterinary or water inputs
Mkonga Iginyu	Kisiwani, Same MGR 0–10 km	Grazing shortage; cattle theft; no veterinary service/health provision; youth unemployment; Slump followed livestock market collapse.
Kisiwani	Kisiwani, Same MGR 0–10 km	Grazing land shortage; health and veterinary provision poor; donor projects incomplete; community needs involvement in external support programme planning; NGO has divided community
Mheza	Maore, Same MGR 0–12 km	Lack of irrigation furrows; grazing shortage; no pastoralist social services in Nadururu
Maore	Maore, Same MGR 12 km	No *ndiva*; shortages of grazing, firewood and irrigated land; no veterinary services or farm inputs
Mpirani	Maore, Same MGR 15–20 km	Damage to crops by livestock and wildlife; no market for crops; no government dispensary
Misufini	Ndungu, Same MGR 25 km	Irrigation scheme land, lack of control, fees, sisal worker housing; lack of employment; no government dispensary
Ndungu	Ndungu, Same MGR 25–30 km	Conflict of farmers vs Japanese irrigation scheme; rice yields; plantation land dispute; wildlife damage
Makokane	Ndungu, Same MGR 15 km	Water loss to project; diseases; lack of grazing/road/market; wildlife damage
Mkundi Mbaru	Mnazi, Lushoto MGR 10–15 km	Water supply; interethnic conflict; lack of grazing land; poor village leadership
KwemKwazu	Mnazi, Lushoto MGR 10–15 km	Schistosomiasis; donors damaged traditional irrigation; food insecurity; farm inputs; crop pests

quarter, valued at Tanzanian Shillings (TSh) 1,200/litre). Finally several villages expressed an interest in mining gemstones, if possible within MGR. Several villages saw the lack of employment opportunities as a particular problem for their young people who turn to charcoal making, or to theft of livestock, raiding beehives and standing crops, and poaching.

Most villages experience problems of land shortage both for cultivation and especially for grazing because they are situated so close to the reserve (Figure 33.2, previous chapter). Detailed studies on land tenure and use have shown that the lack of water reservoirs (*ndiva*) and the need for extension and reinforcement of the traditional irrigation furrows are major constraints on the availability of cultivable land for many villagers, especially women and young people. Shortage of land for grazing (a direct result of exclusion from MGR), and lack of support from MGR over identifying areas away from farms that could provide good fod-

priorities
Regain land, dam, store and perennial crops lost to MGR; support to primary school; repair water scheme; livestock facilities; repair four collapsed bridges
Regain gazetted land; construct *ndiva* (dams), livestock services; youth employment from MGR; completion of village projects started by the Outreach Programme/World Vision
Complete *ndivas* and donor projects (dispensary, school, water); land for pastoral Maasai/Pare; veterinary services; employment, agricultural inputs; women's projects
Irrigation development; *ndiva*; livestock services; water supply, school, maize mill for Nadururu; youth employment; medicines for dispensary, MGR boundary revision
Support for *ndiva* construction; grazing land; expansion of irrigable land; human and livestock drugs; support for technical secondary school construction; youth employment
Government dispensary; cooperative to organise agricultural inputs; repair irrigation furrows repeatedly destroyed by heavy rain/floods
Housing and wage for plantation workers; land for cultivation and water supply, government dispensary, farmer control over irrigation scheme and fees
Farmers' association control over rice project; expel contentious extension workers; secure market for farmers' rice; control wildlife in rice farms
Regain grazing from MGR; farmers association control of rice project; pump and borehole for domestic and livestock water supply; youth/women's projects
Rehabilitate water supply; water for pastoralist Maasai and Sambaa stock; village leadership and cohesion; crop and grazing land; repair road and bridge
Repair irrigation system/drinking water supply; gem mining equipment; food aid to aged; cotton technical and market support; eradicate schistosomiasis

der, has driven a considerable proportion of villagers to express dissatisfaction over the current extent of the reserve, and over their relations with it. Table 34.2 shows several villages are currently seeking alterations to the MGR boundary.

After the PRA exercise a sample of 39 agricultural and agro-pastoral households in Kisiwani and Mkonga Iginyu (representing a cross-section of socio-economic activities and status) were selected for intensive study of resource use. At the beginning of the study 33% wanted MGR to be degazetted for farming, and a further 10% wanted degazetting for grazing. Many villagers feel that the new boundaries cleared recently as part of the reserve's rehabilitation have brought the boundaries closer to their villages. The villagers expressed a desire to see the old boundary restored in the hope of reducing the effect of wild animals on farms to levels experienced in the past. Villagers felt that access by wildlife to water facilities inside the reserve would greatly reduce crop loss to marauding animals. A final detailed PRA discussion at Kisiwani with this study sample of 39 households and with village leaders ended with half the villagers expressing a wish to see MGR retained (51%), but with the return of land most recently expropriated from villagers (in the 1990s). They also stipulated MGR should provide water to wildlife and that the community should co-manage MGR affairs. By contrast 90% expressed the desire to see nearby forest catchment reserves retained in view of their widely recognised environmental benefits.

Resource use and management by Pare people in Kisiwani and Mkonga Iginyu

The prime need of this group is for water and irrigable or cultivable land. The reserve has little land suitable for irrigated agriculture and few claims are made on the land inside the reserve for cultivation. Besides land and water management there are also important needs for fuelwood and wild foods met in part from the reserve.

Land and water management

Customary and present natural resource tenure, distribution, use and management systems were investigated through detailed key informant interviews, focal group meetings with traditional past and present land and water distributors, through repeated meetings with sampled households and at times a wider representation. Detailed investigation was carried out on use of land, irrigation water, fuelwood, indigenous trees, honey and beehive trade, wild vegetables and fruits.

Land ownership systems among the Pare are described by Kimambo (1969), Omari (1992), Tenga (1979) and confirmed in recent field work (Kiwasila 1996b). In the past the first person to clear an area became its *de facto* owner, and land was passed from father to son with new plots being sought as families expanded. Women did not have full land ownership rights. However this is changing as more and more fathers are allocating land to daughters. Women heads of household reported less land (mean = 3.4 acres, n = 8, sd = 2.75) than did men (mean = 10.3 acres, n = 31, sd = 9.26).

In the days of Pare kingdoms and during the colonial era the chief nominated a land and water distributor *(Mgawa)* with responsibility for organising public land allocation meetings, ruling on land and distribution of irrigation water, and negotiating individual problem cases as they arose. The traditional systems have evolved and sometimes been weakened through successive administrations, and as both the pressure on cultivable (particularly irrigable) land and the complexity of negotiating access have intensified (Kimambo 1969). There is now a mixture of clan land, state land, and privately owned land that may be lent or leased out. The function of the *Mgawa* is now held by the village government land allocation committee. Land is an expensive commodity: in general, half an acre of irrigated land is rented at TSh 40,000 per season while rain fed land is rented at TSh 10,000 per season (Kiwasila 1996b). Litigation over land issues dominates court time and villagers' concerns in the villages investigated. Nonetheless the land allocation committees provide a major component of the social structure of decision making over resource use and should be taken into account in planning consultation, negotiation and participation over the long-term management of Mkomazi resources.

Water management is a matter of daily concern. Present day water management systems have again evolved from customary systems established during precolonial times when the *Mgawa* and his assistants distributed water as well as land. Water is divided between mountain side and plains farms, and again between different blocks of farm land within each of those zones. In theory the system operated such that water in any given drainage line was retained by mountainside farms three days a week, and was allowed to flow to the plains three days a week, where it was stored in *ndiva*. On the seventh day, it was directed to the Mkomazi village charco dam on the edge of the plains for use by livestock and wildlife as well as watering the farms of the *Mgawa* and his assistants.

The *Wagawa* were responsible for inspecting farms to check plots were properly cleared and planted, and to ensure water furrows were kept clear. Water management required a sophisticated level of cooperative management, organisation and supervision. Cooperative work parties (*Msaragambo*) constructed and maintained *ndivas* and water channels. Penalties for water theft and other contraventions were established and enforced by *Mpambuo*, a traditional institution for consfication of property which could then be redeemed.

As with land management, this system has come under the combined stresses of growing demand and of changing political structures. The village government has taken over the function of distributing irrigation water through a production committee of *Wagawa*, but this is perennially under-resourced and overloaded with administrative responsibilities. Negotiation and decisions between mountainside and plains farms have become more difficult as they now come under separate wards and leadership. Kisiwani has grown to the extent that Mkonga Iginyu has budded off as a separate village with its own administration, and the two receive water from the same river (Nakombo) as well as sharing some *ndiva* and irrigation furrows. Water allocation from the shared furrows is meant to alternate between communities after every three days, but coordination and cooperation are weak.

There is a strong perception of water shortages (whether due to climatic downturn, deforestation affecting catchment and run-off, and/or a spiralling demand). Water from irrigated plots used to flow back into the Nakombo waterline and went on to supply Gonja Maore 18 km away from Kisiwani. Now the Nakombo waterline dries up so that water barely flows to the outer sub-villages of Kisiwani. Some farm plots in the sub-villages are left dry (in Njiro, and the far end of Bama and Kandema in Mkonga Iginyu) as is the *ndiva* for wildlife and livestock.

There are allegations of corruption over water allocation, including sales of allocation of water quotas, and that *Mpambuo* is misused over such offenses. Nonetheless, the water management system persists in many villages around Mkomazi. It is, despite problems, another structure in place that is essential to any future consideration of resource management planning.

Use of wild resources

The two main uses of wild resources are wood as fuel and wild vegetables for food. Use and importance of these resources was initially identified during PRA. Their use was quantified during subsequent household questionnaire, 24 hour dietary recall and market surveys. The intensive sample of households was selected to allow comparisons between Kisiwani and Mkonga Iginyu, and also between different socioeconomic categories. However, it was also constrained by logistical problems and by the willingness of different high income households to be involved. The sample of 39 households was made up as shown in Table 34.3.

Taking the intensive sample of 39 households as a whole, 90% depended on farming as their major activity and just under half (40%) supplemented family

Table 34.3 Wealth categories of sample households defined by villagers during PRA, on criteria of housing, assets and occupation (Kiwasila PhD thesis in prep.).

wealth category	Kisiwani	Mkonga Ijinyu	total
poor	15	13	28
medium/wealthy	6	5	11
total	21	18	39

Table 34.4 Top fuelwood species recorded in household surveys, Kisiwani and Mkonga Iginyu Villages 1996. Local names in brackets.

species	questionnaire rank	multi-round rank
Acacia mellifera	5	1
Acacia bussei	1	2
Grewia bicolor	2	3
Wattle species (Nghwati)	–	4
Senna siamea (planted)	–	5
Acacia senegal (Mnoa)	–	7
(Mtalati)	–	8
Balanites species (Mkongori)	–	9
(Mhungululu)	–	10
Balanites species (Ngonge)	–	11
Acacia nilotica	–	12
Cedrela odorata (Mti Kunuka, exotic)	–	13
Dichrostachys cinerea	–	14
Acacia species (Msharanu)	–	15
Acacia albida	3	–
Combretum exalatum	4	–

income with wages from manual labour. One-third (33%) supplement income by selling wild foods (mostly vegetables). 39% sell charcoal (including 36% who actually make as well as selling it). 3% of the sample identify charcoal making (along with farming during good rains) as their major means of livelihood. Appendix 2 (at the end of this chapter) lists 107 wild plants used as food or fuel and indicates preferred species.

Fuelwood species used and preferred by respondents.

Questionnaires in several villages at the beginning of the main study showed species preferred for fuel, and systematic follow-up twice a month counted and identified to species level pieces of fuelwood present at 34 households. Table 34.4 lists the top species ranked according to stated preference and to frequency of occurrence in the multi-round fuelwood sample.

Differences arise mainly from the location of the intensive sample households, which reflects species available at those sites. For some species, botanical names could not be obtained from the herbarium. Further analysis showed that *Acacia bussei* (a hardwood giving long lasting heat) is widely exploited and used by all socio-economic groups. The species is generally available close to and inside MGR, at some distance from most households. *Acacia mellifera* (available only in Njiro sub-village of Kisiwani) is exploited by almost all households in the sub-village.

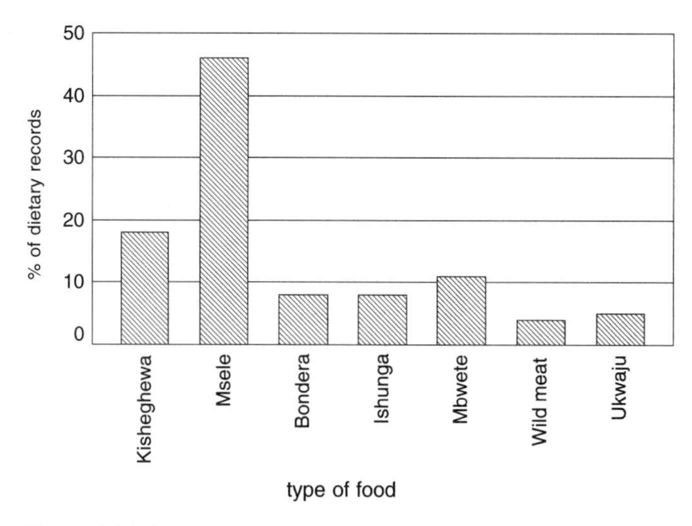

Figure 34.1 Percentage contribution of main wild foods to all wild food items eaten in the systematic sample of 373 meal days of Pare agropastoralists during March 1996–March 1997.

Wealthy households may have access to electricity and kerosene cooking stoves and have lesser fuelwood needs.

Frequency of wild plant use: occurrence in household meals.

Favoured wild vegetables are shown in Figure 34.1. Their use seemed to be widely

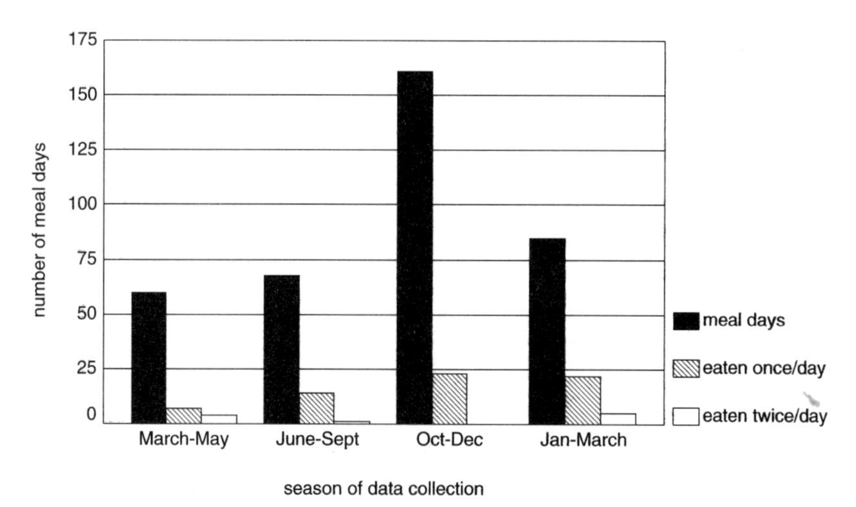

Figure 34.2 Meal days with wild vegetables out of 373 meal days during March 1996–March 1997

popular as 92% of the 39 households studied reported gathering and using wild vegetables as food. Consumption of wild vegetables differed little between different socio-economic groups in the sample. However, the frequency of use varied with 69% of the sample using wild vegetables on a weekly basis, and 15% eating them daily.

Multi-round surveys of the 39 sample households with observations taken during several consecutive repeats made it possible to get some measure of frequency of use of different wild foods relative to the total number of meals cooked. Figure 34.2 shows the mean frequency of use of the most common wild foods over the study. Overall, wild foods were used in 20% of 373 meal-days.

The economic importance of wild foods—market data.

Market surveys carried out on 28 separate market days, evenly distributed by season, monitored the extent to which wild plants were sold for food in addition to opportunist gathering and consumption. Table 34.4 and Figure 34.3 show the relative quantities and frequencies of different plants sold in the market throughout the study. A total of 1,110.5 kg of wild foods were sold at Kisiwani Market on the 28 days sampled between March 1996 and January 1997.

The supply of different wild vegetables varies in different ways with the availability of moisture. *Kisheghewa* for example (*Solanum* sp.) occurs in wet areas and is common on irrigation farms during the dry (hungry) season, while *Mbwete* (*Celosia trigyna*) and *Bondera* (*Amaranthus* sp.), are both prolific weeds wide-

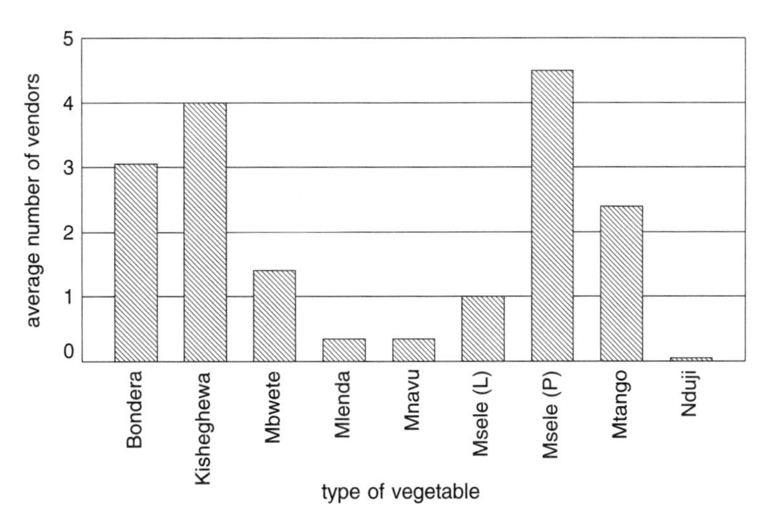

Figure 34.3 The average number of vendors offering each vegetable for sale at each market during 28 market days March 1996–January 1997. Msele (L) = leaves, Msele (P) = powder.

Table 34.5 Average weight (kg) of wild produce sold per market day at Kisiwani market, March 1996–January 1997.

month	number of days	*Amaranthus* sp.	*Solanum nigrum*	*Solanum nigrum*	*Celosia trigyna*	*Corchorus olitorius*	*Zanthoxylum holtziana*	*Zanthoxylum holtziana*	*Ipomea* sp.	
		Bondera	Kishegewa	Mnavu	Mbwete	Mlenda	Msele leaves	Msele powder	Mtango	Nduji fruit/veg.
March	4	47.1	0.0	0.0	11.7	0.7	41.4	66.4	63.3	0.0
April	2	23.8	4.3	0.9	1.1	0.0	0.4	39.0	3.6	0.6
May	4	79.9	13.8	13.4	9.8	3.8	30.8	40.2	7.1	0.0
June	2	13.8	13.2	1.2	2.1	1.0	0.0	21.8	18.5	0.0
July	3	30.9	12.9	0.0	4.9	0.0	0.0	24.5	16.5	0.0
August	2	2.0	28.3	0.0	0.9	0.0	0.0	20.2	20.4	0.0
September	1	0.0	9.0	0.0	0.0	0.0	0.0	13.7	5.6	0.0
October	3	22.5	23.6	0.0	2.8	1.4	0.8	31.5	8.2	0.0
November	2	5.3	28.2	0.0	2.0	0.0	0.0	20.2	3.9	0.0
December	2	26.6	14.3	2.9	3.2	0.0	27.7	36.9	7.2	0.0
January	3	0.6	18.4	0.0	16.0	0.0	15.6	20.8	6.3	0.0
mean	–	23.0	15.1	1.7	4.9	0.6	10.6	30.5	14.6	0.1
std. dev.	–	24.0	9.1	4.0	5.2	1.1	15.6	14.8	17.2	0.2
std. error	–	7.2	2.7	1.2	1.6	0.3	4.7	4.5	5.2	0.1

spread during the rains. Certain species predominate in market sales, particularly *Zanthoxylum (Fagara) holtziana.*

During the extended drought at the end of 1996, the distinctively thorny *Zanthoxylum (Fagara)* tree provided most of the vegetable relish when the few other wild and cultivated vegetables remaining on-farm were largely lost to wild animals dispersing from MGR. *Zanthoxylum* leaves are dried and ground to a powder which can be stored and later reconstituted with water to make a much-used relish. The powder is sold in the market by the dessert spoonful (TSh 20 per spoonful) and retailed door to door at TSh 30–50 per spoonful. Few villages have access to *Zanthoxylum* trees on their own farm land, most trees being found in Vumari (and in one case in Maore). There are none at Kisiwani where detailed studies were done (Table 34.5). In most markets, the *Zanthoxylum* sold is harvested illegally within Mkomazi. *Zanthoxylum* from Mkomazi and from the Mkomazi side of Vumari is considered to be better than that from neighbouring mountain villages or from Handeni. During the drought, much of *Zanthoxylum* sold was brought to Kisiwani from Vumari village because the danger from wild animals in and around MGR deterred harvesters. The cost of the powder reflects the risks involved. From both consumption data and market data it is clear that *Zanthoxylum* is important to villagers throughout the year and that people will take risks to maintain the supply. The tree is not farmed as it is thought to reproduce vegetatively from underground roots and is seen as difficult to plant and establish.

Importance to individuals and households of wild food sales

Analysis of who sells wild vegetables showed a predominance of women and especially school age girls. Table 34.6 shows 36% of the vendors were children. 26% were women of 50 years of age and above. Children used their earnings to buy school materials. Middle aged women vendors (though few) earn more from wild vegetables. Most of these were selling *Zanthoxylum* powder. As the 1996–97 drought grew worse with animals dispersing onto village lands, increasing numbers of men took to harvesting *Zanthoxylum* both for their wives and to sell to women vendors. Fear of wild animals prevented harvesting of *Zanthoxylum* by women. Elderly women whose income earning activity banked on *Zanthoxylum* were completely deprived of any earnings at this period. Wild food sales thus represent a strategy for those with few other options.

Income data are always suspect, as there is every incentive to under report. Nonetheless the data here are felt to be as accurate as it is possible to get, given that the researcher was well known to and accepted by vendors and buyers in the market. Produce for sale was monitored by direct counts and weighing at the market, and sales recorded by direct observation and discussion at the time. Cross checking ensured that items sold earlier in the day were reported and included. Over the 28 market days monitored, a total of TSh 349,942 were reported earned, mostly by women and children. Average earnings reported per market day, per trader was TSh 27 for women and 14 for men, though particular individuals and age/sex classes earned much more than this average as others did not attend all markets. Even small amounts are significant especially for school children and old women who have no other means of income, and also seasonally for men. The district council charge a levy on items sold at the market, and the scale of this levy suggests minimum earnings from wild vegetable sales are significantly higher than the figures reported. Wild vegetable sellers are charged TSh 10 to 100 per category and per amount of wild vegetable sold. A high levy (TSh 100) is charged on *Zanthoxylum*. Those with small piles (mostly school children and the elderly) pay a levy of TSh 0–30. These sums could clearly not be paid from the reported earnings. Women vendors in some cases walked long distances to sell small quantities and used their small earnings to buy food and other family necessities. Some, after walking 10km to market, said they earned as little as TSh 100 per day (c. 10 pence), on which they had to pay TSh 30 as a market levy. From the total income reported earned during the study period 32 % (TSh 111,981) went to the district council as levy from wild vegetables. Assuming that in reality vendors would not tolerate more than 10% of their earnings going on levy, we estimate that a more likely figure for total earnings over the 28 market days is around a million shillings, with individuals earning averages of around TSh 50–100 per market day and an order of magnitude more than this for regular traders. Wild vegetables contribute not only to the welfare of the villagers through consumption and by income generation but also to the village government and district council. Another small market at Kisiwani also collects a levy which is retained by the village

Table 34.6 Age, sex and income of vendors of wild plant foods (28 market days, March 1996–January 1997).

	number of traders		total income (TSh)		average income (TSh)		grand total
age	women	men	women	men	women	men	(TSh)
0–9	15	0	5,280	–	352	–	5,280
10–19	156	18	74,356	2,780	477	154	77,136
20–29	41	0	49,560	–	1,209	–	49,560
30–39	70	1	67,615	150	966	150	67,765
40–49	44	4	60,626	8,450	1,378	2,113	69,076
50–59	78	7	64,365	970	825	139	65,335
60–69	26	0	10,395	–	400	–	10,395
70–79	19	1	5,385	10	283	10	5,395
total	449	31	337,582	12,360			349,942

government. Funds at both levels are used to pay the village executive secretary (by the district council) and village volunteers (by the village government).

Charcoal-making is potentially much more lucrative than sale of wild foods but is also physically very hard work. Sales depend on transport and roadside outlets and the whim of forest officers who often confiscate charcoal bags as most vendors cannot afford a licence.

Conclusion

This paper has summarised patterns of natural resource use by reserve adjacent farming communities. PRA survey established a number of concerns relating to natural resource use in general and to MGR in particular. Communities around MGR are all preoccupied with access to land whether for grazing, irrigated or rainfed crops, or for gathered resources. Cultivators around Mkomazi have long-established social institutions for regulating access to and allocation of land and water resources. Although affected by political and economic change, and under-mined by the limited authority village government and councils of elders can exert, these institutions continue to operate on an everyday basis, and form a framework within which management issues concerning the reserve could be negotiated on a more participatory basis.

Current management of the reserve is seen as bringing little benefit to villages and as having dealt a series of blows to the wider economy of reserve adjacent communities. Consultation would be seen as a benefit in itself, as the farmers ex-press both the need to maintain use of reserve resources for gathered resources, and the desire to have some input to decision-making and management. Mkomazi Game Reserve is not the only development to have impinged on the lives and livelihoods of reserve adjacent farmers. Many of the villages have experienced donor interventions whether agricultural schemes, educational or health oriented. Each scheme tends to act in isolation without reference to inevitable knock on effects, and most generate their own case specific problems and disillusionments. In parallel with the social institutions that manage land and water, established farmers' associations come into conflict with the management structures imposed by such projects. Corruption and poor completion of externally financed projects add to the frustrations that villagers already experience. Development projects associated with the reserve are as prone as any to these problems.

Developments inside and outside the reserve thus interact to produce unplanned results, ethnic tensions and local conflicts that rebound on local institutions and district authorities. Villages are already negotiating directly with MGR authori-ties; pursuing law cases; and in many places there is a continuing if low key use of MGR resources. Gathered resources are common in the diet and gathered fuelwood is the main energy source for all households. Collection and sale of wild resources is a last resort strategy for the poor. Sales of wild vegetables also contribute income to village and local government. However, the broader survey makes even more clear the need to look at the future of the reserve as one piece in a jigsaw of interacting developments. Conservation concerns stress population growth and pressure on the reserve as undermining biodiversity and long term sustainability. The authors would see a more pressing danger in continuing uncoordinated, short term, *ad hoc* schemes that are pursued without consultation, negotiation, or reference to the wider framework of resource use and change in Mkomazi. Community participation with transparency, evaluation and auditing, and providing people with real benefits, must be the fundamental condition for conservation with development here as elsewhere.

Acknowledgements and disclaimer

This work was funded by the Department for International Development. The authors also wish to acknowledge the support of the Royal Anthropological Insitute's Emslie Horniman Fund, the Parkes Foundation, the Kathleen and Margery Elliot Trust, the Central Research Fund of London University and the Graduate School of University College London. The authors are grateful for the help and cooperation of the Institute of Resource Assessment, the University of Dar es Salaam, the Commission for Science and Technology of Tanzania and the Department of Wildlife. We are also indebted to the many members of local and District government and the countless other Tanzanians who live and work near to Mkomazi Game Reserve for their help, cooperation, welcome and warm hospitality.

The Department for International Development (DFID) is the British Government Department which supports programmes and projects to promote overseas development. It provides funding for economic and social research to inform development policy and practice. DFID funds supported this study and the preparation of the summary of findings. DFID distributes the report to bring the research to the attention of policy-makers and practitioners. However, the views and opinions expressed in the document do not reflect DFID's official policies or practices, but are those of the authors alone.

References

Homewood, K., Kiwasila, H.L. & Brockington, D. (1997) *Conservation with development? The case of Mkomazi, Tanzania.* Final Research report to DFID-ESCOR. Department for International Development, London.

Kauzeni, A.S. & Kiwasila, H.L. (1994) *Serengeti Regional Conservation Strategy: a Socio-Economic Study. Final Report submitted to the Serengeti Regional Conservation Strategy.* Institute of Resource Assessment, University of Dar es Salaam.

Kimambo, I. (1991) *Penetration and Protest in Tanzania.* James Currey, London.

Kimambo, I. (1969) *A History of Tanzania*. East African Publishing House, Nairobi.

Kiwasila, H.L. (in prep.) PhD Thesis. University College London.

Kiwasila, H.L. (1994) *A reconnaissance socio-economic study on villagers adjacent to Mkomazi Game Reserve*. Working Paper. Anthropology Department, University College London.

Kiwasila, H.L. (1996a) *Patterns of Natural Resource use and implications for local communities adjacent to Mkomazi Game Reserve in Tanzania*. PhD Upgrading Report and pilot study report. Anthropology Department, University College London.

Kiwasila, H.L. (1996b) *A Village Profile of 11 Mkomazi adjacent Villages*. PRA Report Working paper. Anthropology Department, University College London.

Omari, C.K. (1992) Some notes on self-help programmes and the Pare people and their impact on social development. In: Forster G. *et al.* (eds.) *Peasantry: Economy in Crisis*. Avebury Aldershot, Brookfield USA.

Tenga, R. (1977) *Land Law and Peasantry in Pare District. A Historical Analysis*. PhD Thesis. University of Dar es Salaam.

URT (1993) *Report of the Presidential Commission Inquiry into Land Matters. Vol II. Selected Land Disputes and Recommendation*. Ministry of Land Housing and Urban Development, Dar es Salaam. Available from SIAS, Uppsala, Sweden.

Werkgroup Tanzania (1990) *Religion in Pare Kinship Formation*. Tiburg, Netherlands.

Appendix 1: PRA background

Participatory Rural Appraisal (PRA) seeks to make development related research as relevant as possible and to make local people agents rather than subjects of research. PRA focuses on speed and relevance rather than long term, systematic research.

PRA tools used in MGR

A range of PRA techniques were used to obtain baseline and historical information on use of NTFPs and their importance to villagers (Chambers 1992, Poffenberger *et al.* 1992, IIED 1995). These included:

- *Discussion groups with key informants* including people with in depth local knowledge on key issues in resource use and management: water distributors, former traditional leaders, former civil servants from the colonial period, past and present village leaders. It also included representatives of special interest groups (youth, women and men beehive owners, pastoral representatives); current civil servants from agricultural, livestock, community development, and forestry extension sectors.
- *Historical transects* highlighting former patterns of residence in relation to the establishment of MGR, and famine coping strategies.
- *Preference ranking matrices* were used to rank preferred species for both wild edible products and fuelwood, and problems/development needs perceived by villagers as directly or potentially related to MGR.
- *Wealth ranking* (Grandin 1988) used the community's own socio-economic definition of wealth to assign households to high, medium and low income categories for analysing differences in natural resource use, food security and

Table 34.7 Wealth class criteria: Kisiwani and Mkonga Iginyu farming households.

	high	medium	low
farm	>11 acres rainfed >1 acre irrigated	6–10 acres rainfed 0.5–1 acre irrigated	none/some rainfed rents/shares irrigated
livestock	>20 cattle small stock	some cattle small stock	few or no livestock
house	brick/cement block walls; plaster; iron roof; cement floor	mud brick, brick or cement block walls; iron or thatch roof	no house owned, or house in poor repair
other occupation	business e.g. grain mill, transport, lodging house etc.	small scale trade (e.g. grain, fruit, veg.) or business (e.g. tea or food vendor)	paid manual work (e.g. charcoal making, farm labour)

expenditure. The criteria (see Table 30.7) included landholdings (irrigated/rainfed), number of livestock, type of housing, remittances, ownership of a business or investment in assets; schooling of children in high level education; type of clothing.

- *Problem tree analysis* asked participants to list problems in resource use, and to put forward potential solutions in the context of past and present experience.
- *Seasonal calendar of village activities through the year* documented seasonal patterns of food stress, effects of rainfall and pest populations on crop yields, availability of wild resources and fuelwood collection patterns.
- *Transect walks* to protected forests, farm plots, irrigation systems, fuelwood collection sites, livestock and wildlife drinking sites, areas where wild fruits and vegetables are procured, areas important for honeybee foraging etc.

PRA meetings held 1995–97

1995 (19 meetings)

VILLAGES AND SUBVILLAGES: Mkonga Iginyu, Kisiwani, Njiro, Maore, Nadururu, Mpirani, Misufini, Ndungu, Makokane, Kwemkazu, Kiwanja, Mrigirigi, Mkundi Mtae, Vumari.
GROUPS: Pare, Sambaa and Maasai/Parakuyo villagers and key informants, village elders, village and ward government, women, youth groups, farmers, agro-pastoralists, salt makers, development project staff, clinic and school staff, District government staff.

1996–97 (17 meetings)

VILLAGES: Kisiwani, Mkonga Iginyu.
GROUPS: Villagers, village elders, women, youth groups, agropastoralists, Pare traditional natural resource managers, key informants, agricultural extension workers, village government, local development advisors, Pare evictees, domestic and irrigation water supply management groups.

Appraisal of methods

PRA can be of great value in turning up issues, processes and linkages of importance in community development. It can also be misleading, given that it relies on rapid methods of data collection and that this inevitably limits the possibilities for cross checking and validation which longer term research should allow. Ultimately PRA work is as good as the practitioner: an informed, perceptive, rigorous researcher can use PRA to great effect, those with little experience of the area and issues, or working to lower standards, may generate poor quality or misleading information. Communities are not homogenous groups with consensus views, and some subgroups may be in a better position to put forward their views than are others, whether in public meetings or more subtly in monopolising access to the research team. In particular there is a problem in dealing with sensitive issues such as income or wealth, or issues of access to resources, where these are contested and a source of conflict within the community. It may not be straightforward to identify the main groups involved—high profile confrontations commonly mask the vested interests of silent third parties. Secondly, it is unlikely that outsiders asking about sensitive issues, and present for only a short period of time, will be given accurate information directly. Thirdly there is no way, other than long term research, to establish the extent to which the groups and individuals involved in the PRA are representative of the wider community. Finally, intensive crosschecking is necessary with other sources such as archival records and contemporary historical accounts.

In the present study PRA was used in conjunction with long term systematic research over a three year period. The first author is Tanzanian and has 10 years' experience of using PRA in different rural Tanzanian areas, on research issues in natural resource use, conservation and development. Long experience and local knowledge help ensure use of PRA in an informed and appropriate way. The long term household study ensured in-depth cross-checking.

References

Chambers, R. (1992) *Rural appraisal: rapid, relaxed and particpatory.* Institute of Development Studies Discussion Paper 311 (especially pp. 35-7)

Grandin, B. (1988) *Wealth ranking in smallholder communities: a field manual.* London. Intermediate Technology Group

Poffenberger, M., McGean, B., Ravindranath, N., & Gadgil, M. (1992) *Field methods manual. Diagnostic tools for supporting joint forest management systems. Joint Forest Management Support Programme.* Society for the Promotion of Wasteland Management, New Delhi.

IIED (1995) *Critical reflections from practice.* PLA Notes Volume 24. IIED, London.

Appendix 2: wild food and fuel plants

Wild plants preferred for food or fuel around Mkomazi Game Reserve (107 species). Identified by F.M. Mbago and Suleiman, Senior Herbarium Technician, Dar es Salaam.

family name	botanical name [1]	local name	use of plant [2]
Agavaceae	*Sansevieria bagamoyoensis*	Mkongepori	CL
Amaranthaceae	*Alternanthera sessilis*	Kilemba cha Bwana	EL **(FWF)**
Amaranthaceae	*Amaranthus spinosus*	Bwache (Sambaa)	EL **(FWF)**
Amaranthaceae	*Amaranthus cruentus*	Buuza	EL **(FWF)**
Amaranthaceae	*Amaranthus dubius*	Sungumsanga	EL
Amaranthaceae	*Amaranthus* sp.	Bondera	**(FWF)**
Amaranthaceae	*Celosia trigyna*	Mbwete	EL **(FWF)**
Anacardiaceae	*Lannea alata*	Mgarito	FR
Anacardiaceae	*Lannea stuhlmannii*	Msighe	M, EB
Anacardiaceae	*Sorindeia madagascariensis*	Mkunguma	FR, F
Apocynaceae	*Saba comorensis* var. *florida*	Ivungo	FR, R
Balanitaceae	*Balanites aegyptica* *	Mkizingo, Mkonga	F, U, LF **(GFW)**
Balanitaceae	*Balanites aegyptica* *	Mkonga	U, F, FR
Balanitaceae	*Balanites aegyptica* *	Mkizingo	U, FR
Bignoniaceae	*Kigelia aethiopium*	Myegea (Swahili) Mlegea (Pare)	Y, BF
Boraginaceae	*Cordia sinensis*	Mhololo	F, FR **(GFW)**
Boraginaceae	*Cordia* sp.	Isiborabagosi	F, M **(GFW)**
Boraginaceae	*Trichodesma zeylanicum*	Ishahangoswe	IFS
Burseraceae	*Commiphora africana*	Msusu	CL, M, G
Burseraceae	*Commiphora caerulea*	Isongolanyiko	F, U
Burseraceae	*Commiphora merkeri*	Msiga	U
Burseraceae	*Commiphora mollis*	Bambara	SC, CG, EF
Burseraceae	*Commiphora pteleifolia*	Isume (Pare)	M
Caesalpiniaceae	*Afzelia cuanzensis*	Mkokola	U, M
Caesalpiniaceae	*Delonix elata*	Mwerange	CL, SH, M
Caesalpiniaceae	*Senna septemtrionalis*	Mheti	F
Caesalpiniaceae	*Tamarindus indica*	Ukwaju	FR
Capparidaceae	*Boscia mossambicensis*	Mtero	F, FR
Capparidaceae	*Maerua holstii*	Mdamwai	FR
Capparidaceae	*Maerua triphylla*	Mdudu-Mbuzi	F, LF, EL
Capparidaceae	*Ritchiea capparoides*	Mdudu-mbuzi	LF, ER
Combretaceae	*Combretum exalatum*	Mzuru	F, P, M **(GFW)**
Combretaceae	*Combretum molle*	Mwamamjiru (Pare)	F, G
Combretaceae	*Paniculatum* sp.	Mgatuu	P
Combretaceae	*Terminalia sericea*	Muruku	F, P
Commelinaceae	*Commelina benghalensis*	Ikongwe	LF, IFS
Commelinaceae	*Commelina diffusa*	Itonge	IFS, EL
Compositae	*Bidens pilosa*	Kimbala, Mashonanguo (Pare)	EL, M
Compositae	*Brachylaena huillensis*	Mvovo	F
Compositae	*Galinsoga parviflora*	Mngeeza	EL
Compositae	*Sonchus bipotini*	Ishunga	M
Compositae	*Sonchus oleraceus*	Isikio	EL, LF
Convulvulaceae	*Ipomea* sp.	Mtango	FR
Convolvulaceae	*Ipomea eriocarpa*	Mfundofundo	IFS

Convolvulaceae	*Merremia* sp.	Mndoo, Busewe	EL
Cucurbitaceae	*Cucumis dipsaceus*	Kakoko, Kakokouganga	EL
Euphorbiaceae	*Acalypha* sp.	Shekizeu	EL
Euphorbiaceae	*Euphorbia bussei*	Lumere	RB, NT
Euphorbiaceae	*Euphorbia nyikae* var. *nyikae*	Msosongo	R, NT, SB
Euphorbiaceae	*Synadenium* sp.	Mgi, Mugi (Pare)	M
Gramineae	*Coix lachryma-jobi*	Itatabiri	O
Hernandiaceae	*Gyrocarpus americanus*	Mdendei	CP
Labiatae	*Ocimum basilicum*	Nyenye, Kivumbasi	BB, M
Labiatae	*Solum kitivuense*	Wanyika	EL
Loganiaceae	*Strychnos mitis*	Mtisi	F **(GFW)**
Malvaceae	*Azanza garckeana*	Mtakataka	FR
Malvaceae	*Hibiscus* spp.	Ngararuge/Mkokoro	FR
Malvaceae	*Hibiscus sabdariffa*	Rusela	FR, M
Meliaceae	*Trichilia emetica*	Mkoromaji	FR
Meliaceae	*Ekebergia capensis*	Mkalambato	F, U, RF
Mimosaceae	*Acacia albida*	Mkababu	FR, F
Mimosaceae	*Acacia bussei*	Gulela	F, LF **(GFW)**
Mimosaceae	*Albizia harveyi*	Mnyaa	F
Mimosaceae	*Acacia kirkii* var. *kirkii*	Mwerera	F, CH, M
Mimosaceae	*Acacia mellifera*	Kalanyika	F, CH
Mimosaceae	*Acacia nilotica*	Nzameli	F, LF **(GFW)**
Mimosaceae	*Acacia zanzibarica*	Geranghunga	F **(GFW)**
Mimosaceae	*Dichrostachys cinerea*	Mkame	F, LF, M **(GFW)**
Moraceae	*Ficus exasperata*	Msasa (Swahili)	WP, M
Moraceae	*Ficus glumosa*	Kishago	FR, SH, BF
Moraceae	*Ficus sur*	Ihuu	FR
Moraceae	*Ficus thonningi*	Mvumo Mdogo (Pare)	SH, FR
Moraceae	*Ficus sanzibarica* ssp. *sanzibarica*	Mvumo Mkubwa (Pare)	SH, FR
Moraceae	*Milicia excelsa*	Mvule	T
Moraceae	*Morus alba*	Mtelia	EL
Moringaceae	*Moringa oleifera*	Mlongelonge	EL, F, SH, CL
Nyctaginaceae	*Boerhavia erecta*	Yogwe	IFS
Olacaceae	*Ximenia caffra*	Mtundutwa	FR, F, M
Papilionaceae	*Dalbergia melanoxylon*	Mwingo	F, CH, C **(GFW)**
Papilionaceae	*Lonchocarpus capposa*	Kiveti	F, CH
Papilionaceae	*Neourautanenia* cf. *imitis*	Kahawa soya	CF
Papilionaceae	*Ormocarpum* sp.	Mghara or mghaa	F
Passifloraceae	*Adenia cissampeloides*	Iyore (Pare)	FS
Polygonaceae	*Oxygonum sinuatum*	Nkhong'o (Pare), Mbigiri (Swahili)	EL
Rhamnaceae	*Ziziphus mucronata*	Isanzulakimbughu	FR, F, M
Rhizophoraceae	*Cassipourea* sp.	Mdelelo	U, F
Rubiaceae	*Vagueria infausta*	Mdaria	FR, M
Rutaceae	*Zanthoxylum (Fagara) holtziana*	Msele, Mlungulungu	EL, M **(FWF)**
Salvadoraceae	*Dobera loranthifolia*	Msiga	F, U, FR
Salvadoraceae	*Salvadora persica* var. *persica* *	Mkayo	FR, U, M
Salvadoraceae	*Salvadora persica* var. *persica* *	Mtasara/Mkakayo	FR, U, M
Sapindaceae	*Pappea capensis*	Mtundawe	FR, F
Sapotaceae	*Pachystela brevipes*	Msambia	FR, F
Sapotaceae	*Vittellaliopsis kirkii*	Msara	FR, SH
Solanaceae	*Solanum nigrum*	Kisheghea, Kisegeju, Kisheghewa	EL **(FWF)**

Solanaceae	*Physalis minima*	Mnavu jike	EL **(FWF)**
Sterculiaceae	*Sterculia africana*	Muoja	R
Sterculiaceae	*Sterculia appendiculata*	Mfune	EN, T
Thymelacaceae	*Gnidia goerzeana*	Mshombori	R, P
Tiliaceae	*Corchorus olitorius*	Mlenda	EL **(FWF)**
Tiliaceae	*Grewia bicolor*	Mlawa Ng'ombe	F, FR **(GFW)**
Tiliaceae	*Grewia forbesii* *	Mlawa-Mkore	FR, M
Tiliaceae	*Grewia forbesii* *	Mvongovongo (Pare)	FR
Tiliaceae	*Grewia monticola* *	Mangura	FR, M
Tiliaceae	*Grewia monticola* *	Mlawa Mnyindo	FR, F, M
Zingiberaceae	*Afromomum angustifolium*	Litungululu	FR

[1] Different forms of species marked * are distinguished by name and/or use.

[2] Key to uses of plants:

BB bait for bee swarms
BF bee forage
C carving
CF coffee
CG chewing gum
CH charcoal
CL fencing, corralling livestock
CP children play with seeds
EB edible bark
EL edible leaves
EN edible nuts
ER edible roots
F firewood
FR fruit
FS fertilises soil
G glue
IFS indicates fertile soil
IR indicates rain
LF livestock feed
M medicine
NT night torch
O ornamental
P poles, building
R ropes
RB rhino bait
RF roofing
SB smoking bees
SC scent
SH shade
T timber
U utensils for cooking and eating, tooth brush
WP washing pots
Y yeast

FWF favoured wild food
GFW good fuel wood

Note: The following food plant has not been fully identified: Nduji.

SECTION III

Management

Sections I and II have detailed, respectively, patterns of biodiversity within Mkomazi and of resource utilisation by local communities within and outside the reserve. Section II highlighted apparent conflicts between biodiversity conservation and utilisation interests. In general, the resource demands of increasing numbers of humans who expect an improving quality of life are leading to a reduced and less sutainable biodiversity through increased migration of pastoralists, cultivators and hunters into many protected areas.

The *Wildlife Policy of Tanzania* advocates the development of Tanzania's protected area network as part of a strategy addressing the simultaneous objectives of conserving biodiversity and meeting local and national development needs. The management challenges presented by the policy are considerable. Decision-makers and planners are simultaneously faced with the complexity of the problem and the inadequacy of available information.

The chapter by Packer considers the broader issues involved in managing resources to achieve multiple and often conflicting goals, high-lighting particular challenges for managing Mkomazi. The final chapter of the book, by Mbano, introduces the process of management planning that is currently in place and how it has resulted in the draft General Management Plan of the Mkomazi/Umba Game Reserves. The plan is a first step towards addressing conservation and development objectives for Mkomazi in the context of the Wildlife Policy.

CHAPTER 35

Management for biodiversity in Mkomazi

Michael J. Packer

Management for biodiversity implies taking actions that are designed to achieve the long-term maintenance of biodiversity. The process requires: the definition of management objectives, within a particular policy context; the identification of management options, taking account of potentially conflicting interests in the natural resources; and the development and implementation of a strategic plan, in which actions are placed in a monitoring framework. The process is driven by knowledge of patterns of biodiversity and of ecological, environmental and anthropogenic driving factors and processes.

This chapter outlines the policy context of management for biodiversity in Tanzania, considers the conservation and utilisation importance of biodiversity, and asks how, given new knowledge of the biodiversity of Mkomazi, feasible management approaches can be identified, highlighting particular management challenges and opportunities.

Policy contexts

Tanzania is a Party to the Convention on Biological Diversity (CBD), which makes it clear that use of biodiversity must be on a sustainable basis: that is, current use must not lead to long-term decline. Several Articles of the CBD are particularly relevant to efforts to manage for biodiversity, most notably Articles 8 and 10 that focus, respectively, on conservation and sustainable use of biodiversity. Together these Articles emphasise: management and protection of biological diversity within and outside a system of protected areas; rehabilitation and restoration of degraded ecosystems; avoidance or minimisation of adverse impacts on biodiversity arising from its use and from other threatening processes and activities; and customary use of biological resources in accordance with traditional cultural practices that are compatible with conservation or sustainable use.

The CBD has influenced Tanzania's Wildlife Policy (MNRT 1998), which is cast in a framework of environmental sustainability and socio-economic transformation. Within Tanzania's wildlife protected area (PA) network, which the policy seeks to maintain and develop, game reserves are identified as providing particular opportunities for pursuing economic and development aims: "a flexible approach to collecting revenue from harvests of biological natural products" within game reserves is to be adopted; and the status and functions of game reserves are to be reviewed with a view to possible management changes that will lead to a community-based conservation approach.

Biodiversity: meaning, importance and use

Biodiversity is most frequently used to mean the 'variety of life', from genes to ecosystems. In attempting to inventory biodiversity it is necessary to focus on one aspect of this broad view. For various reasons, ecological science uses species diversity or 'species richness' as a meaningful surrogate of biodiversity. Much of the research within Mkomazi has focused on determining species presence and patterns of species richness. Within a particular area, different species interact with each other to form communities, and interactions between communities 'bind' them together as an ecosystem. The 'normal' structure of a natural ecosystem is determined by particular types and strengths of interactions between species and with the physical environment. These interactions constitute the ecological processes by which an ecosystem functions; that is, remains intact and productive.

Efforts to conserve biodiversity have focused on protection of species and the maintenance of ecological processes: but why is biodiversity important? Biodiversity is valued in a number of ways, which are usefully classified into use, passive use, and non-use values. The *use value* of biodiversity relates to its being a key resource base for many subsistence and economic purposes, providing many 'goods': food, medicine, genetic resources, building and industrial materials, recreational exploitation, tourism. The *passive use value* of biodiversity concerns the ecological 'services' that it provides: the means to produce 'goods', atmospheric, hydrological and climatic 'regulation', nutrient cycling, soil formation and maintenance, pest control, pollination. Biodiversity also has *non-use value*, which concerns its intrinsic, aesthetic (pleasure that people derive directly from biodiversity) and existence (knowledge that biodiversity exists even if never encountered) values. Fundamentally, the maintenance of biodiversity is essential for the normal functioning of ecosystems, and so the continued provision of goods and services upon which increasing human populations depend.

Current economic valuation of biodiversity focuses only on 'use values' and tends to promote short-term, consumptive exploitation, which generally has a negative impact on species and ecosystems and threatens long-term productivity. Efforts to promote sustainable use (entailing a long-term view) must acknowledge that current patterns of use are, on the whole, destructive. Exploited ecosystems are at increasing risk of being destabilised through increased rates of use due to human population growth, environmental change and unpredictable ecological processes. Efforts to conserve biodiversity must therefore seek to limit rates of consumptive use (a politically unpopular policy option), as well as ensure that local communities obtain economic benefit from biodiversity conservation.

Management for biodiversity

Management planning is a knowledge-driven process. Relevant knowledge may comprise: information about particular phenomena (such as patterns of species and factors influencing them); scientific principles and concepts; or less well-tested ideas and theories. Limited knowledge and conflicting interests make management planning a difficult task. The task, however, is one that can be aided by the use of a formal decision theory approach, which may involve quantitative or qualitative methods.

Decision support for management planning

The primary task in any decision-making process is to define clear management objectives. Determining objectives can be challenging, especially for complex ecosystems such as Mkomazi, which tend to behave in unpredictable ways in response to perturbation. Single species approaches to conservation tend to give rise to straight-forward objectives, such as minimising risks of extinction. Ecosystem approaches are likely to give rise to complex or more vaguely phrased objectives that are a challenge to analyse in a decision theory framework. This is particularly true of 'conservation with development' approaches to management.

The next step is to define possible management options. This step is directed by policy and draws on all available knowledge of a particular ecosystem (biodiversity patterns, ecosystem function, driving factors and processes, and the temporal and spatial scales at which different processes operate), as well as ecological and social scientific concepts that are relevant. Resources (time and finance) available for implementation will constrain those management options which are feasible, possibly ruling out sophisticated options that might, nevertheless, have a high chance of achieving the management objective.

Management options need to be defined in terms of variables that describe the state of the ecosystem and that can be monitored to measure any changes. Deciding which options to implement involves predicting changes in the variables that might arise from each management action. The predictability (level of certainty, risk, uncertainty and chaos) of a particular management action is determined by the extent of our knowledge of the ecosystem, and will influence whether that action is selected as an option for further consideration. Thus the baseline research findings of the Mkomazi Ecological Research Programme can be used critically to provide meaningful 'state variables' (such as canopy arthropod diversity or grass species diversity) for defining options, as well as to provide an understanding of important ecosystem processes such as fire.

Where there is a lack of 'hard' data to make sound judgements, subjective *expert* opinion can be of use: ecosystem management tends to involve more qualitative processes like ecological risk and multi-criteria assessments, which rely more on expert knowledge. Traditional knowledge about how ecosystems function, acquired in relation to hunting, gathering and cultivation practices (often over many hundreds of years of experience), may provide useful management insights.

Adaptive management

Selecting between alternative options for managing ecosystems that are highly dynamic and thereby less predictable is problematic. Under these circumstances, an adaptive management process can be used to guide management intervention. In this approach, alternative management strategies (effectively different models of how the ecosystem works) are defined. Analysis of each strategy gives rise to a set of management actions with predicted consequences on the ecosystem. The strategies and their associated predictions are, in a sense, hypotheses that can be 'tested' through implementation. A critical step is monitoring the effects of implementation. This provides data to help assess a particular strategy and so constitutes the essential feedback loop. Modifications to the strategies and associated management actions can be made in light of the new information.

Management for biodiversity in Mkomazi

Research in Mkomazi has gone a long way towards assessing the biodiversity of the reserve, as the many chapters of Section I clearly show. Coupling this information with current ecological and conservation thinking provides the scientific inputs to the management planning process.

Ecology of semi-arid savanna

Ecological research is just beginning to understand the relationships between biodiversity, ecological processes and ecosystem structure and function. The *stability* of an ecosystem arises from its ability to maintain structure and/or function during disturbance (termed *resistance*), and to recover following disturbance (termed *resilience*). These properties are positively related to species diversity (McCann *et al.* 1998, McGrady-Steed *et al.* 1997): the existence of many weak, trophic interactions between species (although for a particular ecosystem a number of interactions may be strong) serves to dampen population fluctuations, reduce variability in ecological processes and so reduce species loss.

Spatially heterogeneous and very dynamic ecosystems, such as semi-arid savanna, tend to have several alternative stable states, adding to their resilience. When the ecosystem is functioning normally, unusual perturbations (such as unusually high rainfall or fire or intense megaherbivore activity) may move its state away from one stable equilibrium towards another. When the ecosystem is degraded in some way and so not functioning normally, these kinds of perturbations may push the state of an ecosystem away from all possible equilibria and towards an ecological *limit*, risking destabilisation of the entire ecosystem. The risks of this happening are greater if the ecosystem is already experiencing other perturbations, such as climate extremes, excessive anthropogenic exploitation, and pollution.

Evidence suggests that semi-arid savannas evolve through boom and bust cycles, driven by fire, herbivory, episodic heavy rains and drought (El Niño Southern Oscillation and longer-term events). Recent work also suggests a remarkable comparative resilience to perturbation, in which the capacity for 'regeneration' is maintained for relatively long periods (Dublin 1995, Leuthold 1996). Despite this apparent resilience, it is well observed that human use of low productivity areas like Mkomazi leads to important changes in biodiversity. These changes *might* be reversible but this will depend on the state of the ecosystem and the extent and density of anthropogenic disturbance. Local anthropogenic disturbance that does not mimic natural disturbances, in terms of nature, intensity, frequency, spatial patterning, and variability, is likely to degrade ecosystems in ways that may not be apparent for long periods of time.

Management options

The task of defining management objectives and actions for Mkomazi is influenced by ecological, socio-economic and political forces. As outlined in Chapter 36, the Wildlife Division has established the basis of a general management plan for Mkomazi in which the overall management objective is to enhance the conservation status of the reserve, essentially through improved interaction between reserve-adjacent communities and management processes, and through efforts to manage for sustainable use. Brief consideration is given here, in light of these objectives, to some management options that might be appropriate to a management situation lying somewhere between the extreme possibilities of uncontrolled access and total exclusion.

Consumptive use of resources

Mkomazi's reserve-adjacent communities incur opportunity costs associated with a lack of access to reserve resources. An apparently conservative estimate of the livestock opportunity costs is £150,000 per year (see Chapters 33 and 37). Opportunity costs associated with cultivation and gathering of resources may also be significant. Reducing these opportunity costs by allowing restricted utilisation of reserve resources might be considered as a management option.

Leaving aside the challenge of deciding to whom access would be granted, planning this option requires knowledge of which resources (such as grazing land, wood, bush meat, medicines) are most important in supporting or enhancing local livelihoods, where they are located, when they are available and what 'demand' exists for them. Control of access and utilisation would need to be carefully managed, especially where demand is high. Effective control might be achieved by the use of existing local resource management systems (see Chapters 33 and 34), probably following adaptation to ensure that access was spatially and temporally restricted so as to minimise negative ecosystem impacts.

There are various kinds of negative effects on ecosystems of multiple-use options. Consumptive use of wild species has detectable negative impacts on the population dynamics of those species. In some cases these impacts may be enough to cause the extinction of species in the area and threaten the stability of the ecosystem. Other more general impacts include certain, even though possibly not long-term, habitat loss and fragmentation (a significant factor in loss of biodiversity), possible reduced habitat heterogeneity, and likely disruption of ecological processes, which, even if temporary, may reduce ecosystem stability and productivity. The chance of negative impacts increases as the spatial extent, frequency and intensity of exploitation increases. Multiple use of Mkomazi will also increase the chances of human-wildlife conflicts, which is a major risk factor for the persistence of species.

Applying current conservation theory (Hanski 1997) to Mkomazi suggests that multiple use of the reserve is not likely to be a sustainable option. The habitat loss and fragmentation that would result, even if temporarily, from increasing pastoral and cultivator access will tend to lead to a reduction in biodiversity. The consequences of this for ecosystem stability and productivity are difficult to predict but it is safe to assume that in the short term, at least, there will be measurable effects. It is likely that long-term negative impacts will arise from patterns of utilisation like those observed prior to eviction of pastoralists in 1988.

Another way of reducing opportunity costs is to develop tourism within the reserve (see below). Tourist income could be earned directly through provision of market goods and services, and/or could be dispersed within the community as financial compensation for opportunity costs. It is increasingly acknowledged, however, that where local populations are to have a vested interest, in terms of consumptive use, in management for biodiversity there must be "constructive engagement with existing economy", amounting to enhancement rather than replacement of local livelihoods (Brown 1998).

Simulating and controlling ecological processes

Whether local community use of the reserve is permitted or not, consideration has to be made of the possible need to manage ecological processes. Disturbances at various spatial and temporal scales, such as the falling of a forest tree, the grazing of a herd of zebra or the flooding of a watershed resulting from an El Niño Southern Oscillation (ENSO) event, are normal in every ecosystem on Earth.

Especially dynamic ecosystems, like semi-arid savanna, tend to experience a wide range of disturbances at a variety of scales and intensities. In these cases, the disturbance events shape the ecosystem and its associated biodiversity. The temporal patterning and intensity of rainfall, for instance, is critically important to the dynamics of the Mkomazi ecosystem. Variably periodic extreme rainfall events, like that experienced by Mkomazi during the 1996–98 ENSO event, when a prolonged period of unusually intense drought was followed by exceptionally heavy rainfall for many months, undoubtedly shape ecosystem dynamics.

Key, potentially manageable disturbance factors and processes in Mkomazi are fire and herbivory. Fire is both a natural and an anthropogenic phenomenon in Mkomazi. The ecosystem is adapted to fire regimes that are characterised by their frequency, spatial patterning and intensity of events, much like many other savanna ecosystems (Braithwaite 1996). Significant changes in fire regime, such as very frequent and/or spatially extensive fires, or fires of greater than usual intensity, can have negative impacts on ecosystem structure and function.

The main issue in managing fire is to ensure that 'hot' burns, which are largely dependent on the amount of readily-combustible biomass (itself related to the interval between fires), do not occur. Avoiding accumulated biomass in the grassland and bushland areas probably requires annual burning. Also important, however, is to determine how frequent and what the spatial patterning of 'cool' burning should be, as well as to ensure that any controlled fires are controllable. Management options involving fire are considered in Chapter 9.

Temporal and spatial patterns of herbivory also play a significant role in shaping Mkomazi's habitat. The variety of wild grazers and browsers, from dikdik, through impala, gerenuk and oryx to elephant, influence the dynamics of different aspects of the ecosystem in different ways and to different extents. Cattle grazing tends not to mimic natural processes. Controlling wild herbivore populations (preventing over- or under-population) is usually considered as an option only in special ecological or species-specific circumstances, such as elephant culling to reduce habitat destruction and intense anti-poaching activity to prevent rhino loss. Management of wild herbivores usually implies protection but may be related to other management options, such as water management and tourism (see below).

Manipulating water availability

Most of the water holes in Mkomazi are artificial. They each hold water well into the dry season with the effect that populations of many animal species are kept within the reserve and concentrated during a period of particularly low productivity. Without the water the individuals might die but would certainly be forced to migrate to other water sources, possibly into Tsavo West National Park or towards Lake Jipe (possibly increasing wildlife-human conflicts). One consequence of maintaining water availability is that rates of herbivory increase in the area surrounding the water hole. The immediate vicinity of water holes is also heavily trampled, although this effect is very local.

The cost of maintaining water availability, and therefore animal species populations, for a period in the dry season is usually high but might be considered necessary for development of tourism (see below). The ecological impacts of this option are difficult to predict. The increased, relatively local herbivory experienced annually during a non-productive period might be intense, with the possibility of significant, negative longer-term impacts on ecosystem processes.

Ecosystem enhancement

Restoring or rehabilitating an ecosystem assumes that it is in a degraded state. The state of dynamic and heterogeneous ecosystems is a challenge to assess. Where there is a history of human modification of these ecosystems, it is usually possible to identify ways of enhancing them.

Re-introductions of locally extinct species, such as black rhino and wild dog, are controversial actions. Such actions, involving endangered species, need to be evaluated in the context of international policy and priorities, not least because of high costs. The ecological role, impacts and advisability of re-introductions also need close analysis. Release of captive-bred wild dogs, for instance, should only be attempted if it is clear that a population is viable. Many factors threaten the viability of a wild dog population in Mkomazi, including disease, limited access to resources, and risk of wildlife-human interaction and persecution. The fact that wild dog are present in neighbouring Tsavo but have not yet re-established in Mkomazi suggests that currently there are constraints on population viability.

Tourism

The development of consumptive and non-consumptive tourism is often seen as the answer to resolving conflicts over biodiversity conservation and use: economic opportunities for reserve-adjacent communities, and perhaps even income from some kind of tourist levy, provide compensation for opportunity costs of exclusion from the reserve. Tourism can be perceived as providing little local benefit and even as degrading local livelihoods. Clearly, any consideration of tourism as a management option needs to fully engage local communities.

Sustainable, consumptive tourism is almost certainly not feasible in Mkomazi because of the reserve's size and the population sizes of trophy species. Non-consumptive tourism (such as game viewing) might be feasible but would need to be of a type and on a scale appropriate to the size and shape of the reserve, and compatible with its ecological integrity. Development of specialist, low impact wildlife tourism, such as bird-watching and even butterfly safaris, is likely to be the most viable tourist option. This kind of tourism has a niche market, and it would require relatively modest infra-structural development of the reserve. It is likely that gross income from tourism within Mkomazi itself would be modest. The added value of including low impact tourism in Mkomazi within specialist safaris to several protected areas might be significant.

Conclusion

Mkomazi is spatially and temporally very heterogeneous, with a great diversity of habitat types resulting, largely, from marked physiographic and environmental variability. This variability gives rise to an especially rich floral diversity compared with other, similar semi-arid savanna reserves in the region (Chapter 5). Vegetation diversity is the template against which species-rich arthropod (Chapter 10) and avian (Chapters 26 and 27) communities have evolved in the area, and which supports a relatively high large mammal diversity (Chapter 31). Mkomazi's contribution to the persistence of semi-arid savanna floral and faunal diversity is significant and is made greater as increasing rates of habitat loss, fragmentation and environmental change disrupt these ecosystems.

The conservation significance of Mkomazi is enhanced by is geographical position. Mkomazi's biodiversity and its proximity to Tsavo West National Park in Kenya makes it an integral part of the greater Tsavo ecosystem. The spatial relationship of Mkomazi and Tsavo is important for both protected areas: Mkomazi probably provides important wet season resources for mammal populations in the Tsavo ecosystem, not to mention a 'buffer' between human settlement and the National Park, while in Tsavo, which similarly buffers Mkomazi, relatively stable mammal populations act as 'source' populations for Mkomazi.

Short-term impacts of anthropogenic activity (primarily pastoral use and associated fire) on the vegetation composition and large mammal diversity of Mkomazi are obvious. Whether these changes persist in the long term or have long-term negative impacts on ecosystem stability and productivity depends on the frequency, spatial patterning and intensity of activities. Increasing human populations and improving technology pose a threat to the sustainability of low productivity ecosystems through changes in rates of biodiversity use.

The needs and development aspirations of growing local populations cannot be ignored in any consideration of management of natural resources. The opportunity costs to reserve-adjacent communities of the protection of Mkomazi can be significant, to the extent that local livelihoods cannot provide a reasonable quality of life (see Chapter 33). Expecting to meet these needs and aspirations through recent modes of exploitation of Mkomazi's biological resources is almost certainly unrealistic. Continued use of biological resources, of the type seen during the 1980s, leads to changes in ecosystem structure and function that threaten the already low productivity.

Balancing the need to conserve biodiversity with the need to use biological resources requires compromise. In preparing the general management plan for Mkomazi (outlined in Chapter 36), the Wildlife Division has taken a bold step for biodiversity. A wise manager would place this step in an adaptive management context, monitoring ecosystem and socio-economic indicators to assess progress towards the management objectives, and adjusting the management process as necessary.

Acknowledgements

The Darwin Initiative for the Survival of Species provided financial support. Thanks to Susan Canney and Malcolm Coe for comments on a draft of this chapter.

References

Braithwaite, R. (1996) Biodiversity and fire in the savanna landscape. In Solbrig, O., Medina, E. & Silva, J. (eds.) *Biodiversity and Savanna Ecosystem Processes: a Global Perspective*. Ecological Studies 121. Springer, Berlin. pp. 121-141.

Brown, D. (1998) *Participatory biodiversity conservation—rethinking the strategy in the low tourist potential areas of tropical Africa*. Natural Resource Perspectives No. 33, Overseas Development Institute, London.

Dublin, H. (1995) Vegetation dynamics in the Serengeti-Mara ecosystem: the role of elephants, fire and other factors. In Sinclair, A. & Arcese, P. (eds.) *Serengeti II: dynamics, management and conservation of an ecosystem*. Chicago University Press, Chicago. pp. 71-90.

Hanski, I.A. (1997) Metapopulation dynamics: from concepts and observations to predictive models. In Hanski, I.A. & Gilpin, M.E. (eds.) *Metapopulation biology: ecology, genetics and evolution*. Academic Press. pp. 69-91.

Leuthold, W. (1996) Recovery of woody vegetation in Tsavo East National Park, Kenya, 1970-1994. *African Journal of Ecology* 34: 101-112.

McCann, K., Hastings, A. & Huxel, G. (1998) Weak trophic interactions and the balance of nature. *Nature* 395: 794-798.

McGrady-Steed, J., Harris, P.M. & Morin, P.J. (1997) Biodiversity regulates ecosystem predictability. *Nature* 390: 162-165.

Ministry of Natural Resources and Tourism (MNRT) (1998) *The Wildlife Policy of Tanzania*. MNRT, Tanzania.

CHAPTER 36

The status and future management of Mkomazi Game Reserve

Bakari N. N. Mbano

Background

The Mkomazi Game Reserve was established under the British colonial administration in 1951. From 1951 to 1972, MGR was administered by the then Game Department. However, with the coming of the Government's Decentralisation policy of 1972, the western part of MGR which is situated in Same District fell under the administration of Kilimanjaro region. The eastern part of MGR (Umba GR) which is in Lushoto District was taken over by the Tanga Regional administration. The boundaries of the two parts of MGR (Mkomazi GR and Umba GR, known collectively as MUGR) are described in the Wildlife Conservation Act No. 12 of 1974. From 1988 to the present MGR has been administered by the Wildlife Division of the Ministry of Natural Resources and Tourism.

The changes of administration have affected the conservation and management of MGR in different ways. Phase one administration (1951–1972) invested substantially on infrastructure development such as roads, game outposts and dams to store water which was a limiting factor as far as drinking water for animals is concerned. The availability of water for wildlife in these dams also attracted livestock.

The low priority in planning, budgetary limitations, high poaching and poor management in the phase two administration resulted in serious deterioration in terms of both wild animal numbers and vegetation condition, including the disappearance of the black rhino *(Diceros bicornis)* from the reserve. Extended drought and overgrazing by livestock resulted in siltation of dams.

The Wildlife Division, which is currently managing the reserve, does not have enough funds to manage the reserve effectively. Assistance has been sought from different donors/conservationists.

1. *The George Adamson Wildlife Preservation Trust (GAWPT)* In 1988, the GAWPT was requested by the Wildlife Division to reintroduce the endangered species (black

rhinoceros and the African wild dog) into the MGR. After the construction of the sanctuary in 1995, four rhinos were reintroduced in the reserve in October 1997 from Addo Elephant National Park in South Africa. The rhinos are being kept in an electrical fenced sanctuary at Kisima before they are released into the reserve. 25 wild dog puppies were captured from Simanjiro areas and are kept in a breeding station at Kisima. The dogs at Kisima are kept in three dens and have started reproducing.
2. *The Mkomazi Ecological Research Programme* This was a joint programme between the Wildlife Division and the Royal Geographical Society (with IBG). The aim of the programme was to collect ecological data which could assist in planning the management of the reserve.

MUGR at present is not generating any revenue. Its day to day functioning depends entirely on government funding and other game reserves in Tanzania. The funds provided by the government are in most cases inadequate to meet the development and recurrent budgets of the reserve. Nevertheless, the government is trying to revamp the dilapidated infrastructure in the MUGR. For example, a two family staff house has been constructed at Njiro game post. The roads in the reserve have regularly been graded to enhance anti-poaching operations and other activities. An information centre (funded by Friends of Conservation) is under construction at the reserve headquarters at Zange. In 1997 the government provided 25 million Tanzanian shillings for the rehabilitation of the damaged Dindira Dam. Furthermore, the government has been supporting village development projects in the surrounding villages such as the construction of a secondary school at Kisiwani village, a livestock watering dam at Same-Mkai area and a laboratory at Same Secondary School.

The future of MUGR

Mkomazi Game Reserve was among the first five priority Game Reserves in need of General Management Plans (GMPs) to guide their management activities. The other four are Ugalla, Moyowosi/Kigosi, Rungwa and Saadani Game Reserves. The GMPs for Ugalla and Rungwa game reserves were prepared during the time of a USAID funded project, Planning and Assessment for Wildlife Management (PAWM) project.

The PAWM project which started in 1991 ended in June 1996, leaving Mkomazi, Moyowosi/Kigosi and Saadani Game Reserves without General Management Plans. As resource use conflicts in and around MUGR intensified the Wildlife Division in 1997 requested assistance from USAID to develop the MUGR GMP. Having realised that Mkomazi GR has stakeholders with different interests to defend and that resource use conflicts are intensifying, the preparation of MUGR GMP had to take into account the interests of all stakeholders. This made the preparation of the GMP very expensive. To ensure involvement and effective participation of all stakeholders the preparation of GMT was done in a series of workshops in three phases.

1. All current MUGR stakeholders were invited to attend a consultative workshop in Same town. The workshop was proceeded by an excursion to selected areas in and around Mkomazi. This enabled the participants to acquaint themselves with the local environment and key issues in the future conservation and management of the reserve.
2. A workshop of stakeholders was conducted using Logical Framework Approach and a combination of participatory techniques, to ensure that consensus on issues was reached and the interests of each group were taken into consideration. Participants suggested a number of activities to be carried out inside and outside of the reserve, to ensure that the conservation goal of Mkomazi is attained. This was followed by a workshop by a group of experts. The main function of this workshop was to use the information collected during the excursion in the first workshop, and other technical information, to come up with statements of MUGR management zones and environmental impacts of the proposed activities in each zone.
3. Wildlife staff prepared the draft GMP based on the information collected in all the workshops. The draft GMP was tabled at the reviewers workshop, which besides the stakeholders also included selected scholars and other experts. Those who could not attend were requested to send their comments in writing.

Highlights of the GMP

1. *Sectoral policies* The preparation of the MUGR GMP took cognisance of the Wildlife Policy (1998) and other relevant sectoral policies operating in and around MUGR such as, the Agricultural and Livestock Policy (draft 1997), Tourism Policy (draft 1997), National Land Policy (1994) and National Environmental Policy (draft 1996).
2. *Management issues* The GMP identified 23 management issues as revealed by the MUGR stakeholders. Issues identified are related to wildlife–human interactions, MUGR management–local people interactions, natural resources management, administration and operations, and wildlife policy and laws.
3. *Management objectives* The main objective (of MUGR) is to enhance the conservation of MUGR. Specific objectives of MUGR are to enhance reserve relationship with adjacent communities, management of natural resources sustainably, streamline reserve administration and management operations, promote visitor use and development and review policy and laws.
4. *Management strategies* These were carefully developed to ensure that they will enhance the attainment of management objectives. These are: boundary conflict resolving, ensuring community involvement and participation in conservation of natural resources, managing natural resources sustainably, developing & man-

aging visitor use facilities, administering resources and staff, protecting and managing social-cultural resources, facilitating preparation and review of policy and laws, and zoning scheme. Studies and plans were also identified and recommended as strategies to assist in carrying out proper management and appropriate decisions. These include: developing community conservation programmes, preparing rhino management, road network and construction, and tourism plans. Livestock numbers and dynamics, water sources, impact of fencing conflict areas, and alternative ways of disposing/using wild dogs were among the studies recommended.

5. *Management Zones* Seven zones were identified. Identification of zones was based on the protection of resource values of MUGR, solving the existing conflicts and attainment of management objectives. These are:

(a) Administrative Use Zone: areas which will be used for general management and administration of the reserve.

(b) Wilderness Zone: areas on top of mountain hills with outstanding resource values which need high level of protection. The areas have been selected because they are fragile and prone to environmental perturbations.

(c) Restoration Zone: an area demarcated for restoration of the endangered black rhinoceros. This area is an appropriate natural habitat for rhino, facilities for reintroduction of rhino already exist and it is centrally placed in the reserve, therefore providing high security to the rhino.

(d) General Tourism Use Zone: the area west of Tanga/Kilimanjaro regional border. It will be used for photographic tourism only. The area was selected as a tourism zone because existing high outstanding resource values to attract tourists, high visibility, can form a tourist circuit and link the coastal area with the northern parks, and has previously been used for the same purpose.

(e) Sport Hunting Zone: the eastern side of the Tanga/Kilimanjaro regional boundary (Umba Game Reserve). Criteria used for zoning are: the potential to generate revenue through hunting, its contribution to increase anti-poaching surveillance in the area, and availability of hunting species like oryx, gerenuk, buffalo, waterbuck, eland and cat species such as lions and leopards.

(f) Cultural Use Zone: areas/sites in MUGR where local people associate with their cultural and traditional values, rituals and collect local medicines. These areas have been previously used by local people through a permission from MUGR management. Use of resources in these areas will be regulated by MUGR management.

(g) Community Based Conservation Zone: areas outside and contiguous with MUGR. According to the Wildlife Policy, 1998, these areas will be gazetted as wildlife management areas. In these areas local communities will manage and utilise wildlife resources for their own benefits. Areas were selected based on the level of existing and potential wildlife–human resource use conflict. The strategy is intended to reduce/minimise the existing conflicts.

In describing the boundaries of each zone, descriptive features will be used. Maps will be drawn and provided to visitors and MUGR personnel to assist in the identification of zonal boundaries on the ground.

Evaluation of selected strategic actions to the environment

Specific environmental topics were identified based on the professional judgement and vast resource knowledge of the interdisciplinary planning team. Rationale for identification was based on priorities of stakeholders, identified management problems, defined management objectives, and protection of MUGR significant resource values. The same rationale was used to select major actions to be discussed. Actions include: allowing prescribed early burning in the reserve, reintroducing black rhinoceros, existing African hunting dog project, rehabilitating and constructing new dams, prohibiting all permanent physical development within the reserve, establishing camping and picnic sites, allowing hiking, night game drive, collecting medicines and performing rituals in the reserve, fencing parts of MUGR, and making Ibaya camp a tourist facility.

The rhino project

Environmental impact analysis showed high initial cost of the rhino project, and that, patrols, monitoring and maintenance of the electric fence will add cost to the management. This puts the viability of the project at stake. Some revenue is expected from tourism and conservation donor agencies. These might cushion the management cost. A rhino management plan has also been recommended to show the duration of the project and how negative impacts will be mitigated.

The wild dog project

The wild dog project has potential interests with the surrounding local communities. Environmental impact analysis indicated high management costs in feeding, regular medical treatments, maintenance of cleanliness and regular patrols on dens. The major impact is on after release effect.

Conclusion

Mutual agreement on the future management of MUGR, has been reached in the planning process of MUGR by the majority of stakeholders. Most of the stakeholders' interests have been accommodated and the Government hopes that human/wildlife conflict will be minimal and conservation of MUGR will be enhanced. In order to maintain the good relation initiated during the planning phase, the implementation phase, involvement and participation of the local communities surrounding the reserve will be, highly encouraged. We urge the donor community and all wildlife conservation agencies to support MUGR conservation efforts, by assisting the implementation of the General Management Plan.

References

Ministry of Natural Resources and Tourism (1997) *The Mkomazi/Umba Game Reserves General Management Plan (draft)*. Dar es Salaam, Tanzania.

Ministry of Natural Resources and Tourism (1998) *The Wildlife Policy of Tanzania*. Dar es Salaam, Tanzania.

Government of Tanzania (1974) *The Wildlife Conservation Act No. 12 of 1974*. Dar es Salaam, Tanzania.

CHAPTER 37

Discussion and conclusions

Malcolm Coe

When the Department of Wildlife requested the Royal Geographical Society to undertake a study of biodiversity in the Mkomazi Game Reserve, we little realised what an area of great biological and ecological interest we were about to investigate. We were given a very free hand to achieve our aim of completing an inventory of the area's biodiversity, in order that the information, which is contained in this volume, could be used by the Department of Wildlife to plan for the area's future management and utilisation. It has recently been implied (Sanşom 1998) that the eviction of pastoralists from the MGR was in some way a prerequisite to the Royal Geographical Society participating in this programme. At no time was this matter ever discussed or our views sought, nor would it have been appropriate for the Department of Wildlife to have done so. Indeed it is true to say that neither during the stages of either planning or execution were members of the MERP consulted about current or future management policy or its execution. This does not necessarily mean to say that as a group of ecologists we would not like to be able to provide some input into the development of management programmes—but over many years of experience in Africa we have learnt that sensitivities about interference in administrative procedures is something that we all need to be continuously aware of. In the contemporary corridors of conservation there is often still a powerful atmosphere of 'father knows best'.

The Mkomazi Game Reserve is a comparatively small area of semi-arid savanna lying on the Tanzania-Kenya border. Its surface is virtually waterless for much of the year which has made both it and the Taru Desert across the Kenya border a formidable barrier to local human husbandry and exploration. Those wishing to live or travel here have tended to avoid it and crossed the area either further south via the Usambara and Pare Mountains or along the Yatta Plateau and the Galana River to the north. Until the Game Department (Department of Wildlife) constructed dams or increased the holding capacity of natural pans in the late 1950s and early 1960s, water was only available at the Umba River in the east, the springs of Kifukua, the Ngurunga pot holes and a number of small springs in gullies, whose water production was and still is entirely dependent on a low level of disturbance to woody vegetation on upper hill slopes and summits. Small water holes that only last a few weeks after the rains have finished are scattered throughout the area and have been almost entirely created by the activity of elephants on the sites of old *Macrotermes* mounds. Indeed Ayeni (1975a, 1975b & 1979) concluded that over 80% of these transient water sources in Tsavo East were formed through the agency of elephant activity. It seems reasonable to assume that excavation and natural maintenance of these water holes has fallen dramatically since poaching has reduced the elephant population in the greater Tsavo ecosystem by over 80% since the early 1980s.

During the rainy season there is abundant forage and water but black cotton soils in the valley bottoms render many of these areas impassable to both man and beast. To a large extent this lack of water and the presence of tsetse flies *(Glossina brevipalpis, G. longipennis* and *G. pallidipes*) have always rendered the MGR, except for the west to south-central fringe, unsuitable for domestic stock. It is only in the last 30 or so years that the Maasai have extended their traditional range eastwards[1] and have taken to entering this area (Anstey 1998[2]) from the west and the south for seasonal grazing.

The MGR slopes downwards from south to north and west to east under the topographic influence of the Usambara and Pare Mountains. These mountains which are composed of the intensely folded metamorphic rocks of the Mozambique Belt have their origins in the upper Pre-Cambrian (c. 570 million years before present) and were probably elevated above the surface during the mid-Tertiary period (c. 30 million years before present) (Morgan 1973). Since their elevation, these massifs have been intensely eroded, depositing extensive colluvial material on the lower slopes and deep accumulations of hygroscopic clay or black cotton soil (vertisols) in basins between outlying hills. Many of the smaller hills have been completely eroded, their presence now discernible only through the extensive circular or ovoid patches of bright red oxidised soils which remain on the surface.

Thus we have seen that the habitats of the MGR are to a large degree controlled by the important physical factors of elevation, slope, soil type and drainage. The major habitat types are here classified according to their plant species composi-

---

[1] The Maasai are thought to have entered Kenya from the north about 500 years ago (Ogot 1981). In the pre-colonial era there was almost continual warfare between the Maasai in the west and the Wakamba and the Galla in the east, to which we must add the pressure from advancing Somali herders from the north-east (Coe 1985). Indeed it is the historical belief of the Maasai that Engai (Heaven or God) (Krapf 1860) gave them all the cattle in the world, which justified their philosophy of stealing the cattle of their neighbours and annexing their grazing lands. Some of the Maasai war parties of up to 1,000 warriors (*moran*) penetrated as far as the ancient sea port of Malindi in 1867 (New 1873).

[2] David Anstey, who knew this area well as the Senior Warden of the Tanganyika/Tanzania Game Division during the 1950s and 60s, kindly made extracts from his field notes available to us.

tion and their location in relation to local physiography. Current studies being undertaken by Susan Canney will generate a vegetation map, the divisions of which will be based upon the analysis of species composition samples measured in the field and plotted on our GIS grid with the help of both aerial photographs and satellite images. Bearing in mind however that the boundaries of most of the habitat divisions tend to grade into one another, a very fine sub-division of habitats will probably be of more value in the laboratory than in the field. Field staff of the Department of Wildlife will probably find the descriptions in this volume (Chapter 7) of value in looking at the effects of both long and short term disturbance on the local biota. The current plant species (taxa) in the MGR number up to 1,300 with a predicted eventual total in excess of 1,500, which comprises 15% of the 10,000 species predicted for the whole of Tanzania by Polhill (1968). This high level of plant species richness in relation to its area, is together with the high level of local physiographic diversity, a major determinant of this region's great ecological importance.

There is however also an important temporal component to both habitat distribution and composition, since the unpredictable climates of these savanna environments almost certainly have a profound influence on plant size and longevity. Glover (pers. comm. 1974) suggested from his observations of the woody flora of the Tsavo National Park, Kenya that few of the tree species lived longer than 40 years. This observation seemed to be corroborated when we cut sections of *Acacia* and *Commiphora* trees which had been knocked over by elephants in the Tsavo East NP in the mid-1970s, suggesting that most (if not all) of the trees being felled were probably 'time expired', their roots less secure and that they were at the end of their active growing phase. The apparent periodicity of drought, subsequent heavy rainfall (Laws 1969) and the recurrent damage to trees by elephants (Caughley 1976), converting mature woodland to open grassland, seem to be a significant part of the complex interaction between large herbivorous mammals and the savanna environment, which has been modelled by Wijngaarden (1985) in Tsavo. However the high levels of elephant damage to woody habitats in Tsavo at this time, when elephant densities reached nearly 1.5 km^2, largely occurred as a result of elephants being concentrated in the Park through human population increase in surrounding areas (Corfield 1973).

Bearing in mind that both woodland and elephants turn over their biomass at a rate of between 3 and 5% per annum (Coe 1990), while perennial grasslands turn over about 33% of their biomass per annum, we may postulate that these comparatively short-lived savanna woodlands exist as part of an irregular grassland-woodland cycle which has been driven in the distant past by drought, natural fires and the megaherbivores. Thus in the woodland phase the nutrients are immobilised in the trees while during the grassland phase the same materials are cycled through the herbivorous mammals. We find therefore that woodlands (and forests) carry a high standing crop biomass of woody material and a low standing crop biomass of large herbivores, while the grasslands bear a low standing crop biomass of grass but may carry a very high standing crop of vertebrate grazers. It is perhaps no coincidence that there is a remarkable similarity between these cycles and the almost eternal wandering of herders with their domestic stock, who utilised the local resources until they were exhausted and then move on in search of new grazing lands or take over the lands and stock of their neighbours (Coe 1990). We note that the comparatively recent history of pastoralists in eastern Africa (c. 1,000–2,000 years before present) has almost continuously been one of southward movement rather than northwards. In one of the most important ecological contributions to this debate, Sinclair & Fryxell (1985) have pointed out that there is a remarkable similarity between the migrations of wild large herbivorous mammals and the pastoralists. Today, sadly, unless we can create more conservation areas or pastoralists can be helped to change their way of life, unsustainable at present human densities in many areas, the outlook for the richest surviving large mammalian fauna in the world is, to say the least, bleak.

During the course of our studies, the MGR has received a great deal of adverse publicity as a result of a long court case, in which a small group of Maasai pastoralists have attempted to secure land-rights in the reserve. This litigation has been extensively supported by a number of organisations based in the United Kingdom (including African Initiatives, Pilot Light and Survival International; the first two named having been funded by Comic Relief). It has been claimed that as a result of the eviction of a small group of Maasai from the MGR by the Government of Tanzania, the village of Kisiwani's[1] cattle market could have lost an income of £150,000 a year (or even as high as £500,000) (see Brockington & Homewood, Chapter 33). Assuming that local cattle would fetch about £25 each in the market, this would represent the sale of 6,000 animals each year. Bearing in mind the near lack of permanent water in the MGR, it seems reasonable to suggest that these animals were being carried on no more than 50% of the total area (in fact probably a lot less—perhaps no more than 15–25%) or 1,850km^2. Assuming the unit weight of these cattle is 180 kg (Coe *et al.* 1976) we may predict that this represents 3.2 animals/km^2 or an offtake of of 576 kg/km^2, though on only 20% of the area it is elevated to 1,440 kg/km^2. This is though only the average annual offtake and does not reflect the whole population, which at a cattle production/biomass ratio of 0.15 or 15% per annum the actual cattle biomass needed to support such an offtake would be 21.3 animals/km^2 or 3,836 kg/km^2. To put this in perspective, if we assume that the rainfall in this area of the MGR is 400mm then we may predict that the Optimal Carrying Capacity (Coe 1976) for *all* the large herbivores would be 2,266 kg/km^2 or 59% of the minimum predicted domestic stock biomass. Thus

[1] This village lies at the foot of the northern slope of the South Pare Mountain and was formerly a busy trading centre until the collapse of the sisal market, the closure of local plantations and the re-routing of the main road to the northern side of the mountain.

we observe that if the domestic stock exceeds the carrying capacity by 1,570 kg/km^2 the wild herbivores would be under unacceptable pressure. These savannas appear green and lush with the advent of adequate rain but for much of the year they are parched. It is frequently suggested that that all that is needed is to excavate more water holes in the MGR but this is only a short-term solution since during periods of drought the primary cause of herbivore mortality is starvation not a lack of water (Foster & Coe 1968).

Although natural fires do occur in savanna environments (Trollope 1984) it is those generated by humans that have the most profound effect on the indigenous flora and fauna. In MGR the commonest origins of these fires are from pastoralists to stimulate post-fire grass growth towards the end of the dry season by burning off the dried-grass standing crop. Since the removal of the small number of illegally resident pastoralists from the reserve, our observations since 1993 suggest that the number of these fires has been reduced. However, they are still quite common in the western segments of MGR from Pangaro, east to Ngurunga and northwards to the Kavateta dam where the standing crop biomass of indigenous large herbivorous mammals is greater than elsewhere. Over the last 15 years rising human numbers in the reserve's surroundings[1] have led to settlers moving into the hills (Zange) on and within the reserve's western boundary, for subsistence agriculture as well as felling in hill-summit forests for poles and charcoal production. It is fires generated by these activities that have almost certainly had a much more severe and recent effect on the vegetation of the Mbono valley and its surrounding hill slopes than that of fires generated by pastoralists, who are largely restricted to the valley bottoms. These hill-fires are pushed over the major ridges and down into the valleys by katabatic winds. The combined effect of fire sweeping both up and down the valley walls has led to the loss of most of the woodland in the more productive areas of MGR. Herein lies the problem. If an income is to be generated from MGR which can be partly used to benefit the local human communities, then these west and west-central habitats must be preserved: their weekend and longer-term visitors represent the only potential source of income other than the limited resources of the Department of Wildlife and the very considerable input from Tony and Lucy Fitzjohn at Kisima and the George Adamson Wildlife Preservation Trust.

Ferguson (Chapter 9) studied the effects of fire on the savanna vegetation of the western MGR and its influence on both vegetation composition and plant survival. In particular he identifies capparaceous shrubs such as the *Cadaba* species which regenerate from stem buds at or just below ground level after fire, the so called 'sprouters' of Bond and van Wilgen (1996). It is also apparent that many plant families and genera produce tubers, some of which are often massive (those

[1] The population of Same town has doubled in little more than 10 years (see Kiwasila and Homewood, Chapter 34).

of *Pyrenacantha malvifolia* may be up to 1.5 m in diameter), which may either lie on the surface or are subterranean. Although these plants quickly regenerate after fire it is suggested here (see Coe, Abdallah & Mboya on phenology, Chapter 8) that this adaptation evolved as an anti-drought adaptation, which has pre-adapted them to resist both natural and anthropogenically generated fire regimes. The fact that a number of these large subterranean tubers are a source of famine food for the Pare people makes them additionally interesting. The thickets of the small mimosaceous tree *Dichrostachys cinerea* ssp. *cinerea* are becoming very extensive in valley bottoms in the west-central areas of the MGR, indicating that this sub-species produces shoots just below the surface following the removal of aerial material by fire. Subsequently these thickets allow a number of species to become established in the centre of the thickets where they are protected from fire. Ultimately the thicket will die off at the centre but continue to spread, in the presence of fire, as separate foci (see Coe *et al.* on habitats, Chapter 7). Fire, whether natural or otherwise, adds an important element of physical and habitat diversity to a savanna environment. However, the effect of these perturbations depends almost entirely on its frequency: unless tree seedlings are able to grow to a height where they are out of reach of subsequent fires, local woody plant diversity will decrease quite rapidly[1].

It is worth noting here that Russell-Smith and his colleagues (Chapter 10) conclude from their arthropod sampling in ground and herbaceous vegetation that there were more species in pit-fall and sweep net samples from burnt than unburnt plots, although this conclusion was for species richness rather than species diversity (numbers of species in relation to their individual relative abundance). This is quite understandable, bearing in mind the almost two dimensional nature of burnt ground and the greater 'catchability' of the predominantly detritivorous fauna and their predators. What future studies should concentrate on is the degree of species (or morpho-species) overlap between these plots, compared with the very low level of overlap between tree genera (*Acacia* and *Commiphora*) observed by McGavin (Chapter 22). The very rich arthropod fauna of the MGR is undoubtedly a consequence of physiographic diversity through its profound influence on habitat diversity. George McGavin concludes that although we would encounter more species in a rain forest of similar area, there are more individuals on the MGR woody vegetation. He recognises that many more tree genera need to be sampled, but based on our current measure of species overlap, information from the ground

---

[1] Nick McWilliam, Raphael Abdallah and Emmanuel Mboya measured the heights of all woody plants on the lower slope of the Simba Plot. In 500m^2 they recorded 816 plants, attributed to 50 taxa (29 genera). Apart from a single specimen of *Adansonia digitata* (15 m), the average height of these plants was 0.9 m, indicating the profound effect that fire has on vegetation stature, although the survival of so many species indicates what a large number of common species are fire strongly adapted. We hope to be able to get back to this plot to observe their survival.

arthropod samples and the number of morpho-species being trapped that are almost certainly new to science, it is quite probable that the total arthropod fauna of the MGR is possibly as high as 90,000 species.

Staggering though these estimates may seem, the minute area, the habitats and micro-habitats that have currently been sampled and the fact that species accumulation curves plotted for the better sampled arthropod groups show no sign of levelling off, suggests that such estimates are 'not far off the mark'. The estimate of 650 beetle morpho-species (Chapters 10 and 14) is high but probably way below the real total, although very few beetle groups have been studied in detail in these savannas. Kingston (1977) recorded at least 102 dung beetle taxa (Family: Scarabaeidae) in a sample area of 4,100 km^2 in the Tsavo East National Park (or 22% of the total area), which is close to that of the MGR (3,250km^2). The predicted butterfly species total is 418 (Chapter 17) in addition to the very abundant moth fauna, which except for the macro-species, has hardly been studied in these savanna habitats. We can however estimate a potential total for the Lepidoptera by examining the ratio of butterfly species to that of the estimated lepidopterous total for southern Africa, where these arthropods are much better known than in east and central Africa. Scholtz & Holm (1985) estimate that there are 10–12,000 species of Lepidoptera in southern Africa, while the number of butterfly species is 632 (Ferrar 1989). Thus the ratio of moths to butterflies is 17.4:1, allowing us to suggest that the total number of Lepidoptera in the MGR is at least 7,300 species; though I suspect that with the very large number of unnamed micro-lepidoptera being observed the true figure is much higher. If we were able in the future to examine the arthropod groups that have been studied in the MGR in relation to those about which we know very little, McGavin's estimates do not appear quite so improbable. Indeed the results of the very hard work of everyone who has contributed (to what is possibly the most detailed study of arthropods in any African savanna) give a strong indication of the biotic riches that still await discovery—unless of course it has all been destroyed before we have ever finished cataloguing its wealth.

The MGR has a far less spectacular vertebrate (more obviously mammalian) fauna than we might expect to find in large National Parks like Serengeti, Tsavo, Amboseli or the Kruger. Yet its geographical position and undoubted flowering plant and invertebrate wealth indicate that the reason for extending a high degree of protection to an area of savanna wilderness cannnot be measured solely in terms of its wildebeest, elephant or lion populations. Although we must accept that these mammals are the primary generators of badly needed foreign currency for all of these countries, who are struggling to raise the standard of living of their people, despite hideous demographic problems which their governments seem unwilling or unable to come to terms with.

The fish fauna of the MGR is only of any importance on the Umba River and its adjacent pools where a small population of fish were reported by our entomologists. Bearing in mind the arid nature of the MGR we were unable to devote the financial resources necessary for a fish survey. However, the Umba River rises in that renowned centre of endemicity, the Usambara Mountains, and it is highly likely that some unusual species are be present—especially I suspect amongst the cichlids, cyprinodonts and cyrinids.

The amphibians and reptiles have been studied by our colleagues from South Africa (Chapters 24 and 25). The uncertain rainfall in this area meant that routine collection of amphibians was somewhat limited, although Mike Cherry managed to collect 14 anuran (frog and toad) species, representitive of four families. All are pretty widespread except for *Kassina somalica* (= *K. senegalensis,* Schiøtz 1975), which exhibits a sahelian arid distribution from southern Somalia to northern Tanzania. The potential species total for the anurans could easily be between 40 and 60 species (Coe & Stone 1995). A notable absence from the amphibians recorded in Mkomazi are the apodan (or caecilian) species (Amphibia: Gymnophiona) which are recorded in moderately mesic upland environments in Tanzania. There appears to be a series of adjacent ancient centres of caecilian radiation in the Teita Hills just over the border in Kenya, the Usambara and Uluguru Mountains, Tanzania where a total of up to at least six species are found. It seems highly likely that these interesting amphibians will be found in the threatened summit forests of the MGR, notably Hafino, Igire/Ibaya, Kinondo, Kisima and Maji Kunanua, where their level of isolation may well have been responsible for the evolution of endemic species or sub-species.

Of the 49 reptile species 58% are considered to be distinctive representitives of the Somalia-Masai phytochorian, or more especially its *Acacia-Commiphora* woodland component. The very rich chamaeleon fauna of Usambara (9 species) (Rodgers & Homewood 1982) and other Eastern Arc Mountains (c. 19 species) (Broadley & Howell 1991) are not represented in the MGR, except for the Beardless Pigmy Chamaeleon (*brachyurus*) and the Flap-necked Chamaeleon (*Chamaeleo dilepis*). The first named is distributed from south-eastern Tanzania, northern Mozambique and Malawi. A Pigmy Chamaeleon was collected below Ibaya camp in burnt capparaceous scrub, which was identified as *Rampholeon kerstenii kerstenii* by the Natural History Museum (Accession Number BMNH 1994: 399). This discovery was of some interest for it represents yet another Somalian form whose sub-species are recorded from southern Somalia to northern Tanzania. Most of the other reptiles were expected but it is clear that the snakes are under-represented. The sub-order Sauria are also a fairly rich group, although representatives of the upland forest-woodland habitats are probably rather low. The spectacular Gecko *Holodactylus africanus* was reported by Drewes (1971) to occur in Mkomazi.

The birds are one of the only animal groups that may be expected to yield a fairly accurate indication of its species richness and in this case we are not disappointed (see Lack, Baker & Baker, Chapters 26 and 27). The current species total for the MGR is 399 although there are a number of other species which require confirmation before our ornithologists are willing to include them in the MGR list. I suspect that the final total will be closer to 450 than 400. It is interesting to note that the total recorded by Lack, Leuthold & Smeenk (1980) in the Tsavo East National Park (but clearly representative of much of Tsavo West) was close to 460 species, yet despite their proximity to the MGR[1] we note that out of a joint total of close to 500 species in about 25,000 km^2, 100 of them were observed only in Tsavo, while 37 have only been recorded in Mkomazi. Although some of these species may well be recorded in the adjacent area in the future, it is clear that there are real differences as well as similarities between the two areas. It is worth noting that since Lack *et al.* (1980) made their observations, there have been fairly profound habitat changes in Tsavo East following the reduction in the elephant population of up to 80% by poaching and the resulting reduced pressure on woodland environments.

We may also note that bird species quite common (even very common) in Tsavo are uncommon or rare in Mkomazi. This is particularly well illustrated by the Fiscal Shrike and the Superb Starling which are common in Tsavo but are replaced in the northern parts of Mkomazi by the Long-tailed Fiscal and Hildebrandt's Startling respectively. One of the most interesting features of the MGR check list is the addition of six bird species to the Tanzanian avifauna, all of which (Violet Wood Hoopoe, Friedmann's Lark, Somali Long-billed Crombec, Yellow-vented Eremomela, Three-streaked Tchagra and Shelleys's Starling are representitives of the somalian arid avifauna. It is clear that much still needs to be done to understand the reasons for species changes between Tsavo and Mkomazi, and it is only if the MGR remains relatively free from excessive anthropogenic disturbance that it will be possible.

The study of small mammals carried out by Morley (Chapter 30) showed a number of parallels with the invertebrates. Although it was only possible to sample a very small area, it was demonstrated that a number of rare or even new rodent and/or shrew species are found in the MGR and especially in the summit forests. The discovery of a skull of the eastern tree hyrax (*Dendrohyrax validus*) by the author was of considerable interest, for this species is limited in its distribution to montane uplands and islands in north-eastern Tanzania (*D.v. validus*, Mount Meru and Kilimanjaro; *D.v. terricola*, Pare, Usambara, and Uluguru Mountains; *D.v. neumanni*, Zanzibar and Pemba (Kingdon 1971)) and is another illustration of the importance of the isolated montane massifs within the MGR. Kingdon (pers. comm.) visited the Igire/Ibaya summit forests during our studies and heard the calls of a bush baby which may well prove to be one of the new forms that have been described in Tanzania very recently.

---

[1] The Tsavo East National Park, Kenya and the Mkomazi Game Reserve are separated by about 110 km of bushland.

The large mammal fauna is not very obvious in the MGR although, provided the artificial dams are maintained, the more open habitats of the western localities of the reserve support healthy populations of plains game (see Eltringham *et al.*, Chapter 31). The elephant populations have been subjected to a high level of poaching in the 1980s, although between the late 1960s (Parker 1969) and the present time (TWCM 1991), their numbers have changed far less than in the Tsavo National Park. This is largely because very few of the elephant are permanent residents[1], but enter the area during the rains to take advantage of areas of forage that are unavailable to most large herbivores in the long dry season. The most noteworthy of the large mammals are lesser kudu (*Tragelaphus imberbis*) and gerenuk (*Litocranius walleri*) in the *Acacia-Commiphora* woodland. Their presence represents fast-dwindling populations of these mammals, as their habitats are being rapidly altered or destroyed by expanding human populations and their domestic stock. Among many interesting features of the Mkomazi mammal fauna is the large and healthy population of the diminutive antelope, the steinbuck (*Raphicerus campestris*), and frequent observations of melanistic serval cats (*Felis serval*) in the Viteo and Mbono Valleys.

It is extremely difficult to predict the way forward in terms of future planning priorities, but the recent consultations between the Department of Wildlife and the local people in the Kisiwani and Same areas demonstrate the manner in which the problems faced by rapidly dwindling conservation areas may be overcome (see Mbano, Chapter 36). It is clear that the biotic wealth of the MGR suggests it should be given a high degree of protection from any activity that may reduce its biotic diversity. This of course does not mean that the people who live around the reserve have no right to expect benefits from this unique area of Sahelian savanna. However, conservation areas like the MGR are only likely to provide an income that can generate obvious local social benefits.

It is my view, and that of many of my colleagues, that the scenery, plants, insects and the birds of the MGR are all features that can be developed as potential tourist attractions both for overseas visitors and the large number of Tanzanian residents for whom weekend and short do-it-yourself holiday facilities are a large untapped resource. To these developments we should add the potential of the new Ibaya facilities as a focus for the development of simple educational and research accommodation. One of these very simple ideas was highlighted by the enormous interest expressed by many of our visitors in the small garden we established, where we were able to persuade many local and attractive species to flower for most of the year.

If we are to be able to preserve areas like the MGR, it is vital that the developed world should continue to provide funds for both the development and

[1] It is possible that a small population of elephant has been resident for most of the year in and around the valleys on the eastern side of Maji Kunanua.

maintenance of the natural biota of this wilderness and the welfare of the local people. Without this understanding and mutual co-operation there will be no future for either man or the rich local flora and fauna. I feel sure that those who avow the cause of human welfare against the preservation of the natural environment, as if these are diametrically opposed aims, will come to realise it is neither a logical or a profitable strategy to spend charitable funds from the developed world on interfering in the internal affairs of sovereign African governments. Hopefully the information contained in this monograph will provide the stimulus that is necessary to ensure a sustainable financial and administrativly secure future for this magical landscape.

References

Anstey, D. (1998) In Litt.

Ayeni, J.S.O. (1975a) Periodicity of African wildlife at waterholes in Tsavo East National Park, Kenya. *Bulletin of Animal Health and Production in Africa* 23(2): 131-137.

Ayeni, J.S.O. (1975b) utilisation of water holes in Tsavo National Park (East). *East African Wildlife Journal* 13: 305-23.

Ayeni, J.S.O. (1979) Big game utilisation of natural mineral licks. In *Wildlife Management in Savannah Woodland.* (eds. S.S. Ajayim & L.B. Halstead). Taylor and Francis, London. pp. 85-95.

Bond, W.J. & van Wilgen, B.W. (1996) *Fire and Plants.* Chapman and Hall, London.

Broadley, D.G. & Howell, K.M. (1991) A check list of the reptiles of Tanzania, with synoptic keys. *Syntarsus* 1: 1-70.

Caughley, G. (1976) The elephant problem-an alternative hypothesis. *East African Wildlife Journal* 14: 265-83.

Coe, M.J., Cumming, D.H. & Phillipson, J. (1976) Biomass and production of large herbivores in relation and rainfall and primary production. *Oecologia* (Berl.) 22: 41-54.

Coe, M. (1985) *Islands in the Bush: A Natural History of the Kora National Reserve, Kenya.* George Philip, London.

Coe, M. (1990) The conservation and management of semi-arid rangelands and their animal resources. In: Goudie, A.S. (ed.) *Techniques for Desert Reclamation.* John Wiley, Chichester. pp. 219-249.

Coe, M. & Stone, G. (eds.) (1995) *Mkomazi Research Programme 1993-97: Progress Report, July 1995.* Royal Geographical Society, London.

Corfield, T.F. (1973) Elephant mortality in Tsavo National Park. *East African Wildlife Journal* 11: 339-68.

Drewes, R.C. (1971) Notes on the distribution of *Holodactylus africanus* Boettger. *Journal of the East African Natural History Society and National Museum* 28(126): 1-3.

Ferrar, A.A. (1989) The role of Red Data books in conserving biodiversity in Southern Africa. In: Huntley, B.J. (ed.) *Biotic Diversity in Southern Africa: Concepts and Conservation*. Oxford University Press, Cape Town. pp. 136-147.

Foster, J.B. & Coe, M.J. (1968) The biomass of game animals in Nairobi National Park, 1960-66. *Journal of Zoology, London* 155: 413-425.

Kingston, T.J. (1977) *Natural Manuring by Elephants in the Tsavo National Park, Kenya*. Unpublished D.Phil. Thesis. University of Oxford.

Krapf, J.L. (1860) *Travels, Researches and Missionary Labours in East Africa*. Trübner, London.

Laws, R.M. (1969) Aspects of reproduction in the African Elephant, *Loxodonta africana*. *Journal of Reproduction and Fertility, Supplement 6*: 193-217.

Morgan, W.T.W. (1973) *East Africa: Geographies for Advanced Study*. Longman, London.

New, C. (1871) *Life, Wanderings and Labours in Eastern Africa*. Frank Cass, London.

Ogot, B.A. (1981) *Historical Dictionary of Kenya*. (1981) African Historical Dictionaries 29. The Scarecrow Press, Inc., Metuchen, N.J.

Parker, A.C. (1969) *Results of two elephant harvests in the Mkomazi (Game Reserve) accompanied by considerable background information extracted from previous research and synthesis efforts*. Wildlife Services Limited, Nairobi.

Polhill, R.M. (1968) Tanzania. In: Hedberg, I. & Hedberg, O. (eds.) *Conservation of Vegetation in Africa South of the Sahara*. Acta Phytogeographica Suecica 54: 166-178.

Rodgers, W.A. & Homewood, K.M. (1982) Species richness and endemism in the Usambara Mountain forests, Tanzania. *Biological Journal of the Linnean Society* 18: 197-242.

Sansom, M. (1998) The Sahel continues. *Geographical* 70(6): 75.

Schiøtz, A. (1975) *The tree frogs of Eastern Africa*. Steenstrupia, Copenhagen.

Scholtz, C.H. & Holm, E. (1985) *Insects of Southern Africa*. Butterworth, Durban.

Sinclair, A.R.E. & Fryxell, J.M. (1985) *Canadian Journal of Zoology* 63: 987-994.

TWCM (1991) *Wildlife Census: Mkomazi 1991*. Tanzanian Wildlife Conservation Monitoring, Arusha.

Trollope, W.S.W. (1984) Fire in savanna. In: Booysen, P. de V. & Tainton, N.M. (eds.) *Ecological effects of fire in South African ecosystems*. Springer-Verlag, Berlin. pp. 149-175.

Wijngaarden, W. (1985) *Elephants-Trees-Grass-Grazers: Relationships between climate, soils, vegetation and large herbivores in a semi-arid savanna ecosystem (Tsavo, Kenya)*. International Institute for Aerospace Survey & Earth Sciences, Netherlands. ITC Publication 4.

Mkomazi Mind and Memory Maps

Jonathan Kingdon

An essay written for exhibitions of the Artists in Residence's work in the British Council, Dar es Salaam, June 1997 and in the RGS, London, October 1998.

This essay is a collection of personal notes designed to provide visitors with some background to what we, the artists, have tried to express.

This exhibition sets out to accomplish several very different objectives. Examine first how these may relate to the work that is on show. The Mkomazi research project was initiated by the Tanzanian Department of Wildlife, the Royal Geographical Society and the Oxford University Zoology Department to study the ecology of one of Tanzania's most important and unique bioreserves.

The British Council supported the idea of having two artists join the scientific team, one Professor Elias Jengo from Dar es Salaam, the other, myself, based in Oxford.

The integration of artists into scientific enterprises has a venerable past (particularly for the Royal Geographical Society) but it should be remembered that, before cameras were invented, the prime role for artists was to record the fauna, flora and topography. If the quirks and foibles of 17th and 18th century artists showed through that was essentially 'cultural contamination'. In that pre-photographic age the artist's eye was meant to be as detached and exact as a camera lens. From today's perspective the cultural and personal styles of such expedition artists become as revealing about their times as the objects they recorded. Few expedition artists try to compete with the camera today. It is the artists' freshness or originality of vision (even their 'quirks and foibles') that are valued now. While there is still a role for the topographic craftsman neither of us has gone down that route. Instead we have agreed on the title 'Mkomazi Mind and Memory Maps' in order to signpost our preoccupation with what the mind and memory makes of a place and the experience of that place. Use of the word map is both a geographic reference and a metaphor for charting our thoughts.

Both of us have made notes and sketches in the field but we have painted our pictures back in our studios. Elias has pinpointed symbolic moments, counterpointing people with the landscape, its fauna and flora (and with other people).

In my own principal canvasses, which I call the 'transect series', I have elaborated images out of eight literal outline maps which 'sample' different areas and scales of the Mkomazi landscape. Each map has associations with a season and a time of day or night. The dominant colours link with other associations, events, animals or plants. It is here that personal memories and experiences come in. For example, one two metre panel maps the two Ndea Hills, the Vitewini ridge and the Ngurunga gorge. It commemorates watching the dawn come up from the top of a giant rock near the base of Ndea Hill. Bush-fires from the previous day were still smouldering in the semi-darkness. Suddenly some small birds, red and yellow barbets, clambered up a tree beside the rock and as the first rays of light broke the horizon they sang a sweet and complex syncopated chorus. The male's disproportionately large head, its red plumage fluffed and pulsing, bobbed feverishly as it stomped and bucked about on its perch, wracked by the sheer energy of its performance. The precise music of the chorus, its extraordinary volume and the flamboyance of the barbets' display brought back memories of many other dawns in the Tanganyika of my youth. Other flocks answered from different directions—then the sun was up and the day took over. The colours of this map are those of the barbet, of fires and dawn, it commemorates my memory of a very particular experience in a very precise place and it asserts that there are seasons when very small organisms can pervade the atmosphere of a place—like us, they have their moments.

A predominantly green map of the whole Pare Mountain chain suggests that time towards the end of the wet season when trees and grass are full of caterpillars and the air pulses with the monotonous calls of cuckoos. The emerald and diederic cuckoos are particularly common at Mkomazi and their barred iridescent plumage is a good match for the colours of the vegetation during the caterpillar season. For me the cuckoos quite literally colour my perception of the landscape. We often describe animals as being 'well camouflaged'—for both painters and biologists such a flip description seems quite inadequate to the wonder of it. A host of different physiological and anatomical processes have been co-ordinated to evolve colours and shapes that 'fit' the animal into its landscape. 'Appearances' are important (for prey animals it can mean the difference between survival and extinction). Although external appearances are only a tiny fraction of any living animal's being, they account for most of what we can portray or visualise about them. Painting is also about appearances and most of a good picture's message lies in the wordless associations, the hidden meanings behind the choice of subject or theme. You have to learn to read a map but once you can interpret all those lines and symbols a more substantial world opens up to you.

One appeal of maps is their lofty viewpoint—the emancipation that they offer from earthbound gravity and the reminders of our own physical smallness and frailty. With maps we are no longer hedged in by the boundaries of horizons. We see the structure of organisms differently from above. The endless battle of animals and plants against gravity disappears; we no longer see the props, struts, trunks, limbs and towers that rear living bodies up from their ultimate fate—collapse. Linearity and verticality become blobs and a pattern of blobs resolves into vegetation, termitaries, herds, towns and crowds, all unified into a pattern where flow seems to be the main dynamic. The flow of streams and rivers cutting into the flows, cracks and bucklings of a restless earth. The flow of traffic, from elephants and cattle-paths to highways and bush-fires. Above it all the flow of air, sometimes laden with rain or dust, sometimes lazily filling valleys with mist or sculpting mountains with clouds. Migrating flocks of birds and butterflies also join these aerial flows. Fragments of their passage, a broken wing or feather, may float down to earth losing the complexity of dynamic three dimensional structures but their exquisite details are also flat mappings of a life that once fluttered across this broad land, mementos of hidden tempos.

When the mind is as airy, nomadic and gravity-free as the migrant, miniature patterns can become landmarks as significant as the monumental masses of mountain ranges.

Some of my Mkomazi paintings are intended to serve as mental and emotional route-maps into some associations between big landscapes and the small details that are such an important part of the scientist's work in this extraordinary place. I see a number of links between what the scientists are doing, their personal experiences of the place and those of artists.

How we record and inventory the geography and resources of a place and how we visualise it or remember it tend to be two very different things. The first often begins with a map and annotated lists of phenomena or species. Memories are less substantial, even when helped by sketches or photographs. Personal experience of a place is illuminated by all the small details, accidental encounters and vagaries of weather or schedule. We also see partly through the eyes and experiences of guides, companions, documents and historical references—we try to visualize what forces of geology and evolution shaped this land and its inhabitants.

Both artists and scientists may travel into strange territories, exploring the lives of animals and plants with histories more astonishing that anything invented by science fiction but it is important to remember that our enterprises are pursued in the very practical climate of a late twentieth century world. Here the uniqueness of Tanzania's extraordinary biological diversity and the evolutionary significance of its geography need to be and are becoming more widely known. Future benefits for Tanzania lie in both tourism and the building of a scientific and educational infrastructure of global significance.

In a newly global world that is hungry for information the various members of the Mkomazi research programme are significant contributors to the process of making both Mkomazi and Tanzania better known abroad.

When a metropolitan club sets off to reach and study some distant global landmark its members start a social process that may eventually impart fame and significance to some small stretch of the earth's surface. What were once very local names become familiar, even household words all over the world.

The Royal Geographical Society is just such a club and its meetings, expeditions and publications have, for more than a hundred years, advertised what an interesting planet it is that we inhabit. Among its many expeditions have been historic journeys to the Poles, Everest and the sources of the Nile.

Today the focus is on a frontier land in Tanzania known as Mkomazi. It is a frontier land in two senses, one trivial the other profound. The Tanzania-Kenya border was very recently drawn but there is another much older frontier where a very ancient and once very large block of equatorial mountains meets the lowlands of north-east Africa. For tens of millions of years the former have been moistened by ocean winds while the latter have been dry. This has been a wet/dry, high/low frontier for so long that its animals and plants have also adapted to both extremes. Now the RGS and Tanzania's Department of Wildlife have discovered such biological richness in Mkomazi it can no longer be seen as straddling a line between two extremes—the frontier itself is a complex of habitats which may help explain how we have come up with some of the highest measures of biodiversity ever recorded on earth. Single trees have been found to shelter 70,000 *dudus* (arthropods) and an average of 1,600 *dudus* inhabits one square metre of Mkomazi scrub. Over 400 species of birds have been recorded and the lists of flora grow and grow!

This abundance and natural diversity probably reflects the great age, the relative stability and the dense, narrow stratification of Mkomazi's ecosystems. Each year this immensely old chain keeps on eroding into the dry lowlands around it, the last remnants mere koppies of hard rock while the heights have montane rain forest and lush banana gardens. In between lies every permutation of altitude and vegetation. Desert roses and Hamets bloom in the rubble of well-drained screes while cycads and lichens are nursed by mist and dew on the upper crags. Every nook and cranny of this ancient landscape is inhabited, a living fabric embroiders the great vistas of plains and mountains. From elephants and baobabs to earwigs and fungi each animal and plant pursues its own unique life history. Much of their lives, like those of strangers in a city, can only be guessed at but the vitality and variety of natural life in Mkomazi dwarfs the human bustle of a Manchester or a Moscow and its riches will outlive both these cities, after all it has inhabitants whose lineages have been present for tens of millions of years. Or will they?

All the Mkomazi team have been impressed by the steady erosion of Mkomazi's complexity by the influence of people—fire, felling and flocks. Every year the fires burn deeper into the woods and thickets, shade and moisture go as more trees fall and voracious cattle and goats cut back the remaining herbage. Perhaps we have had the privilege of glimpsing a vanishing world. If so it is a terrible reflection on the quality of our civilisation that we must burn, kill and consume the biological riches of our planet. In this sense Mkomazi is the world.

One dimension of this exhibition that has not been touched upon is the shared identity of the two artists as Makerereans. This does not mean we come from some distant planet, Mercury or Mars but that we have both been participants in one of the most distinctive, dynamic but least-known art movements in modern Africa. This exhibition is probably best looked at as the product of an only partly familiar culture. Makerere is the name of a university in Uganda which, when we were there together, was the University of East Africa. In this unlikely setting a very distinctive School of Art flourished which continues to be a major cultural influence in Africa.

To be a Makererean at the beginning of the 1960s was to belong simultaneously to a very local indigenous institution and to be the fervent believer in a global Uhuru movement. The optimism of boom-time coincided with the dismantling of Europe's colonial empires and had replaced the pessimism and austerity of wartime.

For a short while independence meant a respect for all cultures and acceptance of the many faces of cultural and personal expression. It was in this tolerant climate that open minds, civil rights and self-discovery flourished. Elias and I shared that global atmosphere as well as the more local one of love for your own back yard and family folk. We both helped set up Tanganyika's Independence Exhibition in Dar es Salaam in 1961 where we both showed work. Now, 36 years later, this exhibition has offered us the opportunity to reaffirm our delight in the diversity, beauty, importance and interest to be found in one precious corner of our birth place.

In our work we also challenge the assumption that race, class or nationality are overriding contexts for art. Just as we learnt from scientists and technicians on the Makerere campus, the biologists and geographers have directed our eyes and minds to many wonderful details of the Mkomazi environment, asking endless questions and revealing some truly amazing stories. As at Makerere, we too, seek to show the scientists how to tease out meanings from the landscape and express it in our own way, without words or graphs. If maps are the ultimate scientific artefact, we have appropriated a much older concept of the chart to explore our own thoughts about a memorable place.

We have also left our mark in Mkomazi. While our first collaborations (as inexperienced teacher and student) were nearly 40 years ago we felt we should commemorate the place and the project that brought us together again. With permission from Mkomazi's chief warden we trekked up to a gully known as the Amphitheatre, where erosion had exposed great rocks with flat surfaces and an overhanging roof. We took raw earth oxides, red, yellow, white and carbon black. We emulsified them with eggs, gum and water and started to apply them with local *mswaki* tooth brushes (so inefficient we soon abandoned them for Same decorators brushes). Our subjects were drawn from our own observations, enthusiasms and experience of the place but also from the objectives of the project. Fauna, flora, interrelationships, natural processes, we wanted to imply them all, past and present. To unify the whole each of us left his images unfinished. We exchanged places so that Elias could complete my frontal lion portrait while I added roots to his baobab. We say our eggs cannot now be unscrambled! During a future wet season the rocks we painted will slither into the gully below; before that the hyraxes will have over-painted our images with their urine. Even so, for a while there will be a record, like fallen feathers from a migrant bird, that in remote Ibaya Valley two Makerereans joined a team of scientists whose task was to proclaim to a wider world that Mkomazi, one of Tanzania's most precious assets, is a place worth celebrating.

The Mkomazi Ecological Research Programme

Malcolm Coe

On 11th October 1988 the Director of Wildlife, Mr F.M.R. Lwezaula of the Ministry of Lands, Natural Resources & Tourism in Tanzania wrote to the Royal Geographical Society (RGS) in London with a request to consider undertaking an ecological inventory study of the Mkomazi Game Reserve. The view was to provide information for the Department of Wildlife to use in planning the area's future management and utilisation. This invitation grew from a similar study conducted in Kenya between 1982 and 1985 as the Kora Research Project, a joint research venture between the RGS and the National Museums of Kenya. The Society's Director, Dr John Hemming, wrote to the Department of Wildlife to suggest that I should engage in preliminary discussions with them.

Our initial aims followed a common pattern for all contemporary projects by the Royal Geographical Society (with The Institute of British Geographers) (RGS-IBG): to provide research and training opportunities for overseas and local scientists. Conducting this sort of research has become prohibitively expensive and can only be undertaken if it is possible to attract suitable funding. Few bodies today are willing to finance the infrastructure needed to support broad-based field studies, for you not only need to get into the field but provide living quarters, water and power for the scientists, often in very remote localities. Fortunately a number of well known multi-nationals still have the imagination and foresight to contribute to such programmes of environmental research. Oddly enough once we have managed to establish a safe and reliable base, the relative cost of subsistence is comparatively low and contributes in no small measure to the local economy in terms of purchasing local supplies, food and employment.

Once a base has been established we must then be able to count on the participation of a large group of experienced scientists who are willing to assist the studies without consultancy fees or other remuneration—other than the opportunity to engage in research in the tropics and to interact with overseas colleagues who often lack even the basic necessities for conducting field work. Such studies provide all participants with the opportunity to contribute to the understanding of increasingly endangered environments, and ultimately to contribute to programmes of conservation-oriented sustainable development. Those most commonly contributing to such co-operative ventures today are academics who can get away to work in the field—usually during a vacation.

Between January 1990 and June 1993 Dr Malcolm Coe and Nigel Winser, the Deputy Director of the RGS-IBG, visited Dar es Salaam, Arusha and Mkomazi on five occasions. We were able to discuss the logistics of operating in the Mkomazi Game Reserve, co-operation with local institutions including the University of Dar es Salaam, the Tropical Pesticides Research Institute in Arusha and the College of African Wildlife Management at Mweka, and to visit potential local sponsors. Throughout these early phases we were ably supported by the British Council who, throughout the Programme, have helped with our transport, provided facilities for our scientific meetings in their superb Conference Hall, acted as a communications centre and allowed us to use their computers and office facilities when we were in Dar es Salaam. It is no exaggeration to say that without them the Programme would never have gone beyond the planning stage. The British High Commission was always ready to help us in our dealings with Government Departments and to act as a centre for the entertainment of both our scientists and visitors. British Airways, BP (Tanzania), CMC (Land Rover) and the Sheraton Hotel (Dar es Salaam) were all stalwart allies during our frequent visits to Tanzania. In the UK we were given massive support and encouragement by British Airways, Friends of Conservation and Land Rover; while the offices of the RGS-IBG remained the centre of our fund-raising operations.

The Department of Wildlife provided considerable logistic support and seconded Paul Marenga to the team as Programme Manager from their Planning and Assessment for Wildlife Management unit, working between Dar es Salaam and Mkomazi. Having decided on an outline Programme of research, the new Director of Wildlife, Costa Mlay, and I signed a Memorandum of Understanding in Dar es Salaam on 18th July 1991. This document was revised when Mr M.A. Ndolanga took over as Director of Wildlife and a new Memorandum was signed in Mkomazi on 2nd August 1993. Bakari Mbano, who had been a very considerate and helpful liaison officer for most of the time we worked in Mkomazi, took over as Director of Wildlife in 1995.

After the signing of the first Memorandum, the RGS-IBG set up Programme Planning Committee, initially with Lord Chorley in the Chair and later Professor Grenville Lucas, to co-ordinate the UK aspects of the scientific programme, its finance and administration. Initially the Committee was fortunate in being able to communicate directly with Tony Fitzjohn who had established a camp in Mkomazi as part of a programme of rehabilitating the reserve and its infrastructure—efforts which were funded by the George Adamson Wildlife Preservation Trust, who later paid generously for the refurbishment of the borehole at Ibaya and who contributed towards the publication of this book.

The aim of the study was to describe the habitats of the Mkomazi Game Reserve in both floral and faunal terms, in order to stimulate the generation of models which will identify factors responsible for their observed patterns of distribution, abundance and species diversity. It was important for us to assist the Department of Wildlife wherever possible in integrating our observations on the comparative ecology of areas that are still in a relatively natural state with those that have been affected by anthropogenic activities such as grazing and uncontrolled burning, into a Management Plan for the area's future utilisation. Additionally we were anxious that we should attempt to provide baseline data, against which the effects of disturbance and change in surrounding areas might be understood.

Members of the Mkomazi Ecological Research Programme

Administration and logistics

Programme Directors
Dr Malcolm Coe, Science Director
Bakari Mbano, Director, Department of Wildlife
M.A. Ndolanga and Costa Mlay, former Directors, Department of Wildlife

Department of Wildlife, Dar es Salaam
Paul Marenga, Programme Manager

Field Directors
Hugh Watson (1993)
Tim Morgan (1994–97)

RGS-IBG, London
Nigel Winser, Deputy Director
Venetia Simonds, Helen Lawrenson, Joanne Lyas, Alex Walters

Scientific Field Assistants
Angus Jackson, RGS-IBG
Nicholas McWilliam, RGS-IBG

Rangers, Department of Wildlife
Daniel Mafunde
Japhet Taiko (Max)
Maneno Myinga
Hamisi Ayuby
Johnson Kaanankira
Elias Venance

Ibaya Camp Staff
Omari Abi di
Mathias Mark
Omari Mohamed
Ramadhani Othman
Firimini Saidi
Mashaka Yahaya
Yona Zawadi

Scientific Programme

Climate
Angus Jackson, RGS-IBG
Nicholas McWilliam, Oxford
Tim Morgan, RGS-IBG

Geochemistry
Dr Peter Abrahams, University of Wales
Dr Rob Bowell, University of Wales

Geographical information systems and mapping
Susan Canney, Oxford
Kent Cassells, DICE
Fay Hercod, RGS-IBG
Nicholas McWilliam, Oxford
Dr Mike Packer, Oxford
Dr Shaun Russell, DICE
Azmillen bin Ramlee, RGS-IBG

Vegetation/habitats and taxonomic botany
Raphael Abdallah, TPRI

Dr Malcolm Coe, Oxford
Julie Cox, NRI
Yvette Kalema, UDSM & TPRI
Jafari Kidegesho, CAWM & DICE
Baker Masaruli, CAWM & DICE
Emmanuel Mboya, TPRI
Leonard Mwsumbi, UDSM
Dr William Mziray, TPRI
Dr Roger Polhill, RBG Kew
Dr Kaj Volleson, RBG Kew

Entomology
Jonathan Davies, Natural History Museum, London
Elias Kihumo, TPRI
Daniel Mafunde, Department of Wildlife
Ramhadani Makusi, TPRI
Dr George McGavin, Oxford
Bruno Nyundo, UDSM
Dr Mark Ritchie, NRI
Dr Hamish Robertson, South African Museum, Cape Town
Dr Tony Russell-Smith, NRI
Linsey Stapley, University of Cambridge
Dr Graham Stone, Oxford
Dr Simon van Noort, South African Museum, Cape Town
Dr Pat Willmer, University of St Andrews

Herpetology
Dr Mike Cherry, University of Stellenbosch
Dr Alex Flemming, National Museum, Bloemfontein
Melissa Stander, University of Stellenbosch

Birds
Neil and Liz Baker, Tanzanian Bird Atlas Project, Moshi
Dr Peter Cotgreave, Save British Science, London
Dr Peter Lack, British Trust for Ornithology
Dr James Matee, TPRI, Arusha
Paul Strecker, Angus Jackson, Ryan Mellor, Caroline Peciuch, Joanne Taylor, Ben Underwood, St Peter's College Oxford

Mammals
Dr Keith Eltringham, University of Cambridge
Deborah Epperson, University of Florida
Ken Ferguson
Jonathan Kingdon, Oxford
Dr Larry Harris, University of Florida
Paul Marenga, Department of Wildlife
Nicholas McWilliam, Oxford
Ian Cooksey, William Dixon, Nigel Raine, Chris Sheldrick, Oxford
Bartholomeo Maganga, Moses Malekia, Kiping'ot Ngurumwa, CAWM

Human communities around MGR
Dr Dan Brockington, UCL
Professor Katherine Homewood, UCL
Hilda Kiwasila, UCL & UDSM

Medical Advisor
Dr Liz Tayler

Artists in residence
Professor Elias Jengo
Jonathan Kingdon, Oxford

Visiting photographers
Chris Caldicott
Tom Craig

CAWM—College of African Wildlife Management at Mweka. DICE—Durrell Institute for Conservation Ecology, University of Kent. NRI—Natural Resources Institute. Oxford—University of Oxford. RBG—Royal Botanic Gardens, Kew. RGS-IBG—Royal Geographical Society with The Institute of British Geographers. TPRI—Tropical Pesticides Research Institute, Arusha. UCL—University College, London. UDSM—University of Dar es Salaam

Acknowledgements

Malcolm Coe

Since the Royal Geographical Society was founded in 1830 it has played a pre-eminent role in encouraging and sponsoring field research and exploration throughout the world. Over the last 50 years these activities have increasingly become joint academic ventures with the host countries, with the aim of providing field experience and excitement for all our participants. In spite of the many generous benefactors who have helped the Society in establishing a healthy endowment, its own research programmes must rely almost exclusively on outside funds from Corporate Sponsors, Trusts and Research Councils. Members of our Field Research Programmes are drawn from Academic Institutions, our Fellowship and members of the public who give their time and expertise on a purely voluntary basis to ensure that we may make a significant contribution to our understanding of the world in which we all live and continue to disturb and damage.

The Mkomazi Ecological Research Programme (MERP) has been no exception to the above philosophy, except that it has become increasingly difficult to obtain adequate funding for such long-term studies: unless ecological and biodiversity information can be collected over a reasonably extended period it is impossible to begin to understand the influence of environmental variables, whose periodicity lies in decades rather than years, on the local flora and fauna. Without the help of all the staff of the RGS-IBG, the University of Oxford, the Ministry of Natural Resources and Tourism and above all the officers of the Department of Wildlife in Dar es Salaam we would never have succeeded in getting such a large and effective team into the field.

The Director of the Royal Geographical Society, Dr John Hemming, played a central role in all our planning. Leading the Expeditions Office team at the Royal Geographical Society was the Deputy Director, Nigel Winser, whose enthusiasm and contacts have almost always been able to find a way around our frequent logistic and financial crises. Indeed it is true to say that without his knowledge and foresight, many RGS-IBG projects would probably never get into the field at all. His involvement in MERP was not only at the London planning end but he also found the time, often at the expense of his family, to accompany me to Dar es Salaam, Arusha, Moshi and the Mkomazi Game Reserve itself. Nigel was ably supported throughout the planning and execution of MERP by our London administrator Venetia Simonds, his secretary Helen Lawrenson and more recently by Jo Lyas. Shane Winser and her staff in the Expedition Advisory Centre have always been ready to help us with contacts, and in the case of Fay Hercod even in the field at Ibaya where she contributed to the early stages of our GIS programme.

The fact that we were able to spend such an extended period in the field has been almost entirely due to the generous funding provided by our major sponsors. At the very beginning of our planning, Friends of Conservation (FoC) came forward and offered to fund the refurbishment of the camp at Ibaya and the provision of new living and working facilities, including the handsome laboratory and an Information Centre at the Zange Gate. The Baring Foundation funded a preliminary field survey in 1993 and during these early stages logistic support was provided by Abercrombie & Kent's office in Arusha. The British Council were magnificent at all stages of the Programme, with successive Directors Robert Sykes and Rosemary Hilhorst and their assistants Peter Llewellyn and Sharon Crowther helping us in ways that were, and continue to be, way beyond the normal call of duty. They helped with insurance and the licensing of our vehicles, acted as a centre for the receipt of and onward transmission of important messages, organised and funded several of our meetings in Dar es Salaam and provided the premier point of liaison with Tanzanian Government Departments. The British Council staff became and remain some of our closest friends in Tanzania. British Airways flew most of us to and from Dar es Salaam over a period of six years, while BP (Tanzania) fuelled the handsome TDI vehicles generously provided by Land Rover (UK) and serviced by CMC Land Rover (Tanzania). The Sheraton Hotel (Dar es Salaam) continues to support us with their Going Green scheme and arranged a sumptuous reception when our Patron HRH The Duke of Kent visited the Programme in 1996. The British High Commissioners Roger Westbrook and Alan Montgomery provided valuable logistic support throughout the Programme and entertained our members and collaborators on numerous occasions.

The MERP Planning Committee met regularly under the chairmanship of Lord Chorley and Professor Grenville Lucas, while up-date meetings for our UK contributors and overseas visitors were accommodated with their usual efficiency in the Society's headquarters. The fact that the Society still shoulders our financial deficit, largely brought about as a consequence of the failure of one of our early sponsors, is a striking measure of the Society's recognition that ecological and conservation studies are at the very forefront of contemporary geography. The logistic support provided by the Department of Zoology at the University of Oxford in the earlier phases of MERP, particularly through Professor Sir Richard Southwood, was an important factor in getting such an active scientific team into the field in Tanzania.

In Tanzania our many friends have made our work so much easier than it might otherwise have been. Four Directors (1989–98) of the Division of Wildlife (F.M.R. Lwezaula, Costa Mlay, M.A. Ndolanga and Bakari Mbano) all played a significant role in guiding us through the process of running a Programme over 500 km away from their Headquarters. In the early stages of our operation they loaned us a radio that would enable us to communicate with them and the Flying Doctor Service at AMREF. Paul Marenga joined us as a Programme Manager at an early stage of our work and spent a great deal of time travelling between Ibaya and Dar es Salaam to deal with supplies, customs and other matters concerning official communication with government departments. In the Mkomazi Game Reserve we received regular advice from the successive Wardens Messrs. H.E. Mungure and Issai Swai. Since the completion of our work, Paul Marenga has been posted to Mkomazi as its new Project Manager. In all the best regulated families there are differences of opinion, but bearing in mind the period that MERP maintained a presence in the MGR, I am still impressed by the goodwill that was generated by almost everyone we dealt with.

During the course of our studies we were delighted to be able to welcome a number of students from the University of Dar es Salaam, the College of African Wildlife Management (Mweka), the Durrrell Institute of Conservation and Ecology (University of Kent) and the University of Oxford who carried out their own projects or took part in on-going MERP investigations. David Manyanza, the Principal at Mweka remains a good friend who provided support at many points during our time in the MGR. Finally but by no means least is the Tropical Pesticides Research Institute (TPRI) in Arusha, which was a staunch supporter of our work from the very outset of our studies. In particular the Director Dr W.F. Mosha and Dr William Mziray of the National Herbarium assisted us in many ways in our aim of contributing to their plant and insect collections. Without the help of Raphael Abdallah, Elias Kihumo, Ramahdani Makusi and Emmanuel Mboya our field work would have been much less impressive than it was.

I know that my colleagues in MERP would not forgive me if I did not mention the great bonds of friendship that we forged with the people of Kisiwani and Same town, for at every stage of our work they were ready to help and to share their incredible knowledge of the local bush. Above all it was their love of the Mkomazi wilderness that especially impressed us, for some visitors imagine that our eccentric passion for the *bundu* is not shared by local villagers. In this case their enthusiasm was quickly transmitted to the whole team, so that an emergency at midnight would be dealt with by a veritable army of volunteers.

During the long process of refurbishing our accommodation and building the kitchen, staff dining room, store, generator room and the handsome FoC Ibaya Conservation Centre, we were able to obtain all our building materials, hardware and furniture locally and if not in stock, they would be obtained in a few days. Our building programme was entirely carried out by our friends Firmini Saidi,

Omari 'Mbili' Mohamed, Ramadhani Othman and Mashaka Yahaya. Much of our equipment was maintained by the incomparable *fundis* Firimini and Omari Mbili, but when this proved impossible we could almost always find a local *fundi* who could put it right as we all watched in ignorance and admiration. Our culinary activities were conducted by the incomparable Mathias Mark assisted by Omari 'Moja' Abdi and Yona Zawadi from whose kitchen bread would emerge from our purpose-built oven, of a quality that would rival anything available in Dar es Salaam or Oxford.

The Savanna Hotel provided some of the best samosas and chapatis in town, while the market not only provided all our perishable supplies but showed a level of friendship that makes much of our own materialist world a poor second by comparison. The late Mr Kamwala's (Esso) Service Station in Same was very efficient in managing our fuel supplies while the ever helpful BP Office in Moshi was able to ensure that the supply was maintained. Also in Same the Mother Superior and Sisters at the Catholic Mission provided a haven of peace and quiet, where their accommodation, gardens and excellent Tanzanian cuisine enabled us to operate in Mkomazi before the Ibaya buildings had been renovated and cleared of several hundred kilograms of accumulated bat dung. Later the Mission generously allowed us to hire their splendid Hall for our local Report Meeting in 1995 and whatever it lacked in the sophistication of Dar es Salaam it made up with pure Tanzanian countryside atmosphere. It would be invidious to recall single groups, people or places but I cannot resist the temptation to suggest that one of our great Same highlights was Max's (alias Japhet Taiko) wedding in the Pentecostal Church.

Our stay at Ibaya was strengthened and in many respects only possible because of the invaluable assistance that we received from our own staff and the Department of Wildlife Rangers (who included Danial Madundi, Japhet (Max) Taiko, Maneno Myinga, Hamisi Ayuby, Johnson Kaanankira and Elias Venance) who were seconded to Ibaya Camp and MERP. Our contribution to the knowledge of the ecology of Mkomazi was only possible because of the help we received from them. It is also of considerable satisfaction that we were able to make such a significant financial contribution to the local economy while we were at Ibaya, something that it is very easy to forget. In thanking the people who were responsible for the success of our work in the MGR we are all deeply indebted to Tim 'Timbo' Morgan who was 'Head of the Ibaya Family' for over three years. Even as I write he is back tying up a few last-minute details in Same, where he was almost instantly joined by the many members of his extended family. Without Tim we would have accomplished little or nothing and we are all deeply indebted to him for his dedication to the cause, supported by little more than his pension.

Asante sana

Patron

His Royal Highness the Duke of Kent

Corporate Benefactors

Friends of Conservation—launch sponsor
British Airways
BP Tanzania
CMC Land Rover (Tanzania)
British Council
Land Rover
Sheraton Hotel (Dar es Salaam)

Sponsors

Abercrombie & Kent
Appledore Trust
Booker plc
William A. Cadbury Charitable Trust
Clothworkers Foundation
Baring Foundation
GreenCard Trust
River Island Clothing Company
George Adamson Wildlife Preservation Trust

Research partners

College of African Wildlife Management at Mweka (Tanzania)
Commonwealth Science Council (South Africa)
Darwin Initiative, Department of the Environment (UK)
Durrell Institute of Conservation & Ecology, University of Kent (UK)
National Herbarium of Tanzania, Arusha (Tanzania)
Natural Resources Institute, Chatham (UK)
Department for International Development (UK)
Royal Botanic Gardens, Kew (UK)
Royal Society, London (UK)
Tropical Pesticides Research Institute, Arusha (Tanzania)
University of Cambridge, Department of Zoology (UK)
University of Oxford, Department of Zoology (UK)
Natural History Museum (UK)

RGS-IBG MERP Committee Members

Lord Chorley (RGS, Chairman)
Professor Gren Lucas (Royal Botanic Gardens Kew, Chairman)
Dr Jeff Burley (University of Oxford)
Mrs Jorie Butler Kent (Friends of Conservation)
Dr Neil Chalmers (Natural History Museum)
Dr Malcolm Coe (Programme Director)
Ian Haines (Overseas Development Adminsitration)
Dr John Hemming (Royal Geographical Society)
Stuart Innes (Foreign & Commonwealth Office)
Len Mole (Royal Society)
Stephen Walters (British Council)
Dr Peter Warren (Royal Society)
Nigel Winser (Royal Geographical Society)

Support & assistance—organisations

Apgar Fund, Magdalen College, Oxford
Blackwells Science
Britain-Tanzania Society
British High Commission, Dar es Salaam
Cheesemans Ecological Safaris
COLAB
COSTECH
CPL Colour Processing
East African Wildlife Society
Fuji
Long Distance Walking Association, Wessex Group
Next Day Printing, Cardiff
Oxford University Exploration Club
Silva Compasses
Varley-Gradwell Fellowship

Support & assistance—individuals

David Anstey
John Boshe
Laurent Certelet
Sir John Chapple
Jacques Chevasson
Hugh Coe
Unity Coe
Edward de Courcy-Bryant
Sharon Crowther
Sandy Evans
Albert Fernandes
Tony and Lucy Fitzjohn
Lisa Gillett
Abdul Haji and family
Rosemary Hilhorst
Dr Kim Howell
Roger Ivens
Lord Jellicoe
Peter Kangwa
Hatim Karimjee
Mrs Jorie Butler Kent
Jonathan Knocker
Peter Llewellyn
Rose Lugembe
Dr David Manyanza
Ali Mchumo
Barry Melbourne Webb
Costa Mlay
Alan Montgomery
Dr F.W. Moshi
H.E. Mungure
Dr William Mziray
Mr M.A. Ndolanga
Mr Nguli
Hon. Jumar Hamad Omar
John Palmer
Dr Roger Polhill
Diane Pilot
Dr A.A. Shareef
Ian Stewart
Robert Sykes
Kathaleen Stephenson
Issai Swai
Mama Tumbo
Professor David Warrell
Roger Westbrook
Lord Nicholas Windsor
Shane Winser
Mrs Miriam Zachariah

Friends of Conservation

Friends of Conservation (FOC) was founded in 1982 by its Chairman Jorie Butler Kent and her husband Geoffrey Kent to conserve wildlife and habitat in Kenya's Masai Mara National Reserve and other areas of the world. In 1988, His Royal Highness The Prince of Wales honoured FOC by becoming its Patron.

The organisation's programmes fall into three categories: Community Conservation and Education, Wildlife Conservation and Research and Eco-Tourism. FOC's Masai Mara Community Conservation and Education Programme (MMCC&EP) has been on the cutting edge of projects of this kind since its inception in 1989. Projects such as this are now widely seen to be the future of successful conservation in developing countries. The MMCC&EP works with Masai Mara area residents, both adults in the communities and trade centres and children in the schools. Education is a key component of this work, which is developing with the Maasai an economic return from their preservation of natural habitat and wildlife on their communal lands. When environmental and economic goals are united in this way, maintaining natural resources becomes a natural way of life.

FOC Chairman Mrs Jorie Butler Kent and HRH The Duke of Kent unveiling the plaque at the FOC Ibaya Research Centre in Mkomazi, March 1996, with (left to right) MERP co-director Dr Malcolm Coe; British High Commissioner to Tanzania Mr Alan Montgomery; and Lord Nicholas Windsor.

Equally important, FOC encourages visitors to east Africa to understand and respect local people and customs as well as wildlife and habitat. Ecosystems and wildlife continue to benefit from FOC conservation and research projects, which include such far reaching work as anti-poaching and wildlife veterinary support, as well as scientific research.

FOC is an international organisation with offices in the USA, the UK and Kenya. Supporters include governmental agencies, foundations and trusts, members of the travel industry and media, corporations and private individuals.

In 1995 FOC was the launch sponsor of the Mkomazi Ecological Research Programme and donated £30,000 to the work of the programme. The donation was used to renovate and equip the Ibaya Research Centre. The centre, situated 20 km into the reserve, provided the Programme's international team of visiting scientists with all the facilities and support needed for their fieldwork.

Friends of Conservation is continuing its support to the Mkomazi Ecological Research Programme through the funding of this book.

Mrs Jorie Butler Kent with Lord Jellicoe, former President of RGS-IBG, at the launch of the Mkomazi Ecological Research Programme.

For more information about Friends of Conservation please contact one of the offices listed below:

Friends of Conservation UK, Riverbank House, 1 Putney Bridge Approach, London SW6 3JD, UK. Tel. (020) 7731 7803, Fax (020) 7731 8213.

Friends of Conservation USA, 1520 Kensington Road, Oak Brook, IL 60523-2141, USA. Tel. (630) 954 3388, Fax (630) 954 1016.

Friends of Conservation Kenya, PO Box 74901, Nairobi, Kenya. Tel. (2) 442 075, Fax (2) 442 048.

The George Adamson Wildlife Preservation Trust

Since 1988, the George Adamson Wildlife Preservation Trust and its sister organisations in America, the Netherlands and Germany have worked together to provide the capital and running costs for the Mkomazi Project which is being undertaken by the Government of Tanzania in the Mkomazi Game Reserve. Under Tony Fitzjohn, who reports to the Director of Wildlife, the management and rehabilitation of the reserve has taken many forms including the cutting of roads and airstrips, desilting of dams, anti-poaching patrols, liaison and assistance to local communities through outreach programmes, and the reintroduction of endangered species, in particular the wild dog *(Lycaon pictus lupinus)* and the black rhino (*Diceros bicornis minor*).

The Trusts are funded by small groups of dedicated individuals alongside the generous backing of corporate sponsors and charitable organisations around the world to perpetuate the survival of wild and natural areas. Since the Mkomazi work began, a policy of hands-on conservation has been followed, using limited resources to the best possible effect. The project is dedicated to restoring the ecosystem in order that the reserve may be self-sustaining in the long term, a process seen as the driving force behind the Mkomazi Project.

In late 1997, four (Kenyan) black rhino were translocated from South Africa to the Mkomazi Game Reserve where they have settled well into a highly secure, purpose-built sanctuary. The establishment of a viable breeding programme will require further animals to be translocated. With patience, appropriate funding and the co-operation of all interested parties, this important work can continue and become a major success story.

The Trusts are proud of their contribution to the vital initiatives provided by the Project and are delighted to be associated with the work of the Department of Wildlife and the Royal Geographical Society (with IBG) in the Mkomazi Game Reserve.

For further information please contact:

Georgina Mortimer, George Adamson Wildlife Preservation Trust, 16a Park View Road, London N3 2JB, UK. Tel. (020) 8343 4246, Fax 020 8343 4048, e-mail *mortimer@btinternet.com*

Authors' addresses

Names, addresses and e-mail addresses of authors. Authors can also be contacted c/o Mkomazi Office, Royal Geographical Society (with IBG), 1 Kensington Gore, London SW7 2AR, UK.

Raphael Abdallah Consultant Botanist (formerly at TPRI), PO Box 25562, Dar es Salaam, Tanzania

Dr Peter W. Abrahams Institute of Geography and Earth Sciences, University of Wales, Aberystwyth, Ceredigion SY23 3DB, Wales, UK *pwa@aber.ac.uk*

Neil E. Baker & Elizabeth M. Baker Tanzanian Bird Atlas Project, PO Box 9601, Moshi, Tanzania *tzbird@habari.co.tz*

Michael F. Bates National Museum, PO Box 266, Bloemfontein 9300, South Africa

Dr Robert J. Bowell Steffen, Robertson and Kirsten (UK) Ltd., 9–10 Windsor Place, Summit House, Cardiff CF1 3BX, UK

Dr Dan Brockington Department of Geography, University of Cambridge, Downing Place, Cambridge CB2 3EN, UK *db261@cam.ac.uk*

Susan M. Canney Department of Zoology, University of Oxford, South Parks Road, Oxford OX1 3PS, UK *susan.canney@zoo.ox.ac.uk*

Dr Michael I. Cherry Department of Zoology, University of Stellenbosch, Private Bag X1, Matieland 7602, South Africa *Mic@land.sun.ac.za*

Dr Malcolm Coe St. Peter's College, Oxford OX1 2DL, UK. *Address for correspondence*: 39 Sandfield Road, Oxford OX3 7RN, UK

Dr Stephen G. Compton Department of Biology, University of Leeds, Leeds LS2 9JT, UK *pab6sgc@west-01.novell.leeds.ac.uk*

Ian Cooksey Magdalen College, Oxford OX1 4AU, UK *stcookseyi@tiffin.kingston.sch.uk*

Dr Peter Cotgreave Save British Science, 29 Tavistock Square, London WC1H 9EZ, UK *peter.cotgreave@ucl.ac.uk*

Jonathan G. Davies Cresswell Associates – Environmental Consultants, The Old Convent, Beeches Green, Stroud, Gloucestershire GL5 4AD, UK *cresswell.associates@dial.pipex.com*

William Dixon Department of Zoology, University of Oxford, South Parks Road, Oxford OX1 3PS, UK *william.dixon@zoo.ox.ac.uk*

Dr Keith Eltringham Department of Zoology, University of Cambridge, Downing Street, Cambridge CB2 3EJ, UK *ske1000@cus.cam.ac.uk*

Ken Ferguson 113 Mayfield Road, Edinburgh EH9 3AJ, UK

Dr Alexander F. Flemming Department of Zoology, University of Stellenbosch, Matieland 7602, South Africa *AFF@land.sun.ac.za*

Professor Katherine M. Homewood Department of Anthropology, University College London, Gower Street, London WC1E 6BT, UK *k.homewood@ucl.ac.uk*

Elias Kihumo Insect Museum, Tropical Pesticides Research, PO Box 3024, Arusha, Tanzania *tpri@marie.gn.apc.org*

Jonathan Kingdon Research Associate, Department of Zoology, University of Oxford, South Parks Road, Oxford OX1 3PS, UK

Hildegarda L. Kiwasila Institute of Resource Assessment, University of Dar es Salaam, PO Box 35097, Dar es Salaam, Tanzania *wrdp@udsm.ac.tz*

Dr Peter Lack British Trust for Ornithology, The Nunnery, Thetford, Norfolk IP24 2PU, UK *peter.lack@bto.org*

Dr Jason G.H. Londt Natal Museum, Private Bag 9070, Pietermaritzburg, 3200, South Africa *jason@mail.tcs.co.za*

Jo Lyas Royal Geographical Society (with IBG), 1 Kensington Gore, London SW7 2AR, UK *j.lyas@rgs.org*

Daniel Mafunde Mkomazi Game Reserve, PO Box 376, Same, Tanzania

Ramadhani Makusi (formerly at TPRI) PO Box 11734, Arusha, Tanzania

Dr Mervyn W. Mansell Biosystematics Division, ARC – Plant Protection Research Institute, Private Bag X134, Pretoria, 0001, South Africa *vrehmwm@plant5.agric.za*

James J. Matee Tropical Pesticides Research Institute, PO Box 3024, Arusha, Tanzania *tpri@marie.gn.apc.org*

Bakari N.N. Mbano Director, Department of Wildlife, Ministry of Natural Resources & Tourism, PO Box 1994, Dar es Salaam, Tanzania *wildlife-division@twiga.com*

Emmanuel Mboya Tropical Pesticides Research Institute, PO Box 3024, Arusha, Tanzania *tpri@marie.gn.apc.org*

Dr George C. McGavin Oxford University Museum of Natural History, Parks Road, Oxford OX1 3PW, UK *george.mcgavin@oum.ox.ac.uk*

Nicholas C. McWilliam Linacre College, Oxford OX1 3JA, UK *nicholas.mcwilliam@bas.ac.uk*

Robert C. Morley 9 Shinecroft Cottages, Rye Lane, Otford, Kent TN14 5NA, UK

Bruno Nyundo Department of Zoology and Marine Biology, University of Dar es Salaam, Tanzania

Dr Michael J. Packer Department of Zoology, University of Oxford, South Parks Road, Oxford OX1 3PS, UK *mike.packer@zoo.ox.ac.uk*

Dr John C. Poynton The Natural History Museum, Cromwell Road, London SW7 5BD, UK

Nigel Raine Department of Zoology, University of Oxford, South Parks Road, Oxford OX1 3PS, UK *nigel.raine@zoo.ox.ac.uk*

Dr Hamish G. Robertson Life Sciences Division, South African Museum, PO Box 61, Cape Town 8000, South Africa *hroberts@samuseum.ac.za*

Dr Alex Rowe Institute of Cell, Animal and Population Biology, University of Edinburgh, King's Buildings, Edinburgh EH9 3JT, UK *Alex.Rowe@ed.ac.uk*

Dr Tony Russell-Smith Natural Resources Institute, Central Avenue, Chatham Marine, Chatham, Kent ME4 4TB, UK *tony.russell-smith@nri.org*

Chris Sheldrick Hertford College, Oxford OX1 3BW, UK

Melissa J. Stander Department of Zoology, University of Stellenbosch, Private Bag X1, Matieland 7602, South Africa

Dr Linsey Stapley Elsevier Science London, 68 Hills Road, Cambridge CB2 1LA, UK *Linsey.Stapley@current-trends.com*

Dr Graham N. Stone Institute of Cell, Animal & Population Biology, University of Edinburgh, King's Buildings, Edinburgh EH9 3JT, UK *Graham.Stone@ed.ac.uk*

Dr Simon van Noort Division of Life Sciences, South African Museum, PO Box 61, Cape Town, 8000, South Africa *svannoort@samuseum.ac.za*

Dr Martin H. Villet Department of Zoology & Entomology, Rhodes University, Grahamstown, 6140, South Africa *zomv@giraffe.ru.ac.za*

Dr Kaj Vollesen Royal Botanic Gardens, Kew, Surrey TW9 3AB, UK *K.Vollesen@rbgkew.org.uk*

Dr Pat Willmer School of Biology, University of St Andrews, Bute Building, St Andrews, Fife KY16 9TS, UK *pgw@st-and.ac.uk*